D0936711

Introduction to Plant Physiology

Introduction to Plant Physiology

Fourth Edition

William G. Hopkins

and

Norman P. A. Hüner

The University of Western Ontario

WILEY

John Wiley & Sons, Inc.

VICE PRESIDENT AND EXECUTIVE PUBLISHER	Kaye Pace
SENIOR ACQUISITIONS EDITOR	Kevin Witt
PRODUCTION SERVICES MANAGER	Dorothy Sinclair
PRODUCTION EDITOR	Janet Foxman
CREATIVE DIRECTOR	Harry Nolan
SENIOR DESIGNER	Kevin Murphy
EDITORIAL ASSISTANT	Alissa Etrheim
SENIOR MEDIA EDITOR	Linda Muriello
PRODUCTION SERVICES	Katie Boilard/Pine Tree Composition
COVER DESIGN	David Levy
COVER IMAGE	©Mark Baigent/Alamy

This book was set in 10/12 Janson Text by Laserwords Private Limited, Chennai, India and printed and bound by Courier/Kendallville. The cover was printed by Courier/Kendallville.

This book is printed on acid-free paper. ∞

Copyright © 2009 John Wiley & Sons, Inc. All rights reserved. No part of this publication may be reproduced, stored in a retrieval system or transmitted in any form or by any means, electronic, mechanical, photocopying, recording, scanning or otherwise, except as permitted under Sections 107 or 108 of the 1976 United States Copyright Act, without either the prior written permission of the Publisher, or authorization through payment of the appropriate per-copy fee to the Copyright Clearance Center, Inc., 222 Rosewood Drive, Danvers, MA 01923, website *www.copyright.com*. Requests to the Publisher for permission should be addressed to the Permissions Department, John Wiley & Sons, Inc., 111 River Street, Hoboken, NJ 07030-5774, (201) 748-6011, fax (201) 748-6008, website *www.wiley.com/go/permissions*.

To order books or for customer service, please call 1-800-CALL WILEY (225-5945).

Library of Congress Cataloging-in-Publication Data:

Hopkins, William G.
  Introduction to plant physiology / William G. Hopkins and Norman P. A. Hüner. –4th ed.
    p. cm.
  Includes index.
  ISBN 978-0-470-24766-2 (cloth)
  1. Plant physiology.   I. Hüner, Norman P. A. II. Title.
  QK711.2.H67 2008
  571.2–dc22

                                                2008023261

Printed in the United States of America

10 9 8 7 6 5 4 3

Preface

When the first edition of this text appeared thirteen years ago, its writing was guided by several of objectives.

- The text should be suited for a semester course for undergraduate students encountering the subject of plant physiology for the first time. It was assumed that the student would have completed a first course in botany or biology with a strong botanical component. The book should provide a broad framework for those interested in pursuing advanced study in plant physiology, but it should also provide the general understanding of plant function necessary for students of ecology or agriculture.

- In keeping with the above objective, the text should focus on fundamental principles of how plants work while attempting to balance the demands of biochemistry and molecular biology on the one hand, and traditional "whole-plant" physiology on the other.

- The text should be interesting and readable. It should include some history so the student appreciates how we arrived at our current understanding. It should also point to future directions and challenges in the field.

- The sheer breadth of plant physiology and the rapidly expanding volume of literature in the field make it impossible to include all of the relevant material in an entry-level text. Consequently, the text must be selective and focused on those topics that form the core of the discipline. At the same time, the student should be introduced to the significance of physiology in the role that plants play in the larger world outside the laboratory.

While we have made every effort to retain the readability and overall approach of previous editions; we have also introduced a number of significant changes in this fourth edition. Those changes include:

- For this edition the illustration program has been completely revised. Some figures have been deleted, others have been revised, and many new figures have been introduced. With the help of the publisher, we have also introduced color into the illustrations. The use of color improves the clarity of the figures, draws attention to important elements in the figure, and helps students visualize the relationships between the figure and the concepts described in the text. At the same time, we are mindful of costs and hope that this has been done in a way that does not add significant cost to the student.

- The number of complex chemical structures in many figures has been reduced and biosynthetic pathways have been simplified in order to provide greater emphasis on fundamental principles.

- We have removed the traditional introductory chapter on Cells, Tissues, and Organs and distributed some of this information in chapters to which it pertains directly.

- The list of references at the end of each chapter has been updated throughout the new edition.

- All life depends on energy and water. Unlike previous editions, the fourth edition begins with four chapters that focus on the properties of water, osmosis, water potential, and plant–water relations, followed by a series of eight chapters dealing with bioenergetics, primary plant metabolism, and plant productivity.

- A major change in this edition is the presence of three new chapters (13, 14, and 15). Using the basic information and concepts developed in chapters 1 to 12, these chapters focus on the inherent plasticity of plants to respond to environmental change on various time scales. This includes a discussion of abiotic and biotic stress, plant acclimation to stress, and finally, long-term, heritable adaptations to environmental stress.

- We have revised the treatment of hormones because many instructors have told us that the separate treatment of each hormone fits their syllabus better. The coverage of each hormone concludes with a general description of the current status of receptors and signal transduction pathways.

- A new chapter focuses on the molecular genetics of flower and fruit development.

- A Glossary has been created for the new edition.

William G. Hopkins
Norman P. A. Hüner
London, Ontario
April 2008

To the Student

This is a book about how plants work. It is about the questions that plant physiologists ask and how they go about seeking answers to those questions. Most of all, this book is about how plants do the things they do in their everyday life.

The well-known conservationist John Muir once wrote: *When we try to pick out anything by itself, we find it hitched to everything in the universe.* Muir might well have been referring to the writing of a plant physiology textbook. The scope of plant physiology as a science is very broad, ranging from biophysics and molecular genetics to environmental physiology and agronomy. Photosynthetic metabolism not only provides carbon and energy for the growing plant, but also determines the capacity of the plant to withstand environmental stress. The growth and development of roots, stems, leaves, and flowers are regulated by a host of interacting factors such as light, temperature, hormones, nutrition, and carbon metabolism. As a matter of practical necessity more than scientific reality, we have treated many of these topics in separate chapters. To get the most out of this book, we suggest you be aware of these limitations as you read and think about how various mechanisms are integrated to form a functional plant.

Plant physiology is also a very active field of study and new revelations about how plants work are reported in the literature almost daily. Many models and explanations contained in this book may have been revised by the time the book appears on the market. If you find a particular topic interesting and wish to learn more about it, the listed publications at the end of each chapter are your gateway into the relevant research literature. You can learn what has happened since this book was written by seeking out reviews and opinions published in the more recent editions of those same journals.

In spite of its presumed objectivity, science ultimately relies on the interpretation of experimental results by scientists—interpretations that are often found to be inadequate and filled with uncertainty. However, as results and observations accumulate, interpretations are refined and the degree of uncertainty diminishes. This is the nature of scientific discovery and the source of the real excitement of doing science. In this book, we have attempted to convey some sense of this scientific process.

We hope that, through this book, we are able to share with you some of our own fascination with the excitement, mystery, and challenge of learning about plant physiology.

William G. Hopkins
Norman P. A. Hüner

Contents

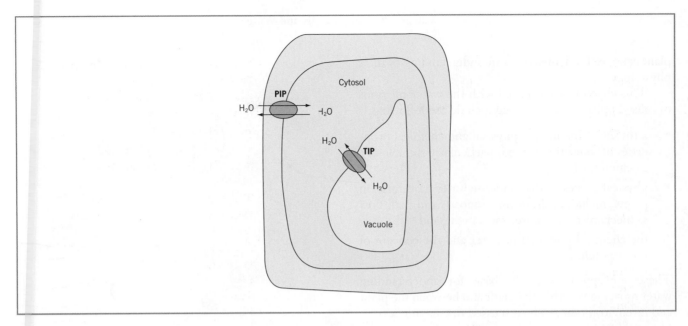

1

Plant Cells and Water

Without water, life as we know it could not exist. Water is the most abundant constituent of most organisms. The actual water content will vary according to tissue and cell type and it is dependent to some extent on environmental and physiological conditions, but water typically accounts for more than 70 percent by weight of non-woody plant parts. The water content of plants is in a continual state of flux, depending on the level of metabolic activity, the water status of the surrounding air and soil, and a host of other factors. Although certain desiccation-tolerant plants may experience water contents of only 20 percent and dry seeds may contain as little as 5 percent water, both are metabolically inactive, and resumption of significant metabolic activity is possible only after the water content has been restored to normal levels.

Water fills a number of important roles in the physiology of plants; roles for which it is uniquely suited because of its physical and chemical properties. The **thermal properties** of water ensure that it is in the liquid state over the range of temperatures at which most biological reactions occur. This is important because most of these reactions can occur only in an aqueous medium. The thermal properties of water also contribute to temperature regulation, helping to ensure that plants do not cool down or heat up too rapidly. Water also has excellent **solvent properties**, making it a suitable medium for the uptake and distribution of mineral nutrients and other solutes required for growth. Many of the **biochemical reactions** that characterize life, such as oxidation, reduction, condensation, and hydrolysis, occur in water and water is itself either a reactant or a product in a large number of those reactions. The **transparency** of water to visible light enables sunlight to penetrate the aqueous medium of cells where it can be used to power photosynthesis or control development.

Water in land plants is part of a very dynamic system. Plants that are actively carrying out photosynthesis experience substantial water loss, largely through evaporation from the leaf surfaces. Equally large quantities of water must therefore be taken up from the soil and moved through the plant in order to satisfy deficiencies that develop in the leaves. For example, it is estimated that the turnover of water in plants due to photosynthesis and transpiration is about 10^{11} tonnes per year.

This constant flow of water through plants is a matter of considerable significance to their growth and survival. The uptake of water by cells generates a pressure known as **turgor**; in the absence of any skeletal system, plants must maintain cell turgor in order to remain erect. As will be shown in later chapters, the uptake of water by cells is also the driving force for cell enlargement. Few plants can survive desiccation. There is no doubt that the water relations of plants and plant cells are fundamental to an understanding of their physiology.

This chapter is concerned with the water relations of cells. Topics to be addressed include the following:

- a review of the unique physical and chemical properties of water that make it particularly suitable as a medium for life,

- physical processes that underlie water movement in plants, including diffusion, osmosis, and bulk flow as mechanisms for water movement, and

- the chemical potential of water and the concept of water potential.

These concepts provide the basis for understanding water movement within the plant and between the plant and its environment, to be discussed in Chapter 2.

1.1 WATER HAS UNIQUE PHYSICAL AND CHEMICAL PROPERTIES

The key to understanding many of the unique properties of water is found in the structure of the water molecule and the strong intermolecular attractions that result from that structure. Water consists of an oxygen atom covalently bonded to two hydrogen atoms (Figure 1.1). The oxygen atom is strongly **electronegative**, which means that it has a tendency to attract electrons. One consequence of this strong electronegativity is that, in the water molecule, the oxygen tends to draw electrons away from the hydrogen. The shared electrons that make up the O—H bond are, on the average, closer to the oxygen nucleus than to hydrogen. As a consequence, the oxygen atom carries a *partial negative charge*, and a corresponding *partial positive charge* is shared between the two hydrogen atoms. This asymmetric electron

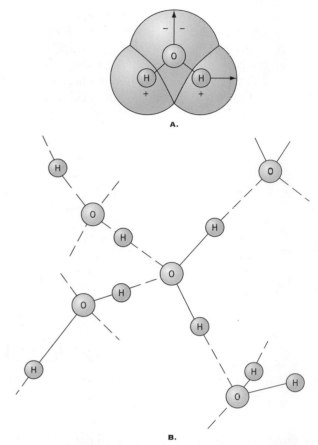

FIGURE 1.1 (*A*) **Schematic structure of a water molecule. (*B*) The hydrogen bond (dashed line) results from the electrostatic attraction between the partial positive charge on one molecule and the partial negative charge on the next.**

distribution makes water a **polar molecule**. Overall, water remains a neutral molecule, but the separation of partial negative and positive charges generates a strong mutual (electrical) attraction between adjacent water

TABLE **1.1**   **Some physical properties of water compared with other molecules of similar molecular size. Because thermal properties are defined on an energy-per-unit mass basis, values are given in units of joules per gram**

	Molecular mass (Da)	Specific heat (J/g/°C)	Melting point (°C)	Heat of fusion (J/g)	Boiling point (°C)	Heat of vaporization (J/g)
Water	18	4.2	0	335	100	2452
Hydrogen sulphide	34	—	−86	70	−61	—
Ammonia	17	5.0	−77	452	−33	1234
Carbon dioxide	44	—	−57	180	−78	301
Methane	16	—	−182	58	−164	556
Ethane	30	—	−183	96	−88	523
Methanol	32	2.6	−94	100	65	1226
Ethanol	46	2.4	−117	109	78	878

molecules or between water and other polar molecules. This attraction is called **hydrogen bonding** (Figure 1.1). The energy of the hydrogen bond is about $20\,kJ\,mol^{-1}$. The hydrogen bond is thus weaker than either covalent or ionic bonds, which typically measure several hundred $kJ\,mol^{-1}$, but stronger than the short-range, transient attractions known as Van der Waals forces (about $4\,kJ\,mol^{-1}$). Hydrogen bonding is largely responsible for the many unique properties of water, compared with other molecules of similar molecular size (Table 1.1).

In addition to interactions between water molecules, hydrogen bonding also accounts for attractions between water and other molecules or surfaces. Hydrogen bonding, for example, is the basis for hydration shells that form around biologically important macromolecules such as proteins, nucleic acids, and carbohydrates. These layers of tightly bound and highly oriented water molecules are often referred to as **bound water**. It has been estimated that bound water may account for as much as 30 percent by weight of hydrated protein molecules. Bound water is important to the stability of protein molecules. Bound water "cushions" protein, preventing the molecules from approaching close enough to form aggregates large enough to precipitate.

Hydrogen bonding, although characteristic of water, is not limited to water. It arises wherever hydrogen is found between electronegative centers. This includes alcohols, which can form hydrogen bonds because of the—OH group, and macromolecules such as proteins and nucleic acids where hydrogen bonds between amino ($-NH_2$) and carbonyl ($\overset{|}{\underset{|}{C}} = O$) groups help to stabilize structure.

1.2 THE THERMAL PROPERTIES OF WATER ARE BIOLOGICALLY IMPORTANT

Perhaps the single most important property of water is that it is a liquid over the range of temperatures most compatible with life. Boiling and melting points are generally related to molecular size, such that changes of state for smaller molecules occur at lower temperatures than for larger molecules. On the basis of size alone, water might be expected to exist primarily in the vapor state at temperatures encountered over most of the earth. However, both the melting and boiling points of water are higher than expected when compared with other molecules of similar size, especially ammonia (NH_3) and methane (CH_4) (Table 1.1). Molecules such as ammonia and the hydrocarbons (methane and ethane) are associated only through weak Van der Waals forces and relatively little energy is required to change their state. Note, however, that the introduction of oxygen raises the boiling points of both methanol (CH_3—OH)

and ethanol ($CH_3\,CH_2\,OH$) to temperatures much closer to that of water. This is because the presence of oxygen introduces polarity and the opportunity to form hydrogen bonds.

1.2.1 WATER EXHIBITS A UNIQUE THERMAL CAPACITY

The term **specific heat**[1] is used to describe the thermal capacity of a substance or the amount of energy that can be absorbed for a given temperature rise. The specific heat of water is $4.184\,J\,g^{-1}\,°C^{-1}$, higher than that of any other substance except liquid ammonia (Table 1.1). Because of its highly ordered structure, liquid water also has a high **thermal conductivity**. This means that it rapidly conducts heat away from the point of application. The combination of high specific heat and thermal conductivity enables water to absorb and redistribute large amounts of heat energy without correspondingly large increases in temperature. For plant tissues that consist largely of water, this property provides for an exceptionally high degree of temperature stability. Localized overheating in a cell due to the heat of biochemical reactions is largely prevented because the heat may be quickly dissipated throughout the cell. In addition, large amounts of heat can be exchanged between cells and their environment without extreme variation in the internal temperature of the cell.

1.2.2 WATER EXHIBITS A HIGH HEAT OF FUSION AND HEAT OF VAPORIZATION

Energy is required to cause changes in the state of any substance, such as from solid to liquid or liquid to gas, without a change in temperature. The energy required to convert a substance from the solid to the liquid state is known as the **heat of fusion**. The heat of fusion for water is $335\,J\,g^{-1}$, which means that $335\,J$ of energy are required to convert 1 gram of ice to 1 gram of liquid water at $0°C$ (Table 1.1). Expressed on a molar basis, the heat of fusion of water is $6.0\,kJ\,mol^{-1}$ (18 g of water per mole $\times$ $335\,J\,g^{-1}$). The heat of fusion of water is one of the highest known, second only to ammonia. The high heat of fusion of water is attributable to the large amount of energy necessary to overcome the

---

[1]Specific heat is defined as the amount of energy required to raise the temperature of one gram of substance by $1°C$ (usually at $20°C$). The specific heat of water is the basis for the definition of a quantity of energy called the *calorie*. The specific heat of water was therefore assigned the value of 1.0 calorie. In accordance with the International System of Units (Système Internationale d'Unites, or SI), the preferred unit for energy is the *joule* (J). 1 calorie = 4.184 joules.

strong intermolecular forces associated with hydrogen bonding.

The density of ice is another important property. At 0°C, the density of ice is less than that of liquid water. Thus water, unlike other substances, reaches its maximum density in the liquid state (near 4°C), rather than as a solid. This occurs because molecules in the liquid state are able to pack more tightly than in the highly ordered crystalline state of ice. Consequently, ice floats on the surface of lakes and ponds rather than sinking to the bottom where it might remain year-round. This is extremely important to the survival of aquatic organisms of all kinds.

Just as hydrogen bonding increases the amount of energy required to melt ice, it also increases the energy required to evaporate water. The **heat of vaporization** of water, or the energy required to convert one mole of liquid water to one mole of water vapor, is about 44 kJ mol^{-1} at 25°C. Because this energy must be absorbed from its surroundings, the heat of vaporization accounts for the pronounced cooling effect associated with evaporation. Evaporation from the moist surface cools the surface because the most energetic molecules escape the surface, leaving behind the lower-energy (hence, cooler) molecules. As a result, plants may undergo substantial heat loss as water evaporates from the surfaces of leaf cells. Such heat loss is an important mechanism for temperature regulation in the leaves of terrestrial plants that are often exposed to intense sunlight.

1.3 WATER IS THE UNIVERSAL SOLVENT

The excellent solvent properties of water are due to the highly polar character of the water molecule. Water has the ability to partially neutralize electrical attractions between charged solute molecules or ions by surrounding the ion or molecule with one or more layers of oriented water molecules, called a **hydration shell**. Hydration shells encourage solvation by reducing the probability that ions can recombine and form crystal structures (Figure 1.2).

The polarity of molecules can be measured by a quantity known as the **dielectric constant**. Water has one of the highest known dielectric constants (Table 1.2). The dielectric constants of alcohols are somewhat lower, and those of nonpolar organic liquids such as benzene and hexane are very low. Water is thus an excellent solvent for charged ions or molecules, which dissolve very poorly in nonpolar organic liquids. Many of the solutes of importance to plants are charged. On the other hand, the low dielectric constants of nonpolar molecules helps to explain why charged solutes do not readily cross the predominantly nonpolar, hydrophobic lipid regions of cellular membranes.

1.4 POLARITY OF WATER MOLECULES RESULTS IN COHESION AND ADHESION

The strong mutual attraction between water molecules resulting from hydrogen bonding is also known as **cohesion**. One consequence of cohesion is that water has an exceptionally high **surface tension**, which is most evident at interfaces between water and air. Surface tension arises because the cohesive force between water molecules is much stronger than interactions between water and air. The result is that water molecules at the surface are constantly being pulled into the bulk water (Figure 1.3). The surface thus tends to contract and behaves much in the manner of an elastic membrane. A

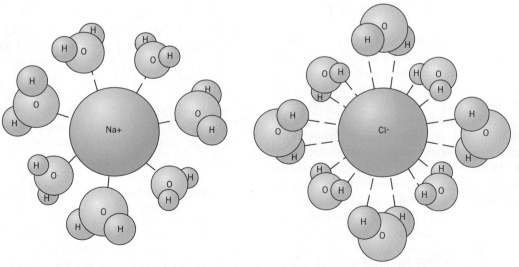

FIGURE 1.2 **Solvent properties of water. The orientation of water molecules around the sodium and chloride ions screens the local electrical fields around each ion. The screening effect reduces the probability of the ions reuniting to form a crystalline structure.**

TABLE 1.2   Dielectric constants for some common solvents at 25°C

Water	78.4
Methanol	33.6
Ethanol	24.3
Benzene	2.3
Hexane	1.9

high surface tension is the reason water drops tend to be spherical or that a water surface will support the weight of small insects.

Cohesion is directly responsible for the unusually high **tensile strength** of water. Tensile strength is the maximum tension that an uninterrupted column of any material can withstand without breaking. High tensile strength is normally associated with metals but, under the appropriate conditions, water columns are also capable of withstanding extraordinarily high tensions—on the order of 30 megapascals (MPa).[2]

The same forces that attract water molecules to each other will also attract water to solid surfaces, a process known as **adhesion**. Adhesion is an important factor in the capillary rise of water in small-diameter conduits.

The combined properties of cohesion, adhesion, and tensile strength help to explain why water rises in capillary tubes and are exceptionally important in maintaining the continuity of water columns in plants. Cohesion, adhesion, and tensile strength will be discussed in greater detail in Chapter 2, when evaporative water loss from plants and water movement in the xylem are examined.

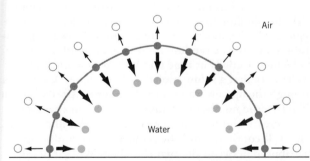

FIGURE 1.3 **Schematic demonstration of surface tension in a water drop. Intermolecular attractions between neighboring water molecules (heavy arrows) are greater than attractions between water and air (light arrows), thus tending to pull water molecules at the surface into the bulk water.**

[2]The pascal (Pa), equal to a force of 1 newton per square meter, is the standard SI unit for pressure.

$$MPa = Pa \times 10^6.$$

1.5 WATER MOVEMENT MAY BE GOVERNED BY DIFFUSION OR BY BULK FLOW

One objective of plant physiology is to understand the dynamics of water as it flows into and out of cells or from the soil, through the plant, into the atmosphere. Movement of substances from one region to another is commonly referred to as **translocation**. Mechanisms for translocation may be classified as either *active* or *passive*, depending on whether metabolic energy is expended in the process. It is sometimes difficult to distinguish between active and passive transport, but the translocation of water is clearly a passive process. Although in the past many scientists argued for an active component, the evidence indicates that water movement in plants may be indirectly dependent upon on expenditure of metabolic energy. Passive movement of most substances can be accounted for by one of two physical processes: either **bulk flow** or **diffusion**. In the case of water, a special case of diffusion known as **osmosis** must also be taken into account.

1.5.1 BULK FLOW IS DRIVEN BY HYDROSTATIC PRESSURE

Movement of materials by *bulk flow* (or *mass flow*) is pressure-driven. Bulk flow occurs when an external force, such as gravity or pressure, is applied. As a result, all of the molecules of the substance move in a mass. Movement of water by bulk flow is a part of our everyday experience. Water in a stream flows in response to the hydrostatic pressure established by gravity. It flows from the faucet in the home or workplace because of pressure generated by gravity acting on standing columns of water in the municipal water tower. Bulk flow also accounts for some water movement in plants, such as through the conducting cells of xylem tissue or the movement of water into roots. In Chapter 9, we discuss how bulk flow is a major component of the most widely accepted hypothesis for transport of solutes through the vascular tissue.

1.5.2 FICK'S FIRST LAW DESCRIBES THE PROCESS OF DIFFUSION

Like bulk flow, *diffusion* is also a part of our everyday experience. When a small amount of sugar is placed in a cup of hot drink, the sweetness soon becomes dispersed throughout the cup. The scent of perfume from a bottle opened in the corner of a room will soon become uniformly distributed throughout the air. If the drink is not stirred and there are no mass movements of air in the room, the distribution of these substances occurs by diffusion. Diffusion can be interpreted as a *directed movement* from a region of a high concentration to a region of lower concentration, but it is accomplished through the *random thermal motion* of individual molecules (Figure 1.4). Thus, while bulk flow is pressure-driven, diffusion is driven principally by concentration differences. Diffusion is a significant factor in the uptake and distribution of water, gases, and solutes throughout the plant. In particular, diffusion is an important factor in the supply of carbon dioxide for photosynthesis as well as the loss of water vapor from leaves.

The process of diffusion was first examined quantitatively by A. Fick. **Fick's first law**, formulated in 1855, forms the basis for the modern-day quantitative description of the process.

$$\mathcal{J} = -D \cdot A \cdot \Delta C \cdot l^{-1} \qquad (1.1)$$

$\mathcal{J}$ is the **flux** or the amount of material crossing a unit area per unit time (for example, mol m^{-2} s^{-1}). D is the **diffusion coefficient**, a proportionality constant that is a function of the diffusing molecule and the medium through which it travels. A and l are the *cross-sectional area* and the *length of the diffusion path*, respectively. The term ΔC represents the difference in concentration between the two regions, also known as the **concentration gradient**. ΔC is the driving force for simple diffusion. In the particular case of gaseous diffusion, it is more convenient to use the difference in density (gm m^{-3}) or vapor pressure (KPa, kilopascal) in place of concentration. The negative sign in Fick's law accounts for the fact that diffusion is toward the *lower* concentration or vapor pressure. In summary, Fick's law tells us that the rate of diffusion is directly proportional to the cross-sectional area of the diffusion path and to the concentration or vapor pressure gradient, and it is inversely proportional to the length of the diffusion path.

1.6 OSMOSIS IS THE DIFFUSION OF WATER ACROSS A SELECTIVELY PERMEABLE MEMBRANE

Fick's law is most readily applicable to the diffusion of solutes and gases. In the general model illustrated in Figure 1.4, for example, the diffusing molecules could be glucose dissolved in water, carbon dioxide dissolved in water, or carbon dioxide in air. We note that the diffusion of solute molecules from chamber A to chamber B from the time t_0 to t_1 does not affect the volume in either chamber, as indicated by no difference in the height of the liquid in the columns of chambers A and B (Figure 1.4). While Fick's law theoretically applies to the diffusion of *solvent* molecules as well, it can at first be difficult to imagine a situation in which diffusion of solvent molecules could occur. Consider what would happen if, for example, water were added to one of the chambers in Figure 1.4. As soon as the water level in the

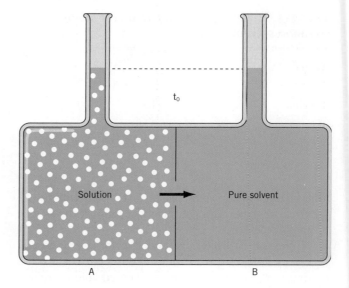

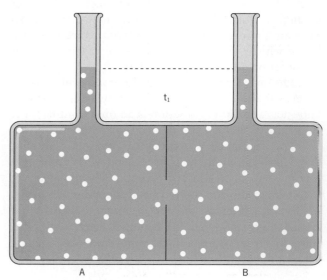

FIGURE 1.4 **Diffusion in solutions is usually associated with the directed movement of a solute molecule from a region of high concentration to a region of lower concentration, due to the random thermal motion of the solute molecules. Initially at time 0 (t_0), there is a much higher probability that a solute molecule in chamber A will pass through the open window into chamber B. After a certain time (t_1), the number of solute molecules in chamber B will increase and the number in chamber A will decrease. This will continue until the molecules are uniformly distributed between the two chambers. At that point, the probability of solute molecules passing between the chambers in either direction will be equal and *net diffusion* will cease. Note, the dotted line indicates that there is no change in volume in either chamber A or B, due to the diffusion of the solute molecule.**

first chamber reached the *open* window, it would flow over into the second chamber—an example of bulk flow.

Alternatively, suppose that we separated chambers A and B by a selectively permeable membrane that allowed solvent molecules to pass freely between the two chambers but not the solute molecules (Figure 1.5). At time t_0, the heights of the liquids in the columns of chambers A and B are the same, indicating comparable volumes. However, after time t_1, the height of the liquid in the column of chamber A has increased while the height of the liquid in the column of chamber B has decreased, indicating an increased volume in chamber A and a decreased volume in chamber B. The difference in the height of the columns (Δh) is a measure of the difference in volume between the two chambers separated by a selectively permeable membrane. The increased volume in chamber A is due to the diffusion of solvent (water) from chamber B to chamber A. Diffusion of water, a process known as *osmosis*, will occur only when the two chambers are separated from one another by a **selectively permeable** membrane. A selectively permeable membrane allows virtually free passage of water and certain small molecules, but restricts the movement of large solute molecules. Thus, all cellular membranes are selectively permeable. Osmosis, then, is simply a special case of diffusion through a selectively permeable membrane.

1.6.1 PLANT CELLS CONTAIN AN ARRAY OF SELECTIVELY PERMEABLE MEMBRANES

Although plants, like all multicellular organisms, exhibit a wide variation in cellular morphology and function, these disparate cells are, in fact, remarkably alike. All cells are built according to a common basic plan and at least start out with the same fundamental structures. In its simplest form, a cell is an aqueous solution of chemicals called **protoplasm** surrounded by a **plasma membrane**. The membrane and the protoplasm it contains are collectively referred to as a **protoplast**. Of course, all of the components that make up protoplasm have important roles to play in the life of a cell, but the plasma membrane is particularly significant because it represents the boundary between the living and non-living worlds. The plasma membrane is also selectively permeable, which means that it allows some materials to pass through but not others. The plasma membrane thus not only physically defines the limits of a cell; it also controls the exchange of material and serves to maintain essential differences between the cell and its environment. The plant protoplast is, in turn, surrounded by a **cell wall**. The cell wall defines the shape of the cell and, through adhesion to the walls of adjacent cells, provides support for the plant as a whole.

In an electron micrograph (an image seen through the electron microscope), membranes are a singularly prominent feature (Figure 1.6). In addition to the plasma membrane, other membranes are found throughout the

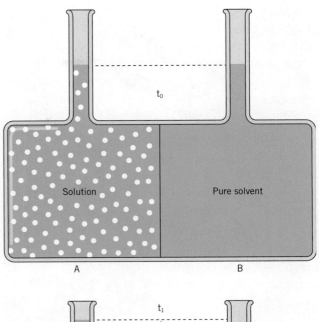

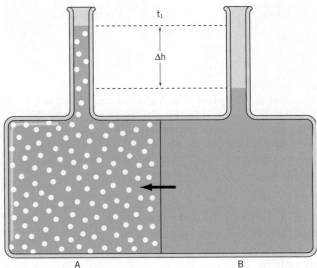

Figure 1.5 Osmosis is the directed movement of the solvent molecule (usually water) across a selectively permeable membrane. Chamber A is separated from chamber B by a selectively permeable membrane. The selectively permeable membrane allows the free movement of the solvent (water) molecules between chambers A and B, but restricts the movement of the solute molecules. At time zero (t_0), all the solute molecules are retained in chamber A and chambers A and B exhibit identical volumes, as indicated by the broken line. After a certain time t_1, all solute molecules are still retained in chamber A, but the volume of chamber A has increased while the volume in chamber B has decreased due to the diffusion of water across the selectively permeable membrane from chamber B to chamber A. This change in volume is represented by Δh.

protoplast where they form a variety of subcellular structures called **organelles** ("little organs"). Organelles serve to compartmentalize major activities within the cell. For example, photosynthesis (Chapters 7 and 8) is localized to the chloroplasts whereas respiration

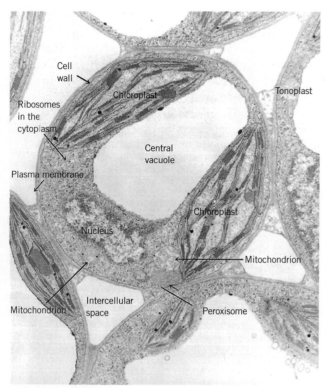

FIGURE 1.6 **The plant cell. A mature mesophyll cell from a *Coleus* leaf, as seen in the electron microscope. Note the prominent, large central vacuole surrounded by the tonoplast, chloroplasts, and mitochondria. (Electron micrograph by Wm. P. Wergin, courtesy of E. H. Newcomb, University of Wisconsin–Madison)**

(Chapter 10) is localized to the mitochondria. The central vacuole and its surrounding tonoplast membrane is critical in the regulation of the osmotic properties of the cytoplasm.

1.6.2 OSMOSIS IN PLANT CELLS IS INDIRECTLY ENERGY DEPENDENT

Water, like any other substance, will only move down an energy gradient—that is, when there is a difference in the energy of water in two parts of a system. In the case of the Figure 1.5, water initially moves from chamber B to chamber A because the energy of pure water in chamber B is greater than the energy of the water in the solution in chamber A. Net movement of water stops when there is no longer an energy gradient across the selectively permeable membrane. Why is the energy of pure water in chamber B of Figure 1.5 greater than the energy of the water in the solution present in chamber A? The energy content of water, like any substance, is most easily described in terms of its chemical potential. **Chemical potential (μ)** is defined as the free energy per mole of that substance and is a measure of the capacity of a substance to react or move. *The rule is that osmosis occurs*

only when there is a difference in the chemical potential $(\Delta\mu)$ *of water on two sides of a selectively permeable membrane.* In other words, osmosis occurs only when the molar free energy of water, that is, the chemical potential of water (μ_w) on one side of a selectively permeable membrane, exceeds the molar free energy or chemical potential of water on the other side of the same selectively permeable membrane.

It appears that the dissolution of a solute in water in some way affects the chemical potential and hence the free energy of the solvent water molecules. Why is this so? This effect is due to the fact that increasing solute concentration in an aqueous solution decreases the mole fraction of water in the solution. The mole fraction of water (X_w) in a solution can be represented as

$$X_w = w/w + s \qquad (1.2)$$

where w represents the moles of H_2O in a given volume and s the moles of solute in the same volume of solution. Thus, as one increases s at a constant value of w, the mole fraction of water, X_w, decreases. Remember that the concentration of pure water is 55.5 moles L^{-1}. The chemical potential of water (μ_w) is related to the mole fraction of water (X_w) according to the following equation

$$\mu_w = \mu_w{}^* + RT \ln X_w \qquad (1.3)$$

where $\mu_w{}^*$ is the chemical potential of pure water under standard temperature and pressure, R is the universal gas constant, and T is the absolute temperature. This equation tells us that as the mole fraction of water decreases, the chemical potential and hence the molar free energy of water decreases. Therefore, the higher the solute concentration of an aqueous solution, the lower the chemical potential of the solvent water. Since energy flows "downhill" spontaneously, water flows or diffuses spontaneously from chamber B to chamber A due to the difference in the chemical potential of water between the two chambers separated by a selectively permeable membrane (Figure 1.5).

Plant cells control the movement of water in and out of cells by altering the solute concentration of the cytosol relative to the solution external to the cell. One way to accomplish this is to regulate transport of ions across the cell membrane into or out of the cell. For example, root cells can take up nitrate ions NO_3^- from the soil by active transport to create a NO_3^- ion gradient across the cell membrane (Figure 1.7) such that

$$(\mu_{nitrate})_{in} > (\mu_{nitrate})_{out} \qquad (1.4)$$

where $\mu_{nitrate}$ is the chemical potential of NO_3^-. Therefore, it follows that

$$\Delta\mu_{nitrate} = (\mu_{nitrate})_{in} - (\mu_{nitrate})_{out} \qquad (1.5)$$

Since $\Delta\mu_{nitrate} > 0$, the uptake of NO_3^- is an active transport process that requires an input of energy. Concomitantly, the increase in cytosolic NO_3^- decreases the mole fraction of water in the cytosol relative to the mole

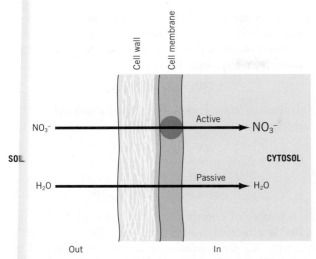

FIGURE 1.7 Water movement in plant cells requires solute gradients. For example, soil nitrate (NO_3^-) is *actively transported* across the selectively permeable cell membrane into plant root cells. Thus, the establishment of a solute gradient such that chemical potential of nitrate is higher in the cytosol than in the soil water requires an input of metabolic energy. The higher concentration of nitrate in the cytosol than the soil water also establishes a gradient for the chemical potential of water such that the chemical potential of the soil water is greater than the chemical potential of cytosolic water. As a consequence, water *diffuses passively* from the soil across the cell membrane into the cell.

fraction of water in the soil water. Therefore, it follows that

$$\Delta\mu_w = (\mu_w)_{in} - (\mu_w)_{out} \qquad (1.6)$$

Since $\Delta\mu_w < 0$, water will diffuse spontaneously into the cell (Figure 1.7).

The chemical potential of water (μ_w) is particularly useful in the study of cellular and plant water relations because it defines the amount of work that can be done by water in one location, for example, a cell or vacuole, compared with pure water at atmospheric pressure and the same temperature. More importantly, a difference in μ_w between two locations means that water is not in equilibrium and there will be a tendency for water to flow toward the location with the lower value. At this point, then, it can be said that the driving force for water movement is a gradient in chemical potential. Clearly, the tendency for water to flow from a location where the μ_w is high to a location where the μ_w is lower will occur spontaneously and does not require energy because $\Delta\mu_w < 0$ (Figure 1.7). However, the only way plant cells can regulate the movement of water is through the establishment of solute concentration gradients across the selectively permeable cell membrane; that is, the movement of water may be coupled to solute transport (Figure 1.7). Although the flow of water is *passive* ($\Delta\mu_w < 0$), the movement of solutes requires

active transport, which is energy dependent ($\Delta\mu_{nitrate} > 0$). Thus, osmosis is indirectly dependent on energy through the requirement of solute concentration gradients that alter the chemical potential of water. Since the cell membrane is permeable only to water (Figure 1.7), only the water diffuses freely across the cell membrane. The process of regulating the water status of a plant cell by the accumulation of solutes is known as **osmotic adjustment**, which is an energy-requiring process.

1.6.3 THE CHEMICAL POTENTIAL OF WATER HAS AN OSMOTIC AS WELL AS A PRESSURE COMPONENT

Osmosis can be easily demonstrated using a device known as an **osmometer**, constructed by closing off the open end of a thistle tube with a selectively permeable membrane (Figure 1.8). If the tube is then filled with a sugar solution and inverted in a volume of pure water, the volume of solution in the tube will increase over time. The increase in volume is due to a net diffusion of water across the membrane into the solution.

The increase in the volume of the solution will continue until the hydrostatic pressure developed in the

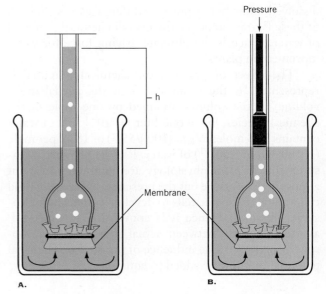

FIGURE 1.8 A demonstration of hydrostatic pressure. A selectively permeable membrane is stretched across the end of a thistle tube containing a sucrose solution and the tube is inverted in a container of pure water. Initially, water will diffuse across the membrane in response to a chemical potential gradient. Diffusion will continue until the force tending to drive water into the tube is balanced by (*A*) the force generated by the hydrostatic head (h) in the tube or (*B*) the pressure applied by the piston. When the two forces are balanced, the system has achieved equilibrium and no further net movement of water will occur.

tube is sufficient to balance the force driving the water into the solution (Figure 1.8A). Alternatively, the tube could be fitted with a piston that would allow us to measure the amount of force required to *just prevent* any increase in the volume of solution (Figure 1.8B). This force, measured in units of pressure (force per unit area), is known as **osmotic pressure** (symbol = π; pi). The magnitude of the osmotic pressure that develops is a function of solute concentration, at least in dilute solutions. It is useful to note that an isolated solution cannot have an osmotic pressure. It has only the *potential* to manifest a pressure when placed in an osmometer. For this reason, we say that the solution has an **osmotic potential** (symbol = Ψ_S). It is convention to define osmotic potential as the negative of the osmotic pressure, since they are equal but opposite forces ($\Psi_S = -\pi$).

Whereas simple diffusion is driven entirely by differences in mole fractions of the solvent, it is apparent from Figure 1.8 that pressure is also a factor in determining both the direction and rate of water movement in an osmometer. Sufficient pressure applied to the piston (Figure 1.8B) will prevent further net movement of water into the solution. If additional pressure were applied, we might expect the net movement of water to reverse its direction and instead flow out of the solution. Thus, osmosis is driven not only by the mole fraction of dissolved solute but by pressure differences as well. Both of these factors influence the overall chemical potential of water, which is the ultimate driving force for water movement in plants.

The effect of pressure on chemical potential is represented by the value **VP**. V is the partial molal volume, or the volume occupied by one mole of the chemical species. Since one liter ($= 10^{-3}$ m^3) of water contains 55.5 moles, $V_w = (1000/55.5)$ or 18 ml per mole ($= 18000$ mm^3 mol^{-1}) of water. P is the pressure. Measurements in plant physiology are commonly made at atmospheric pressure but the presence of relatively rigid cell walls allows plant cells to develop significant hydrostatic pressures. Hence it is convention to express P as the difference between actual pressure and atmospheric pressure. The influence of pressure on chemical potential can now be added to equation 1.3:

$$\mu_w = \mu_w{}^* + RT \ln X_w + V_w P \qquad (1.7)$$

The chemical potential of water may also be influenced by electrical potential and gravitational field. In spite of its strong dipole nature, the net electrical charge for water is zero and so the electrical term can be ignored. This is why we talk about the *chemical* potential of water rather than its *electrochemical* potential. While the gravitational term may be large and must be considered where water movement in tall trees is concerned, it is not significant at the cellular level. Where water

movement involves heights of 5 to 10 meters or less, the gravitational term is commonly omitted.

Since $\mu_w = \mu_w{}^*$ under very dilute conditions, it follows from Equation 1.7 that

$$PV_w = -RT \ln X_w \qquad (1.8)$$

Substituting π for P, we obtain the formal definition of osmotic pressure (Equation 1.10):

$$\pi V_w = -RT \ln X_w \qquad (1.9)$$

or

$$\pi = -RT \ln X_w V_w{}^{-1} \qquad (1.10)$$

As a matter of convenience, the value of X_w for pure water is arbitrarily designated as unity ($= 1$). Consequently, X_w *for a solution is always less than 1.* Thus, equation 1.10 tells us that addition of solute decreases the value of X_w, which in turn increases the osmotic pressure, π. By substituting equation 1.9 in equation 1.7, we can now arrive at an expression relating the chemical potential of water, hydrostatic pressure and osmotic pressure:

$$\mu_w = \mu_w{}^* + V_w(P - \pi) \qquad (1.11)$$

According to equation 1.11, the extent to which the chemical potential of water in a solution (μ_w) differs from that of pure water under conditions of standard temperature and pressure (μ_w^*) is a function of an osmotic component (π) and a pressure component (P).

Osmotic pressure or osmotic potential is one of the four **colligative properties** of a solution. Since the magnitude of the osmotic pressure or osmotic potential exhibited by any solution is dependent on the *number* of particles present in solution, osmotic pressure (π) and therefore osmotic potential (Ψ_S) is a colligative property. Thus, a solution of 0.5 M CaCl$_2$ (Equation 1.12) will exhibit a greater osmotic pressure than a solution of 0.5 M NaCl (Equation 1.13) because the former solution will exhibit twice as many Cl$^-$ ions present as the latter at the same concentration.

$$CaCl_2 \rightarrow Ca^{2+} + 2Cl^- \qquad (1.12)$$
$$NaCl \rightarrow Na^+ + Cl^- \qquad (1.13)$$

The other three colligative properties of solutions are freezing point depression, boiling point elevation, and the lowering of the vapor pressure of the solvent.

Because π is directly proportional to the solute molal concentration (moles of solute in 1 L of solvent), π can be used to calculate the molecular mass of the solute molecule. According to the **Ideal Gas Law**,

$$PV = nRT \qquad (1.14)$$

where P is the pressure, V is the volume, n is the moles of the gaseous molecule, R is the universal gas constant, and T is the absolute temperature. Thus, for an ideal, dilute solution

$$\pi V = nRT \qquad (1.15)$$

or

$$\pi = nRT\,V^{-1} \qquad (1.16)$$

where π is the osmotic pressure of the solution, V is the volume, n is the moles of solute in the solution, R is the universal gas constant, and T is the absolute temperature. Since n = mass (m)/molecular mass (M), it follows that

$$\pi = m\,M^{-1}RT\,V^{-1} \qquad (1.17)$$

Thus, by knowing the mass of the solute (m) dissolved in one liter (V^{-1}), one can measure π experimentally for this solution and subsequently calculate the molecular mass of the solute. Furthermore, if one knows the molecular mass of the solute, equation 1.17 allows one to calculate the osmotic pressure (π) and hence the osmotic potential of a solution. Note that the solute must be perfectly soluble in water in order to use this method to determine either the molecular mass of the solute or the osmotic pressure of the solution.

1.7 HYDROSTATIC PRESSURE AND OSMOTIC PRESSURE ARE TWO COMPONENTS OF WATER POTENTIAL

It is usually more convenient to measure relative values than it is to measure absolute values. The absolute chemical potential of water in solutions is one of those quantities that is not conveniently measured. However, equation 1.11 canbe rearranged as:

$$(\mu_{w} - \mu_{w}{}^{*})V_{w}^{-1} = P - \pi \qquad (1.18)$$

Although the value of $(\mu_{w} - \mu_{w}^{*})$ is more easily measured, the task of plant physiologists was simplified even further when, in 1960, R. O. Slatyer and S. A. Taylor introduced the concept of **water potential** (symbolized by the Greek uppercase psi, Ψ). Water potential is proportional to $(\mu_{w} - \mu_{w}^{*})$ and can be defined as:

$$\Psi = (\mu_{w} - \mu_{w}{}^{*})V_{w}^{-1} = P - \pi \qquad (1.19)$$

or simply:

$$\Psi = P - \pi \qquad (1.20)$$

where P is the hydrostatic pressure and π is the osmotic pressure.

The concept of water potential has been widely accepted by plant physiologists because it avoids the difficulties inherent in measuring chemical potential. Instead, it enables experimenters to predict the behavior of water on the basis of two easily measured quantities, P and π. It also makes it possible to express water potential in units of pressure (pascals), which is more relevant to soil-plant-atmosphere systems than units of energy (joules). This distinction is not trivial. In practice it is far easier to measure pressure changes than it is to measure the energy required to affect water movement. Finally,

we can restate the driving force for water movement as the *water potential gradient*; that is, water will move from a region of high water potential to a region of lower water potential. As we shall see, however, water potentials are usually negative. This means that water moves from a region of less negative water potential to a region where the water potential is more negative.

In the same way that the chemical potential of water in a solution is measured against that of pure water, water potentials of solutions are also measured against a reference. For water potential, the reference state is arbitrarily taken as pure water at atmospheric pressure. Under these conditions there is neither hydrostatic pressure nor dissolved solutes; that is, both P and π are zero. According to equation 1.20, the value of Ψ for pure water is therefore also zero.

1.8 WATER POTENTIAL IS THE SUM OF ITS COMPONENT POTENTIALS

Water potential may be also be defined as the sum of its component potentials:[3]

$$\Psi = \Psi_{P} + \Psi_{S} \qquad (1.21)$$

The symbol Ψ_{P} represents the **pressure potential**. It is identical to P and represents the hydrostatic pressure in excess of ambient atmospheric pressure. The term Ψ_{S} represents the *osmotic potential*. Note the change in sign ($\pi = -\Psi_{S}$). As pointed out earlier, osmotic potential is equal to osmotic pressure but carries a negative sign. Osmotic potential is also called **solute potential** (hence the designation Ψ_{S}) because it is the contribution due to dissolved solute. The term *osmotic* (or *solute*) *potential* is preferred over *osmotic pressure* because it is more properly a property of the solution.

We can see from equation 1.21 that hydrostatic pressure and osmotic potential are the principal factors contributing to water potential. A third component, the **matric potential** (M), is often included in the equation for water potential. Matric potential is a result of the adsorption of water to solid surfaces. It is particularly important in the early stages of water uptake by dry seeds (called **imbibition**) and when considering water held in soils (Chapter 3). There is also a matric component in cells, but its contribution to water potential is relatively small compared with solute component. It is also difficult to distinguish the matric component from osmotic potential. Consequently,

---

[3] Students reading further in the literature will find that a variety of conventions, names, and symbols have been used to describe the components of water potential. These differences are for the most part superficial, but careful reading is required to avoid confusion.

matric potential may be excluded for purposes of the present discussion. We will return to matric potential when we discuss soil water in Chapter 3.

Returning to equation 1.21, we can see that an increase in hydrostatic pressure or osmotic potential will increase water potential while a decrease in the pressure or osmotic potential (more negative) lowers it. We can use these changes to explain what happened earlier in our example of the osmometer (Figure 1.8). The dissolved sucrose generated an osmotic potential in the thistle tube, thereby lowering the water potential of the solution ($\Psi < 0$) compared with the pure water ($\Psi = 0$) on the other side of the membrane. Water thus diffused across the membrane into the solution. As the volume of the solution increased, a hydrostatic pressure developed in the thistle tube. When the positive hydrostatic pressure was sufficient to offset the negative osmotic potential, the water potential of the solution was reduced to zero. At that point $\Psi = 0$ on both sides of the membrane and there was no further *net* movement of water. It is also interesting to note that where volume in the osmometer is permitted to increase, the osmotic potential will decrease. This is because along with the volume change accompanying diffusion of water into the solution, the mole fraction of water would also increase. In effect, the solute concentration decreases due to dilution. At equilibrium, then, the osmotic potential of the solution would be higher (i.e., less negative) than at the beginning. In the end the pressure required to balance osmotic potential is less than what would have been required had the volume increase not occurred. Because the contribution of Ψ_S to water potential is always negative, water will, at constant pressure, always move from the solution with the higher (less negative) osmotic potential to the solution with the lower (more negative) osmotic potential.

We can now ask what contributes to the osmotic and pressure potentials, and thus water potential, in plant cells. The osmotic potential of most plant cells is due primarily to the contents of the large central vacuole. With the exception of meristematic and certain other highly specialized cells, cell vacuoles contain on the order of 50 to 80 percent of the cellular water and a variety of dissolved solutes. These may include sugars, inorganic salts, organic acids, and anthocyanin pigments. Most of the remaining cellular water is located in the cell wall spaces, while the cytoplasm accounts for as little as 5 to 10 percent. Methods for determining the osmotic potential of cells and tissues do not generally discriminate between the cytoplasmic and vacuolar contributions—the result is an average of the two. The osmotic potential of a parenchyma cell is typically in the range of -0.1 to -0.3 MPa, the largest part of which is due to dissolved salts in the vacuole.

In a laboratory osmometer, pressure (Ψ_P) can be estimated as the difference between atmospheric pressure (0.1 MPa) and the hydrostatic pressure generated by the height of the water column. In cells, the pressure component arises from the force exerted outwardly against the cell walls by the expanding protoplast. This is known as **turgor pressure**. An equal but opposite inward pressure, called **wall pressure**, is exerted by the cell wall. A cell experiencing turgor pressure is said to be **turgid**. A cell that experiences water loss to the point where turgor pressure is reduced to zero is said to be **flaccid**. Instruments are available for measuring P directly in large algal cells, but in higher plants it is usually calculated as the difference between water potential and osmotic potential. In nonwoody herbaceous plants, turgor pressure is almost solely responsible for maintaining an erect habit. Indeed, one of the first outward signs of water deficit in plants is the wilting of leaves due to loss of turgor in the leaf cells.

1.9 DYNAMIC FLUX OF H₂O IS ASSOCIATED WITH CHANGES IN WATER POTENTIAL

The water status of plant cells is constantly changing as the cells adjust to fluctuations in the water content of the environment or to changes in metabolic state. **Incipient plasmolysis** is the condition in which the protoplast just fills the cell volume. At incipient plasmolysis, the protoplast exerts no pressure against the wall but neither is it withdrawn from the wall. Consequently, turgor pressure (Ψ_P) is zero and the water potential of the cell (Ψ_{cell}) is equal to its osmotic potential (Ψ_S). When the cell is bathed by a **hypotonic**[4] solution such as pure water ($\Psi = 0$), water will enter the cell as it moves down the water potential gradient. This causes simultaneously a small dilution of the vacuolar contents (with a corresponding increase in osmotic potential) and the generation of a turgor pressure. Net movement of water into the cell will cease when the osmotic potential of the cell is balanced by its turgor pressure and, by equation 1.21, the water potential of the cell is therefore also zero. When the cell is bathed by a **hypertonic** solution, which has a more negative osmotic potential than the cell, the water potential gradient favors loss of water from the cell. The protoplast then shrinks away from the cell wall, a condition known as **plasmolysis**. Continued removal of water concentrates the vacuolar contents, further lowering the osmotic potential. Turgor

---

[4]A solution with a lower solute content than a cell or another solution and, hence, less negative osmotic potential, is referred to as *hypotonic*. A *hypertonic* solution has a higher solute content and more negative osmotic potential. A solution with an equivalent osmotic potential is known as **isotonic**.

presure remains at zero and the water potential of the cell is determined solely by its osmotic potential. In either situation described above, the water potential of the cell is determined as the algebraic sum of the turgor pressure and osmotic potential (Equation 1.21).

In addition to water movement between cells and their environment, diffusion down a water potential gradient can also account for water movement between cells (Figure 1.9). Individual cells in a series may experience different values for Ψ_S and Ψ_P, depending on the specific circumstances of each cell. Nonetheless, water will flow through the series of cells so long as a continuous gradient in water potential is maintained.

The phenomena of plasmolysis and wilting are superficially the same, but there are some important differences. Plasmolysis can be studied in the laboratory simply by subjecting tissues to hypertonic solutions and observing protoplast volume changes under the microscope. As plasmolysis progresses, protoplast volume progressively decreases, and the protoplast pulls away from the cell wall. The void between the outer protoplast surface (the plasma membrane) and the cell wall will become filled with external solution, which readily penetrates the cell wall. For this reason, plasmolysis does not normally give rise to a significant negative pressure (or tension) on the protoplast. Plasmolysis remains essentially a laboratory phenomenon and, with the possible exception of conditions of extreme water stress or saline environments, seldom occurs in nature. Wilting, on the other hand, is the typical response to dehydration in air under natural conditions. Because of its extreme surface tension, water in the small pores of the cell wall resists the entry of air and the collapsing protoplast maintains contact with the cell wall. This tends to pull the wall inward and substantial negative pressures may develop. The water potential of wilted cells becomes even more negative as it is the sum of the *negative* osmotic potential plus the *negative* pressure potential.

1.10 AQUAPORINS FACILITATE THE CELLULAR MOVEMENT OF WATER

Porins are a class of membrane proteins that belong to a large family of proteins called **major intrinsic proteins (MIPs)** that are found in the cell membranes of all living organisms including plants, microorganisms, and animals. Porin-type channels are nonselective cation channels which are characterized by a β-pleated sheet protein structure. In plants, porins are generally restricted to the outer membranes of mitochondria and chloroplasts and account for their highly permeable properties. In contrast, **aquaporins** are membrane protein channels or pores controlling the selective movement of water primarily. This is indicated by the fact that the presence of aquaporins does not affect the electrical conductance of a membrane which indicates that small ions such as H^+ are not conducted by these membrane channels. The results of genome sequencing indicate that *Arabidopsis thaliana* exhibits 35 aquaporin gene homologs and 33 homologues have been detected in the rice genome. Thus, plants exhibit a multiplicity of aquaporin isoforms which, in part, reflects the multiplicity of internal membrane types in which these channels are localized within a plant cell (Figure 1.10A). For example, 13 homologues have been detected in plant plasma membranes of Arabidopsis and are designated PIPs for *p*lasma membrane *i*ntrinsic *p*roteins whereas the tonoplast membrane, which surrounds the inner vacuole of plant cells, exhibits 10 homologues and are designated TIPs.

The structure of aquaporins is highly conserved between plants, microbes, and animals. Four subunits (tetramers) associate to form a single plant aquaporin. Each subunit has a molecular mass of 23 to 31 kDa and spans the membrane with six α-helices joined by

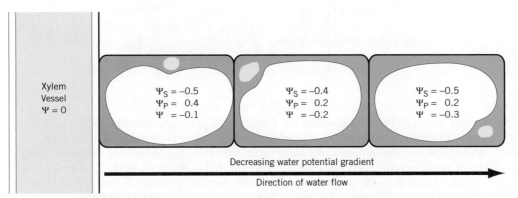

Xylem Vessel $\Psi = 0$

$\Psi_S = -0.5$
$\Psi_P = 0.4$
$\Psi = -0.1$

$\Psi_S = -0.4$
$\Psi_P = 0.2$
$\Psi = -0.2$

$\Psi_S = -0.5$
$\Psi_P = 0.2$
$\Psi = -0.3$

Decreasing water potential gradient

Direction of water flow

FIGURE 1.9 **Diagram illustrating the contributions of osmotic potential (Ψ_S), turgor pressure (Ψ_p), and water potential (Ψ) to water movement between cells. The direction of water movement is determined solely by the value of the water potential in adjacent cells.**

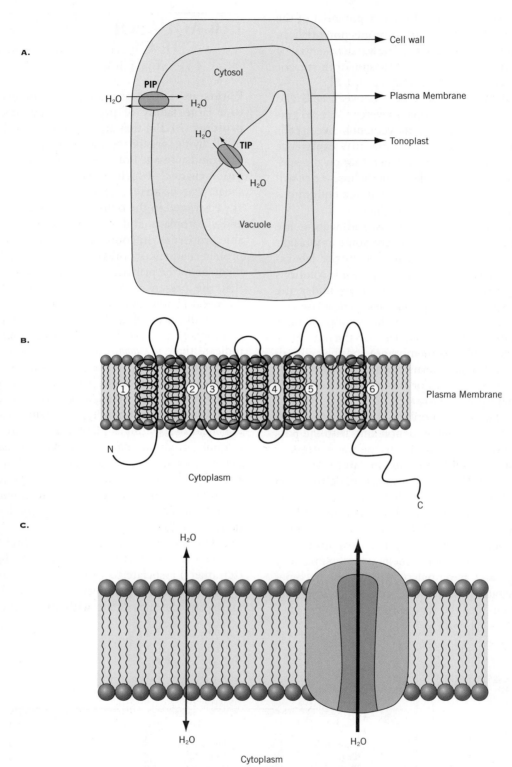

A.

Cell wall

Cytosol

PIP

H_2O → → H_2O

H_2O

TIP

H_2O

Vacuole

Plasma Membrane

Tonoplast

B.

N

Cytoplasm

① ② ③ ④ ⑤ ⑥

C

Plasma Membrane

C.

H_2O

H_2O

H_2O

Cytoplasm

FIGURE 1.10 (*A*) **Plant aquaporins are found in the plasma membrane (PIPs) as well as the tonoplast membrane (TIPs) of the vacuole to regulate cellular water flow.** (*B*) **Each of the four subunits of a plant aquaporin spans the membrane six times. Both the N-terminus (N) and C-terminus (C) are located in the cytoplasm.** (*C*) **Aquaporins enhance water flow (thick arrow) and because they are gated allow the cell to regulate cellular flow compared to water flow through pure lipid bilayers (thin arrow). The interior of the aquaporin channel contains hydrophilic amino acids, which interact with water, whereas the exterior of the protein channel consists of hydrophobic amino acids, which interact with the lipid fatty acids of the membrane bilayer.**

5 intervening loops (Figure 1.10B) in contrast to the β-pleated sheet structure of porins. These protein subunits fold within the membrane such that the hydrophobic amino acids are on the outer side of the pore and interact with the hydrophobic fatty acids of the lipid bilayer whereas the hydrophilic amino acids are in the inner side of the pore and interact with water molecules as they move through the pore from one side of the membrane to the other. Aquaporins can be open or closed to regulate the movement of water across the membrane. **Gating** is the term used to describe this regulated opening and closing of these protein channels. Gating through PIPs can be controlled by cytoplasmic pH, the concentration of divalent cations such as Ca^{2+} as well as by aquaporin protein phosphorylation.

As discussed above, lipid bilayers are quite permeable to water. Thus, there are two possible pathways for the movement of water across a membrane—one pathway through the lipid bilayer itself and the second pathway through an aquaporin (Figure 1.10C). If this is so, why does a cell require any aquaporins at all? The results of recent experiments clearly indicate that the rate of water movement through a lipid bilayer with aquaporins is faster than a membrane that contains lipids only. Thus, the presence of aquaporins provides a low resistance pathway for the movement of water across a membrane. Furthermore, since aquaporins are gated, this provides greater control for the movement of water intracellularly as well as intercellularly. For example, the water permeability of the tonoplast is two orders of magnitude (10^2 times) greater than that of the plasma membrane. Thus, this allows the vacuole to replenish or buffer the cytoplasm with water when the cell is exposed to hypertonic conditions. Thus, aquaporins are important in regulating the osmotic properties of plant cells. This process is called **osmoregulation.**

1.11 TWO-COMPONENT SENSING/SIGNALLING SYSTEMS ARE INVOLVED IN OSMOREGULATION

Plants, green algae, fungi, as well as prokaryotic organisms (bacteria, cyanobacteria) have cell walls and thus are sensitive to turgor pressure. These organisms must be initially able to sense and then subsequently respond accordingly at the physiological, biochemical, and genetic level in order to ensure daily survival as well as seasonal changes in their aqueous environment. These organisms have evolved sophisticated sensing and signaling strategies to respond to changes in their abiotic environment. This strategy generally employs, first, a **sensing mechanism**, which can usually detect changes in specific environmental parameters (e.g., temperature, pressure, light quality, irradiance). A change

in an environmental parameter will usually activate a specific sensor that will generate a specific intracellular signal. Second, the sensor is usually linked to or coupled to a **transducing mechanism**, which propagates the signal generated by the sensor and subsequently elicits a specific cellular response. **Two-component systems** are examples of such sensory signaling mechanisms and were first discovered in prokaryotes. All two-component systems consist of a transmembrane **sensor protein** located in the cell membrane and a cytoplasmic protein called the **response regulator** (Figure 1.11). The sensor protein detects an environmental signal and transmits this signal to the response regulator located in the cytoplasm. The response regulator mediates the cellular response to the external signal by DNA binding or other regulatory functions that provide transcriptional control over one or more target genes. Thus, response regulators are typically DNA-binding proteins or transcriptional factors.

An **osmosensor** is a device that is able either to detect changes in the chemical potential of extracellular water ($\Delta\mu_{H2O}$) directly or to detect the mechanical perturbations to the cell membrane as a consequence of changes in turgor pressure due to the differences in osmotic potential (Ψ_S) between the external aqueous environment and the cytosol. In the heterotrophic bacterium, *Escherichia coli*, a two-component system made up of a sensor, **EnvZ**, and a response regulator, **OmpR**, enable this bacterium to sense changes in external osmolarity. EnvZ and OmpR form an osmosensing/signal transduction pathway that responds to ($\Delta\mu_{H2O}$) by regulating the relative transcription of genes encoding two pore proteins, **OmpF** and **OmpC**. Both proteins form pores in the outer membrane that control cell membrane permeability. Although the total amount of both OmpF and OmpC remains relatively constant in response to $\Delta\mu_{H2O}$, the relative proportions of OmpF : OmpC changes with changes in $\Delta\mu_{H2O}$. Under conditions of low external osmolarity the expression of the larger OmpF pores are favored, whereas under high external osmolarity the expression of the smaller OmpC pores are favored. As a consequence, the permeability of the bacterial cell membrane is altered.

Several two-component sensing/signal transduction pathways have been reported in both plants and fungi. In the yeast, *Saccharomyces cerevisiae*, the histidine kinase **SLN1** acts as a transmembrane osmosensor. The involvement of a two-component, histidine kinase osmosensing system (**ATHK1**) has been described recently in *Arabidopsis thaliana*. Transforming yeast cells, in which the *SLN1* gene was inactivated, with the plant *ATHK1* gene restored the ability of the yeast cells to sense and transduce a signal of changes in external osmolarity ($\Delta\mu_{H2O}$). The accumulation of *ATHK1* mRNA in *Arabidopsis* is tissue specific with highest levels observed in root tissue. Furthermore, the high

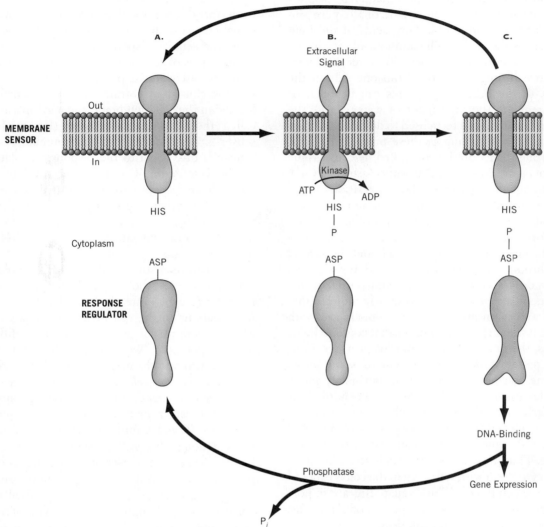

FIGURE 1.11 **A general model illustrating two-component sensory signaling.** (*A*) Two-component systems contain a transmembrane-bound protein sensor that exhibits an extracellular and intracellular domain. The intracellular domain exhibits a histidine kinase enzyme activity, which phosphorylates a specific histidine residue (HIS) within the intracellular domain of the sensor. The intracellular domain of the sensor is capable of interacting with a cytoplasmic response regulator protein, which may be a transcriptional activator and thus has DNA-binding activity. (*B*) The stimulation of the transmembrane sensor by an extracellular signal causes a conformational change in the extracellular domain of the membrane sensor, which activates the histidine kinase activity to autophosphorylate the intracellular domain of the sensor using ATP. (*C*) Subsequently, the sensory information flows from a membrane-bound sensor to a response regulator located in the cytoplasm. The phosphate group on the sensor histidine is transferred to an aspartyl group (ASP) on the response regulator. This relaxes the conformational of the transmembrane sensor but activates the DNA-binding activity of the response regulator. The binding of the phosphorylated response regulator to DNA leads to an alteration in the expression of a specific gene or gene families in response to the extracellular signal. The response regulator also exhibits protein phosphatase activity which dephosphorylates the ASP group of the response regulator and inhibits its DNA-binding activity.

external salt concentrations stimulate the expression of *ATHK1* in the roots of *Arabidopsis*. The high level of expression of *ATHK1* in root tissue combined with its sensitivity to external salt concentrations suggest that the histidine kinase *ATHK1*, is part of a two-component signal transduction system efficient in osmosensing.

SUMMARY

Water has numerous chemical and physical properties that make it particularly suitable as a medium in which life can occur. Most of these properties are the result of the tendency of water molecules to form hydrogen bonds. At the cellular level, water moves primarily by osmosis, in response to a chemical potential gradient across a selectively permeable membrane. The movement of water can be predicted on the basis of water potential. Water potential is a particularly useful concept because it can be calculated from two readily measured quantities: pressure and osmotic potential. Plants derive mechanical support from the turgidity of cells, due at least in part to the high structural strength of cell walls. Aquaporins are gated, membrane protein channels that regulate the permeability of cell membranes to water. Two-component sensing/signalling mechanisms play an important role in the ability of plant cells to osmoregulate in reponse to changes in external water potential.

CHAPTER REVIEW

1. What is a hydrogen bond and how does it help to explain many of the unique physical and chemical properties of water?

2. Describe osmosis as a special case of diffusion. Distinguish between osmotic *pressure* and osmotic *potential*.

3. Understand the concept of water potential and how it is related to the chemical potential of water. In what way does the concept of water potential help the plant physiologist explain water movement?

4. Explain why osmosis is, indirectly, an energy-dependent process in plants.

5. Show how one can use osmotic potential to determine molecular mass.

6. Can you suggest an important role for turgor in the plant?

7. Is it osmotic potential or turgor pressure that has the more significant role in regulating water potential of plant cells?

8. Estimate the values of Ψ_{cell}, Ψ_S, and Ψ_P for a tissue that neither gains nor loses weight when equilibrated with a 0.4 molal mannitol solution and in which, when placed in a 0.6 molal mannitol solution, 50 percent of the cells are plasmolyzed.

9. What is an aquaporin? What role do they play in cell membrane permeability to water?

10. What is an osmosensor? Describe how a two-component sensing/signalling system contributes to osmosensing.

FURTHER READING

Borstlap, A. C. 2002. Early diversification of plant aquaporins. *Trends in Plant Science* 7: 529–530.

Buchanan, B. B., W. Gruissem, R. L. Jones. 2000. *Biochemistry and Molecular Biology of Plants*. American Society of Plant Physiologists Rockville, Maryland.

Evert, R. F. 2006. *Esau's Plant Anatomy*. Hoboken, NJ: John Wiley & Sons, Inc.

Hoch, J. A., T. J. Silhavy. 1995. *Two-Component Signal Transduction*. Washington, D. C. ASM Press.

Maurel, C., L. Verdoucq, D.-T. Luu, V. Santoni. 2008. Plant aquaporins: membrane channels with multiple integrated functions. *Annual Review of Plant Biology* 59: 595–624.

Nobel, P. S. 2005. *Physicochemical and Environmental Plant Physiology*. Elsevier Science & Technology Burlington, MA.

Pollack, G. H. 2001. *Cells, Gels and the Engines of Life*. Seattle: Ebner and Sons Publishers.

Urao, T., K. Yamaguchi-Shinozaki, K. Shinozaki. 2000. Two-component systems in plant signal transduction. *Trends in Plant Science* 5: 67–74.

Wood, J. M. 1999. Osmosensing by bacteria: Signals and membrane-based sensors. *Microbiology and Molecular Biology Reviews* 63: 230–262.

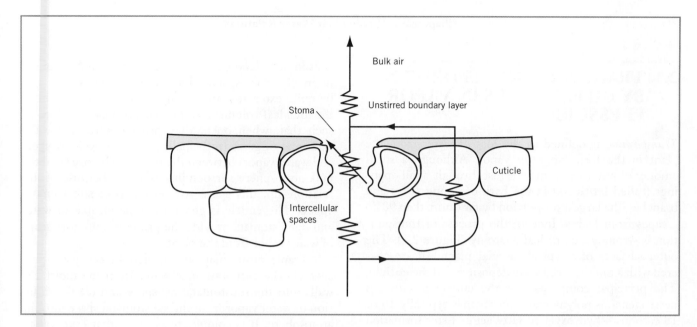

2

Whole Plant Water Relations

The dominant process in water relations of the whole plant is the absorption of large quantities of water from the soil, its translocation through the plant, and its eventual loss to the surrounding atmosphere as water vapor. Of all the water absorbed by plants, less than 5 percent is actually retained for growth and even less is used biochemically. The balance passes through the plant to be lost as water vapor, a phenomenon known as **transpiration**. Nowhere is transpiration more evident than in crop plants, where several hundred kilograms of water may be required to produce each kilogram of dry matter and excessive transpiration can lead to significant reductions in productivity.

The quantitative importance of transpiration has been indicated by a variety of studies over the years. In his classic 1938 physiology textbook, E. C. Miller reported that a single maize plant might transpire as much as 200 liters of water over its lifetime— approximately 100 times its own body weight. Extrapolated to a field of maize plants, this volume of water is sufficient to cover the field to a depth of 38 cm over the course of a growing season. A single, 14.5 m open-grown silver maple tree may lose as much as 225 liters of water per hour. In a deciduous forest, such as that found in the southern Appalachians of the United States, one-third of the annual precipitation will be absorbed by plants, only to be returned to the atmosphere as vapor.

Whether there is any positive advantage to be gained by transpiration is a point for discussion, but the potential for such massive amounts of water loss clearly has profound implications for the growth, productivity, and even survival of plants. Were it not for transpiration, for example, a single rainfall might well provide sufficient water to grow a crop. As it is, the failure of plants to grow because of water deficits produced by transpiration is a principal cause of economic loss and crop failure across the world. Thus, on both theoretical and practical grounds, transpiration is without doubt a process of considerable importance.

This chapter will examine the phenomena of transpiration and water movement through plants. The principal topics to be addressed include

- the process of transpiration and the role of vapor pressure differences in directing the exchange of water between leaves and the atmosphere,

- the role of environmental factors, in particular temperature and humidity, in regulating the rate of transpirational water loss,

- the anatomy of the water-conducting system in plants, and how plants are able to maintain standing columns of water to the height of the tallest trees, and

- water in the soil and how water is taken up by roots to meet the demands of water loss at the other end.

2.1 TRANSPIRATION IS DRIVEN BY DIFFERENCES IN VAPOR PRESSURE

Transpiration is defined as the loss of water from the plant in the form of water vapor. Although a small amount of water vapor may be lost through small openings (called **lenticels**) in the bark of young twigs and branches, the largest proportion by far (more than 90%) escapes from leaves. Indeed, the process of transpiration is strongly tied to leaf anatomy (Figure 2.1). The outer surfaces of a typical vascular plant leaf are covered with a multilayered waxy deposit called the **cuticle**. The principal component of the cuticle is **cutin**, a heterogeneous polymer of long-chain—typically 16 or 18 carbons—hydroxylated fatty acids. Ester formation between the hydroxyl and carboxyl groups of neighboring fatty acids forms cross-links, establishing an extensive polymeric network.

The cutin network is embedded in a matrix of cuticular **waxes**, which are complex mixtures of long-chain (up to 37 carbon atoms) saturated hydrocarbons, alcohols, aldehydes, and ketones. Because cuticular waxes are very hydrophobic, they offer extremely high resistance to diffusion of both liquid water and water vapor from the underlying cells. The cuticle thus serves to restrict evaporation of water directly from the outer surfaces of leaf epidermal cells and protects both the epidermal and underlying mesophyll cells from potentially lethal desiccation.

The integrity of the epidermis and the overlying cuticle is occasionally interrupted by small pores called **stomata** (sing. **stoma**). Each pore is surrounded by a pair of specialized cells, called **guard cells**. These guard cells function as hydraulically operated valves that control the size of the pore (Chapter 8). The interior of the leaf is comprised of photosynthetic **mesophyll** cells. The somewhat loose arrangement of mesophyll

cells in most leaves creates an interconnected system of intercellular air spaces. This system of air spaces may be quite extensive, accounting for up to 70 percent of the total leaf volume in some cases. Stomata are located such that, when open, they provide a route for the exchange of gases (principally carbon dioxide, oxygen, and water vapor) between the internal air space and the bulk atmosphere surrounding the leaf. Because of this relationship, this space is referred to as **substomatal space**. The cuticle is generally impermeable to water and open stomata provide the primary route for escape of water vapor from the plant.

Transpiration may be considered a two-stage process: (1) the evaporation of water from the moist cell walls into the substomatal air space and (2) the diffusion of water vapor from the substomatal space into the atmosphere. It is commonly assumed that evaporation occurs primarily at the surfaces of those mesophyll cells that border the substomatal air spaces. However, several investigators have proposed a more restricted view, suggesting instead that most of the water evaporates from the inner surfaces of epidermal cells in the immediate vicinity of the stomata. Known as **peristomal evaporation**, this view is based on numerous reports indicating the presence of cuticle layers on *mesophyll cell walls*. In addition, mathematical modeling of diffusion in substomatal cavities has predicted that as much as 75 percent of all evaporation occurs in the immediate vicinity of the stomata. The importance of peristomal evaporation versus evaporation from mesophyll surfaces generally remains to be established by direct experiment. Whether the evaporation occurs principally at the mesophyll or epidermal cell surfaces is an interesting problem, reminding us that physiological processes are often not as straightforward as they may first appear.

The diffusion of water vapor from the substomatal space into the atmosphere is relatively straightforward. Once the water vapor has left the cell surfaces, it diffuses through the substomatal space and exits the leaf through the stomatal pore. Diffusion of water vapor through the stomatal pores, known as **stomatal transpiration**, accounts for 90 to 95 percent of the water loss from leaves. The remaining 5 to 10 percent is accounted for by **cuticular transpiration**. Although the cuticle is composed of waxes and other hydrophobic substances and is generally impermeable to water, small quantities of water vapor can pass through. The contribution of cuticular transpiration to leaf water loss varies considerably between species. It is to some extent dependent on the thickness of the cuticle. Thicker cuticles are characteristic of plants growing in full sun or dry habitats, while cuticles are generally thinner on the leaves of plants growing in shaded or moist habitats. Cuticular transpiration may become more significant, particularly for leaves with thin cuticles, under dry conditions when stomatal transpiration is prevented by closure of the stomata.

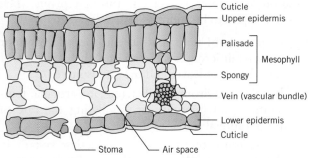

FIGURE 2.1 Diagrammatic representation of a typical mesomorphic leaf (*Acer* sp.) shown in cross-section. Note especially the presence of a cuticle covering the outer surfaces of both the upper and lower epidermis. Note also the extensive intercellular spaces with access to the ambient air through the open stomata.

2.2 THE DRIVING FORCE OF TRANSPIRATION IS DIFFERENCES IN VAPOR PRESSURE

In Chapter 1, it was shown that water movement is determined by differences in water potential. It then can be assumed that the driving force for transpiration is the difference in water potential between the substomatal air space and the external atmosphere. However, because the problem is now concerned with the diffusion of water vapor rather than liquid water, it will be more convenient to think in terms of vapor systems. Consider what happens, for example, when a volume of pure water is introduced into a closed chamber (Figure 2.2). Initially the more energetic water molecules will escape into the air space, filling that space with water vapor. Some of those water molecules will then begin to condense into the liquid phase. Eventually water in the chamber will reach a dynamic equilibrium; the rate of evaporation will be balanced by the rate of condensation. The air space will then contain the maximum amount of water vapor that it can hold at that temperature. In other words, *at equilibrium the gas phase will be saturated with water vapor*. The concentration of water molecules in a vapor phase may be expressed as the vapor mass per unit volume (g m^{-3}), called **vapor density**. Alternatively, the concentration may be expressed in terms of the pressure exerted by the water vapor molecules against the fluid

surface and walls of the chamber. This is called **vapor pressure** (symbol = e). With an appropriate equation, vapor density and vapor pressure are interconvertible. However, because we are now accustomed to dealing with the components of water potential in pressure units, it will be more consistent for us to use vapor pressure (expressed as kilopascals, kPa) in our discussion. We can then say that when a gas phase has reached equilibrium and is saturated with water vapor, the system will have achieved its **saturation vapor pressure**.

The vapor pressure over a solution at atmospheric pressure is influenced by both solute concentration and temperature. As was previously discussed with respect to water potential (Chapter 1), the effect of solute concentration on vapor pressure may be expressed in terms of the mole fraction of water molecules. This relationship is given by a form of Raoult's law, which states:

$$e = X_i \, e^o \qquad (2.1)$$

where e is vapor pressure of the solution, X_i is the mole fraction of water (= number of water molecules/number of water molecules + number of solute molecules), and e^o is the saturation vapor pressure over pure solvent.

The actual reduction in vapor pressure due to solute turns out to be quite small. This is because even in relatively concentrated solutions the mole fraction of solvent remains large. Consider, for example, a 0.5 molal solution, which is approximately the concentration of vacuolar sap in a typical plant cell. A 0.5 molal solution

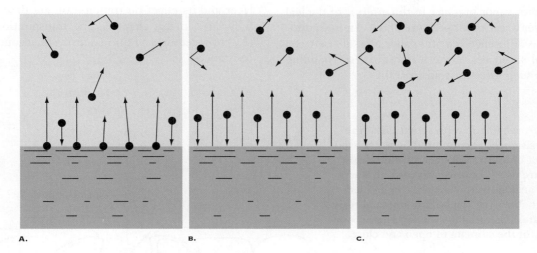

A.   B.   C.

FIGURE 2.2 **Vapor pressure in a closed container. Initially (*A*), more molecules escape from the water surface than condense, filling the air space with water vapor molecules. The vaporous water molecules exert pressure—vapor pressure—against the walls of the chamber and the water surface. At equilibrium (*B*) the rate of condensation equals evaporation and the air is saturated with water vapor. The vapor pressure when the air is saturated is known as the saturation vapor pressure. At higher temperature (*C*), a higher proportion of water molecules have sufficient energy to escape. Both the concentration of water molecules in the vapor phase and the saturation vapor pressure are correspondingly higher.**

contains 1/2 mole of solute dissolved in 1,000 grams (55.5 mol) of water. The mole fraction of water in a 0.5 molal solution is therefore $55.5/(55.5 + 0.5) = 0.991$. According to equation 2.1, the saturation vapor pressure of a half-molal solution would be reduced by less than 1 percent compared with pure water.

Temperature, on the other hand, has a significant effect on vapor pressure. This is due to the effect of temperature on the average kinetic energy of the water molecules. As the temperature of a volume of water or an aqueous solution increases, the proportion of molecules with sufficient energy to escape the fluid surface also increases. This in turn will increase the concentration of water molecules in the vapor phase and, consequently, the equilibrium vapor pressure. An increase in temperature of about 12°C will nearly double the saturation vapor pressure.

According to Fick's law of diffusion (Chapter 1), molecules will diffuse from a region of high concentration to a region of low concentration, or, down a concentration gradient. Because vapor pressure is proportional to vapor concentration, water vapor will also diffuse down a vapor pressure gradient, that is, from a region of high vapor pressure to a region of lower vapor pressure. In principle, we can assume that the substomatal air space of a leaf is normally saturated or very nearly saturated with water vapor. This is because the mesophyll cells that border the air space present a large, exposed surface area for evaporation of water. On the other hand, the atmosphere that surrounds the leaf is usually unsaturated and may often have a very low water content. These circumstances create a gradient between the high water vapor pressure in the interior of the leaf and the lower water vapor pressure of the external atmosphere. This difference in water vapor pressure between the internal air spaces of the leaf and the surrounding air is the driving force for transpiration.

2.3 THE RATE OF TRANSPIRATION IS INFLUENCED BY ENVIRONMENTAL FACTORS

The rate of transpiration will naturally be influenced by factors such as *humidity* and *temperature*, and *wind speed*, which influence the rate of water vapor diffusion between the substomatal air chamber and the ambient atmosphere. Fick's law of diffusion tells us that the rate of diffusion is proportional to the difference in concentration of the diffusing substance. It therefore follows that the rate of transpiration will be governed in large measure by the *magnitude* of the vapor pressure difference between the leaf (e_{leaf}) and the surrounding air (e_{air}). In other words,

$$T \propto e_{leaf} - e_{air} \qquad (2.2)$$

where T is the rate of transpiration. At the same time, the escape of water vapor from the leaf is controlled to a considerable extent by resistances encountered by the diffusing water molecules both within the leaf and in the surrounding atmosphere (Figure 2.3). Resistance is encountered by the vapor molecules as they pass through the intercellular spaces, which are already saturated with water vapor, and the stomatal pores. Note that the symbol for stomatal resistance indicates that it is variable, to account for the fact that stomata may at any time be fully open, partially open, or closed. Additional resistance is encountered by the **boundary layer**, a layer of undisturbed air on the surface of the leaf. The boundary layer and its effect on tranpiration are described more fully later in this chapter. The transpiration equation (2.2) now requires additional terms to account for these resistances.

$$T = \frac{e_{leaf} - e_{air}}{r_{air} + r_{leaf}} \qquad (2.3)$$

FIGURE 2.3 **A schematic representation of the principal resistances encountered by water vapor diffusing out of a leaf. Symbols for electrical resistance are used because resistance to water diffusion is analogous to resistance in an electrical circuit. Note the symbol for stomatal resistance indicates it is variable—taking into account the capacity of the stomata to open and close.**

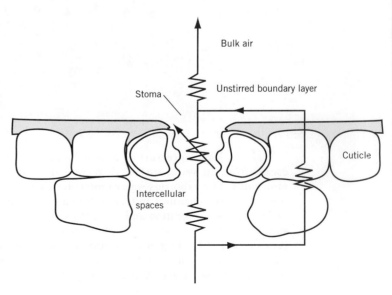

where r_{air} and r_{leaf} are resistances due to the air and the leaf respectively. Equation 2.3 tells us that the rate of transpiration is proportional to the difference in vapor pressure between the leaf and the atmosphere divided by the sum of resistances encountered in the air and the leaf.

2.3.1 WHAT ARE THE EFFECTS OF HUMIDITY?

Humidity is the actual water content of air, which, as noted earlier, may be expressed either as vapor density (g m^{-3}) or vapor pressure (kPa). In practice, however, it is more useful to express water content as the **relative humidity (RH)**. Relative humidity is the ratio of the actual water content of air to the maximum amount of water that can be held by air at that temperature. Expressed another way, relative humidity is the ratio of the actual vapor pressure to the saturation vapor pressure. Relative humidity is most commonly expressed as RH × 100, or **percent relative humidity**. The effects of humidity and temperature on the vapor pressure of air are illustrated in Table 2.1. Air at 50 percent RH by definition contains one-half the amount of water possible at saturation. Its vapor pressure is therefore one-half the saturation vapor pressure. Note also that a 10°C rise in temperature nearly doubles the saturation vapor pressure. Relative humidity and temperature also have a significant effect on the water potential of air (Table 2.2).

As indicated earlier, the vapor pressure of the substomatal leaf spaces is probably close to saturation most of the time. Even in a rapidly transpiring leaf the relative humidity would probably be greater than 95 percent and the resulting water potential would be close to zero (Table 2.2). Under these conditions, the vapor pressure in the substomatal space will be the saturation vapor pressure at the leaf temperature. The vapor pressure of atmospheric air, on the other hand, depends on both the relative humidity of the air and its temperature. Humidity and temperature thus have the potential to modify the magnitude of the vapor pressure gradient ($e_{leaf} - e_{air}$), which, in turn, will influence the rate of transpiration.

TABLE 2.1   Water Vapor Pressure (kPa) in air as a function of temperature and varying degrees of saturation. Air is saturated with water vapor at 100% relative humidity (RH)

Temperature (°C)	Relative Humidity				
	100%	80%	50%	20%	10%
30	4.24	3.40	2.12	0.85	0.42
20	2.34	1.87	1.17	0.47	0.23
10	1.23	0.98	0.61	0.24	0.12

TABLE 2.2   Some values for water potential (Ψ) as a function of relative humidity (RH) at 20°C

RH(%)	Ψ (MPa)
100	0
95	−6.9
90	−14.2
50	−93.5
20	−217.1

Water potential is calculated from the following relationship:
$\Psi = 1.06\,T \log (RH/100)$

2.3.2 WHAT IS THE EFFECTS OF TEMPERATURE?

Temperature modulates transpiration rate through its effect on vapor pressure, which in turn affects the vapor pressure gradient as illustrated by the three examples in Table 2.3. In the first example (A), assuming an ambient temperature of 10°C and a relative humidity of 50 percent, the leaf-to-air vapor pressure gradient is 0.61 kPa. This might be a typical situation in the early morning hours. As the sun comes up, the air temperature will increase. A 10°C increase in temperature (Table 2.3B), assuming the water content of the atmosphere remains constant, will increase the leaf-to-air vapor pressure gradient and, consequently, the potential for transpiration, by a factor of almost 3. Note that in this example it is assumed that leaf temperature is in equilibrium with the atmosphere. This is not always the case. A leaf exposed to full sun may actually reach temperatures 5°C to 10°C higher than that of the ambient air. Under these circumstances, the vapor pressure gradient may increase as much as

TABLE 2.3   The effect of temperature and relative humidity on leaf-to-air vapor pressure gradient. In this example it is assumed that the water content of the atmosphere remains constant

Leaf	Atmosphere	$e_{leaf} - e_{air}$
(A)		
T = 10°C	T = 10°C	
e = 1.23 kPa	e = 0.61 kPa	0.61 kPa
RH = 100%	RH = 50%	
(B)		
T = 20°C	T = 20°C	
e = 2.34 kPa	e = 0.61 kPa	1.73 kPa
RH = 100%	RH = 26%	
(C)		
T = 30°C	T = 20°C	
e = 4.24 kPa	e = 0.61 kPa	3.63 kPa
RH = 100%	RH = 26%	

sixfold (Table 2.3C). As long as the stomata remain open and a vapor pressure gradient exists between the leaf and the atmosphere, water vapor will diffuse out of the leaf. This means transpiration may occur even when the relative humidity of the atmosphere is 100 percent. This is often the case in tropical jungles where leaf temperature and, consequently, saturation vapor pressure is higher than the surrounding atmosphere. Because the atmosphere is already saturated, the water vapor condenses upon exiting the leaf, thereby giving substance to the popular image of the steaming jungle.

2.3.3 WHAT IS THE EFFECT OF WIND?

Wind speed has a marked effect on transpiration because it modifies the effective length of the diffusion path for exiting water molecules. This is due to the existence of the boundary layer introduced earlier (Figure 2.3). Before reaching the bulk air, water vapor molecules exiting the leaf must diffuse not only through the thickness of the epidermal layer (i.e., the guard cells), but also through the boundary layer. The thickness of the boundary layer thus adds to the length of the diffusion path. According to Fick's law, this added length will decrease the rate of diffusion and, hence, the rate of transpiration.

The thickness of the boundary layer is primarily a function of leaf size and shape, the presence of leaf hairs (trichomes), and wind speed. The calculated thickness of the boundary layer as a function of wind speed over a typical small leaf is illustrated in Figure 2.4. With increasing wind speed, the thickness of the boundary layer and, consequently, the length of the diffusion path decreases. In accordance with Fick's law, the vapor pressure gradient steepens and, all other factors being equal, the rate of transpiration increases. This relationship holds truest at lower wind speeds, however. As wind speed increases it tends to cool the leaf and may cause sufficient desiccation to close the stomata. Either one of these factors tends to lower the rate of transpiration.

High wind speeds will therefore have less of an effect on transpiration rate than expected on the basis of their effect on boundary layer thickness alone.

Boundary layer thickness can also be influenced by a variety of plant factors. Boundary layers are thicker over larger leaves and leaf shape may influence the wind pattern. Leaf pubescence, or surface hairs, helps to maintain the boundary layer, and thus reduce transpiration, by breaking up the air movement over the leaf.

Given our discussion of transpiration, it is clear that plant–water relations reflects the acquisition of water from the soil through the constant loss of water through the leaves. This apparent conundrum may lead one to question whether transpiration offers any positive advantage to plants. This question is addressed in Box 2.1

2.4 WATER CONDUCTION OCCURS VIA TRACHEARY ELEMENTS

The distinguishing feature of vascular plants is the presence of **vascular tissues**, the **xylem** and **phloem**, which conduct water and nutrients between the various organs. Vascular tissues begin differentiating a few millimeters from the root and shoot apical meristems and extend as a continuous system into other organs such as branches, leaves, flowers, and fruits. In organs such as leaves, the larger veins subdivide into smaller and smaller veins such that no photosynthetic leaf cell is more than a few cells removed from a small vein ending. Xylem tissue is responsible for the transport of water, dissolved minerals, and, on occasion, small organic molecules upward through the plant from the root through the stem to the aerial organs. Phloem, on the other hand, is responsible primarily for the translocation of organic materials from sites of synthesis to storage sites or sites of metabolic demand (Chapter 9).

Xylem consists of **fibers, parenchyma cells**, and **tracheary elements** (Figure 2.5). Fibers are very elongated cells with thickened secondary walls. Their principal function is to provide structural support for the plant. Parenchyma cells provide for storage as well as the lateral translocation of solutes. The tracheary elements include both **tracheids** and **vessel elements** (Figure 2.5). Tracheary elements are the most highly specialized of the xylem cells and are the principal water-conducting cells. Tracheids and vessels are both elongated cells with heavy, often sculptured, secondary cell walls. Their most distinctive feature, however, is that when mature and functioning, both tracheids and vessels form an interconnected network of nonliving cells, devoid of all protoplasm. The hollow, tubular nature of these cells together with their extensive interconnections facilitates the rapid and efficient transport of large volumes of water throughout the plant.

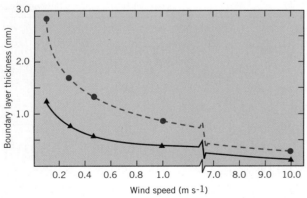

FIGURE 2.4 **The impact of wind speed on calculated boundary layer thickness for leaves 1.0 cm (triangles) or 5.0 cm (circles) wide. A wind speed of 0.28 m s^{-1} = 1 km hr^{-1}. (Plotted from the data of Nobel, 1991.)**

BOX 2.1
WHY TRANSPIRATION?

Our discussion of transpiration in this chapter has focused on the mechanism of water loss and the role of transpiration in the ascent of sap—a sort of operational approach to the problem. We cannot fail to be impressed by the amount of water that must be made available to a plant in order to support transpiration and the possible consequences of such water loss to plant survival. Transpiration often results in water deficits and desiccation injury, especially when high temperature and low humidity favor transpiration but the soil is deficient of water. This raises an interesting and often controversial question: Is there any positive advantage in transpiration to be gained by the plant? It has been argued that transpiration is required to bring about the ascent of sap, that it increases nutrient absorption, and that it assists in the cooling of leaves. It has also been argued that transpiration is little more than a "necessary evil."

Transpiration does speed up the movement of xylem sap, but it seems unlikely that this is an essential requirement. The growth of cells alone would cause a slow ascent of xylem sap, even in the absence of transpiration. Transpiration serves only to increase the rate and quantity of water moved and there is no evidence that the higher rates are beneficial. Another argument is that, because mineral nutrients absorbed by the roots move largely in the xylem sap, transpiration may benefit nutrient distribution. It is true that minerals in the xylem sap will be carried along with a rapidly moving transpiration stream, but the rate-limiting step in nutrient supply is more likely to be the rate at which the nutrients are absorbed by the roots and delivered to the xylem. Moreover, experiments with radioactive tracers have shown that minerals continue to circulate within the plant in the absence of transpiration.

Because transpiration involves the evaporation of water, it can assume a significant role in the cooling of leaves. This is illustrated by the energy budget for a typical mesophyte leaf shown in the accompanying table. Because leaves are heavily pigmented, they absorb large amounts of direct solar radiation. Some of this absorbed solar radiation will not be utilized in photochemical reactions, such as photosynthesis, but will instead account for a significant heat gain by the leaves. Leaves also exchange infrared energy with their surroundings, both absorbing and radiating infrared.

Overall a leaf will radiate more infrared energy than it gains, leaving a negative net infrared exchange. This leaves a net radiation gain in this example of $370\,W\,m^{-2}$, which must be dissipated by other means. One way of dissipating the heat load is by evaporation of water from the leaf surface, or transpiration. The latent heat of vaporization of water is $44\,kJ\,mol^{-1}$ and a typical mesophyte leaf might transpire at the rate of about 4 mmol of water m^{-2} per second. The heat energy consumed by transpiration may be calculated as: $(4 \times 10^{-3}\,mol\,m^{-2}\,s^{-1})\,(44 \times 10^{-3}\,J\,mol^{-1}) = 176\,J\,m^{-2}\,s^{-1} = 176\,W\,m^{-2}$. In this example, transpiration can thus account for dissipation of approximately one-half of the net radiation balance. Dissipation of the remaining heat is probably accounted for by convection from the leaf to the surrounding air.

The energy budget for a typical mesophyte leaf.

Energy gain	$W\,m^{-2}$
a. Absorbed solar radiation	+605
b. Net infrared exchange	−235
c. Net radiation balance (a + b)	+370
Energy loss	
d. Loss by transpiration	−176
e. Loss by convection	−194
Net =	−370

Data from Nobel, 1991.

One argument raised against a significant role for transpiration is that there is seldom any clear correlation between transpiration and plant growth. While some plants may develop more slowly at high humidity, many are able to complete their life cycle without apparent harm under conditions such as 100 percent relative humidity, where transpiration is minimal. Under such circumstances, the supply of water and nutrients is clearly adequate. If the leaf were not cooled by evaporative water loss, other processes such as convection might remove sufficient energy to prevent the leaf reaching lethal temperatures.

In Chapter 8, it is argued that the evolutionary function of stomata is to ensure an adequate supply of carbon dioxide for photosynthesis. It has been suggested that transpiration is simply an unfortunate consequence of this function; that is, a structure that is efficient for the diffusive uptake of carbon dioxide is equally efficient for the outward diffusion of water vapor. According to this view, leaf structure represents a compromise between the need to restrict desiccation of leaf cells while at the same time maintaining access to atmospheric carbon dioxide.

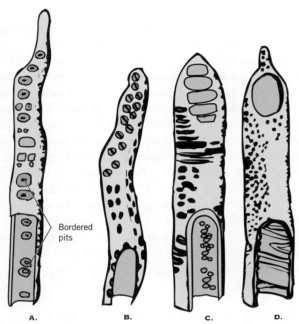

FIGURE 2.5 **Tracheids from (*A*) spring wood of white pine (*Pinus*) and (*B*) oak (*Quercus*). Vessel elements from (*C*) Magnolia and (*D*) basswood (*Tilia*). Only short tip sections are shown. (From T. E. Weier et al., *Botany*, 6th ed. New York, Wiley, 1982, Figure 7.18. Used by permission of the authors.)**

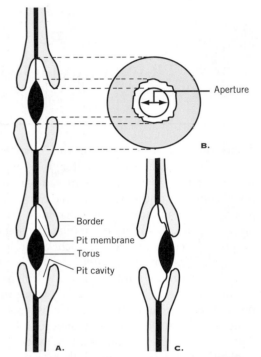

FIGURE 2.6 **Diagram of bordered pit pairs. (*A*) Two bordered pit pairs in the wall between two xylem tracheids, in side view. (*B*) Surface view. (*C*) A pressure differential—lower to the right—pushes the torus against the border, thereby sealing the pit and preventing water movement. (From K. Esau, *Anatomy of Seed Plants*, New York, Wiley, 1977. Reprinted by permission.)**

Tracheids are single cells with diameters in the range of 10 to 50 μm. They are typically less than 1 cm in length, although in some species they may reach lengths up to 3 cm. Tracheids also have thickened secondary walls composed mainly of cellulose, hemicellulose, and lignin. Because of the high lignin content, secondary walls are less permeable to water than are the primary walls of growing cells. On the other hand, the additional strength of secondary walls helps to prevent the cells from collapsing under the extreme negative pressure that may develop in the actively transpiring plants. Although their principal function is to conduct water, the thickened secondary walls of tracheids also contribute to the structural support of the plant.

The movement of water between tracheids is facilitated by interruptions, known as **pit pairs**, in the secondary wall (Figure 2.6). During the development of tracheids, regions that are to become pit pairs avoid the deposition of secondary wall material. This leaves only the middle lamella and primary walls to separate the hollow core, or **lumen**, of one cell from that of the adjacent cell. The combined middle lamella and primary wall is known as the **pit membrane**. Pit membranes are not solid but have openings (about 0.3 μm diameter) that permit the relatively free passage of water and solutes. The origin of these openings is not clear, but they are thought to represent regions where cytoplasmic strands (**plasmodesmata**) once penetrated the cell walls when the cells were still alive. **Bordered pit pairs** have secondary wall projections over the pit area and a swollen

central region of the pit membrane called the **torus**. When pressure is unequal in adjacent vascular elements, such as when one contains an air bubble, the torus is drawn toward the element with the lower pressure. Pressure of the torus against the borders seals off the pit as shown in Figure 2.6.

Successive tracheids commonly overlap at their tapered ends. As a result, tracheids line up in files running longitudinally through the plant. Water moves between adjacent tracheids (either vertically or laterally) through the pit pairs in those regions where overlap occurs. The movement of water is no doubt facilitated by the openings in the pit membranes.

Vessels are very long tracheary elements made up of individual units, known as **vessel members**, which are arranged end-to-end in longitudinal series. At maturity, the end walls of the vessel members have dissolved away, leaving openings called **perforation plates**. As the cell that is to become a vessel member develops, the future perforation area becomes thickened due to swelling of the middle lamella (Figure 2.7). The thickened area contains little cellulose, consisting almost entirely of noncellulosic polysaccharides. As the cell matures and the cytoplasm begins to break down, the unprotected site of the perforation is attacked by hydrolytic enzymes and dissolved away. The rest of the wall area has been

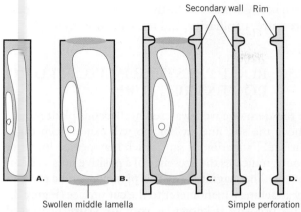

Secondary wall   Rim

Swollen middle lamella   Simple perforation

FIGURE 2.7 **The development of a vessel member. (*A*) Meristematic cell. (*B*) Swelling of the middle lamella in the region of a future perforation plate. (*C*) Secondary wall deposition except over area of future perforation. (*D*) Mature vessel member. The primary wall and middle lamella have dissolved away and the protoplast has disappeared. (From K. Esau, *Anatomy of Seed Plants*, 1st ed., New York, Wiley, 1960. Reprinted by permission.)**

covered with lignified secondary wall materials and is protected from degradation. In some cases, the resulting perforation will encompass virtually the entire end wall, leaving only a ring of secondary wall to mark the junction between two successive vessel members (Figure 2.7). In other cases, the plate may be multiperforate. If the perforations are elongate and parallel, the pattern is called **scalariform**. An irregular, netlike pattern is called **reticulate**. The perforations generally allow for a relatively free flow of water between successive vessel members. There are no perforation plates at the ends of vessels (i.e., the last vessel member in a sequence), but water is able to move laterally from one vessel to the next due to the presence of pit pairs similar to those found in tracheids.

The size of vessels is highly variable, although they are generally larger than tracheids. In maples (*Acer* sps.), for example, vessels range from 40 µm to 60 µm in diameter, while in some species of oak (*Quercus* sps.) diameters may range up to 300 µm to 500 µm. (The large-diameter vessels account for the ring porus character of spring wood in woody species.) The length of vessels in maple is generally 4 cm or less, but some may reach lengths of 30 cm. In oak, on the other hand, vessel lengths up to 10 m have been recorded. However, because of extensive branching of the vascular system and the large number of lateral connections between overlapping tracheary elements, the xylem constitutes a single continuous, interconnected system of water-conducting conduits between the extremes of the plant—from the tip of the longest root to the outermost margins of the highest leaf.

Vessels are considered evolutionarily more advanced than tracheids. For example, xylem tissue

in the gymnosperms, considered evolutionarily more primitive than the angiosperms, consists entirely of tracheids. Although tracheids do occur in angiosperms, the bulk of the water is conducted in vessels. Also because of their larger size, vessels are considerably more efficient than tracheids when it comes to conducting water. An empirical equation relating flow rate to the size of conduits was developed in the nineteenth century by the French scientist Jean L. M. Poiseuille. Poiseuille showed that when a fluid is pressure-driven, the volume flow rate (J_v) is a function of the viscosity of the liquid (η), the difference in pressure or pressure drop (ΔP), and the radius of the conduit:

$$J_v = \Delta P \, \pi r^4 / 8\eta \qquad (2.4)$$

Equation 2.4 applies to water movement in the xylem tracheary elements because, as will be shown below, it is driven by a difference in pressure between the soil and the leaves. The important point to note, then, is that the volume flow rate is *directly proportional to the fourth power of the radius*. The impact of this relationship can be seen by comparing the relative volume flow rates for a 40-µm-diameter (r = 20 µm) tracheid and a 200-µm-diameter (r = 100 µm) vessel. Although the relative diameter of the vessel is 5 times that of the tracheid, its relative volume flow rate will be 625 (i.e., 5^4) times that of the tracheid. The high rate of flow in the larger vessels occurs because the flow rate of water is not uniform across the conduit. The flow rate of molecules near the conduit wall is reduced by friction, due to adhesive forces between the water and the conduit wall. As the diameter of the conduit increases, the proportion of molecules near the wall and consequently subject to these frictional forces will decrease. Put another way, the faster-moving molecules in the center of the conduit constitute a larger proportion of the population and the overall rate of flow increases accordingly.

2.5 THE ASCENT OF XYLEM SAP IS EXPLAINED BY COMBINING TRANSPIRATION WITH THE COHESIVE FORCES OF WATER

The tallest-standing trees are generally found growing in the rainforests along the Pacific coast of the northwestern United States and southwestern British Columbia. The best known are the redwoods (*Sequoia sempervirens*) of northern California, some of which exceed 110 m in height. Individual specimens of Douglas fir (*Pseudotsuga menziesii*) have been reported in excess of 100 m and

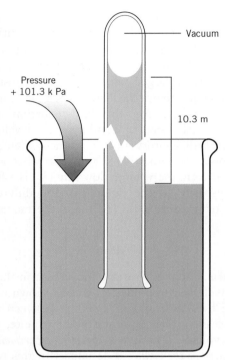

FIGURE 2.8 Atmospheric pressure can support a water column to a maximum height of 10.3 m.

a Sitka spruce (*Picea sitchensis*) measuring 95 m has been located in the Carmanah Valley of Vancouver Island. In Australia, there have been reports of *Eucalyptus* trees measuring more than 130 m in height.

The forces required to move water to such heights are substantial. Were we able to devise a sufficiently long tube closed at one end, fill it with water, and invert it as shown in Figure 2.8, we would find that atmospheric pressure (ca. 101 kPa at sea level) would support a column of water approximately 10.3 m in height. To push the water column any higher would require a correspondingly greater pressure acting on the open surface. Clearly, elevating water to the height of the tallest trees would require a force 10 to 15 times greater than atmospheric, or 1.0 to 1.5 MPa. This force would be equivalent to the pressure at the base of a standing column of water 100 m to 150 m high.

But even a force of this magnitude would not be sufficient. In addition to the force of gravity, water moving through the plant will encounter a certain amount of resistance inherent in the structure of the conducting tissues—irregular wall surfaces, perforation plates, and so forth. We can assume that a force at least equal to that required to support the column would be necessary to overcome these resistances. In that case, a force on the order of 2.0 to 3.0 MPa would be required to move water from ground level to the top of the tallest known trees. How can such a force be generated? This is a question that has long held the interest of plant physiologists and over the years a number of theories

have been advanced. The three most prominent are **root pressure**, **capillarity**, and the **cohesion theory**.

2.5.1 ROOT PRESSURE IS RELATED TO ROOT STRUCTURE

If the stem of a well-watered herbaceous plant is cut off above the soil line, xylem sap will exude from the cut surface. Exudation of sap, which may persist for several hours, indicates the presence of a positive pressure in the xylem. The magnitude of this pressure can be measured by attaching a manometer to the cut surface (Figure 2.9). This pressure is known as **root pressure** because the forces that give rise to the exudation originate in the root.

Root pressure has its basis in the structure of roots and the active uptake of mineral salts from the soil. The xylem vessels are located in the central core of a root, the region known as the **stele**. Surrounding the stele is a layer of cells known as the **endodermis**. In most roots, the radial and transverse walls of the endodermal cells develop characteristic thickenings called the **Casparian band** (Figure 2.10). The Casparian band is principally composed of **suberin**, a complex mixture of *hydrophobic*, long-chain fatty acids, and alcohols. These hydrophobic molecules impregnate the cell wall, filling in the spaces between the cellulose microfibrils as well as the intercellular spaces between the cells. Because it is both space-filling and hydrophobic, the Casparian band presents an effective barrier to the movement of water through the apoplastic space of the endodermis. The result is that water can move into or out of the stele only by first passing through the membranes of the endodermal cells and then through the plasmodesmatal connections.

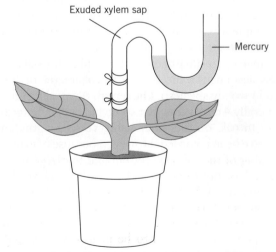

FIGURE 2.9 A simple manometer for measuring root pressure. Root pressure can be calculated from the height of the mercury in the glass tube.

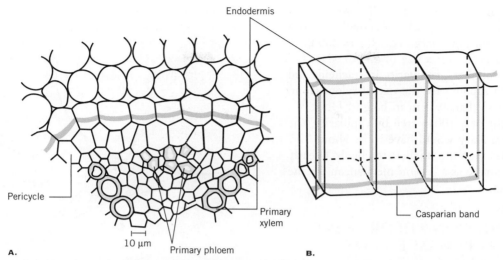

FIGURE 2.10 Suberin deposits (Casparian strip) in the walls of root endodermal cells. (A) Cross section. (B) Three-dimensional view. Suberin deposits in the radial walls establish a barrier to movement of water and salts in the apoplast of the endodermis. (From K. Esau, *Anatomy of Seed Plants***, New York, Wiley, 1977. Reprinted by permission.)**

As roots take up mineral ions from the soil, the ions are transported into the stele where they are actively deposited in the xylem vessels. The accumulation of ions in the xylem lowers the osmotic potential and, consequently, the water potential of the xylem sap. In response to the lowered water potential, water follows, also passing from the cortical cells into the stele through the membranes of the endodermal cells. Since the Casparian band prevents the free return of water to the cortex, a positive hydrostatic pressure is established in the xylem vessels. In a sense, the root may be thought of as a simple osmometer (refer to Figure 1.8) in which the endodermis constitutes the differentially permeable membrane, the ions accumulated in the xylem represent the dissolved solute, and the xylem vessels are the vertical tube. So long as the root continues to accumulate ions in the xylem, water will continue to rise in the vessels or exude from the surface when the xylem vessels are severed.

The question to be answered at this point is whether root pressure can account for the rise of sap in a tree. The answer is probably no, for several reasons. To begin with, xylem sap is not as a rule very concentrated and measured root pressures are relatively low. Values in the range of 0.1 to 0.5 MPa are common, which are no more than 16 percent of that required to move water to the top of the tallest trees. In addition, root pressure has not been detected in all species. Finally, it has been clearly established that during periods of active transpiration, when water movement through the xylem would be expected to be most rapid, the xylem is under **tension** (i.e., *negative* pressure). Root pressure clearly cannot serve as the mechanism for the ascent of sap in all cases. However, root pressure could serve

to fill vessels in small, herbaceous plants and in some woody species in the spring when sap moves up to the developing buds.

2.5.2 WATER RISE BY CAPILLARITY IS DUE TO ADHESION AND SURFACE TENSION

If a glass capillary tube (i.e., a tube of small diameter) is inserted into a volume of water, water will rise in the tube to some level above the surface of the surrounding bulk water. This phenomenon is called **capillary rise**, or simply **capillarity**. Capillary rise is due to the interaction of several forces. These include **adhesion** between water and polar groups along the capillary wall, **surface tension** (due to cohesive forces between water molecules), and the force of gravity acting on the water column. Adhesive forces attract water molecules to polar groups along the surface of the tube. When these water-to-wall forces are strong, as they are between water and glass tubes or the inner surfaces of tracheary elements, the walls are said to be *wettable*. As water flows upward along the wall, strong cohesive forces between the water molecules act to pull the bulk water up the lumen of the tube. This will continue until these lifting forces are balanced by the downward force of gravity acting on the water column.

The following equation can be used to calculate the rise of water in a capillary tube where h is the

$$h = 1.49 \times 10^{-5}\,\mathrm{m^2}/r\,\mathrm{m} \qquad (2.5)$$

rise in meters and r is the radius of the capillary tube in meters (m). Clearly, the calculated rise of water in a capillary tube is inversely proportional to the radius of the tube. In a large tracheid or small vessel, with a diameter of 50 μm (r = 25 μm), water will rise to a height of about 0.6 m. For a large vessel (r = 200 μm), capillarity would account for a rise of only 0.08 m. On the basis of these numbers, capillarity in tracheids and small vessels might account for the rise of xylem sap in small plants, say less than 0.75 m in height. However, to reach the height of a 100 m tree by capillarity, the diameter of the capillary would have to be about 0.15 μm—much smaller than the smallest tracheids. Clearly capillarity is inadequate as a *general* mechanism for the ascent of xylem sap.

2.5.3 THE COHESION THEORY BEST EXPLAINS THE ASCENT OF XYLEM SAP

The most widely accepted theory for movement of water through plants is known as the **cohesion theory**. This theory depends on there being a continuous column of water from the tips of the roots through the stem and into the mesophyll cells of the leaf. The theory is generally credited to H. H. Dixon, who gave the first detailed account of it in 1914.

2.5.3.1 *What is the driving force?*   According to the cohesion-tension theory, the driving force for water movement in the xylem is provided by evaporation of water from the leaf and the tension or negative pressure that results. Water covers the surfaces of the mesophyll cells as a thin film, adhering to cellulose and other hydrophillic surfaces. As water evaporates from this film, the air–liquid interface retreats into the small spaces between cellulose microfibrils and the angular junctions between adjacent cells. This creates very small curved surfaces or microscopic menisci (Figure 2.11). As the radii of these menisci progressively decrease, surface tension at the air–water interface generates an increasingly negative pressure, which in turn tends to draw more liquid water toward the surface. Because the water column is continuous, this negative pressure, or tension, is transmitted through the column all the way to the soil. As a result, water is literally pulled up through the plant from the roots to the surface of the mesophyll cells in the leaf.

The cohesion theory raises two very important questions: (1) Is the xylem sap of a rapidly transpiring plant under tension? (2) How is the integrity of very tall water columns maintained? In the absence of any direct evidence, the answers to these questions provide substantial indirect support for the theory.

Except in certain circumstances, such as when root pressure is active, pressures in the xylem are rarely positive. On the other hand, several lines of evidence support the conclusion that xylem water is instead under significant tension. First, if one listens carefully when

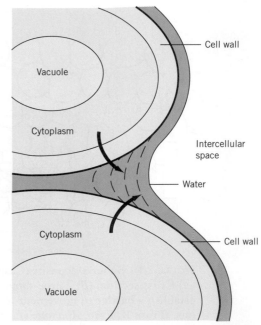

FIGURE 2.11 Tension (negative pressure) in the water column. Evaporation into the leaf spaces causes the water–air interface (dashed lines) to retreat into the spaces between and at the junctions of leaf mesophyll cells. As the water retreats, the resulting surface tension pulls water from the adjacent cells. Because the water column is continuous, this tension is transmitted through the column, ultimately to the roots and soil water.

the xylem of a rapidly transpiring plant is severed, it is sometimes possible to hear the sound of air being drawn rapidly into the wound. If severed beneath the surface of a dye solution, the dye will be very rapidly taken up into tracheary elements in the immediate vicinity of the wound. A second line of evidence involves sensitive measuring devices, called *dendrographs*, which can be used to measure small changes in the diameter of woody stems. Diameters decrease significantly during periods of active transpiration. This will happen because the stem is slightly elastic and the tension in the water column pulls the tracheary walls inward. In the evening, when transpiration declines, the tension is released and stem diameter recovers. Moreover, the shrinkage occurs first in the upper part of the tree, closest to the transpiring leaves, when transpiration begins in the morning. Only later does it show up in the lower part. This observation has been confirmed by experiments in which a localized pulse of heat is generated in the xylem and the flow of heat away from that point is monitored. Flow rates indicate that tensions are greater near the top of the tree. A variety of other experiments have demonstrated that rapidly transpiring shoots are able to pull a column of mercury to heights greater than can be accounted for by atmospheric pressure alone.

Direct measurement of tension in xylem vessels was made possible by the introduction of the **pressure bomb** technique by P. F. Scholander. If the xylem solution is under tension, it will withdraw from the cut surface but can be forced back to the surface by increasing the pressure in the chamber. With such a device, Scholander and others have measured tensions on the order of −0.5 to −2.5 MPa in rapidly transpiring temperate-zone trees.

Finally, it has been observed that water potentials near the bottom of the tree are less negative than water potentials higher in the crown. Such a pressure drop between the bottom and the top of the crown is consistent with a tension resulting from forces originating at the top of the tree. The weight of evidence clearly supports the hypothesis that the xylem water column is literally pulled up the tree in response to transpiration.

2.5.3.2 How is the integrity of the water column maintained?

The ability to resist breakage of the water column is a function of the **tensile strength** of the water column. Tensile strength is a measure of the maximum tension a material can withstand before breaking. Tensile strength is expressed as force per unit area, where the area for the purpose of our discussion is the cross-sectional area of the water column. Tensile strength is yet another property of water attributable to the strong intermolecular cohesive forces, or hydrogen bonding, between the water molecules. The tensile strength of water (or any fluid, for that matter) is not easily measured—a column of water does not lend itself to testing in the same way as a steel bar or a copper wire. The tensile strength of water will also depend on the diameter of the conduit, the properties of the conduit wall, and the presence of any dissolved gases or solute. Still, a number of ingenious approaches have been developed to measure the tensile strength of water with fairly consistent results. It is now generally accepted that pure water, free of dissolved gas, is able to withstand tensions as low as −25 to −30 MPa at 20°C. This is approximately 10 percent of the tensile strength of copper, and 10 times greater than the −2.5 to −3.0 MPa required to pull an uninterrupted water column to the top of the tallest trees. As noted above, tensions in the xylem are more typically in the range of −0.5 to −2.5 MPa for temperate deciduous trees such as maple (*Acer* sps.), but may sometimes be as low as −10 Mpa.

Because xylem water is under tension, it must remain in the liquid state well below its vapor pressure—recall that the vapor pressure of water at 20°C and 100% RH is 2.3 kPa or 0.0023 MPa (Table 2.1). A water column under tension is therefore physically unstable. Physicists call this condition a *metastable state*, a state in which change is ready to occur but does not occur in the absence of an external stimulus. Stability can be achieved

in a water column under tension by introducing a vapor phase. Water molecules in the vapor phase have very low cohesion, which allows the vapor to expand rapidly, thus causing the column to rupture and relieve the tension. How might a vapor phase be introduced to the xylem column? Xylem water contains several dissolved gases, including carbon dioxide, oxygen, and nitrogen. When the water column is under tension, there is a tendency for these gases to come out of solution. Submicroscopic bubbles first form at the interface between the water and the walls of the tracheid or vessel, probably in small, hydrophobic crevices or pores in the walls. These small bubbles may redissolve or they may coalesce and expand rapidly to fill the conduit. This process of rapid formation of bubbles in the xylem is called **cavitation** (L. *cavus*, hollow). The resulting large gas bubble forms an obstruction, called an **embolism** (Gr. *embolus*, stopper), in the conduit. The implications of embolisms with respect to the cohesion theory are quite serious, because a conduit containing an embolism is no longer available to conduct water. Indeed, the potential for frequent cavitation in the xylem was raised as a principal objection to the cohesion theory when it was initially proposed. In order to satisfy these objections, it was necessary to determine just how vulnerable the xylem was to cavitation.

Early attempts to relate cavitation to tensions developed in the xylem were largely inconclusive. There were no satisfactory methods for observing cavitation in the xylem itself and model systems, employing glass tubes, did not necessarily duplicate the interface conditions present in plant tissues. This all changed in 1966 when J. A. Milburn and R. P. C. Johnson introduced an acoustic method for detecting cavitation in plants. In laboratory experiments with glass tubes, the rapid relaxation of tension that follows cavitation produces a shockwave that can be heard as an audible click. Milburn and Johnson found that similar clicks could be "heard" in plant tissue by using sensitive microphones and amplifiers. Each click is believed to represent formation of an embolism in a single vessel element.

Milburn and Johnson studied cavitation in water-stressed leaves of castor bean (*Ricinus communis*). Water stress was introduced by detaching the leaf from the plant and permitting it to wilt. As the leaf wilted, the number of clicks occurring in the petiole was recorded. A total of 3,000 clicks were detected, which is approximately equal to the number of vessels that might be expected in such a petiole. Cavitation could be prevented by adding water to the severed end of the petiole. Various methods that either increased or decreased transpiration from the leaf resulted in a corresponding increase or decrease in the number of clicks. These results indicate a reasonably straightforward relationship between cavitation and tension in the xylem, which appears to support the cohesion theory. They further suggest that cavitation is readily induced by water stress, a condition that herbaceous plants might be expected to encounter on a daily basis. A long-term study of cavitation in a stand of sugar maple (*Acer saccharum*) has been conducted using a method that measures changes in **hydraulic conductance**. In its simplest form, conductance is the inverse of resistance. Hydraulic conductance is therefore a measure of the total capacity of the tissue to conduct water. The acoustic method is limited to counting the number and frequency of cavitations. The hydraulic method, on the other hand, assesses the impact of the resulting embolisms on the capacity of the tissue to transport water.

During the summer growing season, embolisms appeared to be confined to the main trunk and reduced hydraulic conductance by 31 percent. During the winter, loss of conductance in the main trunk increased to 60 percent, while some twigs suffered a 100 percent loss! A decline in conductance during the summer months is no doubt attributable to water stress, as it is in herbaceous plants. The rise in embolisms during the winter is probably related to freeze–thaw cycles. The solubility of gases is very low in ice; when tissue freezes, gas is forced out of solution. During a thaw, these small bubbles will expand and nucleate cavitation.

Problems related to cavitation are not limited to mature trees. Newly planted seedlings often experience water stress due to poor root–soil contact and may be vulnerable to cavitation. A recent study found that seedlings of western hemlock (*Tsuga heterophylla*) experienced water stress and declining xylem pressure potential when planted out. The resulting cavitation and embolism formation in the tracheids caused a decline in hydraulic conductance in the seedling. If the decline in hydraulic conductance is severe enough, it can lead to defoliation or death of the seedling.

Clearly, the effect of cavitation and embolisms on long-term survival of plants would be disastrous if there were not means for their removal or for minimizing their effects. The principal mechanism for minimizing the effect of embolisms is a structural one. The embolism is simply contained within a single tracheid or vessel member. In those tracheary elements with bordered pit pairs, the embolism is contained by the structure of the pit membrane (Figure 2.12A). A difference in pressure between the vessel member containing the embolism and the adjacent water-filled vessel causes the torus to press against the pit border, thus preventing the bubble from being pulled through. At the same time, surface tension prevents the bubble from squeezing through the small openings in the perforation plates between successive vessel members

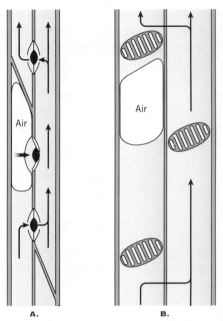

FIGURE 2.12 **Diagram to illustrate how water flow bypasses embolisms in tracheids and vessels. In tracheids (*A*), the pressure differential resulting from an embolism causes the torus to seal off bordered pits lining the affected tracheary element. In vessels (*B*), the bubble may expand through perforation plates, but will eventually be stopped by an imperforate end wall. In both tracheids and vessels, surface tension prevents the air bubbles from squeezing through small pits or capillary pores in the side walls. Water, however, continues to move around the blockage by flowing laterally into adjacent conducting elements.**

(Figure 2.12B). Water, however, will continue to flow laterally through available pits, thus detouring around the blocked element by moving into adjacent conduits. In addition to bypassing embolisms, plants may also avoid long-term damage by repairing the embolism. This can happen at night, for example, when transpiration is low or absent. Reduced tension in the xylem water permits the gas to simply redissolve in the xylem solution. An alternative explanation, particularly in herbaceous species, is that air may be forced back into solution on a nightly basis by positive root pressure.

The repair of embolisms in taller, woody species is not so easily explained. As noted above, sugar maples do recover from freezing-induced embolisms in the spring. Maples and many other tree species also exhibit positive xylem pressures in the spring. These pressures account for the springtime sap flow and could also serve to reestablish the continuity of the water column. Springtime recovery has also been documented for grapevines, although this is apparently due to relatively high root pressures at this time of the year. It might be possible that woody plants in general develop higher-than-normal root pressures in the spring in order to overcome winter damage. Finally, most woody species produce new secondary xylem each spring. This new xylem tissue is laid down before the buds break and may meet the hydraulic conductance needs of the plant, replacing older, nonfunctional xylem.

It is clear, even on the basis of the relatively few studies that have been completed, that xylem is particularly vulnerable to cavitation and embolism. If herbaceous plants were unable to recover on a nightly basis or woody species were unable to recover in the spring, their growth and ultimate survival would be severely compromised. This might lead one to question why plants have not evolved mechanisms to lessen their effects. Interestingly, M. H. Zimmermann turned this question around, proposing instead that cell walls might actually be designed to cavitate. This statement may appear to contradict an earlier observation that surface tension prevents the passage of embolisms through perforations in xylem conduits, but it does not. Where a water-filled conduit under tension is separated from an air volume (at atmospheric pressure) by a porous wall, a concave meniscus (the "lifting force") will form in the pore to balance the negative pressure. As the pressure differential increases, the radius of the meniscus will decrease. At a "suitable" tension in the xylem, the meniscus will reach a radius less than the diameter of the pore (a fraction of a μm in diameter) and will be pulled through the pore into the water-filled conduit. The resulting bubble nucleates a cavitation, which relieves the tension, reduces the pressure differential, and prevents the further entry of air. Results consistent with Zimmermann's hypothesis have been obtained through a comparison of the pressure differences required to embolize sugar maple stems with diameters of the pores in the pit membranes.

Accordingly, Zimmermann's **designed leakage** hypothesis represents a kind of safety valve. Plants appear to be constructed in such a way as to allow cavitation when water potentials reach critically low levels, yet still allow the "damage" to be repaired when conditions improve. Water stress and, consequently, cavitations would be expected to appear first in leaves and smaller branches. These localized cavitations would serve the additional advantage of cutting off peripheral structures while preserving the integrity of the main stem or trunk during extended dry periods.

Roots are even more vulnerable to cavitation than shoots, which could benefit the whole plant during periods of drought. Complete cavitation of the xylem in the smallest roots, for example, would isolate those roots from drying soil, reduce hydraulic conductance, and ultimately reduce transpiration rates. This would help buffer the water status of the stem until the drought eased and the cavitated conduits were refilled or new growth replaced the damaged roots.

2.6 WATER LOSS DUE TO TRANSPIRATION MUST BE REPLENISHED

Uptake of water from the soil by the roots replenishes the water lost as a consequence of leaf transpiration. This establishes an integrated flow of water from the soil, through the plant, and into the atmosphere, referred to as the **soil-plant-atmosphere continuum**. The concept of a soil-plant-atmosphere continuum reinforces the observation that plants do not exist in isolation, but are very interdependent with their environment.

2.6.1 SOIL IS A COMPLEX MEDIUM

In order to understand interactions between roots and soil water, a review of the nature of soils would be helpful. Soil is a very complex medium, consisting of a solid phase comprised of inorganic rock particles and organic material, a soil solution containing dissolved solutes, and a gas phase generally in equilibrium with the atmosphere. The inorganic solid phase of soils is derived from parent rock that is degraded by weathering processes to produce particles of varying size (Table 2.4). In addition to the solid, liquid, and gas phases, soils also contain organic material in varying stages of decomposition as well as algae, bacteria, fungi, earthworms, and various other organisms.

The clay particles in a soil combine to form complex aggregates that, in combination with sand and silt, determine the structure of a soil. Soil structure in turn affects the porosity of a soil and, ultimately, its water retention and aeration. **Porosity**, or pore space, refers to the interconnected channels between irregularly shaped soil particles. Pore space typically occupies approximately 40 percent to 60 percent of a soil by volume. Two major categories of pores—**large pores** and

TABLE 2.4 **Classification of soil particles and some of their properties. A mixture of 40 percent sand, 40 percent silt, and 20 percent clay is known as a loam soil. A sandy soil contains less than 15 percent silt and clay, while a clay soil contains more than 40 percent clay particles**

Particle Class	Particle (mm) Size	Water Retention	Aeration
Coarse sand	2.00–0.2	poor	excellent
Sand	0.20–0.02		
Silt	0.02–0.002	good	good
Clay	less than 0.002	excellent	poor

capillary pores—are recognized. Although there is no sharp line of demarcation between large pores and capillary pores—the shape of the pore is also a determining factor—water is not readily held in pores larger than 10 to 60 μm diameter. When a soil is freshly watered, such as by rain or irrigation, the water will percolate down through the pore space until it has displaced most, if not all, of the air. The soil is then *saturated* with water. Water will drain freely from the large pore space due to gravity. The water that remains after free (gravity) drainage is completed is held in the capillary pores. At this point, the water in the soil is said to be at **field capacity**. Under natural conditions, it might require two to three days for a loam soil to come to field capacity following a heavy rainfall. The relative proportions of large and capillary pore space in a soil can be estimated by determining the water contents of the soil when freshly watered and at field capacity. Water content, expressed as the weight of water per unit weight of dry soil, may be determined by drying the soil at 105°C.

It should not be surprising that a sandy soil, with its coarse particles, will have a relatively high proportion of large pores. A sandy soil will therefore drain rapidly, has a relatively low field capacity, and is well-aerated (Table 2.4). The pore space of a clay soil, on the other hand, consists largely of capillary pores. Clay soils hold correspondingly larger quantities of water and are poorly aerated. A loam soil represents a compromise, balancing water retention against aeration for optimal plant growth.

The water held by soil at or below field capacity is found in capillary channels and the interstitial spaces between contacting soil particles, much as it is in the cell walls and intercellular spaces of mesophyll cells in a leaf. Soil water is therefore also subject to the same forces of surface tension found in the capillary spaces of mesophyll tissues, as discussed earlier. Consequently, soil water at or below field capacity will be under tension and its water potential will be negative. As the water content of the soil decreases, either by evaporation from the soil surface or because it is taken up by the roots, the air–water interface will retreat into the capillary spaces between the soil particles. Because water adheres strongly to the soil particles, the radius of the meniscus decreases and pressure becomes increasingly negative. In principle, water movement in the soil is primarily pressure-driven in the same manner that capillary tension in the mesophyll cells of a leaf draws water from the xylem column. As water is removed from the soil by a root, tensions in the soil water will draw more bulk water toward the root. If there is an abundance of water in the soil, these pressure differences may draw water from some distance.

Except in highly saline soils, the solute concentration of soil water is relatively low—on the order of 10^{-3} M—and soil water potential is determined principally by the negative pressure potential. As might be expected, the uptake of water by roots occurs because of a water potential gradient between the soil and the root. Thus as the soil dries and its water potential declines, plants may experience difficulty extracting water from the soil rapidly enough to balance losses by transpiration. Under such conditions, plants will lose turgor and wilt. If transpiration is reduced or prevented for a period of time (such as at night, or by covering the plant with a plastic bag), water uptake may catch up, turgor will be restored, and the plants will recover. Eventually, however, a point can be reached where the water content of the soil is so low that, even should all water loss by transpiration be prevented, the plant is unable to extract sufficient water from the soil and the loss of turgor is permanent. The soil water content at this point, measured as a percentage of soil dry weight, is known as the **permanent wilting percentage**. The actual value of the permanent wilting percentage varies between soil types; it is relatively low (in the range of 1 to 2 percent) for sand and high (20 to 30 percent) for clay. Loam soils fall between these two extremes, depending on the relative proportion of sand and clay. Regardless of soil type, however, the water potential of the soil at the permanent wilting percentage is relatively uniform at about −1.5 MPa. Although there are some exceptions to the rule, most plants are unable to extract significant amounts of water when the soil water potential falls below −1.5 MPa. In a sense, field capacity may be considered a property of the soil, while the permanent wilting percentage is a property of the plant. The water content of the soil between field capacity and the permanent wilting percentage is considered **available water**, or water that is available for uptake by plants. The range of available water is relatively high in silty loam soils, somewhat less in clay, and relatively low in sand. Not all water in this range is uniformly available, however. In a drying soil, plants will begin to show signs of water stress and reduced growth long before the soil water potential reaches the permanent wilting percentage.

2.7 ROOTS ABSORB AND TRANSPORT WATER

Roots have four important functions. Roots (1) anchor the plant in the soil; (2) provide a place for storage of carbohydrates and other organic molecules; (3) are a site of synthesis for important molecules such as alkaloids and some hormones; and (4) absorb and transport upward to the stem virtually all the water and minerals taken up by plants.

The effectiveness of roots as absorbing organs is related to the extent of the root system. Over the years, a number of efforts have attempted to establish the true extent of root systems. Such studies involve careful and often tedious procedures for excavating the plant and removing the soil without damaging the roots. More elaborate efforts involved construction of subterranean walkways, called *rhizotrons*, equipped with windows through which the growth and development of roots could be observed. The root systems of several prairie grasses have been examined in different types of soil. By carefully excavating the grasses and washing the roots free of soil, they were able to show massive networks of roots penetrating to a depth of 1.5 m. Using a radioactive tracer technique, it was later found the roots of a single 14-week-old corn plant had penetrated to a depth of more than 6 m and extended horizontally as much as 5 m in all directions. Furthermore, the measurements of the roots of a *single* mature rye plant grown in a box of soil measuring 30 cm × 30 cm × 56 cm deep have been reported. The combined length of all roots was 623 km with an estimated total surface area of 639 m². Many species invest substantially more than 50 percent of their body weight in roots. Roots clearly comprise a large proportion of the plant body, although their importance to plants is less obvious to the casual observer than the more visible shoot system.

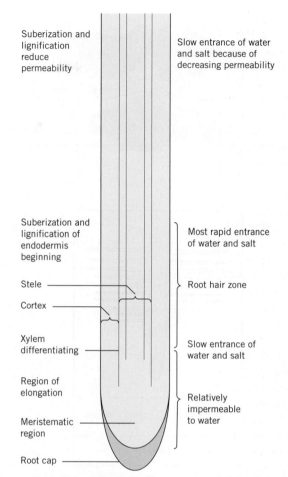

FIGURE 2.13 Diagrammatic illustration of the relationship between differentiation of root tissues and water uptake.

2.8 THE PERMEABILITY OF ROOTS TO WATER VARIES

A large number of anatomical and physiological studies have established that the region of most active water uptake lies near the root tip. Beyond this generalization, the permeability of roots to water varies widely with age, physiological condition, and the water status of the plant. Studies with young roots have shown that the region most active with respect to water uptake starts about 0.5 cm from the tip and may extend down the root as far as 10 cm (Figure 2.13). Little water is absorbed in the meristematic zone itself, presumably because the protoplasm in this zone is dense and there are no differentiated vascular elements to carry the water away. The region over which water appears to be taken up most rapidly corresponds generally with the zone of cell maturation. This is the region where vascular tissue, in particular the xylem, has begun to differentiate. Also in this region the deposition of suberin and lignin in the walls of endodermal cells is only beginning and has not

yet reached the point of offering significant resistance to water movement.

The region of most rapid water uptake also coincides with the region of active root hair development. Root hairs are thin-walled outgrowths of epidermal cells that increase the absorptive surface area and extend the absorptive capacity into larger volumes of soil (Figure 2.14). Depending on the species and environmental conditions, root hairs may reach lengths of 0.1 mm to 10 mm and an average diameter of 10 μm. In some species such as peanut, pecan, and certain conifers, root hairs are rare or absent. More commonly a single root tip may contain as many as 2,500 hairs cm⁻² and may increase the absorbing surface of the root 1.5 to twentyfold. For two reasons, root hairs greatly increase the contact of the root with soil water. First, their small diameter permits root hairs to penetrate capillary spaces not accessible to the root itself (Figure 2.14A). Second, root hairs extend contact into a cylinder of soil whose diameter is twice that of the length of the hair (Figure 2.14B).

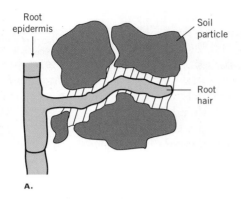

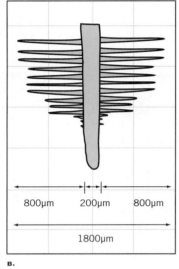

FIGURE 2.14 **Root hairs and water uptake. (A) Root hairs enhance water uptake by their ability to penetrate water-containing capillary spaces between soil particles. (B) Root hairs increase by several times the volume of soil that can be extracted of water by a root. (A from T. E. Weier et al.,** *Botany*, **6th ed., New York, Wiley, 1982, Figure 9.7A. Used by permission of the authors.)**

2.9 RADIAL MOVEMENT OF WATER THROUGH THE ROOT INVOLVES TWO POSSIBLE PATHWAYS

Once water has been absorbed into the root hairs or epidermal cells, it must traverse the cortex in order to reach the xylem elements in the central stele. In principle, the pathway of water through the cortex is relatively straightforward. There appear to be two options: water may flow either past the cells through the **apoplast** of the cortex or from cell to cell through the plasmodesmata of the **symplast**. In practice, however, the two pathways are not separate. Apoplastic water is in constant equilibration with water in the symplast and cell vacuoles. This means that water is constantly being exchanged across both the cell and vacuolar membranes. In effect,

then, water flow through the cortex involves both pathways. The cortex consists of loosely packed cells with numerous intercellular spaces. The apoplast would thus appear to offer the least resistance and probably accounts for a larger proportion of the flow.

In the less mature region near the tip of the root, water will flow directly from the cortex into the developing xylem elements, meeting relatively little resistance along the way. Moving away from the tip toward the more mature regions, water will encounter the endodermis (see section 2.5.1 on root pressure). Suberization of the endodermal cell walls imposes a permeability barrier, forcing water taken up in these regions to pass through the cell membranes. While the endodermis does increase resistance to water flow, it is far from being an absolute barrier. Indeed, under conditions of rapid transpiration the region of most rapid water uptake will shift toward the basal part of the root (Table 2.5).

The resistance offered by the root to water uptake is reflected in the absorption lag commonly observed when water loss by transpiration is compared with absorption by the roots (Figure 2.15). That this lag is due to resistance in the root can be demonstrated experimentally. If the roots of an actively transpiring plant are cut off (under water, of course), there is an immediate increase in the rate of absorption into the xylem. In some species, absorption lag may cause a water deficit in the leaves sufficient to stimulate a temporary closure of the stomata (Figure 2.15). This phenomenon is known as **midday closure**. Closure of the stomata reduces transpiration, allowing absorption of water to catch up and the stomata to then reopen.

Finally, although the absorption of water by roots is believed to be a passive, pressure-driven process, it is nonetheless dependent on respiration in the

TABLE 2.5 **Relative water uptake for different zones along a** *Vicia faba* **root as a function of transpiration rate. The measured xylem tension at the low transpiration rate was −0.13 MPa and −10.25 MPa at the high transpiration rate. When transpiration is low, most of the water uptake occurs near the root tip. When transpiration is high, the resulting increase in tension shifts the region of uptake toward more basal regions**

Water Uptake Zone (cm from apex)	Transpiration Rate	
	Low	High
0–2.5	100	132
2.5–5.0	103	216
5.0–7.5	54	216
7.5–10.0	27	270
10.5–12.5	19	283

From data of Brouwer, 1965.

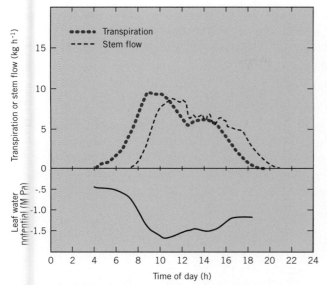

FIGURE 2.15 The absorption lag in a *Larix* (larch) tree. Upper: Absorption of water, measured as the flow of water through the stem, lags about two hours behind transpiration. A transient decline in transpiration rate near midday may occur if the lag is sufficient to create a water deficit and stimulate temporary stomatal closure. Lower: Rapid transpiration causes a decrease in the leaf water potential, which slowly recovers as water moves in to satisfy the deficit. (Adapted from E.-D. Schulze et al., Canopy transpiration and water fluxes in the xylem of *Larix* and *Picea* trees—A comparison of xylem flow, porometer and cuvette measurements, *Oecologia* 66:475–483, 1985. Reprinted by permission of Springer-Verlag.)

root cells. Respiratory inhibitors (such as cyanide or dinitrophenol), high carbon dioxide levels, and low oxygen (anaerobiosis) all stimulate a decrease in the hydraulic conductance of most roots. Anaerobiosis is most commonly encountered by plants in water-logged soils and may lead to extreme wilting in species not specifically adapted to such situations. The exact role of respiration is not clear. The requirement is probably indirect, such as maintaining the cellular integrity and continued elongation of roots. Active uptake of nutrient ions (Chapter 3) may also be a factor. On the other hand, killing roots outright—for example, by immersion in boiling water—dramatically reduces resistance and allows water to be absorbed more rapidly than when the roots were alive.

SUMMARY

Large amounts of water are lost by plants through evaporation from leaf surfaces, a process known as transpiration. Transpiration is driven by differences in water vapor pressure between internal leaf spaces and the ambient air. A variety of factors influence transpiration rate, including temperature, humidity, wind,

and leaf structure. Water is conducted upward through the plant primarily in the xylem, a tube-like system of tracheary elements including tracheids and vessels. The principal driving force for water movement in the xylem is transpiration and the resulting tension in the water column. The water column is maintained because of the high tensile strength of water. Water lost by transpiration is replenished by the absorption of water from the soil through the root system.

CHAPTER REVIEW

1. Explain why transpiration rate tends to be greatest under conditions of low humidity, bright sunlight, and moderate winds.

2. Describe the anatomy of xylem tissue and explain why it is an efficient system for the transport of water through the plant.

3. Trace the path of water from the soil, through the root, stem, and leaf of a plant, and into the atmosphere.

4. Explain how water can be moved to the top of a 100 m tree, but a mechanical pump can lift water no higher than about 10.3 m. What prevents the water column in a tree from breaking? Under what conditions might the water column break, and, if it does break, how is it reestablished?

5. Many farmers have found that fertilizing their fields during excessively dry periods can be counterproductive, as it may significantly damage their crops. Based on your knowledge of the water economy of plants and soils, explain how this could happen.

6. Does transpiration serve any useful function in the plant?

7. Explain the relationships between field capacity, permanent wilting percentage, and available water. Even though permanent wilting percentage is based on soil weight, it is often said to be a property of the plants. Explain why this might be so.

FURTHER READING

Waisel, Y., A. Eshel, U. Kafkafi. 1996. *Plant Roots: The Hidden Half*. 2nd ed. New York: M. Dekker.

Buchanan, B. B., W. Gruissem, R. L. Jones. 2000. *Biochemistry and Molecular Biology of Plants*. Rockville, MD.: American Society of Plant Physiologists.

Evert, R. F. 2006. *Esau's Plant Anatomy*. Hoboken, NJ: John Wiley & Sons, Inc.

Nobel, P. S. 2005. *Physicochemical and Environmental Plant Physiology*. Burlington, MA: Elsevier Science & Technology.

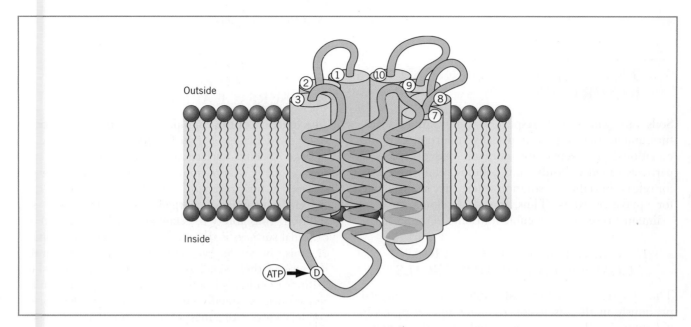

3

Roots, Soils, and Nutrient Uptake

With the exception of carbon and oxygen, which are supplied as carbon dioxide from the air, terrestrial plants generally take up nutrient elements from the soil solution through the root system. The root systems of most plants are surprisingly extensive. Through a combination of primary roots, secondary and tertiary branches, and root hairs, root systems penetrate massive volumes of soil in order to mine the soil for required nutrients and water. Soil is a complex medium. It consists of a solid phase that includes mineral particles derived from parent rock plus organic material in various stages of decomposition, a liquid phase that includes water or the soil solution, gases in equilibrium with the atmosphere, and a variety of microorganisms. The solid phase, in particular the mineral particles, is the primary source of nutrient elements. In the process of weathering various elements are released into the soil solution, which then becomes the immediate source of nutrients for uptake by the plant.

The soil solution, however, is very dilute (total mineral content is on the order of 10^{-3} M) and would quickly become depleted by the roots were it not continually replenished by the release of elements from the solid phase. In one study of phosphorous uptake, for example, it was calculated that the phosphate content of the soil was renewed on the average of 10 times each day. Availability of nutrient elements is not, however, limited to the properties of the soil itself. Access to elements is further enhanced by the continual growth of the very dynamic root system into new, nutrient-rich regions of the soil.

In this chapter we will examine the availability of nutrients in the soil and their uptake by roots. This will include

- the soil as a source of nutrient elements, the colloidal nature of soil, and ion exchange properties that determine the availability of nutrient elements in a form that can be taken up by roots,

- mechanisms of solute transport across membranes, including simple and facilitated diffusion and active transport, the function of membrane proteins as ion channels and carriers, and the role of electrochemical gradients,

- ion traffic into and through the root tissues and the concept of apparent free space, and

- the beneficial role of microorganisms, especially fungi, with respect to nutrient uptake by roots.

The unique situation with respect to uptake and metabolism of nitrogen will be addressed in Chapter 11.

3.1 THE SOIL AS A NUTRIENT RESERVOIR

Soils vary widely with respect to composition, structure, and nutrient supply. Especially important from the nutritional perspective are inorganic and organic soil particles called **colloids**. Soil colloids retain nutrients for release into the soil solution where they are available for uptake by roots. Thus, the soil colloids serve to maintain a reservoir of soluble nutrients in the soil.

3.1.1 COLLOIDS ARE A SIGNIFICANT COMPONENT OF MOST SOILS

The physical structure of soils was introduced previously in the discussion of water uptake by roots (Chapter 2). There it was noted that the mineral component of soils consists predominantly of **sand, silt**, and **clay**, which are differentiated on the basis of particle size (see Chapter 2, Table 2.4). The three components are easily demonstrated by stirring a small quantity of soil into water. The larger particles of sand will settle out almost immediately, leaving a turbid suspension. Over the course of hours or perhaps days, if left undisturbed, the finer particles of *silt* will settle slowly to the bottom as well and the turbidity will in all likelihood disappear. The very small *clay* particles, however, remain in stable suspension and will not settle out, at least not within a reasonable time frame. Clay particles in suspension are not normally visible to the naked eye—they are simply too small. They are, however, small enough to remain suspended. These suspended clay particles can be detected by directing a beam of light through the suspension. The suspended clay particles will scatter the light, causing the path traversed by the light beam to become visible. Particles that are small enough to remain in suspension but too large to go into true solution are called colloids and the light-scattering phenomenon, known as the **Tyndall effect**, is a distinguishing characteristic of colloidal suspensions. A true solution, such as sodium chloride or sucrose in water, on the other hand, will not scatter light. This is because, in a true solution, the solute and solvent constitute a single phase. A colloidal suspension, on the other hand, is a two-phase system. It consists of a solid phase, the colloidal **micelle**, suspended in a liquid phase. Light scattered by the solid phase is responsible for the Tyndall effect.

Clay is not the only soil component that forms colloidal particles. Many soils also contain a colloidal carbonaceous residue, called **humus**. Humus is organic material that has been slowly but incompletely degraded to a colloidal dimension through the action of weathering and microorganisms. In a good loam soil, the colloidal humus content may be substantially greater than the colloidal clay content and make an even greater contribution to the nutrient reservoir.

3.1.2 COLLOIDS PRESENT A LARGE, NEGATIVELY CHARGED SURFACE AREA

The function of the colloidal soil fraction as a nutrient reservoir depends on two factors: (1) colloids present a large specific surface area, and (2) the colloidal surfaces carry a large number of charges. The charged surfaces in turn reversibly bind large numbers of ions, especially positively charged cations from the soil solution. This ability to retain and exchange cations on colloidal surfaces is the single most important property of soils, in so far as plant nutrition is concerned. Because of their small size, one of the distinguishing features of colloids is a high surface area per unit mass, also known as **specific surface area**. Consider, for example, a cube with a mass of one gram that measures 10 mm on a side. The specific surface area of this cube is $600 \, mm^2 \, g^{-1}$ (Figure 3.1). If the cube is then subdivided into particles of colloidal dimensions, say 0.001 mm on a side, the specific surface area increases to 6,000,000 $mm^2 g^{-1}$, a 10,000-fold increase. On a mass basis, then, colloids provide an incredibly large surface area for interaction with mineral elements in the soil solution.

A second important feature of soil colloids, in addition to the large specific surface area, is the large number of charges on the colloidal surfaces. Colloidal clays consist primarily of aluminum silicates (the chemical formula for kaolinite, one of the simplest clays, is $Al_2Si_2O_5(OH)_4$). The predominantly negative charges arise by virtue of ionization of alumina and silica at the

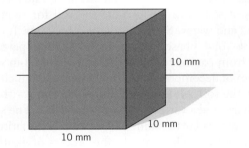

FIGURE 3.1 **Particles of colloidal dimension have a high surface area per unit mass, or specific surface area.**

edges of the clay particle. Because colloidal carbon is derived largely from lignin and carbohydrates, it also carries negative charges arising from exposed carboxyl and hydroxyl groups.

3.1.3 SOIL COLLOIDS REVERSIBLY ADSORB CATIONS FROM THE SOIL SOLUTION

The association of cations (positively charged ions) with negatively charged colloidal surfaces depends on electrostatic interactions; hence, binding affinity varies according to the **lyotropic series**:

$$Al^{3+} > H^+ > Ca^{2+} > Mg^{2+} > K^+ = NH_4^+ > Na^+$$

In this series, aluminum ions have the highest binding affinity and sodium ions the least, reflecting the general rule that trivalent ions (3+) are retained more strongly than divalent ions (2+) and divalent ions more strongly than monovalent ions (1+). Ions, however, are also hydrated, which means that they are surrounded by shells of water molecules, and electrostatic rules are modulated by the *relative hydrated size* of the ion. Since ions of smaller hydrated size can approach the colloidal surfaces more closely, they tend to be more tightly bound. Also, cation adsorption is not an all-or-none phenomenon. The degree of association and ion concentration both decline in a continuous gradient with increasing distance from the surface of the colloid.

Cation adsorption is also *reversible*. Consequently, any ion with a higher affinity (e.g., H^+) is capable of displacing an ion lower in the series (e.g., Ca^{2+}). Alternatively, an ion with a lower affinity can, if provided in sufficient quantity, displace an ion with higher affinity by mass action (Figure 3.2). This process of exchange between adsorbed ions and ions in solution is known as **ion exchange**. The ease of *removal*, or **exchangeability** of an ion, is indicated by the reverse of the lyotropic series shown above. Thus, sodium ions are the most readily exchanged in the series and aluminum ions the least.

Although the immediate source of mineral nutrients for the plant are the ions in the soil solution, the colloidal fraction with its adsorbed ions represents the principal nutrient reservoir. It is important to view the soil as a very dynamic system, with cations in the soil solution freely exchangeable with cations adsorbed to colloidal surfaces. As the soluble nutrients are taken up by the roots from the dilute soil solution, they are continually replaced by exchangeable ions held in the colloidal reservoir. The reservoir is then replenished by ions derived from the weathering of rock particles. In this way, ion exchange at the colloidal surface plays a major role in providing a controlled release of nutrients to the plant. It may not always work to the advantage of the plant, however. One effect of acid

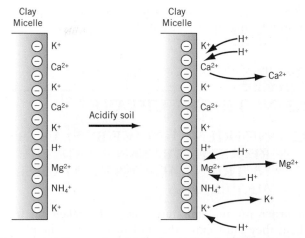

FIGURE 3.2 Ion exchange in the soil. (*A*) Cations are adsorbed to the negatively charged soil particles by electrostatic attraction. (*B*) Acidifying the soil increases the concentration of hydrogen ions in the soil. The additional hydrogen ions have a stronger attraction for the colloidal surface charges and so displace other cations into the soil solution.

rain, for example, appears to be the displacement of cations from the colloidal reservoir due to the high concentration of hydrogen ions. Those solubilized nutrient ions not immediately assimilated by the roots are readily leached out of the soil by the rain and percolating ground water. Both the soil solution and the reservoir of nutrients are depleted more rapidly than they can be replenished and the plants are deprived of an adequate nutrient supply. Under normal circumstances, roots also secrete hydrogen ions, which assist in the uptake of nutrients.

3.1.4 THE ANION EXCHANGE CAPACITY OF SOIL COLLOIDS IS RELATIVELY LOW

The soil colloids are predominantly negatively charged and, consequently, they do not tend to attract negatively charged anions. Although some of the clay minerals do contain cations such as Mg^{2+}, the anion exchange capacity of most soils is generally low. The result is that anions are not held in the soil but tend to be readily leached out by percolating ground water. This situation has important consequences for agricultural practice. Nutrients supplied in the form of anions, in particular nitrogen (NO_3^-), must be provided in large quantity to ensure sufficient uptake by the plants. As a rule, farmers sometimes find they must apply at least twice—sometimes more—the amount of nitrogen actually required to produce a crop. Unfortunately, much of the excess nitrate is leached into the ground water and eventually finds its way into wells or into streams and lakes, where it contributes to problems of eutrophication by stimulating the growth of algae.

3.2 NUTRIENT UPTAKE

3.2.1 NUTRIENT UPTAKE BY PLANTS REQUIRES TRANSPORT OF THE NUTRIENT ACROSS ROOT CELL MEMBRANES

In order for mineral nutrients to be taken up by a plant, they must enter the root by crossing the plasma membranes of root cells. From there they can be transported through the symplast to the interior of the root and eventually find their way into the rest of the plant. Nutrient uptake by roots is therefore fundamentally a cellular problem, governed by the rules of membrane transport. Membrane transport is inherently an abstract subject. That is to say, investigators measure the kinetics of solute movement across various natural and artificial membranes under a variety of circumstances. Models are then constructed that attempt to explain these kinetic patterns in terms of what is currently understood about the composition and architecture of membranes. As our understanding of membrane structure has changed over the years, so have the models that attempt to interpret how solutes cross these membranes. There are, however, three fundamental concepts—**simple diffusion**, **facilitated diffusion**, and **active transport**—that have persevered, largely because they have proven particularly useful in categorizing and interpreting experimental observations. These three concepts now make up the basic language of transport across all membranes of all organisms. These three basic modes of transport are interpreted schematically in Figure 3.3.

3.2.2 SIMPLE DIFFUSION IS A PURELY PHYSICAL PROCESS

According to Fick's law (Chapter 1), the rate at which molecules in solution diffuse from one region to another is a function of their concentration difference. For a membrane-bound cell, Fick's law may be restated as:

$$J = PA(C^o - C^i) \qquad (3.1)$$

where J is the flux, or amount of solute crossing the membrane per unit time. A is the cross-sectional area of the diffusion path, which, in this case, is the area of the cell membrane (in cm^2). P is the permeability coefficient. It measures the velocity (in cm^{-1}) with which the solute crosses that membrane and is specific for a particular membrane-solute combination. Since the membrane barrier is primarily lipid in character, nonpolar solute molecules tend to pass through more rapidly than polar molecules. Membrane lipid bilayers are particularly impermeable to most ions. This is because their charge and high degree of hydration renders ions insoluble in lipids and thus effectively prevents them from entering the hydrocarbon phase of membranes. Synthetic lipid bilayers, or artificial membranes, for example, are some nine orders of magnitude less permeable to smaller ions such as K^+ or Na^+ than to water. The permeability coefficient in Fick's equation thus generally reflects the lipid solubility of diffusing molecules. Few solutes of biological importance are nonpolar and only three (O_2, CO_2, NH_3) appear to traverse membranes by simple diffusion through the lipid bilayer. Water, in spite of its high polarity, also diffuses rapidly through lipid bilayers; this is because water passes through

FIGURE 3.3 **The exchange of ions and solutes across membranes may involve simple diffusion, facilitated diffusion, or active transport.**

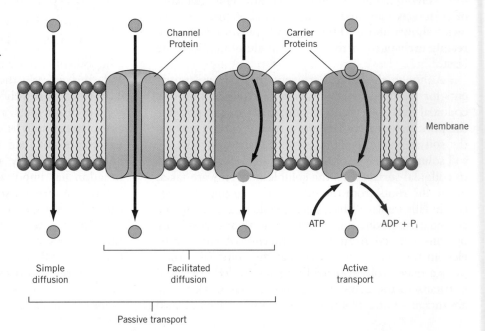

water-selective channels called *aquaporins*. Aquaporins were discussed in Chapter 1.

Transport by diffusion is a passive process, meaning that the transport process does not require a direct input of metabolic energy. The energy for transport by diffusion comes from the concentration or electrochemical gradient of the solute being transported. As a consequence, transport by diffusion does not lead to an accumulation of solute against a concentration or electrochemical gradient.

3.2 3 THE MOVEMENT OF MOST SOLUTES ACROSS MEMBRANES REQUIRES THE PARTICIPATION OF SPECIFIC TRANSPORT PROTEINS

In the 1930s, students of membrane transport recognized that certain ions entered cells far more rapidly than would be expected on the basis of their diffusion through a lipid bilayer. We now know that this is because natural membranes contain a large number of proteins, many of which function as **transport proteins**. Some of these transport proteins *facilitate* the diffusion of solutes, especially charged solutes or ions, into the cell by effectively overcoming the solubility problem. The term **facilitated diffusion** was coined to describe this rapid, assisted diffusion of solutes across the membrane. In facilitated diffusion, as in simple diffusion, the direction of transport is still determined by the concentration gradient (for uncharged solute) or electrochemical gradient (for charged solutes and ions). Facilitated diffusion is also bidirectional and, like simple diffusion, net movement ceases when the rate of movement across the membrane is the same in both directions. Two major classes of transport proteins are known. **Carrier proteins** (also known as **carriers, transporters**, or simply, **porters**) bind the particular solute to be transported, much along the lines of an enzyme–substrate interaction. Binding of the solute normally induces a conformational change in the carrier protein, which delivers the solute to the other side of the membrane. Release of the solute at the other surface of the membrane completes the transport and the protein then reverts to its original conformation, ready to pick up another solute.

Channel proteins are commonly visualized as a charged-lined, water-filled channel that extends across the membrane. Channels are normally identified by the ion species that is able to permeate the channel, which is in turn dependent on the size of the hydrated ion and its charge. Diffusion through a channel is dependent on the *hydrated size* of the ion because the associated water molecules must diffuse along with the ion. The number of ion channels discovered in the membranes of plant cells is increasing. Currently there is solid evidence for K^+, Cl^-, and Ca^{2+} channels, while additional channels

for other inorganic and organic ions are strongly suggested. Channel proteins are frequently **gated**, which means they may be open or closed (Box 3.1). Solutes of an appropriate size and charge may diffuse through only when the channel "gate" is open. Two types of gates are known. An electrically gated channel opens in response to membrane potentials of a particular magnitude. Other channels may open only in the presence of the ion that is to be transported and may be modulated by light, hormones, or other stimuli. The precise mechanism of gated channels is not known, although it is presumed to involve a change in the three-dimensional shape, or conformation, of the protein.

The importance of carriers lies in the selectivity they impart with respect to which solutes are permitted to enter or exit the cell. Channels, on the other hand, appear to be involved wherever large quantities of solute, particularly charged solutes or ions, must cross the membrane rapidly. Whereas a carrier may transport between 10^4 and 10^5 solute molecules per second, a channel may pass on the order of 10^8 ions per second. It should also be stressed that large numbers of channels are not required to satisfy the needs of most cells. The rate of efflux through guard cell K^+ channels during stomatal closure, for example, has been estimated at 10^7 K^+ ions sec^{-1}—a rate that conceivably could be accommodated by a single channel. Many carrier and channel proteins are inducible, which means that they are synthesized by the cell only when there is solute available to be taken up.

3.2.4 ACTIVE TRANSPORT REQUIRES THE EXPENDITURE OF METABOLIC ENERGY

Many transport processes, in addition to being rapid and specific, will lead to an **accumulation** of solute inside the cell. In other words, the transport process will establish significant concentration or electrochemical gradients and will continue to transport solute against those gradients. The transport process involved is known as **active transport**. By definition, active transport is tightly coupled to a metabolic energy source—usually, although not always, hydrolysis of adenosine triphosphate (ATP). In other words, active transport requires an input of energy and does not occur spontaneously. Unlike simple and facilitated diffusion, active transport is also unidirectional—either into *or* out of the cell—and is always mediated by carrier proteins.

Active transport serves to accumulate solutes in the cell when solute concentration in the environment is very low. When used to transport solute out of the cell, active transport serves to maintain a low internal solute concentration. Because active transport systems move solutes against a concentration or electrochemical gradient, they are frequently referred to as **pumps**.

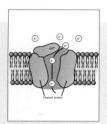

Box 3.1
ELECTROPHYSIO-LOGY—EXPLORING ION CHANNELS

The exchange of ions across cellular membranes is facilitated by the presence of transmembrane proteins referred to as ion channels. Most ion channels are highly specific for one or a limited number of ion species, which can diffuse through an open channel at rates as high as 10^8 s^{-1}.

Channel proteins may exist in two different conformations, referred to as open and closed. In the open conformation, the core of the protein forms a pathway for diffusion of ions through the membrane (Figure 3.4). A channel that can open and close is said to be gated—in the open conformation, the "gate" is open and ions are free to diffuse through the channel. When the gate is closed, the channel is not available for ion diffusion. A number of stimuli, including voltage, light, hormones, and ions themselves, are known to influence the frequency or duration of channel opening. The channel protein is believed to contain a sensor that responds to the appropriate stimulus by changing the conformation of the protein and opening the gate.

Because ions are mobile and carry a charge, their movement across membranes establishes an electrical current. These currents, typically on the

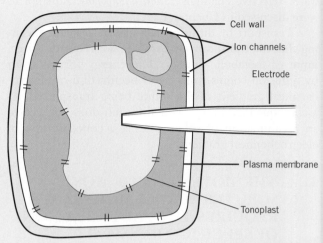

FIGURE 3.5 **Changes in electrical properties of the cell related to ion flow can be measured by inserting a micro-electrode directly into the cell.**

order of picoamperes (pA = 10^{-12} ampere), can be measured using microelectrodes constructed from finely drawn-out glass tubing. The first evidence for gated channels was based on experiments in which the electrode was inserted directly into the cell (Figure 3.5). This method has several limitations. It requires relatively large cells and the results reflect the activities of many different channels of various types at the same time. Moreover, when applied to plant cells, the electrode usually penetrates the vacuolar membrane as well as the plasma membrane, thus summing the behavior of channels in both membranes.

These problems were largely circumvented by development of the patch clamp method that permits the study of single ion channels in selected membranes. In this technique, the tip of the microelectrode is placed in contact with the membrane surrounding an isolated protoplast (a cell from which the cell wall has been removed) (Figure 3.6). A tight seal between the electrode and the membrane is formed by applying a slight suction. The small region of membrane in contact with the electrode is referred to as the "patch." Measurements may be made in this configuration, with the whole cell attached, or, alternatively, the electrode can be pulled away from the cell. In that case the patch then remains attached to the electrode tip and can be bathed in solutions of known composition. Note that the exterior surface of the membrane is in contact with the microelectrode solution (called the "inside-out" configuration). Variations in the technique permit the orientation of the patch to be reversed, with the internal surface facing the electrode solution (the "outside-out" configuration).

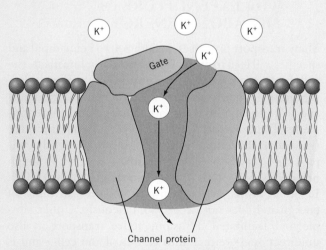

FIGURE 3.4 **A gated membrane channel. Gated channels may be open, in which case ions are permitted to pass through the channel, or closed to ion flow. Opening may be stimulated by changes in membrane potential, the presence of hormones, or the ion itself.**

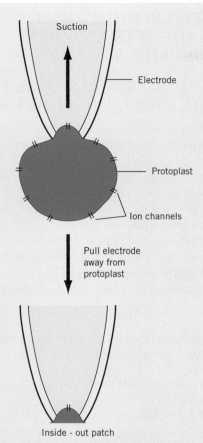

FIGURE 3.6 **Current flow through individual channels can be measured by the patch-clamp technique. A small piece of membrane containing a single channel can be isolated at the tip of a microelectrode. Ions flowing through the channel carry the current, which can be measured with sensitive amplifiers.**

With a sufficiently small electrode tip (ca. 1.0 μm diameter), the patch may contain a single ion channel.

With an appropriate electrical circuit, the experimenter can hold, or "clamp," the potential difference (voltage) across the patch at some predetermined value. At the same time, the circuit will monitor any current that flows through the membrane patch. Typical experimental results are shown in Figure 3.7. Figure 3.7A illustrates a patch-clamp recording for a single K^+ channel, using an inside-out patch of plasma membrane isolated from a stomatal guard cell protoplast. For the lower trace, the voltage was clamped at 0 mV. Note the appearance of one small, transient current pulse. For the upper trace, the voltage was stepped up to +90 mV. The current trace indicates that the channel opened immediately, in response to the voltage stimulus, and remained open for about 50 milliseconds (ms).

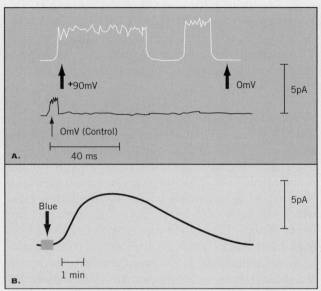

FIGURE 3.7 (*A*) **Current generated by a K^+-selective channel in the plasma membrane of a stomatal guard cell protoplast. Channel opening was stimulated by applying a 90-mV pulse across the membrane. Note that the channel exhibited a transient closure and opening before the voltage pulse was terminated. A 0 mV control is shown for comparison. (Reprinted with permission from J. I. Schroeder et al.,** *Nature* **312:361–362. Copyright 1984, Macmillan Magazines Ltd.) (*B*) Blue light stimulates a current in whole-cell membranes, an indication that membrane ion pumps are activated by blue light. (Reprinted with permission from S. M. Assmann et al.,** *Nature* **318:285–287. Copyright 1985, Macmillan Magazines Ltd.)**

It then spontaneously closed and reopened again for about 12 ms. This is an example of a voltage-gated ion channel. Figure 3.7B is a trace obtained with guard cell protoplasts, using a whole cell configuration. Note the difference in time scale. In this case, a pulse of blue light initiates current flow by activating a blue-light-activated proton pump.

The patch-clamp technique represents a major advance in electrophysiology that is providing insights into the mechanisms of plant membrane transport.

FURTHER READING

Molleman, A. 2003. *An Introductory Guide to Patch Clamp Electrophysiology*. Hoboken, NJ: John Wiley & Sons, Inc.

Volkov, A.G. 2006. *Plant Electrophysiology: Theory and Methods*. Berlin: Springer-Verlag.

3.3 SELECTIVE ACCUMULATION OF IONS BY ROOTS

The selectivity of transport proteins and the ability of cells to accumulate solutes by active transport together enables roots to selectively accumulate nutrient ions from the soil solution. Selective accumulation of ions by roots is illustrated by a typical set of data shown in Table 3.1. Accumulation refers to the observation that the concentrations of some ions inside the cell may reach levels much higher than their concentration in the surrounding medium. This difference in ion concentration is expressed quantitatively by the **accumulation ratio**, which can be defined as the ratio of the concentration inside the cell (C^i) to the concentration outside the cell (C^o). Note that in Table 3.1 the internal concentration of K^+ is more than 1,000 times greater than it is in the bathing medium. In the past, an accumulation ratio greater than 1 has been considered compelling evidence in favor of active transport, since that solute has evidently moved in against a concentration gradient. Conversely, an accumulation ratio less than 1 implies that the solute has been actively *excluded* or *extruded* from the cell. As will be shown below, this is not always the case, especially where charged solutes are involved. When assessing solute uptake by cells, it is especially important to distinguish between uncharged and charged solutes.

Ion uptake is highly selective. Note there is virtually no accumulation of Na^+ by maize roots (Table 3.1) and accumulation ratios for K^+ and NO_3^- are substantially higher than for SO_4^{2-}. The low concentrations of Na^+ in plant cells (unlike animal cells) may result from limited uptake of Na^+ in the first place, but also because Na^+ is actively expelled from most plant cells.

TABLE **3.1**   **The uptake of selected ions by maize roots.**

Ion	Accumulation Ratio		
	C^o (m)	C^i (m)	$[C^i/C^o]$
K^+	0.14	160	1142
Na^+	0.51	0.6	1.18
NO_3^-	0.13	38	292
SO_4^{2-}	0.61	14	23

Maize roots were bathed in nutrient solutions for four days. C^o and C^i are the ion concentrations of the medium and root tissue, respectively. C^i was measured as the concentration of ions in the sap expressed from the roots.
From data of H. Marschner, 1986.

3.4 ELECTROCHEMICAL GRADIENTS AND ION MOVEMENT

3.4.1 IONS MOVE IN RESPONSE TO ELECTROCHEMICAL GRADIENTS

Accumulation ratios for uncharged solutes, such as sugars, are relatively straightforward. It can be assumed that uptake is fundamentally dependent on the difference in concentration on the two sides of the membrane. In other words, for uncharged solutes it is the concentration gradient alone that determines the gradient in **chemical potential** (see Chapter 1). The chemical potential gradient ($\Delta\mu$) can be expressed by the following equation, where μ_i is the chemical potential of the uncharged solute in the cytosol and μ_o is the chemical potential of uncharged solute outside of the cell.

$$\Delta\mu = \mu_i - \mu_o \qquad (3.2)$$

Thus, for the uncharged solute, U, its chemical potential gradient ($\Delta\mu_u$) is represented by

$$\Delta\mu_u = RT \ln[U_i]/[U_o] \qquad (3.3)$$

Equation 3.3 simplifies to

$$\Delta\mu_u = 59 \log[U_i]/[U_o] \qquad (3.4)$$

where U_i is the concentration of U in the cytosol and U_o is the concentration of U outside of the cell. It is a relatively simple matter to measure experimentally internal and external concentrations of the solute and thus calculate the accumulation ratio from which one can calculate the chemical potential gradient. From equations 3.3 and 3.4, it is clear that when $U_i > U_o$, $\Delta\mu$ is a positive value, which indicates that uptake of U must occur by energy-dependent, active transport. On the other hand, when $U_i < U_o$, $\Delta\mu$ is a negative value, which indicates that uptake of U will occur by energy-independent, facilitated diffusion.

With charged solutes, or ions, the situation is more complex and the accumulation ratio is not always a valid indication of passive or active transport. Because ions carry an electrical charge, they will diffuse in response to a gradient in **electrical potential** as well as chemical potential. Positively charged potassium ions, for example, will naturally be attracted to a region with a preponderance of negative charges. Consequently, the movement of ions is determined by a gradient that has two components: one concentration and one electrical. In other words, ions will move in response to an **electrochemical gradient**, and the electrical properties of the cell, or its **transmembrane potential**, must be taken into account.

A transmembrane potential (a voltage or potential difference across a membrane) develops because of an unequal distribution of anionic and cationic charges across the membrane. The cytosol, for example, contains a large number of fixed or nondiffusible charges such as the carboxyl ($-COO^-$) and amino ($-NH_4^+$) groups of proteins. At the same time, cells use energy to actively pump cations, in particular H^+, Ca^{2+}, and Na^+, into the exterior space. The resulting unequal distribution of cations establishes a potential difference, or voltage, across the membrane. The cytosol remains negative relative to the cell wall space, which accumulates the positively charged cations.

A simple example illustrates how a transmembrane potential can influence ion movement into and out of cells (Figure 3.8). In this example, it is assumed that (1) the internal K^+ concentration is high relative to that outside the cell; (2) K^+ can move freely across the membrane, perhaps through K^+ channels; and (3) the internal K^+ concentration is balanced by a number of organic anions restrained within the cell. Under these conditions, it might be expected that K^+ will diffuse out of the cell, driven by its concentration gradient, until the concentrations of K^+ outside and inside the cell are equal. However, as K^+ diffuses out of the cell it leaves behind the nondiffusible anionic charges, thus creating a charge imbalance and thus a voltage (potential) difference across the membrane. The potential that is generated by such a combination of nondiffusible anions and mobile cations is referred to as a **Donnan potential**. The negative charges tend to pull the positively charged potassium ions back onto the cell. As a result, equilibrium is achieved not when the *concentrations* of K^+ are equal on both sides of the membrane, but when the membrane potential difference reaches a value such that the force of the concentration gradient pulling K^+ out of the cell is balanced by the force of the electrical gradient pulling K^+ back into the cell. Under these circumstances, the cell will maintain a high internal K^+ concentration and the accumulation ratio will be greater than unity, yet the movement of potassium ion is solely by passive diffusion. Because the unequal distribution of K^+ at equilibrium results from a Donnan potential, it is an example of **Donnan equilibrium**.

Anion distribution would also be influenced by the membrane potential, but in the opposite direction. Anions would be repelled by the preponderance of internal negative charges and attracted by the preponderance of external positive charges, thus leading to an accumulation ratio less than unity. It is clear from these examples that an accumulation ratio other than unity does not necessarily mean that active transport is involved.

Transmembrane potentials can be measured with a microelectrode made from finely drawn-out glass tubing. With the aid of a microscope, the electrode is inserted into the vacuole of a cell. A reference electrode is placed in the medium surrounding the cell. The difference in potential between the two electrodes can be measured with the aid of a sensitive voltmeter. It is by no means an easy technique, but many experimenters have become quite proficient with it. Potentials measured in this way are commonly in the range of -100 to $-130\,mV$ for young roots and stems, although potentials as high as $-200\,mV$ have been recorded for some algal cells. The cytosol is always negative with respect to the surrounding medium. Such potentials do not require a large charge imbalance. As few as one unbalanced charge in a population of a million ions is sufficient to generate a potential of $100\,mV$.

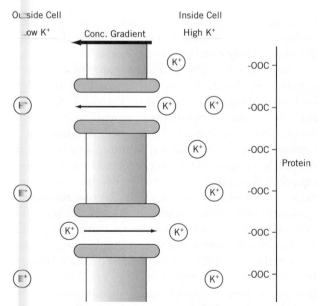

FIGURE 3.8 An electrical potential gradient may drive an apparent accumulation of cations (e.g., K^+) against a concentration gradient. The combination of concentration differences and charge differences constitutes an *electrochemical* gradient. Thus, potassium ions tend to accumulate in cells in response to the large number of fixed charges on proteins and other macromolecules.

3.4.2 THE NERNST EQUATION HELPS TO PREDICT WHETHER AN ION IS EXCHANGED ACTIVELY OR PASSIVELY

To predict the distribution of a charged ion, C, across a membrane, we have to take into account the concentration difference of the ion across the membrane as well as the contribution of electrical potential difference (ΔE_c) across the membrane. Thus, equation 3.3 simplifies to

$$-z\,\Delta E_c = 59 \log[C_i]/[C_o] \qquad (3.5)$$

TABLE 3.2   The uptake of selected ions by roots of pea (*Pisum sativum*) and oat (*Avena sativa*). The Nernst equation was used to predict the internal concentration (C_i) assuming cell membrane potentials of $-110\,mV$ and $-84\,mV$ for pea and oat roots, respectively. The accumulation ratio was calculated on the basis of the measured (or actual) C_i. The symbol E under probable uptake mechanism refers to active exclusion from the root. The symbol U denotes active uptake by the roots.

Ion	C^o	Predicted C^i	Actual C^i	Accumulation Ratio	$\dfrac{\text{Actual}}{\text{Predicted}}$	Probable Uptake Mechanism
			Pea Root			
K^+	1.0	74	75	75	1.01	Diffusion
Na^+	1.0	74	8	8	0.108	E
Ca^{2+}	1.0	5400	1.0	1.0	0.00018	E
NO_3^-	2.0	0.027	28	14	1037	U
$H_2PO_4^-$	1.0	0.014	21	21	1500	U
SO_4^{2-}	0.25	0.000047	9.5	38	202,127	U
			Oat Root			
K^+	1.0	27	66	66	2.4	Diffusion (?)
Na^+	1.0	27	3	3	0.11	E
Ca^{2+}	1.0	700	1.5	1.5	0.0021	E
NO_3^-	2.0	0.076	56	28	741	U
$H_2PO_4^-$	1.0	0.038	17	17	447	U
SO_4^{2-}	0.25	0.000036	2	8	5,555	U

From data of Higinbotham et al., 1967.

where z is the valency or absolute charge of the ion, C, and C_i and C_o are the concentrations of this ion in the cytosol and outside the cell respectively. Equation 3.5, which shows relationship between transmembrane potential gradient and ion distribution across the membrane is called the **Nernst equation** where ΔE_c is the electrical potential difference (also known as the **Nernst potential**). The value of z for a univalent cation, for example, would be 1 while for calcium or magnesium it would be 2. For chloride or nitrate it would be -1 and for sulphate it would be -2.

The Nernst equation is useful because it allows us to make certain predictions about the equilibrium concentrations for ions inside the cell (C_i) when ion transport is due to facilitated diffusion. In order to apply the equation, it is necessary to measure the transmembrane potential and the concentrations of ions both inside and outside the cell. If the actual concentrations deviate significantly from those predicted by the Nernst equation, it may be considered evidence that either active uptake or active expulsion of the ions is involved. In other words, the Nernst equation can be used to determine the probability of whether an ion is actively or passively distributed across the membrane. If the measured internal concentrations are approximately equal to the calculated Nernst value, it can be assumed that the ion has been distributed passively. If the measured

concentration is greater than predicted, active uptake is probably involved, and, if lower, it is likely that the ion is being actively expelled from the cell.

Application of the predictive value of the Nernst equation is illustrated by the experiment of Higinbotham and colleagues shown in Table 3.2. At equilibrium the measured membrane potentials were $-110\,mV$ for pea roots and $-84\,mV$ for oat roots. C^o, the concentration of ions in the external solution, was known and the predicted value of C^i was calculated from the Nernst equation. The ratio of the predicted value to the actual value is a measure of how well the Nernst relationship applies to that particular ion. Note that the accumulation ratios for almost all of the ions are greater than 1.0, indicating some degree of accumulation in the cell. Only in the case of K^+ was the ratio of predicted concentration to actual concentration near 1.0. This indicates that K^+ is near electrochemical equilibrium and was probably accumulated passively, at least in pea roots. There appears to be a possibility of some active accumulation of K^+ by oat roots. Cellular concentrations of both Na^+ and Ca^{2+} are lower than predicted. Since other evidence supports the existence of Na^+ and Ca^{2+} pumps in the membrane, these ions probably entered passively down their electrochemical gradients but were then actively expelled. Internal concentrations for all three anions are much higher than predicted, indicating they are actively taken up by the cells. This is understandable, since energy would be required to overcome the normal transmembrane potential and move negatively charged ions into the predominantly negative environment inside the cell.

Although these results are indicative of active and passive transport, further tests would help to confirm the conclusion in each case. Since active transport requires a direct input of metabolic energy, it is sensitive to oxygen and respiratory poisons. Thus reduced uptake of a particular ion in the absence of oxygen or in the presence of respiratory inhibitors such as cyanide or 2,4-dinitrophenol would be evidence in support of active transport. Even this evidence is not always compelling, however, since effects of inhibitors may indirectly influence nutrient uptake. For example, even transport by diffusion ultimately requires an expenditure of metabolic energy, if only to establish and maintain the organization of membranes and other properties of the cell that make transport possible. It is not always a simple matter to distinguish between direct and indirect involvement of energy. The criteria for active transport usually require that the solute distribution not be in electrochemical equilibrium and that there be a quantitative relationship between energy expended and the amount of solute transported.

Finally, it should be noted that when ion concentrations are known, the Nernst equation can also be used to estimate the transmembrane potential, or Nernst potential, *contributed by that ion*. At steady state, however, calculation of the membrane potential is complicated by the fact that many different ion species, each with a different permeability, are simultaneously crossing the membrane in both directions. The individual contributions of all ion gradients must consequently be taken into account and summed in order to arrive at the overall potential for the cell. In practice, however, K^+, Na^+, and Cl^- are the dominant ions. These ions have the highest permeabilities and concentrations in plant cells and a reasonable estimation of transmembrane potential can be based on these three ions alone.

3.5 ELECTROGENIC PUMPS ARE CRITICAL FOR CELLULAR ACTIVE TRANSPORT

3.5.1 ACTIVE TRANSPORT IS DRIVEN BY ATPASE-PROTON PUMPS

Energy to drive active transport comes chiefly from the hydrolysis of ATP. The energy-transducing membranes of chloroplasts and mitochondria contain large, multiprotein ATPase complexes (Chapters 7, 10). The chloroplast and mitochondrial ATPases, known as F-type ATPases, utilize the energy associated with an electrochemical proton gradient across a membrane to

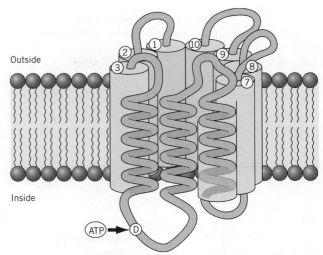

FIGURE 3.9 A model for the plasma membrane H⁺-ATPase. The enzyme is a single chain with 10 hydrophobic, membrane-spanning domains. Only three are shown here as helical coils, while the remaining seven are schematically represented as cylinders. Linking adjacent membrane-spanning domains are hydrophilic loops that project into the cytosol (inside) and the cell wall apoplast (outside). The ATP-binding site is an aspartic acid residue (D) located on the hydrophilic loop between the fourth and fifth membrane-spanning domains. The H⁺-binding site is located in the hydrophobic domain. Hydrolysis of ATP at the binding site is thought to change the conformation of the enzyme, thereby exposing the H⁺-binding site to the outside of the membrane where the H⁺ is released. (Compare with Figure 5.9.)

drive ATP *synthesis*. ATPase complexes are also found in the plasma and vacuolar membranes of cells and possibly other membranes as well. These ATPases are called **ATPase-proton pumps** (or, **H⁺-ATPase**). Known as P-type ATPases, the plasma membrane ATPase-proton pumps are structurally distinct (Figure 3.9) and operate in reverse of the F-type. Rather than synthesizing ATP, the plasma membrane ATPases *hydrolyze* ATP and use the negative free energy to "pump" protons from one side of the membrane to the other against an electrochemical gradient. An ATPase-proton pump thus serves as a proton-translocating carrier protein and the free energy of ATP hydrolysis is conserved in the form of a proton gradient across the membrane. This proton gradient (ΔpH), together with the normal membrane potential ($\Delta \Psi$), contributes to a **proton motive force (pmf)** that tends to move protons back across the membrane (Equation 3.6).

$$pmf = \Delta \Psi - 59 \, \Delta pH \qquad (3.6)$$

It is generally conceded that the proton motive force established by pumping protons across membranes is the primary source of energy for a variety of plant activities. We will revisit this equation and the concept of a proton motive force in our discussion of ATP synthesis in the chloroplast and the mitochondrion (Chapter 5).

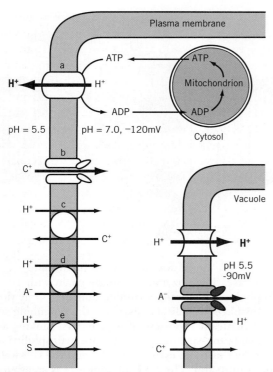

FIGURE 3.10 **Schematic diagram relating the activity of a membrane ATPase-proton pump to solute exchange. The proton pump (*a*) uses the energy of ATP to establish both a proton gradient and a potential difference (negative inside) across the membrane. The energy of the proton gradient may activate an ion channel (*b*), or drive the removal of ions from the cell by an antiport carrier (*c*), or drive the uptake of ions or uncharged solute by a symport carrier (*d*, *e*). Similar pumps and carriers operate across the vacuolar membrane. C^+, cation; A^-, anion; S, uncharged solute.**

Included are activities such as active transport of solutes (cations, anions, amino acids, and sugars), regulation of cytoplasmic pH, stomatal opening and closure, sucrose transport during phloem loading, and hormone-mediated cell elongation.

A schematic model relating ATPase-proton pumps to solute exchange across membranes is shown in Figure 3.10. The ATP required to drive the pump is ultimately derived from oxidative phosphorylation in the mitochondria. The proton-translocating protein is shown extending across the plasma membrane, with its ATP binding site on the cytosolic side. Hydrolysis of the ATP results in the translocation of one or more protons from the cytosol to the surrounding apoplastic cell wall space.

There are several particularly interesting consequences of the ATPase-proton pump. First, a single ion species is translocated in one direction. This form of transport is consequently known as a **uniport** system. Second, because the ion transported carries a charge, an electrochemical gradient is established across the membrane. In other words, the ATPase-proton pump is **electrogenic**—it contributes directly to the negative

potential difference across the plasma membrane. In fact, the electrogenic proton pump is a major factor in the membrane potential of most plant cells. From equation 3.5, a tenfold difference in proton concentration (one pH unit) at 25°C contributes 59 mV to the potential. Since the proton gradient across the plasma membrane is normally on the order of 1.5 to 2 pH units, it can account for approximately 90 to 120 mV of the total membrane potential. Third, since the ions translocated are protons, the ATPase-proton pump establishes a proton gradient as well as an electrical gradient across the membrane. Energy stored in the resulting electrochemical proton gradient (or the *proton motive force*) can then be coupled to cellular work in accordance with Mitchell's *chemiosmotic hypothesis* (see Chapter 5). Indeed, this is an excellent demonstration of how chemiosmotic coupling is not restricted to ATP synthesis in chloroplasts and mitochondria but can be used to perform other kinds of work elsewhere in the cell.

Several ways of coupling the electrochemical proton gradient to solute movement across the membrane are illustrated in Figure 3.10. In the first case, the electrogenic pump contributes to the charge-dependent uptake of cations through ion-specific channels (Figure 3.10) (b)). Second, the return of protons to the cytosol can be coupled with the transport of other solute molecules at the same time, or **cotransport**. Transport of both ions is mediated by the same carrier protein and the movement of the second solute is obligatorily coupled to the inward flux of protons down their electrochemical gradient. If the second ion moves in opposite direction from the proton, the method of cotransport is referred to more specifically as **antiport**. In Figure 3.10 (c), for example, proton flux into the cell is shown coupled to the efflux of other cations out of the cell. Here the energy of the proton electrochemical gradient is used to maintain low internal concentrations of specific cations. Any cations that do chance to "leak" into the cell, no doubt passively through ion channels, are thus pumped out against their electrochemical gradient. If the two solutes move in the same direction at the same time, the method of cotransport is referred to as **symport**. Two examples are shown in Figure 3.10 (d, e). In the first example, proton flux into the cell is coupled with the uptake of anions (A^-) against their electrochemical gradient. In the second example of symport, the proton gradient can be used to power the uptake of uncharged solutes (S), such as sugars. All three examples of cotransport are forms of active transport mediated by specific carrier proteins.

3.5.2 THE ATPASE-PROTON PUMPS OF PLASMA MEMBRANES AND VACUOLAR MEMBRANES ARE DIFFERENT

Much of the pioneering experimental work on membrane ATPase has been conducted with small, spherical vesicles obtained from isolated cellular membranes. When membranes are disrupted, the pieces naturally seal off to form vesicles because of their strongly hydrophobic nature. While the technique is relatively straightforward, the preparation of vesicles from a single membrane source, and thus containing a single type of ATPase, presents some difficulties. Contamination by chloroplast ATPase can be avoided by isolating the membranes from dark-grown, etiolated tissue, while mitochondria can usually be separated from other membranes by differential centrifugation. It is more difficult, however, to separate plasma membranes from other cellular membranes such as the vacuolar membrane and, consequently, many of the early studies were characterized by inconsistent results from different laboratories. These inconsistencies were resolved when it became clear that the membrane preparations often contained two types of electrogenic ATPase-proton pumps: one associated with the plasma membrane and one with the vacuolar membrane. Improved techniques have enabled at least partial separation of the two membranes by density gradient centrifugation and it is now possible to characterize their respective ATPases. The plasma membrane–type proton-pumping ATPase is characteristically inhibited by vanadate ion (VO_3^-) but is generally insensitive to other anions such as NO_3^-. Vanadate competes with phosphate for binding sites, indicating that ATP transfers a phosphate group to the ATPase protein (Figure 3.9). The resulting energy-rich **phosphoenzyme** then undergoes a conformational change that exposes the proton-binding site to the outside. Evidence thus far indicates that a single proton is translocated for each ATP hydrolyzed.

The vacuolar-type ATPase-proton pump (or V-type) differs from the plasma membrane–type in several ways. It is, for example, insensitive to vanadate but strongly inhibited by nitrate. In this respect it is similar to mitochondrial, or F-type, ATPase, which is also insensitive to vanadate. Should the preparation be contaminated with any mitochondrial ATPase, however, its activity can be blocked by including oligomycin or azide in the assay medium. Both inhibit mitochondrial ATPase without affecting the activity of the tonoplast-type. Structurally, the vacuolar-type ATPase is also more similar to the mitochondrial F-type ATPase than to the plasma membrane ATPase. Like the F-type, the V-type can be separated into a complex of hydrophobic subunits embedded in the membrane (analogous to F_0) and a complex of soluble, hydrophilic subunits (analogous to F_1). Although the soluble complex contains an ATP-binding site, insensitivity of the V-type ATPase to vanadate suggests that it does not form a phosphorylated intermediate. The vacuolar version also appears to differ in that it transports two protons for each molecule of ATP that is hydrolyzed. The function of the vacuolar ATPase is

to pump protons from the cytosol into the vacuole, thus accounting for the fact that the potential of the vacuole is more positive than the cytosol by some 20 to 30 mV (Figure 3.10). In extreme cases, large pH gradients can be maintained across the vacuolar membrane and the vacuolar sap may become quite acidic. For example, the pH of lemon juice (which is predominantly vacuolar sap) is normally about 2.5. Like the plasma membrane pump the vacuolar pump is also electrogenic, except that the accumulated protons serve to reduce the potential of the vacuole relative to the cytosol. The resulting potential difference serves to drive anions (e.g., Cl^- or malate) into the vacuole, which is less negative than the cytosol. The electrochemical proton gradient can also be used to drive cations (e.g., K^+ or Ca^{2+}) into the vacuole by an antiport carrier. Both of these activities make important contributions to the turgor changes that drive stomatal guard cell movement and the specialized motor cells that control nyctinastic responses (Chapter 23).

In addition to the H^+-ATPases, some membranes, such as the plasma membranes, chloroplast envelope, the endoplasmic reticulum, and vacuolar membrane, also contain calcium-pumping ATPases (**Ca^{2+}-ATPases**). Ca^{2+}-ATPases couple the hydrolysis of ATP with the translocation of Ca^{2+} across the membrane. In the case of the plasma membrane, the calcium is pumped out of the cytosol. This serves to keep the cytosolic Ca^{2+} concentration low, which is necessary in order to avoid precipitating phosphates and to keep Ca^{2+}-dependent signaling pathways operating properly (Chapters 16 and 17).

3.5.3 K^+ EXCHANGE IS MEDIATED BY TWO CLASSES OF TRANSPORT PROTEINS

Since the 1950s, the uptake of K^+ into cells—especially root cells—has been studied more thoroughly than any other ion species. Much of the early work was carried out by Emanuel Epstein, who pioneered the use of $^{86}Rb^+$, a radioactive K^+ analog, to follow K^+ uptake in low-salt roots. Epstein was also the first to treat ion transporters as enzymes and to analyze their data using the methods of classical enzyme kinetics. An analysis of the initial rate of K^+ absorption at different external concentrations showed that K^+ absorption is biphasic. On the basis of these results, Epstein proposed that there were two types of K^+ transport systems in plant cells: a high-affinity transport system (HAT) that is active at low K^+ concentrations ($\leq 200\,\mu M$), and a low-affinity transport system (LAT) that is active at high K^+ concentrations. Such transport systems have now been identified for myriad macronutrient ions including Ca^{2+}, NO_3^-, SO_4^{2-} and PO_4^{2-}. A consistent feature of all HATs is they mediate a slow rate of ion uptake when the external concentrations of the ion are low. This means that HATs exhibit a low capacity for ion uptake. However, although their capacity for ion uptake is low, their efficiency for ion uptake is very high because of their high binding affinity for the specific ion. Thus, the uptake kinetics for HATs exhibit nonlinear, saturation kinetics. In contrast, although LATs exhibit low affinity for ion uptake in a linear concentration dependence, the capacity of LATs for uptake is high. The ion concentration which typically induces a transition from LATs to HATs is about 1 mM for most macronutrient ions. Biophysical and molecular genetic studies have confirmed the existence of multiple transporters with different substrate affinities. Mosts physiological, biochemical, and molecular studies of macronutrient ion transport have focused on the characterization of HATs. However, given that agricultural soil concentrations of K^+, NO_3^-, and NH_4^+ may exceed 1 mM, LATs may also be important in maintaining plant productivity under field conditions.

Experiments using microelectrodes that measure cytoplasmic K^+ concentration and membrane potential simultaneously in single root cells have confirmed that when the external concentration is low, there is a strong K^+ electrochemical gradient across the plasma membrane. In order for K^+ to move into the cell under those conditions, the high-affinity uptake system must be driven by an active transport mechanism. This high-affinity uptake system is probably a H^+-ATPase-linked K^+-H^+ symporter. Patch-clamp studies, on the other hand, have established that the low-affinity uptake system involves channels that move K^+ either into (K_{in}^+ channels) or out (K_{out}^+ channels) of the cell. While the focus tends to be on K_{in}^+ channels, channels that allow K^+ to move out of the cell are also important in controlling osmotic adjustment, maintaining stomatal function, and driving nyctinastic movements of leaves (Chapter 23).

K^+ transporters are highly efficient and, consequently, their relative abundance in the cell membranes is very low. This makes it difficult to isolate and purify transporters by traditional biochemical techniques. Some genes that encode K^+ transporters and other plant transport proteins have been identified by cloning the genes in transport-deficient mutants of yeast (*Saccharomyces cerevisiae*). A yeast mutant deficient of a high-affinity K^+ transporter, for example, will grow on a medium with a high K^+ concentration, but not on a low-concentration medium. However, transformation of the yeast with *KAT1* (a gene encoding a high-affinity K^+ transporter from the guard cells and vascular tissue of *Arabidopsis*), will restore the capacity of the yeast to grow on low K^+. Experiments of this type have led to the identification of at least a dozen putative K^+

transporters as well as transporters for sugars, amino acids, NH_4^+, and SO_4^{2-}.

3.6 CELLULAR ION UPTAKE PROCESSES ARE INTERACTIVE

Deprivation of the macronutrients nitrogen (N), phosphorus (P), potassium (K), and sulfur (S) can be a limiting factor for plant growth and survival under natural conditions. How do plant root cells sense changes in soil nutrient macronutrient availability and how is this signal transduced by the root cell to elicit an appropriate response at the molecular, biochemical, and physiological levels to adjust the rate of growth to match the macronutrient availability? Although the cellular networks involved in the sensing and signalling of limitations in macronutrient resource availability are not well defined, it is clear that the deficiency in one nutrient can affect the uptake of another nutrient. For example, a decrease in SO_4^{2-} availability can disrupt nitrogen metabolism resulting in the accumulation of NO_3^- in the leaves. Similarly, NH_4^+ availability is known at affect K^+ uptake. Numerous molecular studies indicate that K^+ limitation represses the expression of NO_3^- transporters but stimulates the expression of NH_4^+ transporters. PO_4^{2-} deprivation is one of the most studied phenomena associated with nutrient uptake. Recent gene micro-array studies indicate that phosphorus deficiency in plants causes a plethora of transcriptional responses, including the up-regulation of SO_4^{2-} transporters as well as iron transporters. PII proteins (Chapter 11) are key regulators of cellular NO_3^- supply to ensure a balance between C- and N-metabolism. Thus, the mechanisms controlling the uptake of these macronutrients nutrients do not occur necessarily independently of one another but rather are interactive. The complex, interactive effects of macronutrient deprivation on gene transcription and cellular physiology indicate that nutrient uptake is best described as a complex network of pathways. As a consequence, there must be molecular **cross talk** between the different ion uptake mechanisms to counteract potential imbalances caused by deficiencies in nutrients.

3.7 ROOT ARCHITECTURE IS IMPORTANT TO MAXIMIZE ION UPTAKE

Even though the uptake of ions by roots is essentially a cellular problem, the organization of roots at the tissue level cannot be totally ignored. The organization and architecture of roots are such that they can absorb some mineral salts without them ever entering a cell.

3.7.1 A FIRST STEP IN MINERAL UPTAKE BY ROOTS IS DIFFUSION INTO THE APPARENT FREE SPACE

Most nutrient uptake studies are carried out with "intact" tissues, such as excised barley roots. A typical pattern for the uptake of Ca^{2+} by low-salt barley roots is illustrated by the kinetic diagram in Figure 3.11. Note that initially, usually within the first few minutes, uptake of Ca^{2+} is very rapid. It then settles into a slow but steady accumulation over time. If at some point the roots are transferred to a large volume of solution lacking calcium, Ca^{2+} will be lost from the root into the bathing solution as shown by the dashed lines. When the bathing solution is distilled water, the quantity of ions lost is usually less than the quantity taken up during the initial rapid phase. If the roots are then transferred from distilled water to a bathing solution containing another cation, say Mg^{2+}, an additional quantity of Ca^{2+} will be lost from the tissue. If volumes of the bathing solutions are sufficiently large, the total quantity of Ca^{2+} lost from the tissue will approximately equal the quantity taken up during the initial rapid phase.

The kinetics of calcium uptake and release in this experiment can be interpreted as follows. Assume that

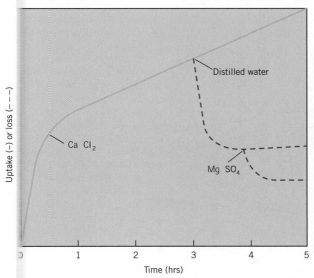

FIGURE 3.11 Typical kinetics for the uptake of Ca^{2+} into roots. When low-salt roots are placed in a solution of calcium chloride, an initial rapid uptake is followed by a slower but steady accumulation of calcium ion. If the roots are then transferred to a large volume of distilled water, some of the calcium diffuses out of the roots. Transfer to a strong magnesium solution releases additional calcium into the medium. The total amount of calcium released is equivalent to the amount taken up by free diffusion during the initial rapid phase.

there is a fraction of the root tissue volume, called **apparent free space (AFS)**, that is not separated from the environment by a membrane or other diffusion barrier. Because there are no barriers, Ca^{2+} in the surrounding medium would have access to the apparent free space by simple diffusion. When root tissue is immersed in the calcium solution, Ca^{2+} will rapidly diffuse into the AFS until the Ca^{2+} concentration in the AFS reaches equilibrium with the bathing solution. This accounts for the initial, rapid uptake of calcium. Thereafter, calcium ions are more slowly but steadily transported across the cell membrane and accumulated by the tissue. When the roots are transferred to distilled water, some of the Ca^{2+} ions present in the free space are free to diffuse back into the surrounding solution—and they will do so until equilibrium is again reached. The Ca^{2+} taken up by the cells, having already been transported across the cell membrane, is not free to diffuse back and remains in the cells.

A further loss of calcium when the roots are transferred to the solution containing magnesium ions is taken as evidence that the tissue behaves as a cation exchange material. That is to say, the AFS matrix, primarily cell wall components, is negatively charged and holds some cations by electrostatic attraction just as soil colloids do. These adsorbed ions are not free to diffuse out of the tissue into distilled water, but can be displaced by other cations, such as magnesium in the example given above. Thus, apparent free space describes that portion of the root tissue that is accessible by free diffusion and includes ions restrained electrostatically due to charges that line the space.

A variety of techniques have been developed to measure the root volume given over to AFS. In principle, the volume of the AFS can be estimated by the following experiment. A sample of roots weighing 1.0 g was immersed in a solution containing 20 μmoles ml^{-1} potassium sulfate (K_2SO_4). The roots were then removed from the solution, blotted to remove excess solution, and placed in a large volume of distilled water. It was found that 4.5 μmoles of sulphate were released from the roots into the distilled water. If it is assumed that sulphate in the AFS was in equilibrium with the external solution, that is, 20 μmoles ml^{-1}, then the volume occupied by the sulphate in the tissue can be calculated: 4.5 μmole/20 μmoles ml^{-1} = 0.22 ml. Thus, the volume of root tissue freely accessible by diffusion is 0.22 ml. By further assuming that 1 g of root tissue occupies approximately 1 ml of volume, the proportion of tissue freely accessible by diffusion is approximately 22 percent by volume. Estimates for the volume occupied by AFS do vary, depending on the species, conditions under which the roots were grown, whether the measurements are corrected for surface films, and so forth. Still, most measured values for AFS tend to fall in the 10 to 25 percent range.

3.7.2 APPARENT FREE SPACE IS EQUIVALENT TO THE APOPLAST OF THE ROOT EPIDERMAL AND CORTICAL CELLS

Exactly what constitutes AFS in a root? If it is assumed that AFS is the volume of the root that is accessible by free diffusion, then it probably consists of the cell walls and intercellular spaces (equivalent to the apoplastic space) of the epidermis and cortex. These are the regions of the root that can be entered *without crossing a membrane*. In most cases there is a strong correlation between the calculated volume of AFS and the calculated volume of cell walls in the cortex of the root. Furthermore, the cation exchange capacity of the AFS can be traced to the carboxyl groups (—COO⁻) associated with the galacturonic acid residues in the cell wall pectic compounds.

Almost certainly, the AFS stops at the endodermis where, in most roots, the radial and transverse walls develop characteristic thickenings called the **Casparian band** (see Chapter 2, Figure 2.10). The Casparian band is principally composed of a complex mixture of hydrophobic, long-chain fatty acids and alcohols called **suberin**. These hydrophobic substances impregnate the cell wall, filling in the spaces between the cellulose microfibrils. They are, in addition, strongly attached to the plasma membrane of the endodermal cells. The hydrophobic, space-filling nature of the Casparian band along with its attachment to the membrane greatly reduces the possibility that ions or small hydrophilic molecules can pass between the cortex and stele without first entering the symplast. This means, of course, that they must pass through the plasma membrane of the cortical or endodermal cells and are, consequently, subject to all of the control and selectivity normally associated with membranes.

3.8 THE RADIAL PATH OF ION MOVEMENT THROUGH ROOTS

3.8.1 IONS ENTERING THE STELE MUST FIRST BE TRANSPORTED FROM THE APPARENT FREE SPACE INTO THE SYMPLAST

Rapid distribution of nutrient ions throughout the plant is accomplished in the xylem vessels. In order to reach these conducting tissues, which are located in the central core, or **stele**, of the root, the ions must move in a radial path through the root. The path these ions must follow is diagrammed in Figure 3.12. For these purposes we may consider the root as consisting of three principal regions. The outermost region consists of the root epidermis (often referred to as the **rhizodermis**) and the cortical cells. The innermost region consists of vascular tissues—the vessel elements and associated parenchyma cells—which are of particular

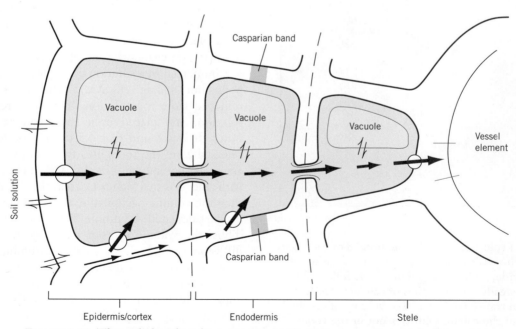

FIGURE 3.12 **The radial paths of ion movement through a root. Arrows indicate the alternative paths that may be taken by nutrient ions as they move from the soil solution into the vascular elements in the stele. Arrows with circles indicate active transport of ions across plasma membranes.**

interest for our discussion. Separating the two is the endodermis with its suberized Casparian band.

Ion uptake begins with free diffusion into the apparent free space. As noted in the previous section, the apparent free space is equivalent to the apoplast outside the endodermis and the Casparian band effectively prevents further apoplastic diffusion through the endodermis into the stele. Hence the only possible route for ions to pass through the endodermis is to enter the symplast by some carrier- or channel-mediated transport at the cell membrane. This may occur either on the outer tangential wall of the endodermal cell itself or through any of the epidermal or cortical cells. Regardless of which cell takes up the ions, symplastic connections (i.e., plasmodesmata) facilitate their passive movement from cell to cell until they arrive at a xylem parenchyma cell in the stele. At this point the ions may be unloaded into the xylem vessels for long distance transport to the leaves and other organs.

3.8.2 IONS ARE ACTIVELY SECRETED INTO THE XYLEM APOPLAST

With the exception of the very tip of the root where the young xylem vessel elements are still maturing, functional xylem is part of the apoplast. The interconnected vessel elements are devoid of cytoplasm and consist only of nonliving tubes filled with an aqueous solution. Release of ions into the xylem thus requires a transfer from the symplast into the apoplast. At one time, it was thought that this transfer was simply a passive leakage, but it is now clear that ions are actively secreted from xylem parenchyma cells. Although there is some conflicting evidence, ion concentration in the apoplast of the stele is generally much higher than in the surrounding cortex. This suggests that ions are being accumulated in the xylem against a concentration gradient, presumably by an energy-dependent, carrier-mediated process. It is also interesting to speculate, in this regard, that the Casparian band also functions to prevent loss of ions from the stele by blocking their diffusion down a concentration gradient.

In addition to working uphill against a concentration gradient, delivery of ions into the xylem vessels is sensitive to metabolic inhibitors such as carbonyl-cyanide-*m*-chlorophenylhydrazone (CCCP), which uncouples ATP formation. It is interesting that ion transport into the xylem is also sensitive to cycloheximide, an inhibitor of protein synthesis, but uptake into the root, at least initially, is not affected. Two plant hormones (abscisic acid and cytokinin) have a similar effect. Whether inhibitors of protein synthesis and hormones are affecting symplastic transport through the endodermis or unloading of ions from the endodermis into the xylem is not certain, but these results at least raise the possibility that ion release into

the vessels is a different kind of process than ion uptake by the roots.

3.8.3 EMERGING SECONDARY ROOTS MAY CONTRIBUTE TO THE UPTAKE OF SOME SOLUTES

The possibility remains that a limited portion of ion uptake may be accomplished entirely through the apoplast, at least in some roots. More basal endodermal cells—the distance from the tip is variable, but measured in centimeters—are characterized by additional suberin deposits that cover the entire radial and inner tangential wall surfaces. This would seem to present an additional barrier to apoplastic flow. However, in *some* plants, a small number of endodermal cells, called **passage cells**, remain unsuberized. Passage cells might represent a major point of entry for solutes into the stele.

Apoplastic continuity between the cortex and stele may also be established at the point of lateral root formation. One series of experiments, for example, followed the path of fluorescent dyes into the vascular tissues and shoots of corn (*Zea mays*) and broad bean (*Vicia faba*) seedlings. These dyes were chosen because they cannot be taken up by cells and thus are normally confined to the apoplast. The point of dye entry was traced to recently emerged secondary roots. These branch roots arise in the **pericycle**, a layer of cells immediately *inside* the endodermis. The emergence of the root primordia through the endodermis disrupts the continuity of the Casparian band and establishes, at least temporarily, the apoplastic continuity required to allow diffusion of the dye into the vascular tissue. Continuity of the apoplast through passage cells and secondary roots has been cited to explain increased calcium uptake in certain regions of corn roots. It may also help to account for the fact that a plant appears to contain virtually every element that is found in its environment, even those not known to be essential or not accumulated by plant cells.

The uptake of ions is not uniform along the length of the root. As shown in Table 3.3, uptake of calcium is highest in the apical 3 cm of the root while potassium is taken up in roughly equivalent amounts along the first 15 cm. Moreover, most of what is taken up in the tip (almost two-thirds of the calcium and three-fourths of the potassium) remains in the root. The proportion of ions translocated to the shoot increases with increasing distance from the tip. It is also interesting that when calcium is taken up further along the root (12 to 15 cm from the tip), it is translocated to the shoot but not to the tip. Clearly, although substantial progress has been made in several laboratories, the transport of ions through roots and into the xylem remains a complex and challenging field of study.

TABLE 3.3   Uptake and translocation of potassium and calcium as a function of position along a corn root.

Zone of application[1]	Ion	Total Uptake[2]	Percent Retained	Percent Translocated to:	
				Root Tip	Shoot
0–3	K^+	15.3	75	—	25
	Ca^{2+}	6.3	63	—	37
6–9	K^+	22.7	17	19	64
	Ca^{2+}	3.8	42	—	58
12–15	K^+	19.5	10	10	80
	Ca^{2+}	2.8	14	—	86

[1] Distance from root tip, cm.
[2] Uptake expressed as microequivalents per 24 hours.
Based on data of H. Marschner and C. Richter, 1973, Z. Pflenzenernaehr, Bodenkd, 135:1–15.

3.9 ROOT-MICROBE INTERACTIONS

The influence of living roots extends well beyond the immediate root surface into a region of the soil defined as the **rhizosphere**. A principal manifestation of this influence is the numerous associations that develop between roots and soil microorganisms, especially bacteria and fungi. Root-microbe associations can at times be quite complex and may involve invasion of the host root by the microorganism. Alternatively the microorganism may remain free-living in the soil. In either case, the association may prove beneficial to the plant or it may be pathogenic and cause injury.

3.9.1 BACTERIA OTHER THAN NITROGEN FIXERS CONTRIBUTE TO NUTRIENT UPTAKE BY ROOTS

Plant roots generally support large populations of bacteria, principally because of the large supply of energy-rich nutrients provided by the growing root system. The immediate environment of the roots is so favorable to bacterial growth that the bacterial population in the rhizosphere may exceed that in the surrounding bulk soil by as much as 50 percent. Nutrients provided by the roots are comprised largely of amino acids and soluble amides, reducing sugars, and other low-molecular-weight compounds. These compounds may either leak from the cells (a nonmetabolic process) or be actively secreted into the apoplastic space from whence they readily diffuse into the surrounding rhizosphere.

Dominant among the root secretions are the mucilages: polysaccharides secreted by Golgi vesicles in cells near the growing tip. Secretion of mucilage appears to be restricted to cells such as root cap cells, young epidermal cells, and root hairs where sec-ondary walls have yet to form. Secretion of mucilage in the more basal regions of the root appears to be restricted by development of secondary walls. The mucilage is rapidly invaded by soil bacteria that contribute their own metabolic products, including mucopolysaccharides of the bacterial capsule. In addition, mucilage also attracts colloidal mineral and organic matter from the soil. The resulting mixture of root secretions, living and dead bacteria, and colloidal soil particles is commonly referred to as **mucigel**.

There is no doubt that bacteria are intimately involved in the nitrogen nutrition of plants. Both invasive and free-living nitrogen-fixing bacteria, known since the late nineteenth century, are the primary source of nitrogen for plants. In addition, other soil bacteria convert ammonium nitrogen to nitrate. But to what extent do the bacteria influence other aspects of plant nutrition? In the previous chapter we pointed out that phosphorous is sparingly soluble in most soils and, in natural ecosystems, is often the limiting nutrient. There is some evidence that soil bacteria can assist in making phosphorous available by solubilizing the water-insoluble forms. It is considered unlikely, however, that this represents a major source of phosphorous for plants, especially in light of the extensive fungal associations described in the next section. Bacteria can, however, enhance nutrient uptake other than by simply making nutrients more available. One way is to influence the growth and morphology of roots. One of the more striking examples is the formation of **proteoid** roots. This is a phenomenon of localized, intense lateral root production observed originally in the Proteaceae, a family of tropical trees and shrubs. (The Proteaceae includes the genus *Macadamea*, the source of the popular macadamia nut.) Proteoid roots have now been found in several other families. Their induction has been traced to localized aggregations of bacteria in the mucigel. The larger number of lateral roots allows a more intensive mining of the soils for poorly mobile nutrients, such as phosphorous. In addition, proteoid roots are generally found near the soil surface where they can take advantage of nutrients leached out of the litter. The mechanism for proteoid root induction has not been determined, but could be related to the production of a plant hormone (indoleacetic acid) by the bacteria.

3.9.2 MYCORRHIZAE ARE FUNGI THAT INCREASE THE VOLUME OF THE NUTRIENT DEPLETION ZONE AROUND ROOTS

Perhaps the most widespread—and from the nutritional perspective, more significant—associations between plants and microorganisms are those formed between roots and a wide variety of soil fungi. A root infected with a fungus is called a **mycorrhiza** (literally, fungus root). Mycorrhizae are a form of **mutualism**, an association in which both partners derive benefit. The significance of mycorrhizae is reflected in the observation that more than 80 percent of plants studied, including virtually all plant species of economic importance, form mycorrhizal associations. Two major forms of mycorrhizae are known: **ectotrophic** and **endotrophic**. The ectotrophic form, also known as **ectomycorrhizae**, is restricted to a few families consisting largely of temperate trees and shrubs, such as pines (Pinaceae) and beech (Fagaceae). Ectomycorrhizae are typically short, highly branched, and ensheathed by a tightly interwoven mantle of fungal hyphae. The fungus also penetrates the intercellular or apoplastic space of the root cortex, forming an *intercellular* network called a **Hartig net**. Endotrophic mycorrhizae, or **endomycorrhizae**, are found in some species of virtually every angiosperm family and most gymnosperms as well (except the Pinaceae). Unlike the ectomycorrhizae, the hyphae of endomycorrhizae develop extensively within cortical cells of the host roots.

The most common type of endomycorrhiza, found in the majority of the world's vegetation, is the **vesicular-arbuscular mycorrhiza (VAM)**. The hyphae of VAM grow between and into root cortical cells, where they form highly branched "treelike" structures called **arbuscules** (meaning dwarf tree). Each branch of the arbuscule is surrounded by the plasma membrane of the host cell. Thus, while the hyphae do penetrate the host cell wall, they do not actually invade the protoplast. The arbuscule serves to increase contact surface area between the hypha and the cell by two to three times. At the same time, it apparently influences the host cell, which may increase its cytoplasmic volume by as much as 20 to 25 percent. Less frequently, VAMs form large ellipsoid **vesicles** either between or within the host cells. The presence of arbuscules and vesicles provides a large surface for the exchange of nutrients between the host plant and the invading fungus. Although VAMs do not form a well-defined sheath around the root, the hyphae, like those of the ectomycorrhizae, do effectively extend the rhizosphere by growing outward into the surrounding soil.

Mycorrhizae were originally discovered by the nineteenth-century German botanist A. B. Frank, who concluded, on the basis of experiments conducted with beech seedlings, that mycorrhizal inoculation stimulated seedling growth. Although not universally accepted in the beginning, these results have been amply confirmed by more modern studies. Numerous studies with pine and other tree seedlings in the United States, Australia, and the former Soviet Union have demonstrated 30 to 150 percent increases in dry weight of tree seedlings infected with mycorrhizae when compared with noninfected controls. Similar results have been obtained in studies with agricultural plants such as maize (Figure 3.13). In one experiment, for example, the dry weight of VAM-infected *Lavendula* plants increased 8.5 times over noninfected controls. The primary cause of mycorrhizal-enhanced growth appears to be enhanced uptake of nutrients, especially phosphorous. In a classic experiment, Hatch demonstrated in 1937 that infected pine seedlings absorbed two to three times more nitrogen, potassium, and phosphorous. Coupled

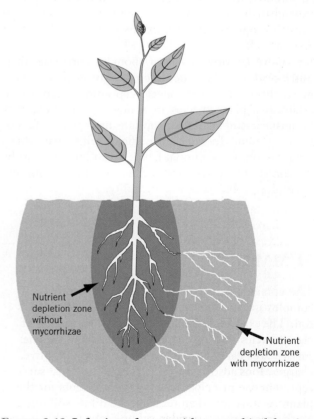

FIGURE 3.13 Infection of roots with mycorrhizal fungi extends the nutrient depletion zone for a plant. The nutrient depletion zone is the zone from which nutrients are drawn by the root system.

with enhanced nutrient uptake is the observation that mycorrhiza-induced growth responses are more pronounced in nutrient-deficient soils. VAM infection, for example, can be effectively eliminated by supplying the plant with readily available phosphorous. With a surplus of phosphorous fertilizer, uninoculated plants will grow as well as those inoculated with mycorrhizal fungi.

The beneficial role of mycorrhizae, particularly with respect to the uptake of phosphorous, appears to be related to the **nutrient depletion zone** that surrounds the root. This zone defines the limits of the soil from which the root is able to readily extract nutrient elements. Additional nutrients can be made available only by extension of the root into new regions of the soil or by diffusion of nutrients from the bulk soil into the depletion zone. The extent of the depletion zone varies from one nutrient element to another, depending on the solubility and mobility of the element in the soil solution. The depletion zone for nitrogen, for example, extends some distance from the root because nitrate is readily soluble and highly mobile. Phosphorous, on the other hand, is less soluble and relatively immobile in soils and, consequently, the depletion zone for phosphorous is correspondingly smaller. Mycorrhizal fungi assist in the uptake of phosphorous by extending their mycelia beyond the phosphorous depletion zone (Figure 3.13). Apparently, mycorrhizal plants find it advantageous to expend their carbon resources supporting mycorrhizal growth as opposed to more extensive growth of the root system itself.

As we continue to learn about mycorrhizae, their nutritional role becomes increasingly evident. Many mycorrhizae are host species–specific. Attempts to establish a plant species in a new environment may be unsuccessful if the appropriate mycorrhizal fungus is not present. Inoculation of fields with mycorrhizal fungi is now an additional factor taken into account by the forest and agricultural industries when attempting to resolve problems of soil infertility.

SUMMARY

The uptake of nutrient salts by plants involves a complex interaction between plant roots and the soil. The colloidal component of soil, consisting of clay particles and humus, presents a highly specific surface area carrying numerous, primarily negative, charges. Ions adsorbed to the charged colloidal surfaces represent the principal reservoir of nutrients for the plant. As ions are taken up from the dilute soil solution by the roots, they are replaced by exchangeable ions from the colloidal reservoir.

For nutrients to be taken up by a plant, they must be transported across the cell membrane into the root cell—thus making nutrient uptake fundamentally a cellular problem. Solutes may cross a membrane by simple diffusion, facilitated diffusion, or active transport. Facilitated diffusion and active transport are mediated by channel and carrier proteins—proteins that span the lipid bilayer. Only active transport achieves accumulation of ions against an electrochemical gradient. Active transport requires a source of metabolic energy, normally in the form of ATP. The free movement of water across membranes has long been an enigma, although it now appears to be a special case of facilitated diffusion through water-selective channels called aquaporins. The uptake of nutrients by most plants is enhanced by association of the roots with soil microorganisms, especially fungi. Fungal-root associations (mycorrhizae) benefit the plant by significantly increasing the volume of soil accessible to the roots.

CHAPTER REVIEW

1. Distinguish between simple diffusion, facilitated diffusion, and active transport. Which of these three mechanisms would most probably account for:

   (a) entry of a small lipid-soluble solute;
   (b) extrusion of sodium ions leaked into a cell;
   (c) rapid entry of a neutral hydrophilic sugar;
   (d) accumulation of potassium ions?

2. The Casparian band was encountered earlier with regard to root pressure and again in this chapter with regard to ion uptake. What is the Casparian band and how does it produce these effects?

3. Trace the pathway taken by a potassium ion from the point where it enters the root to a leaf epidermal cell.

4. What are channel proteins and what role do they play in nutrient uptake?

5. What is the Nernst equation and what does it tell us about ion transport?

6. How does the concept of molecular cross talk pertain to plant cellular ion transport?

7. Describe the concept of apparent free space. What role does apparent free space play in the uptake of nutrient ions?

8. Why is an accumulation ratio greater than 1.0 not necessarily an indication that active transport is involved?

9. How do mycorrhizae assist a plant in the uptake of nutrient elements?

10. Describe the colloidal properties of soil. How do the properties of colloids help to ensure the availability of nutrient elements in the soil?

FURTHER READING

Brito, D. T, H. J. Kronzucker. 2006. Futile cycling at the plasmamembrane: A hallmark of low affinity nutrient transport. *Trends in Plant Science* 11: 529–534.

Buchanan, B. B., W. Gruissem, R. L. Jones. 2000. *Biochemistry and Molecular Biology of Plants*. Rockville, MD: American Society of Plant Physiologists.

Cooper, J. E., J. R. Rao. 2006. *Molecular Approaches to Soil, Rhizosphere and Plant Micro-organism Analysis*. Cambridge: CABI Publishing.

Demidchik, V., R. J. Davenport, M. Tester. 2002. Nonselective cation channels in plants. *Annual Review of Plant Biology* 53: 67–107.

Epstein, E., A. J. Bloom. 2005. *Mineral Nutrition of Plants: Principles and Perspectives*. Sunderland: Sinauer Associates Inc.

Evert, R. F. 2006. *Esau's Plant Anatomy*. Hoboken, NJ: John Wiley & Sons, Inc.

Nobel, P. S. 2005. *Physicochemical and Environmental Plant Physiology*. Burlington, MA: Elsevier Science & Technology.

Peterson, R. L. 2004. *Mycorrhizas: Anatomy and Cell Biology*. Ottawa: NRC Press.

Schachtman, D. P., R. Shin. 2007. Nutrient sensing: NPKS. *Annual Review of Plant Biology* 58: 47–69.

Waisel, Y., A. Eshel, U. Kafkafi. 1996. *Plant Roots: The Hidden Half*. 2nd ed. New York: M. Dekker.

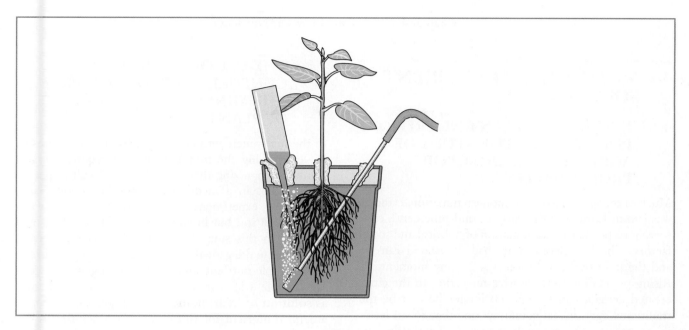

4

Plants and Inorganic Nutrients

Unlike **heterotrophic** organisms, which depend for their existence on energy-rich organic molecules previously synthesized by other organisms, plants must survive in an entirely inorganic environment. As **autotrophic** organisms, plants must take in carbon dioxide from the atmosphere and water and mineral nutrients from the soil, and from these simple, inorganic components, make all of the complex molecules of a living organism. Since plants stand at the bottom of the food chain, mineral nutrients assimilated by plants eventually find their way into the matter that makes up all animals, including humans. All organisms must continually draw material substance from their environment in order to maintain their metabolism, growth, and development. The means for making these materials available to the organism is the subject of plant **nutrition**.

Plant nutrition is traditionally treated as two separate topics: organic nutrition and inorganic nutrition. Organic nutrition focuses on the production of carbon compounds, specifically the incorporation of carbon, hydrogen, and oxygen via photosynthesis, while inorganic nutrition is concerned primarily with the acquisition of mineral elements from the soil. Photosynthesis and the acquisition of mineral ions from the soil are so interdependent, however, that this distinction between organic and inorganic nutrition is more a matter of convenience than real. Nevertheless, because the acquisition and assimilation of carbon are addressed in Chapters 7 and 8, this chapter will focus on the acquisition of mineral elements and the role of those elements in plant metabolism.

This chapter will examine the nutritional requirements of plants that are satisfied by mineral elements. This will include

- methods employed in the study of mineral nutrition,

- the concept of essential and beneficial elements and the distinction between macronutrients and micronutrients,

- a general discussion of the metabolic roles of the 14 essential mineral elements, the concept of critical and deficient concentration, and symptoms associated with deficiencies of the mineral elements, and

- a brief discussion of micronutrient toxicity.

The soil as a nutrient reservoir and mechanisms for mineral uptake by roots was covered in Chapter 3.

4.1 METHODS AND NUTRIENT SOLUTIONS

4.1.1 INTEREST IN PLANT NUTRITION IS ROOTED IN THE STUDY OF AGRICULTURE AND CROP PRODUCTIVITY

Much of the groundwork for modern nutritional studies was laid in Europe in the early to mid-nineteenth century, in response to a combination of political and social factors. The Napoleonic wars had devastated Europe and the industrial revolution was gaining momentum. Rising populations and massive migration to the cities created demands that could no longer be met by the traditional agricultural economy, one that relied heavily on the use of organic manures. Greater efficiency in agriculture was required and this was not possible without a more thorough understanding of plant nutrition.

One of the first to make significant progress in the study of plant nutrition was N. T. de Saussure (1767–1845), who studied both photosynthesis and the absorption of nutrient elements with the same careful, quantitative methods. De Saussure conducted some of the first elemental analyses of plant material and introduced the concept that some, but not necessarily all, of the elements found might be indispensable, or *essential*, to plant growth. De Saussure's ideas concerning the importance of elements derived from the soil generated considerable debate at the time, but received support from the work of C. S. Sprengel (1787–1859), working in Germany, and Jean-Baptiste Boussingault in France. Sprengel introduced the idea that soils might be unproductive if deficient in but one single element necessary for plant growth, and Boussingault stressed quantitative relationships between the effects of fertilizer and nutrient uptake on crop yields. Boussingault is also credited with providing the first evidence that legumes had the unique capacity to assimilate atmospheric nitrogen, a finding that was later confirmed by the discovery of the nitrogen-fixing role of bacteria in root nodules.

By the middle of the nineteenth century, many pieces of the nutritional puzzle were beginning to fall into place. In 1860, Julius Sachs, a prominent German botanist, demonstrated for the first time that plants could be grown to maturity in defined nutrient solutions in the complete absence of soil. J. B. Lawes and J. H. Gilbert, working at Rothamsted in England, had successfully converted insoluble rock phosphate to soluble phosphate (called **superphosphate**), and by the end of the century the agricultural use of NPK (nitrogen, phosphorous, and potassium) fertilizers was well established in Europe.

4.1.2 THE USE OF HYDROPONIC CULTURE HELPED TO DEFINE THE MINERAL REQUIREMENTS OF PLANTS

In the mid-nineteenth century, J. Sachs was interested in determining the minimal nutrient requirements of plants. Recognizing that it would be difficult to pursue such studies in a medium as complex as soil, Sachs devised an experimental system such that the roots grew not in soil but in an aqueous solution of mineral salts. With this simplified system, Sachs was able to demonstrate the growth of plants to maturity on a relatively simple nutrient solution containing six inorganic salts (Table 4.1). Variations on Sachs's system, known as **solution** or **hydroponic culture** (growing plants in a defined nutrient solution), have remained to this day the principal experimental system for study of plant nutrient requirements. Hydroponic culture is also now used extensively in North America for the year-round commercial production of vegetables such as lettuce, tomato, sweet peppers, and seedless cucumber.

The nutrient solution devised by Sachs contributed a total of nine mineral nutrients (K, N, P, Ca, S, Na, Cl, Fe, Mg). Carbon, hydrogen, and oxygen were excluded from this total because they were provided in the form of carbon dioxide and water and were not considered mineral elements. It was at least another half century before the need for additional mineral nutrients was demonstrated. There was no magic to the success of Sachs's experiments. Many of the mineral nutrients used by plants are required in very low amounts and Sachs unknowingly provided these nutrients as impurities in the salts and water he used to make up his nutrient solution. Analytical techniques have now improved to the point where it is possible to detect mineral contents several orders of magnitude lower than was possible in Sachs's time. Most mineral elements are now measured by either atomic absorption spectrometry or atomic emission spectrometry. These techniques involve vaporization of the elements at temperatures

TABLE 4.1   **The composition of Sachs's nutrient solution (1860) used for solution culture of plants.**

Salt	Formula	Approximate Concentration (mM)
Potassium nitrate	KNO_3	9.9
Calcium phosphate	$Ca_3(PO_4)_2$	1.6
Magnesium sulfate	$MgSO_4 \cdot 7H_2O$	2.0
Calcium sulfate	$CaSO_4$	3.7
Sodium chloride	$NaCl$	4.3
Iron sulfate	$FeSO_4$	trace

in excess of several thousand degrees. In the vaporous state, the element will either absorb or emit light at very narrow wavelength bands. The wavelength of light absorbed or emitted is characteristic of a particular element and the quantity of absorbed or emitted energy is proportional to the concentration of the element in the sample. In this way, concentrations as low as 10^{-3} g ml^{-1} for some elements can be measured in samples of plant tissue, soil, or nutrient solutions within a few minutes.

Aside from the commercial applications of hydroponic plant culture, a great deal of plant physiology and other botanical research is conducted with plants grown under controlled environments. This may include relatively simple greenhouses or complex growth rooms in which temperature and lighting are carefully regulated. Plant nutrient supply must also be regulated, and over the years a large number of nutrient solutions have been formulated for this purpose. Most modern formulations are based on a solution originally developed by D. R. Hoagland, a pioneer in the study of plant mineral nutrition. Individual investigators may introduce minor modifications to the composition of the nutrient solution in order to accommodate specific needs. Such formulations are commonly referred to as **modified Hoagland's solutions** (Tables 4.2, 4.3).

The concentration of minerals in most nutrient solutions is many times greater than that normally found in soils. An excess is necessary in order to maintain a continual supply of nutrients as they are taken up by the roots. The nutrient concentration of the soil solution, on the other hand, is relatively low but is continually replenished by nutrients adsorbed on the soil particles (Chapter 3).

TABLE 4.2   The composition of a typical one-half strength "modified" Hoagland's nutrient solution, showing the nutrient salts used and their approximate millimolar (mM) concentrations.

		Concentration (mM)
Calcium nitrate	Ca(NO)₃	2.5
Potassium phosphate	KH₂PO₄	0.5
Potassium nitrate	KNO₃	2.5
Magnesium sulfate	MgSO₄	1.0
Zinc sulfate	ZnSO₄	0.00039
Manganous sulfate	MnSO₄	0.0046
Copper sulfate	CuSO₄	0.00016
Boric acid	H₃BO₃	0.0234
Molybdic acid	MoO₃	0.000051
Iron sequestrene	Fe	0.179

TABLE 4.3   The quantity of each nutrient element in modified Hoagland's nutrient solution.

Element	Mg/L
Calcium	103
Nitrogen	105
Potassium	118
Sulfur	33
Magnesium	25
Phosphorous	15
Iron	10
Boron	0.25
Manganese	0.25
Zinc	0.025
Copper	0.01
Molybdenum	0.0052

4.1.3 MODERN TECHNIQUES OVERCOME INHERENT DISADVANTAGES OF SIMPLE SOLUTION CULTURE

In the simplest form of solution culture, a seedling is supported in the lid of a container, with its roots free to grow in the nutrient solution (Figure 4.1). Note that the solution must be aerated in order to obtain optimal root growth and nutrient uptake. A solution that is not aerated becomes depleted of oxygen, a condition known as **anoxia**. Anoxia inhibits the respiration of root cells and, because nutrient uptake requires energy, reduces nutrient uptake. The container in which the plants are grown is usually painted black or wrapped with an opaque material in order to keep out light. The purpose

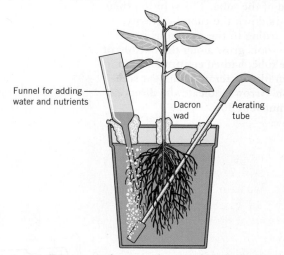

FIGURE 4.1 Diagram of a typical setup for nutrient solution culture. (From Epstein, E. 1972. *Mineral Nutrition of Plants: Principles and Perspectives*. New York: Wiley. Reprinted by permission)

of excluding light is to reduce the growth of algae that would compete with the plants for nutrients or possibly produce toxic byproducts.

There are some disadvantages in the use of a simple solution culture to study the nutrient requirements of plants. The major problems are a selective depletion of ions and associated changes in the pH of the solution that occur as the roots continue to absorb nutrients. Plants maintained in pure solution culture will continue to grow vigorously only if the nutrient solution is replenished on a regular basis. In order to avoid such problems, some investigators grow the plants in a *nonnutritive* medium such as acid-washed quartz sand, *perlite*, or *vermiculite*.[1] Plants can then be watered by daily application of fresh nutrient solution from the top of the medium (a technique called **slop culture**) or by slowly dripping onto the culture from a reservoir (**drip culture**). Alternatively, the nutrient culture can be **subirrigated**. In this case, the nutrient solution is alternately pumped into the culture from below and then allowed to drain out. This fill-and-empty process is repeated on a regular basis and serves both to replenish the nutrient solution and to *aerate* the roots. Most commercial hydroponic operations now utilize some variation of the **nutrient film** technique in which the roots are continuously bathed with a thin film of recirculating nutrient solution (Figure 4.2). The advantage of the nutrient film technique is that it not only provides for good aeration of the roots and nutrient uptake; it also allows the pH and nutrient content of the solution to be continuously monitored and adjusted.

These methods overcome some of the problems inherent in pure solution culture, but may not be suitable for many laboratory experiments. This is because no medium is truly nonnutritive. Any medium, even the glass, plastic, or ceramic containers used in solution culture, may provide some nutrients at very low levels. For example, soft (sodium silicate) glass provides sodium, hard (borosilicate) glasses provide boron, and plastics might provide chloride or fluoride, and so forth. Water used to prepare nutrient solutions must be carefully distilled, avoiding, wherever possible, metallic components in the distillation apparatus.

---

[1]Vermiculite is a silicate mineral of the mica family. It expands on heating to produce a lightweight product that has high water retention and is commonly used as a mulch in seed beds. Perlite is a coarsely ground glassy volcanic rock. Both vermiculite and perlite are effectively inert substances that provide no plant nutrients.

FIGURE 4.2 **The nutrient film technique for hydroponic plant production. Plants are grown in a tube or trough placed on a slight incline. A pump (P) circulates nutrient solution from a reservoir to the elevated end of the tube. The solution then flows down the tube by gravity, returning to the reservoir. Inset: the roots grow along the bottom of the tube, bathed continuously in a thin film of aerated nutrient solution. Arrows indicate the direction of nutrient flow.**

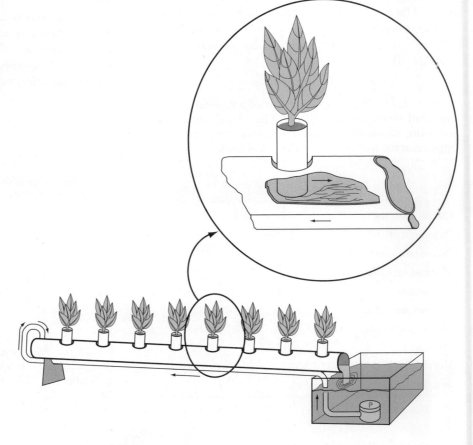

4.2 THE ESSENTIAL NUTRIENT ELEMENTS

4.2.1 SEVENTEEN ELEMENTS ARE DEEMED TO BE ESSENTIAL FOR PLANT GROWTH AND DEVELOPMENT

Most plants require a relatively small number of nutrient elements in order to successfully complete their life cycle. Those that are required are deemed to be **essential nutrient elements**. Essentiality is based primarily on two criteria formulated by E. Epstein in 1972. According to Epstein, an element is considered essential if ( ) *in its absence the plant is unable to complete a normal life cycle*, or (2) *that element is part of some essential plant constituent or metabolite*. By the first criterion, if a plant is unable to produce viable seed when deprived of that element, then that element is deemed essential. By the second criterion, an element such as magnesium would be considered essential because it is a constituent of the chlorophyll molecule and chlorophyll is essential for photosynthesis. Similarly, chlorine is essential because it is a necessary factor in the photosynthetic oxidation of water. Most elements satisfy both criteria, although either one alone is usually considered sufficient.

Although the criteria for essentiality are quite clear, it is not always easy to demonstrate that an element is or is not essential. D. Arnon and P. Stout had earlier suggested a third criterion: they suggested that an essential element must act directly in the metabolism of the plant and not simply to correct an unfavorable microbial or chemical condition in the nutrient medium. The use of solution cultures, from which the element in question has been omitted, has largely circumvented the need to apply this third criterion. On the other hand, some plants may form viable seeds even though a particular element has been excluded from the nutrient solution and other symptoms of deficiency are evident. In such cases there may be present in the seed, or contaminating the nutrient solution, a quantity of the element sufficient to moderate the deficiency and allow seed formation. It is assumed that in the complete absence of a nutrient the deficiency symptoms would be severe enough to kill the plant before viable seed could be formed. If required, this can be confirmed by careful purification of nutrient salts and exclusion of atmospheric contaminants. Where a sufficient quantity of the element may be carried within the seed, essentiality can be confirmed by growing several successive generations from seed that was itself produced in the absence of that element. This is usually sufficient to reduce the concentration of that element in the seed to the deficient range.

It is generally agreed, based on these criteria, that only 17 elements are essential for the growth of all higher plants (Table 4.4).

4.2.2 THE ESSENTIAL NUTRIENTS ARE GENERALLY CLASSED AS EITHER MACRONUTRIENTS OR MICRONUTRIENTS

The essential elements are traditionally segregated into two categories: (1) the so-called **macronutrients** and (2) the **trace elements** or **micronutrients**. The distinction between macro- and micronutrients simply reflects the relative concentrations found in tissue or required in nutrient solutions (Table 4.4) and does not infer importance relative to the nutritional needs of the plant. The first nine elements in Table 4.4 are called macronutrients because they are required in large amounts (in excess of 10 mmole kg^{-1} of dry weight). The macronutrients are largely, but not exclusively, involved in the structure of molecules, which to some extent accounts for the need for large quantities. The remaining eight essential elements are considered micronutrients. Micronutrients are required in relatively small quantities (less than 10 mmole kg^{-1} of dry weight) and serve catalytic and regulatory roles such as enzyme activators. Some macronutrients, calcium and magnesium for example, serve as regulators in addition to their structural role.

4.2.3 DETERMINING ESSENTIALITY OF MICRONUTRIENTS PRESENTS SPECIAL PROBLEMS

The essentiality of micronutrients is particularly difficult to establish because they are required in such small quantities. Most micronutrient requirements are fully satisfied by concentrations in the range of 0.1 to 1.0 µg L^{-1}—amounts that are readily obtained from impurities in water or macronutrient salts, the containers in which the plants are grown, and contamination by atmospheric dust. A micronutrient may be required at concentrations below detectable limits, so it is far easier to establish that a micronutrient *is* essential than that it *is not*.

As micronutrients go, iron is usually supplied at relatively high concentrations. This is necessary because availability of iron is very sensitive to pH and other soil conditions. At a pH above 7, iron tends to form insoluble iron hydroxides and calcium complexes. In acidic solution, iron reacts with aluminum to form insoluble complexes. In both cases, iron readily precipitates out of solution and, consequently, is frequently deficient in natural situations. For these reasons, the need for iron as an essential plant nutrient was established early in the study of plant nutrition. The need for other micronutrients, however, was not recognized until salts of sufficient purity became available in the early part of the twentieth century. In some cases, what had been for some time recognized as a plant disease turned out to be a nutrient deficiency. In 1922, for example, J. S. McHargue demonstrated that the disorder known as *gray speck of oats* was actually caused by a manganese deficiency, and the

TABLE 4.4   **The essential nutrient elements of higher plants and their concentrations considered adequate for normal growth.**

Element	Chemical Symbol	Available Form	Concentration in Dry Matter (mmol/kg)
Macronutrients			
Hydrogen	H	H_2O	60,000
Carbon	C	CO_2	40,000
Oxygen	O	O_2, CO_2	30,000
Nitrogen	N	NO_3^-, NH_4^+	1,000
Potassium	K	K^+	250
Calcium	Ca	Ca^{2+}	125
Magnesium	Mg	Mg^{2+}	80
Phosphorous	P	HPO_4^-, HPO_4^{2-}	60
Sulfur	S	SO_4^{2-}	30
Micronutrients			
Chlorine	Cl	Cl^-	3.0
Boron	B	BO_3^{3-}	2.0
Iron	Fe	Fe^{2+}, Fe^{3+}	2.0
Manganese	Mn	Mn^{2+}	1.0
Zinc	Zn	Zn^{2+}	0.3
Copper	Cu	Cu^{2+}	0.1
Nickel	Ni	Ni^{2+}	0.05
Molybdenum	Mo	Mo_4^{2-}	0.001

following year Katherine Warington showed that boron was required for several legume species. By 1939, the need for zinc, copper, and molybdenum had also been clearly established. In each case, the nutrient deficiency was found to cause a well-known disorder previously thought to be a disease. Chlorine was not added to the list until 1954, although its essential nature was suggested nearly one hundred years earlier. The need for chlorine became evident in the course of experiments to determine whether cobalt was required for tomato (*Lycopersicum esculentum*). T. C. Broyer and his coworkers had purified their nutrient salts by methods that removed not only cobalt but halides (including chlorine) as well. Plants grown in solutions prepared from these purified salts developed browning and necrosis of the leaves. The symptoms could be avoided by supplementing the nutrient solution with cobalt chloride. Subsequent investigation, however, established that it was the deficiency of chloride rather than cobalt that gave rise to the symptoms.

There is now mounting evidence that nickel should be added to the list of essential elements. Nickel is an essential component of urease, an enzyme widely distributed in plants, microorganisms, and some marine invertebrates. Urease catalyzes the hydrolysis of urea into NH_3 and CO_2 and is thought to play an important role in mobilization of nitrogenous compound in plants. In 1987, P. H. Brown and his colleagues showed that

nickel depletion led to the formation of nonviable seed in barley (*Hordeum vulgare*). The addition of nickel would bring to 17 the total number of nutrient elements essential for higher plants.

4.3 BENEFICIAL ELEMENTS

In addition to the 17 essential elements listed in Table 4.4, some plants appear to have additional requirements. However, because these have not been shown to be requirements of higher plants generally, they are excluded from the list of essential elements. They are referred to instead as **beneficial elements**. If these elements are essential to all plants, they are required by most at concentrations well below what can be reliably detected by present analytical techniques. The definition of *beneficial* currently applies primarily to sodium, silicon, selenium, and cobalt. With time, and as experimental methods improve, one or more of these beneficial elements may be added to the list of essential elements.

4.3.1 SODIUM IS AN ESSENTIAL MICRONUTRIENT FOR C4 PLANTS

A **sodium** requirement was first demonstrated for the bladder salt-bush (*Atriplex vesicaria*), a perennial pasture species of arid inland areas of Australia. By carefully purifying the water, recrystallizing the nutrient salts, and using sodium-free vessels, P. F. Brownell and C. J. Wood were able to reduce the sodium content of the final culture medium to less than 1.6 μg L^{-1}. Plants grown in the depleted solution showed reduced growth, **chlorosis** (yellowing due to loss of chlorophyll), and **necrosis** (dead tissue) of the leaves. Based on a survey of 32 species of plants, it was concluded that sodium is generally essential as a micronutrient for plants utilizing specifically the C4 photosynthetic pathway, but not for most C3 plants (see Chapter 8 and 15).

4.3.2 SILICON MAY BE BENEFICIAL FOR A VARIETY OF SPECIES

Given the high content of silicon dioxide in normal soils, it should not be surprising that many plants take up appreciable quantities of silicon. Silicon may comprise 1 to 2 percent of the dry matter of maize (*Zea mays*) and other grasses and as much as 16 percent of the scouring rush (*Equisetum arvense*), yet experiments have generally failed to demonstrate that silicon is essential for most other plants.[2] The ubiquitous presence of silicon in glass, nutrient salts, and atmospheric dust makes it especially difficult to exclude silicon from nutrient experiments. However, there are numerous reports of beneficial effects of silicon in a variety of species. Silicon seems to be particularly beneficial to grasses, where it accumulates in the cell walls, especially of epidermal cells and may play a role in fending off fungal infections or preventing **lodging**, a condition in which stems are bent over by heavy winds or rain.

4.3.3 COBALT IS REQUIRED BY NITROGEN-FIXING BACTERIA

Cobalt is essential for the growth of legumes, which are host to symbiotic nitrogen-fixing bacteria (Chapter 11). In this case, the requirement can be traced to the needs of the nitrogen-fixing bacterium rather than the host plant. A similar cobalt requirement has been demonstrated for the free-living nitrogen-fixing bacteria, including the cyanobacteria. In addition, when legumes are provided with fixed nitrogen such as nitrate, a cobalt requirement cannot be demonstrated.

4.3.4 SOME PLANTS TOLERATE HIGH CONCENTRATIONS OF SELENIUM

Selenium salts tend to accumulate in poorly drained, arid regions of the western plains of North America. Although selenium is generally toxic to most plants, cer-

tain members of the legume genus *Astragalus* (milk-vetch or poison-vetch) are known to tolerate high concentrations of selenium (up to 0.5 percent dry weight) and are found only on soils containing relatively high concentrations of selenium. Such concentrations of selenium would be toxic to most other plants. At one time it was thought that selenium might be essential to these "accumulator species," but there is no definitive supporting evidence.

Selenium accumulators are of considerable importance to ranchers, however, as they are among a diverse group of plants known as "loco weeds." The high selenium content in these plants causes a sickness known as alkali poisoning or "blind-staggers" in grazing animals.

4.4 NUTRIENT FUNCTIONS AND DEFICIENCY SYMPTOMS

The essential elements are essential because they have specific metabolic functions in plants. When they are absent, plants will exhibit characteristic deficiency symptoms that, in most cases, are related to one or more of those functions.

Some students of plant mineral nutrition prefer to classify the macro- and micronutrients along functional lines. For example, elements such as carbon, hydrogen, and oxygen have a predominantly structural role—they are the stuff of which molecules are made—while others appear to be predominantly involved in regulatory roles, such as maintaining ion balance and activating enzymes. Other investigators have proposed more complicated schemes with up to four categories of biochemical function. Unfortunately, any attempt to categorize the nutrient elements in this way runs into difficulty because the same element often fills both structural and nonstructural roles. Magnesium, for example, is an essential component of the chlorophyll molecule but also serves as a cofactor for many enzymes, including ATPases and others involved in critical energy-transfer reactions. Calcium is an important constituent of cell walls where its role is largely structural, but there are also Ca-ATPases and calcium is implicated as a second messenger in hormone responses and photomorphogenesis (Chapters 16, 17, 22). Regardless of how they are classified, it is clear that these elements are essential because they satisfy specific metabolic requirements of the plant. When those requirements are not met or are only partially met, the plant will exhibit characteristic deficiency symptoms that, if severe enough, result in death.

4.4.1 A PLANT'S REQUIREMENT FOR A PARTICULAR ELEMENT IS DEFINED IN TERMS OF CRITICAL CONCENTRATION

Typically, when the supply of an essential element becomes limiting, growth is reduced. The concentration

---

[2] Grazing animals appear to have adapted to the high silicon content of grasses. The teeth of grazing animals (such as cows and horses) grow continuously, compensating for the wear caused largely by silicon. On the other hand, the teeth of browsing animals such as deer, whose diets contain little grass, do not continue to grow.

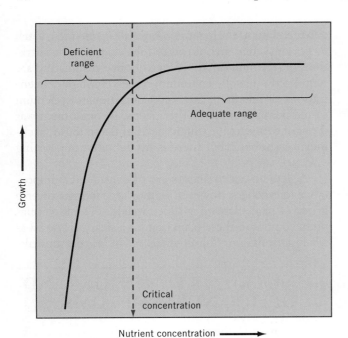

FIGURE 4.3 **Generalized plot of growth as a function of nutrient concentration in tissue. The critical concentration is that concentration giving a 10 percent reduction in growth. At concentrations less than critical, the nutrient is said to be deficient. Greater concentrations are considered adequate.**

of that nutrient, *measured in the tissue*, just below the level that gives maximum growth, is defined as the **critical concentration**. This concept is illustrated in Figure 4.3. At concentrations above the critical concentration, additional increments in nutrient content will have no beneficial effect on growth and the nutrient content is said to be **adequate**. At concentrations below the critical concentration, the nutrient content becomes **deficient** and growth falls off sharply. In other words, at tissue nutrient levels below the critical concentration, that nutrient becomes *limiting to growth*.

When nutrient levels exceed the critical concentration, that nutrient is, with one qualification, no longer limiting. The qualification is that at sufficiently high tissue levels, virtually all nutrients become toxic. Toxic levels are seldom achieved with the macronutrients, but are common in the case of the micronutrients (Section 4.5). Normal concentrations of copper, for example, are in the range of 4 to $15 \, \mu g \, g^{-1}$ of tissue dry weight (dwt). Deficiency occurs at concentrations below $4 \mu g$ but most plants are severely damaged by concentrations in excess of $20 \mu g \, g^{-1}$ dwt. Boron is toxic above $75 \mu g \, g^{-1}$ dwt and zinc above $200 \mu g \, g^{-1}$ dwt.

Since each element has one or more specific structural or functional roles in the plant, in the absence of that element the plant will be expected to exhibit certain morphological or biochemical symptoms of that deficiency. In some cases the deficiency symptoms will clearly reflect the functional role of that element. One

example is yellowing, or **chlorosis**, which is characteristic of several nutrient deficiencies. In the case of magnesium deficiency, for example, the plant turns yellow because it is unable to synthesize the green pigment chlorophyll. In other cases, the relationship between deficiency symptoms and the functional role of the element may not always be so straightforward. Moreover, deficiency symptoms for some elements are not always consistent between one plant and the next. Nonetheless, for each element there are certain generalizations that can be made with respect to deficiency symptoms.

Deficiency symptoms also depend in part on the mobility of the element in the plant. Where elements are mobilized within the plant and exported to young developing tissues, deficiency symptoms tend to appear first in older tissues. Other elements are relatively immobile—once located in a tissue they are not readily mobilized for use elsewhere. In this case, the deficiency symptoms tend to appear first in the younger tissues.

In this section we will review the functional roles of the essential elements and, in general terms, morphological and biochemical abnormalities that result from their deficiencies. Carbon, hydrogen, and oxygen will be excluded from this discussion as they are required for the structural backbone of all organic molecules. A deficiency of carbon, consequently, leads quickly to starvation of the plant, while a deficiency of water leads to desiccation. Instead, our discussion will be limited to deficiencies of those essential elements taken in from the soil solution.

4.4.2 NITROGEN IS A CONSTITUENT OF MANY CRITICAL MACROMOLECULES

Although the atmosphere is approximately 80 percent nitrogen, only certain prokaryote species—bacteria and cyanobacteria—can utilize gaseous nitrogen directly. The special problems of nitrogen availability and metabolism are addressed in Chapter 11. To summarize, most plants absorb nitrogen from the soil solution primarily as inorganic nitrate ion (NO_3^-) and, in a few cases, as ammonium (NH_4^+) ion. Once in the plant, NO_3^- must be reduced to NH_4^+ before it can be incorporated into amino acids, proteins, and other nitrogenous organic molecules. Nitrogen is most often limiting in agricultural situations. Many plants, such as maize (*Zea mays*), are known as "heavy feeders" and require heavy applications of nitrogen fertilizer. The manufacture and distribution of nitrogen fertilizers for agriculture are, in both energy and financial terms, an extremely costly process.

Nitrogen is a constituent of many important molecules, including proteins, nucleic acids, certain hormones (e.g., indole-3-acetic acid; cytokinin), and chlorophyll. It should not be surprising, then, that the most overt symptoms of nitrogen deficiency are a slow, stunted growth and a general chlorosis of the leaves. Nitrogen is very mobile in the plant. As the older leaves yellow and die, the nitrogen is mobilized, largely in the form of soluble amines and amides, and exported from the older leaves to the younger, more rapidly developing leaves. Thus the symptoms of nitrogen deficiency generally appear first in the older leaves and do not occur in the younger leaves until the deficiency becomes severe. At this point, the older leaves will turn completely yellow or brown and fall off the plant. Conditions of nitrogen stress will also lead to an accumulation of anthocyanin pigments in many species, contributing a purplish color to the stems, petioles, and the underside of leaves. The precise cause of anthocyanin accumulation in nitrogen-starved plants is not known. It may be related to an overproduction of carbon structures that, in the absence of nitrogen, cannot be utilized to make amino acids and other nitrogen-containing compounds.

Excess nitrogen normally stimulates abundant growth of the shoot system, favoring a high shoot/root ratio, and will often delay the onset of flowering in agricultural and horticultural crops. Similarly, a deficiency of nitrogen reduces shoot growth and stimulates early flowering.

4.4.3 PHOSPHOROUS IS PART OF THE NUCLEIC ACID BACKBONE AND HAS A CENTRAL FUNCTION IN INTERMEDIARY METABOLISM

Phosphorous is available in the soil solution primarily as forms of the polyprotic phosphoric acid (H_3PO_4). A polyprotic acid contains more than one proton, each with a different dissociation constant. Soil pH thus assumes a major role in the availability of phosphorous. At a soil pH less than 6.8, the predominant form of phosphorous and the form most readily taken up by roots is the monovalent *orthophosphate* anion ($H_2PO_4^-$). Between pH 6.8 and pH 7.2, the predominant form is HPO_4^{2-}, which is less readily absorbed by the roots. In alkaline soils (pH greater than 7.2), the predominant form is the trivalent PO_4^{3-}, which is essentially not available for uptake by plants. The actual concentration of soluble phosphorous in most soils is relatively low—on the order of $1\,\mu M$—because of several factors. One factor is the propensity of phosphorous to form insoluble complexes. At neutral pH, for example, phosphorous tends to form insoluble complexes with aluminum and iron, while in basic soils calcium and magnesium complexes will precipitate the phosphorous. Because insoluble phosphates are only very slowly released into the soil solution, phosphorous is always limited in highly calcareous soils.

Substantial amounts of phosphorous may also be bound up in organic forms, which are not available for uptake by plants. Organic phosphorous must first be converted to an inorganic form by the action of soil microorganisms, or through the action of phosphatase enzymes released by the roots, before it is available for uptake. In addition, plants must compete with the soil microflora for the small amounts of phosphorous that are available. For these sorts of reasons, phosphorous, rather than nitrogen, is most commonly the limiting element in natural ecosystems. One of the more successful strategies developed by plants for increasing the uptake of phosphorous is the formation of intimate associations between roots and soil fungi, called mycorrhiza. Mycorrhizal associations were discussed in Chapter 3.

In the plant, phosphorous is found largely as phosphate esters—including the sugar-phosphates, which play such an important role in photosynthesis and intermediary metabolism. Other important phosphate esters are the nucleotides that make up DNA and RNA as well as the phospholipids present in membranes. Phosphorous in the form of nucleotides such as ATP and ADP, as well as inorganic phosphate (P_i), phosphorylated sugars, and phosphorylated organic acids also plays an integral role in the energy metabolism of cells.

The most characteristic manifestation of phosphorous deficiency is an intense green coloration of the leaves. In the extreme, the leaves may become malformed and exhibit necrotic spots. In some cases, the blue and purple anthocyanin pigments also accumulate, giving the leaves a dark greenish-purple color. Like nitrogen, phosphorous is readily mobilized and redistributed in the plant, leading to the rapid senescence and death of the older leaves. The stems are usually shortened and slender and the yield of fruits and seeds is markedly reduced.

An excess of phosphorous has the opposite effect of nitrogen in that it preferentially stimulates growth of roots over shoots, thus reducing the shoot/root ratio. Fertilizers with a high phosphorous content, such as bone meal, are often applied when transplanting perennial plants in order to encourage establishment of a strong root system.

4.4.4 POTASSIUM ACTIVATES ENZYMES AND FUNCTIONS IN OSMOREGULATION

Potassium (K^+) is the most abundant cellular cation and so is required in large amounts by most plants. In agricultural practice, potassium is usually provided as potash (potassium carbonate, K_2CO_3). Potassium is frequently deficient in sandy soils because of its high solubility and the ease with which K^+ leaches out of sandy soils. Potassium is an activator for a number of enzymes, most notably those involved in photosynthesis and respiration. Starch and protein synthesis are also affected by potassium deficiency. Potassium serves an important function in regulating the osmotic potential of cells (Chapter 3). As an osmoregulator, potassium is a principal factor in plant movements, such as the opening and closure of stomatal guard cells (Chapter 8) and the sleep movements, or daily changes in the orientation of leaves (Chapter 23). Because it is highly mobile, potassium also serves to balance the charge of both diffusible and nondiffusible anions.

Unlike other macronutrients, potassium is not structurally bound in the plant, but like nitrogen and phosphorous is highly mobile. Deficiency symptoms first appear in older leaves, which characteristically develop mottling or chlorosis, followed by necrotic lesions (spots of dead tissue) at the leaf margins. In monocotyledonous plants, especially maize and other cereals, the necrotic lesions begin at the older tips of the leaves and gradually progress along the margins to the younger cells near the leaf base. Stems are shortened and weakened and susceptibility to root-rotting fungi is increased. The result is that potassium-deficient plants are easily lodged.

4.4.5 SULFUR IS AN IMPORTANT CONSTITUENT OF PROTEINS, COENZYMES, AND VITAMINS

Several forms of sulfur are found in most soils, including iron sulfides and elemental sulfur. Sulfur is taken up by plants, however, as the divalent sulfate anion (SO_4^{2-}). Sulfur deficiency is not a common problem because there are numerous microorganisms capable of oxidizing sulfides or decomposing organic sulfur compounds. In addition, heavy consumption of fossil fuels in industry as well as natural phenomena such as geysers, hot sulfur springs, and volcanos together contribute large amounts of sulfur oxides (SO_2 and SO_3) to the atmosphere. Indeed it is often difficult to demonstrate sulfur deficiencies in greenhouses in industrial areas because of the high concentrations of airborne sulfur.

Sulfur is particularly important in the structure of proteins where disulphide bonds (—S—S—) between neighboring cysteine and methionine residues contribute to the tertiary structure, or folding. Sulfur is also a constituent of the vitamins thiamine and biotin and of coenzyme A, an important component in respiration and fatty acid metabolism. In the form of iron-sulfur proteins, such as ferredoxin, it is important in electron transfer reactions of photosynthesis and nitrogen fixation. The sulfur-containing **thiocyanates** and **isothiocyanates** (also known as mustard oils) are responsible for the pungent flavors of mustards, cabbages, turnips, horseradish, and other plants of the family Brassicaceae. Because of the presence of mustard oils, many species of the Brassicaceae prove fatal to livestock that graze on them. Mustard oils also appear to serve as a defense against insect herbivory.

Sulfur deficiency, like nitrogen, results in a generalized chlorosis of the leaf, including the tissues surrounding the vascular bundles. This is due to reduced protein synthesis rather than a direct impairment of chlorophyll synthesis. However, chlorophyll is stabilized by binding to protein in the chloroplast membranes. With impaired protein synthesis, the ability to form stable chlorophyll-protein complexes is also impaired. Unlike nitrogen, however, sulfur is not readily mobilized in most species and the symptoms tend to occur initially in the younger leaves.

4.4.6 CALCIUM IS IMPORTANT IN CELL DIVISION, CELL ADHESION, AND AS A SECOND MESSENGER

Calcium is taken up as the divalent cation (Ca^{2+}). Calcium is abundant in most soils and is seldom deficient under natural conditions. Calcium is important to dividing cells for two reasons. It plays a role in the mitotic spindle during cell division and it forms calcium pectates in the middle lamella of the cell plate that forms between daughter cells. It is also required for the physical integrity and normal functioning of membranes and, more recently, has been implicated as a second messenger in a variety of hormonal and environmental responses. As a second messenger involved in protein phosphorylation, Ca^{2+} is an important factor in regulating the activities of a number of enzymes.

Because of its role in dividing cells, calcium deficiency symptoms characteristically appear in the meristematic regions where cell division is occurring and new cell walls are being laid down. Young leaves are typically deformed and necrotic and, in extreme cases, death of the meristem ensues. In solution cultures, calcium deficiency results in poor root growth. The roots are discolored and may feel "slippery" to the touch because of the deterioration of the middle lamella. Calcium is relatively immobile and the symptoms typically appear in the youngest tissues first.

4.4.7 MAGNESIUM IS A CONSTITUENT OF THE CHLOROPHYLL MOLECULE AND AN IMPORTANT REGULATOR OF ENZYME REACTION

Like calcium, magnesium is also taken up as the divalent cation (Mg^{2+}). Magnesium is generally less abundant in soils than calcium but is required by plants in relatively large amounts. Magnesium deficiencies are most likely in strongly acid, sandy soils. Magnesium has several important functions in the plant. By far the largest proportion is found in the porphyrin moiety of the chlorophyll molecule, but it is also required to stabilize ribosome structure and is involved as an activator for numerous critical enzymes. It is critical to reactions involving ATP, where it serves to link the ATP molecule to the active site of the enzyme. Mg^{2+} is also an activator for both ribulosebisphosphate carboxylase and phosphoenolpyruvate carboxylase, two critical enzymes in photosynthetic carbon fixation (Chapters 8 and 15).

The first and most pronounced symptom of magnesium deficiency is chlorosis due to a breakdown of chlorophyll in the lamina of the leaf that lie between the veins. Chloroplasts in the region of the veins are for some reason less susceptible to magnesium deficiency and retain their chlorophyll much longer. Magnesium is also quite mobile. It is readily withdrawn from the older leaves and transported to the younger leaves that are more actively growing and synthesizing chlorophyll. Consequently, chlorosis due to Mg^{2+} deficiency is, at least initially, most pronounced in the older leaves.

4.4.8 IRON IS REQUIRED FOR CHLOROPHYLL SYNTHESIS AND ELECTRON TRANSFER REACTIONS

Of all the micronutrients, iron is required by plants in the largest amounts (it is considered a macronutrient by some). Iron may be taken up as either the ferric (Fe^{3+}) or ferrous (Fe^{2+}) ion, although the latter is more common due to its greater solubility.[3] The importance of iron is related to two important functions in the plant. It is part of the catalytic group for many redox enzymes and it is required for the synthesis of chlorophyll. Important redox enzymes include the heme-containing cytochromes and non-heme iron-sulfur proteins (e.g., Rieske proteins, ferredoxin, and photosystem I) involved in photosynthesis (Chapter 7), respiration (Chapter 10) and nitrogen fixation (Chapter 11). During the course of electron transfer the iron moiety is reversibly reduced

from the ferric to the ferrous state. Iron is also a constituent of several oxidase enzymes, such as catalase and peroxidase.

Iron is not a constituent of the chlorophyll molecule itself and its precise role in chlorophyll synthesis remains somewhat of a mystery. There is, for example, no definitive evidence that any of the enzymes involved in chlorophyll synthesis are iron-dependent. Instead, the iron requirement may be related to a more general need for iron in the synthesis of the chloroplast constituents, especially the electron transport proteins. Iron deficiencies invariably lead to a simultaneous loss of chlorophyll and degeneration of chloroplast structure. Chlorosis appears first in the interveinal regions of the youngest leaves, because the mobility of iron in the plant is very low and it is not easily withdrawn from the older leaves. Chlorosis may progress to the veins and, if the deficiency is severe enough, the very small leaves may actually turn white.

Iron deficiencies are common because of the propensity of Fe^{3+} to form insoluble hydrous oxides ($Fe_2O_3 \cdot 3H_2O$) at biological pH. This problem is particularly severe in neutral or alkaline calcareous soils. On the other hand, iron is very soluble in strongly acidic soils and iron toxicity due to excess iron uptake can result. The problem of iron deficiency can usually be overcome by providing chelated iron, either directly to the soil or as a foliar spray. A **chelate** (from the Greek, *chele* or claw) is a stable complex formed between a metal ion and an organic molecule, called a chelating agent or **ligand**. The ligand and the metal ion share electron pairs, forming a **coordinate bond**. Because chelating agents have a rather high affinity for most metal ions, formation of the complex reduces the possibility for formation of insoluble precipitates. At the same time, the metal can be easily withdrawn from the chelate for uptake by the plant. One of the more common synthetic chelating agents is the sodium salt of **ethylenediaminetetraacetic acid (EDTA)** (Figure 4.4), known commercially as *versene* or *sequestrene*. EDTA and similar commercially available chelating agents, however, are not highly specific and will bind a range of cations, including iron, copper,

---

[3] In the scientific literature, particularly that body of literature dealing with iron uptake by organisms, the ferric form of iron is also referred to as Fe(III) (*iron-three*). By the same convention, ferrous iron is referred to as FE(II) (*iron-two*).

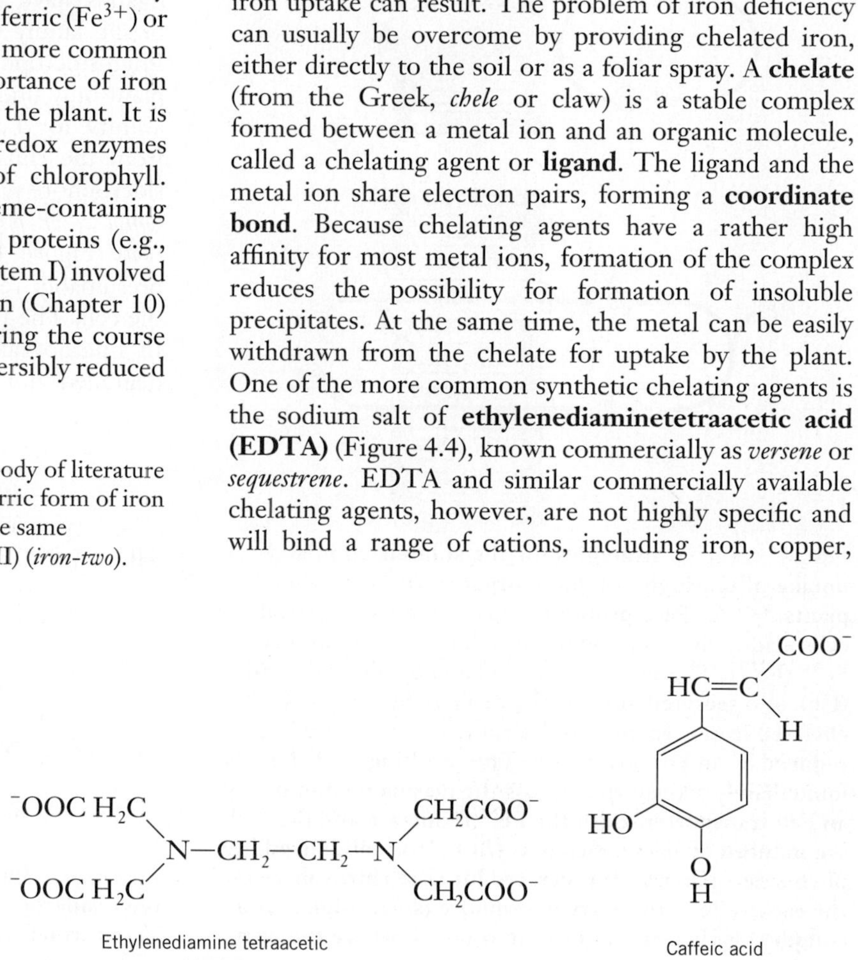

FIGURE 4.4 Examples of organic acids that function as chelating agents. Ethylenediaminetetraacetic acid (EDTA) is a synthetic acid in common commercial use. Complexed with iron, it is sold under the trade name Versenate. Caffeic acid is one of several naturally occurring phenolic acids that may be secreted by roots.

Ethylenediamine tetraacetic acid (EDTA)

Caffeic acid

zinc, manganese, and calcium. Natural chelating agents, including porphyrins (as in hemoglobin, cytochromes, and chlorophyll) and a variety of organic and phenolic acids, are far more specific for iron.

The importance of iron in plant nutrition is highlighted by the strategies plants have developed for uptake under conditions of iron stress. Iron deficiency induces several morphological and biochemical changes in the roots of dicots and nongraminaceous monocots. These

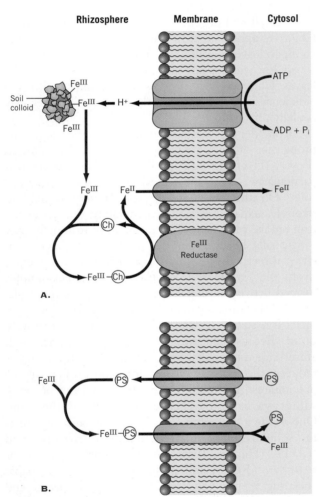

A.

B.

FIGURE 4.5 **Two strategies for the solubilization and uptake of sparingly soluble inorganic iron by higher plants.** (*A*) ATPase proton pumps in the root cortical cells acidify the rhizosphere, which helps to solubilize as Fe^{3+} (Fe^{III}). The Fe^{3+} is then chelated by phenolic acids (Ch), also secreted into the rhizosphere by the roots. The chelated iron is carried to the root surface where it is reduced by an Fe^{III} reductase. The resulting Fe^{2+} (Fe^{II}) is immediately transported across the plasma membrane by an Fe^{II} transporter. Both the Fe^{III} reductase and the Fe^{II} are induced by iron deficiency. (*B*) Fe^{3+} is solubilized by phytosiderophores (PS) secreted into the rhizosphere by the root. The entire ferrisiderophore (siderophore-iron complex) is then taken into the root cell where the iron is subsequently released.

include the formation of specialized transfer cells in the root epidermis, enhanced proton secretion into the soil surrounding the roots, and the release of strong ligands, such as **caffeic acid** (Figure 4.4), by the roots. Simultaneously, there is an induction of reducing enzymes in the plasma membrane of the root epidermal cells. Acidification of the rhizosphere encourages chelation of the Fe^{3+} with caffeic acid, which then moves to the root surface where the iron is reduced to Fe^{2+} at the plasma membrane (Figure 4.5A). Reduction to Fe^{2+} causes the ligand to release the iron, which is immediately taken up by the plant before it has the opportunity to form insoluble precipitates.

A second strategy for iron uptake by organisms involves the synthesis and release *by the organism* of low-molecular-weight, iron-binding ligands called **siderophores** (Gr. iron-bearers). Most of our knowledge of siderophores comes from studies with aerobic microorganisms (bacteria, fungi, and algae), where they were first discovered and have been studied most extensively. More recently, however, it has been discovered that siderophores are also released by the roots of higher plants (Figure 4.6). Known as **phytosiderophores**, to distinguish them from ligands of microbial origin, these highly specific iron-binding ligands have thus far been found only in members of the family Gramineae, including the cereal grains. Phytosiderophores are synthesized and released by the plant only under conditions of iron stress, have a high affinity for Fe^{3+}, and very effectively scavenge iron from the rhizosphere. The distinctive feature of the siderophore system is that *the entire iron-phytosiderophore complex, or ferrisiderophore, is then reabsorbed into the roots* (Figure 4.5B). Once inside the root, the iron is presumably reduced to Fe^{2+} and released for use by the cell. The fate of the phytosiderophore is unknown. In microorganisms, siderophores may be chemically degraded and metabolized or, alternatively, the same

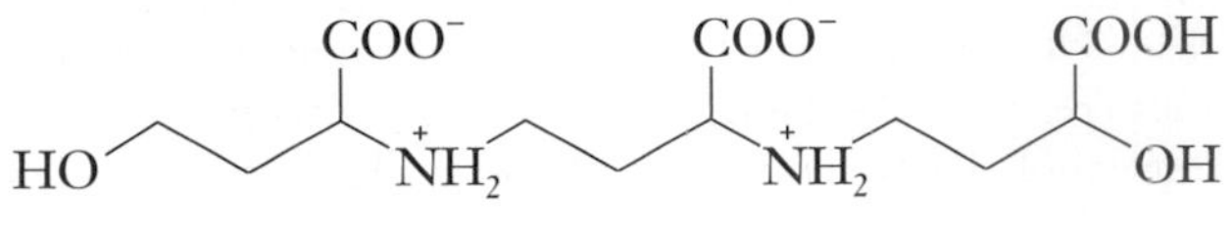

Avenic acid (AA)

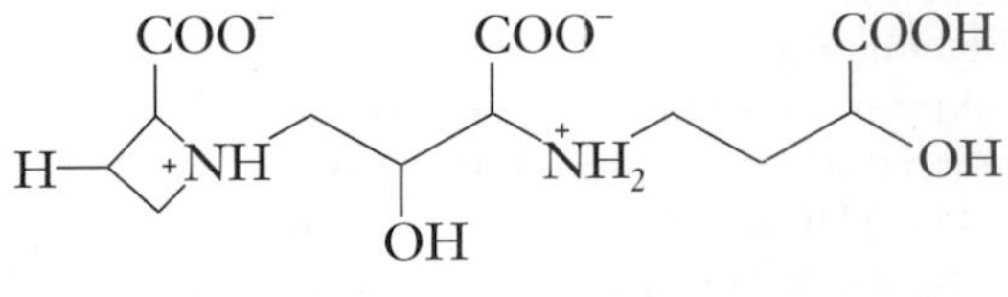

Muginec acid (MA)

FIGURE 4.6 **Phytosiderophores. The structures of two phytosiderophores released by the roots of higher plants. Ferric iron forms coordinate bonds with the nitrogen and carboxyl groups.**

molecule may again be secreted by the cell in order to pick up more iron.

The study of phytosiderophores is a relatively young field and, although substantial progress has been made in recent years, there is still much to be learned. It is not yet known, for example, how widespread the use of phytosiderophores is and the nature of the ferrisiderophore transport system has not been demonstrated in plants. One thing is clear: in those plants that use them, phytosiderophores are an important and effective strategy for supplying iron to the plant under conditions of iron stress.

4.4.9 BORON APPEARS TO HAVE A ROLE IN CELL DIVISION AND ELONGATION AND CONTRIBUTES TO THE STRUCTURAL INTEGRITY OF THE CELL WALL

In aqueous solution, boron is present as boric acid, or H_3BO_3. At physiological pH (<8), it is found predominantly in the undissociated form, which is preferred for uptake by roots. With respect to its biochemical and physiological role, boron is perhaps the least understood of all the micronutrients. There is, for example, no solid evidence for involvement of boron with specific enzymes, either structurally or as an activator. Indeed, most of what we know about the role of boron is based entirely on studies of what happens to plants when boron is withheld.

A substantial proportion of the total borate content of cells is found in the cell wall. This is apparently because borate has a propensity to form stable esters with cell wall saccharides that have adjacent hydroxyl groups. This so-called *cis*-diol configuration is characteristic of some common cell wall polysaccharides, such as mannose and its derivatives. Glucose, fructose, and galactose, on the other hand, do not have this configuration and so do not bind boron. The primary walls of boron-deficient cells exhibit marked structural abnormalities, suggesting that boron is required for the structural integrity of the cell wall.

Other responses to boron deficiency point toward a role in cell division and elongation. One of the most rapid responses to boron deficiency, for example, is an inhibition of both cell division and elongation in primary and secondary roots. This gives the roots a stubby and bushy appearance. Cell division in the shoot apex and young leaves is also inhibited, followed by necrosis of the meristem. In addition, boron is known to stimulate pollen tube germination and elongation. It is not known how boron is involved in cell growth, but both hormone and nucleic acid metabolism have been implicated. Inhibition of cell division and elongation is accompanied by an increased activity of enzymes that oxidize the hormone indole-3-acetic acid and a decrease in RNA content (possibly through impaired synthesis of uracil, an RNA precursor).

In addition to the effects on shoot meristems noted above, common symptoms of boron deficiency include shortened internodes, giving the plant a bushy or rosette appearance, and enlarged stems, leading to the disorder known as "stem crack" in celery. In storage roots such as sugar beets, the disorder known as "heart rot" is due to the death of dividing cells in the growing region because of boron deficiency.

4.4.10 COPPER IS A NECESSARY COFACTOR FOR OXIDATIVE ENZYMES

In well-aerated soils, copper is generally available to the plant as the divalent cupric ion, Cu^{2+}. Cu^{2+} readily forms a chelate with humic acids in the organic fraction of the soil and may be involved in providing copper to the surface of the root. In wet soils with little oxygen, Cu^{2+} is readily reduced to the cuprous form, Cu^+, which is unstable. As a plant nutrient, copper seems to function primarily as a cofactor for a variety of oxidative enzymes. These include the photosynthetic electron carrier plastocyanin; cytochrome oxidase, which is the final oxidase enzyme in mitochondrial respiration; and ascorbic acid oxidase. The browning of freshly cut apple and potato surfaces is due to the activity of copper-containing **polyphenoloxidases** (or phenolase). **Superoxide dismutase (SOD)**, which detoxifies superoxide radicals (O_2^-), is another important copper enzyme.

Common disorders due to copper deficiency are generally stunted growth, distortion of young leaves and, particularly in citrus trees, a loss of young leaves referred to as "summer dieback."

4.4.11 ZINC IS AN ACTIVATOR OF NUMEROUS ENZYMES

Zinc is taken up by roots as the divalent cation Zn^{2+}. Zinc is an activator of a large number of enzymes, including **alcohol dehydrogenase (ADH)**, which catalyzes the reduction of acetaldehyde to ethanol; **carbonic anhydrase (CA)**, which catalyzes the hydration of carbon dioxide to bicarbonate; and, copper SOD which detoxifies O_2^-. However, there is general agreement that disorders associated with zinc deficiency reflect disturbances in the metabolism of the auxin hormone indole-3-acetic acid. Typically, zinc-deficient plants have shortened internodes and smaller leaves (e.g., "little leaf" disorder of fruit trees). The precise role of zinc in auxin metabolism remains obscure, but auxin levels in zinc-deficient plants are known to decline before the overt symptoms of zinc deficiency appear. Furthermore, restoration of the zinc supply is followed by a rapid increase in hormone level and then resumption of growth. Available evidence supports the view that zinc is required for synthesis of the hormone precursor tryptophan.

4.4.12 MANGANESE IS AN ENZYME COFACTOR AS WELL AS PART OF THE OXYGEN-EVOLVING COMPLEX IN THE CHLOROPLAST

Manganese is absorbed and transported within the plant mainly as the divalent cation Mn^{2+}. Manganese is required as a cofactor for a number of enzymes, particularly decarboxylase and dehydrogenase enzymes, which play a critical role in the respiratory carbon cycle. Interestingly, manganese can often substitute for magnesium in reactions involving, for example, ATP. However, the best known and most studied function of manganese is in photosynthetic oxygen evolution (Chapter 7). In the form of a **manganoprotein**, manganese is part of the oxygen-evolving complex associated with photosystem II, where it accumulates charges during the oxidation of water.

Manganese deficiency can be widespread in some areas, depending on soil conditions, weather, and crop species. Deficiency is aggravated by low soil pH (<6) and high organic content. Manganese deficiency is responsible for "gray speck" of cereal grains, a disorder characterized by the appearance of greenish-gray, oval-shaped spots on the basal regions of young leaves. It may cause extreme chlorosis between the leaf veins as well as discoloration and deformities in legume seeds.

4.4.13 MOLYBDENUM IS A KEY COMPONENT OF NITROGEN METABOLISM

Although molybdenum is a metal, its properties more closely resemble those of the nonmetals. In aqueous solution it occurs mainly as the molybdate ion $M_oO_4^{2-}$. Molybdenum requirements are among the lowest of all known micronutrients and appear to be primarily related to its role in nitrogen metabolism. Among the several enzymes found to require molybdenum are **dinitrogenase** and **nitrate reductase**. The molybdenum requirement of a plant thus depends to some extent on the mode of nitrogen supply (Chapter 11). Dinitrogenase is the enzyme used by prokaryotes, including those in symbiotic association with higher plants, to reduce atmospheric nitrogen. Nitrate reductase is found in roots and leaves where it catalyzes the reduction of nitrate to nitrite, a necessary first step in the incorporation of nitrogen into amino acids and other metabolites.

In plants such as legumes, which depend on nitrogen fixation, molybdenum deficiency gives rise to symptoms of nitrogen deficiency. When nitrogen supplies are adequate, a deficiency of molybdenum shows up as a classic disorder known as "whiptail" in which the young leaves are twisted and deformed. The same plants may exhibit interveinal chlorosis and necrosis along the veins of older leaves. Like many of the micronutrients, molybdenum deficiency is highly species dependent—it is particularly widespread for legumes, members of the family Brassicaceae, and for maize. Molybdenum deficiency is aggravated in acid soils with a high content of iron precipitates, which strongly adsorb the molybdate ion.

4.4.14 CHLORINE HAS A ROLE IN PHOTOSYNTHETIC OXYGEN EVOLUTION AND CHARGE BALANCE ACROSS CELLULAR MEMBRANES

Chloride ion (Cl^-) is ubiquitous in nature and highly soluble. It is thus rarely, if ever, deficient. Deficiencies normally can be shown only in very carefully controlled solution culture experiments. Along with manganese, chloride is required for the oxygen-evolving reactions of photosynthesis (Chapter 7). Cl^- is a highly mobile anion with two principal functions: it is both a major counter-ion to diffusible cations, thus maintaining electrical neutrality across membranes, and one of the principal osmotically active solutes in the vacuole. Chloride ion also appears to be required for cell division in both leaves and shoots. Chloride is readily taken up and most plants accumulate chloride ion far in excess of their minimal requirements. Plants deprived of chloride tend to exhibit reduced growth, wilting of the leaf tips, and a general chlorosis.

4.4.15 THE ROLE OF NICKEL IS NOT CLEAR

Nickel is a relatively recent addition to the list of essential nutrient elements. Nickel is an abundant metallic element and is readily absorbed by roots. It is ubiquitous in plant tissues, usually in the range of 0.05 to 5.0 mg Kg^{-1} dry weight. One of the principal difficulties encountered in attempting to establish a role for nickel is its extremely low requirement. It has been estimated that the quantity of nickel needed by a plant to complete one life cycle is approximately 200 ng, a requirement that can be met by the initial nickel content of the seed in most cases. In order to establish a nickel deficiency, it is necessary to undertake extensive purification of the nutrient salts and then grow several successive generations to seed in nickel-deficient solutions. The strongest evidence in favor of essential status for nickel is based on studies with legumes and cereal grains. In one such study of barley (*Hordeum vulgare*), the critical nickel concentration for seed germination was found to be 90 ng g^{-1} of seed dry weight. By growing plants for three generations in the absence of nickel, the nickel content of the seed could be reduced to 7.0 ng g^{-1} dry weight. Germination of these seeds was less than 12 percent. When the plants were grown for the same number of generations in nutrient solution supplemented with 0.6 μM or 1.0 μM nickel, seed germination was 57 and 95 percent, respectively. In other studies, nickel deficiencies have led to depressed seedling vigor, chlorosis, and necrotic lesions in leaves.

The basis for a nickel requirement by plants is not clear, but it may be related to mobilization of nitrogen during seed germination. Nickel is known to be a component of two enzymes; urease and hydrogenase. Urease catalyzes the hydrolysis of urea into NH_3 and CO_2 and is found widely through the plant kingdom. Urease from jack bean seeds (*Canavalia ensiformis*) was in fact the first protein to be crystallized by J. B. Sumner in 1926. One of the principal effects of nickel deficiency in soybean (*Glycine max*) is decreased urease activity in the leaves, although the metabolic significance of urease is not yet clear.

Free urea is rarely, if ever, detected in plant tissue, but it is formed by the action of the enzyme arginase on arginine and its structural analog, canavanine (Chapter 11). Canavanine, a nonprotein amino acid, is abundant in the seeds of some plant groups, such as jack bean, but its concentration diminishes rapidly upon germination. Arginine is also abundant in seeds and both amino acids could function as stored nitrogen that is readily mobilized during seed germination. If this view should be proven valid, then urease, and thus nickel as well, would play an important role in the mobilization of nitrogen during germination and early seedling growth.

A common form of mobile nitrogen in some legumes is a family of urea-based compounds known as **ureides**, such as *allantoic acid* or *citrulline* (Chapter 11). Ureides are formed in root nodules during nitrogen fixation and transported via the xylem throughout the host plant. Ureides are also formed in senescing leaves and transported out to the developing seeds for storage. The breakdown of ureides produces urea, which accumulates to toxic levels in Ni-deficient plants. Furthermore, the metabolism of purine bases (adenine and guanine) in all plants also produces ureides. It seems reasonable to assume that most, if not all, plants have a requirement for urease and nickel.

Hydrogenase is another important enzyme in some nitrogen-fixing plants. Hydrogenase is responsible for recovering hydrogen for use in the nitrogen-fixing process (Chapter 11). A deficiency of nickel leads to depressed levels of hydrogenase activity in the nodules of soybean, which in turn would be expected to depress the efficiency of nitrogen fixation.

4.5 TOXICITY OF MICRONUTRIENTS

As a group, the micronutrient elements are an excellent example of the dangers of excess. Most have a rather narrow adequate range and become toxic at relatively low concentrations. **Critical toxicity levels**, defined as the tissue concentration that gives a 10 percent reduction in dry matter, vary widely between the several micronutrients as well as between plant species. As noted earlier, critical concentrations for copper, boron, and zinc are on the order of 20, 75, and 200 μg g^{-1} dry weight, respectively. On the other hand, critical toxicity levels for manganese vary from 200 μg g^{-1} dry weight for corn, to 600 μg g^{-1} for soybean, and 5300 μg g^{-1} for sunflower. Toxicity symptoms are often difficult to decipher because an excess of one nutrient may induce deficiencies of other nutrients. For example, the classic symptom of manganese toxicity, which often occurs in waterlogged soils, is the appearance of brown spots due to deposition of MnO_2 surrounded by chlorotic veins. But excess manganese may also induce deficiencies of iron, magnesium, and calcium. Manganese competes with both iron and magnesium for uptake and with magnesium for binding to enzymes. Manganese also inhibits calcium translocation into the shoot apex, causing a disorder known as "crinkle leaf." Thus the dominant symptoms of manganese toxicity may actually be the symptoms of iron, magnesium, and/or calcium deficiency.

Excess micronutrients typically inhibit root growth, not because the roots are more sensitive than shoots but because roots are the first organ to accumulate the nutrient. This is particularly true of both copper and zinc. Copper toxicity is of increasing concern in vineyards and orchards due to long-term use of copper-containing fungicides as well as urban and industrial pollution. Zinc toxicity can be a problem in acid soils or when sewage sludge is used to fertilize crops.

In spite of the apparent toxicity of micronutrients, many plant species have developed the capacity to tolerate extraordinarily high concentrations. For example, most plants are severely injured by nickel concentrations in excess of 5 μg g^{-1} dry weight, but species of the genus *Alyssum* can tolerate levels in excess of 10 000 μg g^{-1} dry weight.

SUMMARY

Plants are autotrophic organisms, taking their entire nutritional needs from the inorganic environment. Plants require carbon, hydrogen, and oxygen, plus 14 other naturally occurring elements that are taken from the soil. These 17 elements are considered essential because it has been demonstrated that in their absence all plants are unable to complete a normal life cycle. Essential elements may be considered either macronutrients or micronutrients, depending on the quantity normally required. Micronutrients are normally required in concentrations less than 10 mmole/kg of dry weight.

Each essential element has a role to play in the biochemistry and physiology of the plant and its absence is characterized by one or more deficiency symptoms, commonly related to that role. Additional elements may be considered beneficial because they satisfy special requirements for particular plants. Essential elements, especially micronutrients, may be toxic when present in excess amounts.

CHAPTER REVIEW

1. Explain the difference between autotrophic and heterotrophic nutrition.

2. What is meant by essentiality? What is the difference between an essential element and a beneficial element? Describe the steps you would go through in order to determine the essentiality or nonessentiality of an element for a higher plant.

3. List the 17 elements that are essential for the growth of all higher plants. Be able to identify one or more principal structural or metabolic roles for each essential element.

4. Deficiencies of iron, magnesium, and nitrogen all cause chlorosis. Iron chlorosis develops only between the veins of young leaves while chlorosis due to both magnesium and nitrogen deficiencies develops more generally in older leaves. Explain these differences. Why does each deficiency lead to chlorosis and why are the patterns different?

5. For what reasons might a soil rich in calcium supply too little phosphorous for plant growth?

6. What is a chelating agent? Explain how chelating agents help to maintain iron availability in nutrient cultures and soils.

7. What is meant by critical toxicity level? Which elements are most likely to be both essential and toxic to plants?

8. There are currently 17 elements known to be essential for higher plants. Is it possible that other elements might be added to this list in the future? Explain your answer.

FURTHER READING

Blevins, D. G, K. M. Lukaszewski. 1999. Boron and plant structure and function. *Annual Review of Plant Biology* 49: 481–500.

Broadley, M. R., P. J. White. 2005. *Plant Nutritional Genomics.* Oxford, U.K.: Blackwell Publishing.

Buchanan, B. B., W. Gruissem, R. L. Jones. 2000. *Biochemistry and Molecular Biology of Plants.* Rockville, MD: American Society of Plant Physiologists.

Epstein, E., A. J. Bloom. 2005. *Mineral Nutrition of Plants: Principles and Perspectives.* Sunderland: Sinauer Associates Inc.

Kochïan, L. V., O. A. Hoekenga, M. A. Piñeros. 2004. How do crop plants tolerate acid soils? Mechanisms of aluminum tolerance and phosphate efficiency. *Annual Review of Plant Biology* 55: 459–493.

Salt, D. E. 2008. Ionomics and the study of the plant ionome. *Annual Review of Plant Biology* 59: 709–733.

Schachtman, D. P., R. Shin. 2007. Nutrient sensing: NPKS. *Annual Review of Plant Biology* 58: 47–69.

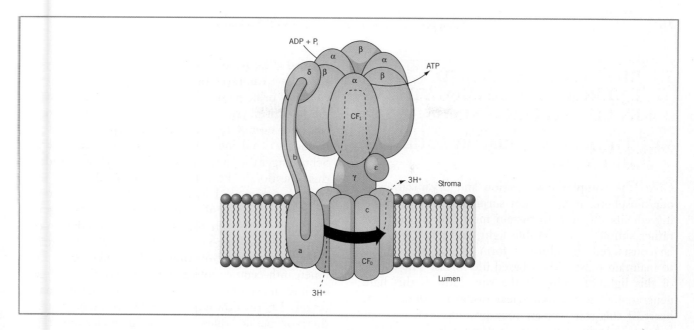

5

Bioenergetics and ATP Synthesis

A unique characteristic of planet earth is the presence of life. On a grand scale, one can consider the entire earth's biosphere to be an enormous, exquisite, but complex energy-transforming system consisting of myriad countless organisms. In addition to water (Chapters 1–4) energy is an absolute requirement for the maintenance and replication of life regardless of its form. Each organism plays a specific role in this teeming web of carbon-based life forms. Regardless of whether we examine the scale of biology at the community, individual, cellular, or molecular level, organization is the very essence of life and yet it is constantly under attack. At the cellular level, proteins, nucleic acids, and other molecules that make up the cell are continually subject to breakdown by hydrolysis. Membranes leak solutes to the environment. Everything on earth, cells and environment alike, is subject to persistent oxidation. Still, all around us we see biological organisms extracting materials from their environment and using them to maintain their organization or to build new, complex structures. Energy to build and preserve order in the face of a constantly deteriorating environment is a fundamental need of all organisms. Two strategies have evolved to satisfy this need. One is **photosynthesis**—the **photoautotrophic** lifestyle—which traps energy from the sun to build complex structures out of simple inorganic substances. By contrast, organisms that live by the alternative lifestyle, **chemoheterotropic**, require a constant intake of organic substances from their environment, from which they can extract their necessary energy through **respiration**. But even many of these substances trace their origins back to photosynthesis. In the end, most life on earth is powered by energy from the sun through photosynthesis. This chapter is concerned with the basic principles of **bioenergetics**—the study of energy transformations in living organisms.

The principal topics to be covered are

- thermodynamic laws and the concepts of free energy and entropy,

- free energy and its relationship with chemical equilibria, illustrating how displacement of a reaction from equilibrium can be used to drive vital reactions,

- oxidation–reduction reactions, showing how they also are involved in biological energy transformations, and

- the chemiosmotic model for synthesis of adenosine triphosphate (ATP), a key mediator of biological energy metabolism.

5.1 BIOENERGETICS AND ENERGY TRANSFORMATIONS IN LIVING ORGANISMS

5.1.1 THE SUN IS A PRIMARY SOURCE OF ENERGY

Given the complex composition and organization of our biosphere, it may seem surprising that the basic ingredients required to sustain most life on earth are rather simple: water, visible light, and air. Light may be considered the ultimate form of energy required to maintain most carbon-based life forms. The source of this light of course is the sun. How is this light generated? The thermonuclear fusion reactions in the heart of this star convert four protons ($4H^+$) to one helium (He) atom, which has an atomic weight of 4.0026. However, since each H^+ has an atomic weight of 1.0079, the expected atomic weight of He should be 4.0316. Clearly, we are missing 0.0290 gm-atoms of mass, which represents a mere 0.72 percent of the total mass of $4H^+$! Einstein showed us that there is a very important relationship between energy and mass:

$$E = mc^2 \tag{5.1}$$

where E is energy, m is mass, and c is the speed of light. Thus, the missing mass (m) of 0.0290 gm-atoms during the conversion of $4H^+$ to He is converted into energy (E) in the form of electromagnetic radiation. A small portion of this electromagnetic energy is in the form of visible light (Chapter 6), which reaches the earth's surface after a trip of about 160 million km. Since the speed of light is 300,000 km s^{-1}, each photon of visible light generated by the sun requires 8.88s to reach the earth's surface. This means that any image of the sun that we detect on earth can never be an original image, but rather, an image of the sun that is 8.88s old!

Air provides the basic elements for all living organisms: C, N, and O. C in the air is in the form of CO_2 (about 0.035%) and the N in air is in the form of N_2 (about 80%). However, most living organisms cannot directly utilize either CO_2 or N_2 as a source of C and N, respectively. Water provides the solvent necessary for enzyme catalysis and formation of biological membranes (Chapters 1 and 2). In Chapters 7 to 11 we will address the role played by plants in utilizing light energy to transform C and N into forms required not only to sustain plant life but all forms of life. However, it is essential that we establish some understanding of the basic principles that govern energy transformations in biological systems, that is, bioenergetics.

5.1.2 WHAT IS BIOENERGETICS?

In 1944, the Nobel Laureate in physics, Erwin Schrödinger, published an intriguing little monograph entitled *What Is Life?* In this book, this famous physicist attempts to unravel the basis of life on physical and chemical principles. Schrödinger simply asked whether the laws of physics and chemistry can account for the complex events that take place within "the spatial boundary of a living organism" through space and time. Schrödinger used a thermodynamic approach to address this question. The term *thermo*dynamics and much of its language and mathematics reflect an historical interest to discover the fundamental laws that govern heat flow. Although the study of thermodynamics is now concerned with energy flow in a more general sense, the science of **thermodynamics** arose from nineteenth-century interests in the workings of steam engines or why heat was evolved when boring cannon barrels. Energy flow is governed by certain fundamental thermodynamic rules. A general understanding of thermodynamic principles is necessary because these principles provide the quantitative framework for understanding energy transformations in biology. In addition to energy transformations, thermodynamics also helps to describe the capacity of a system to do work. Work may be defined in several different ways. The physicist defines work as displacement against a force: the sliding of an object against friction or rolling a boulder uphill, for example. The chemist, on the other hand, views work in terms of pressure and volume. For example, work must be done to overcome the force of atmospheric pressure when the volume of a gas increases. In biology the concept of work is applied more broadly, embracing a variety of work functions against a wide spectrum of forces encountered in cells and organisms. In addition to mechanical work such as muscular activity, the biologist is concerned with such diverse activities as chemical syntheses, the movement of solute against electrochemical gradients, osmosis, and ecosystem dynamics. These and a host of other essential activities of living things can all be described in thermodynamic terms.

Bioenergetics is the application of thermodynamic laws to the study of energy transformations in biological systems. The energetics of cellular processes can be related to chemical equilibrium and oxidation–reduction potentials of chemical reactions. Whether at the level of molecules, cells, or ecosystems, the flow of energy is central to the maintenance of life. A basic understanding of energy flow is therefore essential to grasping the true beauty, significance, and complexity of biology. The field of study concerned with the flow of energy through living organisms is called bioenergetics. It is important to note that although the laws of bioenergetics provide critical insight into the driving forces that govern cellular processes, bioenergetics does not provide any insights into the biochemical reaction mechanisms underlying these processes. For the past several decades, a central focus of bioenergetics has been to unravel the complexities of energy transformations in photosynthesis and respiration and understand how that energy is used to drive energy-requiring reactions such as ATP synthesis and accumulation of ions across membranes.

The purpose of this section is to facilitate an understanding of the laws that govern biological energy transformations by assembling, in the simplest form possible, some basic thermodynamic principles. Thus, a complete understanding of cellular physiology requires an integration of bioenergetics with biochemical reaction mechanisms which is a pervading theme of this textbook. The interested student will find a more comprehensive treatment of thermodynamics and bioenergetics in the excellent monograph by D. G. Nicholls and S. J. Stuart (2002). This publication provided the basis for the discussion that follows.

5.1.3 THE FIRST LAW OF THERMODYNAMICS REFERS TO ENERGY CONSERVATION

Biological energy transformations are based on two thermodynamic laws. The first law, commonly known as the **law of conservation of energy**, states that the energy of the universe is constant. This is not a difficult concept to comprehend—it means simply that there is a fixed amount of energy and, while it may be moved about or changed in form, it can all be accounted for somewhere. More to the point, energy is never "lost" in a reaction—an apparent decrease in one form of energy will be balanced by an increase in some other form of energy. In one of the examples mentioned earlier, some of the energy expended in displacing an object appears as work while some appears as heat generated due to friction. In the same way, some of the chemical energy released in the combustion of glucose will also be found as heat in the environment, while some will be found as bond energy in the product molecules, CO_2 and water.

5.1.4 THE SECOND LAW OF THERMODYNAMICS REFERS TO ENTROPY AND DISORDER

As biologists we are concerned above all with how much work can be done. But, as suggested earlier, not all energy is available to do work. This brings us to the second law of thermodynamics and the concept of **entropy**. Because it involves the concept of entropy (S), the second law is a bit more difficult to comprehend. What is entropy? Entropy has been variously described as a measure of randomness, disorder, or chaos.

However, since entropy is a thermodynamic concept, it is useful to describe it in terms of thermal energy. Temperature is defined as the mean molecular kinetic energy of matter. Thus, any molecular system not at absolute zero ($-273°C$, or $0°K$) contains a certain amount of thermal energy—energy in the form of the vibration and rotation of its constituent molecules as well as their translation through space. This quantity of thermal energy and temperature go hand-in-hand: as the quantity of energy increases or decreases, so does temperature. Because temperature cannot be held constant when this energy is given up, it is said to be "isothermally unavailable" (Gr. *isos*, equal). Quantitatively, isothermally unavailable energy is given by the term TS, where T is the absolute temperature and S is entropy.

Since isothermally unavailable energy, and consequently, entropy, are related to the energy of molecular motion, it follows that the more molecules are free to move about, that is, the more random or less ordered or chaotic the system, the greater will be their entropy. By this same argument it also follows that at absolute zero, a state in which all molecular motion ceases, entropy is also zero. Consider a familiar example: the combustion of glucose. The highly ordered structure of a glucose molecule imposes certain constraints on the movement of the constituent carbon atoms. In the form of six individual carbon dioxide molecules, however, those same atoms are far less constrained. They are individually free to rotate and tumble through space. With respect to carbon, the glucose molecules and carbon dioxide molecules each have entropy, but the product carbon dioxide molecules are less ordered, their freedom of movement is greater, and so is their entropy. It is important to note, however, that for any given reaction, the entropy of all reactants and all products must be taken into consideration. For a system not at rest, the natural tendency is for entropy to increase, that is, for systems to become increasingly chaotic. Equation 5.2 shows the relationship between entropy, S, and disorder, D, where k is Boltzmann's constant ($1.3800662 \times 10^{-23}$ J $\cdot °K^{-1}$). Thus, as disorder increases so does entropy.

$$S = k \log D \qquad (5.2)$$

This tendency was summarized by R. J. Clausius: *the entropy of the universe tends toward a maximum*. Clausius's dictum is one way of stating the second law of thermodynamics. The primary characteristic of all life is complex order which appears to contradict the second law of thermodynamics. However, this is not the case since all living organisms exhibit a finite lifetime which can vary from hours to days to years to centuries depending on the species. Since order (1/D) is the inverse of disorder (D), Equation 5.2 becomes

$$-S = k \log(1/D) \qquad (5.3)$$

Thus, life is negative entropy ($-S$) !

As physiologists, however, our concern with entropy is primarily that it represents energy that is *not available* to do work. In this context, the second law can then be restated as: *the capacity of an isolated system to do work continually decreases*. In other words, it is never possible to utilize all of the energy of a system to do work.

5.1.5 THE ABILITY TO DO WORK IS DEPENDENT ON THE AVAILABILITY OF FREE ENERGY

From the above discussion, it is apparent that some energy will be available under isothermal conditions and is, consequently, available to do work. This energy is called **Gibbs free energy** in honor of J. W. Gibbs, the nineteenth-century physical chemist who introduced the concept. Free energy (G) is related to TS in the following way:

$$H = G + TS \qquad (5.4)$$

H is the total heat energy (also called **enthalpy**), including any work that might be done. H is comprised of isothermally unavailable TS *plus* G. Equation 5.4 thus identifies two kinds of energy: free energy, which is available to do work, and entropy, which is not. Except in a limited number of situations, it is free energy, the energy available to do work, that is of greatest interest to the biologist. Equation 5.4 also suggests a corollary of the second law: *the free energy of the universe tends toward a minimum*.

It is neither convenient nor relevant to measure absolute energies (either G or S), but *changes* (designated by the symbol Δ) in energy during the course of a reaction can usually be measured with little difficulty as, for example, heat gain or loss, or work. Equation 5.4 can be restated as follows:

$$\Delta G = \Delta H - T\Delta S \qquad (5.5)$$

Changes in free energy can tell us much about a reaction. It can tell us, for example, the feasibility of a reaction actually taking place and the quantity of work that might be done if it does take place. Feasibility is indicated by the sign of ΔG. If the sign of ΔG is negative (i.e., $\Delta G < 0$), the reaction is considered *spontaneous*, meaning that it will proceed without an input of energy. Since the free energy of the products is less than the reactants, reactions with a negative ΔG are sometimes known as **exergonic**, or energy yielding. If, on the other hand, ΔG is positive, an input of energy is required for the reaction to occur. The oxidation of glucose is an example of a reaction with a negative ΔG. Once an activation barrier (see Chapter 8 for details) is overcome, glucose will spontaneously oxidize to form CO_2 and water. Despite the existence of large quantities of CO_2 and water in the atmosphere, however, they are not known to spontaneously recombine to form glucose! This is because the equilibrium constant favors the formation of CO_2 and H_2O and the ΔG for glucose formation is positive. Reactions with a positive ΔG (i.e., $\Delta G > 0$) are known as **endergonic**, or energy consuming. Equation 5.5 also tells us that any change in free energy

is associated with a change in entropy. This means that there is always a price to pay in any reaction involved in an energy transformation. The price paid is in the inevitable increase in entropy. Therefore, no energy transformation process can be 100 percent efficient with respect to the retention of free energy.

The magnitude of free energy changes is very much a function of the particular set of conditions for that reaction. For that reason it is convenient to compare the free energy changes of reactions under standard reaction conditions. In biochemistry the **standard free energy change**, $\Delta G^{\circ\prime}$, defines the free energy change of a reaction that occurs at physiological pH (pH = 7.0) under conditions where both reactants and products are at unit concentration (1 M).

5.1.6 FREE ENERGY IS RELATED TO CHEMICAL EQUILIBRIA

Under appropriate conditions, all chemical reactions will achieve a state of equilibrium, at which there will be no further *net* change in the concentrations of reactants and products. There is a fairly straightforward relationship between free energy and chemical equilibria. This relationship, which is central to an understanding of bioenergetics, is illustrated diagrammatically in Figure 5.1. In a reaction where the reactant A is converted to product B, K is the equilibrium mass-action ratio—the ratio of concentration of products to the concentration of reactants when the reaction has come to equilibrium. Thus

$$K_{eq} = [B]_{eq}/[A]_{eq} \qquad (5.6)$$

In Figure 5.1 the slope of the line represents the change in free energy (ΔG) when a small amount of the reactant A is converted to the product B. Several useful points can be drawn from this diagram.

1. At equilibrium, the slope of the line is zero. Consequently, *when reactions are at equilibrium, $\Delta G = 0$ and no useful work can be accomplished.*

2. The further the mass-action ratio is displaced from equilibrium (K_{eq}), the greater the free energy change for conversion of the same small amount of A to B. The free energy change for a reaction is a function of its displacement from equilibrium. Therefore, *the further a reaction is poised away from equilibrium, the more free energy is available as the reaction proceeds toward equilibrium.*

3. As A approaches equilibrium, ΔG is negative and free energy is available to do work. However, as the reaction proceeds past equilibrium toward B, ΔG is becomes positive and energy must be supplied. *A system can do work as it moves toward equilibrium.* Note that if the reaction were initiated with pure B, the direction of the arrows would be reversed and work could be done as B approached equilibrium.

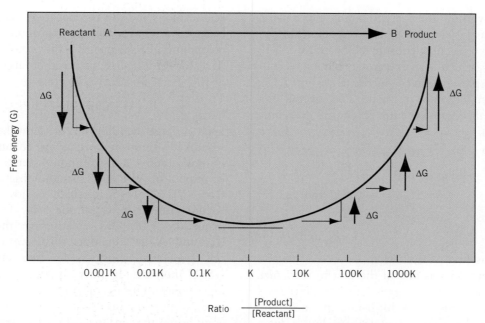

FIGURE 5.1 **The free energy of a reaction is a function of its displacement from equilibrium. K is the mass-action ratio when the reaction is at equilibrium. Vertical arrows indicate the slope of the free energy curve, or change in free energy, as reactant is converted to product. Note that the free energy change at equilibrium is zero and that the magnitude of the free energy change, indicated by the length of the arrow, increases as the reaction moves away from equilibrium toward pure reactant or pure product. A downward arrow indicates a negative free energy change; an upward arrow indicates a positive free energy change. (Redrawn from D. G. Nicholls and S. J. Ferguson,** *Bioenergetics,* **New York: Academic Press, 2002. Reprinted by permission.)**

The relationship between the free energy change (ΔG) and the equilibrium constant (K_{eq}) can be expressed quantitatively as:

$$\Delta G = \Delta G^{\circ\prime} + RT \ln K_{eq} \tag{5.7}$$

However, at equilibrium, $\Delta G = 0$ and Equation 5.7 can be rearranged to give

$$\Delta G^{\circ\prime} = -RT \ln K_{eq} = -2.3RT \log K_{eq} \tag{5.8}$$

Furthermore, the actual free energy change (ΔG) of a reaction *not* at equilibrium is given by:

$$\Delta G = \Delta G^{\circ\prime} + 2.3RT \log \Gamma \tag{5.9}$$

where R is the universal gas constant, T is the absolute temperature, and Γ equals the observed (i.e., nonequilibrium) mass-action ratio. Equation 5.8 can then be substituted in Equation 5.9 and rearranged to give:

$$\Delta G = -2.3RT \log(K_{eq} / \Gamma) \tag{5.10}$$

Equation 5.10 reinforces the observation that the value of ΔG is a function of the degree to which a reaction is displaced from equilibrium. When $\Gamma = K_{eq}$, the reaction is at equilibrium and $\Delta G = 0$ and no useful work can be done. When Γ is less than K_{eq}, $\Delta G < 0$ and the reaction will occur spontaneously with the release of energy which can be used to perform useful work. However, when Γ is greater than K_{eq}, $\Delta G > 0$ and the reaction

will not occur spontaneously, but rather, requires an input of energy to proceed.

5.2 ENERGY TRANSFORMATIONS AND COUPLED REACTIONS

5.2.1 FREE ENERGY OF ATP IS ASSOCIATED WITH COUPLED PHOSPHATE TRANSFER REACTIONS

Biological organisms overcome metabolic restrictions imposed by thermodynamically unfavorable reactions that exhibit a positive ΔG by *coupling* them or linking them to reactions with a negative ΔG. Thus, the free energy released by the exergonic reaction provides the necessary energy to ensure that the endergonic reaction proceeds. However, in order that this system of coupled reactions will occur spontaneously, *the net free energy change must always be negative*. Thus, the concept of such coupled reactions was critical to our understanding of how life, an overall endergonic process, is maintained without defying the laws of thermodynamics. The notion of biological coupled reactions was unknown to Schrödinger in 1944 to explain the complex events that take place within "the spatial boundary of a living organism" through space and time. For example, in glycolysis, the hydrolysis of adenosine triphosphate (ATP) (Equation 5.11), which is an exergonic reaction, is coupled to the phosphorylation of glucose (Equation 5.12), which is an endergonic reaction. The overall net reaction proceeds because the net reaction is exergonic (Equation 5.13).

$$ATP \rightarrow ADP + Pi \qquad \Delta G^{\circ\prime} = -32.2 \, kJ/mol \quad (5.11)$$

$$Glucose + Pi \rightarrow Glucose\text{-}6\text{-}P$$
$$\Delta G^{\circ\prime} = +13.8 \, kJ/mol \quad (5.12)$$

$$ATP + Glucose \rightarrow Glucose\text{-}6\text{-}P + ADP$$
$$\Delta G^{\circ\prime} = -18.4 \, kJ/mol \quad (5.13)$$

When ATP was first isolated from extracts of muscle in 1929, F. Lipmann was one of the first to recognize its significant role in the energy metabolism of the cell (Figure 5.2). He called it a **high-energy molecule** and designated the terminal phosphate bonds by a squiggle symbol ($\sim$). Actually, the designation of ATP as a high-energy molecule is somewhat misleading because it implies that ATP is in some way a unique molecule. It is not. The two terminal phosphate bonds of the ATP molecule are normal, covalent anhydride bonds. Hydrolysis of an anhydride is accompanied by a favorable increase in entropy, largely due to resonance stabilization of the product molecules: the electrons have one additional bond through which they can resonate. In addition, both the ADP and P_i products are

acidic anions and the two negatively charged products, because of mutual charge repulsion, do not readily recombine. Consequently, the equilibrium constant (K_{eq}) for ATP hydrolysis is rather large—on the order of 10^5. Thus, for ATP hydrolysis

$$K_{eq} = [ADP][Pi]/[ATP] \approx 10^5 \qquad (5.14)$$

Such a large equilibrium constant helps to explain why ATP is so important in cellular metabolism. Photosynthesis and respiration serve to maintain a large pool of ATP such that the observed mass-action ratio (Γ) can be maintained as low as 10^{-4}. This is nine orders of magnitude away from equilibrium (Equation 5.14). By substituting these values for K_{eq} and A in Equation 5.10, we can see that a cell generates considerable free energy ($\Delta G \approx -56 \, kJ/mol$) simply through its capacity to maintain high concentrations of ATP. Thus, it is the cell's capacity to maintain the mass-action ratio so far from equilibrium that enables ATP to function as an energy store.

It is important to emphasize that it is the extent of displacement of G from equilibrium that defines the capacity of a reactant to do work, rather than any intrinsic property of the molecule itself. In the words of D. G. Nicholls, if the glucose-6-phosphate reaction ($\Delta G^{\circ\prime} = 13.8 \, kJ/mol$) "were maintained ten orders of magnitude away from equilibrium, then glucose-6-phosphate would be just as capable of doing work in the cell as is ATP. Conversely, the Pacific

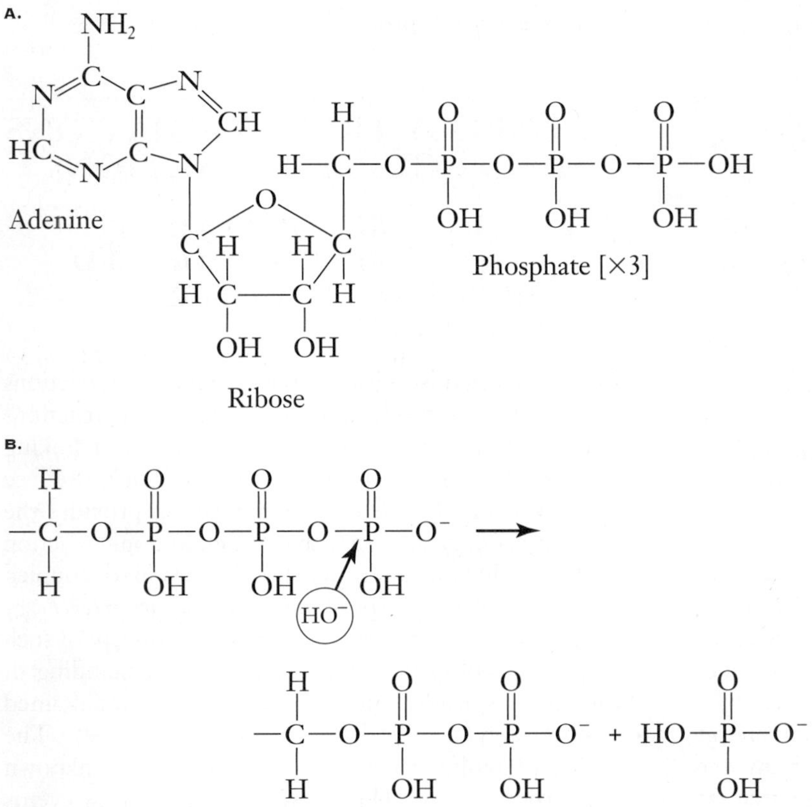

A.

Adenine

Ribose

Phosphate [×3]

B.

FIGURE 5.2 **Adenosine triphosphate (ATP).** (*A*) **The ATP molecule consists of adenine (a nitrogenous base), ribose (a sugar), and three terminal phosphate groups. (*B*) Hydrolysis of ATP yields ADP (adenosine diphosphate) plus an inorganic phosphate molecule.**

Ocean could be filled with an equilibrium mixture of ATP, ADP and P_i but the ATP would have no capacity to do work" (Nicholls and Ferguson, 2002). Of course, it goes without saying that the biochemistry of the cell is structured so as to use ATP, not glucose-6-phosphate, in this capacity.

The relationship between free energy and the capacity to do work is not restricted to chemical reactions. Any system not at equilibrium has a capacity to do work, such as an unequal distribution of solute molecules across a membrane (see Chapter 3). The maintenance of such a steady-state but nonequilibrium condition is called **homeostasis**. Indeed, homeostasis that reflects the capacity to avoid equilibrium in spite of changing environmental conditions is an essential characteristic of all living organisms. When $\Delta G = 0$, no useful work can be done and life ceases to exist.

5.2.2 FREE ENERGY CHANGES ARE ASSOCIATED WITH COUPLED OXIDATION–REDUCTION REACTIONS

Photosynthesis and respiration, which we will discuss in more detail in Chapters 7 and 10, are electrochemical phenomena. Each operates as a sequence of oxidation–reduction reactions in which electrons are transferred from one component to another. Thus, the oxidation of one component is linked or coupled to the reduction of the next component. Such coupled electron transfer reactions are known as *red*uction–*ox*idation or *redox* reactions. This can be illustrated in the following general way. Any compound, A, in its reduced form (A_{red}) becomes oxidized (A_{ox}) when it gives up an electron (e^-) (Equation 5.15). Similarly, any compound, B, in its oxidized form (B_{ox}), becomes reduced (B_{red}) when it accepts an electron (Equation 5.16). When these two reactions are coupled, the net effect is the transfer of an electron from compound A to compound B (Equation 5.17) with A_{red} being the **reductant** and B_{ox} being the **oxidant**.

$$A_{red} \rightarrow A_{ox} + e^- \qquad (5.15)$$

$$B_{ox} + e^- \rightarrow B_{red} \qquad (5.16)$$

$$A_{red} + B_{ox} \rightarrow A_{ox} + B_{red} \qquad (5.17)$$

As a specific biological example, consider the net reduction of three-phosphoglyceric acid (PGA) to glyceraldehyde-3-P (GAP) by the reduced form of nicotinamide adenine dinucleotide phosphate (NADPH):

$$PGA + NADPH + H^+ \rightleftharpoons GAP + NADP^+ \qquad (5.18)$$

Redox reactions may be conveniently dissected into two half-reactions involving the donation and acceptance of electrons. Thus the reduction of PGA to GAP may be considered as the two half-reactions:

$$NADPH \rightleftharpoons NADP^+ + H^+ + 2e^- \qquad (5.19)$$

$$PGA + 2e^- + H^+ \rightleftharpoons GAP \qquad (5.20)$$

Thus, the oxidation of NADPH to $NADP^+$ is coupled to the reduction of PGA to GAP. A reduced/oxidized pair such as $NADPH/NADP^+$ is known as a **redox couple**. Note that oxidation–reduction reactions often involve the transfer of protons. The positively charged protons balance the negative charge of the acquired electrons, which maintains electroneutrality. The involvement of protons indicates that the redox reaction is pH sensitive! The structures of some principal redox compounds involved in biological electron transport are shown in Figure 5.3.

Since each of the half-reactions described above, as well as the net reaction, is reversible, their free energies could be described on the basis of chemical equilibria. However, it is not clear how to treat the electrons, which have no independent existence. Moreover, our interest in redox couples is more in their tendency to accept electrons from or donate electrons to another couple, a tendency known as **redox potential**. Redox potentials allow the feasibility and direction of electron transfers between components in a complex system to be predicted. Indeed, in order to understand electron flow in photosynthesis and respiration it is necessary to have a working understanding of redox potential and how it is applied.

The direction of electron transfer between redox couples can be predicted by comparing their midpoint potentials (E_m). Thermodynamically spontaneous electron transfer will proceed from couples with the more negative (less positive) redox potential to those with the less negative (more positive) redox potential. The energy-transducing membranes of bacteria, mitochondria, and chloroplasts all contain electron-transport systems involving a number of electron carriers with different midpoint redox potentials (Table 5.1).

TABLE **5.1** **Midpoint redox potentials for a selection of redox couples involved in photosynthesis and respiration.**

Reductant/Oxidant	E_m (mV)
Ferredoxin red/ox	−430
$H_2/2H^+$	−420
$NADH + H^+/NAD^+$	−320
$NADPH + H^+/NADP^+$	−320
Succinate/fumarate	+30
Ubiquinone red/ox	+40
Cyt c^{2+}/Cyt c^{3+}	+220
$2H_2O/O_2 + 4H^+$	+820

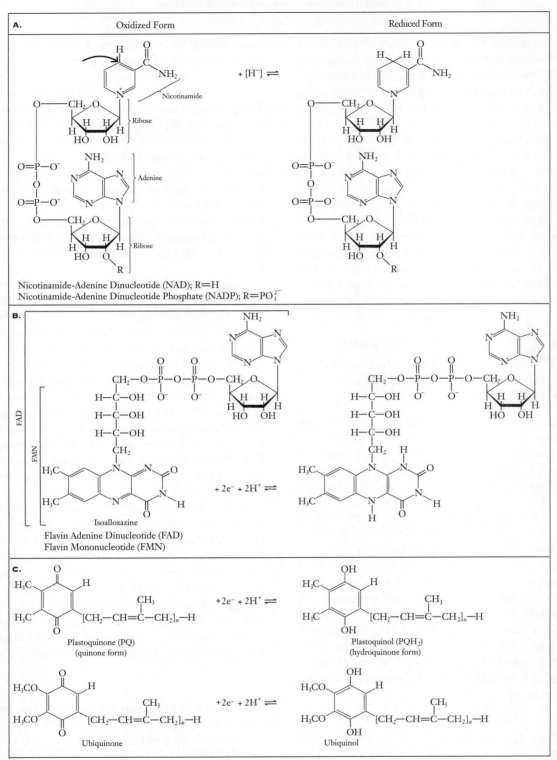

FIGURE 5.3 The chemical structures of some common biological redox agents in oxidized and reduced states. (*A*) Nicotinamide adenine dinucleotide (NAD) and nicotinamide adenine dinucleotide phosphate (NADP). Note that only the nicotinamide ring is changed by the reaction. The nicotinamide ring accepts two electrons but only one proton. Arrow indicates where the electrons are added to the nicotinamide ring. (*B*) Flavin adenine dinucleotide (FAD) consists of adenosine (adenine plus ribose) and riboflavin (ribitol plus isoalloxazine). Flavin mononucleotide (FMN) consists of riboflavin alone. Reduction occurs on the isoalloxazine moiety, which accepts two electrons and two protons. (*C*) Quinones. A quinone ring is attached to a hydrocarbon chain composed of five-carbon isoprene units. The value of *n* is usually 9 for plastoquinone, found in chloroplast thylakoid membranes, and 10 for ubiquinone, found in the inner membrane of mitochondria. Reduction of the quinone ring is a two-step reaction. The transfer of one electron produces the partially reduced, negatively charged semiquinone (not shown). Addition of a second electron plus two protons yields the fully reduced hydroquinone form.

In addition to allowing us to predict the direction of electron transfer, redox potentials also permit the calculation of Gibbs free energy changes for electron-transfer reactions. This can be done using the following relationship:

$$\Delta G^{\circ\prime} = -n\mathrm{F}\Delta E_m \quad (5.21)$$

where n is the number of electrons transferred and F is the Faraday constant (96 500 coulombs mol^{-1}). Biological electron transfers may involve either single electrons or pairs, but energy calculations are almost always based on $n = 2$. ΔE_m is the redox interval through which the electrons are transferred and is determined as

$$\Delta E_m = E_m(\text{acceptor}) - E_m(\text{donor}) \quad (5.22)$$

Thus, for a coupled transfer of electrons from water (the donor) to $NADP^+$ (the acceptor) as occurs in photosynthetic electron transport in chloroplasts, $\Delta E_m = (-320) - (+820) = -1140\,\mathrm{mV} = -1.14\,\mathrm{V}$. Substituting this in Equation 5.21, the value of $\Delta G^{\circ\prime}$ for a two-electron transfer from H_2O to $NADP^+$ is $+220\,\mathrm{kJ}$ mol^{-1}. Note that the sign of $\Delta G^{\circ\prime}$ is positive, indicating that this electron transfer will not occur spontaneously. In photosynthesis, light energy is used to drive this coupled endergonic reaction. In contrast, mitochondria transfer electrons from NADH to O_2. For this coupled electron-transfer reaction, the value of ΔE_m is $+1.14\,\mathrm{V}$ and consequently $\Delta G^{\circ\prime}$ is $-220\,\mathrm{kJ}\,\mathrm{mol}^{-1}$, indicating that this electron transfer in mitochondria is exergonic and thus will occur spontaneously. It is important to note that the molecular mechanisms by which these electrons are transferred through complex processes such as photosynthesis, respiration, and nitrogen assimilation are not in the purview of bioenergetics. The details of the underlying mechanisms will be addressed later in Chapters 7, 8, 10, and 11, respectively.

5.3 ENERGY TRANSDUCTION AND THE CHEMIOSMOTIC SYNTHESIS OF ATP

It has been known for many years that the three principal energy-transducing membrane systems (in bacteria, chloroplasts, and mitochondria) were able to link electron transport with the synthesis of ATP. The mechanism, however, was not understood until Peter Mitchell proposed his **chemiosmotic hypothesis** in 1961. Although not readily accepted by many biochemists in the beginning, Mitchell's hypothesis is now firmly supported by experimental results. In honor of his pioneering work, Mitchell was awarded the Nobel prize for chemistry in 1978.

Mitchell's hypothesis is based on two fundamental requirements. First, energy-transducing membranes are impermeable to H^+. Second, electron carriers are organized asymmetrically in the membrane. The result is

that, in addition to transporting electrons, some carriers also serve to translocate protons across the membrane against a proton gradient. The effect of these **proton pumps** is to conserve some of the free energy of electron transport as an unequal or nonequilibrium distribution of protons, or ΔpH, across the membrane.

5.3.1 CHLOROPLASTS AND MITOCHONDRIA EXHIBIT SPECIFIC COMPARTMENTS

In plants, chloroplasts and mitochondria are the main energy-transducing organelles. The biochemical mechanism by which ATP is synthesized is directly related to the specific compartments that exist in each of these organelles. The structure (Figure 5.4) and development or **biogenesis** of chloroplasts have been studied extensively (Box 5.1). The number of chloroplasts per cell varies from species to species. For example, the unicellular green alga, *Chlamydomonas reinhardtii*, exhibits a single chloroplast while a single mesophyll cell in the leaves of many terrestrial plants can exhibit in excess of 200 chloroplasts.

Within a mature chloroplast, we recognize four major structural regions or compartments: (1) a pair of outer limiting membranes, collectively known as the **envelope**, (2) an unstructured background matrix or **stroma**, (3) a highly structured internal system of membranes, called **thylakoids**, and (4) the intrathylakoid space, or **lumen** (Figure 5.7A). The envelope defines the outer limits of the organelle. These membranes are 5.0 to 7.5 nm thick and are separated by a 10 nm **intermembrane space**. Because the inner envelope membrane

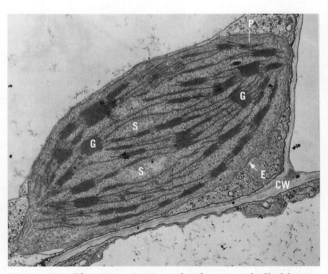

FIGURE 5.4 **Electron micrograph of a mesophyll chloroplast of maize (*Zea mays*). S, stroma; G, granum; P, peripheral reticulum; E, envelope membrane; CW, cell wall.**

BOX 5.1

PLASTID BIOGENESIS

Plastids are a family of double membrane-bound, semi-autonomous organelles common to plant cells. Plastids arise from small, vesicular organelles called **proplastids** (0.2 to 1.0 μm) (Figure 5.5), which are carried from one generation to the next through the embryo and maintained in the undifferentiated state in the dividing cells of plant meristems. Plastids are prokaryotic in origin and arose by **endosymbiosis**. This theory states that sometime during plant evolution, an ancestral eukaryote host cell engulfed a photosynthetic bacterium which established a stable association within the host. As a consequence, plastids contain their own DNA in the form of double-stranded circular molecules very similar to prokaryotic DNA, and replicate by division of existing plastids. Thus, plastids are considered to be semiautonomous and are inherited maternally in most

flowering plants but are paternally inherited in gymnosperms. Several categories of plastids are found in plants and are named according to their color. The most prominent plastids in the leaves of plants are the green photosynthetic **chloroplasts**, which synthesize chlorophyll. **Leucoplasts** are colorless plastids that synthesize volatile compounds called monoterpenes present in essential oils. **Amyloplasts** are unpigmented plastids specialized for the biosynthesis and storage of starch and are prominent organelles in plant storage organs. **Chromoplasts** are pigmented plastids that are responsible for the colors for many fruit (tomatoes, apples, oranges) and flowers (tulips, daffodils, marigolds). Their color is a consequence of the particular combination of carotenes and xanthophylls which they accumulate.

During expansion in the dark, leaves are either white or pale yellow. Such leaves are called **etiolated** leaves. In the development of etiolated leaves, proplastids are converted to **etioplasts** (Figure 5.6). The etioplast is not considered an intermediate stage in the normal development of a chloroplast but rather a specialized plastid present in etiolated leaves. Etioplasts lack chlorophyll but accumulate the colorless chlorophyll precursor, **protochlorophyllide**. A prominent structural feature of etioplasts is a highly ordered, paracrystalline structure called the **prolamellar body (PLB)**. When exposed to light, the protochlorophyll is converted to chlorophyll and the prolamellar body undergoes a reorganization to form the internal thylakoid membranes of the chloroplast.

The assembly of thylakoid membranes during normal chloroplast biogenesis is extremely complex. In angiosperms, chloroplast biogenesis is strictly light-dependent but not in gymnosperms and some green algae. However, regardless of the species, chloroplast biogensis requires coordination between gene expression in the nucleus as well as in the plastids. In addition, plastid protein synthesis has to be coordinated with cytosolic synthesis of nuclear encoded photosynthetic proteins and their subsequent import from the cytosol to the chloroplast. For example, the genes that encode light harvesting polypeptides of photosystem I and photosystem II (Chapter 7) are encoded by the nucleus and synthesized in the cytosol while proteins that form the reaction centers of these same photosystems are encoded by chloroplast DNA and synthesized in the stroma. Elucidation of the mechanisms and regulation of protein import into cellular organelles continues to be a major area of research. The formation of thylakoid membranes during the greening process occurs via the sequential appearance of PS I, followed by PS II, intersystem electron transport components, and lastly the assembly of the light harvesting complexes associated with PS I and PS II. Maximum rates of CO_2 assimilation and O_2 evolution occur once the biogenesis of the thylakoid membranes is complete.

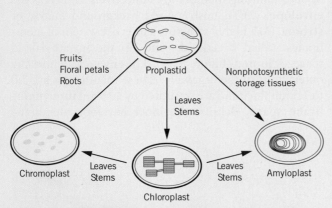

FIGURE 5.5 Plastids. A diagram illustrating some of the interrelationships between various types of plastids. Colorless plastids, that is, plastids without pigments or any prominent internal structure, are called leucoplasts. Leucoplasts in nonphotosynthetic tissues are often the sites of starch accumulation, in which case they are called amyloplasts. In photosynthetic tissues, chloroplasts may become amyloplasts by accumulating excess photosynthetic product in the form of starch granules. When the starch is ultimately degraded, the plastid may once again resume its photosynthetic function. Chloroplasts in maturing fruits and senescing leaves are commonly converted to chromoplasts due to the simultaneous loss of chlorophyll and accumulation of yellow or reddish-orange carotenoid pigments. The characteristic colors of tomato fruit and many autumn leaves are due to such chromoplasts. Chromoplasts may also form in nonphotosynthetic tissues such as carrot roots. The pigments in chromoplasts are often so concentrated that they form a crystalline deposit.

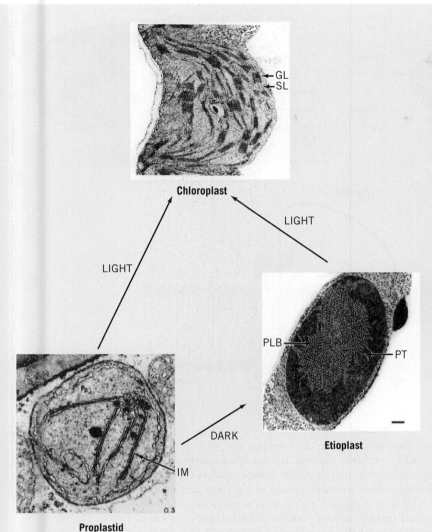

Proplastid

Chloroplast

Etioplast

FIGURE 5.6 **Chloroplast development in higher plants. All chloroplasts are derived from proplastids, which exist as a simple, double membrane vesicle with invaginations of the inner membrane (IM). Upon exposure to light, chlorophyll biosynthesis is initiated with the concomitant differentiation of the thylakoid membranes into granal lamellae (GL; stacks) separated by stromal lamellae (SL; unstacked). If seedlings are initially grown in the dark, the proplastid develops into an etioplast. The etioplast is characterized by the presence of a prolamellar body (PLB) and prothylakoids (PT) extending out from the prolamellar bodies. Upon transfer of the dark grown etiolated seedlings to the light, the prolamellar body in the etioplast disperses and thylakoid membranes develop, producing a normal chloroplast.**

In addition to the conversion of proplastids to chloroplasts, light also induces the conversion of etioplasts into chloroplasts in etiolated plants. This process usually takes approximately 12 to 24 hours (Figure 5.5). Upon exposure to light, carotenoid biosynthesis is stimulated and chlorophyll biosynthesis is initiated by the conversion of protochlorophyllide to chlorophyllide *a* by the enzyme **NADPH:protochlorophyllide oxidoreductase (POR)**, the major protein present in the PLB (Figure 5.6). Thus, exposure of etiolated plants to light causes their leaves to turn green in a process referred to as **de-etiolation or greening**. The photoreduction of protochlorophyllide is associated with the disorganization of the PLB, which leads to the production and assembly of normal, functional thylakoid membranes.

FURTHER READING

Baker, N. R., J. Barber. 1984. *Chloroplast Biogenesis. Topics in Photosynthesis*: *Vol. 5*. Amsterdam: Elsevier Press.

Biswal, U. C., B. Biswal, M. K. Raval. 2004. *Chloroplast Biogensis: From Plastid to Gerontoplast*. Dordrecht: Kluwer Academic Publishers.

is selectively permeable, the envelope also serves to isolate the chloroplast and regulate the exchange of metabolites between the chloroplast and the cytosol that surrounds it. Experiments with spinach chloroplasts have shown that the intermembrane space is freely accessible to metabolites in the cytoplasm. Thus it appears that the outer envelope membrane offers little by way of a permeability barrier. It is left to the inner envelope membrane to regulate the flow of molecular traffic between the chloroplast and cytoplasm.

The envelope encloses the stroma, a predominantly protein solution. The stroma contains all of the

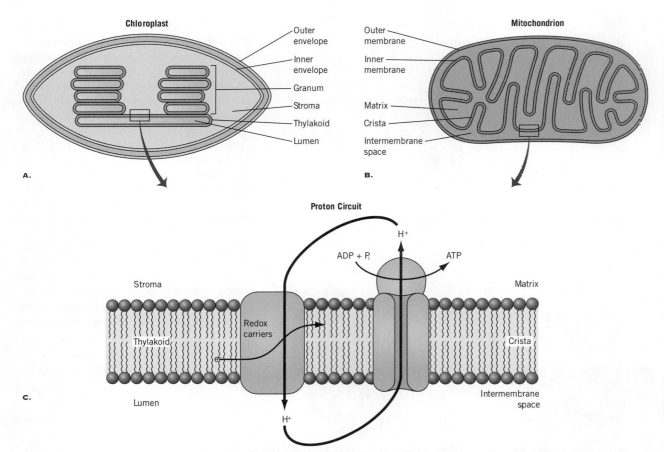

FIGURE 5.7 Using various compartments within chloroplasts and mitochondria, electron transport is coupled to ATP synthesis through the establishment of a proton circuit. (*A*) A typical chloroplast illustrating the major compartments (stroma, thylakoid membrane, lumen) involved in the establishment of a proton circuit for the chemiosmotic synthesis of ATP in chloroplasts. (*B*) A typical mitochondrion illustrating the major compartments (matrix, cristae, intermembrane space) involved in the establishment of a proton circuit for the chemiosmotic synthesis of ATP in mitochondria. (*C*) A model illustrating the similarity in the mechanism by which electron transport is coupled to the establishment of a pH gradient, which is, in turn, consumed in the synthesis of ATP in both chloroplasts and mitochondria. However, the biochemical composition of the redox carriers involved in photosynthetic and respiratory electron transport are quite different.

enzymes responsible for photosynthetic carbon reduction, including **ribulose-1,5-bisphosphate carboxylase/oxygenase**, generally referred to by the acronym **rubisco** (Chapter 8). Rubisco, which accounts for fully half of the total chloroplast protein, is no doubt the world's single most abundant protein. In addition to rubisco and other enzymes involved in carbon reduction, the stroma contains enzymes for a variety of other metabolic pathways as well as DNA, RNA, and the necessary machinery for transcription and translation.

Embedded within the stroma is a complex system of membranes often referred to as **lamellae** (Figure 5.7A). This system is composed of individual pairs of parallel membranes that appear to be joined at the end, a configuration that in cross-section gives the membranes the appearance of a flattened sack, or **thylakoid**

(Gr., sacklike). In some regions adjacent thylakoids appear to be closely appressed, giving rise to membrane stacks known as **grana**. The thylakoids found within a region of membrane stacking are called **grana thylakoids**. Some thylakoids, quite often every second one, extend beyond the grana stacks into the stroma as single, nonappressed thylakoids. These **stroma thylakoids** most often continue into another grana stack, thus providing a network of interconnections between grana. While the organization of thylakoids into stacked and unstacked regions is typical, it is by no means universal. One particularly striking example is the chloroplast in cells that surround the vascular bundles in C4 photosynthetic plants and that have no grana stacks (Chapter 15). Here the thylakoids form long, unpaired arrays extending almost the entire diameter of the chloroplast. The thylakoid membranes contain the chlorophyll and carotenoid pigments and are the site of the light-dependent, energy-conserving reactions of photosynthesis.

The interior space of the thylakoid is known as the **lumen**. The lumen is the site of water oxidation and, consequently, the source of oxygen evolved in photosynthesis. Otherwise it functions primarily as a reservoir for protons that are pumped across the thylakoid membrane during electron transport and that are used to drive ATP synthesis. It is generally assumed that the thylakoids represent a single, continuous network of membranes. This means, of course, that the lumen of each thylakoid seen in cross-section also represents but a small part of a single, continuous system.

Plant mitochondria are typically spherical or short rods approximately 0.5 μm in diameter and up to 2 μm in length. The number of mitochondria per cell is variable but generally relates to the overall metabolic activity of the cell. In one study of rapidly growing sycamore cells in culture, 250 mitochondria per cell was reported. The mitochondria accounted for about 0.7 percent of total cell volume and contained 6 to 7 percent of the total cell protein. In other metabolically more active cells, such as secretory and transfer cells, the number is even higher and may exceed a thousand per cell! Mitochondria, like the other energy-transducing organelle, the chloroplast, are organized into several ultrastructural compartments with distinct metabolic functions: an **outer membrane**, an **inner membrane**, the **intermembrane space**, and the **matrix** (Figure 5.7B). The composition and properties of the two membranes are very different. The outer membrane, which has few enzymatic functions, is rich in lipids and contains relatively little protein. The inner membrane, on the other hand, contains over 70 percent protein on a dry-weight basis. The outer membrane is also highly permeable to most metabolites. It contains large channel-forming proteins, known as **porins** (Chapter 1), which allow essentially free passage of molecules and ions with a molecular mass of 10000 Da or less. The permeability of the inner membrane, like that of the chloroplast, is far more selective—it is freely permeable only to a few small molecules such as water, O_2, and CO_2. Like the chloroplast thylakoid membrane, the permeability of the mitochondrial inner membrane to protons is particularly low, a significant factor with respect to its role in ATP synthesis.

The inner membrane of the mitochondrion is extensively infolded. These invaginations form a dense network of internal membranes called **cristae**. The typical biology textbook picture of cristae is based on animal mitochondria where the infoldings are essentially lamellar and form platelike extensions into the matrix. The common pattern in plant mitochondria is less regular, forming a system of tubes and sacs (Figure 5.8). The unstructured interior of the mitochondrion, or matrix, is an aqueous phase consisting of 40 to 50 percent protein by weight. Much of this protein is comprised of enzymes

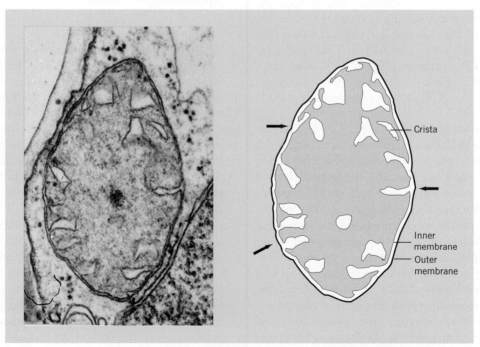

FIGURE 5.8 The mitochondrion. Left: An electron micrograph of a mitochondrion from a maize (*Zea mays*) leaf cell (×10,000). Right: A diagram of the mitochondrion shown in the electron micrograph, to illustrate the essential features. Note that the cristae are continuous with the inner membrane (arrows).

involved in carbon metabolism, but the matrix also contains a mitochondrial genome, including DNA, RNA, ribosomes, and the necessary machinery for transcribing genes and synthesizing protein.

5.3.2 CHLOROPLASTS AND MITOCHONDRIA SYNTHESIZE ATP BY CHEMIOSMOSIS

In chloroplasts the protons are pumped across the thylakoid membrane, from the stroma into the lumen. The difference in proton concentration (ΔpH) across the membrane may be quite large—as much as three or four orders of magnitude. Since protons carry a positive charge, a ΔpH also contributes to an electrical potential gradient across the membrane. A trans-thylakoid ΔpH of 3.5, for example, establishes a potential difference of 207 mV at 25°C. Together the membrane potential difference ($\Delta \Psi$) and the proton gradient (ΔpH) constitute a **proton motive force (pmf)**.

$$\text{pmf} = -59\Delta pH + \Delta \Psi \qquad (5.23)$$

In order to pump protons into the lumen against a proton motive force of this magnitude, a large amount of energy is required. This energy is provided by the negative ΔG generated by photosynthetic intersystem electron transport.

The direction of the proton motive force also favors the return of protons to the stroma, but the low proton conductance of the thylakoid membrane will not allow the protons to simply diffuse back. In fact, the return of protons to the stroma is restricted to highly specific, protein-lined channels that extend through the membrane and that are a part of the ATP synthesizing enzyme, **ATP synthase** (Figure 5.9). This large (400 kDa) multisubunit complex, also known as **coupling factor** or CF_0—CF_1, consists of two multipeptide complexes.

A hydrophobic complex called CF_0 is largely embedded in the thylakoid membrane. Attached to the CF_0 on the stroma side is a hydrophilic complex called CF_1. The CF_1 complex contains the active site for ATP synthesis, while the CF_0 forms an H^+ channel across the membrane, channeling the energy of the electrochemical proton gradient toward the active site of the enzyme. When the electron-transport complexes and the ATP-synthesizing complex are both operating, a proton circuit is established in chloroplasts as well as mitochondria (Figure 5.7C). In chloroplasts, the photosynthetic electron-transport complex pumps the protons from the stroma into the lumen and thus establishes the proton gradient. At the same time, the ATP synthase allows the protons to return to the stroma. Some of the free energy of electron transport is initially conserved in the proton gradient. As the energy-rich proton gradient collapses through the

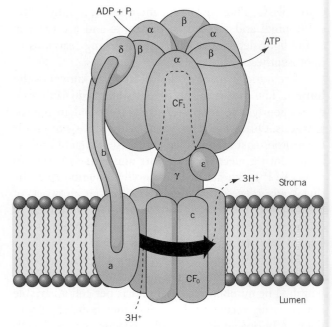

FIGURE 5.9 **A model of the chloroplast ATP synthase.** CF_0 **is an integral membrane protein that forms a proton channel through the thylakoid membrane.** CF_1 **is attached to the stromal side of** CF_0 **and contains the active site for ATP synthesis.** CF_1 **consists of five different subunits with a stoichiometry of** $\alpha_3, \beta_3, \gamma, \delta, \varepsilon$**.** CF_0 **consists of four different subunits with a proposed stoichiometry of a, b, b', c_{10}. The proton channel lies at the interface between subunit a and the ring of c subunits. The pmf rotates the ring of c subunits, which causes the** CF_1 **to rotate and convert ADP and** P_i **to ATP. Large arrow indicates the direction of rotation.**

CF_0-CF_1 complex, that conserved energy is available to drive the synthesis of ATP. The light-dependent synthesis of ATP by chemiosmosis in the chloroplast is called **photophosphorylation**. In mitochondria, the respiratory electron-transport complex pumps protons from the matrix to the intermembrane space (IMS) to establish a proton gradient. The potential energy of this proton gradient is consumed by the mitochondrial ATP synthase to synthesize ATP (Figure 5.7C). The chemiosmotic synthesis of ATP by mitochondria is called **oxidative phosphorylation**.

An essential element of Mitchell's chemiosmotic hypothesis is the reversibility of the ATP synthase reaction. This means that under appropriate conditions the CF_0-CF_1 and other similar complexes can use the negative free energy of ATP hydrolysis to *establish* a proton gradient. For example, both the plasma membrane and tonoplast contain **ATPase proton pumps**, which pump protons out of the cell or the vacuole, as the case may be (Chapter 3). The energy of ATP is thus conserved in the form of a proton gradient that may then be coupled to various forms of cellular work. ATPase proton pumps are a principal means of utilizing ATP to provide energy for the transport of other ions and small solute molecules across cellular membranes (Chapter 3).

Bioenergetics is a fundamental science. Its study is challenging, but this discussion has, of necessity, been restricted to general principles. With this brief background in mind, however, we can now proceed to a discussion of energy conservation through the light-dependent and light-independent reactions of photosynthesis and subsequently examine how plant cells unlock the stored chemical energy needed for growth, development, and the maintenance of homeostasis through the processes of glycolysis and respiration. In addition, we have seen that bioenergetic principles were important in our previous discussion of osmosis and water relations (Chapters 1 and 2) as well as ion transport associated with nutrient uptake (Chapter 3).

SUMMARY

The application of thermodynamic laws to the study of energy flow through living organisms is called bioenergetics. There are two forms of energy, one that is available to do work (free energy) and one that is not (entropy). In a biochemical system such as living organisms, free energy is related to chemical equilibrium. The further a reaction is held away from equilibrium, the more work can be done. Cells utilize this principle to link or couple energy-yielding reactions with energy-consuming reactions. In fact, life exists because cells are able to avoid equilibrium, that is, avoid maximum entropy. Most energy-exchange reactions in the cell are mediated by phosphorylated intermediates, especially ATP and related molecules. ATP is useful in this regard because it has a large equilibrium constant and is highly mobile within the cell. Because ATP is turned over rapidly, it is maintained far from equilibrium. By exploiting specific compartments in chloroplasts (stroma vs. lumen) and mitochondria (matrix vs. intermembrane space), the chemiosmotic synthesis of ATP is linked or coupled to an energy-rich proton gradient (a nonequilibrium proton distribution) across the energy-transducing thylakoid of the chloroplast or the inner membrane (cristae) of the mitochondria. The free energy of electron transport is used to establish the proton gradient and ATP is synthesized as the protons return through transmembrane ATP synthesizing complex.

CHAPTER REVIEW

1. The second law of thermodynamics states that the free energy of the universe tends toward a minimum or that entropy tends toward a maximum. This idea is sometimes referred to as **entropic doom**. Explain what is meant by entropic doom.

2. British writer C. P. Snow has written that understanding the second law of thermodynamics is as much a mark of the literate individual as having read a work of Shakespeare. Can you offer an explanation of what he means by this?

3. Explain the relationship between free energy and chemical equilibria.

4. Explain the relationship between free energy and redox potential.

5. What is the major role of coupled reactions in biology? Give an example of a coupled phosphate transfer reaction and a coupled redox reaction.

6. Compare the compartmentation required for the chemiosmotic synthesis of ATP through photophosphorylation and oxidative phosphorylation.

FURTHER READING

Nicholls, D. G. S., S. J. Ferguson. 2002. *Bioenergetics 2*. New York: Academic Press.

Schrödinger, E. 2000. *What Is Life?* Cambridge: Cambridge University Press.

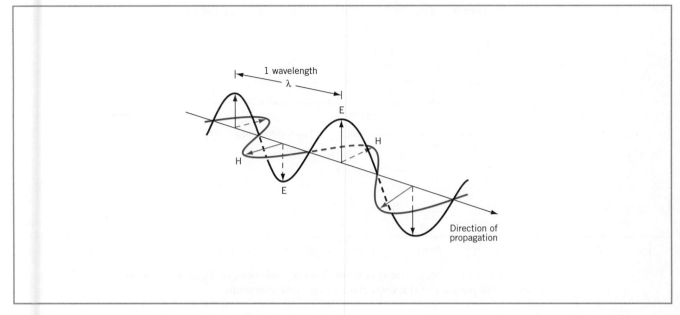

6

The Dual Role of Sunlight: Energy and Information

Sunlight satisfies two very important needs of biological organisms: energy and information. On the one hand, radiant energy from the sun maintains the planet's surface temperature in a range suitable for life and, through the process of photosynthesis, is the ultimate source of energy that sustains most of life in our biosphere. Radiation, primarily in the form of light, also provides critical information about the environment—information that is used by plants to regulate movement, trigger developmental events, and mark the passage of time. The importance of light in the life of green plants is reflected in the study of **photobiology**, which encompasses not only phenomena such as **photosynthesis** (which reflects the role of sunlight as an energy source) but also phenomena such as **photomorphogenesis** and **photoperiodism**, where sunlight provides the necessary information for proper plant development and the measurement of daylength, respectively.

In order to fully appreciate the pervasive importance of light to plants, it is necessary to understand something of the physical nature of light and the molecules with which light interacts in plants. In this chapter, we will

- explore the physical nature of light and how light interacts with matter,

- discuss some of the terminology used in describing light and methods for measuring it,

- discuss briefly the characteristics of light in the natural environment of plants, and

- review the principal pigments and pigment systems found in plants.

The various ways in which light is used by plants to power photosynthesis and regulate development will be discussed throughout many of the subsequent chapters.

6.1 THE PHYSICAL NATURE OF LIGHT

6.1.1 LIGHT IS ELECTROMAGNETIC ENERGY, WHICH EXISTS IN TWO FORMS

What is light? As Johnson recognized more than 200 years ago, "we all know what light is, but it is not easy to *tell* what it is." The simplest answer is that light is a form of radiant energy, a narrow band of energy within the continuous **electromagnetic spectrum**

93

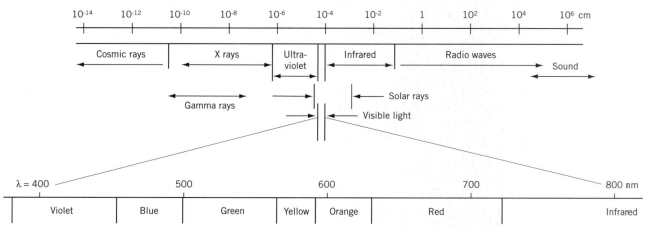

FIGURE 6.1 **The electromagnetic spectrum. Visible radiation, or light, represents only a very small portion of the total electromagnetic spectrum.**

of radiation emitted by the sun (Figure 6.1). The term "light" describes that portion of the electromagnetic spectrum that causes the physiological sensation of vision in humans. In other words, light is defined by the range of wavelengths—between 400 and approximately 700 nanometers—capable of stimulating the receptors located in the retina of the human eye. Strictly speaking, those regions of the spectrum we perceive as red, green, or blue are called light, whereas the ultraviolet and infrared regions of the spectrum, which our eyes cannot detect (although they may have significant biological effects), are referred to as ultraviolet or infrared **radiation**, respectively. While the following discussion will focus on light, it is understood that the principles involved apply to radiant energy in the broader sense.

Like other forms of energy, light is a bit of an enigma and is difficult to define. It is more easily described not by what it is but by how it interacts with matter. Physicists of the late nineteenth and early twentieth centuries resolved that light has attributes of both continuous waves and discrete particles. Both of these attributes are important in understanding the biological role of light.

6.1.2 LIGHT CAN BE CHARACTERIZED AS A WAVE PHENOMENON

The propagation of light through space is characterized by regular and repetitive changes, or waves, in its electrical and magnetic properties. Electromagnetic radiation actually consists of two waves—one electrical and one magnetic—that oscillate at 90° to each other and to the direction of propagation (Figure 6.2). The wave properties of light may be characterized by either **wavelength** or **frequency**. The distance in space between wave crests is known as the wavelength and is represented by the Greek letter lambda (λ). Biologists commonly express wavelengths in units of *nanometers* (nm), where $1\,nm = 10^{-9}$ m. Frequency, represented by the Greek letter nu (ν), is the number of wave crests, or cycles, passing a point in space in one second. Frequency is thus

related to wavelength in the following way:

$$\nu = c/\lambda \qquad (6.1)$$

where c is the speed of light (3×10^8 m s^{-1}). Biologists most commonly use wavelength to describe light and other forms of radiation, although frequency is useful in certain situations. Wavelengths of primary interest to photobiologists fall into three distinct ranges: ultraviolet, visible, and infrared (Table 6.1).

6.1.3 LIGHT CAN BE CHARACTERIZED AS A STREAM OF DISCRETE PARTICLES

When light is emitted from a source or interacts with matter, it behaves as though its energy is divided into discrete units or particles called **photons**. The energy carried by a photon is called a **quantum** (pl. *quanta*), to reflect the fact that the energy can be quantized, that is, it can be divided into multiple units.

The energy carried by a photon (E_q) is related to wavelength and frequency in accordance with the following relationship:

$$E_q = hc/\lambda = h\nu \qquad (6.2)$$

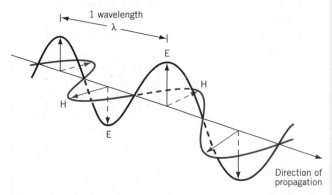

FIGURE 6.2 **Wave nature of light. Electric vectors (E) and magnetic vectors (H) oscillate at 90° to each other.**

TABLE **6.1**   **Radiation of Principal Interest to Biologists.**

Color	Wavelength Range (nm)	Average Energy (kJ mol^{-1} photons)
Ultraviolet	*100–400*	
UV-C	100–280	471
UV-B	280–320	399
UV-A	320–400	332
Visible	*400–740*	
Violet	400–425	290
Blue	425–490	274
Green	490–550	230
Yellow	550–585	212
Orange	585–640	196
Red	640–700	181
Far-red	700–740	166
Infrared	*longer than 740*	85

where h is a proportionality constant, called Planck's constant. The value of h is 6.62×10^{-34} J s photon^{-1}. Accordingly, *the quantum energy of radiation is inversely proportional to its wavelength or directly proportional to its frequency.* The symbol hν (pronounced "h nu") is commonly used to represent a photon in figures and diagrams.

Since both h and c are constants, the energy of a photon is easily calculated for any wavelength of interest. The following example illustrates a calculation of the energy content of red light, with a representative wavelength of 660 nm (6.6×10^{-7} m).

$$E_q = (6.62 \times 10^{-34} \text{J s photon}^{-1}) \\ (3 \times 10^8 \text{m s}^{-1})/6.6 \times 10^{-7} \text{m} \qquad (6.3)$$

Solving for E_q:

$$E_q = 3.01 \times 10^{-19} \text{J photon}^{-1} \qquad (6.4)$$

For blue light, with a representative wavelength of 435 nm (4.35×10^{-7} m),

$$E_q = (6.62 \times 10^{-34} \text{J s photon}^{-1}) \\ (3 \times 10^8 \text{m s}^{-1})/4.35 \times 10^{-7} \text{m} \qquad (6.5)$$

Again, solving for E_q:

$$E_q = 4.56 \times 10^{-19} \text{J photon}^{-1} \qquad (6.6)$$

As the above numbers indicate, the energy content of a single photon is a very small number. However, the *Einstein-Stark law of photochemical equivalence* states that one photon can interact with only one electron. Thus, in any irreversible photochemical reaction, the energy of one photon may be used to convert one molecule of reactant A to one molecule of product B.

$$A + h\nu \rightarrow B \qquad (6.7)$$

Since one mole of any substance contains Avogadro's number (N) of molecules (N = 6.023×10^{23} molecules mol^{-1}), to convert one mole of reactant A to one mole of product B would require N number of photons. Thus, for practical purposes it is convenient to multiply the energy of a single photon by Avogadro's number, which gives the value of energy for a mole of photons. The energy carried by a mole of photons of red light, for example, is 181,292 J mol^{-1}, or 181 kJ mol^{-1} (Table 6.1). The energy carried by a mole of photons of blue light is correspondingly 274 kJ mol^{-1}. The concept of a mole of photons is more useful than dealing with individual photons. For example, as will become apparent in the following section, the law of photochemical equivalence states that a mole of photons of a particular wavelength would be required to excite a mole of pigment molecules.

6.1.4 LIGHT ENERGY CAN INTERACT WITH MATTER

For light to be used by plants, it must first be absorbed. The absorption of light by any molecule is a photophysical event involving internal electronic transitions (Figure 6.3A). The **Gotthaus-Draper principle** tells us that only light that is absorbed can be active in a photochemical process. In contrast to photophysical events, photochemistry refers to any chemical reaction which utilizes absorbed light to convert reactants to products, that is, any light-dependent reaction (Equation 6.7). Therefore, any photobiological phenomenon requires the participation of a molecule that absorbs light. Such a molecule may be defined as a **pigment**. Plants contain a variety of pigments that are prominent visual features and important physiological components of virtually all plants. The characteristic green color of leaves, for example, is due to a family of pigments known as the chlorophylls. Chlorophyll absorbs the light energy used in photosynthesis. The pleasing colors of floral petals

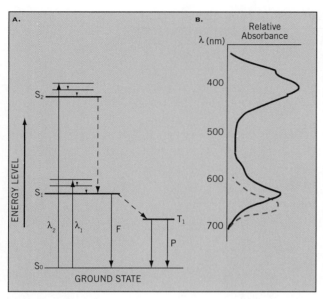

FIGURE 6.3 **The absorption light by a molecule. (*A*) An energy level diagram depicting the various possible transitions when light is absorbed. A nonexcited molecule is said to be in the ground state (S_0). Upon absorption of light of wavelength of either λ_1 or λ_2, a molecule can undergo an electronic transition (solid arrows) to a singlet excited state represented by either S_1 or S_2, respectively. Within each singlet excited state exist various internal energy states representing vibrational and rotational states (smaller horizontal lines). Dashed arrows represent radiationless decay through which energy is given up primarily as heat. Fluorescence (F) is the emission of light from the lowest excited singlet state. T_1 represents the metastable excited triplet state. Energy from the triplet excited state may be lost by radiationless decay or by delayed emission of light known as phosphorescence (P). The triplet state is sufficiently long-lived to allow for photochemical reactions to occur. (*B*) An absorption spectrum (solid line) is a graph in which absorbance is plotted as a function of wavelength. Peaks or absorption bands correspond to the principal excitation levels illustrated in the energy diagram. Also shown is a fluorescence emission spectrum (dashed line) that corresponds to the emission of the absorbed energy as light from the lowest excited singlet state (S_1).**

are due to the anthocyanin pigments that serve to attract insects as pollen vectors. Other pigments, such as phytochrome, are present in quantities too small to be visible but nonetheless serve important roles in plant morphogenesis. These and other important plant pigments will be described later in this chapter.

What actually happens when a pigment molecule absorbs light? Absorption of light by a pigment molecule is a rapid, photophysical, electronic event, occurring within a femtosecond (fs = 10^{-15} s). In accordance with the First Law of Thermodynamics (Chapter 5), the energy of the absorbed photon is transferred to an electron in the pigment molecule during that extremely short period of time. The energy of the electron is thus

elevated from a low energy level, the **ground state**, to a higher energy level known as the **excited**, or **singlet**, state. This change in energy level is illustrated graphically in Figure 6.3. Like photons, the energy states of electrons are also *quantized*, that is, an electron can exist in only one of a series of discrete energy levels. A photon can be absorbed only if its energy content matches the energy required to raise the energy of the electron to one of the higher, allowable energy states.

In the same way that quanta cannot be subdivided, electrons cannot be partially excited. Although, according to the Einstein-Stark law of photochemical equivalence, a single photon can excite only one electron, complex pigment molecules, such as chlorophyll, will have many different electrons, each of which may absorb a photon of a different energy level and, consequently, different wavelength. Moreover, each singlet excited state in which an electron may exist may be subdivided into a variety of smaller but discrete internal energy levels called vibrational and rotational levels. This broadens even further the number of photons that may be absorbed (Figure 6.3). Pigment molecules such as chlorophyll, when exposed to white light, will thus exhibit many different excited states at one time.

An excited molecule has a very short lifetime (on the order of a nanosecond, or 10^{-9} s) and, in the absence of any chemical interaction with other molecules in its environment, it must rid itself of any excess energy and return to the ground state. Dissipation of excess energy may be accomplished in several ways.

1. **Thermal deactivation** occurs when a molecule loses excitation energy as heat (Figure 6.3A). The electron will very quickly drop or *relax* to the lowest excited singlet state. The excess energy is given off as heat to its environment. If the electron then returns to the ground state, that energy will also be dissipated as heat.

2. **Fluorescence** is the emission of a photon of light as an electron relaxes from the first singlet excited to ground state. Since the rate of relaxation through fluorescence is much slower than the rate of relaxation through thermal deactivation, fluorescence emission occurs only as a consequence of relaxation from the first excited singlet state (Figure 6.3A). Consequently, the emitted photon has a lower energy content, and, in accordance with Equation 6.1, a longer wavelength, than the exciting photon. In the case of the photosynthetic pigment chlorophyll, for example, peak fluorescent emission falls to the long-wavelength side of the red absorption band (Figure 6.3B). This is true regardless of whether the pigment was excited with blue light (450 nm, 262 kJ mol^{-1}) or red light (660 nm, 181 kJ mol^{-1}). For pigments such as chlorophyll in solution, a return

to the ground state by emission of red light is often the only available option.

3. Energy may be transferred between pigment molecules by what is known as **inductive resonance** or **radiationless transfer**. Such transfers will occur with high efficiency, but require that the pigment molecules are very close together and that the fluorescent emission band of the donor molecule overlaps the absorption band of the recipient. Inductive resonance accounts for much of the transfer of energy between pigment molecules in the chloroplast (Chapter 7).

4. The molecule may revert to another type of excited state, called the **triplet** state (Figure 6.3A). The difference between singlet and triplet states, related to the spin of the valence electrons, is not so important here. It is sufficient to know that the triplet state is more stable than the singlet state—it is considered a metastable state. The longer lifetime of the metastable triplet state (on the order of 10^{-3} s) is sufficient to allow for photochemical reactions to occur. This could take the form of an oxidation reaction in which the energetic electron is actually given up to an acceptor molecule. When this occurs, the pigment is said to be **photooxidized** and the acceptor molecule becomes reduced.

6.15 HOW DOES ONE ILLUSTRATE THE EFFICIENCY OF LIGHT ABSORPTION AND ITS PHYSIOLOGICAL EFFECTS?

Figure 6.4B illustrates a graph whereby the absorption of light by a pigment (in this case, chlorophyll) measured as relative absorbance is plotted as a function of wavelength. The resulting graph is known as an **absorption spectrum** which emphasizes the correspondence between possible excitation states of the molecule and the principal bands in the absorption spectrum. The energy level diagram in Figure 6.3B represents, in a broad sense, absorption of light by chlorophyll. In this case λ_1 would represent red light and λ_2 would represent blue light.

An absorption spectrum is in effect a probability statement. The height of the absorption curve (or width, as presented in Figure 6.4B) at any given wavelength reflects the probability by which light of that energy level will be absorbed. More importantly, an absorption spectrum is like a fingerprint of the molecule. Every light-absorbing molecule has a unique absorption spectrum that is often a key to its identification. For example, there are several different variations of the green pigment chlorophyll. The pattern of the absorption spectrum for each variation generally resembles that shown in Figure 6.4B, yet each variation of chlorophyll

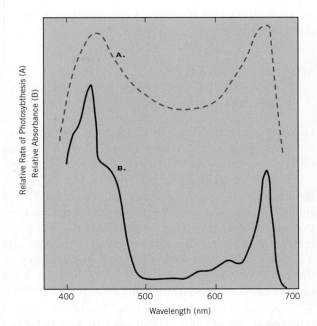

FIGURE 6.4 **A typical action spectrum for leaf photosynthesis (*A*) compared with the absorption spectrum (*B*) of a pigment extract from a leaf containing primarily chlorophyll. The action spectrum has peaks in the blue and red regions of the visible spectrum that correspond to the principal absorption peaks for the pigment extract.**

differs with respect to the precise wavelengths at which maximum absorbance occurs.

Because light must first be absorbed in order to be effective in a physiological process, it follows that there must be a pigment that absorbs the effective light. One of the first tasks facing a photobiologist when studying a light-dependent response is to identify the responsible pigment. One important piece of information is called an **action spectrum**. An action spectrum is a graph that shows the effectiveness of light in inducing a particular process plotted as a function of wavelength. The underlying assumption is that light most efficiently absorbed by the responsible pigment will also be most effective in driving the response. In other words, the action spectrum for a light-dependent response should closely resemble the absorption spectrum of the pigment or pigments that absorb the effective light. A comparison of an action spectrum with the absorption spectra of suspected pigments can therefore provide useful clues to the identity of the pigment responsible for a photosensitive process. As an example, a typical action spectrum for photosynthesis in a green plant is shown in Figure 6.4A. It is compared with the absorption spectrum for a leaf extract that contains primarily chlorophyll and some carotenoid (Figure 6.4B). Note that the action spectrum has pronounced peaks in the red and blue regions of the spectrum and that these action maxima correspond to the absorption maxima for chlorophyll. This is part of the evidence that identifies a role for chlorophyll in photosynthesis.

6.1.6 ACCURATE MEASUREMENT OF LIGHT IS IMPORTANT IN PHOTOBIOLOGY

Given the manifold ways in which light can influence the physiology and development of plants, it should not be too surprising that proper measurement and description of light and light sources has become a significant component of many laboratory and field studies. Many experiments are now conducted in controlled environment-rooms or chambers that allow the researcher to control light, temperature, and humidity. To permit others to interpret the experiments or repeat them in their own laboratories, it is essential that light sources and conditions be fully and accurately described. The spectral distribution of light emitted from fluorescent lamps is very different from that emitted from tungsten lamps or from natural skylight. Because of these and many other factors, each light source may have demonstrably different effects on plant development and behavior. Even natural light changes in quality from dawn through midday to dusk, or between shady and sunny habitats or cloudy and open skies. Understanding photobiology thus requires an understanding of how light is measured and what those measurements mean. Also required is a consistent terminology that is understood by everyone working in the field.

There are three parameters of primary concern when describing light. The first is **light quantity**—how much light has the plant received? The second is the composition of light with respect to wavelength, known as **light quality, spectral composition**, or **spectral energy distribution (SED)**. The third factor is **timing**. What are the duration and periodicity of the light treatment?

The measure of light quantity most widely accepted by plant photobiologists is based on the concept of **fluence**. Fluence is defined as the quantity of radiant energy falling on a small sphere, divided by the cross-section of the sphere. Since light is a form of energy that can be emitted or absorbed as discrete packets or photons, fluence can be expressed in terms of either the number of photons or quanta (in moles, mol) or the amount of energy (in joules, J). **Photon fluence** (units = mol m^{-2}) refers to the total number of photons incident on the sphere while **energy fluence** (units = J m^{-2}) refers to the total amount of energy incident on the sphere. The corresponding rate terms are **photon fluence rate** (units = mol m^{-2} s^{-1}) and **energy fluence rate** (units = J m^{-2} s^{-1} or W m^{-2}). The term **irradiance** is frequently used interchangeably with energy fluence rate, although in principle the two are not equivalent. Irradiance refers to the flux of energy on a flat surface rather than a sphere.

Many instruments for measuring radiation actually measure total energy, including energy outside the visible portion of the spectrum, such as infrared, which is not directly relevant to photobiological processes. In order to avoid such complications, instruments are now commercially available that are limited to that portion of the spectrum between 400 nm and 700 nm. This range of light is broadly defined as **photosynthetically active radiation (PAR)**. Thus photon fluence rates expressed as mol photons m^{-2} s^{-1} PAR, or energy fluence rates, expressed as Watts (W) m^{-2} PAR, are widely accepted for routine laboratory work in plant photobiology. The only serious limitation to PAR measurements is that they exclude light in the 700 to 750 nm range—light which, although inactive in higher plant photosynthesis, plays a significant role in regulating plant development (Chapter 16).

The term "light quality" refers to spectral composition and is usually defined by an emission or incidence

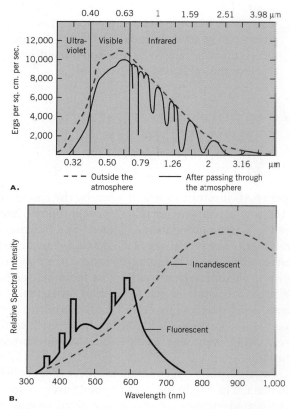

FIGURE 6.5 Spectral energy distribution of sunlight (*A*) compared with fluorescent and incandescent light (*B*). Note the difference in wavelength scale between A and B. About 20 percent of the incoming energy from the sun, particularly in the infrared region, is absorbed by atmospheric gases (primarily CO$_2$ and H$_2$O). The solar spectrum was drawn from measurements made in clear weather from an observatory in Australia by C. W. Allen. (From H. H. Lamb, *Climate: Present, Past and Future*, London: Methuen & Co., 1972.)

spectrum. SED is measured with a spectroradiometer, an instrument capable of measuring fluence rate over narrow-wavelength bands. Depending on the instrument, either **spectral photon fluence rate** (units = mol photons m^{-2} s^{-1} nm^{-1}) or **spectral energy fluence rate** (units = W m^{-2} nm^{-1}) is plotted against wavelength. In practice, spectroradiometers are also equipped with flat surface detectors that measure **spectral irradiance** (W m^{-2} nm^{-1}).

SED can vary depending on the nature of the light source and a number of other factors (Figure 6.5). The SED of natural sunlight, for example, can vary depending on the quality of the atmosphere, cloud cover, and the time of day (Figure 6.5A). The SED of artificial sources, such as incandescent and fluorescent lamps, are significantly different from natural light (Figure 6.5B). Fluorescent light has a relatively high emission in the blue but drops off sharply in the red. Incandescent light, on the other hand, contains relatively little blue light but high emissions in the far-red and infrared.

6.2 THE NATURAL RADIATION ENVIRONMENT

A relatively small proportion of the radiation originating in the sun reaches the earth's atmosphere and even less actually reaches the surface (Figure 6.5A). However, both the quantity and spectral distribution of radiant energy that reaches (or fails to reach) earth may have a significant impact on the physiology of the plant. As well, radiant energy is central to several problems of more immediate and profound consequences for man.

Significant amounts of infrared radiation are absorbed by the water vapor and carbon dioxide and other gases present in the earth's atmosphere (Figure 6.5A), giving rise to a phenomenon known as the **greenhouse effect** (Figure 6.6). Although public awareness of the greenhouse effect has increased markedly in recent years, it is not a phenomenon restricted to the late twentieth century. Indeed, the greenhouse effect has been with us since the beginnings of life on earth. Without it, life as we know it would not be possible. Infrared radiation is of low frequency (or long wavelength) and therefore low energy. Its principal effect is to increase vibrational activity in molecules—that is, heat. Absorption of infrared by atmospheric water vapor and carbon dioxide creates a "thermal blanket" that helps to prevent extreme variations in temperature such as occur on the lunar surface, where these gases are absent. Similar, although less extreme, temperature variations are characteristic of dry, desert regions on earth where high daytime temperatures alternate with very cool nights.

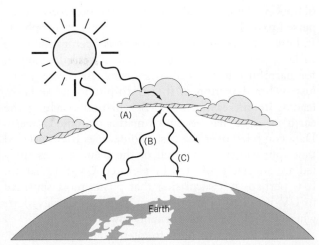

FIGURE 6.6 Diagram representing the greenhouse effect. Radiation from the sun warms both the atmosphere and the earth. (*A*) The earth then reradiates infrared (heat) back into the atmosphere. (*B*) Here the infrared radiation is either reflected back to earth or absorbed by atmospheric gases, such as CO$_2$, H$_2$O vapor, and methanol, thus preventing its escape. (*C*) Some of the trapped infrared is reradiated back to earth, giving rise to increased temperatures.

Public concern about the greenhouse effect arises from evidence that, since the beginnings of the industrial revolution, our prodigious consumption of fossil fuels has contributed to a steady increase in atmospheric carbon dioxide and other so-called "greenhouse gases." Many believe that continued release of carbon dioxide will lead to greater heat retention in the atmosphere and global warming. This could result in partial melting of polar ice caps with extensive flooding of low-lying land areas and major shifts in plant biodiversity and agricultural productivity. A scenario commonly proposed is that higher carbon dioxide levels will stimulate photosynthesis and increase the amount of plant material on earth. However, this is an overly simplistic view of the effects of CO$_2$ concentrations on photosynthesis. Increases in both global temperatures and global CO$_2$ concentrations can have negative effects on photosynthetic rates.

At the other end of the spectrum, ultraviolet radiation is characterized by short wavelength, high frequency, and high energy levels (Table 6.1). Absorption of ultraviolet radiation creates highly reactive molecules, often causing the ejection of an electron, or ionization of the molecule. Such ionizations usually have deleterious effects on organisms. A principal action of UV-C (about 254 nm), for example, is to induce thiamine dimers (hence, mutations) in deoxyribonucleic acid. In the natural environment, UV-induced mutation is not normally a major problem because little far-ultraviolet radiation reaches the surface. Virtually all of the UV-C and most of the UV-B is absorbed by ozone (O_3) and aerosols (dispersed particles of solids or liquids) in the stratosphere. If, however, the atmospheric ozone concentration were to be lowered, there would be an increased potential for harmful effects to all organisms. In recent years, just such a depletion of the stratospheric ozone layer, leading to increases in UV-B radiation reaching the earth's surface, has become a matter of some concern. Data compiled over the past two decades have revealed that approximately one-half of the plant species studied are adversely affected by elevated UV-B radiation. It is perhaps not surprising that plants most sensitive to UV-B radiation are those native to lower elevations where UV-B fluxes are normally low.

With respect to both frequency and energy level, visible light falls between UV and infrared radiation. Absorption of visible light raises the energy level of valence electrons of the absorbing molecule and thus has the potential for initiating useful photochemical reactions. Moreover, the fluence rate and spectral quality of visible light are constantly changing, often predictably, throughout the day or season. These variations convey information about the environment—information that the plant can use to its advantage.

The two most significant changes in visible light on a daily basis are seen in the fluence rate and in spectral distribution. Typically at midday under full sun, the fluence rate approaches 2000 μmol m^{-2} s^{-1}. At twilight, just before the sun sets below the horizon, the fluence rate will have dropped to the order of 10 μmol m^{-2} s^{-1} or less. During the period known as dusk, fluence rate falls rapidly—by as much as one order of magnitude every 10 minutes.

Falling light levels at end of day are accompanied by shifts in spectral quality (see Chapter 24). Normal daylight consists of direct sunlight and diffuse skylight. Diffuse skylight is enriched with blue wavelengths because the shorter wavelengths are preferentially scattered by moisture droplets, dust, and other components of the atmosphere. Consequently, normal daylight is enriched with blue (hence, blue skies!). At twilight, often defined as a solar elevation of $10°$ or less from the horizon, a combination of scattering and refraction of the sun's rays as they enter the earth's atmosphere at a low angle enriches the light with longer red and far-red wavelengths. This is because, at twilight, the path traversed by sunlight through the atmosphere to an observer on earth may be up to 50 times longer than it is when the sun is directly overhead. Much of the violet and blue light is thus scattered out of the line of sight, leaving predominantly the longer red and orange to reach the observer.

Atmospheric factors, such as clouds and air pollution, also influence the spectral distribution of sunlight. Cloud cover reduces irradiance and increases the proportion of scattered (i.e., blue) light. Airborne pollutants will cause scattering, but will also absorb certain wavelengths. Plants growing under a canopy must cope with severe reduction in red and blue light as it is filtered through the chlorophyll-containing leaves above, or with **sunflecks**—spots of direct sunlight that suddenly appear through an opening in the canopy. These sudden changes in irradiance may have a significant impact on the photosynthetic capacity of a plant (Chapter 14).

It is clear that plants are exposed to an ever-changing light environment. Many of these changes, such as cloud cover, are unpredictable but others such as daily changes in fluence rate and spectral energy distribution occur with great regularity. The more regular changes convey precise information about the momentary status of the environment as well as impending changes (see Chapters 13 and 14). It is perhaps not surprising that plants have evolved sophisticated means for interpreting this information as a matter of survival.

6.3 PHOTORECEPTORS ABSORB LIGHT FOR USE IN A PHYSIOLOGICAL PROCESS

Photoreceptors are defined as pigment molecules that process the energy and informational content of light into a form that can be used by the plant. A pigment that contains protein as an integral part of the molecule is known as a **chromoprotein**. Thus, photoreceptors typically are chromoproteins. The **chromophore** (Gr. *phoros*, bearing) is that portion of the chromoprotein molecule responsible for absorbing light and hence, color. The protein portion of a chromoprotein molecule is called the **apoprotein**. The complete molecule, or **holochrome**, consists of the chromophore plus the protein. The principal photoreceptors found in plants are described here. Their roles in various physiological processes will be discussed in detail in later chapters.

6.3.1 CHLOROPHYLLS ARE PRIMARILY RESPONSIBLE FOR HARVESTING LIGHT ENERGY FOR PHOTOSYNTHESIS

As noted earlier, chlorophyll is the pigment primarily responsible for harvesting light energy used in photosynthesis. The chlorophyll molecule consists of two parts, a **porphyrin** head and a long hydrocarbon, or **phytol** tail (Figure 6.7). A porphyrin is a cyclic tetrapyrrole, made up of four nitrogen-containing **pyrrole rings** arranged in a cyclic fashion. Porphyrins are ubiquitous in living organisms and include the heme group found in mammalian hemoglobin and the

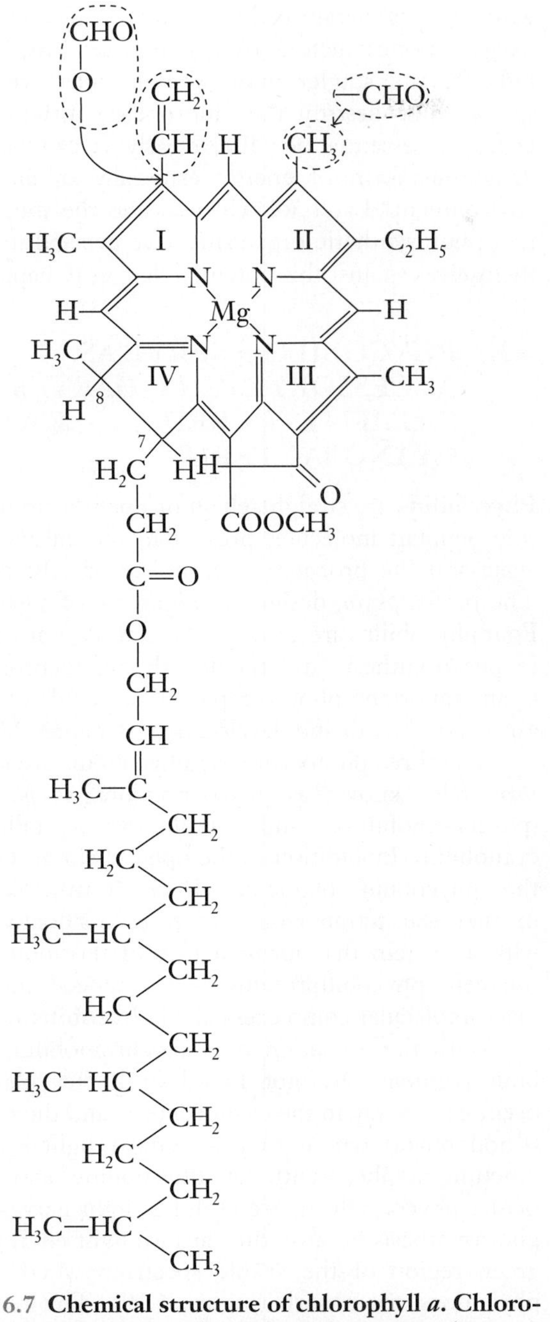

FIGURE 6.7 **Chemical structure of chlorophyll *a*. Chlorophyll *b* is similar except that a formyl group replaces the methyl group on ring II. Chlorophyll *c* is similar to chlorophyll *a* except that it lacks the long hydrocarbon tail. Chlorophyll *d* is similar to chlorophyll *a* except that a —O—CHO group is substituted on ring I as shown.**

photosynthetic and respiratory pigments, cytochromes (Chapters 7, 10). Esterified to ring IV of the porphyrin in chlorophyll is a 20-carbon alcohol, phytol. This long, lipid-soluble hydrocarbon tail is a derivative of the 5-carbon **isoprene**. Isoprene is the precursor to a variety of important molecules, including other pigments (the carotenes), hormones (the gibberellins), and steroids (Chapter 19).

Completing the chlorophyll molecule is a magnesium ion (Mg^{2+}) chelated to the four nitrogen atoms in the center of the ring. Loss of the magnesium ion from chlorophyll results in the formation of a nongreen product, **pheophytin**. Pheophytin is readily formed during extraction under acidic conditions, but small amounts are also found naturally in the chloroplast where it serves as an early electron acceptor (Chapter 7).

Four species of chlorophyll, designated chlorophyll *a*, *b*, *c*, and *d*, are known. The chemical structure of chlorophyll *a*, the primary photosynthetic pigment in all higher plants, algae, and the cyanobacteria, is shown in Figure 6.7. Chlorophyll *b* is similar except that a formyl group (—CHO) substitutes for the methyl group on ring II. Chlorophyll *b* is found in virtually all higher plants and green algae, although viable mutants deficient of chlorophyll *b* are known. The principal difference between chlorophyll *a* and chlorophyll *c* (found in the diatoms, dinoflagellates, and brown algae) is that chlorophyll *c* lacks the phytol tail. Finally, chlorophyll *d*, found only in the red algae, is similar to chlorophyll *a* except that a (—O—CHO) group replaces the (—CH=CH$_2$) group on ring I.

When grown in the dark, angiosperm seedlings do not accumulate chlorophyll (Chapter 5, Box 1). Their yellow color is primarily due to the presence of carotenoids. Dark-grown seedlings do, however, accumulate significant amounts of **protochlorophyll a**, the immediate precursor to chlorophyll *a*. The chemical structure of protochlorophyll differs from chlorophyll only by the presence of a double bond between carbons 7 and 8 in ring IV (Figure 6.7). The reduction of this bond is catalyzed by the enzyme **NADPH:protochlorophyll oxidoreductase**. In angiosperms this reaction requires light, but in gymnosperms and most algae chlorophyll can be synthesized in the dark. There is a general consensus among investigators that chlorophyll *b* is synthesized from chlorophyll *a*.

Note that the respective chlorophylls exhibit generally a similar shape to their absorption spectra in organic solvents, but exhibit absorption maxima at distinctly different wavelengths, in both the blue and the red regions of the spectrum (Figure 6.8). These shifts in the absorbance maxima illustrate that subtle chemical changes in the porphyrin ring of chlorophyll (Figure 6.7) have significant effects on the absorption properties of this pigment. This is evidence that it is the porphyrin ring of chlorophyll that actually absorbs the light and not the phytol tail. Note also that chlorophyll does not absorb strongly in the green region of the visible light spectrum (490–550 nm). The strong absorbance in the blue and red and transmittance in the green is what gives chlorophyll its characteristic green color.

The presence of the long hydrocarbon phytol exerts a dominant effect on the solubility of chlorophyll, rendering it virtually insoluble in water. In the plant,

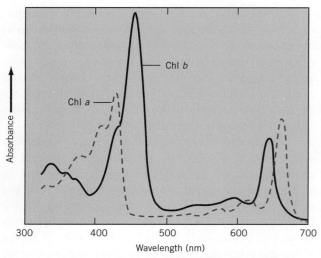

FIGURE 6.8 Absorption spectra of chlorophyll *a* (broken line) and chlorophyll *b* (solid line) in acetone.

with O_2 to generate to highly dangerous and reactive oxygen species such as oxygen free radicals, hydroxyl radicals, and singlet oxygen. These reactive oxygen species may destroy the chloroplast. Although light energy is essential for life, clearly it can be a very dangerous form of energy especially in an aerobic environment! Later we will examine the mechanisms that photosynthetic organisms have evolved to protect themselves against this potential danger (Chapter 14).

6.3.2 PHYCOBILINS SERVE AS ACCESSORY LIGHT-HARVESTING PIGMENTS IN RED ALGAE AND CYANOBACTERIA

Phycobilins are straight-chain or open-chain tetrapyrrole pigment molecules present in the eukaryotic red algae and the prokaryotic cyanobacteria (Figure 6.9). The prefix, *phyco*, designates pigments of algal origin. Four phycobilins are known. Three of these are involved in photosynthesis and the fourth, phytochromobilin, is an important photoreceptor that regulates various aspects of growth and development (Chapter 22).

The three photosynthetic phycobilins are *phycoerythrin* (also known as phycoerythrobilin), *phycocyanin* (phycocyanobilin), and *allophycocyanin* (allophycocyanobilin). In addition to the open-chain tetrapyrrole, the phycobilin pigments differ from chlorophyll in that the tetrapyrrole group is covalently linked with a protein that forms a part of the molecule. In the cell, phycobiliproteins are organized into large macromolecular complexes called **phycobilisomes.**

With the exception of phytochromobilin, phycobilin pigments are not found in higher plants but occur exclusively in the cyanobacteria and the red algae (Rhodophyta) where they assume a light-harvesting function in photosynthesis. Phycobilins, and in particular phycoerythrin, are useful as light harvesters for photosynthesis because they absorb light energy in the green region of the visible spectrum where chlorophyll does not absorb (Figure 6.10). The red algae, for example, appear almost black because the chlorophyll and phycoerythrin together absorb almost all of the visible radiation for use in photosynthesis (compare Figure 6.10 with Figure 6.8).

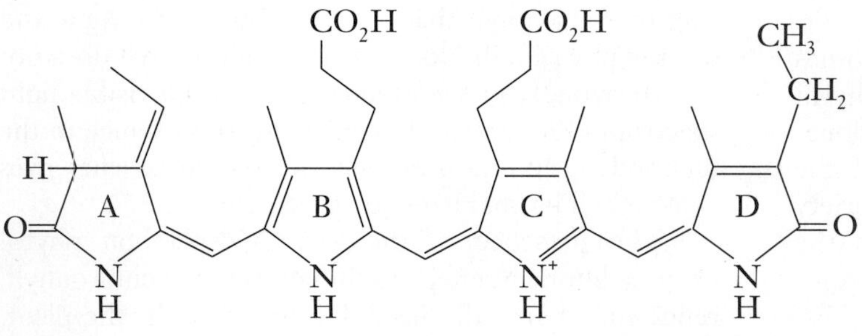

chlorophyll is found exclusively in the lipid domain of the chloroplast membranes, where it forms noncovalent associations with hydrophobic proteins. Only an extremely small percentage of the chlorophyll found *in vivo* is ever free chlorophyll, that is, not bound to proteins. The absorption spectra of these chlorophyll-protein complexes are markedly different from that of free pigment in solution. For example, chlorophyll *a*-protein absorbs primarily in the region of 675 nm as opposed to 663 nm for chlorophyll *a* in acetone. Conjugation of chlorophylls with protein in the membrane is important for three reasons. One is that it helps to maintain the pigment molecules in the precise relationship required for efficient absorption and energy transfer. A second reason is that it provides each pigment with a unique environment that in turn gives each molecule a slightly different absorption maximum. These slight absorbance differences are an important factor in the orderly transfer of energy through the pigment bed toward the reaction center where photochemical conversion actually occurs (Chapter 7). Third, the presence of excess free chlorophyll would photosensitize plants, which would lead to the destruction of chloroplast structure. Chlorophyll in the unbound state is less efficient in photosynthetic inductive energy transfer but reacts more efficiently

FIGURE 6.9 The open-chain tetrapyrrole chromophore of phycocyanin. Compare with the cyclic tetrapyrrole group in the chlorophyll molecule (Figure 6.7).

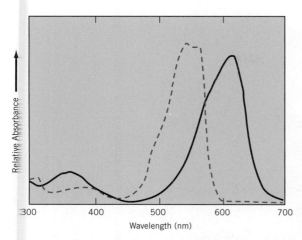

FIGURE 6.10 **Absorption spectra of phycocyanin (solid line) and phycoerythrin (broken line) in dilute buffer. Compare with the absorption spectra of chlorophyll (Figure 6.8). Note that the phycobilins, phycoerythrin in particular, absorb strongly in the 500–600 nm range where chlorophyll absorption is minimal.**

The fourth phycobiliprotein, of particular significance to higher plants, is **phytochrome**, a receptor that plays an important role in many photomorphogenic phenomena. Its chromophore structure and absorption spectrum are similar to that of allophycocyanin. Phytochrome (literally, plant pigment) is unique because it exists in two forms that are photoreversible. The form P660 (or Pr) absorbs maximally at 660 nm. However, absorption of 660 nm light converts the pigment to a second, far-red-absorbing form P735 (or Pfr). Absorption of far-red light by Pfr converts it back to the red-absorbing form. Pfr is believed to be an active form of the pigment that is capable of initiating a wide range of morphogenetic responses. Phytochrome will be discussed in more detail in Chapter 22.

6.3.3 CAROTENOIDS ACCOUNT FOR THE AUTUMN COLORS

Carotenoids comprise a family of orange and yellow pigments present in most photosynthetic organisms. Found in large quantity in roots of carrot and tomato fruit, carotenoid pigments are also prominent in green leaves. In the fall of the year, the chlorophyll pigments are degraded and the more stable carotenoid pigments account for the brilliant orange and yellow colors so characteristic of autumn foliage.

Carotenoid pigments are C_{40} terpenoids biosynthetically derived from the isoprenoid pathway described in Chapter 19. Because the carotenoids are predominantly hydrocarbons, they are lipid soluble and found either in the chloroplast membranes or in specialized plastids called **chromoplasts**. The concentration of pigment in chromoplasts may reach very high levels, to the

extent that the pigment actually forms crystals. The carotenoid family of pigments includes **carotenes** and **xanthophylls** (Figure 6.11). Carotenes are predominantly orange or red-orange pigments. β-carotene is the major carotenoid in algae and higher plants. Note that in β-carotene and α-carotene (a minor form), both ends of the molecule are cyclized. Other forms, such as γ-carotene, found in the green photosynthetic bacteria, have only one end cyclized. Lycopene, the principal pigment of tomato fruit, has both ends open. The yellow carotenoids, xanthophylls, are oxygenated carotenes. Lutein and zeaxanthin, for example, are hydroxylated forms of α-carotene and β-carotene, respectively.

Like chlorophyll, β-carotene in the chloroplast is complexed with protein. β-carotene, which absorbs strongly in the blue region of the visible spectrum (Figure 6.12), is known to quench both the triplet excited chlorophyll as well as the highly reactive singlet excited oxygen, which can be generated by the reaction of triplet chlorophyll with ground state oxygen. Thus, β-carotene protects chlorophyll from photooxidation.

6.3.4 CRYPTOCHROME AND PHOTOTROPIN ARE PHOTORECEPTORS SENSITIVE TO BLUE LIGHT AND UV-A RADIATION

A wide range of plant responses to blue light and UV-A radiation have been known or suspected for a long time. **Cryptochrome** was the name given initially to the blue light/UV-A photoreceptor because blue light responses appeared to be prevalent in cryptograms, an old primary division of plants which do not exhibit true flowers and seeds and included ferns, mosses, algae and fungi. In addition, the molecular nature of the blue light photoreceptor remained unknown, and thus, cryptic (secret or hidden) for many years. However, recent research has established that cryptochromes are found throughout the plant kingdom. The action spectrum for cryptochrome exhibits two peaks, one in the UVA region (320-400 nm) and one in the blue region of the visible spectrum (400-500 nm). The chromophore for cryptochrome is a flavin. The three most common flavins are riboflavin (Figure 6.13) and its two nucleotide derivatives, flavin mononucleotide (FMN) and flavin adenine dinucleotide (FAD). The flavins may occur free or complexed with protein, in which case they are called flavoprotein. However, those flavins that function as photoreceptors probably constitute a very small portion of a much larger pool. Both FMN and FAD, for example, are important cofactors in cellular oxidation–reduction reactions (Chapter 5).

Arabidopsis has two genes for cryptochrome (*CRY1* and *CRY 2*) whereas tomato has at least three genes that encode this photoreceptor. Mosses and ferns exhibit

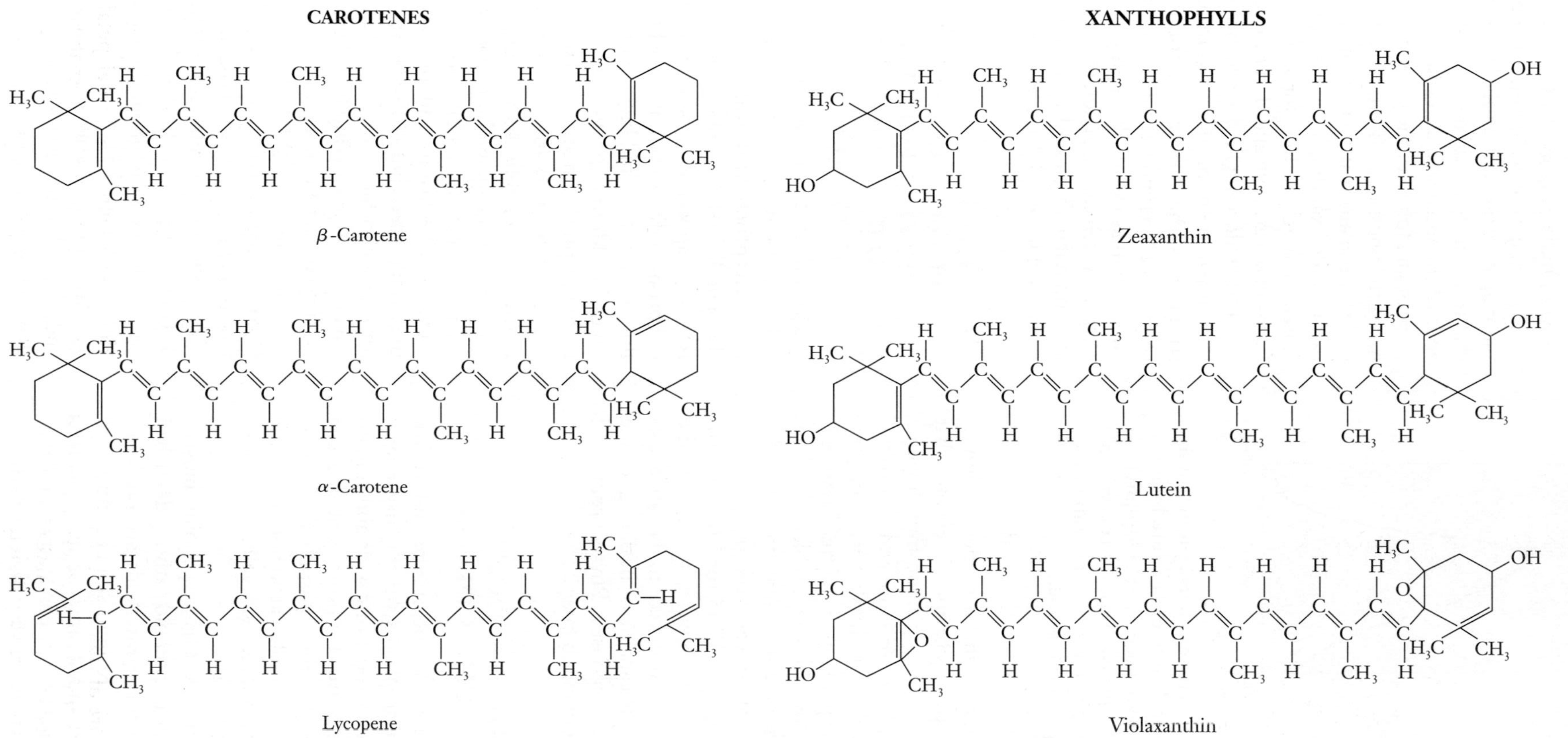

CAROTENES

β-Carotene

α-Carotene

Lycopene

XANTHOPHYLLS

Zeaxanthin

Lutein

Violaxanthin

FIGURE 6.11 The chemical structures of representative carotenes and xanthophylls. The principal distinction between the two is that xanthophylls contain oxygen and carotenes do not. Carotenes are generally orange while xanthophylls are yellow.

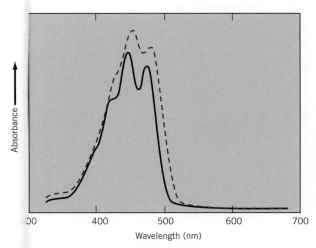

FIGURE 6.12 **Absorption spectra of α-carotene (solid line) and β-carotene (broken line).**

two and five genes for cryptochrome, respectively. Interestingly, the sequence of the CRY1 protein is similar to **photolyase**, a unique class of flavoproteins that use blue light to stimulate repair of UV-induced damage to microbial DNA. Photolyases contain two chromophores; one a flavin (FAD) and one a **pterin** (Figure 6.13). Although the precise nature of the CRY1 chromophores remains to be determined, it appears that one is FAD and the second is likely to be a pterin. Cryptochromes are cytoplasmic proteins with a mass of about 75 kDa and, together with phytochrome, mediate photomorphogenic responses such as photoperiod-dependent

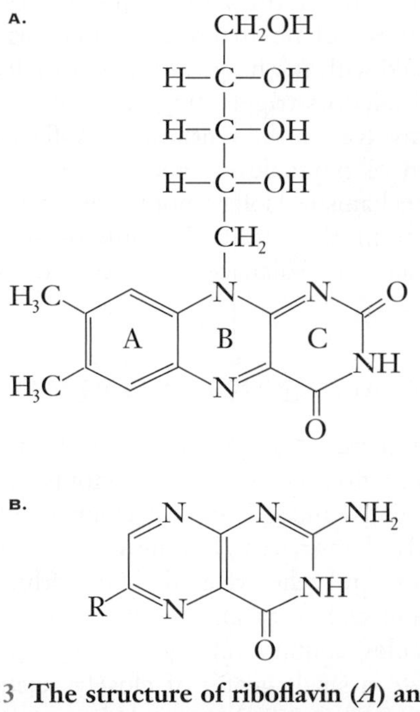

FIGURE 6.13 **The structure of riboflavin (*A*) and pterin (*B*). Note the similarity between the pterin structure and the B and C rings of riboflavin. See Chapter 5 for the structures of riboflavin derivatives, FMN and FAD.**

control of flowering, stimulation of leaf expansion and the inhibition of stem elongation.

Phototropins (PHOT) are a second class of blue light photoreceptor that was first discovered in the late 1980s. Arabidopsis exhibits two phototropin genes designated *PHOT1* and *PHOT2*. Like cryptochrome, phototropin is also a flavoprotein with two FMN molecules as chromophores. The molecular mass of this photoreceptor is about 120 kDa and is localized to the plasmamembrane. Phototropins are not involved in photomorphogenic responses. Analyses of mutants deficient in either *PHOT1* or *PHOT2* indicate that these two genes exhibit partial overlapping roles in the regulation of phototropism. Furthermore, phototropins play important roles in optimizing photosynthetic efficiency of plants such as the regulation of stomatal opening for CO_2 gas exchange as well as chloroplast avoidance movement to protect the photosynthetic apparatus from photoinhibition due to exposure to excess light (Chapter 13). The biochemical nature and physiological roles of these blue light/UV-A photoreceptors are discussed in more detail in Chapter 22.

6.3.5 UV-B RADIATION MAY ACT AS A DEVELOPMENTAL SIGNAL

More recently, a small number of responses, such as anthocyanin synthesis in young milo seedlings (*Sorghum vulgare*) and suspension cultures of parsley or carrot cells have been described with an action spectrum peak near 290 nm and no action at wavelengths longer than about 350 nm. These findings would seem to indicate the presence of one or more UV-B (280–320 nm) receptors in plants, although the nature of the photoreceptors has yet to be identified with certainty.

The impact of ultraviolet radiation, especially UV-B, on plants is receiving increasing attention because of concerns about the thinning of the atmospheric ozone layer. A reduction in the ozone layer results in an increase in UV-B radiation, specifically between 290 and 314 nm, which can cause damage to nucleic acids, proteins, and the photosynthetic apparatus and lead to shorter plants and reduced biomass. The UV-B receptor also appears to modulate responses to phytochrome in some systems. It has yet to be identified.

6.3.6 FLAVONOIDS PROVIDE THE MYRIAD FLOWER COLORS AND ACT AS A NATURAL SUNSCREEN

Although the plant world is predominantly green, it is the brilliant colors of floral petals, fruits, bracts, and occasionally leaves that most attracts humans and a variety of other animals to plants. These various shades of scarlet, pink, purple, and blue are due to the presence of

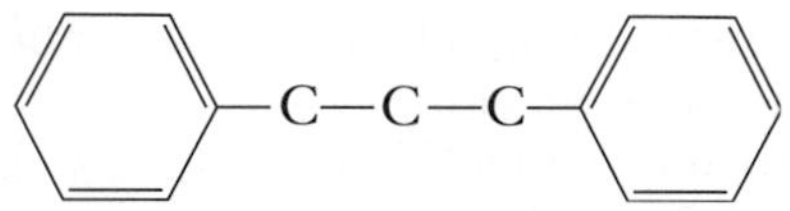

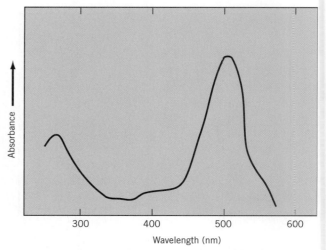

FIGURE 6.14 **Flavonoids are phenylpropane derivatives with a basic C$_6$—C$_3$—C$_6$ composition.**

FIGURE 6.15 **Absorption spectrum of the anthocyanin, pelargonin.**

pigments known as **anthocyanins**. Anthocyanins belong to a larger group of compounds known as **flavonoids**. Other classes of flavonoids (e.g., chalcones and aurones) contribute to the yellow colors of some flowers. Yet others (the flavones) are responsible for the whiteness of floral petals that, without them, might appear translucent. The flavonoids are readily isolated and, because of their brilliant colors, have been known since antiquity as a source of dyes. Consequently, the flavonoids have been extensively studied since the beginnings of modern organic chemistry and their chemistry is well known. The biosynthesis of flavonoids is discussed in Chapter 28.

Flavonoids are phenylpropane derivatives with a basic C$_6$—C$_3$—C$_6$ composition (Figure 6.14). The most strongly colored of the flavonoids are the anthocyanidins and anthocyanins. Anthocyanins are the glycoside derivatives of anthocyanidins. Unlike chlorophyll, the anthocyanins are water-soluble pigments and are found predominantly in the vacuolar sap. They are readily extracted into weakly acidic solution. The color of anthocyanins is sensitive to pH: both anthocyanidins and anthocyanins are natural indicator dyes. For example, the color of cyanidin changes from red (acid) to violet (neutral) to blue (alkaline). The deep violet extract of boiled red cabbage will turn a definitely unappetizing blue-green if boiled in alkaline water!

Anthocyanins in leaves such as *Coleus* and red-leaved cultivars of maple (*Acer* sps.) are found in the vacuoles of the epidermal cells, where they appear to mask the chlorophylls. However, the anthocyanins absorb strongly between 475 nm and 560 nm while transmitting both blue and red light. Consequently, the presence of anthocyanins does not interfere with photosynthesis in the chloroplasts of the underlying mesophyll cells.

Virtually all flavonoids absorb strongly in the UV-B region of the spectrum (Figure 6.15). Since these compounds also occur in leaves, one possible function of the flavonoids is thought to be protection of the underlying leaf tissues from damage due to ultraviolet radiation. Thus, the accumulation of UV-B absorbing flavonoids acts as a natural sunscreen for plants, green algae, and cyanobacteria. As flower pigments, the flavonoids attract insect pollinators. Many insects can detect ultraviolet light and thus can perceive patterns contributed by the colorless flavonoids as well as the colored patterns visible to humans. The synthesis of anthocyanins is stimulated by light, both UV and visible, as well as by nutrient stress

(especially nitrogen and phosphorous deficiencies) and low temperature.

At least one group of flavonoids, the **isoflavonoids**, have become known for their antimicrobial activities. Isoflavonoids are one of several classes of chemicals of differing chemical structures, known as **phytoalexins**, that help to limit the spread of bacterial and fungal infections in plants. Phytoalexins are generally absent or present in very low concentrations, but are rapidly synthesized following invasion by bacterial and fungal pathogens. The details of phytoalexin metabolism are not yet clear. Apparently a variety of small polysaccharides, glycoproteins and proteins of fungal or bacterial origin, serve as **elicitors** that stimulate the plant to begin synthesis of phytoalexins. Studies with soybean cells infected with the fungus *Phytophthora* indicate that the fungal elicitors trigger transcription of mRNA for enzymes involved in the synthesis of isoflavonoids. The production of phytoalexins appears to be a common defense mechanism. Isoflavonoids are the predominant phytoalexin in the family Leguminoseae, but other families, such as Solanaceae, appear to use terpene derivatives.

6.3.7 BETACYANINS AND BEETS

The prominent red pigments of beet root and *Bougainvillea* flowers are not flavonoids (as was long believed), but a more complex group of glycosylated compounds known as **betalains** or **betacyanins**. Betacyanins and the related betaxanthins (yellow) are distinguished from anthocyanins by the fact that the molecules contain nitrogen. They appear to be restricted to a small group of closely related families in the order Chenopodiales, including the goosefoot, cactus, and portulaca families, which are not known to produce anthocyanins.

SUMMARY

Sunlight provides plants with energy to drive photosynthesis and critical information about the environment. Light is a form of electromagnetic energy that has attributes of continuous waves and discrete particles. The energy of a particle of light (a quantum) is inversely proportional to its wavelength.

Light is absorbed by pigments, and pigments that absorb physiologically useful light are called photoreceptors. All pigments have a characteristic absorption spectrum that describes the efficiency of light absorption as a function of wavelength. Because only light that is absorbed by pigments can be effective in a physiological or biochemical process, a comparison of absorption spectra with the action spectrum for a process helps to identify the responsible pigment. When light is absorbed, the pigment becomes excited, or unstable. The excess energy must be dissipated as heat, reemitted as light, or used in a photochemical reaction, thus allowing the pigment to return to its stable, ground state.

Regular and predictable changes in fluence rate and spectral energy distribution provide plants with information about the momentary status of their environment as well as impending changes. The biochemical characteristics of the principal plant pigments of physiological interest are described.

CHAPTER REVIEW

1. Although, as Samuel Johnson said, it is not easy to tell what light is, what is it? Describe the various parameters of light and how it can be measured.

2. Describe the relationship between an absorption spectrum and an action spectrum. Of what significance is an action spectrum to the plant physiologist?

3. When is a pigment a photoreceptor? Make a list of the major plant pigments and identify one or more principal functions of each.

4. Chlorophylls and carotenoids are found predominantly in cellular membranes while anthocyanins are located in vacuoles. What does this distribution tell you about the chemistry of these pigments?

5. Assume you are writing a paper in which you report the effects of artificial light on the growth and photosynthesis of plants. How would you describe the light environment so that a reader could attempt to repeat your experiments in his/her own laboratory?

6. Describe how light energy is absorbed and dissipated by a pigment.

FURTHER READING

Batschauer, A. (ed.). 2003. *Photoreceptors and Light Signaling*. Cambridge: Royal Society of Chemistry.

Bova, B. 2001. *The Story of Light*. Naperville IL: Sourcebooks Inc.

Briggs, W. L., J. L. Supdich (eds.). 2005. *Handbook of Photosensory Receptors*. Weinheim Germany: Wiley-VCH.

Clegg, B. 2001. *Light Years. An Exploration of Mankind's Enduring Fascination with Light*. London: Piatkus Publishers Ltd.

Goodwin, T. W. (ed.). 1988. *Plant Pigments*. London and New York: Academic Press.

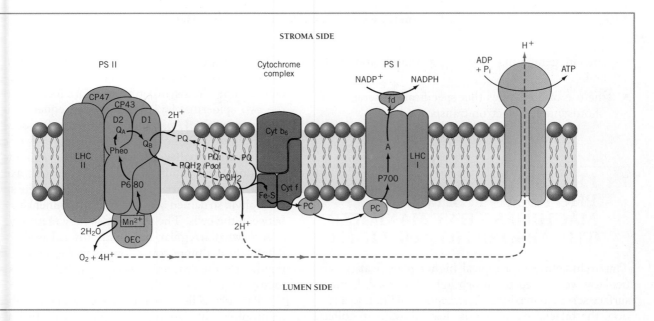

7

Energy Conservation in Photosynthesis: Harvesting Sunlight

Photosynthesis is the fundamental basis of competitive success in green plants and the principal organ of photosynthesis in higher plants is the leaf. From the delicate, pastel hues of early spring through the brilliant reds and oranges of autumn, foliage leaves are certainly one of the dominant features of terrestrial plants. The biologist's interest in leaves, however, goes far beyond their aesthetic quality. Biologists are interested in the structure of organs and how those structures are adapted to carry out effectively certain physiological and biochemical functions. Leaves provide an excellent demonstration of this structure–function relationship. While some leaves may be modified for special purposes (for example, tendrils, spines, and floral parts), the *primary* function of leaves remains photosynthesis. In order to absorb light efficiently, a typical leaf presents a large surface area at approximately right angles to the incoming sunlight. From this perspective, the leaf may be viewed as a *photosynthetic machine*—superbly engineered to carry out photosynthesis efficiently in an extremely hostile environment.

Photosynthesis occurs not only in eukaryotic organisms such as green plants and green algae but also in prokaryotic organisms such as cyanobacteria and certain groups of bacteria. In higher plants and green algae

the reactions of photosynthesis occur in the chloroplast, which is, quite simply, an incredible thermodynamic machine. The chloroplast traps the radiant energy of sunlight and conserves some of it in a stable chemical form. The reactions that accomplish these energy transformations are identified as the **light-dependent reactions** of photosynthesis. Energy generated by the light-dependent reactions is subsequently used to reduce inorganic carbon dioxide to organic carbon in the form of sugars. Both the carbon and the energy conserved in those sugars are then used to build the order and structure that distinguishes living organisms from their inorganic surroundings.

The focus of this chapter is the organization of leaves with respect to the exploitation of light as the primary source of energy and its conversion to the stable, chemical forms of ATP and NADPH by the chloroplast. We will discuss

- the structure of terrestrial plant leaves with respect to the interception of light,

- photosynthesis as the reduction of carbon dioxide to carbohydrate,

- the photosynthetic electron transport chain, its organization in the thylakoid membrane, and its role in generating reducing potential and ATP, and

- the use of herbicides that specifically interact with photosynthetic electron transport.

7.1 LEAVES ARE PHOTOSYNTHETIC MACHINES THAT MAXIMIZE THE ABSORPTION OF LIGHT

The architecture of a typical higher plant leaf is particularly well suited to absorb light. Its broad, laminar surface serves to maximize interception of light. In addition, the bifacial nature of the leaf allows it to collect incident light on the upper surface and diffuse (both scattered and reflected) light on the lower surface. Gross morphology is not, however, the only factor enhancing interception of light—internal cellular arrangements also play an important role.

The anatomy of a typical dicotyledonous mesomorphic leaf is shown in Figure 7.1A,C. The leaf is sheathed with an upper and lower **epidermis**. The exposed surfaces of the epidermal cells are coated with a cuticle. The photosynthetic tissues are located between the two epidermal layers and are consequently identified as **mesophyll** (*meso*, middle; *phyll*, leaf) tissues. The upper photosynthetic tissue generally consists of one to three layers of **palisade** mesophyll cells. Palisade cells are elongated, cylindrical cells with the long axis perpendicular to the surface of the leaf. Below is the **spongy** mesophyll, so named because of the prominent air spaces between the cells. The shape of spongy mesophyll cells is somewhat irregular but tends toward isodiametric. The plan of a monocotyledonous leaf is similar except that it lacks the distinction between palisade and spongy mesophyll (Figure 7.1B,D).

Palisade cells generally have larger numbers of chloroplasts than spongy mesophyll cells. In leaves of *Camellia*, for example, the chlorophyll concentration of the palisade cells is 1.5 to 2.5 times that of the spongy mesophyll cells. The higher number of chloroplasts in the palisade cells no doubt reflects an adaptation to the higher fluence rates for photosynthetically active light generally incident on the upper surfaces of the leaf.

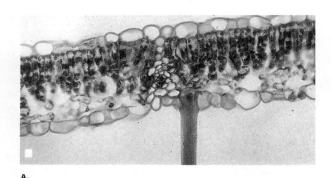

A.

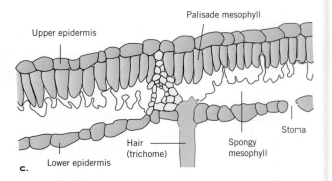

C.

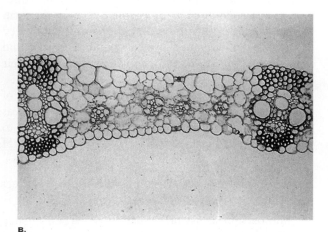

B.

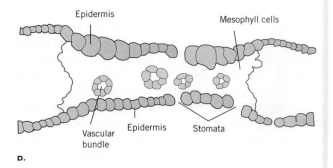

D.

FIGURE 7.1 **The structure of leaves shown in cross-section.** (*A,C*) A dicotyledonous leaf *Acer* sp. (*B,D*) A monocotyledonous leaf (*Zea mays*), showing a section between two major veins. (A,B: From T. E. Weier et al., *Botany*, 6th ed. New York, Wiley, 1982. Used by permission of authors.)

In spite of the relatively large number of chloroplasts in the palisade layers of a dicotyledonous leaf, there is a significant proportion of the cell volume that does not contain chloroplasts. Because the absorbing pigments are confined to the chloroplast, a substantial amount of light may thus pass through the first cell layer without being absorbed. This has been called the **sieve effect**. Multiple layers of photosynthetic cells is one way of increasing the probability that photons passing through the first layer of cells will be intercepted by successive layers (Figure 7.2).

The impact of the sieve effect on the efficiency of light absorption is to some extent balanced by factors that change the direction of the light path within the leaf. Light may first of all be **reflected** off the many surfaces associated with leaf cells. Second, light that is not reflected but passes between the aqueous volume of mesophyll cells and the air spaces that surround them (especially in the spongy mesophyll) will be bent by **refraction**. Third, light may be *scattered* when it strikes particles or structures with diameters comparable to its

wavelengths. In the leaf cell, for example, both mitochondria and the grana structures within chloroplasts have dimensions (500–1000 nm) similar to the wavelengths active in photosynthesis. Both organelles will scatter light. These three factors—reflection, refraction, and scattering—combine to increase the effective path length as light passes through the leaf. The longer light path increases the probability that any given photon will be absorbed by a chlorophyll molecule before it can escape from the leaf (Figure 7.2).

Careful studies of the optical properties of leaves have shown that, in spite of their scattering properties, palisade cells do not appear to absorb as much light as might be expected. That is to say that the palisade cells have a lower than expected *efficiency* of light attenuation. This is apparently because they also act to some extent as a **light guide**. Some of the incident light is channeled through the intercellular spaces between the palisade cells in much the same way that light is transmitted by an optical fiber (Figure 7.2). It is probable that photosynthesis in the uppermost palisade layer is frequently light saturated. Any excess light would be wasteful and could, in fact, give rise to photoinhibition and other harmful effects that we will discuss in more detail later in this chapter. Thus, the increased transmission of light to the lower cell layers resulting from both scattering and the light-guide effect would no doubt be advantageous by contributing to a more efficient allocation of photosynthetic energy throughout the leaf.

Not all leaves are designed like the "typical" dicotyledonous mesomorphic leaf described above. Leaves may be modified in many ways to fit particular environmental situations. Pine leaves (or needles), for example, are more circular in cross-section. Their capacity for light interception has been compromised in favor of a reduced surface-to-volume ratio, a modification that helps to combat desiccation when exposed to dry winter air. In other cases, such as dry land or desert species, the leaves are much thicker in order to provide for storage of water. In extreme cases, such as the cacti, the leaves have been reduced to thorns and the stem has taken over the dual functions of water storage and photosynthesis. These and other modifications to leaf morphology will be discussed more fully in Chapter 14 and 15.

Within the leaf mesophyll cells of plants, the chloroplast is the organelle that transforms light energy into ATP and NADPH to convert CO_2 to sugars. The structure of a typical chloroplast was discussed in Chapter 5. ATP is synthesized by chemiosmosis, whereas NADPH is the product of coupled electron transfer reactions in the chloroplast thylakoid membranes. The enzymatic reactions involved in the conversion of CO_2 to sugars takes place in the chloroplast stroma (Chapter 8).

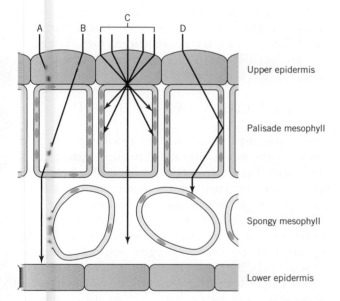

FIGURE 7.2 A simplified diagram illustrating how the optical properties of leaves help to redistribute incoming light and maximize interception by chlorophyll. (A) Photon strikes a chloroplast and is absorbed by chlorophyll. (B) The sieve effect—a photon passes through the first layer of mesophyll cells without being absorbed. It may be absorbed in the next layer of cells or pass through the leaf to be absorbed by another leaf below. (C) The planoconvex nature of epidermal cells creates a lens effect, redirecting incoming light to chloroplasts along the lateral walls of the palisade cells. (D) The light-guide effect. Because the refractive index of cells is greater than that of air, light reflected at the cell–air interfaces may be channeled through the palisade layer(s) to the spongy mesophyll below.

7.2 PHOTOSYNTHESIS IS AN OXIDATION–REDUCTION PROCESS

Although it may not be obvious at first glance, photosynthesis is fundamentally an oxidation–reduction reaction. This can be seen by examining the summary equation for photosynthesis:

$$6CO_2 + 12H_2O \rightarrow C_6H_{12}O_6 + 6O_2 + 6H_2O \quad (7.1)$$

Here photosynthesis is shown as a reaction between CO_2 and water to produce glucose, a six-carbon carbohydrate or hexose. Although glucose is not the first product of photosynthesis, it is a common form of accumulated carbohydrate and provides a convenient basis for discussion. Note that equal molar quantities of CO_2 and O_2 are consumed and evolved, respectively. This is convenient for the experimenter since it means that photosynthesis can be measured in the laboratory as either the uptake of CO_2 or the evolution of O_2. However, it is important to note that the ratio of CO_2 fixed/O_2 evolved = 1 only under conditions where photorespiration is suppressed. We will discuss photorespiration and its impact on photosynthesis in more detail in Chapter 8. For simplicity, we can reduce equation 7.1 to

$$CO_2 + 2H_2O \rightarrow (CH_2O) + O_2 + H_2O \quad (7.2)$$

where the term (CH_2O) represents the basic building block of carbohydrate. Equation 7.2 can be interpreted as a simple redox reaction, that is, a reduction of CO_2 to carbohydrate, where H_2O is the reductant and CO_2 is the oxidant. But it might also be interpreted as a hydration of carbon (e.g., carbo*hydrate*), as it was in early studies of photosynthesis. How do we know that it is one and not the other? And why is it necessary to write the equation with two molecules of water as reactant (and one as product) when one would appear to suffice? These questions can best be answered by reviewing some of the early studies on photosynthesis (see Box 7.1. Historical Perspective—The Discovery of Photosynthesis).

One of the earliest clues to the redox nature of photosynthesis was provided by studies of C. B. van Niel in the 1920s. As a microbiologist, van Niel was interested in the photosynthetic sulfur bacteria that use hydrogen sulfide (H_2S) as a reductant in place of water. Consequently, unlike algae and higher plants, the photosynthetic sulfur bacteria do not evolve oxygen. Instead, they deposit elemental sulfur according to the following equation:

$$CO_2 + 2H_2S \rightarrow (CH_2O) + 2S + H_2O \quad (7.3)$$

The reaction in equation 7.3 can also be written as two partial reactions:

$$2H_2S \rightarrow 4e^- + 4H^+ + 2S \quad (7.4)$$

$$CO_2 + 4e^- + 4H^+ \rightarrow (CH_2O) + H_2O \quad (7.5)$$

Equations 7.4 and 7.5 describe photosynthesis in the purple sulfur bacteria as a straightforward oxidation–reduction reaction. C. B. van Niel adopted a comparative biochemistry approach and argued that the mechanisms for **oxygenic** (i.e., oxygen-evolving) photosynthesis in green plants and **anoxygenic** (i.e., non-oxygen-evolving) photosynthesis in the sulfur bacteria both followed the general plan:

$$2H_2A + CO_2 \rightarrow 2A + (CH_2O) + H_2O \quad (7.6)$$

In this equation, A can represent either oxygen or sulfur, depending on the type of photosynthetic organism. According to equation 7.6, the O_2 released in oxygenic photosynthesis would be derived from the reductant, water. Correct stoichiometry would therefore require the participation of four electrons and hence two molecules of water.

A second important clue was provided by R. Hill who, in 1939, was first to demonstrate the partial reactions of photosynthesis in *isolated* chloroplasts. In Hill's experiments with chloroplasts, artificial electron acceptors, such as ferricyanide, were used. Under these conditions, no CO_2 was consumed and no carbohydrate was produced, but light-driven reduction of the electron acceptors was accompanied by O_2 evolution:

$$4Fe3^+ + 2H_2O \rightarrow 4Fe2^+ + O_2 + 4H^+ \quad (7.7)$$

Hill's experiments confirmed the redox nature of green plant photosynthesis and added further support for the argument that water was the source of evolved oxygen. Direct evidence for the latter point was finally provided by S. Ruben and M. Kamen in the early 1940s. Using either CO_2 or H_2O labeled with ^{18}O, a heavy isotope of oxygen, they showed that the label appeared in the evolved oxygen only when supplied as water ($H_2^{18}O$), not when supplied as $C^{18}O_2$. If the evolved O_2 is derived from water, then two molecules of water must participate in the reduction of each molecule of CO_2.

Based on these results, *photosynthesis can be viewed as a photochemical reduction of CO_2*. The energy of light is used to generate strong reducing equivalents from H_2O—strong enough to reduce CO_2 to carbohydrate. These reducing equivalents are in the form of reduced $NADP^+$ (or, $NADPH + H^+$). Additional energy for carbon reduction is required in the form of ATP, which is also generated at the expense of light. *The principal function of the light-dependent reactions of photosynthesis is therefore to generate the NADPH and ATP required for carbon reduction.* This is accomplished through a series of reactions that constitute the **photosynthetic electron transport chain**.

BOX 7.1

HISTORICAL PERSPECTIVE—THE DISCOVERY OF PHOTOSYNTHESIS

Photosynthesis assumes a role of such dominant proportions in the organization and development of plants, not to mention feeding world populations, it is somewhat surprising that so little was known about the process before the final decades of the eighteenth century. The practice of agriculture was already several thousand years old and practical discussions of crop production had been written at least 2,000 years before. The origins of plant nutrition as a science can be traced as far back as Aristotle and other Greek philosophers, who taught that plants absorbed organic material directly from the soil. This theory, known as the humus theory, prevailed in agricultural circles until the late nineteenth century, long after the principles of photosynthesis had been established.

The first suggestions of photosynthesis appear in the writings of Stephen Hales, an English clergyman and naturalist who is considered "the father of plant physiology." In 1727, Hales surmised that plants obtain a portion of their nutrition from the air and wondered, as well, whether light might also be involved. Hales's insights were remarkably prescient, contrary as they were to the long-established humus theory. However, chemistry had yet to come of age as a science and Hales's ideas were not provable by experiment or by reference to any well-established chemical laws.

Rabinowitch and Govindjee (1969) date the "discovery" of photosynthesis as 1776, the year Joseph Priestly published his two-volume work entitled *Experiments and Observations on Different Kinds of Air*. But as with many other phenomena in science, there was no one moment of discovery. The story gradually fell into place through the cumulative efforts of several clergy, physicians, and chemists over a period of nearly 75 years. J. Priestly (1733–1804) was an English minister whose nonconformist views on religion and politics led to his emigration to the United States in 1794. He was also a scientist engaged in pioneering experiments with gases and is perhaps best known for his discovery of oxygen. Priestly's experiments, begun in 1771 and first published in 1772, led him to observe that air "contaminated" by burning a candle could not support the life of a mouse. He then found that the air could be restored by plants—a sprig of mint was introduced into the contaminated air and "after eight or nine days I found that a mouse lived perfectly in that part of the air in which the sprig of mint had grown" (Priestly, 1772).

Priestly failed to recognize the role of light in his experiments and it was perhaps serendipitous that his laboratory was well enough lighted for the experiments to have succeeded at all. In 1773, Priestly's experiments came to the attention of Jan Ingen-Housz (1730–1799), a physician to the court of Austrian Empress Maria Theresa. During a visit to London, Ingen-Housz heard Priestly's experiments described by the President of the Royal Society. He was intrigued by these experiments and six years later returned to England to conduct experiments. In the course of a single summer, Ingen-Housz performed and had published some 500 experiments on the purification of air! He observed that plants could purify air within hours, not days as observed by Priestly, but only when the *green* parts of plants were exposed to *sunlight*. Together, Priestly and Ingen-Housz had confirmed Hales's guesses made some 52 years earlier.

Although Priestly continued his experiments—in 1781 he agreed with Ingen-Housz on the value of light and green plant parts—neither Priestly nor Ingen-Housz recognized the role of "fixed air," as CO_2 was known at the time. This was left to the Swiss pastor and librarian Jean Senebier (1742–1809). In 1782, Senebier published a three-volume treatise in which he demonstrated that the purification of air by green plants in the light was dependent on the presence of "fixed" air. It is interesting to note that all three scientists had emphasized the purification of air in relation to its capacity to support animal life—plant nutrition was not a central theme. At the same time chemists across Europe, including Priestly in England, Scheele in Germany, and Lavoisier in France, were actively investigating the chemical and physical properties of gases. By 1785 Lavoisier had identified "fixed" air as CO_2 and by 1796 Ingen-Housz had correctly deduced that CO_2 was the source of carbon for plants.

Another important component in the equation of photosynthesis was added by the work of a Geneva chemist, N. T. de Saussure (1767–1845). It was de Saussure who first approached photosynthesis in a sound, quantitative fashion. In his book *Recherches Chimiques sur la Végétation* (1804) he showed that the weight of organic matter plus oxygen formed by photosynthesis was substantially larger than the weight of CO_2 consumed. He thus concluded that the additional weight was provided by water as a reactant. The equation for photosynthesis, using the new language of chemistry founded by Lavoisier, could now be written:

$$CO_2 + H_2O \rightarrow O_2 + \text{organic matter}$$

Finally, it remained for a German surgeon, Julius Mayer (1814–1878), to clarify the energy relationships of photosynthesis. In 1845, he correctly deduced, for the first time, that the energy used by plants and animals in their metabolism is derived from the energy of the sun and that it is transformed by photosynthesis from the radiant to the chemical form. Thus by the middle of the nineteenth century the general outline of photosynthesis was complete. Despite the importance of the process, however, it would be almost another century before the structural and chemical details of photosynthesis would yield to modern methods of microscopic and radiochemical analysis.

FURTHER READING

Priestly, J. 1772. Observations on different kinds of air. *Philosophical Transactions of the Royal Society of London* 62:166–170.

Rabinowitch, E. & Govindjee. 1969. *Photosynthesis*. New York: Wiley.

7.3 PHOTOSYNTHETIC ELECTRON TRANSPORT

7.3.1 PHOTOSYSTEMS ARE MAJOR COMPONENTS OF THE PHOTOSYNTHETIC ELECTRON TRANSPORT CHAIN

The key to the photosynthetic electron transport chain is the presence of two large, multimolecular, pigment-protein complexes known as **photosystem I (PSI)** and **photosystem II (PSII)** (Figure 7.3) PSI consists of 18 distinct subunits whereas PSII consists of 31 individual subunits! These two photosystems operate in series linked by a third multiprotein aggregate called the **cytochrome complex**. Overall, the effect of the chain is to extract low-energy electrons from water and, using light energy trapped by chlorophyll, raise the energy level of those electrons to produce a strong reductant NADPH (see Box 7.2: The Case for Two Photosystems).

The composition, organization, and function of the photosynthetic electron transport chain have been an area of active study and rapid progress in recent years. This interest has led to the development of a variety of experimental methods for the study of PSI, PSII, and other large-membrane protein aggregates. Most significant among these are techniques for the removal of the complexes from the thylakoid membranes by first solubilizing the membrane with a range of detergents. The different photosystems or classes of molecular aggregates can then be separated from each other by centrifugation. If the detergents and the conditions under which the treatments are carried out are carefully selected, not only can complexes be isolated but also individual complexes can be further subdivided into smaller aggregates that retain varying parts of the overall activity. These purified complexes or subunits may then be analyzed for their composition with respect to pigments, protein, or other components or assayed for their capacity to carry out specific photochemical or electron transport reactions. Most recently this approach has led to the crystallization of PSII and PSI reaction centers. By exposing these crystals to X-rays and analyzing the resulting diffraction patterns, scientists have been able to determine the precise three-dimensional location of all the pigment molecules and redox components of PSI and PSII reaction centers. The Nobel Prize in chemistry was awarded to Diesenhoffer, Michel, and Huber in 1987 for the first successful crystallization and X-ray diffraction of bacterial reaction centers. This represents the second Nobel Prize given for research in photosynthesis. See Chapter 8 for research that led to the first Nobel Prize in photosynthesis.

Such fractionation studies have revealed that PSI and PSII each contain several different proteins together with a collection of chlorophyll and carotenoid molecules that absorb photons. The bulk of the chlorophyll in the photosystem functions as **antenna chlorophyll** (Figure 7.4). The association of chlorophyll

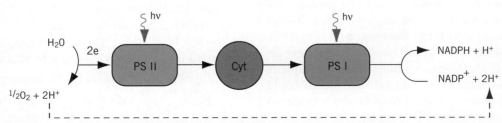

FIGURE 7.3 A linear representation of the photosynthetic electron-transport chain. A sequential arrangement of the three multimolecular membrane complexes extracts low-energy electrons from water and, using light energy, produces a strong reductant, NADPH + H⁺.

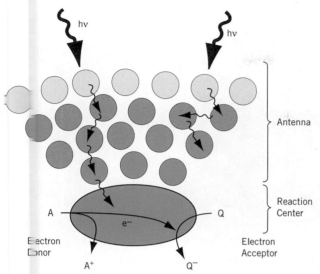

FIGURE 7.4 **A photosystem contains antenna and a reaction center. Antenna chlorophyll molecules absorb incoming photons and transfer the excitation energy to the reaction center where the photochemical oxidation–reduction reactions occur.**

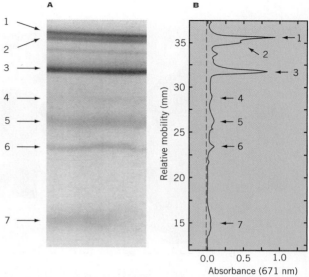

FIGURE 7.5 **Separation of thylakoid chlorophyll-protein complexes by nondenaturing polyacrylamide gel electrophoresis.** (*A*) **In the presence of specific detergents, chlorophyll-protein complexes are removed from thylakoid membranes structurally and functionally intact. These pigment-protein-detergent complexes are charged and thus will migrate when an electric field is applied. The porous matrix through which the electric field is applied is a polyacrylamide gel. Thus, the physical separation of protein complexes through a polyacrylamide gel matrix by applying an electric field is called** *polyacrylamide gel electrophoresis.* **Since these protein complexes still have chlorophyll bound to them, you can watch the chlorophyll-protein complexes separate through the gel matrix according to their molecular mass right in front of your eyes! The largest complexes remain at the top of the gel and the smallest complexes near the bottom of the gel. The illustration shows such an electrophoretic separation of pigment-protein complexes from thylakoid membranes of** *Arabidopsis thaliana* **with the arrow indicating the direction of migration. Typically, seven "green bands" can be resolved. Bands 1 through 6 are individual chlorophyll-protein complexes associated with PSI and PSII. The band exhibiting the greatest migration (band 7) is free pigment.** (*B*) **The relative amount represented by each green band can be quantified by scanning the gel in a spectrophotometer. The peak areas provide an estimate of the relative abundance of each chlorophyll-protein complex. Clearly, green bands 1, 2, and 3 are the most abundant pigment-protein complexes in this particular thylakoid sample. Biochemical and spectroscopic analyses of each pigment-protein complex indicates that band 1 contains chlorophyll a and represents light harvesting complex (LHCI) associated with the core antenna complex (CP1) of PSI. Band 2 is CP1 without its associated LCHI. Band 3 contains chlorophyll a as well as chlorophyll b and represents the trimeric form of the light harvesting complex (LHCII) associated with PSII. Band 4 and 6 represent the dimeric and monomeric forms of LHCII. Band 5 is a chlorophyll a pigment-protein complex designated as CPa and contains the core antenna of PSII (CP47, CP43) associated with the PSII reaction center, P680.**

with specific proteins forms a number of different **chlorophyll-protein (CP) complexes**, which can be separated by gel electrophoresis and identified on the basis of their molecular mass and absorption spectrum (Figure 7.5). The core antenna for photosystem II, for example, consists of two chlorophyll-proteins (CP) known as CP43 and CP47 (Figure 7.6). These two CP complexes each contain 20 to 25 molecules of chlorophyll *a*. The core antenna chlorophyll *a* absorb light but do not participate directly in photochemical reactions. However, protein-bound antenna chlorophylls lie very close together such that excitation energy can easily pass between adjacent pigment molecules by inductive resonance or radiationless energy transfer (Chapter 6). The energy of absorbed photons thus migrates through the antenna complex, passing from one chlorophyll molecule to another until it eventually arrives at the **reaction center** (Figure 7.4).

Each reaction center consists of a unique chlorophyll *a* molecule that is thought to be present as a dimer. This **reaction center chlorophyll** plus associated proteins and redox carriers are directly involved in light-driven redox reactions. The reaction center chlorophyll is, in effect, an energy sink—it is the longest-wavelength, thus the lowest-energy-absorbing chlorophyll in the complex. Because the reaction center chlorophyll *a* is the site of the primary photochemical redox reaction, it is here that light energy is actually converted to chemical energy. The reaction center chlorophyll *a* of PSI and PSII are designated as P700 and P680, respectively. These designations identify the reaction center chlorophyll *a*, or pigment (P), with an absorbance maximum at either 700 nm (PSI) or 680 nm (PSII).

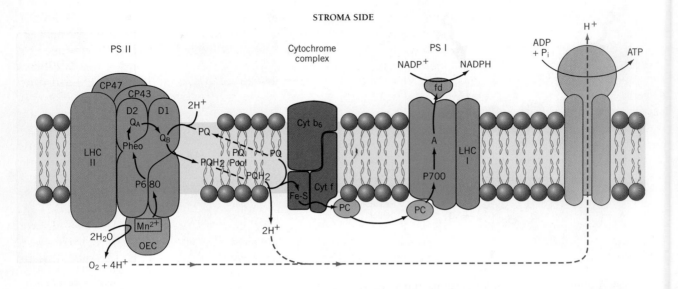

FIGURE 7.6 **The organization of the photosynthetic electron transport system in the thylakoid membrane. See text for details.**

Tightly associated with the reaction centers, P680 and P700, are core antenna complexes. CP47 and CP43 are the core antenna of PSII whereas CP1 is the core complex of PSI (Figure 7.6). Also shown in Figure 7.6 are two additional chlorophyll-protein complexes, depicted in close association with PSII and PSI—light-harvesting complex II (LHCII) and light-harvesting complex I (LHCI), respectively. LHCII is associated with PSII and LHCI is associated with PSI. As their names imply, the light-harvesting complexes function as extended antenna systems for harvesting additional light energy. LHCI and LHCII together contain as much as 70 percent of the total chloroplast pigment, including virtually all of the chlorophyll *b*. LHCI is relatively small, has a chlorophyll *a/b* ratio of about 4/1, and appears rather tightly bound to the core photosystem. LHCII, on the other hand, contains 50 to 60 percent of the total chlorophyll and, with a chlorophyll *a/b* ratio of about 1.2, most of the chlorophyll *b*. LHCII also contains most of the xanthophyll. The function of the light-harvesting complexes and the core antenna are to absorb light and transfer this energy to the reaction centers (Figure 7.4).

The principal advantage of associating a single reaction center with a large number of light harvesting and core antenna chlorophyll molecules is to increase efficiency in the collection and utilization of light energy. Even in bright sunlight it is unlikely that an individual chlorophyll molecule would be struck by a photon more than a few times every second. Since events at the reaction center occur within a microsecond time scale, any reaction center that depended on a single molecule of chlorophyll for its light energy would no doubt lie idle much of the time. Thus, the advantage of a photosystem is that while the reaction center is

busy processing one photon, other photons are being intercepted by the antenna molecules and funneled to the reaction center. This increases the probability that as soon as the reaction center is free, more excitation energy is immediately available. The efficiency of energy transfer through the light harvesting complexes and the core antenna complexes to the reaction is very high—only about 10 percent of the energy is lost. Thus, it is important to appreciate that LCHI and LHCII are not necessarily absolute requirements for photosynthetic electron transport under light saturated conditions, that is, under conditions when light is not limiting. Rather, the light harvesting complexes enhance photosynthetic efficiency under low light, that is, under conditions where light limits photosynthesis. In fact, photosynthetic organisms modulate the structure and function of the light harvesting complexes in response to changes irradiance. This will be discussed in more detail in Chapter 14. In addition, LHCII has an important role in the dynamic regulation of energy distribution between the photosystems which will be discussed in more detail in Chapter 13.

A schematic of the photosynthetic electron transport chain depicting the arrangement of PSI, PSII, and the cytochrome b_6/f complex in the thylakoid membrane is presented in Figure 7.6. A fourth complex—the CF_0-CF_1 coupling factor or ATP synthase—is also shown. All four complexes are membrane-spanning, integral membrane proteins with a substantial portion of their structure buried in the hydrophobic lipid bilayer. Note that the orientation of the complex and their individual constituents is not random—specific polypeptide regions will be oriented toward the stroma or lumen respectively. Such a **vectorial** arrangement of proteins is characteristic of all energy-transducing membranes, if not all membrane proteins, and is an essential element of their capacity to conserve energy through chemiosmosis (Chapter 5). One particularly significant consequence of this arrangement is the directed movement of protons between the stroma and the thylakoid lumen as shown in Figure 7.6. Although PQ reduction and its concomitant protonation occurs on the stromal side of the thylakoid membrane, the oxidation of PQH_2 by the cytochrome b_5/f complex (Cyt b_6/f) requires the diffusion of PQH_2 from the stromal side to the lumen side of the thylakoid membrane. It is this arrangement that gives rise to the proton gradient necessary for ATP synthesis. This aspect of the electron transport chain will be revisited later. Another consequence of the vectorial arrangement is that the oxidation of water and reduction of $NADP^+$ occur on opposite sides of the thylakoid membrane. Water is oxidized and protons accumulate on the lumen side of the membrane where they contribute to the gradient, which drives ATP synthesis. However, both NADPH and ATP are produced in the stroma where they are used in the carbon reduction cycle (Chapter 8) or other chloroplast activities (Chapter 11).

7.3.2 PHOTOSYSTEM II OXIDIZES WATER TO PRODUCE OXYGEN

Electron transport actually begins with the arrival of excitation energy at the photosystem II reaction center chlorophyll, P680, which is located near the lumenal side of the reaction center. As illustrated in Figure 7.7, this excitation energy is required to change the redox potential of P680 from $+0.8\,eV$ to about $-0.4\,eV$ for P680*, the excited form of P680. As a consequence of this initial endergonic excitation process, P680* can rapidly (within picoseconds, 10^{-12} s) transfer electrons exergonically to **pheophytin (Pheo)**. Pheophytin,

considered the primary electron acceptor in PSII, is a form of chlorophyll *a* in which the magnesium ion has been replaced by two hydrogens. Since this initial oxidation of P680 is light dependent, this is called a **photooxidation** event, which results in the formation of $P680^+$ and $Pheo^-$, a **charge separation**. Note that the energy of one photon results in the release of one electron, which is consistent with the Einstein-Stokes law (Chapter 6). This charge separation effectively stores light energy as redox potential energy and represents the actual conversion of light energy to chemical energy. It is essential that this charge separation be stabilized by the rapid movement of the electron from P680 at the lumen side of the PSII reaction center to an electron acceptor molecule localized at the stromal side of the PSII reaction center (see discussion below). If the electron were permitted to recombine with $P680^+$, there would be no forward movement of electrons, the energy would be wasted, and ultimately, carbon could not be reduced.

The role of the reaction proteins, D1 and D2, is to bind and to orient specific redox carriers of the PSII reaction center in such a way as to decrease the probability of charge recombination between $P680^+$ and $Pheo^-$. How does this happen? First, within picoseconds, pheophytin passes one electron on to a quinone electron acceptor called Q_A, resulting in the formation of $[P680^+\ Pheo\ Q_A^-]$. Now the PSII reaction center is considered to be "closed," that is, it is unable to undergo another photo-oxidation event (Figure 7.8). On a slower time scale of microseconds, the electron is passed from Q_A to **plastoquinone(PQ)**, resulting in the formation of $[P680^+\ Pheo\ Q_A]$. PQ is a quinone (see Figure 5.3C) that binds transiently to a binding site (Q_B) that is on the stromal side of the D1 reaction center protein (Figure 7.6). The reduction of PQ to **plastoquinol (PQH_2)** decreases its affinity for the

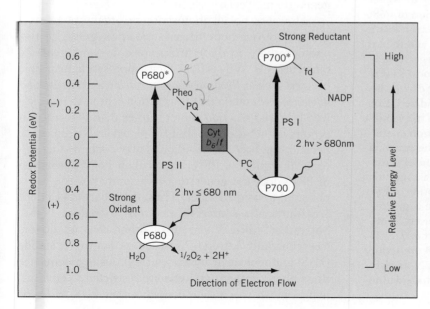

FIGURE 7.7 The Z-scheme for photosynthetic electron transport. The redox components are arranged according to their approximate midpoint redox potentials (E_m, Chapter 2). The vertical direction indicates a change in energy level (ΔG, Chapter 5). The horizontal direction indicates electron flow. The net effect of the process is to use the energy of light to generate a strong reductant, reduced ferredoxin (fd) from the low-energy electrons of water. The downhill transfer of electrons between P680* and P700 represents a negative free energy change. Some of this energy is used to establish a proton gradient, which in turn drives ATP synthesis. Indicated redox potentials are only approximate.

binding site on the D1 polypeptide. The plastoquinol is thus released from the reaction center, to be replaced by another molecule of PQ. PQH$_2$ diffuses from the Q$_B$ site and becomes part of the PQ pool present in the thylakoid membrane. Since PQ requires two electrons to become fully reduced to PQH$_2$, reduction at the Q$_B$ site is considered a **two-electron gate** (Figure 7.8).

Second, the initial charge separation is further stabilized because P680$^+$ is a very strong oxidant (perhaps the strongest known in biological systems) and is able to "extract" electrons from water. Thus P680$^+$ is rapidly reduced, again within picoseconds, to P680, resulting in the formation of [P680 Pheo Q$_A$]. Now the PSII reaction center is said to be "open," that is, it is ready to receive another excitation (Figure 7.8). On a bright, sunny summer day, the photon flux to which a leaf can be exposed may reach 2000 µmol photons m^{-2} s^{-1}. This

means that about 10^{19} charge separations per second may occur over a leaf surface area of 1 cm^2!

The electrons that reduce P680$^+$ are most immediately supplied by a cluster of four manganese ions associated with a small complex of proteins called the **oxygen-evolving complex (OEC)**. As the name implies, the OEC is responsible for the splitting (oxidation) of water and the consequent evolution of molecular oxygen. The OEC is located on the lumen side of the thylakoid membrane. The OEC is bound to the D1 and D2 proteins of the PSII reaction center and functions to stabilize the manganese cluster. It also binds Cl$^-$ which is necessary for the water-splitting function.

$$2H_2O \rightarrow O_2 + 4H^+ + 4e \qquad (7.8)$$

According to equation 7.8, the oxidation of two moles of water generates one mole of oxygen, four moles of protons, and four moles of electrons. It has been determined that only one PSII reaction center and OEC is involved in the release of a single oxygen molecule. Thus, in order to complete the oxidation of two water molecules, expel four protons, and produce a single molecule of O$_2$, the PSII reaction center must be "closed" and then "opened" four times (Figure 7.8). This means that PSII must utilize the energy of four photons in order to evolve one molecule of O$_2$. Experiments in which electron transport was driven by extremely short flashes of light—short enough to excite essentially one electron at a time—have demonstrated that the OEC has the capacity to store charges. Each excitation of P680 is followed by withdrawal of one electron from the manganese cluster, which stores the residual positive charge. When four positive charges have accumulated, the complex oxidizes two molecules of water and releases the product oxygen molecule. As a consequence, organisms containing PSII and the OEC exhibit oxygenic photosynthesis, that is, a photosynthetic process that generates molecular oxygen (O$_2$).

However, not all photosynthetic organisms are oxygenic. Photosynthetic bacteria are **anoxygenic**, that is, photosynthesis in these prokaryotes does not generate molecular O$_2$. Why is this? First, photosynthetic purple bacteria such as *Rhodopseudomonas viridis* and *Rhodobacter sphaeroides* contain only one reaction center, the specialized bacterial reaction center, rather than two reaction centers found in chloroplasts of plants and green algae. Second, the bacterial reaction center contains the bacteriochlorophyll a, P870, rather than P680. Excitation of P870 does not generate a sufficiently positive redox potential to oxidize water. Thus, these bacteria can not use water as a source of electrons to reduce P870$^+$ but rather utilize a variety of other electron donors (see equations 7.4 and 7.5) such as hydrogen sulfide (H$_2$S) and molecular hydrogen (H$_2$). As a consequence these photosynthetic bacteria are restricted to environments which specifically contain either H$_2$S or H$_2$ as

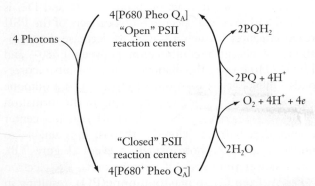

4 Photons

4[P680 Pheo Q$_A$]
"Open" PSII
reaction centers

2PQH$_2$

2PQ + 4H$^+$

O$_2$ + 4H$^+$ + 4e

"Closed" PSII
reaction centers
4[P680$^+$ Pheo Q$_A^-$]

2H$_2$O

FIGURE 7.8 The light-dependent, cyclic "opening" and "closing" on gate of photosystem II reaction centers. The PSII reaction center polypeptides, D1 and D2, bind the following redox components irreversibly: P680, the reaction chlorophyll *a*; pheophytin (Pheo) and Q$_A$, the first stable quinone electron acceptor of PSII reaction centers. The transfer of absorbed light energy to "open" PSII reaction centers causes the photooxidation of P680 and converts it P680$^+$. Subsequently, the electron lost by P680 is transferred very rapidly first to Pheo and then to convert it to Q$_A^-$. This results in a stable charge separation [P680$^+$ Pheo Q$_A^-$]. The PSII reaction is now said to be "closed," that is, the PSII reaction center cannot be photooxidized since P680 is already photooxidized. Excitation of a "closed" reaction center results in irreversible damage. Conversion of a "closed" PSII reaction center to an "open" PSII reaction center requires the concomitant reduction of P680$^+$ through the oxidation of water and the transfer of the electron from Q$_A^-$ to plastoquinone (PQ) bound to the Q$_B$ site on the D1 polypeptide of PSII. Upon the complete reduction, plastoquinone is protonated to form plastoquinol (PQH$_2$) and released from the Q$_B$ site and subsequently reduces the Cyt b_6/*f* complex. The scheme illustrated is balanced for 4 excitations and for the evolution of one molecule of O$_2$. Note that the absorption of *one* photon causes the photooxidation of *one* P680.

a source of reducing power. Since water is generally very abundant in our biosphere, the evolution of PSII allowed oxygenic photosynthetic organisms to survive and reproduce almost anywhere in our biosphere. Thus, the evolution of PSII and its associated OEC was a major factor determining the global distribution of oxygenic photosynthetic organisms that fundamentally changed the development of all life on Earth.

7.3.3 THE CYTOCHROME COMPLEX AND PHOTOSYSTEM I OXIDIZE PLASTOQUINOL

Following its release from PSII, plastoquinol diffuses laterally through the membrane until it encounters a **cytochrome b_6f complex** (Figure 7.6). This is another multiprotein, membrane-spanning redox complex whose principal constituents are cytochrome b_6 (Cyt b_6) and cytochrome f (Cyt f). The cytochrome complex also contains an additional redox component called the **Rieske iron-sulfur (FeS) protein**—iron-binding proteins in which the iron complexes with sulfur residues rather than a heme group as in the case of the cytochromes. Plastoquinol diffuses within the plane of the thylakoid membrane and passes its electrons first to the FeS protein and then to Cyt f. Since the oxidation of plastoquinol is thought to be diffusion limited, this is the slowest step in photosynthetic electron transport and occurs on a time scale of milliseconds (ms). The Rieske FeS protein and the heme of Cyt f are located on the lumenal side of the thylakoid membrane. From Cyt f, the electrons are picked up by a copper-binding protein, **plastocyanin (PC)**. PC is a small peripheral protein that is able to diffuse freely along the lumenal surface of the thylakoid membrane.

In the meantime, a light-driven charge separation similar to that involving P680 has also occurred in the reaction center of PSI. Excitation energy transferred to the reaction center chlorophyll of PSI (P700) is used to change the redox potential of P700 from about $+0.4\,eV$ to about $-0.6\,eV$ for P700*, the excited form of P700 (Figure 7.7). As a consequence of this initial endergonic excitation, P700* is rapidly photooxidized to P700$^+$ by the primary electron acceptor (A) in PSI, a molecule of chlorophyll a (Figure 7.6); the electron is then passed through a quinone and additional FeS centers and finally, on the stroma side of the membrane, to **ferredoxin**. Ferredoxin is another FeS-protein that is soluble in the stroma. Ferredoxin in turn is used to reduce NADP$^+$, a reaction mediated by the enzyme **ferredoxin-NADP$^+$-oxidoreductase**. Finally, the electron deficiency in P700$^+$ is satisfied by withdrawing an electron from reduced PC (Figure 7.6).

The overall effect of the complete electron transport scheme is to establish a continuous flow of electrons between water and NADP$^+$, passing through the two separate photosystems and the intervening cytochrome complex (Figure 7.6). The bioenergetics of this process are illustrated in Figure 7.7. In the overall process, electrons are removed from water, a very weak reductant ($E_m = 0.82\,V$), and elevated to the energy level of ferredoxin, a very strong reductant ($E_m = -0.42\,V$). Ferredoxin in turn reduces NADP$^+$ to NADPH ($E_m = -0.32\,V$). NADPH, also a strong reductant, is a water-soluble, mobile electron carrier that diffuses freely through the stroma where it is used to reduce CO$_2$ in the carbon reduction cycle (Chapter 8). Since two excitations—at PSII and PSI—are required for each electron moved through the entire chain, a substantial amount of energy is put into the system. Based on one 680 nm photon (175 kJ per mol quanta) and one 700 nm photon (171 kJ per mol quanta), 692 kJ are used to excite each mole pair of electrons [$2 \times (175 + 171)$]. Only about 32 percent of that energy is conserved in NADPH ($218\,kJ\,mol^{-1}$).

What happens to the other 68 percent of the energy? An additional portion of the redox free energy of electron transport energy is conserved as ATP. This occurs in part because transfer of electrons between PSII and PSI is energetically downhill—that is, it is accompanied by a negative ΔG (Figure 7.7). In the process of moving electrons between plastoquinone and the cytochrome complex, some of that energy is used to move protons from the stroma side of the membrane to the lumen side. These protons contribute to a proton gradient that can be used to drive ATP synthesis by chemiosmosis (Chapter 5).

The **quantum requirement for oxygen evolution** is defined as the number photons required to evolve one molecule of O$_2$. As discussed above, two excitations, one at PSII and one at PSI, are required to move each electron through noncyclic electron transport from H$_2$O to NADP$^+$. From equation 7.8, the evolution of one molecule of O$_2$ by P680$^+$ generates four electrons that are eventually transferred through PSI to reduce 2NADP$^+$ to 2NADPH. Thus, to transfer four electrons from H$_2$O to NADP$^+$ requires 8 photons. Therefore the minimal theoretical quantum requirement for O$_2$ evolution is 8 photons/molecule of O$_2$ evolved.

Conversely, **quantum yield of oxygen evolution** is the inverse of quantum requirement, that is, number of molecules of O$_2$ evolved per photon absorbed. Since this is also the definition of **photosynthetic efficiency**, the terms quantum yield and photosynthetic efficiency are interchangeable. Consequently, the maximum theoretical quantum yield for oxygen evolution or photosynthetic efficiency of O$_2$ evolution must be 1/8 or 0.125 molecules of O$_2$ evolved/photon absorbed. Photosynthetic efficiency or quantum yield for O$_2$ evolution will vary depending on the environmental conditions to which a plant is exposed. This will be discussed in more detail in Chapters 13 and 14.

7.4 PHOTOPHOSPHORYLATION IS THE LIGHT-DEPENDENT SYNTHESIS OF ATP

In Chapter 5, we examined the bioenergetics of the light-dependent synthesis of ATP. However, by definition, thermodynamics does not provide specific information with respect to kinetics and biochemical mechanism. Here we discuss the molecular basis underlying the chemiosmotic synthesis of ATP in chloroplasts.

The ATP required for carbon reduction and other metabolic activities of the chloroplast is synthesized by photophosphorylation in accordance with Mitchell's chemiosmotic mechanism (Chapter 5). Light-driven production of ATP by chloroplasts is known as **photophosphorylation**. Photophosphorylation is very important because, in addition to using ATP (along with NADPH) for the reduction of CO_2, a continual supply of ATP is required to support a variety of other metabolic activities in the chloroplast. These activities include amino acid, fatty acid, and starch biosynthesis, the synthesis of proteins in the stroma, and the transport of proteins and metabolites across the envelope membranes.

When electron transport is operating according to the scheme shown in Figures 7.6 and 7.7, electrons are continuously supplied from water and withdrawn as NADPH. This flow-through form of electron transport is consequently known as either *noncyclic* or *linear* electron transport. Formation of ATP in association with noncyclic electron transport is known as **noncyclic photophosphorylation**. However, as will be shown later, PSI units and PSII units in the membrane are not physically linked as implied by the Z scheme, but are even segregated into different regions of the thylakoid.

One consequence of this heterogeneous distribution in the membranes is that PSI units may transport electrons independently of PSII, a process known as **cyclic electron transport**. In terrestrial plants, the major pathway for PSI cyclic electron transport is thought to occur via P700 to ferredoxin (fd) which transfers the electrons back to PQ via a recently discovered protein, PGR5 rather than to $NADP^+$. The electron then returns to $P700^+$, passing through the cytochrome b_6/f complex and plastocyanin. Using a genetic approach in *Arabidopsis thaliana*, the gene, *PGR5*, was shown to encode a small thylakoid polypeptide that is essential for PSI cyclic electron transport (Figure 7.9). However, the precise role of PGR5 in the electron transfer process has yet to be elucidated. Since these electrons also pass through PQ and the cytochrome complex, cyclic electron transport will also contribute to the establishment of the pH gradient required to support ATP synthesis, a process known as **cyclic photophosphorylation**. It is thought that cyclic photophosphorylation is a source of ATP required for chloroplast activities over and above that required in the carbon-reduction cycle. Since noncyclic photophosphorylation results in the production of both ATP and NADPH whereas cyclic photophosphorylation does not generate NADPH, switching between cyclic and noncyclic photophosphorylation also represents a mechanism by which the chloroplast can regulate the stromal ATP/NADPH ratios, which is important in the maintenance of chloroplast metabolic activity.

A key to energy conservation in photosynthetic electron transport and the accompanying production of ATP is the light-driven accumulation of protons in the lumen. There are two principal mechanisms that account for this accumulation of protons: the oxidation of water, in which two protons are deposited into the lumen for each

FIGURE 7.9 **Cyclic electron transport. PSI units operating independently of PSII may return electrons from P700 through ferredoxin (fd), and PGR5 to the thylakoid plastoquinone (PQ) pool and the cytochrome b_6/f complex. In cyclic electron transport, the oxidation of PQ by the cytochrome b_6/f complex generates a proton gradient that can be used for ATP synthesis but no NADPH is produced.**

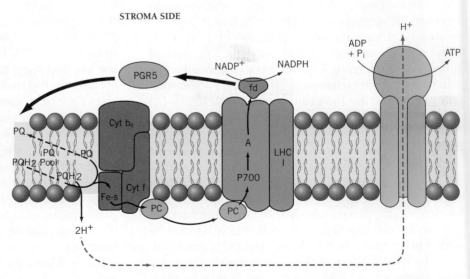

water molecule oxidized, and a PQ-cytochrome proton pump. The energy of the resulting proton gradient is then used to drive ATP synthesis in accordance with Mitchell's chemiosmotic hypothesis (Chapter 5).

The precise mechanism by which protons are moved across the membrane by the cytochrome complex is not yet understood, although several models have been proposed. The most widely accepted model is known as the Q-cycle, based on an original proposal by Mitchell. A simplified version of the Q-cycle during steady-state operation is shown in Figure 7.10. When PQ is reduced by PSII, it binds temporarily to the D1 protein (Q_B) as a semiquinone after it accepts the first electron from Q_A. Subsequently, the Q_B semiquinone is converted to the fully reduced plastoquinol (PQH_2) after it has accepted another electron from Q_A, plus two protons are picked up from the surrounding stroma. PQH_2 dissociates from the PSII complex and diffuses laterally through the membrane until it encounters the lumenal PQH_2 binding site of the cytochrome b_6/f complex. There, two PQH_2 bind sequentially and are reoxidized to PQ through a semiquinone intermediate (PQH_2, not shown) by the combined action of the Rieske FeS-protein and the low reduction potential form of cytochrome b_6 (LP). Concomitantly, 4 H^+ are transferred to the lumen. One of the PQ molecules returns to the thylakoid PQ pool to be reduced again by PSII, while the other PQ molecule is transferred to the

stromal binding site of the cytochrome b_6/f complex where it becomes reduced by the high reduction potential form of cytochrome b_6 (HP) and is protonated using 2 H^+ from the stroma. This PQH_2 molecule is then released from the stromal binding site and recycled into the thylakoid PQH_2 pool.

Thus, for *each pair* of electrons passing from plastoquinone through the Rieske FeS-center and cytochrome f to plastocyanin, *four* protons are translocated from the stroma into the lumen of the thylakoid. If this scheme is correct, then each pair of electrons passing through noncyclic electron transport from water to $NADP^+$ contributes six protons to the gradient—four from the Q-cycle (Equation 7.9) plus two from water oxidation (Equation 7.10).

$$2PQH_2 \rightarrow PQ + PQH_2 + 4H^+ + 2e \qquad (7.9)$$

therefore $4H^+/2e$

$$H_2O \rightarrow \tfrac{1}{2}O_2 + 2H^+ + 2e \qquad (7.10)$$

therefore $2H^+/2e$

For cyclic electron transport, the number of protons transferred per pair of electrons would be four.

Since it is generally agreed that three protons must be transported through the CF_0-CF_1 for each ATP synthesized ($3H^+$/ATP), a pair of electrons passing through noncyclic electron transport would be expected to yield two ATP molecules for every NADPH

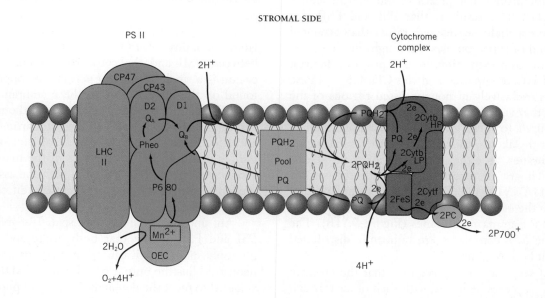

FIGURE 7.10 **The Q-cycle, a model for coupling electron transport from plastoquinol (PQH_2) to the cytochrome b_6/f complex (Cyt b_6/f) with the translocation of protons across the thylakoid membrane. Six protons are translocated for each pair of electrons that passes through the electron transport chain. Fe, Rieske FeS-center; Cyt f, cytochrome f of the cytochrome b_6/f complex; Cyt b_{LP}, low reduction potential form of cytochrome b_6 of the cytochrome b_6/f complex; Cyt b_{HP}, high reduction potential form of cytochrome b_6 of the cytochrome b_6/f complex; PC, plastocyanin.**

produced (2ATP/NADPH). The precise stoichiometry, however, is difficult to determine, in part because of uncertainty with regard to the relative proportions of cyclic and noncyclic photophosphorylation occurring at any specific moment in time.

According to the chemiosmotic theory,

$$pmf = -59\Delta pH + \Delta\Psi \qquad (7.11)$$

As protons accumulate in the lumen relative to the stroma, divalent magnesium ions (Mg^{2+}) released from the thylakoid membrane accumulate in the chloroplast stroma. This minimizes the difference in electrical charge between the stroma and the lumen. Thus, in chloroplasts, the ΔpH (i.e., the H^+ concentration gradient) is the major factor that contributes to chloroplastic pmf, whereas $\Delta\Psi$ contributes minimally.

7.5 LATERAL HETEROGENEITY IS THE UNEQUAL DISTRIBUTION OF THYLAKOID COMPLEXES

In addition to the vectorial arrangement of electron transport components across the membrane that accounts for the simultaneous electron transfer within the thylakoid membrane and active transport of protons from the stroma to the lumen, there is also a distinct **lateral heterogeneity** with respect to their distribution of the major protein complexes within the thylakoids (Figure 7.11). The result is that PSI and PSII, for example, are spatially segregated, rather than arranged as some kind of supercomplex that might be suggested by the static representation in the previous figures. The PSI/LHCI complexes and the CF_0-CF_1 ATPase are located exclusively in nonappressed regions of the thylakoid; that is, those regions where the membranes are not paired to form grana. These regions include the stroma thylakoids, the margins of the grana stacks, and membranes at either end of the grana stacks, all of which are in direct contact with the stroma (Figure 7.11A). Virtually all of the PSII complexes and LHCII, on the other hand, are located in the appressed regions of the grana membranes (Figure 7.11B). The cytochrome b_6/f complexes are uniformly distributed throughout both regions.

Spatial segregation also requires that the electron transport complexes be linked with each other through one or more **mobile carriers** that can deliver electrons between complexes. These carriers are plastoquinone (PQ), plastocyanin (PC), and ferredoxin. All three are mobile carriers that are not permanently part of any electron transport complex. Plastoquinone is a hydrophobic molecule and is consequently free to diffuse laterally within the lipid matrix of the thylakoid membrane. Its estimated diffusion coefficient is 10^6 cm^{-2} s^{-1}, which means that it could travel more than the diameter of a typical granum in less than one millisecond. The

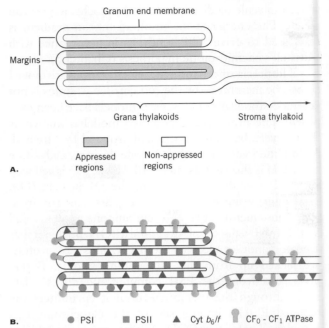

A. Appressed regions | Non-appressed regions

B.   ● PSI   ■ PSII   ▲ Cyt b_6/f CF$_0$ - CF$_1$ ATPase

FIGURE 7.11 Lateral heterogeneity in the thylakoid membrane. (A) Nonappressed membranes of the stroma thylakoids, grana end membranes, and grana margins are exposed to the stroma. Appressed membranes in the interior of grana stacks are not exposed to the stroma. (B) PSII units are located almost exclusively in the appressed regions while PSI and ATP synthase units are located in nonappressed regions. The cytochrome b_6/f complex, plastoquinone, and plastocyanin are uniformly distributed throughout the membrane system.

lateral mobility of PQ allows it to carry electrons between PSII and the cytochrome complex. Plastocyanin is a small (10.5 kDa) peripheral copper-protein found on the lumenal side of the membrane. It readily diffuses along the lumenal surface of the membrane and carries electrons between the cytochrome complex and PSI. Ferredoxin, a small (9 kDa) iron-sulfur protein, is found on the stroma side of the membrane. It receives electrons from PSI and, with the assistance of the ferredoxin-NADP oxidoreductase, reduces NADP$^+$ to NADPH.

An unequal number and spatial segregation of PSI and PSII means that both cyclic and noncyclic photophosphorylation can occur more or less simultaneously. Thus the output of ATP and NADPH can be adjusted to meet the demands not only of photosynthesis but of other biosynthetic energy requirements within the chloroplast (see Chapter 8).

Are granal stacks required for oxygenic photosynthesis? The unequivocal answer to this question is no. Contrary to the illusion created by a transmission electron micrograph of a chloroplast (see Figure 5.4), the stacking of thylakoid membranes into grana is not static but very dynamic. If granal stacks are not required for oxygenic photosynthesis, what regulates their formation and why do they occur in most chloroplasts? First, in vitro experiments with isolated thylakoids showed that decreasing the concentration of monovalent cations such as K^+ and Mg^{2+}, respectively, in the surrounding thylakoid isolation buffer decreases stacking and results in the homogeneous distribution of the photosynthetic electron protein complexes within the thylakoid membrane. This is reversible upon the re-addition of these cations which induces stacking of the thylakoids and the reestablishment of lateral heterogeneity. Later, it was established that the N-terminal domain of the major LHCII polypeptides mediated the stacking process. Wild type barley chloroplasts exhibit typical granal stacks whereas the chloroplasts of the *chlorina f2* mutant of barley, which lacks both chlorophyll b and the major LHCII polypeptides, does not exhibit granal stacks. Proteolytic cleavage of this N-terminal domain of the major LHCII polypeptides also inhibited granal stacking. It was concluded that the presence of cations shields the negative surface charges created by the exposed N-terminal domains of the major LHCII polypeptides

present thylakoid membranes. Thus, unstacking of thylakoid membranes is the result of electrostatic repulsion of the negative surface charges on thylakoid membranes. The possible functions of this remarkable process are still not completely understood. We will return to this subject in Chapter 13 when we discuss the regulation of energy distribution between PSII and PSI.

7.6 CYANOBACTERIA ARE OXYGENIC

In contrast to photosynthetic bacteria, cyanobacteria (Figure 7.12A) are a large and diverse group of prokaryotes which perform oxygenic photosynthesis because they exhibit PSI as well as PSII with its associated OEC and an intersystem electron transport chain comparable to that of eukaryotic photoautotrophs. However, in contrast to the intrinsic, major light har-

FIGURE 7.12 (*A*) An electron micrograph of a typical single cell cyanobacterium, *Synechocystis*. Note that there are no granal stacks. The thylakoid membranes are arranged in concentric rings in the cell cytoplasm. (*B*) A confocal micrograph of the cyanobacterium, *Plectonema boryanum*. This is an example of a filamentous cyanobacterium. Each red circle represents one cell and is visible due to the red fluorescence emanating from chlorophyll a. (*C*) General structure of PSII and its associated phycobilisome in cyanobacteria. Note that the phycobilisome is associated with PSII at the surface of the thylakoid membrane. Phycoerythrin (red) and phycocyanin (dark blue) are the pigments bound to the protein rod structure. The rods are bound to the allophycocyanin proteins (blue circles) which bind the phycobilisome to CP43 and CP47 of PSII.

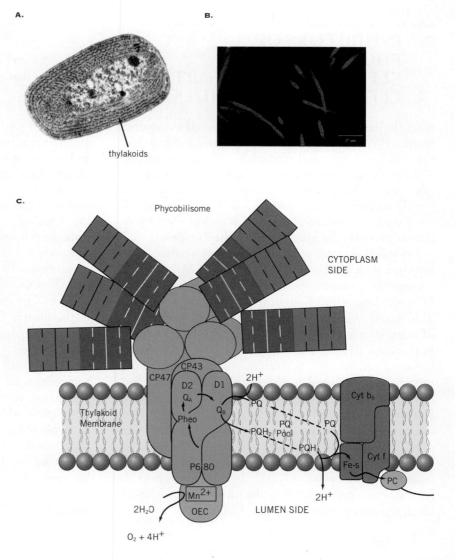

vesting pigment-protein complex found in chloroplast thylakoid membranes of plants and green algae, the light harvesting complex of cyanobacteria is an extrinsic pigment-protein complex called a phycobilisome which is bound to the outer, cytoplasmic surface of cyanobacterial thylakoids (Figure 7.12B). **Phycobilisomes (PBSs)** are rod-shaped chromoproteins called phycobiliproteins which may constitute up to 40 percent of the total cellular protein. The phycobiliproteins usually associated with PBS include **allophycocyanin (AP), phycocyanin (PC), and phycoerythrin (PE).** In addition to PBS, PSII of cyanobacteria include the Chl a core antenna CP47 and CP43 similar to that found in eukaryotic organisms. Cyanobacteria are distinct from chloroplasts because the redox carriers involved in respiratory as well as photosynthetic electron transport are located in the cyanobacterial thylakoid membranes where they share a common PQ pool and a common Cyt b_6f complex. Because PBSs are large, extrinsic pigment-protein complexes, this prevents appression of cyanobacterial thylakoid membranes and the formation of granal stacks characteristic of eukaryotic chloroplasts. This is further evidence that granal stacks are not a prerequisite for oxygenic photosynthesis.

7.7 INHIBITORS OF PHOTOSYNTHETIC ELECTRON TRANSPORT ARE EFFECTIVE HERBICIDES

Since the dawn of agriculture, man has waged war against weeds. Weeds compete with crop species for water, nutrients, and light and ultimately reduce crop yields. Traditional methods of weed control, such as crop rotation, manual hoeing, or tractor-drawn cultivators were largely replaced in the 1940s by labor-saving chemical weed control. Modern agriculture is almost completely dependent upon the intensive use of **herbicides.**

A wide spectrum of herbicides is now available that interfere with a variety of cell functions. Many of the commercially more important herbicides, however, act by interfering with photosynthetic electron transport. Two major classes of such herbicides are **derivatives of urea,** such as monuron and diuron, and the **triazine herbicides,** triazine and simazine (Figure 7.13). Both the urea and triazine herbicides are taken up by the roots and transported to the leaves. There they bind to the Q_B binding site of the D1 protein in PSII (also known as the herbicide-binding protein). The herbicide interferes with the binding of plastoquinone to the same site and thus blocks the transfer of electrons to plastoquinone. Because of its action in blocking electron transport at this point, DCMU is commonly used in laboratory experiments where the investigator wishes to block electron transport between PSII and PSI.

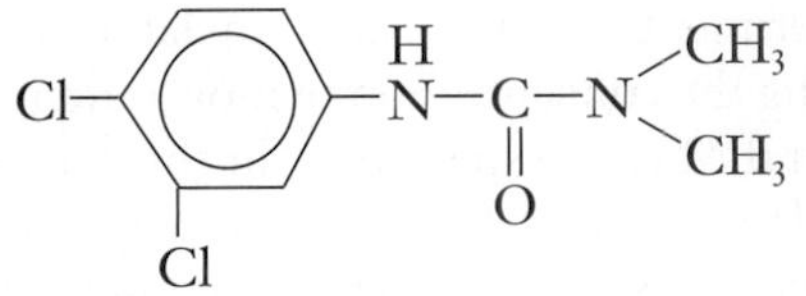

3-(3,4-Dichlorophenyl)-1,1-dimethylurea
(common names: Diuron, DCMU)

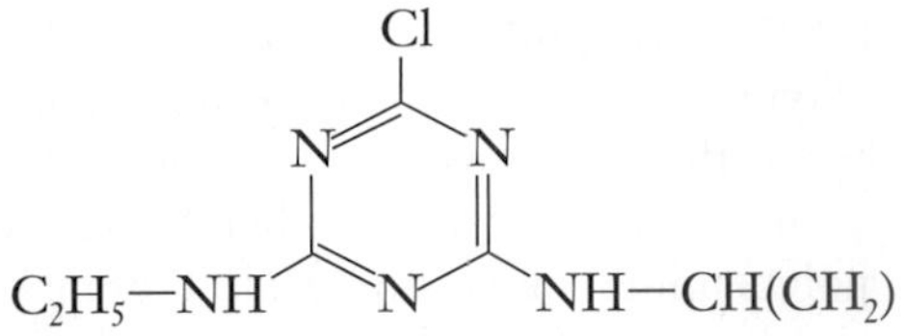

2-Chloro-4-ethylamino-6-
isopropylamino-s-triazine
(common name: Atrazine)

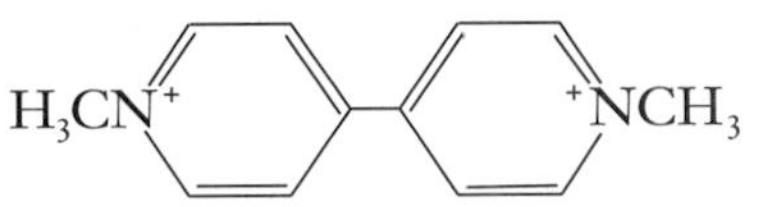

Paraquat
(methyl viologen)

FIGURE 7.13 **The chemical structures of some common herbicides that act by interfering with photosynthesis.**

The triazine herbicides are used extensively to control weeds in cornfields, since corn roots contain an enzyme that degrades the herbicide to an inactive form. Other plants are also resistant. Some, such as cotton, sequester the herbicide in special glands while others avoid taking it up by way of root systems that penetrate deep below the application zones. In many cases, however, weeds have developed triazine-resistant races, or biotypes. In several cases, the resistance has been traced to a single amino acid substitution in the D1 protein. The change in amino acid reduces the affinity of the protein for the herbicide but does not interfere with plastoquinone binding and, consequently, electron transport.

The availability of herbicide-resistant genes together with recombinant DNA technology has stimulated considerable interest in the prospects for developing additional herbicide-resistant crop plants. It is possible, for example, to transfer the gene for the altered D1 protein into crop species and confer resistance to triazine herbicides. This approach will be successful, however, only if weed species do not continue to acquire resistance to the same herbicides through natural evolutionary change.

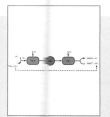

Box 7.2

THE CASE FOR TWO PHOTOSYSTEMS

The photosynthetic unit of oxygenic photosynthetic organisms is organized as two separate **photosystems** that operate in series. While the two-step series formulation, or "Z-scheme," for photosynthesis is general knowledge today, the idea generated considerable excitement when it was first proposed in the early 1960s. The two-step idea was based on a series of experiments conducted during the 1950s, which laid the foundation for significant advances in our understanding of photosynthetic electron transport. The first of these experiments was centered around the concept of quantum efficiency. Information about quantum efficiency is very useful when attempting to understand photochemical processes. Quantum efficiency can be expressed in two ways—either as **quantum yield** or as **quantum requirement**. Quantum yield (ϕ) expresses the efficiency of a process as a ratio of the yield of product to the number of photons absorbed. In photosynthesis, for example, product yield would be measured as the amount of CO_2 taken up or O_2 evolved. Alternatively, the quantum requirement ($1/\phi$) (sometimes referred to as quantum number) tells how many photons are required for every molecule of CO_2 reduced or oxygen evolved. Equation 7.8 identifies that a minimum of four electrons are required for every molecule of CO_2 reduced. In Chapter 5 it was established that one photon is required for each electron excited.

Therefore, the minimum *theoretical* quantum requirement for photosynthesis is four. However, it has been well established *experimentally* that the minimum quantum requirement for photosynthesis is eight to ten photons for every CO_2 reduced. If eight photons are required (it is usual to assume the minimum) for four electrons, then *each* electron must be excited twice! A second line of evidence, again from the laboratory of R. Emerson, was based on attempts to determine the action spectra for photosynthesis in *Chlorella*. Emerson and his colleague C. M. Lewis reported in 1943 that the value of ϕ was remarkably constant over most of the spectrum (Emerson and Lewis, 1943). This would indicate that any photon absorbed by chlorophyll was more or less equally effective in driving photosynthesis. However, there was an unexpected drop in the quantum yield at wavelengths greater than 680 nm, even though chlorophyll still absorbed in that

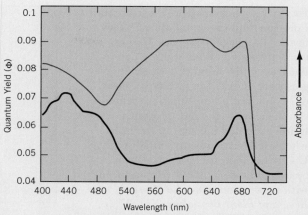

Figure 7.14 **The Emerson "red drop" in the green alga *Chlorella*.** Lower curve: Absorption spectrum of photosynthetic pigments. Upper curve: Action spectrum for quantum yield of photosynthesis. (Redrawn from the data of R. Emerson, C. M. Lewis, *American Journal of Botany* 30:165–178, 1943).

range (Figure 7.14). This puzzling drop in quantum efficiency in the long red portion of the spectrum was called the **red drop**.

In another experiment, Emerson and his colleagues set up two beams of light—one in the region of 650 to 680 nm and the other in the region of 700 to 720 nm. The fluence rates of both beams were adjusted to give equal rates of photosynthesis. Emerson discovered that when the two beams were applied simultaneously, the rate of photosynthesis was *two to three times greater than the sum of the rates* obtained with each beam separately! This phenomenon has become known as the **Emerson enhancement effect** (Figure 7.15). The enhancement effect suggests that photosynthesis involves two photochemical events or systems, one driven by short-wavelength light ($\leq$680 nm) and one driven by long-wavelength light (>680 nm). For optimal photosynthesis to occur, both systems must be driven simultaneously or in rapid succession.

In an attempt to explain conflicting information about the role of cytochromes and redox potential values, R. Hill and Fay Bendall, in 1960, proposed a new model for electron transport. The Hill and Bendall model involved two photochemical acts operating in series—one serving to oxidize the cytochromes and one serving to reduce them (Figure 7.16). The following year, L. Duysens confirmed the Hill and Bendall model, showing that cytochromes were oxidized in the presence of long-wavelength light. The effect could be reversed by short-wavelength light.

FIGURE 7.15 **Schematic to illustrate the Emerson "enhancement effect." Two beams of light (660 nm and 710 nm) were presented either singly (A and B) or in combination (C). Beam energies were adjusted to give equal rates of oxygen evolution. When presented simultaneously, the rate of oxygen evolution exceeded the sum of the rates when each beam was presented singly. Up arrows indicate light on. Down arrows indicate light off. (Reproduced with permission from the *Annual Review of Plant Physiology*, Vol. 22, copyright 1971 by Annual Reviews, Inc.)**

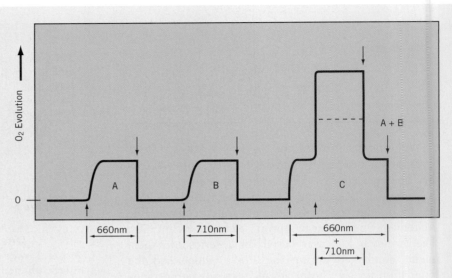

Although the scheme has been significantly modified and considerable detail has been added since it was originally proposed, the Hill and Bendall scheme provided the catalyst that has led to our present understanding of photosynthetic electron transport and oxygen evolution. As a consequence, today the Z-scheme is the prevailing paradigm for photosynthetic electron transport in plants, algae, and cyanobacteria. However, Daniel Arnon spent part of his illustrious scientific career challenging the Z-scheme. Supported with experimental evidence, he maintained until his death in 1995 that PSI and PSII could operate independently of one another and still support CO_2 assimilation. He maintained that PSI can operate in a cyclic mode to generate ATP and PSII could reduce $NADP^+$ directly. Although the reduction of $NADP^+$ by PSII can be shown *in vitro*, the quantum yield for this process appears to be very low and this reaction has never been reported *in vivo*. If PSII could reduce $NADP^+$ directly, one would expect to see CO_2 fixation in the absence of PSI. However, all experiments with mutants of the green alga, *Chlamydomonas reinhardtii*, that lack PSI indicate that these mutants are unable to fix CO_2. Although the consensus is that the Z-scheme reflects an accurate description of photosynthetic electron flow in photosynthetic organisms grown under optimal growth conditions, data continue to accumulate that indicate that the Z-scheme may not explain photosynthetic electron flow during growth under extreme conditions. For recent controversies surrounding the Z-scheme see the papers by Redding and Peltier (1998) as well as by Ivanov et al. (2000).

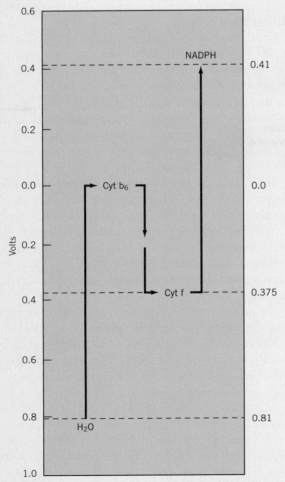

FIGURE 7.16 **The Z scheme as originally proposed by Hill and Bendall. For a current version, see Figure 7.7. (Redrawn from Hill and Bendall, 1960).**

REFERENCES

Arnon, D. I. 1995. Divergent pathways of photosynthetic electron transfer: The autonomous oxygenic and anoxygenic photosystems. *Photosynthesis Research* 46:47–71.

Duysens, L. N. M., J. Amesz, B. M. Kamp. 1961. Two photochemical systems in photosynthesis. *Nature* 190:510–511.

Emerson, R., C. M. Lewis. 1943. The dependence of the quantum yield of *Chlorella* photosynthesis on wavelength of light. *American Journal of Botany* 30:165–178.

Emerson, R., R. Chalmers, C. Cederstrand. 1957. Some factors influencing the long-wave length limit of photosynthesis. *Proceedings of the National Academy of Science USA* 43:133–143.

Hill, R., F. Bendall. 1960. Function of the two cytochrome components in chloroplasts: A working hypothesis. *Nature* 186:136–137.

Ivanov, A. G., Y.-I. Park, E. Miskiewicz, J. A. Raven, N. P. A. Huner, G. Öquist. 2000. Iron stress restricts photosynthetic intersystem electron transport in *Synechococcus* sp. PCC 7942. *FEBS Lett.* 485:173–177.

Redding, K., G. Peltier. 1998. Reexamining the validity of the Z-scheme: Is photosystem I required for oxygenic photosynthesis in *Chlamydomonas*? In: J.-D. Rochaix, M. Goldschmidt-Clermont, S. Merchant (eds.), *The Molecular Biology of Chloroplasts and Mitochondria in Chlamydomonas? Advances in Photosynthesis*, Vol. 7, pp. 349–362. Dordrecht: Kluwer Academic.

Another class of herbicides are the bipyridylium viologen dyes—paraquat (Figure 7.13)—which act by intercepting electrons on the reducing side of PSI. The viologen dyes are auto-oxidizable, immediately reducing oxygen to superoxide. Not only do the viologen dyes interfere with photosynthetic electron transport, but the superoxide they produce causes additional damage by rapidly inactivating chlorophyll and oxidizing chloroplast membrane lipids. Because viologen herbicides are also highly toxic to animals, their use is banned or tightly regulated in many jurisdictions.

Chemical herbicides have become an important management tool for modern agriculture, but their value as a labor-saving device must be carefully weighed against potentially harmful ecological effects. Many of these herbicides are carcinogenic, and thus the potential accumulation of these hazardous compounds in water supplies continues to be a major public concern. In addition, the overuse of herbicides promote herbicide tolerance in weeds, which exacerbates the weed problem in the long term.

the photosynthetic electron transport chain are not distributed homogeneously throughout the thylakoid membranes of eukaryotic chloroplasts but exhibit lateral heterogeneity. However, lateral heterogeneity is dynamic and is a consequence electrostatic shielding of negative surface charges on thylakoid membranes created by the exposed N-terminal domains of the major LHCII polypeptides. Granal stacks are not an absolute requirement for oxygenic photosynthesis. Although cyanobacteria exhibit PSI, PSII and are oxygenic, thylakoids of these prokaryotes do not exhibit granal stacks due to the presence of extrinsic pigment-protein complexes called phycobilisomes. In contrast to eukaryotic photosynthetic organisms and cyanobacteria, photosynthetic bacteria contain only one specialized bacterial reaction center that oxidizes H_2S or H_2 and is incapable of oxidizing H_2O. Several classes of economically important herbicides act by interfering with photosynthetic electron transport.

SUMMARY

The function of the light-dependent reactions of photosynthesis is to generate the ATP and reducing potential (as NADPH) required for subsequent carbon reduction. The electron transport chain in the thylakoid membranes of oxygenic photoautrophs is composed of two photosystems (PSI, PSII) and a cytochrome b_6f complex. The three complexes are linked by plastoquinone and plastocyanin, mobile carriers that freely diffuse within the plane of the membrane. Each photosystem consists of a reaction center, core antenna, and associated light-harvesting (LHC) complexes. Light energy gathered by the antenna and LHC is passed to the reaction center. In the reaction center, electron flow is initiated by a charge separation (photooxidation). As a result, electrons obtained from the oxidation of water are passed through PSII, the cytochrome b_6f complex, and PSI to $NADP^+$. Protons pumped across the membrane between PSII and PSI drive photophosphorylation. The components of

CHAPTER REVIEW

1. ATP formation in chloroplasts is based on the stepwise conservation of energy. Trace the conservation of energy from the initial absorption of light by an antenna chlorophyll molecule to the final formation of a molecule of ATP.

2. Describe the concept of a photosystem and how it is involved in converting light energy to chemical energy.

3. Explain the difference between cyclic and non-cyclic electron transport. How can noncyclic photosynthetic electron transport function if the PSII and PSI units are located in different regions of the thylakoid membrane?

4. How is lateral heterogeneity regulated?

5. Explain the difference between oxygenic photosynthesis and anoxygenic photosynthesis. What role did the evolution of oxygenic photosynthesis play in the global distribution of photosynthetic organisms?

6. What is LHCII and where is it localized? What are phycobilisomes and where are they localized?

7. The herbicide DCMU is commonly used in laboratory investigations of electron transport reactions in isolated chloroplasts. Can you suggest why DCMU might be useful for such studies?

FURTHER READING

Aro, E.-M., B. Andersson. 2001. *Regulation of Photosynthesis. Advances in Photosynthesis and Respiration, Vol. 11.* Dordrecht: Kluwer.

Blankenship, R. E. 2002. *Molecular Mechanisms of Photosynthesis.* Williston: Blackwell Science.

Buchanan, B. B., W. Gruissem, R. L. Jones. 2000. *Biochemistry and Molecular Biology of Plants.* Rockville, MD: American Society of Plant Physiologists.

Merchant, S., M. R. Sawaya. 2005. The light reactions: A guide to recent acquisitions for the picture gallery. *Plant Cell* 17:648–663.

Nelson, N., C. F. Yocum. 2006. Structure and function of photosystems I and II. *Annual Review of Plant Biology* 57:521–565.

Ort, D. T., C. F. Yokum (eds.). 1996. *Oxygenic Photosynthesis: The Light Reactions. Advances in Photosynthesis and Respiration, Vol. 4.* Dordrecht: Kluwer.

Shikanai, T. 2007. Cyclic electron transport around photosystem I: Genetic approaches. *Annual Review of Plant Biology* 58:199–217.

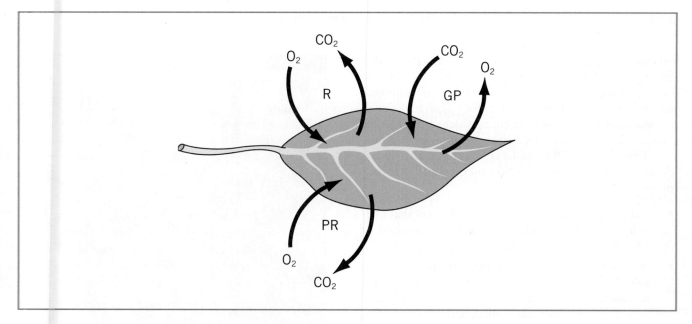

8

Energy Conservation in Photosynthesis: CO_2 Assimilation

Chapter 7 showed how chloroplasts conserve light energy by converting it to reducing potential, in the form of NADPH and ATP. In this chapter, attention is focused on how CO_2 enters the leaf and subsequently is reduced through the utilization of the NADPH and ATP produced by photosynthetic electron transport. Air is the source of CO_2 for photosynthesis. Gas exchange between the leaf and the surrounding air is dependent upon diffusion and is controlled by the opening and closing of special pores called stomata. Stomatal movements are very sensitive to external environmental factors such as light, CO_2, water status, and temperature. The reactions involved in the reduction of CO_2 have been traditionally designated the "dark reactions" of photosynthesis, but this designation is quite misleading, since it implies they can proceed in the absence of light. However, several critical enzymes in the carbon reduction cycle are light activated; in the dark they are either inactive or exhibit low activity. Consequently, carbon reduction cannot occur in the dark, even if energy could be made available from some source other than the photochemical reactions.

Only a few decades ago, knowledge of carbon metabolism was in its infancy and, as is to be expected when opening up new areas of study, understanding of the process was somewhat unsophisticated.

Photosynthesis and respiration, long recognized as the two major divisions of carbon metabolism, were thought to be separate and independent metabolic processes, neatly compartmentalized in the cell. Photosynthesis was localized in the chloroplast while respiration appeared restricted to the cytoplasm and the mitochondrion. The task of photosynthesis was to reduce carbon and store it as sugars or starch. When required, these storage products could be mobilized and exported to the mitochondrion where they were oxidized to satisfy the energy and carbon needs of the cell through respiration. Thus, the relationships between photosynthesis and respiration appeared simple and uncomplicated.

Over the past 40 years, however, knowledge and understanding of carbon metabolism has improved considerably and, with that, so has the apparent complexity. Photosynthetic carbon metabolism can no longer be explained by a single, invariable cycle. It is no longer restricted to just the chloroplast or even to a single cell. In addition to carbon reduction, photosynthetic energy is used to drive nitrogen assimilation, sulphate reduction, and other aspects of intermediary metabolism. The rate of photosynthesis is influenced and even controlled by events occurring outside the chloroplast and elsewhere in the plant. These and other complexities of metabolic integration within the cell and between different parts of the plant are only beginning to become obvious. One thing is certain: a more holistic approach to carbon metabolism is called for. The traditional, compartmentalized vision of independent processes no longer adequately explains carbon metabolism in plants.

This chapter will describe several interrelated variations of photosynthetic carbon metabolism in higher plants, including

- leaf gas exchange through stomatal pores that provide an efficient mechanism for the absorption of CO_2 along a shallow concentration gradient,

- the path of carbon, energetics, and regulation of the photosynthetic carbon reduction cycle, or C3 metabolism—the pathway that all organisms ultimately use to assimilate carbon, and

- photorespiration and how limitations are imposed on carbon assimilation in C3 plants by the photosynthetic carbon oxidation cycle

Later chapters will address carbon partitioning, respiratory carbon metabolism, and factors influencing the distribution of carbon throughout the plant as well as the ecological significance of C4 and CAM (Crassulacean Acid Metabolism) photosynthesis.

8.1 STOMATAL COMPLEX CONTROLS LEAF GAS EXCHANGE AND WATER LOSS

The epidermis of leaves contains pores that provide for the exchange of gases between the internal air spaces and the ambient environment. The opening, or stoma, is bordered by a pair of unique cells called **guard cells** (Figure 8.1). In most cases the guard cells are in turn surrounded by specialized, differentiated epidermal cells called **subsidiary cells**. The stoma, together with its bordering guard cells and subsidiary cells, is referred to as the **stomatal complex**, or **stomatal apparatus**.

The distinguishing feature of the stomatal complex is the pair of guard cells that functions as a hydraulically operated valve. Guard cells take up water and swell to open the pore when CO_2 is required for photosynthesis, and lose water to close the pore when CO_2 is not required or when water stress overrides the photosynthetic needs of the plant. The mechanical, physiological, and biochemical properties of the guard cells have attracted scholars almost since their occurrence was first reported by M. Malpighi in the late seventeenth century. A continuing interest in stomatal movement is understandable, given the foremost importance of stomata in regulating gas exchange and consequent effects on photosynthesis and productivity.

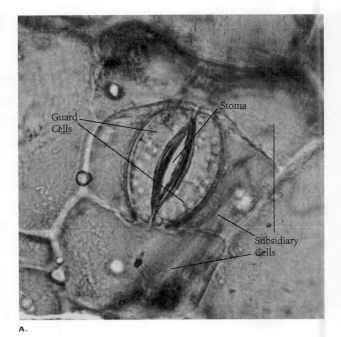

A.

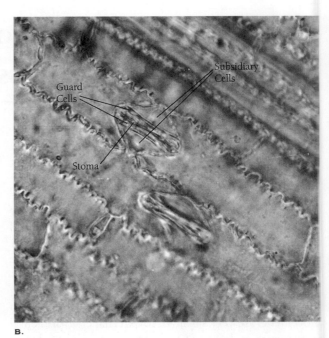

B.

FIGURE 8.1 Stomata. (*A*) **Elliptic type in the lower epidermis of *Zebrina*. In this picture the stoma is open (× 250). (*B*) Graminaceous type from the adaxial surface of maize (Zea mays) leaf. These stomata are closed (× 250).**

More than 90 percent of the CO_2 and water vapor exchanged between a plant and its environment passes through the stomata. Stomata are therefore involved in controlling two very important but competing processes: uptake of CO_2 for photosynthesis and, as discussed in Chapter 2, transpirational water loss. It is important, therefore, to take into account stomatal function when considering photosynthetic productivity and crop yields.

More recently, additional interest in stomatal function has been prompted by recognition that airborne pollutants such as ozone (O_3) and sulphur dioxide (SO_2) also enter the leaf through open stomata.

Stomata are found in the leaves of virtually all higher plants (angiosperms and gymnosperms) and most lower plants (mosses and ferns) with the exception of submerged aquatic plants and the liverworts. In angiosperms and gymnosperms they are found on most aerial parts including nonleafy structures such as floral parts and stems, although they may be nonfunctional in some cases. The frequency and distribution of stomata is quite variable and depends on a number of factors including species, leaf position, ploidy level (the number of chromosome sets), and growth conditions. A frequency in the range of 20 to 400 stomata mm^{-2} of leaf surface is representative, although frequencies of 1000 mm^{-2} or more have been reported. Although there are exceptions to every rule, the leaves of herbaceous monocots such as grasses usually contain stomata on both the adaxial (upper) and abaxial (lower) surfaces with roughly equal frequencies. Stomata occur on both the upper and lower surfaces of herbaceous dicots' leaves, but the frequency is usually lower on the upper surface. Most woody dicots and tree species have stomata only on the lower leaf surface while floating leaves of aquatic plants (e.g., water lily) have stomata only on the upper surface. In most cases the stomata are randomly scattered across the leaf surface, although in monocots with parallel-veined leaves the stomata are arranged in linear arrays between the veins.

The most striking feature of the stomatal complex is the pair of guard cells that border the pore. These specialized epidermal cells have the capacity to undergo reversible turgor changes that in turn regulate the size of the aperture between them. When the guard cells are fully turgid the aperture is open, and when flaccid, the aperture is closed. While there are many variations on the theme, anatomically we recognize two basic types of guard cells: the graminaceous type and the elliptic type (Figure 8.1).

Elliptic or kidney-shaped guard cells are so called because of the elliptic shape of the opening. In surface view, these guard cells resemble a pair of kidney beans with their concave sides opposed. In cross-section the cells are roughly circular in shape, with a **ventral wall** bordering the pit and a **dorsal wall** adjacent to the surrounding epidermal cells (Figure 8.2). The mature guard cell has characteristic wall thickenings, mainly along the outer and inner margins of the ventral wall. These thickenings extend into one or two **ledges** that protect the **throat** of the stoma. In some plants, particularly the gymnosperms and aquatic species, the inner ledge may be small or absent. The outer ledge appears to be an architectural adaptation that helps to prevent the penetration of liquid water from the outside into the substomatal air space, which would otherwise have disastrous consequences for gas exchange.

The graminaceous type of guard cell is largely restricted to members of the Gramineae and certain other monocots (e.g., palms). Often described as dumbbell-shaped, the graminaceous-type guard cells have thin-walled, bulbous ends that contain most of the cell organelles (Figure 8.1). The "handle" of the dumbbell is characterized by walls thickened toward the lumen. The pore in this case is typically an elongated slit. The guard cells are flanked by two prominent subsidiary cells.

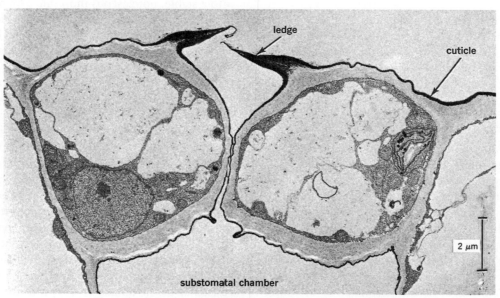

FIGURE 8.2 **Guard cells seen in cross-section. (From K. Esau,** *Anatomy of Seed Plants,* **New York, Wiley, 1977. Reprinted by permission).**

8.2 CO_2 ENTERS THE LEAF BY DIFFUSION

Diffusion of CO_2 into the leaf through the stoma is more efficient than would be predicted on the basis of stomatal area alone. A fully open stomatal pore typically measures 5 to 15 μm wide and about 20 μm long. The combined pore area of open stomata thus amounts to no more than 0.5 to 2 percent of the total area of the leaf. Since leaves contain no active pumps, all of the CO_2 taken into the leaf for photosynthesis must enter by diffusion through these extremely small pores. One might think that diffusion through such a limited area would be extremely restricted, yet it has been calculated that the rate of CO_2 uptake by an actively photosynthesizing leaf may approach 70 percent of the rate over an absorbing surface with an area equivalent to that of the entire leaf! This extraordinarily high diffusive efficiency appears to be related to the special geometry of gaseous diffusion through small pores.

The high efficiency of gaseous diffusion through stomata can be demonstrated experimentally by measuring CO_2 diffusion into a container of CO_2-absorbing agent such as sodium hydroxide. The container is covered with a thin membrane perforated with pores of known dimensions. Diffusion of CO_2 through the membrane can be measured as the amount of carbonate present in the sodium hydroxide solution after, for example, one hour. It was discovered that the rate of CO_2 diffusion through a perforated membrane varies *in proportion to the diameter of the pores, not the area*. How can these results be reconciled with **Fick's law**, which states that

$$\text{Rate of diffusion} = v = D \cdot A(dc/dx) \qquad (8.1)$$

where D is the diffusion coefficient, A is the surface area over which diffusion occurs, and (dc/dx) is the concentration gradient over which diffusion occurs? Clearly, the rate of diffusion is directly proportional to the surface area, A, and the concentration gradient (dc/dx).

The physical explanation for this paradox lies in the pattern of diffusive flow as the gases *enter* and *exit* the stomatal pore. This is illustrated schematically for a stoma in Figure 8.3. Note that in the aperture itself (i.e., in the throat of the stoma) CO_2 molecules can flow only straight through and diffusion is proportional to the cross-sectional area of the throat as predicted by Fick's law of diffusion. But when the gas molecules pass through the aperture into the substomatal cavity, they can "spill over" the edge of the pore. The additional diffusive capacity contributed by spillover is proportional to the amount of edge, or the perimeter of the pore. Because the area of a pore decreases by the square of the radius (r) while perimeter varies directly with the

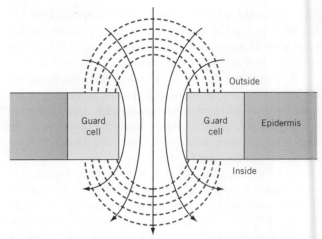

FIGURE 8.3 **The spillover effect for diffusion of CO_2 through a stomatal pore. The dashed lines are isobars, representing regions of equivalent CO_2 partial pressure.**

diameter (2r), the relative contribution of the perimeter effect increases as the pore size decreases. Thus in very small pores (e.g., the size of stomata) the bulk of gas movement is accounted for by diffusion over the perimeter. Even this effect is exaggerated with respect to stomata. Because of their elliptical shape, the ratio of perimeter to area is greater than for circular pores.

How is a high concentration gradient for CO_2 established and maintained? To ensure a constant diffusion of CO_2 from the air into the leaf, the CO_2 concentration within the substomatal cavity and leaf air spaces must be less than the CO_2 concentration in the air above the leaf. This CO_2 concentration gradient (dc/dx) is established because, in the light, chloroplasts continuously *fix* CO_2, that is, chloroplasts within the leaf mesophyll cells continuously convert gaseous CO_2 into a stable, nongaseous molecule 3-phosphoglycerate (PGA) through the reductive pentose phosphate cycle (see below). Thus, this biochemical cycle constantly removes CO_2 from intercellular air spaces of a leaf, thereby ensuring that the internal leaf CO_2 concentration is less than the ambient CO_2 concentrations in the light. In the dark, photosynthesis stops but respiration generates CO_2 such that the internal leaf CO_2 concentrations are greater than the ambient CO_2 concentrations, and thus CO_2 diffuses out of a leaf in the dark. The rate of CO_2 evolution from a leaf in the dark is a measure of the rate of leaf mitochondrial respiration.

The above arguments, of course, represent an ideal situation. In reality, the stomatal pore itself is not the only barrier to gaseous diffusion between the leaf and its environment. A number of other factors—such as unstirred air layers on the leaf surface and the aqueous path between the air space and the chloroplast—offer resistance to the uptake of CO_2 into the leaf and complicate the actual situation. Nonetheless, stomata are remarkably efficient structures. They permit very high rates of CO_2 absorption, without which photosynthesis would be severely limited. This creates a paradox. A system that is efficient for the uptake of CO_2 is also efficient for the loss of water vapor from the internal surfaces of the leaf (Chapter 2). Thus, the principal functional advantage offered by the stomatal apparatus is an ability to conserve water by closing the pore when CO_2 is not required for photosynthesis or when water stress overrides the leaf's photosynthetic needs.

8.3 HOW DO STOMATA OPEN AND CLOSE?

This question may be answered by first asking what mechanical forces are involved in guard cell movement. The driving force for stomatal opening is known to be the osmotic uptake of water by the guard cells and the consequent increase in hydrostatic pressure. The result is a deformation of the opposing cells that increases the size of the opening between them. In the case of elliptic guard cells the thickened walls become concave, while in the dumbbell-shaped cells the handles separate but remain parallel. Stomatal closure follows a loss of water, and the consequent decrease in hydrostatic pressure and relaxation of the guard cell walls.

Deformation of elliptic guard cells during opening is due to the unique structural arrangement of the guard cell walls. In normal cells, bands of cellulose microfibrils encircle the cell at right angles to the long axis of the cell. Studies with polarized light and electron microscopy have demonstrated that the microfibrils in the guard cell walls are oriented in radial fashion, fanning out from the central region of the ventral wall (Figure 8.4). Additional microfibrils are arranged longitudinally within the ventral wall thickenings, crosslinking with the radial bands and restricting expansion along the ventral wall. When the guard cells take up water, expansion follows the path of least resistance—which is to push the relatively thin dorsal walls outward into the neighboring epidermal cells. This causes the cells to arch along the ventral surface and form the stomatal opening. The dumbbell-shaped guard cells of the grasses also depend on the osmotic uptake of water, but operate in a slightly different way. In this case the bulbous ends of the cells push against each other as they swell, driving the central handles apart in parallel and widening the pore between them.

What controls stomatal opening and closure? To answer this question it is necessary instead to ask what regulates the osmotic properties of the guard cells. This question has proven difficult to answer, partly because so many factors seem to be involved and partly because it has been difficult to study guard cell metabolism free of complications introduced by the surrounding epidermal and mesophyll cells. This problem has been partially

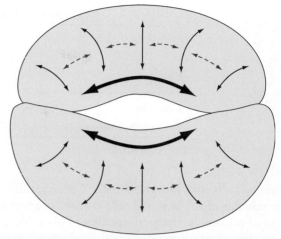

FIGURE 8.4 **The role of microfibrils in guard cell movement. The orientation of the microfibrils (solid arrows) in elliptic guard cells allows expansion of the cells only in the direction shown by the dashed arrows. This causes the cells to buckle and thereby increase the size of the opening between their adjacent walls.**

resolved by studying guard cell behavior in peeled strips of epidermal cells. More recently, techniques for preparation of guard cell protoplasts have become available, making it possible to study guard cell metabolism and ion movement in isolation.

Over the years a variety of mechanisms have been offered to explain changing osmotic concentrations of guard cells. Most have centered on the observation that guard cells normally contain chloroplasts and were assumed to be photosynthetically competent. One way or another it was proposed that an accumulation of photosynthetic product—sugars and other small molecules—contributed directly to the observed osmotic changes in the guard cells. While it is true that most guard cells do have chloroplasts, the number of chloroplasts varies considerably. As well, the guard cells of some species (e.g., some orchids and variegated regions of *Pelargonium*) have no chloroplasts but remain fully functional. Furthermore, investigators have been unable to detect significant levels of Rubisco (the principal carbon-fixing enzyme; see below) in the guard cells of at least 20 species, leading to the conclusion that the carbon-fixing portion of photosynthesis does not operate in guard cells. The conclusion is inescapable: photosynthetic carbon metabolism cannot be invoked as a general mechanism to explain guard cell movement.

In the late 1960s it became evident that K^+ levels are very high in open guard cells and very low in closed guard cells (Table 8.1). A variety of techniques, including electron microprobes and histochemical methods specific for K^+, have confirmed that the K^+ content of closed guard cells is low compared with that of the surrounding subsidiary and epidermal cells. Upon opening, large amounts of K^+ move from the subsidiary and

TABLE 8.1   Potassium content of open and closed guard cells.

Species	K⁺ Content			
	pmol/Guard Cell		mM	
	Open	Closed	Open	Closed
Vicia faba	2.72	0.55	552	112
Commelina communis	3.1	0.4	448	95

Data from MacRobbie, 1987.

epidermal cells into the guard cells. Consequently, an accumulation of K^+ in guard cells is now accepted as a universal process in stomatal opening. This work gave rise to the current hypothesis that the osmotic potential of guard cells and, consequently, the size of the stomatal opening, is determined by the extent of K^+ accumulation in the guard cells.

Although we lack a thorough understanding of the mechanisms involved, available information about guard cell metabolism and stomatal movements is summarized in the general model shown in Figure 8.5. It is widely accepted that accumulation of ions by most plant cells is driven by an ATP-powered proton pump located on the plasma membrane (Chapter 3). Two lines of evidence indicate that K^+ uptake by stomatal guard cells fits this general mechanism. First, the fungal toxin **fusicoccin**, which is known to stimulate active proton extrusion by the pump, stimulates stomatal opening. Second, **vanadate** (VO_3^-), which inhibits the proton pump, also inhibits stomatal opening. This constitutes

reasonably good evidence that proton extrusion is one of the initial events in stomatal opening. By removing positively charged ions, proton extrusion would tend to hyperpolarize the plasma membrane (i.e., lower the electrical potential inside the cell relative to the outside) as well as establish a pH gradient. Hyperpolarization is thought to open K^+ channels in the membrane, which then allows the passive uptake of K^+ in response to the potential difference or charge gradient across the membrane.

In order to maintain electrical neutrality, excess K^+ ion accumulated in the cells must be balanced by a counterion carrying a negative charge. According to the model shown in Figure 8.5, charge balance is achieved partly by balancing K^+ uptake against proton extrusion, partly by an influx of chloride ion (Cl^-), and partly by production within the cell of organic anions such as malate. In most species, malate production probably accounts for the bulk of the required counterion while in others, such as corn (*Zea mays*), as much as 40 percent of the K^+ moving into the cell is accompanied by Cl^-. In those few species whose guard cells lack chloroplast or starch, Cl^- is probably the predominant counterion.

In addition to its role in maintaining charge balance, the accumulation of malate also helps to maintain cellular pH during solute accumulation. Proton extrusion would tend to deplete the intracellular proton concentration and increase cellular pH. However, because malate is an organic anion, each carboxyl group ($-COO^-$) accumulated releases one proton into the cytosol. The synthesis of malate therefore tends to replenish the supply of protons lost by extrusion and maintain cellular pH at normal levels.

The evidence for malate as a counterion is quite strong. To begin with, malate levels in guard cells of open stomata are five to six times that of closed stomata. Second, guard cells contain high levels of the enzyme phosphoenolpyruvate carboxylase (PEPcase), which catalyzes the formation of malate (Figure 8.5). Third, there is a decrease in the starch content of open stomata that correlates with the amount of malate formed. Finally, factors that influence stomatal opening and closure also influence the activity of PEPcase. For example, fusicoccin, which induces stomatal opening, also causes an

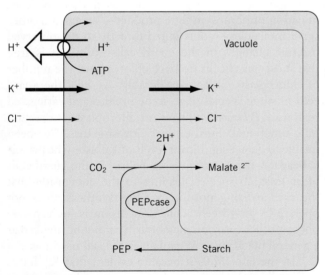

FIGURE 8.5 A simplified model for ion flow associated with the guard cells during stomatal opening. Potassium uptake is driven by an ATPase-proton pump of ions located in the plasma membrane. The accumulation of ions in the vacuole lowers the water potential of the guard cell, thereby stimulating the osmotic uptake of water and increased turgor.

increase in both malate concentration and the activity of PEPcase. Conversely, the plant hormone abscisic acid, which normally induces stomatal closure, antagonizes the effect of fusicoccin. The effect of fusicoccin is to stimulate the phosphorylation of PEPcase, a process well known to activate a variety of enzymes and other proteins in the cell.

The accumulation of K^+, Cl^-, and malate in the vacuoles of the guard cells would lower both the osmotic potential and the water potential of the guard cells. The consequent uptake of water would increase the turgor and cause the stomata to open. At present, this remains a working model for stomatal opening since many of the details have yet to be verified experimentally.

Stomatal closure has not received the same attention that opening has, but it is generally assumed that closure is effected by a simple reversal of the events leading to opening. On the other hand, the rate of closure is often too rapid to be accounted for simply by a passive leakage of ions from the guard cells, leading to the suggestion that other specific metabolic pumps are responsible for actively extruding ions upon closure. One possibility is that signals for stomatal closure stimulate the uptake of Ca^{2+} into the cytosol. Ca^{2+} uptake would depolarize the membrane, thus initiating a chain of events that includes opening anion channels to allow the release of Cl^- and malate. According to this scenario, a loss of anions would further depolarize the membrane, opening K^+ channels and allowing the passive diffusion of K^+ into the adjacent subsidiary and epidermal cells.

What is the source of ATP that powers the guard cell proton pumps? The two most logical sources would be either photosynthesis in the guard cell chloroplasts or cellular respiration. Although most guard cells do contain chloroplasts, they are generally smaller, less abundant, and with fewer thylakoids than those of underlying mesophyll cells. As noted above, guard cell chloroplasts apparently lack the enzymatic machinery for photosynthetic carbon fixation. On the other hand, although ATP production has not been measured directly, indirect evidence indicates that they are capable of using light energy to produce ATP, a process known as **photophosphorylation** (see Chapters 5 and 7). Photosynthesis is probably not the only immediate source of energy, however, since stomatal movement can occur in the dark. An alternative source of energy is cellular respiration. Guard cells do have large numbers of mitochondria and high levels of respiratory enzymes. They may well be able to derive sufficient ATP from the oxidation of carbon through **oxidative phosphosphorylation** (see Chapters 5 and 10). It appears that guard cells have more than adequate capacity to produce, through either respiration or photosynthesis, all the energy necessary to drive stomatal opening.

8.4 STOMATAL MOVEMENTS ARE ALSO CONTROLLED BY EXTERNAL ENVIRONMENTAL FACTORS

The major role of stomata is to allow entry of CO_2 into the leaf for photosynthesis while at the same time preventing excessive water loss. In this sense, they evidently serve a **homeostatic** function; they operate to maintain a constancy of the internal environment of the leaf. It should come as no surprise, then, to find that stomatal movement is regulated by a variety of environmental and internal factors such as light, CO_2 levels, water status of the plant, and temperature. It might be expected, for example, that stomata will open in the light in order to admit CO_2 for photosynthesis or partially close when CO_2 levels are high in order to conserve water while allowing photosynthesis to continue. On the other hand, conditions of extreme water stress should override the plant's immediate photosynthetic needs and lead to closure, protecting the leaf against the potentially more damaging effects of desiccation. In general, these expectations have been verified by direct observation. Each of these factors can theoretically be studied independently under the controlled conditions of the laboratory, but the extent to which they interact under natural conditions makes it far more difficult to study the effects of one relative to another. Moreover, it must be kept in mind that stomatal opening is not an all-or-none phenomenon. At any given time, the extent of stomatal opening and its impact on both photosynthesis and water loss will be determined by the sum of all of these factors and not by any one alone.

8.4.1 LIGHT AND CARBON DIOXIDE REGULATE STOMATAL OPENING

Both light and CO_2 appear to make a substantial contribution to the daily cycle of stomatal movements. Their effects are also tightly coupled, which makes it very difficult to distinguish their relative contributions. In general, low CO_2 concentrations and light stimulate opening while high CO_2 concentrations cause rapid closure even in the light. The response of the stomata is to the *intracellular concentration of CO_2 in the guard cells*. Recall that the outer surfaces of the epidermis, including the guard cells, are covered with the CO_2-impermeable cuticle. Once induced to close by high CO_2 treatment, stomata are not easily forced to open by treatment with CO_2-free air. This is because the closed guard cells remain in equilibrium with the high CO_2 content of the air trapped in the substomatal chamber. Consequently, it is the CO_2 content of the substomatal chamber rather than the ambient atmosphere that is most important in regulating stomatal opening. The actual mechanism by which CO_2 regulates stomatal opening is not understood.

Stomata normally open at dawn. As well, stomata closed by exposure to high CO_2 can be induced to open slowly if placed in the light. Both responses appear to result from two separate effects of light; one indirect and one direct. The indirect effect requires relatively high fluence rates and is usually attributed to a reduction in intercellular CO_2 levels due to photosynthesis in the mesophyll cells. By the same argument, closure of the guard cells in the dark can be attributed to the accumulation of respiratory CO_2 inside the leaf. This interpretation is reinforced by the observation that the action spectrum for moderate to high fluence rates resembles that for photosynthesis with peaks in both the red and blue. Thus it appears that CO_2 is a primary trigger and that, at least in intact leaves, the indirect effect of light may operate through regulation of intercellular CO_2 levels. A significant difficulty with this interpretation, however, is that similar action spectra have been obtained for isolated epidermal peels. Such a result in the absence of an intact leaf argues strongly for an important but yet undefined role of the guard cell chloroplasts.

Perhaps one of the more significant advances to emerge in recent years is the unequivocal demonstration of a direct effect of low-fluence blue light on stomatal opening. If the stomata depended solely on photosynthetically active light, it would likely suffer from two limitations. First, the guard cells would be unable to respond to light levels below the **photosynthetic light compensation point** (i.e., the minimum fluence rate at which photosynthesis exceeds respiration). Second, the system would be prone to extreme oscillations as the rate of photosynthesis fluctuated with rapid changes in PAR. A direct effect of blue light on stomatal opening would seem to circumvent these limitations.

The blue light effect has been demonstrated in a variety of ways. Although stomatal opening is promoted by both red and blue light, it is generally more sensitive to blue light than to red. At low fluence rates, below 15 $\mu mol\ m^{-2}\ s^{-1}$, blue light will cause stomatal opening but red light is ineffective. At higher fluence rates stomatal opening under blue light (which presumably activates both systems) is consistently higher than under red at the same fluence rate. The response of stomata to red light is probably indirect, mediated by the guard cell chloroplasts and involving photosynthetic ATP production. The action spectrum of the blue light response, on the other hand, is typical of other blue light responses and is probably mediated by **cryptochrome**, a putative blue light receptor (Chapter 6). The mode of action of blue light is not certain, but blue light does cause swelling of isolated guard cell protoplasts. This result

indicates that blue light acts directly on the guard cells. Several investigators have reported that blue light activates proton extrusion by the guard cells and stimulates malate biosynthesis; both are prerequisites to stomatal opening.

But what function does the blue light response serve under natural conditions? One interesting and plausible suggestion is that it may have a role in the early morning opening of stomata. Opening can often be observed before sunrise, when fluence rates are much lower than that required to drive photosynthesis. They may also remain open after sunset. The high sensitivity of the blue light response to low fluence rates together with the relatively high proportion of blue light in sunlight at dawn and dusk suggests that the blue light response could function as an effective "light-on" signal. From an ecophysiological standpoint, the blue light response anticipates the need for atmospheric CO_2 and drives stomatal opening in preparation for active photosynthesis. Another possible role is to stimulate rapid stomatal opening in response to sunflecks—the sunfleck itself would be analogous to a blue light pulse—in order to maximize the opportunity for photosynthesis under this particular condition (Chapter 14).

8.4.2 STOMATAL MOVEMENTS FOLLOW ENDOGENOUS RHYTHMS

Many biological processes undergo periodic fluctuations that persist under constant environmental conditions. This phenomenon, known as **endogenous rhythm**, is discussed further in Chapter 25. It was demonstrated that stomatal opening and closure in *Tradescantia* leaves persisted for at least three days, even though the plants were maintained under continuous light. A periodicity of approximately 24 hours was maintained, although the timing of opening or closure could be shifted by a six-hour dark period. Results such as these clearly indicate an involvement of an endogenous circadian rhythm in control of stomatal opening, although it is not clear how the rhythm interacts with other stimuli.

8.5 THE PHOTOSYNTHETIC CARBON REDUCTION (PCR) CYCLE

Now that we understand the processes involved in the control of CO_2 entry into a leaf, we will examine in some detail the biochemical mechanisms by which chloroplasts fix this CO_2 and convert it to stable phosphorylated carbon intermediates.

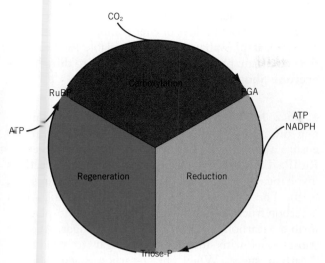

FIGURE 8.6 The three stages of the photosynthetic carbon reduction cycle.

8.5.1 THE PCR CYCLE REDUCES CO₂ TO PRODUCE A THREE-CARBON SUGAR

The pathway by which all photosynthetic eukaryotic organisms ultimately incorporate CO_2 into carbohydrate is known as carbon fixation or the **photosynthetic carbon reduction (PCR) cycle**. It is also referred to as the **Calvin cycle**, in honor of Melvin Calvin, who directed the research effort that elucidated the pathway. Mapping the complex sequence of reactions involving the formation of organic carbon and its conversion to complex carbohydrates represented a major advance in plant biochemistry. For his efforts and those of his associates, Calvin was awarded the Nobel Prize for chemistry in 1961.

The PCR cycle can be divided into three primary stages (Figure 8.6): (1) **carboxylation** which fixes the CO_2 in the presence of the five-carbon acceptor molecule, ribulose bisphosphate (RuBP), and converts it into two molecules of a three-carbon acid; (2) **reduction**, which consumes the ATP and NADPH produced by photosynthetic electron transport to convert the

three-carbon acid to triose phosphate; and (3) **regeneration**, which consumes additional ATP to convert some of the triose phosphate back into RuBP to ensure the capacity for the continuous fixation of CO_2.

8.5.2 THE CARBOXYLATION REACTION FIXES THE CO₂

Calvin's strategy for unraveling the path of carbon in photosynthesis was conceptually very straightforward: identify the first stable organic product formed following uptake of radiolabeled CO_2. In order to achieve this, cultures of the photosynthetic green alga *Chlorella* were first allowed to establish a steady rate of photosynthesis. $^{14}CO_2$ was then introduced and photosynthesis continued for various periods of times before the cells were dropped rapidly into boiling methanol. The hot methanol served two functions: it denatured the enzymes, thus preventing any further metabolism, while at the same time extracting the sugars for subsequent chromatographic analysis. When the time of photosynthesis in the presence of $^{14}CO_2$ was reduced to as little as two seconds, most of the radioactivity was found in a three-carbon acid, **3-phosphoglycerate** (3-PGA). Thus 3-PGA appeared to be the first stable product of photosynthesis. Other sugars that accumulated the label later in time were probably derived from 3-PGA. Because Calvin's group determined that the first product was a three-carbon molecule, the PCR cycle is commonly referred to as the **C3 cycle**. The next step was to determine what molecule served as the acceptor—the molecule to which CO_2 was added in order to make the three-carbon product. Systematic degradation of 3-PGA demonstrated that the ^{14}C label was predominantly in the carboxyl carbon. A two-carbon acceptor molecule would be logical, but the search was long and futile. No two-carbon molecule could be found. Instead, Calvin recognized that the acceptor was the five-carbon keto sugar, **ribulose-1,5-bisphosphate (RuBP)**. This turned out to be the key to the entire puzzle. The reaction is a carboxylation in which CO_2 is added to RuBP, forming a six-carbon intermediate (Figure 8.7). The

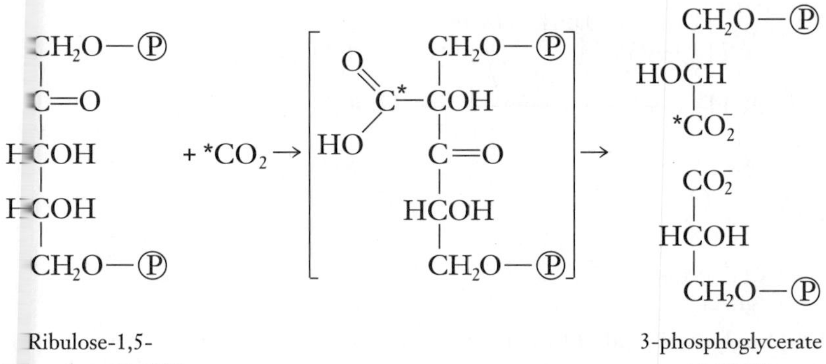

FIGURE 8.7 The carboxylation reaction of the photosynthetic carbon reduction cycle.

$$\begin{array}{c} CH_2O-\text{\textcircled{P}} \\ | \\ C=O \\ | \\ HCOH \\ | \\ HCOH \\ | \\ CH_2O-\text{\textcircled{P}} \end{array} + {}^*CO_2 \rightarrow \left[\begin{array}{c} O \\ \parallel \quad CH_2O-\text{\textcircled{P}} \\ C^*-COH \\ HO \quad | \\ \quad C=O \\ \quad | \\ \quad HCOH \\ \quad | \\ \quad CH_2O-\text{\textcircled{P}} \end{array} \right] \rightarrow \begin{array}{c} CH_2O-\text{\textcircled{P}} \\ | \\ HOCH \\ | \\ {}^*CO_2^- \\ \\ CO_2^- \\ | \\ HCOH \\ | \\ CH_2O-\text{\textcircled{P}} \end{array}$$

Ribulose-1,5-bisphosphate (RuBP)

3-phosphoglycerate (3-PGA)

intermediate, which is transient and unstable, remains bound to the enzyme and is quickly hydrolyzed to **two** molecules of 3-PGA. The carboxylation reaction is catalyzed by the enzyme **ribulose-1,5-bisphosphate carboxylase-oxygenase**, or **Rubisco**. Rubisco is without doubt the most abundant protein in the world, accounting for approximately 50 percent of the soluble protein in most leaves. The enzyme also has a high affinity for CO_2 that, together with its high concentration in the chloroplast stroma, ensures rapid carboxylation at the normally low atmospheric concentrations of CO_2. Thus, the reaction catalyzed by Rubisco maintains the CO_2 concentration gradient (dc/dx) between the internal air spaces of a leaf and the ambient air to ensure a constant supply of this substrate for the PCR cycle.

8.5.3 ATP AND NADPH ARE CONSUMED IN THE PCR CYCLE

The carboxylation reaction, with a ΔG of -35 kJ mol^{-1}, is energetically very favorable. This poses an interesting question. If the equilibrium constant of the reaction favors carboxylation with such a high negative free energy change, where is the need for an input of energy from the light reactions of photosynthesis? Energy is required at two points: first for the reduction of 3-PGA and second for regeneration of the RuBP acceptor molecule. Each of these requirements will be discussed in turn.

8.5.3.1 *Reduction of 3-PGA*   In order for the chloroplast to continue to take up CO_2, two conditions must be met. First, the product molecules (3-PGA) must be continually removed and, second, provisions must be made to maintain an adequate supply of the acceptor molecule (RuBP). Both require energy in the form of ATP and NADPH.

The 3-PGA is removed by *reduction* to the triose phosphate, **glyceraldehyde-3-phosphate**. This is a two-step reaction (Figure 8.8) in which the 3-PGA is first phosphorylated to 1,3-bisphosphoglycerate, which is then reduced to glyceraldehyde-3-phosphate (G3P). Both the ATP and the NADPH required in these two steps are products of the light reactions and together represent one of two sites of energy input. The resulting triose sugar-phosphate, G3P, is available for export to the cytoplasm, probably after conversion to **dihydroxyacetone phosphate (DHAP)** (Chapter 9).

8.5.3.2 *Regeneration of RuBP*   In order to maintain the process of CO_2 reduction, it is necessary to ensure a continuing supply of the acceptor molecule, RuBP. This is accomplished by a series of reactions involving 4-, 5-, 6-, and 7-carbon sugars (Figures 8.9, 8.10). These reactions include the condensation of a 6-carbon fructose-phosphate with a triose-phosphate to form a 5-carbon sugar and a 4-carbon sugar. Another triose joins with the 4-carbon sugar to produce a 7-carbon sugar. When the 7-carbon sugar is combined with a third triose-phosphate, the result is two more 5-carbon sugars. All of the five-carbon sugar can be isomerized to form **ribulose-5-phosphate (Ru5P)**. Ru5P can, in turn, be phosphorylated to regenerate the required ribulose-1,5-bisphosphate.

The net effect of these reactions is to recycle the carbon from five out of every six G3P molecules, thus regenerating three RuBP molecules to replace those used in the earlier carboxylation reactions. The summary reactions shown in Figures 8.9 and 8.10 include three molecules of RuBP on each side of the equation. This is to emphasize that the cycle serves to regenerate the original number of acceptor molecules and maintain a steady-state carbon reduction. Figures 8.9 and 8.10 show that for every three turns of the cycle (i.e., the uptake of three CO_2) there is sufficient carbon to regenerate the required number of acceptor molecules *plus one additional triose phosphate*, which is available for export from the chloroplast. The stoichiometry in Figures 8.9 and 8.10 was chosen to illustrate this point. Six turns of the cycle would regenerate 6 molecules of RuBP, leaving the equivalent of one additional hexose sugar as net product. Twelve turns would generate the equivalent of a sucrose molecule, and so on.

As a general rule it is necessary to show that the required enzymes are present and active before a complex metabolic scheme can be accepted as fact. Calvin's

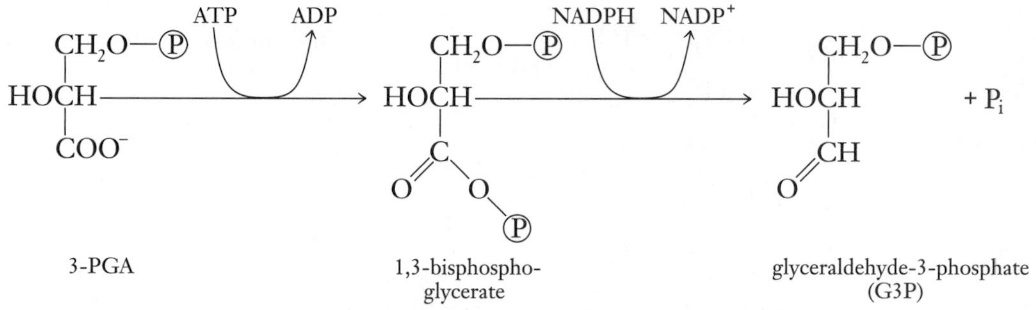

FIGURE 8.8 **Reduction of phosphoglyceric acid (PGA) to glyceraldehyde-3-phosphate (G3P).**

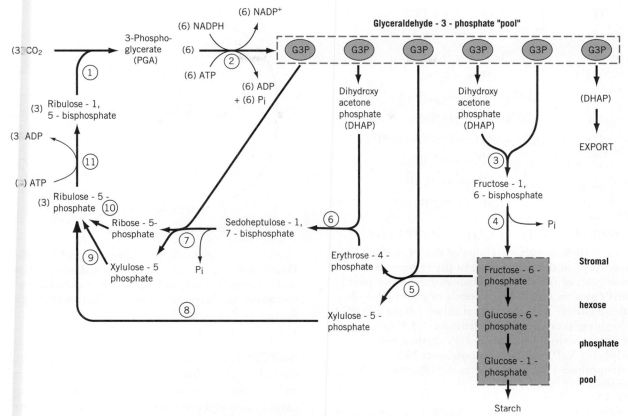

FIGURE 8.9 The photosynthetic carbon reduction (PCR) cycle. Numbers in brackets indicate stoichiometry. Enzymes, indicated by circled numbers are: (1) ribulose-1,5-bisphosphate carboxylase/oxygenase (Rubisco); (2) 3-phosphoglycerate kinase and glyceraldehyde-3-phosphate dehydrogenase; (3) aldolase; (4) fructose-1,6-bisphosphatase; (5) transketolase; (6) aldolase; (7) sedoheptulose-1,7- bisphosphatase; (8, 9) ribulose-5-phosphate epimerase; (10) ribose-5-phosphate isomerase; (11) ribulose-5-phosphate kinase.

PCR cycle has met this criterion since all of the enzymes required by the scheme in Figure 8.9 have now been demonstrated in the stroma. Moreover, all of the reactions have been demonstrated *in vitro*, at rates that would support maximal rates of photosynthesis.

8.5.4 WHAT ARE THE ENERGETICS OF THE PCR CYCLE?

Figure 8.9 shows that for three turns of the cycle, that is, the uptake of 3 molecules of CO_2, a total of 6 molecules of NADPH and 9 molecules of ATP are required. Therefore, the reduction of each molecule of CO_2 requires 2 molecules of NADPH and 3 molecules of ATP for a ratio of ATP/NADPH of 3/2 or 1.5. Since each NADPH stores 2 electrons, we can see that a total of 4 electrons are required to fix each molecule of CO_2. This total represents an energy input of 529 kJ mol^{-1} of CO_2. Oxidation of one mole of hexose would yield about 2817 kJ, or 469 kJ mol^{-1} of CO_2. Thus, the photosynthetic reduction process represents an **energy storage efficiency** of about 88 percent. If we include the energy consumed in the form of the three ATP

per CO_2 (3×31.4 kJ $mol^{-1} = 282$ kJ mol^{-1}) for the regeneration of RuBP, then energy storage efficiency is about 58 percent. An important assumption underlying these simple calculations is that all of the CO_2 fixed by the PCR cycle actually remains fixed in the leaf. Later in this chapter we will see that this assumption does not necessarily hold under all conditions.

8.6 THE PCR CYCLE IS HIGHLY REGULATED

It was originally believed that the PCR cycle did not require a significant level of regulation, in part because early *in vitro* studies of Rubisco suggested a low, and probably rate-limiting, reactivity for this critical enzyme. (Its *in vivo* reactivity is now known to be much higher, although it may still be rate limiting.) In addition, plants were widely believed to be opportunistic and would use available light, water, and CO_2 to conduct photosynthesis at maximum rates. However, it is now recognized that photosynthesis does not operate in isolation and an

$$\text{(1)} \quad 3 \text{ RuBP} + 3 \text{ CO}_2 \longrightarrow 6 \text{ PGA}$$

$$\text{(2)} \quad 6 \text{ PGA} \longrightarrow 5 \text{ G3P} + \text{G3P}$$

$$\text{(3)} \quad 2 \text{ G3P} \longrightarrow \text{FBP}$$

$$\text{(4)} \quad \text{FBP} \longrightarrow \text{F6P}$$

$$\text{(5)} \quad \text{G3P} + \text{F6P} \longrightarrow \text{E4P} + \text{Xu5P}$$

$$\text{(6)} \quad \text{G3P} + \text{E4P} \longrightarrow \text{SBP}$$

$$\text{(7)} \quad \text{G3P} + \text{SBP} \longrightarrow \text{R5P} + \text{Xu5P}$$

$$\text{(8, 9)} \quad 2 \text{ Xu5P} \longrightarrow 2 \text{ R5P}$$

$$\text{(10)} \quad \text{R5P} \longrightarrow \text{RuBP}$$

$$\text{(11)} \quad 2 \text{ R5P} \longrightarrow 2 \text{ RuBP}$$

$$\text{SUM} = 3 \text{ RuBP} + 3 \text{ CO}_2 \rightarrow 3 \text{ RuBP} + \text{G3P}$$

FIGURE 8.10 Summary reactions of the PCR cycle. Three turns of the cycle result in the regeneration of 3 molecules of the acceptor ribulose-1,5-bisphosphate (RuBP) plus an additional molecule of glyceraldehyde-3-phosphate (G-3-P). Additional abbreviations are: PGA, 3-phosphoglyceric acid; FBP, fructose-1,6-bisphosphate; F6P, fructose-6-phosphate; E-4-P, erythrose-4-phosphate; XuP, xylulose-5-phosphate; SBP, sedoheptulose-1,7-bisphosphate; R-5-P, ribulose-5-phosphate.

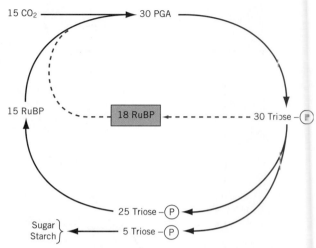

FIGURE 8.11 Autocatalytic properties of PCR cycle. When required, carbon can be retained within the PCR cycle (dashed arrows) to build up the amount of receptor molecules and increase the rate of photosynthesis.

unregulated photosynthetic machinery is incompatible with an orderly and integrated metabolism. Changing levels of intermediates between light and dark periods and competing demands for light energy and carbon with other cellular needs (nitrate reduction, for example) demand some degree of regulation. The most effective control is, of course, at the level of enzyme activities. Molecular biology combined with classical enzyme kinetics (see Box 8.1) and structural information obtained through protein crystallization has begun to elucidate the sophisticated nature of photosynthetic enzyme regulation. A principal factor in the regulation of the PCR cycle is, perhaps not surprisingly, light.

8.6.1 THE REGENERATION OF RuBP IS AUTOCATALYTIC

The rate of carbon reduction is partly dependent on the availability of an adequate pool of acceptor molecules, CO₂ and RuBP. The PCR cycle can utilize newly fixed carbon to increase the size of this pool, when necessary, through the **autocatalytic** regeneration of RuBP. During the night, when photosynthesis is shut down and carbon is required for other metabolic activities, the concentrations of intermediates in the cycle (including RuBP) will fall to low levels. Consequently, when photosynthesis starts up again, the rate could be severely limited by the availability of RuBP, the CO₂ acceptor molecule. Normally the extra carbon taken in through

the PCR cycle is accumulated as starch or exported from the chloroplast. However, the PCR cycle has the potential to augment supplies of acceptor by retaining that extra carbon and diverting it toward generating increasing amounts of RuBP instead (Figure 8.11). In this way the amount of acceptor can be quickly built up within the chloroplast to the level needed to support rapid photosynthesis. Only after the level of RuBP has been built up to adequate levels will carbon be withdrawn for storage or export. The time required to build up the necessary levels of PCR cycle intermediates in the transition from dark to light is called the **photosynthetic induction time**. No other sequence of photosynthetic reactions has this capacity, which may help to explain why *all* photosynthetic organisms ultimately rely on the C3 cycle for carbon reduction. How autocatalysis is regulated is not altogether clear. However, the most effective control would be to enhance the activities of enzymes favoring recycling over those leading to starch synthesis or export of product.

8.6.2 RUBISCO ACTIVITY IS REGULATED INDIRECTLY BY LIGHT

Rubisco activity declines rapidly to zero when the light is turned off and is regained only slowly when the light is once again turned on. Light activation is apparently indirect and involves complex interactions between Mg²⁺ fluxes across the thylakoid, CO₂ activation, chloroplast pH changes, and an activating protein.

As noted in the previous chapter, light-driven electron transport leads to a net movement of protons into the lumen of the thylakoids. The movement of protons across the thylakoid membrane generates a proton gradient equivalent to 3.0 pH units and an increase in the pH of the stroma from around pH 5.0 in the dark to about pH 8.0 in the light. *In vitro*, Rubisco is generally more active at pH 8.0 than at pH 5.0. The Mg^{2+} requirement for Rubisco activity was noted some years ago. Light also brings about an increase in the free Mg^{2+} of the stroma as it moves out of the lumen to compensate for the proton flux in the opposite direction.

Work in the laboratory of G. H. Lorimer, again using isolated Rubisco *in vitro*, has shown that Rubisco uses CO_2 not only as a substrate but also as an activator. The activating CO_2 must bind to an activating site, called the allosteric site, that is separate and distinct from the substrate-binding site (see Box 8.1). Based on these *in vitro* studies, Lorimer and Miziorko proposed a model for *in vivo* activation that takes into account all three factors: CO_2, Mg^{2+}, and pH. According to this model, the CO_2 first reacts with an ε-amino group of a lysine residue in the allosteric site, forming what is known as a **carbamate** (Figure 8.12). Carbamate formation requires the release of two protons and, consequently, would be favored by increasing pH. The Mg^{2+} then becomes coordinated to the carbamate to form a carbamate-Mg^{2+} complex, which is the active form of the enzyme.

Further experiments, however, indicated that the *in vitro* model could not fully account for the activation of Rubisco in leaves. In particular, measured values for *in vivo* Mg^{2+} and CO_2 concentrations and pH differences were not sufficient to account for more than half the expected activation level. This paradox was resolved by the discovery of an *Arabidopsis* mutant that failed to activate Rubisco in the light, even though the enzyme isolated from the mutant was apparently identical to that isolated from the wildtype. Electrophoretic analysis revealed that the *rca* mutant, as it was called, was missing a soluble chloroplast protein. Subsequent experiments demonstrated that full activation of Rubisco could be restored *in vitro* simply by adding the missing protein to a reaction mixture containing Rubisco, RuBP, and physiological levels of CO_2. This protein has been named **Rubisco activase** to signify its role in promoting light-dependent activation of Rubisco.

Rubisco activase is known to require energy in the form of ATP. The protein has been identified in at least 10 genera of higher plants as well as the green alga *Chlamydomonas*. It is clear that Rubisco activase has a significant and probably ubiquitous role to play in regulating eukaryotic photosynthesis.

8.6.3 OTHER PCR ENZYMES ARE ALSO REGULATED BY LIGHT

Rubisco is not the only PCR cycle enzyme requiring light activation. Studies with algal cells, leaves, and isolated chloroplasts have shown that the activities of at least four other PCR cycle enzymes are also stimulated by light. These include glyceraldehyde-3-phosphate dehydrogenase (G-3-PDH) (reaction 2, Figure 8.9), fructose-1,6-bisphosphatase (FBPase) (reaction 4, Figure 8.9), sedoheptulose-1,7-bisphosphatase (SBPase) (reaction 7, Figure 8.9), and ribulose-5-phosphate kinase (R-5P-K) (reaction 11, Figure 8.9).

The mechanism for light activation is different from that of Rubisco and is best demonstrated in the case of FBPase. Light activation of FBPase can be blocked by the electron transport inhibitor DCMU and agents that selectively modify sulfhydryl groups. On the other hand, the enzyme can be activated in the dark by the reducing agent, dithiothreitol (DTT). It gradually emerged that activation requires the participation of both chloroplast ferredoxin, a product of the light-dependent reactions, and **thioredoxin** (Figure 8.13). Like ferredoxin, thioredoxin is a small (12 kDa) iron-sulphur protein, known to biochemists for its role in the reduction of ribonucleotides to deoxyribonucleotides. It contains two **cysteine** residues in close proximity that undergo reversible reduction–oxidation from the disulphide (—S—S—)

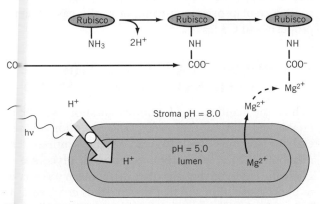

FIGURE 8.12 Light-driven ion fluxes and activation of Rubisco. Activation of Rubisco is facilitated by the increase in stromal pH and Mg^{2+} concentration that accompanies light-driven electron transport.

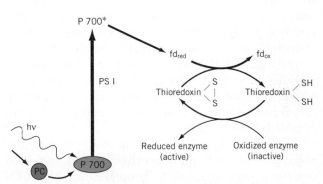

FIGURE 8.13 The ferredoxin/thioredoxin system for light-driven enzyme activation.

state to the sulfhydryl (—SH HS—) state. In the chloroplast, PSI drives the reduction of ferredoxin, which in turn reduces thioredoxin. The reaction is mediated by the enzyme ferredoxin-thioredoxin reductase. Thioredoxin subsequently reduces the appropriate disulphide bond on the target enzyme, resulting in its activation. Subsequent deactivation of the enzymes in the dark is not well understood, but clearly the sulfhydryl groups are in some way reoxidized and the enzymes rendered inactive.

The traditional view of the PCR cycle was that it did not require the direct input of light. These reactions were consequently referred to as the "dark reactions" of photosynthesis. In view of the fact that at least five critical enzymes in the cycle require light activation, such a designation is clearly not appropriate.

8.7 CHLOROPLASTS OF C3 PLANTS ALSO EXHIBIT COMPETING CARBON OXIDATION PROCESSES

The most widely used method for assessing the rate of photosynthesis in whole cells (e.g., algae) or intact plants is to measure gas exchange—either CO_2 uptake or O_2 evolution. This is, at best, a complicated process since there are several different and competing metabolic reactions that contribute to the gas exchange of an algal cell or a higher plant leaf. Cellular (or mitochondrial) respiration **(R)** is an example of opposite gas exchange, since it results in an evolution of CO_2 and uptake of O_2. Historically, it was assumed that mitochondrial-based respiration and chloroplast-based photosynthesis were effectively independent and that their respective contributions to gas exchange could also be assessed independently. (One argument held that photosynthesis could supply the entire energy need of the leaf directly and the mitochondria would consequently "shut down" in the light!). We now know that measuring gas exchange is a far less certain process, complicated in part by oxidative metabolism and the consequent *evolution* of CO_2 directly associated with photosynthetic metabolism (Figure 8.14). Called **photorespiration (PR)**, this process involves the reoxidation of products just previously assimilated in photosynthesis. The photorespiratory pathway involves the activities of at least three different cellular organelles (the chloroplast, the peroxisome, and the mitochondrion) and, because CO_2 is evolved, results in a net loss of carbon from the cell.

The measured CO_2 uptake in the light is termed **apparent** or **net photosynthesis (AP)**, since it represents photosynthetic CO_2 uptake minus the CO_2 evolved from mitochondrial respiration plus photorespiration (Equation 8.2). **True** or **gross photosynthesis**

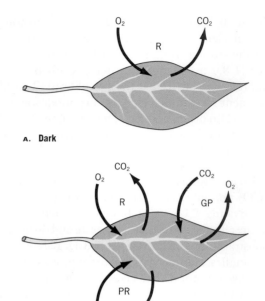

FIGURE 8.14 Gas exchange observed in a C3 leaf in the dark (A) and in the light (B). GP, gross photosynthesis; PR, photorespiration; R, mitochondrial respiration.

(GP) is thus calculated by adding the amount of mitochondrial-respired CO_2 plus photorespired-CO_2 to that taken up in the light (Equation 8.3).

$$AP = GP - (R + PR) \qquad (8.2)$$

$$GP = AP + R + PR \qquad (8.3)$$

Early experiments based on discrimination between carbon isotopes suggested there were both qualitative and quantitative differences between the process of respiration (i.e., CO_2 *evolution*) as it occurred in the dark and in the light. On this basis, CO_2 evolution in the light was called photorespiration. Initially, the concept that light would alter the rate of respiration was, to say the least, controversial. However, biochemical and molecular evidence has firmly established photorespiration as important process contributing to the gas exchange properties of C3 leaves.

8.7.1 RUBISCO CATALYZES THE FIXATION OF BOTH CO₂ AND O₂

While the legitimacy of photorespiration was being established during the 1960s, the attention of several investigators was attracted to the synthesis and metabolism of a two-carbon compound, **glycolate**. It gradually emerged that glycolate metabolism was related to photorespiration and that the enzymes involved were located in peroxisomes and mitochondria as well as the chloroplast. The key to photorespiratory CO_2 evolution and glycolate metabolism is the bifunctional nature of Rubisco. In addition to the **carboxylation** reaction,

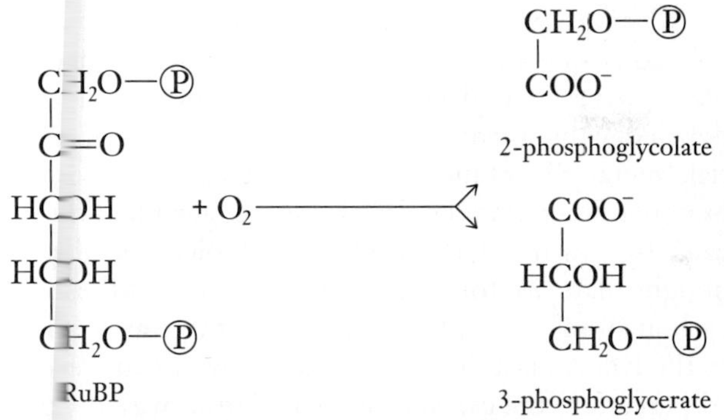

FIGURE 8.15 The RuBP oxygenase reaction.

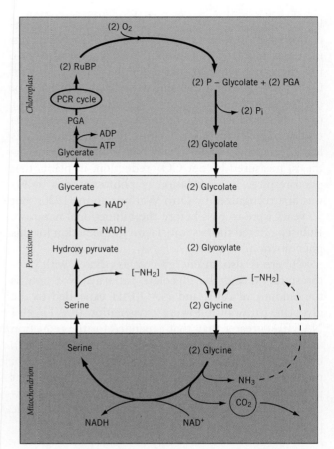

FIGURE 8.16 The photorespiratory glycolate pathway.

Rubisco also catalyzes an **oxygenase** reaction, hence the name ribulose-1,5-bisphosphate carboxylase-oxygenase. With the addition of a molecule of oxygen, RuBP is converted into one molecule of 3-PGA and one molecule of **phosphoglycolate** (Figure 8.15). The phosphoglycolate is subsequently metabolized in a series of reactions in the peroxisome and the mitochondrion that result in the release of a molecule of CO_2 and recovery of the remaining carbon by the PCR cycle (Figure 8.16).

The **C2 glycolate cycle**, also known as the **photosynthetic carbon oxidation (PCO) cycle**, begins with the oxidation of RuBP to 3-PGA and P-glycolate. The 3-PGA is available for further metabolism by the PCR cycle, but the P-glycolate is rapidly dephosphorylated to glycolate in the chloroplast. The glycolate is exported from the chloroplast and diffuses to a **peroxisome**. Taken up by the peroxisome, the glycolate is oxidized to glyoxylate and hydrogen peroxide. The peroxide is broken down by catalase and the glyoxylate undergoes a transamination reaction to form the amino acid glycine. Glycine is then transferred to a mitochondrion where two molecules of glycine (4 carbons) are converted to one molecule of serine (3 carbons) plus one CO_2. *Glycine is thus the immediate source of photorespired CO_2.* The serine then leaves the mitochondrion, returning to a peroxisome where the amino group is given up in a transamination reaction and the product, hydroxypyruvate, is reduced to glycerate. Finally, glycerate is returned to the chloroplast where it is phosphorylated to 3-PGA.

The release of carbon as CO_2 during the conversion of glycine to serine is accompanied by the release of an equivalent amount of nitrogen in the form of ammonia. During active photorespiration, the rate of ammonia release may be substantially greater than the rate of nitrogen assimilation. This nitrogen is not lost, however, as the ammonia is rapidly reassimilated in the chloroplast, using the enzymes of the glutamate synthase cycle (Chapter 11).

The C2 glycolate pathway involves complex interactions between photosynthesis, photorespiration, and various aspects of nitrogen metabolism in at least three different cellular organelles. Much of the supporting evidence comes from labeling studies employing either $^{14}CO_2$ or specific intermediates, or $^{18}O_2$, in which the fate of the label is followed through the various suspected chemical transformations. As with the PCR cycle, all of the enzymes necessary to carry out the C2 glycolate cycle have been demonstrated. The distribution of intermediates between the three organelles, however, is not conclusively established. It is largely inferred from the location of the enzymes. All of the subcellular organelles involved have been isolated and shown to contain the appropriate enzymes.

8.7.2 WHY PHOTORESPIRATION?

In normal air (21% O_2), the rate of photorespiration in sunflower leaves is about 17 percent of gross photosynthesis. Every photorespired CO_2, however, requires an input of two molecules of O_2 (Figure 8.16). The true rate of oxygenation is therefore about 34 percent and the ratio of carboxylation to oxygenation is about 3 to 1 (1.00/0.34). This experimental value agrees with similar values calculated for several species based on the known characteristic of purified Rubisco. The ratio of carboxylation to oxygenation depends, however, on the relative levels of O_2 and CO_2 since both gases compete for binding at the active site on Rubisco. As the concentration of O_2 declines, the relative level of carboxylation increases until, at zero O_2, photorespiration is also zero. On the other hand, increases in the relative level of O_2 (or decrease in CO_2) shifts the balance in favor of oxygenation. An increase in temperature will also favor oxygenation, since as the temperature increases the solubility of gases in water declines, but O_2 solubility is less affected than CO_2. Thus O_2 will inhibit photosynthesis, measured by net CO_2 reduction, in plants that photorespire. The inhibition of photosynthesis by O_2 was first recognized by Otto Warburg in the 1920s, but 50 years were to pass before the bifunctional nature of Rubisco offered the first satisfactory explanation for this phenomenon.

There is also an energy cost associated with photorespiration and the glycolate pathway. Not only is the amount of ATP and NAD(P)H expended in the glycolate pathway following oxygenation (5 ATP + 3 NADPH) greater than that expended for the reduction of one CO_2 in the PCR cycle (3 ATP + 2 NADPH), but *there is also a net loss of carbon*. On the surface, then, photorespiration appears to be a costly and inefficient process with respect to both energy and carbon acquisition. It is logical to ask, as many have, why should the plant indulge in such an apparently wasteful process?

This question is not easily answered, although several ideas have been put forward. One has it that the oxygenase function of Rubisco is inescapable. Rubisco evolved at a time when the atmosphere contained large amounts of CO_2 but little oxygen. Under these conditions, an inability to discriminate between the two gases would have had little significance to the survival of the organism. Both CO_2 and O_2 react with the enzyme at the same active site, and oxygenation requires activation by CO_2 just as carboxylation does. It is believed that oxygen began to accumulate in the atmosphere primarily due to photosynthetic activity, but by the time the atmospheric content of O_2 had increased to significant proportions, the bifunctional nature of the enzyme had been established without recourse. In a sense, C3 plants were the architect of their own problem—generating the oxygen that functions as a competitive inhibitor of carbon reduction. By this view, then, the oxygenase function is an evolutionary "hangover" that has no useful role. However, this is an oversimplified view of photorespiration since photorespiratory mutants of *Arabidopsis* proved to be lethal under certain growth conditions, indicating the essential nature of the photorespiratory pathway in C3 plants. Clearly, any inefficiencies resulting from photorespiration in C3 plants are apparently not severe. There is no evidence that selection pressures have caused evolution of a form of Rubisco with lower affinity for O_2.

While most agree that oxygenation is an unavoidable consequence of evolution, many have argued that plants have capitalized on this apparent evolutionary deficiency by turning it into a useful, if not essential, metabolic sequence. The glycolate pathway, for example, undoubtedly serves a scavenger function. For each two turns of the cycle, two molecules of phosphoglycolate are formed by oxygenation. Of these four carbon atoms, one is lost as CO_2 and three are returned to the chloroplast. The glycolate pathway thus recovers 75 percent of the carbon that would otherwise be lost as glycolate. The salvage role alone may be sufficient justification for the complex glycolate cycle. There is also the possibility that some of the intermediates, serine and glycine, for example, are of use in other biosynthetic pathways, although this possibility is still subject to some debate.

Recently, strong experimental support has been provided for the thesis that photorespiration could also function as a sort of safety valve in situations that require dissipation of excess excitation energy. For example, a significant decline in the photosynthetic capacity of leaves irradiated in the absence of CO_2 and O_2 has been reported. Injury is prevented, however, if sufficient O_2 is present to permit photorespiration to occur. Apparently the O_2 consumed by photorespiration is sufficient to protect the plant from photooxidative damage by permitting continued operation of the electron transport system. This could be of considerable ecological value under conditions of high light and limited CO_2 supply, for example, when the stomata are closed due to moisture stress (Chapter 14). Indeed, photorespiratory mutants of *Arabidopsis* are more sensitive to photoinhibition than their wildtype counterparts.

A claim made frequently in the literature is that crop productivity might be significantly enhanced by inhibiting or genetically eliminating photorespiration. As a result, substantial effort has been expended in the search for chemicals that inhibit the glycolate pathway or selective breeding for low-photorespiratory strains. Others have surveyed large numbers of species in an effort to find a Rubisco with a significantly lower affinity for oxygen. All of these efforts have been unsuccessful, presumably because the basic premise that photorespiration is detrimental to the plant and counterproductive is incorrect. Clearly, success in increasing photosynthesis and improving productivity lies in other directions. For example, a mechanism for concentrating CO_2 in the photosynthetic cells could be one way to suppress photorespiratory loss and improve the overall efficiency of carbon assimilation. That is exactly what has been achieved by C4 and CAM plants and will be discussed further in Chapter 15.

8.7.3 IN ADDITION TO PCR, CHLOROPLASTS EXHIBIT AN OXIDATIVE PENTOSE PHOSPHATE CYCLE

Although the oxidative pentose phosphate cycle (OPPC) is restricted to the cytosol in animals, this pathway is present in both the chloroplast (Figure 8.17) and the cytosol (Chapter 10) in plants. Furthermore, the chloroplastic OPPC shares several intermediates with the PCR pathway and is closely integrated with it (Figure 8.17). The first step in the oxidative pentose phosphate cycle is the oxidation of glucose-6-P (G-6 P) to **6-phosphogluconate** (6-P-gluconate) by the enzyme **glucose-6-phosphate dehydrogenase** (Figure 8.17, reaction 1). The glucose-6-phosphate and fructose-6-phosphate are components of the same stromal hexose phosphate pool that is shared with the RPPC (Figure 8.9). This reaction is highly exergonic ($\Delta G < 0$), and thus is not reversible. As a consequence, this reaction is apparently the rate-determining step for the stromal OPPC. The second reaction in the OPPC involves the oxidation of 6-phosphogluconate to **ribulose-5-phosphate (R-5-P)** by the enzyme **gluconate-6-phosphate dehydrogenase** with the production of one molecule of NADPH and one CO_2 (Figure 8.17, reaction 2).

The simultaneous operation of both the PCR pathway and the OPPC in the stroma would result in the reduction of one molecule of CO_2 to carbohydrate at the expense of three ATP and two NADPH through the PCR pathway. Subsequently, the carbohydrate would be reoxidized to CO_2 by the OPPC yielding two NADPH. Thus, if both metabolic pathways operate simultaneously in the stroma, three ATP would be consumed with no net fixation of CO_2. This would represent **futile cycling** of CO_2 with the net consumption of ATP. This would be terribly wasteful!

How do plants overcome the apparent conundrum created by the presence of both a reductive and an oxidative pentose phosphate cycle in the same compartment? The potential for the futile cycling of CO_2 is overcome by metabolic regulation, which ensures that the key enzymes of the PCR cycle are active only in the light and inactive in the dark. In contrast, the key regulatory enzymes of the OPPC are active only in the dark. Figure 8.13 shows that key regulatory enzymes of the PCR cycle (FBPase, SBPase and Ru-5-P kinase) are converted by light from their inactive to their active forms by reduced thioredoxin through the reducing equivalents generated by photosynthetic electron transport. In contrast to stromal FBPase, SBPase, and Ru-5-P kinase, which are active when their disulfide bonds are reduced by thioredoxin (—S—S— → — SH HS—), the key regulatory enzyme in the OPPC (glucose-6-P dehydrogenase; Figure 8.17, reaction 1) *is active when its internal disulfide bonds are oxidized and inactive when they are reduced by thioredoxin*. As a consequence, Rubisco (Figure 8.12), as well as stromal FBPase, SBPase, and Ru-5-P kinase (Figure 8.13) are in their active states in the light but phosphogluconate dehydrogenase is in the inactive state, whereas in the dark, phosphogluconate dehydrogenase is in its active state and the key enzymes of the PRC pathway are inactive. Thus, this exquisite regulation ensures that photosynthesis results in the net fixation of CO_2 and conversion to carbohydrate and prevents the wasteful consumption of ATP.

The OPPC is thought to be a means to generate NADPH required to drive biosynthetic reactions such as lipid and fatty acid biosynthesis in plant mesophyll cells. The oxidative pentose phosphate cycle represents an important source of pentose phosphate, which serves as a precursor for the **ribose** and **deoxyribose** required in the synthesis of nucleic acids. Another intermediate of the oxidative pentose phosphate pathway with potential significance to plants is the 4-carbon erythrose-4-P, a precursor for the biosynthesis of aromatic amino acids, lignin, and flavonoids. In addition, the Ru-5-P generated by the OPPC in the dark can be converted to RuBP in the light to provide the necessary acceptor molecule to get the RPPC started.

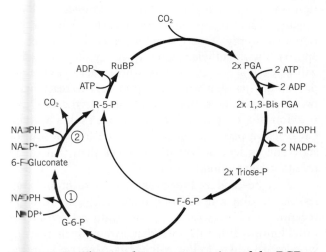

FIGURE 8.17 The simultaneous operation of the PCR cycle and the oxidative pentose phosphate cycle (OPPC) illustrating the potential for the futile cycling of CO_2 in the chloroplast. Reaction 1 is catalyzed by the enzyme glucose-6-P dehydrogenase and reaction 2 by the enzyme phosphogluconate dehydrogenase.

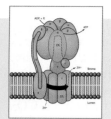

BOX 8.1
ENZYMES

Living cells must carry out an enormous variety of biochemical reactions, yet cells are able to rapidly construct very large and complicated molecules, or regulate the flow of materials through complex metabolic pathways, with unerring precision and accuracy. All this is made possible by **enzymes**. Enzymes are biological catalysts; they facilitate the conversion of *substrate* molecules to product, but are not themselves permanently altered by the reaction. Cells contain thousands of enzymes, each catalyzing a particular reaction.

Enzyme-catalyzed reactions differ from ordinary chemical reactions in four important ways:

1. *High specificity*. Enzymes are capable of recognizing subtle and highly specific differences in substrate and product molecules, to the extent of discriminating between mirror images of the same molecules (called stereoisomers or *enantiomers*) in the same way you do not fit your right hand into your left glove.

2. *High reaction rates*. The rates of enzyme-catalyzed reactions are typically 10^6 to 10^{12} greater than rates of uncatalyzed reactions. Many enzymes are capable of converting thousands of substrate molecules every second.

3. *Mild reaction conditions*. Enzyme reactions typically occur at atmospheric pressure, relatively low temperature, and within a narrow range of pH near neutrality. There are exceptions, such as certain protein-degrading enzymes that operate in vacuoles with a pH near 4.0, or enzymes of thermophilic bacteria that thrive in hot sulfur springs, where temperatures are close to 100°C. Most enzymes, however, enable biological reactions to occur under conditions far milder than those required for most chemical reactions.

4. *Opportunity for regulation*. The presence of a particular enzyme and its amount is regulated by controlled gene expression and protein turnover. In addition, enzyme activity is subject to regulatory control by a variety of activators and inhibitors.

These opportunities for regulation are instrumental in keeping complex and often competing metabolic reactions in balance.

The first step in an enzyme-catalyzed reaction is the reversible binding of a substrate molecule (S) with the enzyme (E) to form an enzyme-substrate complex (ES):

$$E + S \leftrightarrows ES \rightarrow E + P \qquad (8.4)$$

The enzyme-substrate complex then dissociates to release the product molecule (P). The free enzyme is regenerated and is then available to react with another molecule of substrate.

Enzymes are proteins and the site on the protein where the substrate binds and the reaction occurs is called the **active site**. Active sites are usually located in a cleft or pocket in the folded protein, and contain reactive amino acid side chains, such as carboxyl (—COO⁻), amino (—NH₃⁺), or sulfur (—S⁻) groups that position the substrate and participate in the catalysis. The shape and polarity of the active site is largely responsible for the specificity of an enzyme, since the shape and polarity of the substrate molecule must complement or "fit" the geometry of the active site in order for the substrate to gain access and bind to the catalytic groups. Where two or more substrates participate in a common reaction, binding of the first substrate may induce a change in the conformation of the protein, which then allows the second substrate access to the active site.

Enzymes increase the rate of a reaction because they lower the amount of energy, known as the **activation barrier**, required to initiate the reaction. This effect is illustrated by the ball and hill analogy (Figure 8.18A). In order for the ball to roll down the hill, it must first be pushed over the lip of the depression in which it sits. This act increases the potential energy of the ball. When the ball is poised at the very top of the lip, it is in a **transition state**; that is, there is an equal probability that it will fall back into the depression or roll forward and down the hill.

Chemical reactions go through a similar transition state (Figure 8.18B). As reacting molecules come together, they increasingly repel each other and the potential energy of the system increases. If the reactants approach with sufficient kinetic energy, however, they will achieve a transition state where there is an equal probability that they will decompose back to reactants or proceed to products. In the case of an enzyme-catalyzed reaction, the enzyme-substrate complex takes a different reaction pathway—a pathway that has a transition state energy level substantially lower than that of the uncatalyzed reaction (Figure 8.18B).

Enzyme-catalyzed reactions exhibit reaction kinetics that exhibit a hyberbolic relationship between the reaction velocity, v, and the substrate concentration, [S] (Figure 8.19). Enzymes that exhibit such reaction kinetics are said to follow **Michaelis-Menton kinetics**. Michaelis-Menton kinetics are characterized by **substrate saturation**, which reflects the fact the enzyme becomes saturated with substrate with increasing substrate concentration at constant enzyme concentration. The Michaelis-Menton equation describes this relationship mathematically:

$$v = V_{max} [S]/K_m + [S] \qquad (8.5)$$

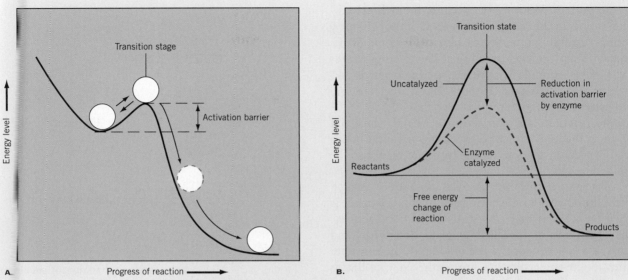

FIGURE 8.18 Enzymes. (*A*) The "ball and hill" analogy of chemical reactions. (*B*) Enzymes reduce the activation barrier, measured as transition state energy, for a reaction.

where v is the initial rate of the reaction, V_{max} is the maximum substrate-saturated rate of the reaction, and the K_m is the substrate concentration that provides the half-maximal substrate-saturated rate of the reaction. The K_m is used as a measure of the *affinity* that an enzyme has for its substrate: a high K_m value implies low affinity of the enzyme for its substrate, whereas a low K_m value implies a high affinity of the enzyme for its substrate.

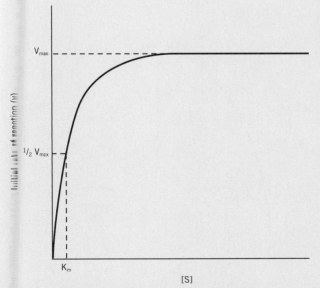

FIGURE 8.19 Plot of initial reaction rate, v, versus substrate concentration [S] for an enzyme-catalyzed reaction. V_{max} is the maximal rate of the reaction under substrate-saturated conditions. K_m is the value of [S] which provides 1/2 V_{max}.

It is important to note that enzymes do not alter the course of a reaction. They do not change the equilibrium between reactants and products, nor do they alter the free energy change (ΔG) for the reaction. (See Chapter 5 for a discussion of free energy changes.) Enzymes change only the rate of a reaction.

Most enzymes are identified by adding the suffix *-ase* to the name of the substrate, often with some indication of the nature of the reaction. For example, α-amyl*ase* digests amylose (starch), malate dehydrogen*ase* oxidizes (that is, removes hydrogen from) malic acid, and phosphoenolpyruvate carboxylase adds carbon dioxide (a carboxyl group) to a molecule of phosphoenolpyruvate.

Many enzymes do not work alone, but require the presence of nonprotein cofactors. Some cofactors, called **coenzymes**, are transiently associated with the protein and are themselves changed in the reaction. Many electron carriers, such as NAD^+ or FAD, for example, serve as coenzymes for many dehydrogenase enzymes (see Chapter 10). They are, in fact, cosubstrates and are reduced to NADH or $FADH_2$ in the reaction. **Prosthetic groups** are nonprotein cofactors more or less permanently associated with the enzyme protein. The heme group of hemoglobin is an example of a tightly bound prosthetic group. Many plant enzymes utilize ions such as iron or calcium as prosthetic groups.

Enzymes and enzyme reactions are sensitive to both temperature and pH. Like most chemical reactions, enzyme reactions have a Q_{10} of about 2, which means that the rate of the reaction doubles for each 10°C rise in temperature. The rate increases with temperature until an optimum is reached, beyond which the rate usually declines sharply. The decline is normally caused by **thermal denaturation**, or unfolding of the enzyme protein. With most enzymes, thermal denaturation occurs in the range of 40 to 45°C, although many enzymes exhibit temperatures closer to 25 or 30°C. Some enzymes exhibit instability at lower temperatures as well. One example is pyruvate, pyrophosphate dikinase (PPDK) (Chapter 15). PPDK is unstable and loses activity at temperatures below about 12 to 15°C. Enzyme reactions are also sensitive to pH, since pH influences the ionization of catalytic groups at the active site. The conformation of the protein may also be modified by pH.

Substrate molecules as well as other related metabolites may not only participate in enzyme catalysis (Equation 8.4) but may also stimulate enzyme activity. This phenomenon is called enzyme **activation** and occurs as a consequence of the binding of the substrate or metabolite molecule to a site on the enzyme that is distinct from the active site. This second alternative binding site on the enzyme is called the **allosteric site**. The binding of the substrate to the allosteric site induces a conformational change in the active site which enhances the rate at which the substrate (S) is converted to product (P). Molecules capable of binding to the allosteric site are called **effector** molecules. Enzymes which exhibit such regulation are called **allosteric enzymes**. Rubisco (8.6.2) is an example of an allosteric enzyme. CO_2 is not only the substrate for the reaction catalyzed by this enzyme, but also activates Rubisco activity by binding to the ε-amino group located in the allosteric site of Rubisco (Figure 8.12).

Conversely, a variety of ions or molecules may combine with an enzyme in such a way that it reduces the catalytic activity of the enzyme. These are known as **inhibitors**. Inhibition of an enzyme may be either **irreversible** or **reversible**. Irreversible inhibitors act by chemically modifying the active site so that the substrate can no longer bind, or by permanently altering the protein in some other way. Reversible inhibitors often have chemical structures that closely resemble the natural substrate. They bind at the active site, but either do not react or react very slowly. For example, the oxidation of succinate to fumarate by the enzyme succinic dehydrogenase is competitively inhibited by malonate, an analog of succinate (Figure 8.20).

Because substrate and inhibitor *compete* with one another for attachment to the active site, this form of inhibitor is known as **competitive inhibition**. Another form of reversible inhibitor, the **noncompetitive inhibitor**, does not compete with the substrate for the active site, but binds elsewhere on the enzyme and, in doing so, restricts access of the substrate to the active site. Alternatively, noncompetitive inhibitors may bind directly to the enzyme-substrate complex, thereby rendering the enzyme catalytically inactive.

Enzymes play a key role in **feedback inhibition**, one of the most common modes for metabolic

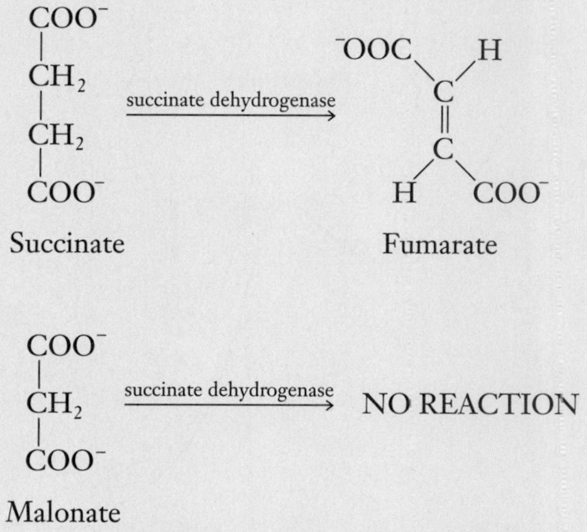

FIGURE 8.20 Malonate, a structural analog of succinate, inhibits the enzyme succinate dehydrogenase. Malonate binds to the enzyme in place of succinate, but does not enter into a reaction.

regulation. Feedback inhibition occurs when the end product of a metabolic pathway controls the activity of an enzyme near the beginning of the pathway. When demand for the product is low, excess product inhibits the activity of a key enzyme in the pathway, thereby reducing the synthesis of product. Once cellular activities have depleted the supply of product, the enzyme is deinhibited and the rate of product formation increases. The enzyme subject to feedback regulation is usually the first one

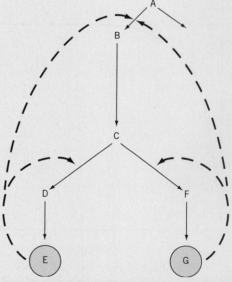

FIGURE 8.21 Feedback inhibition. Excess product inhibits the enzyme that catalyzes a first committed step leading to product formation.

past a metabolic branch point. This is known as the **committed step**. In the example shown in Figure 8.21, reactions A→B, C→D, and C→F all represent committed steps. In this example, an excess of product G would reduce the flow of precursor through the reaction of C → F, thereby diverting more precursor, C, to product E. Alternatively, an excess of both E and G would regulate the conversion of A to B. Feedback regulation is an effective way of coordinating product formation within complex pathways. Many of the enzymes of respiratory metabolism, for example, are subject to feedback regulation, thereby balancing the flow of carbon against the constantly changing energy demands of the cell.

Enzymes are remarkable biological catalysts that both enable and control the enormous variety of biochemical reactions that comprise life.

FURTHER READING

Buchanan, B. B., W. Gruissem, R. L. Jones. 2000, *Biochemistry and Molecular Biology of Plants*. Rockville MD: American Society of Plant Physiologists.

SUMMARY

Photosynthetic gas exchange between the leaf and the air is dependent upon diffusion and is regulated by the opening and closing of specialized epidermal pores called stomata. Stomatal movement is regulated by K^+ levels in the guard cells. Opening and closing of stomata are also sensitive to environmental factors such as CO_2 levels, light, temperature, and the water status of the plant.

The photosynthetic carbon reduction (PCR) cycle occurs in the chloroplast stroma. It is the sequence of reactions all plants use to reduce carbon dioxide to organic carbon. The key enzyme is ribulose-1,5-bisphosphate carboxylase-oxygenase (Rubisco), which catalyzes the addition of a carbon dioxide molecule to an acceptor molecule, ribulose-1,5-bisphosphate (RuBP). The product is two molecules of 3-phosphoglycerate (3-PGA). Energy from the light-dependent reactions is required at two stages: ATP and NADPH for the reduction of 3-PGA and ATP for the regeneration of the acceptor molecule RuBP. The bulk of the cycle involves a series of sugar rearrangements that (1) regenerate RuBP and (2) accumulate excess carbon as 3-carbon sugars. This excess carbon can be stored in the chloroplast in the form of starch or exported from the chloroplast for transport to other parts of the plant.

Photosynthesis, like all other complex metabolic reactions, is subject to regulation. In this case, the primary activator is light. Several key PCR cycle enzymes, including Rubisco, are light activated. This is one way of integrating photosynthesis with other aspects of metabolism, regulating changing levels of intermediates between light and dark periods and competing demands for carbon with other cellular needs.

Plants that utilize the PCR cycle exclusively for carbon fixation also exhibit a competing process of light- and oxygen-dependent carbon dioxide evolution, called photorespiration. The source of carbon dioxide is the photosynthetic carbon oxidation (PCO) cycle. The PCO cycle also begins with Rubisco, which, in the presence of oxygen, catalyzes the oxidation, as well as carboxylation, of RuBP. The product of RuBP oxidation is one molecule of 3-PGA plus one 2-carbon molecule, phosphoglycolate. Phosphoglycolate is subsequently metabolized in a series of reactions that result in the release of carbon dioxide and recovery of the remaining carbon by the PCR cycle. The role of the PCO cycle is not yet clear, although it has been suggested that it helps protect the chloroplast from photo-oxidative damage during periods of moisture stress, when the stomata are closed and the carbon dioxide supply is cut off.

Chloroplasts also exhibit an oxidative pentose phosphate cycle (OPPC) that potentially would lead to the futile cycling of CO_2. This is prevented by the differential light regulation of reductive and oxidative pentose phosphate cycles through the action of thioredoxin.

CHAPTER REVIEW

1. Review the reactions of the photosynthetic carbon reduction cycle and show how:

   (a) product is generated;
   (b) the carbon is recycled to regenerate the acceptor molecule.

2. In what chemical form(s) and where is energy put into the photosynthetic carbon reduction (PCR) cycle? What is the source of this energy?

3. The photosynthetic carbon reduction (PCR) cycle is said to be autocatalytic. What does this mean and of what advantage is it?

4. Describe the photorespiratory pathway. What is the relationship between photorespiration and photosynthesis?

5. Debate the position that the oxygenase function of Rubisco is an evolutionary "hangover."

6. How do plants overcome the potential for futile cycling of CO_2 in the chloroplast?

FURTHER READING

Blankenship, R. E. 2002. *Molecular Mechanisms of Photosynthesis*. London: Blackwell Science.

Buchanan, B. B., Y. Balmer. 2005. Redox regulation: A broadening horizon. *Annual Review of Plant Biology* 56: 187–220.

Buchanan, B. B., W. Gruissem, R. L. Jones. 2000. *Biochemistry and Molecular Biology of Plants*. Rockville MD: American Society of Plant Physiologists.

Leegood, R. C., T. D. Sharkey, S. von Caemmerer. 2000. *Photosynthesis: Physiology and Metabolism. Advances in Photosynthesis*, Vol. 9. Dordrecht: Kluwer.

Spreitzer, R. J., M. E. Salvucci. 2002. Rubisco: Structure, regulatory interactions and possibilities for a better enzyme. *Annual Review of Plant Biology* 53: 449–475.

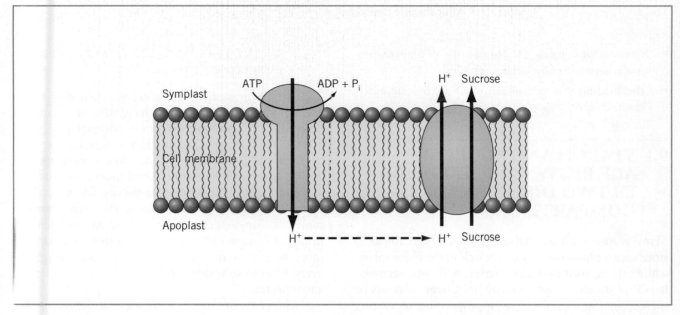

9

Allocation, Translocation, and Partitioning of Photoassimilates

The previous two chapters showed how energy was conserved in the form of carbon compounds, or **photoassimilates**. The primary function of photosynthesis is to provide energy and carbon sufficient to support maintenance and growth not only of the photosynthetic tissues but of the plant as a whole. During daylight hours, photoassimilate generated by the PCR cycle is temporarily accumulated in the leaf as either sucrose in the mesophyll vacuole or starch in the chloroplast stroma. The conversion of photoassimilates to either sucrose or starch is called **carbon allocation**. Although a portion of the carbon assimilated on a daily basis is retained by the leaf to support its continued growth and metabolism, the majority is exported out of the leaf to nonphotosynthetic organs and tissues. There, it is either metabolized directly or placed in storage for retrieval and metabolism at a later time. The transport of photoassimilates over long distances is known as **translocation**. Translocation occurs in the vascular tissue called **phloem**. Phloem translocation is a highly significant process that functions to ensure an efficient distribution of photosynthetic energy and carbon between organs throughout the organism. This is called **carbon partitioning**. Phloem translocation is also important from an agricultural perspective because it plays a significant role in determining productivity,

crop yield, and the effectiveness of applied herbicides and other xenobiotic chemicals.

This chapter is focused on the biosynthesis of the photosynthetic end-products, starch and sucrose, and the structure of the phloem and its function in the translocation and distribution of these important photoassimilates. The principal topics to be covered include

- the biosynthesis of starch and sucrose,

- the allocation of fixed carbon between the starch and sucrose biosynthetic pathways,

- the basis for identifying phloem as the route for translocation of photoassimilates and the nature of substances translocated in the phloem,

- the structure of the phloem tissue, especially the several unique aspects of sieve tube structure and composition,

- the source-sink concept and the significance of sources and sinks to the translocation process,

- the pressure-flow hypothesis for phloem translocation and the processes by which photoassimilates gain entry into the phloem in the leaf and are subsequently removed from the translocation stream at the target organ,

- factors that regulate the distribution of photoassimilates between competing sinks, and

- the loading and translocation of xenobiotic agrochemicals.

9.1 STARCH AND SUCROSE ARE BIOSYNTHESIZED IN TWO DIFFERENT COMPARTMENTS

Many plants, such as soybean, spinach, and tobacco, store excess photoassimilate as starch in the chloroplast, while others, such as wheat, barley, and oats, accumulate little starch but temporarily hold large amounts of sucrose in the vacuole. The appropriation of carbon fixed by the PCR cycle into either starch or sucrose biosynthesis is called **carbon allocation**. This will be discussed further in relation to source sink relationships (see below). The starch and sucrose will later be mobilized to support respiration and other metabolic needs at night or during periods of limited photosynthetic output. Sucrose exported from the leaf cell to nonphotosynthetic tissues may be metabolized immediately, stored temporarily as sucrose in the vacuoles, or converted to starch for longer-term storage in the chloroplasts.

9.1.1 STARCH IS BIOSYNTHESIZED IN THE STROMA

The dominant storage carbohydrate in higher plants is the polysaccharide starch, which exists in two forms (Figure 9.1). **Amylose** is a linear polymer of glucose created by linking adjacent glucose residues between the first and fourth carbons. Amylose is consequently known as an α-(1,4)-glucan. **Amylopectin** is similar to amylose except that occasional α-(1,6) linkages, about every 24 to 30 glucose residues, create a branched molecule. Amylopectin is very similar to glycogen, the principal storage carbohydrate in animals. Glycogen is more highly branched, with one α-(1,6) linkage for every 10 glucose residues compared with one in 30 for amylopectin.

The site of starch synthesis in leaves is the chloroplast. Large deposits of starch are clearly evident in electron micrographs of chloroplasts from C3 plants. In addition, the two principal enzymes involved—**ADPglucose pyrophosphorylase** and **starch synthase**—are found localized in the chloroplast stroma. Starch synthesis in the chloroplast begins with the hexose phosphate pool generated by the PCR cycle (Figure 9.2). Fructose-6-phosphate (F-6-P), one component of the stromal hexose phosphate pool, is converted to glucose-1-phosphate, another component of the stromal hexose phosphate pool, by the two

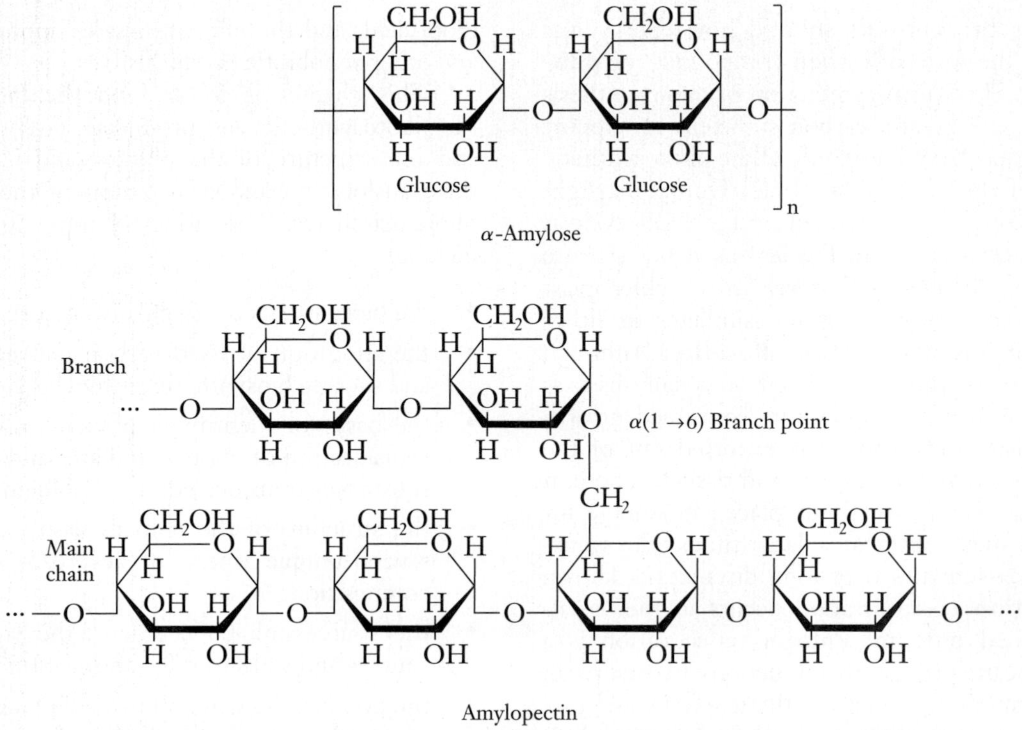

FIGURE 9.1 **The chemical structures of the two forms of starch: amylose and amylopectin. Amylose is a long chain of $\alpha(1\rightarrow4)$-linked glucose residues. Amylopectin is a multibranched polymer of $\alpha(1\rightarrow4)$-linked glucose containing $\alpha(1\rightarrow6)$ branch points.**

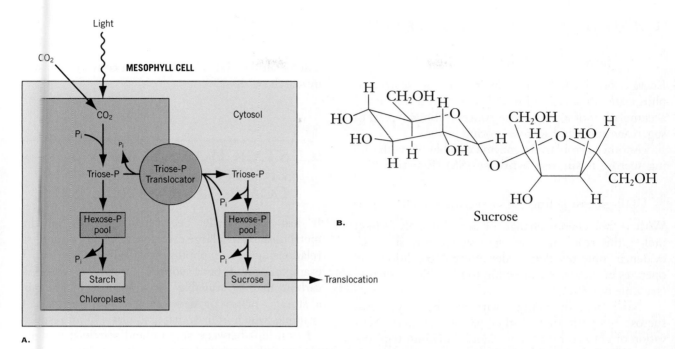

FIGURE 9.2 (*A*) **Allocation of fixed carbon between the chloroplast and the cytosol.**
(*B*) **The structure of sucrose.**

chloroplast enzymes, hexose-phosphate isomerase (Equation 9.1) and phosphoglucomutase (Equation 9.2). The glucose-1-P subsequently reacts with ATP to form ADP-glucose (Equation 9.3). Reaction 9.3 is catalyzed by the enzyme ADP-glucose phosphorylase. ADP-glucose is an activated form of glucose and serves as the immediate precursor for starch synthesis. Starch deposits within the chloroplast stroma are evident as insoluble starch grains. As a consequence, this form of stored carbon is osmotically inactive (Chapter 1) which allows plants to store large amounts of fixed carbon in chloroplasts with minimal influence on the osmotic pressure of the stroma. This prevents the chloroplast membrane from bursting upon the accumulation and storage of fixed carbon as starch.

$$\text{fructose-6-P} \leftrightarrow \text{glucose-6-P} \qquad (9.1)$$

$$\text{glucose-6-P} \leftrightarrow \text{glucose-1-P} \qquad (9.2)$$

$$\text{ATP} + \text{glucose-1-P} \leftrightarrow \text{ADP-glucose} + H_2O + PPi \qquad (9.3)$$

$$PPi + H_2O \leftrightarrow 2Pi \qquad (9.4)$$

Finally, the enzyme starch synthase catalyzes formation of a new α-(1,4) link, adding one more glucose to the elongating chain (Equation 9.5).

$$\text{ADP-glucose} + \alpha\text{-}(1 \rightarrow 4)\text{-glucan} \leftrightarrow \text{ADP} \\ + \alpha\text{-}(1 \rightarrow 4)\text{-glucosyl-glucan} \quad (9.5)$$

Formation of the α-(1,6) branching linkages, giving rise to amylopectin, is catalyzed by the **branching enzyme**, also known as the Q-enzyme.

9.1.2 SUCROSE IS BIOSYNTHESIZED IN THE CYTOSOL

Sucrose is a soluble disaccharide containing a glucose and a fructose residue (Figure 9.2B). It is one of the more abundant natural products that not only plays a vital role in plant life but is also a leading commercial commodity. Sucrose may function as a storage product as it does in sugarbeets or sugarcane, where it is stored in the vacuoles of specialized storage cells. Alternatively, sucrose may be translocated to other, nonphotosynthetic tissues in the plant for direct metabolic use or for conversion to starch. Sucrose is by far the most common form of sugar found in the translocation stream.

The site of sucrose synthesis in the cell was the subject of debate for some time. On the basis of cell fractionation and enzyme localization studies it has now been clearly established that sucrose synthesis occurs exclusively in the cytosol of photosynthetic cells (Figure 9.2). Earlier reports of sucrose synthesis in isolated chloroplasts appear attributable to contamination of the chloroplast preparation with cytosolic enzymes. Moreover, the inner membrane of the chloroplast envelope is impermeable to sucrose, so that if sucrose were synthesized inside the chloroplast it would be unable to exit the chloroplast and enter the translocation stream.

Two routes of sucrose synthesis are possible. The principal pathway for sucrose synthesis in photosynthetic cells is provided by the enzymes **sucrose phosphate synthase** (Equation 9.6) and **sucrose phosphate phosphatase** (Equation 9.7).

$$\text{UDP-glucose} + \text{fructose-6-P} \leftrightarrow \text{sucrose-6-P} + \text{UDP} \tag{9.6}$$

$$\text{sucrose-6-P} + H_2O \leftrightarrow \text{sucrose} + \text{Pi} \tag{9.7}$$

Energy provided by the hydrolysis of sucrose-6-phosphate (about 12.5 kJ mol^{-1}) may play a role in the accumulation of high sucrose concentrations typical of sugarcane and other sucrose-storing plants.

Another cytoplasmic enzyme capable of synthesizing sucrose is **sucrose synthase (SS)** (Equation 9.8):

$$\text{UDP-glucose} + \text{fructose} \leftrightarrow \text{sucrose} + \text{UDP} \tag{9.8}$$

With a free energy change of approximately $+14 \text{ kJ mol}^{-1}$, this reaction is not spontaneous. Most of the evidence indicates that under normal conditions SS operates in the reverse direction to break down sucrose (see Equation 9.11).

Note that, in contrast with starch biosynthesis, sucrose biosynthesis by either pathway requires activation of glucose with the nucleotide **uridine triphosphate (UTP)** rather than ATP:

$$\text{UTP} + \text{glucose-1-P} \leftrightarrow \text{UDP-glucose} + \text{PPi} \tag{9.9}$$

$$\text{PPi} + H_2O \leftrightarrow 2\text{Pi} \tag{9.10}$$

Although sucrose phosphate synthase in some tissues can use ADP-glucose, UDP-glucose is clearly predominant.

Carbon for cytoplasmic sucrose biosynthesis is exported from the chloroplast through a special orthophosphate (P_i)-dependent transporter located in the chloroplast envelope membranes (Figure 9.2). This **P_i/triose phosphate transporter** exchanges P_i and triose phosphate—probably as dihydroxyacetone phosphate (DHAP)—on a one-for-one basis. Once in the cytoplasm, two molecules of triose phosphates (glyceraldehyde-3-phosphate and DHAP) are condensed to form fructose-1,6-bisphosphate. Subsequently, the fructose-1,6-bisphosphate enters the **cytosolic hexose phosphate pool** where it is converted to glucose-1-phosphate as it is in the chloroplast, employing cytoplasmic counterparts of the chloroplastic enzymes. Some of the orthophosphate generated in sucrose synthesis is used to regenerate UTP while the rest can reenter the chloroplast in exchange for triose-P.

Sucrose translocated from the leaf to storage organs such as roots, tuber tissue, and developing seeds is most commonly stored as starch. The conversion of sucrose to starch is generally thought to involve a reversal of the sucrose synthase reaction:

$$\text{Sucrose} + \text{UDP} \rightarrow \text{fructose} + \text{UDP} - \text{glucose} \tag{9.11}$$

Because ADP-glucose is preferred for starch biosynthesis, UDP-glucose is converted to ADP-glucose as shown in (Equation 9.12) and (Equation 9.13):

$$\text{UDP-glucose} + \text{PPi} \leftrightarrow \text{UTP} + \text{glucose-1-P} \tag{9.12}$$

$$\text{ATP} + \text{glucose-1-P} \leftrightarrow \text{ADP-glucose} + H_2O + \text{PPi} \tag{9.13}$$

The resulting ADP-glucose is then converted to starch by starch synthase.

9.2 STARCH AND SUCROSE BIOSYNTHESIS ARE COMPETITIVE PROCESSES

It has traditionally been held that carbohydrate metabolism is to a large extent governed by source-sink relationships. The photosynthetically active leaf, for example, would be a **source**, providing assimilated carbon that is available for transport to the **sink**, a storage organ or developing flower or fruit, for example, which utilizes that assimilate. With respect to relationships between sucrose and starch, it was often observed that removal of a sink, thus reducing demand for photoassimilate, resulted in accumulation of starch in the leaves. This led to the assumption that starch represented little more than excess carbon. There is now good evidence that this assumption is false.

In soybean plants (*Glycine max*), starch accumulation is not related to the length of the photosynthetic period. Plants maintained on a 7-hour light period put a larger proportion of their daily photoassimilate into starch than those maintained on a 14-hour light period, even though the assimilation period is only half as long. Thus it appears that foliar starch accumulation is more closely related to the energy needs of the daily dark period than photosynthetic input. Just how these needs are anticipated by the plant is unknown. However, many species are now known to distribute different proportions of carbon between starch and sucrose in ways apparently unrelated to sink capacity or inherent capacities of isolated chloroplasts to form starch. Carbon distribution thus appears to be a programmed process, implying some measure of control beyond a simple source-sink relationship. Moreover, it is essential that sucrose synthesis be controlled in order to maintain an efficient operation of photosynthesis itself. If the rate of sucrose synthesis should exceed the rate of carbon assimilation, demand for triose-P in the cytoplasm could deplete the pool of PCR cycle intermediates, thereby decreasing the capacity of Calvin cycle enzymes for regeneration of RuBP and seriously inhibiting photosynthesis.

While the enzyme sucrose phosphate synthase (SPS) determines the maximum capacity for sucrose synthesis, it appears that cytosolic fructose-1,6-bisphosphate phosphatase (FBPase) plays the more important role in balancing the allocation of carbon between sucrose and starch synthesis. The highly exergonic reaction (fructose-1,6-bisphosphate $\rightarrow$ fructose-6-phosphate + P_i) occupies a strategic site in the sucrose synthetic pathway—it is the first *irreversible* reaction in the conversion of triose-P to sucrose. Consequently the flow of carbon into sucrose can easily be controlled by regulating the activity of FBPase—similar to regulating the flow of water by opening or closing a valve.

Unlike the chloroplastic FBPase, which is light regulated by thioredoxin, the cytosolic FBPase is not regulated by thioredoxin, but rather is quite sensitive to inhibition by **fructose-2,6-bisphosphate (F-2,6-BP)** (Figure 9.3). F-2,6-BP, an analog of the natural substrate fructose-1,6-bisphosphate, is considered a **regulator metabolite** because it functions as a regulator rather than a substrate (Figure 9.4). F-2,6-BP levels are, in turn, sensitive to a number of interacting factors including the concentration of F-6-P of the cytosolic hexose phosphate pool and the cytosolic triose-P/P_i ratio (Figure 9.4).

Control of sucrose synthesis by F-2,6-BP is essential to ensure a balance between rates of CO_2 assimilation and the allocation of fixed carbon. For example, sucrose export from the cell slows in the light, leading to an accumulation of intermediates such as F-6-P in the cytosolic hexose phosphate pool and triose-P. This causes a shift in allocation in favor of starch. When the consumption of sucrose decreases, sucrose and its precursors (e.g., F-6-P) will accumulate in the leaf cytosolic hexose pool (Figure 9.2). Since F-6-P is also the precursor for F-2,6-BP, levels of the inhibitor will increase as well—leading to an inhibition of FBPase and an accumulation of triose-P. The accumulation of phosphorylated intermediates probably also leads to a decrease in the concentration of P_i. The combined accumulation of triose-P and decrease of P_i will in turn decrease the rate at which triose-P can be exported from the chloroplast through the transporter. The consequent accumulation of triose-P and decrease of orthophosphate in the chloroplast in turn stimulate the synthesis of starch (Figure 9.2). The decrease in stromal P_i leads to a reduction in ATP synthesis (ADP + P_i →

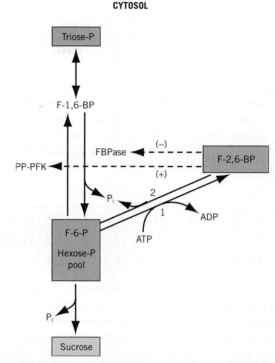

FIGURE 9.4 The synthesis of sucrose is regulated by F-6-P triose-P/P_i ratio. The accumulation of F-6-P in the cytosolic hexose-P pool favors the synthesis of fructose-2,6-bisphosphate (F-2,6-BP) from F-6-P. F-6-P in the cytosolic hexose-P pool is phosphorylated by the enzyme, F-6-P-2-kinase (reaction 1) and converted to F-2,6-BP. Accumulation of this regulatory metabolite inhibits the activity of the enzyme, FBPase, but stimulates the enzyme, pyrophosphate-dependent phosphofructokinase (PP$_i$-PFK), which catalyzes the reverse reaction. The net result is a decreased rate of entry of carbon into the cytosolic hexose-P pool and hence a decreased rate of sucrose synthesis. Low levels of F-6-P favors the breakdown of F-2,6-BP to F-6-P by the enzyme fructose- 2,6-bisphosphatase (reaction 2). This releases cytosolic FBPase from inhibition, and favors the synthesis of F-6-P which, in turn, favors sucrose synthesis.

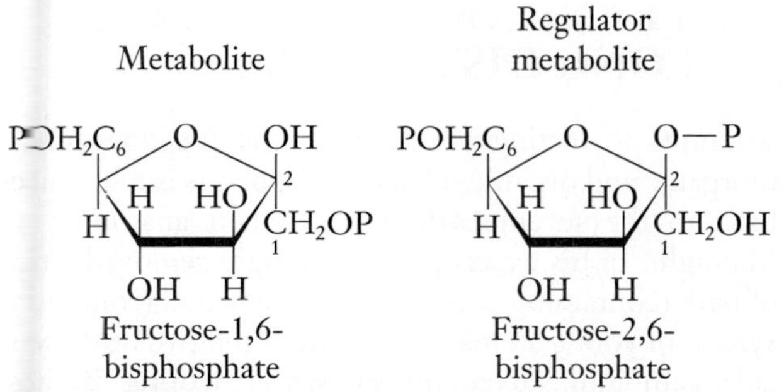

ATP). This causes a buildup of the transthylakoid ΔpH, which in turn inhibits the rate of photosynthetic electron transport through what is called **photosynthetic control**. Photosynthetic control is defined as the regulation of the rate of photosynthetic electron transport by the transthylakoid ΔpH. Hence, this causes a reduction in the rate of photosynthetic O_2 evolution and ultimately the rate of CO_2 assimilation. Plants exhibiting such an inhibition of photosynthesis are said to be **feedback limited**.

Many of the details remain to be worked out, but it is clear that starch synthesis in the chloroplast, triose-P export, and sucrose synthesis in the cytosol are in

Metabolite

Regulator metabolite

POH_2C_6 O OH

H HO CH_2OP

OH H

Fructose-1,6-bisphosphate

POH_2C_6 O O—P

H HO CH_2OH

OH H

Fructose-2,6-bisphosphate

FIGURE 9.3 Comparison of the structures of F-1,6-BP and F-2,6-BP.

delicate balance. The balance is modulated by very subtle changes in the level of triose-P and P_i as well as precise regulation of a number of enzymes and requires intimate communication between the two cellular compartments. Thus, the allocation of fixed carbon to either starch or sucrose biosynthesis in leaf mesophyll cells illustrates that these metabolic pathways actually have a dual role: (1) to provide *energy and carbon* for growth and the maintenance of homeostasis and (2) to provide *information* with respect to the metabolic status in two different compartments.

9.3 FRUCTAN BIOSYNTHESIS IS AN ALTERNATIVE PATHWAY FOR CARBON ALLOCATION

In addition to carbon allocation to sucrose and starch, about 10 percent of terrestrial plant species exhibit the capacity to allocate carbon to water-soluble fructose polymers called **fructans**, which are biosynthesized in the vacuole. Plants capable of forming vacuolar fructans include agronomically important crops such as cereals (wheat, barley, rye), in addition to onion, garlic, leek, Jerusalem artichoke, and chicory. The most common form of fructan in these plants is based on the sequential enzymatic addition of fructose from a donor sucrose molecule to a sucrose acceptor molecule by the enzyme sucrose:sucrose fructosyl transferase (SST). This results in the formation of the trisaccharide, 1-kestose (Equation 9.14), which is composed one of glucosyl unit linked two fructosyl

$$\text{Sucrose} + \text{sucrose} \rightarrow \text{1-kestose} + \text{glucose} \quad (9.14)$$

units. This trisaccharide is extended by the action of an additional vacuolar enzyme, fructan:fructan fructosyl transferase (FFT), which results in the formation of a polymer of fructose in the form of glucosyl-1, 2-fructosyl-1,2-fructrosy-(fructosyl)$_N$ linked in the 1,2-β orientation and where N can vary between 1 (kestose) and 40 (Equation 9.15).

$$\text{1-Kestose} + \text{fructan} \rightarrow \text{glucosyl-1,2-fructosyl-1,}$$
$$\text{2-fructrosyl-(fructosyl)}_N \quad (9.15)$$

Under conditions where the rate of carbon accumulation exceeds the rate of carbon utilization, sucrose accumulates and the enzymes of vacuolar fructan metabolism, SST and FFT, are induced. Thus, it is presumed that increase in cytosolic sucrose concentrations trigger the biosynthesis of fructans. The sucrose accumulated in the cytosol is transported to the vacuole and converted to fructans. Since fructan accumulation can attain levels as high as 40 percent of the dry weight of cereals, the biosynthesis of vacuolar fructans represents an important mechanism for carbon allocation.

TABLE 9.1   **The effects of sucrose accumulation on starch biosynthesis in spinach and the grass, *Lolium temulentum*.**

Plant	Leaf Sucrose (μmol mg^{-1} Chl)	Starch/Sucrose Ratio
Spinach	1	0.2
	5	0.4
	10	0.6
Lolium	5	0.1
	33	0.1
	50	0.1

Data from Pollack et al., 1995.

As discussed above, plant species such as spinach, in which starch is the major storage carbohydrate, the regulatory metabolite, fructose-2,6-bisphosphate, feedback inhibits the export of triose phosphate from the chloroplast stroma in response to increases in cytosolic sucrose concentrations. This stimulates starch biosynthesis through the activation of the enzyme, ADP-glucose pyrophosphorylase and results in an increase in the starch/sucrose ratio (Table 9.1). However, this type of feedback inhibition does not appear to occur in plants such as *Lolium temulentum* that convert sucrose to fructans in the vacuole (Table 9.1). It has been proposed that the apparent insensitivity of chloroplast metabolism to cytosolic sucrose accumulation in fructan-accumulators may represent a selective advantage for grasses which evolved in environments where there were rapid changes in the balance between the supply of and demand for fixed carbon due either shading, cool temperatures, perennial growth habit, or perhaps herbivory. This decreased sensitivity to feedback limited photosynthesis would allow fructan-accumulators to maintain higher rates of CO_2 assimilation due to the maintenance of a greater flux of carbon through the sucrose biosynthetic pathway than starch accumulators.

9.4 PHOTOASSIMILATES ARE TRANSLOCATED OVER LONG DISTANCES

Attempts to distinguish between the translocation of inorganic and organic substances in plants can be traced back to the seventeenth-century plant anatomist M. Malpighi. In his experiments, Malpighi removed a ring of bark (containing phloem) from the wood (containing xylem) of young stems by separating the two at the vascular cambium, a technique known as **girdling**. Because the woody xylem tissue remained intact, water and inorganic nutrients continued to move up to the leaves and the plant was able to survive for some time. Girdled plants, however, developed characteristic swellings of the bark in the region immediately above the girdle (Figure 9.5).

Over the years, this experiment has been repeated and refined to include nonsurgical girdling such as by localized steam-killing or chilling. The characteristic swelling is attributed, in part, to an accumulation of photoassimilate flowing downward, which is blocked from moving further by removal or otherwise interfering with the activity of the phloem. As we now know, the downward stream also contains nitrogenous material and probably hormones that help to stimulate proliferation and enlargement of cells above the blockage. Eventually, of course, the root system will starve from the lack of nutrients and the girdled plant will die.

An analysis of phloem exudate provides more direct evidence in support of the conclusion that photoassimilates are translocated through the phloem. Unfortunately, phloem tissue does not lend itself to analysis as easily as xylem tissue does (described in Chapter 2). This is because the translocating elements in the phloem are, unlike xylem vessels and tracheids, living cells when functional. These cells contain a dense, metabolically active cytoplasm and, because of an inherent sealing action of its cytoplasm, do not exude their contents as readily as do xylem vessels. Moreover, phloem contains numerous parenchyma cells that, while not directly involved in the transport process, do provide contaminating cytoplasm. Cutting the stems of some herbaceous

plants will produce an exudate of largely phloem origin, but in some plants, such as some representatives of the family Cucurbitaceae, the exudate may quickly gel on contact with oxygen, making collection and subsequent analysis difficult. The gelling of phloem exudate is due to the properties of a particular phloem protein, which is described more fully later in this chapter. In spite of these difficulties, however, numerous investigators have successfully completed analyses of phloem exudates obtained by making incisions into the phloem tissue, assisted in part by the development of modern analytical techniques applicable to very small samples.

One intriguing solution to the problem of obtaining the contents of sieve tubes uncontaminated by other cells was provided by insect physiologists studying the nutrition of aphids. Aphids are one of several groups of small insects that feed on plants by inserting a long mouthpart (the stylus) directly into individual sieve tubes. When feeding aphids are anaesthetized with a stream of carbon dioxide and the stylus carefully severed with a razor blade, phloem sap continues to exude from the cut stylus for several days. The aphid technique works well for a number of herbaceous plants and some woody shrubs, but it is restricted to those plants on which the aphids naturally feed. The principal advantage of this technique is that the severed aphid stylet delivers an uncontaminated sieve tube sap. Although the volumes delivered are relatively low, this technique has proven extremely useful in studies of phloem transport. The continued exudation, incidentally, demonstrates that phloem sap is under pressure, an important observation with respect to the proposed mechanism for phloem transport to be discussed later.

The third line of evidence involves the use of radioactive tracers, predominantly ^{14}C and usually fed to a leaf. A typical example is the translocation of photoassimilate in petioles of sugarbeet (*Beta vulgaris*) leaves. In these experiments, attached leaves were allowed to photosynthesize in a closed chamber containing a radioactive carbon source ($^{14}CO_2$). After 10 minutes, the radiolabeled photoassimilate being transported out of the leaf was immobilized by freezing the petiole in liquid nitrogen. Cross sections of the frozen petiole were prepared and placed in contact with X-ray film. The resulting image on the X-ray film, or radioautograph, indicated that the radioactive photoassimilate being translocated out of the leaf was localized exclusively in the phloem (Figure 9.6). Similar experiments have been conducted on a variety of herbaceous and woody plants and with other radioactive nuclides, such as phosphorous and sulphur, with the same conclusion—*the translocation of photoassimilates and other organic compounds over long distances occurs through the phloem tissue.* There are exceptions to this rule, such as when stored sugars are mobilized in the spring of the year and translocated through the xylem to the developing buds (Chapter 2).

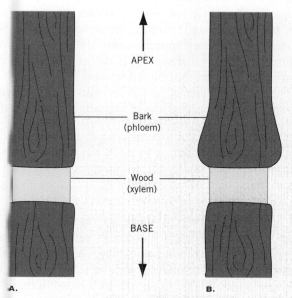

FIGURE 9.5 The results of girdling on woody stems.
(*A*) The phloem tissue can be removed by separating the phloem (the bark) from the xylem (the wood) at the vascular cambium. (*B*) The girdle interrupts the downward flow of nutrients and hormones, resulting in a proliferation of tissue immediately above the girdle.

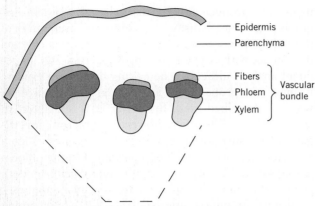

FIGURE 9.6 **Location of radioactivity (gray area) in the phloem of sugarbeet petioles after 10 minutes of photosynthesis in the presence of $^{14}CO_2$.**

9.4.1 WHAT IS THE COMPOSITION OF THE PHOTOASSIMILATE TRANSLOCATED BY THE PHLOEM?

This question may be answered by analyzing the chemical composition of phloem exudate. Phloem sap can be collected from aphid stylets or, alternatively, from some plants by simply making an incision into the bark. If done carefully, to avoid cutting into the underlying xylem, the incision opens the sieve tubes and a relatively pure exudate can be collected in very small microcapillary tubes for subsequent analysis. As might be expected, the chemical composition of phloem exudate is highly variable. It depends on the species, age, and physiological condition of the tissue sampled. Even for a particular sample under uniform conditions, there may be wide variations in the concentrations of particular components between subsequent samples. For example, an analysis of phloem exudate from stems of actively growing castor bean (*Ricinus communis*) (Table 9.2) shows that the exudate contains sugars, protein, amino acids, the organic acid malate, and a variety of inorganic anions and cations. The predominant amino acids are glutamic acid and aspartic acid, which are common forms for the translocation of assimilated nitrogen (Chapter 11). The inorganic anions include phosphate, sulphate, and chloride—nitrate is conspicuously absent—while the predominant cation is potassium. Although not shown in Table 9.1, some plant hormones (auxin, cytokinin, and gibberellin) were also detected, but at very low concentrations. Of course, many of the components identified in phloem exudate—inorganic ions, for example—are cytoplasmic constituents of the translocating cells and do not necessarily represent translocated photoassimilate. Protein found in phloem exudates includes a wide variety of enzymes as well as one predominant protein (called **P-protein**) that is unique to the translocating cells. We will return to a discussion of P-protein later in this chapter.

TABLE 9.2   The chemical composition of phloem exudate from stems of actively growing castor bean (*Ricinus communis*).

Organic	mg 1^{-1}
Sucrose	80–106
Protein	1.45–2.20
Amino acids	5.2
Malic acid	2.0–3.2
Inorganic	**meq 1^{-1}**
Anions (inorganic)	20–30
Cations (inorganic)	74–138
Total dry matter	100–125 mg 1^{-1}

Data from Hall and Baker, 1972.

The principal constituent of phloem exudate in most species is sugar. In castor bean it is sucrose, which comprises approximately 80 percent of the dry matter (Table 9.2). Such a preponderance of sucrose in the translocation stream strongly suggests that it is the predominant form of translocatable photoassimilate. This suggestion has been amply confirmed by labeling experiments. In the example of translocation in sugarbeet petioles described earlier, more than 90 percent of the radioactivity, following 10 minutes of labeling with $^{14}CO_2$, was recovered as sucrose. There are exceptions to this rule—one is the squash family (Cucurbitaceae), where nitrogenous compounds (principally amino acids) are quantitatively more important—but overall sugar, particularly sucrose, accounts for the bulk of the translocated carbon. A survey of over 500 species representing approximately 100 dicotyledonous families confirms that sucrose is almost universal as the dominant sugar in the phloem stream.

A small number of families translocate, in addition to sucrose, oligosaccharides of the raffinose series (raffinose, stachyose, or verbascose) (Figure 9.7). Stachyose, for example, accounts for about 46 percent of the sugars in stem internodes of *Cucurbita maxima*. Yet other families (Oleaceae, Rosaceae) translocate some of their photoassimilates as the sugar alcohols mannitol or sorbitol.

It is interesting to speculate on why sucrose is the preferred vehicle for long-distance translocation of photoassimilate. One possibility is that sucrose, a disaccharide, and its related oligosaccharides are nonreducing sugars. On the other hand, all monosaccharides, including glucose and fructose, are **reducing sugars**. Reducing sugars have a free aldehyde or ketone group that is capable of reducing mild oxidizing agents. Some oligosaccharides, such as sucrose, are *nonreducing* sugars because the acetal link between the subunits is stable and nonreactive in alkaline solution. The exclusive use

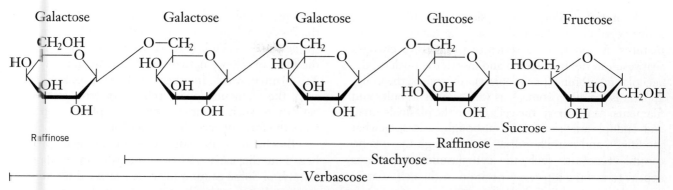

FIGURE 9.7 Sugars of the raffinose series. Raffinose, stachyose, and verbascose consist of sucrose with 1, 2, or 3 galactose units, respectively. All sugars in the raffinose series, including sucrose, are nonreducing sugars.

of nonreducing sugars in the translocation of photoassimilate may be related to this greater chemical stability. Nonreducing sugars are less likely to react with other substances along the way. Indeed, free glucose and fructose, both reducing sugars, are rarely found in phloem exudates. The occasional report of reducing sugars in phloem exudate probably indicates contamination by nonconducting phloem cells, where reducing sugars are readily formed by hydrolysis of sucrose or other oligosaccharides.

A second possible factor is that the β-fructoside linkage between glucose and fructose, a feature of sucrose and other members of the raffinose series, has a relatively high negative free energy of hydrolysis—about -27kJ mol^{-1} compared with about -31kJ mol^{-1} for ATP. Sucrose is thus a small and highly mobile but relatively stable packet of energy, which may account for its "selection" as the principal form of assimilate to be translocated in most plants.

9.5 SIEVE ELEMENTS ARE THE PRINCIPAL CELLULAR CONSTITUENTS OF THE PHLOEM

The distinguishing feature of phloem tissue is the conducting cell called the **sieve element**. Also known as a **sieve tube**, the sieve element is an elongated rank of individual cells, called **sieve-tube members**, arranged end-to-end (Figure 9.8). Unlike xylem tracheary elements, phloem sieve elements lack rigid walls and contain living protoplasts when mature and functional. The protoplasts of contiguous sieve elements are interconnected through specialized **sieve areas** in adjacent walls. Where the pores of the sieve area are relatively large and are found grouped in a specific area, they are known as **sieve plates** (Figure 9.8). Sieve plates are typically found in the end walls of sieve-tube members and provide a high degree of protoplasmic continuity

between consecutive sieve-tube members. Additional pores are found in sieve areas located in lateral walls. These are generally smaller and are not, as a rule, grouped in distinct areas. These sieve areas nonetheless provide cytoplasmic continuity through the lateral walls of adjacent sieve elements.

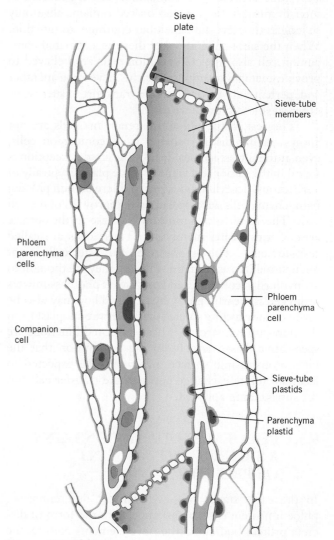

FIGURE 9.8 Phloem tissue from the stem of tobacco.

As noted earlier, mature sieve elements contain active cytoplasm. However, as the sieve element matures, it undergoes a series of progressive changes that result in the breakdown and loss of the nucleus, the vacuolar membrane (or tonoplast), ribosomes, the Golgi apparatus (or dictyosomes), as well as microtubules and filaments. At maturity, the cells retain the plasmalemma, and endoplasmic reticulum (although it is somewhat modified), and mitochondria. Even though there is no central vacuole as such, the cytoplasmic components appear to assume a parietal position in the cell, that is, along the inner wall of the cell.

In addition to sieve elements, phloem tissue also contains a variety of parenchyma cells. Some of these cells are intimately associated with the sieve-tube members and for this reason are called **companion cells**. Companion cells (Figure 9.8) contain a full complement of cytoplasm and cellular organelles. A companion cell is derived from the same mother cell as its associated sieve-tube member and shares numerous cytoplasmic connections with it. The interdependence of the sieve-tube member and companion cell is reflected in their lifetimes—the companion cell remains alive only so long as the sieve-tube member continues to function. When the sieve-tube member dies, its associated companion cell also dies. Companion cells are believed to provide metabolic support for the sieve-tube member and, perhaps, are involved in the transport of sucrose or other sugars into the sieve tube.

The rest of the phloem parenchyma cells are not always readily distinguishable from companion cells, even at the ultrastructural level. The single exception is found in the minor leaf veins of some plants, typically of herbaceous dicotyledonous plants. Here certain phloem parenchyma cells develop extensive ingrowths of the cell wall. The result is a significant increase in the surface area of the plasma membrane. These cells are called **transfer cells**. The precise role of transfer cells is not understood but, as the name implies, they are thought to be involved in collecting and passing on photoassimilates produced in nearby mesophyll cells. They may also be involved in recycling solutes that enter the apoplast from the transpiration stream. These proposed functions are speculative, based largely on the assumption that the high protoplasmic surface area would be expected to facilitate solute exchange between the transfer cell and the surrounding apoplast.

9.5.1 PHLOEM EXUDATE CONTAINS A SIGNIFICANT AMOUNT OF PROTEIN

In the early stages of sieve element differentiation, phloem protein (P-protein) appears in the form of discrete protein bodies. As the sieve elements mature, the P-protein bodies continue to enlarge. At the time the nucleus, vacuole, and other cellular organelles disappear, the P-protein bodies disperse in the cytoplasm. In some species, such as maple (*Acer rubrum*), the P-protein takes the form of a loose network of filaments, ranging from 2 to 20 nm in width. In others, such as tobacco (*Nicotiana* sps.), the filaments appear tubular in cross-section. In yet others, such as some leguminous plants, P-protein takes the form of crystalline inclusions.

Biochemical investigations of phloem proteins began in the early 1970s, principally in exudates of *Cucurbita*. Some caution must be exercised when interpreting these results, however, since phloem exudates contain proteins in addition to P-protein. Using the technique of *sodium dodecyl sulphate polyacrylamide gel electrophoresis* (**SDS-PAGE**) (Chapter 7), a variety of polypeptide subunits with molecular mass values ranging from 15 to 220 kD have been reported. Apparently phloem protein varies widely between species, with respect to both its subunit composition and its chemical properties. One particularly interesting property of phloem protein is its capacity to form a gel. Gelation could be prevented by 2-mercaptoethanol, a reducing agent that prevents formation of intermolecular disulfide (—S—S—) bonds. The effect of reducing agent is fully reversible—removal of the 2-mercaptoethanol allows gelling to proceed. This effect was traced to a single basic protein in the phloem exudate. This protein probably accounts for the propensity of certain phloem exudates, such as from *Cucurbita*, to gel rapidly on exposure to air.

P-protein has been the subject of considerable attention over the years because of its prominence in sieve elements and its propensity to plug the pores in the sieve plates. Still, its role and that of other phloem-specific proteins is not yet clear. P-protein has been implicated in various ways in the transport function of sieve elements. According to some theories, P-protein is considered an active participant in the transport process. At the same time, the presence of P-protein in sieve elements is invoked as an argument against other theories. It is now generally accepted that, *in intact, functioning sieve elements*, P-protein is located principally along the inner wall of the sieve element and does not plug the sieve plate.

The formation of plugs in the sieve plates occurs only when the sieve element is injured. This occurs because the sieve element is normally under *positive* hydrostatic pressure, as evidenced by the continued flow of exudate from aphid stylets. When the pressure is released through injury to the sieve element, the contents, including P-protein, surge toward the site of injury. This results in the accumulation of P-protein, possibly assisted by its gelling properties, as "slime" plugs on the side of the sieve plate away from the pressure release. Thus, it appears that at least one function of P-protein is protective. By sealing off sieve plates in areas where the integrity of the phloem has been breached, P-protein helps to maintain the positive hydrostatic pressure in the phloem and reduce unnecessary loss of translocated photoassimilate.

Another prominent and somewhat controversial feature of sieve elements is the presence of **callose**. Callose, a $\beta1\rightarrow3$-glucan, is related to starch and cellulose. Small amounts of callose are deposited on the surface of the sieve plate or line the pores through which the interconnecting strands of cytoplasm pass between contiguous cells (Figure 9.8). Controversy over the role of callose arises from the frequent observation that callose appears to accumulate in the pores to the extent that it would appear to interfere with translocation. However, it is now known that callose can be synthesized very rapidly (within a matter of seconds) and, similar to P-protein, will accumulate in the sieve area in response to injury. Large amounts of callose also appear to be deposited on the sieve plates of older, nonfunctional sieve elements. In both cases, the function of callose appears to be one of sealing off sieve elements that have been injured or are no longer functional, thus preserving the integrity of the translocating system.

9.6 DIRECTION OF TRANSLOCATION IS DETERMINED BY SOURCE-SINK RELATIONSHIPS

Identification of an organ or tissue as a source or sink depends on the direction of its *net* assimilate transport. An organ or tissue that produces more assimilate than it requires for its own metabolism and growth is a *source*. A source is thus a net exporter or producer of photoassimilate that is, it exports more assimilate than it imports. Mature leaves and other actively photosynthesizing tissues are the predominant sources in most plants. A *sink*, on the other hand, is a net importer or consumer of photoassimilate. Roots, stem tissues, and developing fruits are examples of organs and tissues that normally function as sinks. *The underlying principle of phloem translocation is that photoassimilates are translocated from a source to a sink.* Sink organs may respire the photoassimilate, use it to build cytoplasm and cellular structure, or place it into storage as starch or other carbohydrate.

Any organ, at one time or another in its development, will function as a sink and may undergo a conversion from sink to source. Leaves are an excellent example. In its early stages of development a leaf will function as a sink, drawing photoassimilates from older leaves to support its active metabolism and rapid enlargement. However, as a leaf approaches maximum size and its growth rate slows, its own metabolic demands

diminish and it will gradually switch over to a net exporter. The mature leaf then serves as a source of photoassimilate for sinks elsewhere in the plant. The conversion of a leaf from sink to source is a gradual process, paralleling the progressive maturation of leaf tissue. In simple leaves, for example, the export of photoassimilate from mature regions of the leaf may begin while other regions are still developing and functioning as sinks. In compound leaves, such as ash (*Fraxinus pennsylvanica*) and honeylocust (*Gleditsia triacanthos*), the early maturing basal leaflets may export photoassimilate to the still-developing distal leaflets as well as out of the leaf.

9.7 PHLOEM TRANSLOCATION OCCURS BY MASS TRANSFER

What is the mechanism for assimilate translocation over long distances through the phloem? Any comprehensive theory must take into account a number of factors. These include: (1) the structure of sieve elements, including the presence of active cytoplasm, P-protein, and resistances imposed by sieve plates; (2) observed rapid rates of translocation (50 to 250 cm hr^{-1}) over long distances; (3) translocation in different directions at the same time; (4) the initial transfer of assimilate from leaf mesophyll cells into sieve elements of the leaf minor veins (called **phloem loading**); and (5) final transfer of assimilate out of the sieve elements into target cells (called **phloem unloading**). Phloem loading and unloading will be discussed in the following section.

At various times assimilate transport has been explained in terms of simple diffusion, cytoplasmic streaming, ion pumps operating across the sieve plate, and contractile elements in the transcellular protoplasmic strands. All of these proposals have been largely rejected on both theoretical and experimental grounds.

The most credible and generally accepted model for phloem translocation is one of the earliest. Originally proposed by E. Münch in 1930 but modified by a series of investigators since, the **pressure-flow** hypothesis remains the simplest model and continues to earn widespread support among plant physiologists. The pressure-flow mechanism is based on the mass transfer of solute from source to sink along a hydrostatic (turgor) pressure gradient (Figure 9.9). Translocation of solute in the phloem is closely linked to the flow of water in the transpiration stream and a continuous recirculation of water in the plant (Chapter 2).

Assimilate translocation begins with the loading of sugars into sieve elements at the source. Typically, loading would occur in the minor veins of a leaf, close to a photosynthetic mesophyll or bundle-sheath cell. The increased solute concentration in the sieve element lowers its water potential (Chapter 1) and, consequently,

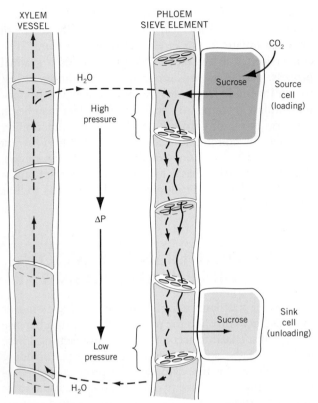

FIGURE 9.9 **A diagram of pressure flow. The loading of sugar into the sieve element adjacent to a source cell causes the osmotic uptake of water from nearby xylem elements. The uptake of water increases the hydrostatic (turgor) pressure in the sieve element. The pressure is lowered at the sink end when sugar is unloaded into the receiver cell and the water returns to the xylem. This pressure differential causes a flow of water from the source region to the sink. Sugar is carried passively along.**

designed to discriminate between energy requirements for the actual movement of assimilate within the sieve elements and the global energy requirements for translocation from source to sink. The results are clear. The effects of low temperature and metabolic inhibitors are either transient or cause disruption of the P-protein and plug the sieve plates. Energy requirements for translocation within the sieve elements are therefore minimal and compatible with the passive character of the pressure-flow hypothesis. The energy requirement indicated in earlier translocation experiments no doubt reflected the needs for loading and unloading the sieve elements.

The principle of pressure flow can be easily demonstrated in the laboratory by connecting two osmometers (Figure 9.10), but a simple physical demonstration does not in itself prove the hypothesis. A number of questions must be answered. First, is the sieve tube under pressure? The prolonged exudation of phloem sap from excised aphid stylets clearly demonstrates that it is. The total volume of exudate may exceed the volume of an individual sieve tube by several thousand times. It is difficult to measure the turgor pressure of individual sieve

is accompanied by the osmotic uptake of water from the nearby xylem. This establishes a higher turgor or hydrostatic pressure (Chapter 1) in the sieve element at the source end. At the same time, sugar is unloaded at the sink end—a root or stem storage cell, for example. The hydrostatic pressure at the sink end is lowered as water leaves the sieve elements and returns to the xylem. So long as assimilates continue to be loaded at the source and unloaded at the sink, this pressure differential will be maintained, water will continue to move in at the source and out at the sink, and assimilate will be carried passively along.

According to the pressure-flow hypothesis, solute translocation in the phloem is fundamentally a passive process; that is, translocation requires no direct input of metabolic energy to make it function. Yet for years it has been observed that translocation of assimilates was sensitive to metabolic inhibitors, temperature, and other conditions, suggesting that metabolic energy was required. More recent experiments, however, have been

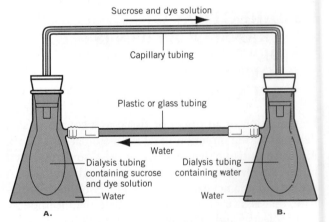

FIGURE 9.10 **A physical model of the pressure-flow hypothesis for translocation in the phloem. Two osmometers are constructed from side-arm flasks and dialysis tubing. Osmometer A (the source) initially contains a concentrated sucrose solution and a dye. Osmometer B (the sink) contains only water. The two osmometers are connected by capillary tubing (the phloem). Water moves into osmometer A by osmosis, generating a hydrostatic pressure that forces water out of osmometer B. Water returns via the tubing (the xylem) that connects the side-arm flasks. As a consequence of the flow of water between the two ends of the system, the sucrose-dye solution flows through the capillary from osmometer A to osmometer B. In the model, the system will come to equilibrium and flow will cease when the sucrose concentration is equal in the two osmometers. In the plant, flow is maintained because sucrose is continually added to the source (A) and withdrawn at the sink (B).**

elements, although a number of attempts have been made over the years. Turgor pressure can be calculated as the difference between sieve tube water potential (Ψ) and osmotic potential (Ψs) (Chapter 1), or it may be measured directly by inserting a small pressure-sensing device, or micromanometer, into the phloem tissue. For example, the turgor pressure in willow saplings can be measured by sealing a closed glass capillary over a severed aphid stylet. Pressure is calculated from the ratio of the compressed and uncompressed air columns in the capillary. As might be expected, values reported in the literature range widely, depending on the method chosen, plant material, the time of day, and physiological status of the subject plant. Whether calculated or measured directly, values of 0.1 MPa to 2.5 MPa are typical.

A second question to be addressed is whether differences in sugar concentration and the turgor pressure drop in the sieve tube are sufficient to account for the measured rates of transport. Sugar concentration is, of course, highly variable, depending on the rate of photosynthesis and the general physiological condition of the plant. However, most studies have confirmed that the sugar content of phloem exudate taken near the source is higher than in exudates taken near sinks. It has been calculated that a pressure drop of about 0.06 MPa m^{-1} would be required for a 10 percent sucrose solution to flow at 100 cm hr^{-1} through a sieve tube with a radius of 12 µm. In these calculations, the resistance offered by sieve plates was taken into account by assuming that (1) the area of the pores in the sieve plate was equal to one-half the area of the sieve tube, (2) there were 60 sieve plates per cm of sieve tube, and (3) the sieve plate pores were not blocked. Assuming that the turgor pressure of sieve tubes in the source regions is typically in the range 1.0 to 1.5 MPa, and that it is zero in the sink (which may not be true), a pressure drop of 0.06 MPa m^{-1} would be sufficient to push a solution through the sieve tubes over a distance of 15 to 25 m. Flow over longer distances could be accomplished if the source sucrose concentration were higher and/or the flow rate were reduced. For example, assimilates can move from the source to the sink, at a velocity of 48 cm hr^{-1}. A pressure drop of 0.2 MPa would be required to achieve this velocity if the sieve plate pores were completely open. From the sucrose concentration in the source and sink, it can be calculated that the actual pressure drop was 0.44 MPa, twice that required! A pressure drop of 0.44 MPa would be sufficient to accomplish a velocity of 48 cm hr^{-1} even if the pores were only 70 to 75 percent open.

Another question that is frequently raised in discussions of the pressure-flow hypothesis is that of bidirectional transport. The translocation of assimilates simultaneously in opposite directions would at first seem incompatible with the pressure-flow hypothesis, but it does occur. Bidirectional transport is first of all

a logical necessity. At any one time, plants will likely have more than one sink being served by the same source—roots for metabolism and storage and developing apical meristems or flowers, for example. It is also easy to demonstrate experimentally the movement of radiolabeled carbon and phosphorous in opposite directions through the same internode or petiole at the same time. This observation might easily be explained by movement through two separate vascular bundles or even through different sieve tubes in the same bundle. As long as the sieve elements are connected to different sinks, the pressure-flow hypothesis does not require that translocation occur in the same direction or even at the same velocity at any one time.

Finally, it has often been argued that sieve elements, because of their structure and composition, offer a substantial resistance to flow and that pressure-flow might not provide sufficient force to overcome this resistance. In this regard it is important to note once again that the sieve plates *in functioning sieve tubes* are not occluded by either P-protein or callose. The presence of viscous cytoplasm and sieve plates undoubtedly imposes some resistance, but a variety of experiments have indicated that the capacity of the phloem to translocate assimilates is not normally a limiting factor in the growth of sinks. The phloem is a flexible system for translocation. It is easily capable of bypassing localized regions of high resistance and the hydrostatic pressure can be adjusted in response to demand at either the source or sink. There are even developmental controls, apparently to ensure that the phloem is adequately "sized" to meet anticipated demand. In wheat, for example, both the number and size of the vascular bundles serving a floral head correlates with the number of flowers. Thus, although there is little, if any, direct proof for the pressure-flow hypothesis, on the balance of evidence it is strongly favored.

9.8 PHLOEM LOADING AND UNLOADING REGULATE TRANSLOCATION AND PARTITIONING

A discussion of phloem translocation is not complete without considering how assimilates are translocated from the photosynthetic mesophyll cells into the sieve elements at the source end (*phloem loading*) or from the sieve elements into the target cells at the sink end (*phloem unloading*).

9.8.1 PHLOEM LOADING CAN OCCUR SYMPLASTICALLY OR APOPLASTICALLY

The path traversed by assimilate from the site of photosynthesis to the sieve element is not long. Most mesophyll cells are within a few tenths of a mm, at most three or four cells' distance from a minor vein ending where loading of assimilate into the **sieve element-companion cell complex (se-cc)** actually occurs.[1] It is generally agreed that sucrose moves from the mesophyll cells to the phloem, probably phloem parenchyma cells, principally by diffusion through the plasmodesmata (i.e., the **symplast**). At this point, the pathway becomes less certain and the subject of some debate. From the phloem parenchyma there are two possible routes into the se-cc complex (Figure 9.11). Sucrose may continue through the symplasm—that is, through plasmodesmata—directly into the se-cc complex. This route is known as the **symplastic pathway**. Alternatively, the sugar may be transported across the mesophyll cell membrane and released into the cell wall solution (i.e., the **apoplasm**). From there it would be taken up across the membrane of the se-cc complex where it enters the long-distance transport stream. This route is known as the **apoplastic pathway**.

The apoplastic model for phloem loading gained favor in the mid-1970s, based largely on studies of translocation in sugarbeet leaves. D. R. Geiger and his coworkers used leaves that had been abraded with carborundum to remove the cuticle, thus improving access to the leaf apoplast. They found that radioactive sucrose appeared in the apoplast following a period of photosynthesis in the presence of $^{14}CO_2$. They also found that exogenously supplied sugar was readily absorbed into the se-cc complex when abraded leaves were bathed in a solution containing ^{14}C-sucrose. These results indicate that sucrose is normally found in the apoplast and can be taken into the sieve elements from the apoplast. Phloem loading in some plants is also inhibited by chemicals such as *p-chloromercuribenzene sulfonic acid* (**PCMBS**) when applied to abraded leaves or leaf disks. PCMBS and certain other sulfhydryl-specific reagents presumably interfere with carrier proteins (Chapter 3) involved in the transport of sucrose across the plasma membranes. Because these reagents do not penetrate the cell membrane, any effect they have must be localized on the apoplastic surface of the membrane.

Sucrose and other sugars are selectively loaded into the se-cc complex against a concentration gradient, which usually implies active transport. In addition, there is an increasingly large body of evidence supporting the existence, in plant cells generally and phloem loading in particular, of a sucrose-uptake mechanism that is both ATP-dependent and linked to the uptake of protons; that is, a sugar-H^+ cotransport (Figure 9.12). This con-

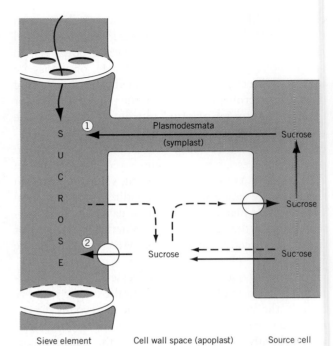

FIGURE 9.11 Loading and retrieval of sugars at the source. Sucrose may be loaded into sieve elements at minor vein endings via one of two pathways. In pathway 1, the symplastic pathway, sugar moves through plasmodesmata that connect the protoplasts of the source cell and the sieve element. In pathway 2, the apoplastic pathway, sugar is released into the cell wall (apoplastic) space, from which it is actively transported across the plasma membrane of the sieve element by sugar-H^+ cotransport. Alternatively, sugar may leak into the apoplast (dashed lines) and be actively retrieved by either the source cell or the sieve element.

clusion is supported by the observation that sugar uptake is accompanied by an increase in pH (i.e., a depletion of protons) or polarity changes in the apoplast. Conversely, if the pH is experimentally increased—that is, protons are removed from the apoplast by infiltrating the apoplast of abraded leaves with a basic buffer—uptake into the phloem cells will be inhibited. Finally, in most cases it can be shown that only sugars taken up into

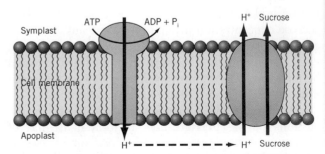

FIGURE 9.12 An illustration of sugar-H^+ cotransport. The energy for sugar uptake may be provided by a plasma membrane ATPase proton pump.

---

[1]It is not technically possible to discriminate between the respective roles of the companion cell and the sieve element in phloem loading and unloading experiments. For this reason, the sieve element and companion cells are considered as a single se-cc complex.

the se-cc complex and translocated in the sieve elements will elicit pH responses. Other sugars, which are not normally taken up by the phloem, elicit no pH changes. In summary, much of the evidence is consistent with an apoplastic pathway for phloem loading, by which sugars first pass from mesophyll or phloem parenchyma cells into the leaf apoplast. The sugar is subsequently taken up into the se-cc complex by sugar-H$^+$ symport, to be translocated out of the source region.

Although the exact nature of the sucrose-H$^+$ symport carrier is not yet worked out, genes (*SUT1, SUC2*) coding for the carrier have been identified and cloned from several species. These include spinach (*Spinacea oleraceae*), potato (*Solanum tuberosum*), *Plantago major*, and *Arabidopsis thaliana*. In several species, expression of the *SUC2* sucrose carrier gene appears limited to the companion cells, while in potato the *SUT1* gene product was located in the plasma membrane of the sieve elements but could not be detected within the companion cells. Such conflicting results might suggest that apoplastic loading occurs differently in different species or, alternatively, that different carriers are active in companion cells and sieve elements.

In spite of the strong evidence in support of an apoplastic pathway for phloem loading, there are other data that support transport through the symplast. Much of the data from sucrose feeding experiments, for example, indicate that a significant proportion—up to 50 percent—of the radioactive sucrose can be detected in *leaf mesophyll cells*. This leaves the experiments open to an alternate interpretation: that sucrose-H$^+$ cotransport exists as a mechanism for retrieving sucrose that has leaked from the photosynthetic cells into the apoplast. Such a retrieval mechanism would not discriminate between sugar that leaked from mesophyll, or possibly the se-cc complex, and sugar that was supplied exogenously by the experimenter. If such leakage does occur, a mechanism for retrieval would serve to prevent unnecessary loss of sugar from the transport stream. Indeed it has been postulated that a leakage-retrieval cycle occurs normally along the entire length of the translocation path.

If the uptake data do reflect retrieval by mesophyll cells rather than phloem loading, one is left with the conclusion that phloem loading occurs via the symplastic pathway. Several laboratories have presented evidence that appears to offer further support for this hypothesis. When mesophyll cells of *Ipomea tricolor* are injected with a fluorescent dye, the dye moves readily into neighboring mesophyll cells and appears in the minor veins within 25 minutes. Since the dye is water-soluble and unable to cross membranes, it is assumed that the dye traveled into the minor veins via the symplastic connection between cells. It is also of interest that much of the data supporting apoplastic loading have come from experiments with one species: sugarbeet. In a recent survey, plants were selected on the basis of whether they had abundant symplastic connections between the se-cc complex and adjacent cells of the minor veins, or whether these cells were *symplastically isolated*, that is, had no symplastic connections. Those plants whose se-cc complexes were symplastically isolated exhibited characteristics of apoplastic loading, while those with abundant symplastic connections exhibited characteristics of symplastic loading.

The concept of symplastic loading does, however, raise some questions. For example, if sugars diffuse freely from the mesophyll into the se-cc complex, they should be equally free to diffuse back into the mesophyll cells. How, then, is it possible for the se-cc complex to accumulate sugars by simple diffusion through the plasmodesmata? Based on studies of phloem loading in *Cucurbita* sps., a **polymer trap** model to account for symplastic loading has been proposed. Species, such as the cucurbits, which have abundant plasmodesmata connections with the se-cc complex and appear to load symplastically, also translocate oligosaccharides in the raffinose series. According to the polymer trap model, sucrose diffuses from the mesophyll or bundle-sheath cells into the companion cells through the connecting plasmodesmata. In the companion cell, the sucrose is converted to an oligosaccharide, such as the tetrasaccharide stachyose, which is too large to diffuse back through the plasmodesmata. The polymer (i.e., stachyose) thus remains "trapped" in the se-cc complex, to be carried away by mass flow.

The symplastic model assumes that the plasmodesmata limit the passage of large molecules, but this may not be the case. Several recent studies of the sucrose transporter gene have indicated that both the transporter protein and its mRNA are able to pass through plasmodesmata between companion cells and sieve elements. If macromolecules can pass through plasmodesmata, it is difficult to imagine why small oligosaccharides cannot. Perhaps plasmodesmata are more than simple tubes allowing solute flux between cells. This is an exciting issue that will no doubt receive considerable attention in the future.

Why there is more than one pathway for phloem loading is not clear. The symplastic pathway appears to have an energetic advantage by avoiding two carrier-dependent membrane transport steps. The observed energy dependence of loading and translocation, however, is more readily explained by the apoplastic model. In the sympastic model, on the other hand, energy is required for the synthesis of oligosaccharides in the companion cells. It has also been suggested that species employing the symplastic pathway are more ancestral or that the apoplastic pathway is an evolutionary adaptation that arose as plants spread from tropical climates into more temperate regions. The new molecular approaches now available will no doubt allow investigators to discriminate between available options. It may be that there is no universal pathway but that the path of phloem loading is family- or species-specific. Given the theoretical and potential practical significance of phloem loading in determining yields, we can expect the investigation and debate to continue.

9.8.2 PHLOEM UNLOADING MAY OCCUR SYMPLASTICALLY OR APOPLASTICALLY

Once assimilate has reached its target sink, it must be unloaded from the se-cc complex into the cells of the sink tissue. In principle, the problem is similar to loading; only the direction varies. In detail there are some significant differences. As with phloem loading, phloem unloading may occur via symplastic or apoplastic routes (see Figure 9.13). The symplastic route (pathway 1) has been described predominantly in young, developing leaves and root tips. Sucrose flows, via interconnecting plasmodesmata, down a concentration gradient from the se-cc complex to sites of metabolism in the sink. The gradient and, consequently, flow into the sink cell is maintained by hydrolyzing the sucrose to glucose and fructose.

There are two possible apoplastic routes, shown as pathways 2 and 3 in Figure 9.13. Pathway 2, which has been studied most extensively in the storage parenchyma cells of sugarcane, involves the release of sucrose from the se-cc complex into the apoplast. Release is insensitive to metabolic inhibitors or PCMBS and therefore does not involve an energy-dependent carrier. Once in the apoplast, sucrose is hydrolyzed by the enzyme **acid invertase**, which is tightly bound to the cell wall and catalyzes the reaction:

$$\text{Sucrose} + H_2O \rightarrow \text{glucose} + \text{fructose} \qquad (9.16)$$

This reaction is essentially irreversible and the hydrolysis products, glucose and fructose, are actively taken up by the sink cell. Once in the cell, they are again combined as sucrose and actively transported into the vacuole for storage. Hydrolysis of sucrose in the apoplast, perhaps combined with the irreversibility of the acid invertase reaction, serves to maintain the gradient and allows the unloading to continue. This pathway seems to be prominent in seeds of maize, sorghum, and pearl millet.

The third pathway for phloem unloading indicates that, at least in legumes, sucrose is unloaded into the apoplast by an energy-dependent carrier. The nature of the carrier has not been conclusively identified, but evidence to date suggests it is probably the same sucrose-H^+ cotransporter described earlier. As with phloem loading, there does not appear to be a universal path for phloem unloading into the developing embryo.

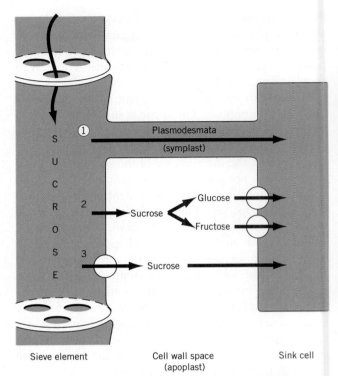

FIGURE 9.13 **Three possible routes for sugar unloading into sink cells. In all three possible routes, a favorable diffusion gradient is maintained by metabolizing the sugar once it enters the sink cell.**

9.9 PHOTOASSIMILATE IS DISTRIBUTED BETWEEN DIFFERENT METABOLIC PATHWAYS AND PLANT ORGANS

Some of the newly fixed carbon or photoassimilate in a source leaf is retained within the leaf, and the rest is distributed to various nonphotosynthetic tissues and organs. This raises several interesting questions. What, for example, determines how much carbon is retained and in what form? What determines how much is exported and to where? What determines how much assimilate, for example, is exported to the roots of a wheat or corn plant and how much is translocated to fill the developing grain? Questions of this sort have been receiving increasing attention of late, because the patterns of distribution, or more to the point, *regulation* of the distribution patterns is highly significant with respect to productivity and yield (Chapter 12). One maize farmer may wish to maximize grain yield while another may require more of the carbon be put into production of vegetative (i.e., leafy) material. Each farmer will assess the **harvest index** (the ratio of usable plant material to total biomass) for the crop in a different way.

The traditional route to improving harvest index has been through breeding and selection. The uncultivated progenitors of modern-day wheat and maize, for example, produced sparse heads with small seeds. Centuries of agricultural selection and, in the last century, careful breeding, have been required to produce the high-yielding wheat and maize varieties in use today. However, the more we learn about the factors regulating carbon distribution and utilization, the greater the prospects for using modern genetic methods to manipulate the harvest index. The distribution of photoassimilate occurs at two levels: **allocation** and **partitioning**. Each of these will be discussed in turn.

9.9.1 PHOTOASSIMILATES MAY BE ALLOCATED TO A VARIETY OF METABOLIC FUNCTIONS IN THE SOURCE OR THE SINK

Allocation refers to the *metabolic* fate of carbon either newly assimilated in the source leaf or delivered to a sink. At the source, there are three principal uses for photoassimilate: leaf metabolism and maintenance of leaf biomass, short-term storage, or export to other parts of the plant.

9.9.1.1 *Leaf metabolism and biomass* Some of the carbon will be allocated to the immediate metabolic needs of the leaf itself. These needs include the maintenance of cell structure, synthesis of additional leaf biomass, and the maintenance of the photosynthetic system itself. Most of this carbon is metabolized through respiration, which provides both the energy and carbon skeletons necessary to support ongoing synthetic activities.

9.9.1.2 *Storage* Under normal light–dark regimes, plants face a dilemma—photosynthesis is restricted to the daylight hours, but a supply of photoassimilate for growth must be maintained over the entire 24 hours. A partial solution to this dilemma is to allocate a portion of the newly fixed carbon for storage in the leaves. Most plants, especially dicots, store the bulk of their carbon as starch, with a smaller amount stored as sucrose. Some, such as barley (*Hordeum vulgare*), sugarcane (*Saccharum spontaneum*), and sugarbeet (*Beta vulgaris*), accumulate little if any starch but store carbon primarily as sucrose in the vacuoles of leaf, stem, or root cells, respectively. Many grasses accumulate fructose polymers called **fructans**. Carbon stored in the leaves serves primarily as a buffer against fluctuations in metabolite levels and is available for reallocation to metabolism when required.

Alternatively, most plants appear to be programmed to maintain a fairly constant rate of translocation and supply to sink tissues. Leaf reserves are therefore available for reallocation to export at night or during periods of stress when photosynthesis is very low. In plants that store both starch and sucrose, there are generally two pools of sucrose, one in the cytoplasm and one in the vacuole. The vacuolar pool, which is larger and turns over more slowly than the cytoplasmic pool, is the first source of sucrose for export at night. Only when the vacuolar pool is depleted will the starch, stored in the chloroplast, be mobilized for export.

9.9.1.3 *Export from the leaf* Normally about half the newly assimilated carbon is allocated for immediate export from the leaf via the phloem. In many plants, a portion of this exported carbon may be stored along the translocation path. As in the leaf, this stored carbohydrate helps to buffer the carbon supply at times when the rate of translocation through the phloem might otherwise be reduced.

Regulating the allocation of photoassimilate is a complex process, involving the interactions of a number of metabolic pathways. Allocation within a source leaf is to a large extent genetically programmed but there is a strong developmental component. Young leaves, for example, retain a large proportion of their newly fixed carbon for growth, but as leaves mature the proportion allocated for export increases. In soybean leaves there are corresponding changes in the activities of enzymes such as acid invertase (Chapter 10) and sucrose synthase (Equation 9.11). The activities of these two degradative enzymes are highest in young, rapidly expanding leaves, which no doubt reflects the need to metabolize sucrose in the early stages of leaf development when the leaf is functioning primarily as a sink.

As a leaf matures and becomes photosynthetically self-sufficient, both its need and capacity to import assimilate decline and the metabolism of the leaf switches over to the synthesis of sucrose for export. There is a corresponding decline in the activities of acid invertase and sucrose synthase and a steady increase in the activity of sucrose phosphate synthase (SPS), a key enzyme in the synthesis of sucrose (Equation 9.6). Because sucrose is the predominant form of translocated carbohydrate and SPS activity is closely correlated with sucrose production, the increase in SPS activity may be a critical factor in determining the transition of the leaf from a sink to a source.

The allocation of photoassimilate between storage and export has been extensively described, but there are few answers to the question of how this allocation is regulated. In most plants, the level of starch fluctuates on a daily basis—increasing during the light period and declining at night. The rate of sucrose export exhibits similar, but less extreme, diurnal fluctuation. The distribution of carbon between starch and sucrose depends primarily on the allocation of triose phosphate between starch synthesis in the chloroplast and sucrose synthesis in the cytoplasm.

Since metabolic regulation of starch and sucrose biosynthesis involves the two key enzymes, fructose-1,6-bisphosphatase (FBPase) and SPS, it is reasonable to expect that factors that influence allocation do so at least in part by influencing the activities of these two enzymes. There are some data to bear out these expectations. In cotton leaves, for example, there is a strong correlation between SPS activity, sucrose content, and export of carbon. All three increase more or less in concert during the photoperiod and drop precipitously at the beginning of the dark period. During the dark period, sucrose content and SPS activity remain low, but the drop in export activity is only transient. The pattern of export recovery during the dark period corresponds very closely with the pattern of starch mobilization. Although there is considerable variation in timing and magnitude, similar diurnal fluctuations in carbon metabolites and enzymes have been found in other species. It thus appears that during periods of active photosynthesis, carbon allocation is largely determined by the activity of SPS. At night, the determining factor appears to be the breakdown of reserve starch.

The single most consistent aspect of source leaf allocation, however, is the generally steady rate of export. Except for transient increases at "dawn" or "dusk," diurnal fluctuations in export are small or nonexistent. Apparently plants are programmed to maintain a steady rate of assimilate translocation over the entire 24-hour period. Whether this program is imposed by photoperiod or some other factor is not known. An understanding of how allocation is regulated in source leaves awaits further investigation.

9.9.2 DISTRIBUTION OF PHOTOASSIMILATES BETWEEN COMPETING SINKS IS DETERMINED BY SINK STRENGTH

The distribution of assimilate between sinks is referred to as **partitioning**. In a vegetative plant, the principal sinks are the meristem and developing leaves at the shoot apex, roots, and nonphotosynthetic stem tissues. With the onset of reproductive growth, the development of flowers, fruits, and seeds creates additional sinks. In general, sinks are competitive and the photoassimilate is partitioned to all active sinks. If the number of sinks is reduced, a correspondingly higher proportion of the photoassimilate is directed to each of the remaining sinks. This is the basis for the common practice of pruning fruit trees to ensure a smaller number of fruit per tree. Partitioning the assimilate among a smaller number of fruit encourages the development of larger, more marketable fruit.

Partitioning of assimilate between competing sinks depends primarily on three factors: the nature of vascular connections between source and sinks, the proximity of the sink to the source, and sink strength. Translocation is clearly facilitated by direct vascular connections between the source leaf and the sink. Each leaf is connected to the main vascular system of the stem by a vascular trace, which diverts from the vascular tissue of the stem into the petiole. Experiments have shown that photoassimilate will move preferentially toward sink leaves above and *in line* (that is, in the same rank) with the source leaf. These sink leaves are most directly connected with the source leaf. Sink leaves not in the same rank, such as those on the opposite side of the stem, are less directly linked; the assimilate must make its way through extensive radial connections between sieve elements.

One of the more significant factors in determining the direction of translocation is **sink strength**. Sink strength is a measure of the capacity of a sink to accumulate metabolites. It is given as the product of sink size and sink activity:

$$\text{Sink strength} = \text{sink size} \times \text{sink activity} \quad (9.17)$$

Sink size is the total mass of the sink (usually as dry weight). **Sink activity** is the rate of uptake, or assimilate intake per unit dry weight of sink per unit time. Differences in sink strength can be measured experimentally, although it is not known exactly what determines sink strength or what causes sink strength to change with time. The rate of phloem unloading is surely a factor, as well as the rate of assimilate uptake by the sink and allocation to metabolism and storage within the sink. Environmental factors (e.g., temperature) and hormones will also have an impact to the extent that they influence the growth and differentiation of the sink tissue.

Photoassimilate from most source leaves is readily translocated in either vertical direction—upward toward the apex or downward toward the roots. All else being equal, however, there is a marked bias in favor of translocation toward the closest sink. In the vegetative plant, photoassimilate from young source leaves near the top of the plant is preferentially translocated toward the stem apex, while older, nonsenescent leaves near the base of the plant preferentially supply the roots. Intermediate leaves may translocate photoassimilate equally in both directions. The direction of translocation is probably related to the magnitude of the hydrostatic pressure gradient in the sieve elements. Given two equivalent sinks at different distances, the sink closest to the source will be served by the steeper pressure gradient. The bias in favor of the shorter translocation distance is sufficient to overcome even sink size.

Because sink strength is closely related to productivity and yield, most studies have been conducted with crop species—in particular the filling of grain in cereals such as wheat (*Triticum aestivum*) and maize (*Zea mays*). Developing grain is a particularly active sink and

Has a major impact on translocation patterns. From the time of anthesis, when the floral parts open to receive pollen, the developing grain becomes the dominant sink. The influence of developing grain on translocation patterns is illustrated by the results shown in Table 9.3. In this experiment the supply of photoassimilate was altered by reducing the supply of carbon dioxide and the dry-weight increase of various plant parts was monitored over the grain-filling period. Reducing photoassimilate supply had virtually no effect on grain weight, which means that a higher proportion of the carbon was translocated to the grain. The difference was made up by an equivalent decrease in the proportion of carbon directed to the roots. Roots and the developing grain are competing sinks. When the supply of photoassimilate is limited, it is preferentially directed toward the sink with the greater strength. The dominant role of developing grain as a sink is also shown by experiments with wheat. When photosynthesis was limited by lowering the light level, the proportion of ^{14}C-photoassimilate from the flag leaf (the leaf directly below the floral head) increased from 49 percent to 71 percent. In this case, however, the difference was made up by an equivalent reduction in the proportion translocated in the lower stem.

The above discussion indicates that sink strength is a significant factor in determining the pattern of translocation, but to suggest that sink strength alone is responsible for the partitioning of assimilate would be to grossly oversimplify the problem. At the very least, assimilate partitioning is a highly integrated system, depending upon interactions between the source leaf, the actively growing sinks, and the translocation path itself. We intuitively expect that such an integrated system will be subject to regulation at one or more points. However, beyond the observation that transport rate generally responds to sink demand—sudden changes in sink activity will cause corresponding changes in transport rate to that sink—relatively little is known about regulation of sink strength and interactions between sink strength and translocation rate.

Two factors that have been implicated in influencing sink strength are cell turgor and hormones. While investigating phloem exudate of castor bean (*Ricinus communis*), it was noted that the act of collecting exudate by making bark incisions, which causes a sudden reduction in the turgor pressure in the sieve elements, gave rise to a marked increase in sucrose loading at the source. Subsequently, through a series of experiments involving artificial manipulation of turgor, it was concluded that phloem loading is dependent on turgor pressure in the sieve elements. Turgor-dependent phloem loading now forms the basis for a relatively simple hypothesis to explain the regulation of transport rate by sink demand. When the se-cc complex is rapidly unloaded at the sink, the reduction in solute concentration causes a corresponding reduction in the hydrostatic pressure, or turgor, at the sink end of the sieve elements (refer to Figure 9.9). This reduced hydrostatic pressure will be transmitted throughout the interconnected system of sieve elements, quickly stimulating increased phloem loading at the source. The resulting increase in solute concentration at the source end of the system would serve to counter the drop in hydrostatic pressure, thus maintaining the pressure gradient and, in accordance with the pressure-flow mechanism, stimulating the flow of assimilate toward the sink. A reduction in sink demand would have the opposite result, leading to a lower rate of solute withdrawal and a higher turgor in the sieve elements. Loading at the source and the hydrostatic pressure gradient would be reduced, thereby lowering the rate of translocation. According to this model, changes in sieve-element turgor would be an important message in the long-distance communication between sinks and sources.

It is not known how the se-cc complex or the mesophyll cells sense changes in turgor. The mechanism by which pressure changes can be translated into changes in sucrose loading is also unknown. However, some experiments have demonstrated that sucrose transport across cell membranes of beet root tissue is turgor regulated, possibly by controlling the activity of an ATPase proton pump located in the plasma membrane.

Plant hormones (see Chapters 18–21) have been implicated in directing long-distance translocation, particularly with regard to redirection of assimilates to new sinks. Hormone-directed transport, however, may be simply an indirect consequence of hormone action. We know that hormones are one of several intrinsic factors involved in regulating the growth and development of organs. Through their influence on the size and metabolic activity of sink organs, hormones will undoubtedly influence sink strength and, as a result, translocation rates. The role of hormones is complicated by the fact that they may, at least in part, be

TABLE 9.3   Patterns of photoassimilate distribution in *Sorghum* plants subjected to high (400 μl l^{-1}) and low (250 μl l^{-1}) concentrations of carbon dioxide. Values are percentage of total dry-weight gain during the grain-filling period. Final grain weight was the same under the two conditions.

	Carbon Dioxide Level	
	High	Low
Grain	71.5	87
Roots	18	4
Other	10.5	9

Based on the data of K. Fischer and G. Wilson, 1975, *Australian Journal of Agricultural Research* 26:11–23.

delivered to new sink organs by the phloem. As well, new sinks often themselves become sources of hormones that may act locally or be translocated to other regions of the plant.

While a role for hormone-directed transport over long distance may be uncertain, there is an accumulating body of evidence that seems to indicate a more direct involvement of hormones in the transfer of solute over short distances. For example, there are a number of reported correlations between the concentration of abscisic acid (ABA; see Chapter 21) and the growth rate of developing fruits. ABA also stimulates the translocation of sugar into the roots of intact bean plants, the uptake of sucrose by sugarbeet root tissue, and the unloading of sucrose into the apoplast of soybean seed-coats and its subsequent uptake into the embryo. There have been conflicting reports on whether ABA stimulates the translocation of ^{14}C-photoassimilate into filling wheat ears. The hormone auxin (IAA; see Chapter 18), on the other hand, inhibits sucrose uptake by sugarbeet roots but stimulates loading in bean leaves. These and other results suggest that loading and unloading may be susceptible to control by hormones.

Although it appears that sink strength is a major factor in determining assimilate distribution, the process of assimilate partitioning remains a complex, highly integrated, and poorly understood phenomenon. Investigators have only begun to address the respective roles of turgor and hormones, while genetic questions and other potential means of regulation have yet to be addressed in any serious way. The regulation of loading, unloading, and source-sink communication should continue to be active and productive areas of research in the future.

9.10 XENOBIOTIC AGROCHEMICALS ARE TRANSLOCATED IN THE PHLOEM

Phloem mobility is of particular interest to the agrochemical industry in producing **xenobiotic** chemicals. The term xenobiotic refers to biologically active molecules that are foreign to an organism. The rate of absorption and translocation of xenobiotic chemicals often determines their effectiveness as herbicides, growth regulators, fungicides, or insecticides. One excellent example is the broad-spectrum herbicide N-(phosphonomethyl)glycine, or **glyphosate**. Glyphosate acts by preventing the synthesis of aromatic amino acids, which in turn blocks the synthesis of protein, auxin hormones, and other important metabolites. Because it is highly mobile in the phloem, glyphosate applied to leaves is rapidly translocated to meristematic

areas or to underground rhizomes for effective control of perennial weeds.

The principal problem with xenobiotics appears to be in gaining entry into the phloem at the minor vein endings in the leaf, that is, phloem loading. Although a few theories have been advanced to explain phloem mobility of xenobiotics, there are relatively few consistent chemical and physical characteristics that describe these molecules. Because xenobiotics are not normally encountered by plants, there are no carriers to mediate their uptake by the cell. Entry is probably by passive diffusion. One consistent characteristic of mobile xenobiotics is their relative level of lipid solubility, or **lipophilicity**, a factor that helps to predict their ability to diffuse through cell membranes.

Efforts to further understand factors controlling the entry of xenobiotic chemicals into plants and their systemic mobility may ultimately lead to advances in our understanding of phloem translocation generally.

SUMMARY

In many plants, the products of photosynthesis may be stored as starch in the chloroplast or exported from the chloroplast to the cytosol where they are converted to sucrose. Storage as starch or export to the cytoplasm are competing processes subject to regulation by subtle changes in the level of triose phosphate and inorganic phosphate (P_i) as well as the regulator metabolite, fructose-2,6-bisphosphate (F-2,6-BP). Other plants such as cereals, store carbon primarily as fructans in the vacuole and exhibit and insensitivity to feedback inhibition of carbon metabolism under conditions where sucrose accumulates in the cytosol.

The long-distance translocation of photoassimilate and other small organic molecules occurs in the phloem tissue. The distinguishing feature of phloem tissue is the conducting tissue called the sieve element or sieve tube. Filled with modified, but active, protoplasm at maturity, sieve tubes are interconnected through perforated end walls called sieve plates.

The direction of long-distance translocation in the phloem is determined largely by source-sink relationships. An organ or tissue that produces more assimilate than it requires for its own metabolism is a source, while a sink is a net importer of assimilate. Sinks include meristems and developing leaves at the apex, nonphotosynthetic stem tissues, roots, and storage organs. Organs such as leaves are commonly sinks in their early stages, but become sources as they mature.

Sugars are translocated in the phloem by mass transfer along a hydrostatic pressure gradient between the source and sink. Loading of sugars into the sieve element–companion cell complex (se-cc) in minor veins of the source is followed by the osmotic uptake of water. The resulting hydrostatic pressure is transmitted throughout the system of sieve elements. Unloading of sugars from the minor veins in the sink maintains the pressure differential that causes mass flow. Phloem loading and unloading may occur through the symplast (plasmodesmata) directly into the se-cc complex. Alternatively, sucrose may be transported across the mesophyll cell membrane into the apoplastic space. From there it would be taken across the membrane of the se-cc complex and enter the long-distance transport stream. There is evidence to support both pathways, but there are a number of issues yet to be resolved.

The distribution of photoassimilate between metabolic pathways and plant organs occurs at two levels: allocation and partitioning. Allocation refers to the immediate metabolic fate of assimilate. It may be allocated to the immediate metabolic needs of the leaf itself and maintenance of leaf biomass, it may be stored for use during nonphotosynthetic periods, or it may be exported from the leaf. Once exported, assimilate will be partitioned between competing sinks. Partitioning is determined by sink strength, which is a combination of sink size and metabolic activity.

CHAPTER REVIEW

1. What factors determine whether the product of the PCR cycle (triose phosphate) will be converted to starch in the chloroplast or sucrose in the cytosol?

2. Distinguish between the roles of F-1,6-BP and F-2,6-BP in the synthesis of sucrose.

3. What is the general structure of a fructan and where does it accumulate?

4. What tissues are removed when a tree is girdled? What causes hypertrophic growth above a girdle wound?

5. Describe the structure of mature phloem tissue. What are its unique features? What kinds of problems do these features raise with respect to phloem translocation?

6. Describe the source-sink concept. To what extent are source-sink relationships involved in determining the direction and rate of translocation in the phloem?

7. Describe the Münch pressure-flow hypothesis and show how it operates to drive translocation in the phloem.

8. How are sugars loaded into the phloem sieve tubes at the source and removed at the sink?

9. Distinguish between allocation and partitioning. What factors determine allocation of carbon within a source leaf? What factors determine partitioning between more than one potential sink?

FURTHER READING

Buchanan, B. B., W. Gruissem, R. L. Jones. 2000. *Biochemistry and Molecular Biology of Plants*. Rockville MD: American Society of Plant Physiologists.

Cairns, A. J., C. J. Pollock, J. A. Gallagher, J. Harrison. 2000. Fructans: synthesis and regulation. In: R. C. Leegood, T. D. Sharkey, S. von Caemmerer, *Advances in Photosynthesis*, Vol. 9, pp. 301–320. Dordrecht: Kluwer.

Foyer, C. H., S. Ferrario-Méry, S. C. Huber. 2000. Regulation of carbon fluxes in the cytosol: Coordination of sucrose synthesis, nitrate reduction and organic acid and amino acid biosynthesis. In: R. C. Leegood, T. D. Sharkey, S. von Caemmerer, *Advances in Photosynthesis*, Vol. 9, pp. 177–203. Dordrecht: Kluwer.

Lough, T. J., W. J. Lucas. 2006. Integrative plant biology: Role of phloem long-distance molecular trafficking. *Annual Review of Plant Biology* 57:203–232.

Trethewey, R. N., A. M. Smith. 2000. Starch metabolism in leaves. In: R. C. Leegood, T. D. Sharkey, S. von Caemmerer, *Advances in Photosynthesis*, Vol. 9, pp. 205–231 Dordrecht: Kluwer.

Turgeon, R. 1996. Phloem loading and plasmodesmata. *Trends in Plant Science* 1: 418–422.

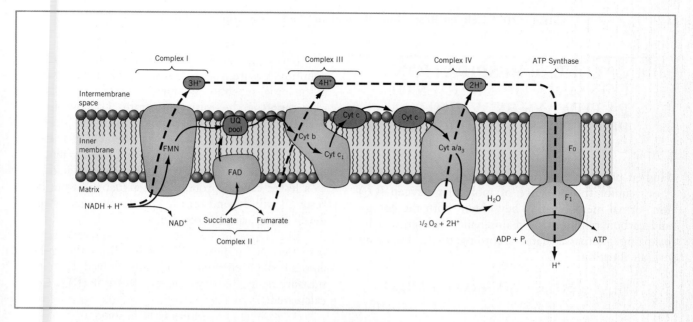

10

Cellular Respiration: Unlocking the Energy Stored in Photoassimilates

The previous four chapters have been devoted to the conservation of light energy as compounds of carbon, or photoassimilates, and factors directing the distribution of those carbon compounds into different plant organs and tissues. Sugars and other photoassimilates represent two important acquisitions by the plant. They represent, first, a highly mobile form of stored photosynthetic energy, and second, a source of carbon skeletons. Through respiration, the plant is able to retrieve the energy in a more useful form, and in the process the sugars are modified to form the carbon skeletons that make up the basic building blocks of cell structure.

This chapter is divided into three principal parts. The first part is devoted to the biochemistry and physiology of cellular respiration. After presenting a brief overview of respiration, the following topics will be discussed:

- pathways and enzymes involved in the degradation of sucrose and starch to hexose sugars,

- the conversion of hexose to pyruvate via the glycolytic pathway and the alternate oxidative pentose phosphate pathways,

- the structure and organization of the mitochondrion, which is the site of oxidative respiratory metabolism,

- the pathway for the complete oxidation of pyruvate to CO_2, known as the citric acid cycle (CAC), the passage of electrons to molecular oxygen via the mitochondrial electron transport chain, and the conservation of energy as reducing potential and ATP,

- several alternative pathways for electron transport that are unique to plants and their possible physiological consequences, and

- the respiration of oils in seeds by first converting fatty acids to hexose sugars via a process known as gluconeogenesis.

In the second part of this chapter, respiration in intact plants and tissues is discussed, showing how environmental factors such as light, temperature, and oxygen availability influence respiration. Finally, the role of respiration in the accumulation of biomass and plant productivity will be briefly examined.

Certain principles introduced earlier (Chapter 5) apply equally well to the present discussion. These include bioenergetics, oxidation and reduction reactions, proton gradients, and the synthesis of adenosine triphosphate (ATP). You may find it helpful at this time to review the appropriate sections of Chapter 5.

10.1 CELLULAR RESPIRATION CONSISTS OF A SERIES OF PATHWAYS BY WHICH PHOTOASSIMILATES ARE OXIDIZED

Higher plants are **aerobic** organisms, which means they require the presence of molecular oxygen (O_2) for normal metabolism. They obtain both the energy and carbon required for maintenance and growth by oxidizing photoassimilates according to the following overall equation:

$$C_6H_{12}O_6 + 6O_2 + 6H_2O \rightarrow 6CO_2 + 12H_2O$$
$$\Delta G^{\circ\prime} = -2869 \text{ kJ mol}^{-1} \qquad (10.1)$$

Note that this equation is written as a reversal of the equation for photosynthesis (Chapter 7, Equation 7.1). The photosynthetic equation is written as the **reduction** of carbon dioxide to hexose sugar, with water as the source of electrons. The equation for respiration, on the other hand, is written as the **oxidation** of hexose to carbon dioxide, with water as a product. Respiration is accompanied by the release of an amount of free energy equivalent to that consumed in the synthesis of the same carbon compounds by photosynthesis. Here the similarity basically ends. Although the two processes overall share the same reactants and products and their energetics are similar, the complex of enzymes involved and the metabolic routes taken are fundamentally different, and they occur in different locations in the cell. Moreover, respiration is a process shared by all living cells in the plant, while photosynthesis is restricted to those cells containing chloroplasts.

Equation 10.1 is written as the direct oxidation of hexose by molecular oxygen, with the consequent release of all of the free energy as heat. Cells do not, of course, oxidize sugars in this way. The release of such a large quantity of energy all at once would literally consume the cells. Instead, the overall process of respiration occurs in three separate but interdependent stages—called **glycolysis**, the **citric acid cycle (CAC)**, and the **respiratory electron transport chain**—comprised of some 50 or more individual reactions in total. The transfer of electrons to oxygen is but the final step in this long and complex process. From the energetic perspective, the function of such a complex process is clear: by breaking the oxidation of hexose down into a series of small, discrete steps, the release of free energy is also controlled so that it can be conserved in metabolically useful forms. Equally important to the cell, as we noted earlier, is the fact that respiration also serves to produce a variety of carbon skeletons that are then used to build other molecules required by the cell. We will return to this point later in the chapter.

The equation for respiration (Equation 10.1) is commonly written with hexose (in particular, glucose) as the initial substrate. In practice, a variety of substrates may serve as the initial substrate. Glucose is itself derived from storage polymers such as starch (a polymer of glucose), fructans (a polymer of fructose), or the disaccharide, sucrose. Other sugars may also be metabolized, as well as lipids, organic acids, and to a lesser extent, protein. The actual substrate being respired will depend on the species or organ, stage of development, or physiological state.

The type of substrate being respired may on occasion be indicated by measuring the relative amounts of O_2 consumed and CO_2 evolved. From these measurements the **respiratory quotient (RQ)** can be calculated:

$$RQ = \frac{\text{moles } CO_2 \text{ evolved}}{\text{moles } O_2 \text{ consumed}} \qquad (10.2)$$

The value of the respiratory quotient is a function of the oxidation state of the substrate being respired. Note that when carbohydrate is being respired (Equation 10.1), the theoretical value of RQ is $6CO_2/6O_2 = 1.0$. Experimental values actually tend to vary in the range 0.97 to 1.17. Because lipids and proteins are more highly reduced than carbohydrate, more oxygen is required to complete their oxidation and the RQ value may be as low as 0.7. On the other hand, organic acids, such as citrate or malate, are more highly oxidized than carbohydrate, less oxygen is required for complete oxidation, and RQ values when organic acids are being respired are typically about 1.3.

While RQ values may provide some useful information, care must be taken when interpreting them. For example, should more than one type of substrate be respired at any one time, the measured RQ will be an average value. Should fermentation be occurring (see below), little or no oxygen will be consumed and an abnormally high RQ may result. Or should either CO_2 or O_2 be trapped in the tissue for any reason, results will be correspondingly misleading. Still, respiratory quotients less than 1 are typical of plants under starvation conditions as lipids and possibly proteins replace carbohydrate as the principal respiratory substrate. Another example of the use of RQ is in germinating seeds. During germination, seeds that store large quantities of lipids will initially exhibit RQ values less than 1. Values will gradually approach 1 as the seedlings consume the lipid reserves and switch over to carbohydrate as the principal respiratory substrate.

The dependence of plant respiration on photosynthesis is illustrated in Figure 10.1. The reduction of CO_2 in the chloroplast leads to the production of fixed carbon in the form of triose phosphates (triose-P), which are represented by the three-carbon phosphorylated intermediates, dihydroxyacetone phosphate and glyceraldehyde-3-phosphate (see Chapter 8). Triose-P

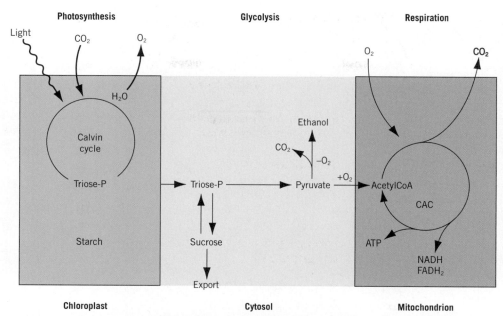

FIGURE 10.1 The metabolic interaction between the chloroplast, cytosol, and mitochondrion in a leaf mesophyll cell. Photosynthesis oxidizes water to O_2 and reduces CO_2 to triose phosphates (triose-P), which represent the fixed carbon substrate for metabolic pathways in the cytosol and the mitochondrion. Triose-P is either stored as starch or exported to the cytosol. In the cytosol, triose-P is either converted to sucrose, which is exported from the mesophyll cell, or is oxidized to pyruvate by the glycolytic pathway. Pyruvate is either oxidized to ethanol in the absence of O_2 (fermentation) or imported to the mitochondrion and completely oxidized to CO_2. Thus, the light energy initially stored as fixed carbon by photosynthesis is eventually converted in the mitochondrion to ATP and reducing power (NADH, $FADH_2$), which is used for growth, development, and the maintenance of cellular homeostasis.

is the carbon intermediate that connects anabolic photosynthetic carbon reduction with respiratory oxidative carbon catabolism. Triose-P generated either directly by the reduction of CO_2 or by the breakdown of starch is exported from the chloroplast to the cytosol, where it is oxidized by the glycolytic pathway to pyruvate. In the presence of molecular oxygen, pyruvate is completely oxidized through mitochondrial respiration to CO_2 with the generation of chemical energy in the form of ATP and NADH used for growth, development, and the maintenance of homeostasis. The metabolic connection between photosynthesis, glycolysis, and respiration not only represents the mechanism by which energy, in the form of fixed carbon, is transferred between three different compartments within a single plant cell, but also represents an "information pathway" connecting the chloroplast with the mitochondrion. This can be illustrated by the fact that inhibition of respiratory ATP synthesis causes PSII reaction centers to become much more sensitive to photoinhibition (Chapter 13). This ability of the chloroplast to respond to events in the mitochondrion is the result of metabolic feedback loops similar to that discussed in Chapters 8 and 9. Clearly, the chloroplast "knows" what is going on in the mitochondrion! Research is only beginning to unravel the precise

nature of these metabolic feedback loops between the mitochondrion, the chloroplast, and the cytosol.

10.2 STARCH MOBILIZATION

There are two distinct pathways for the breakdown, that is, the mobilization of starch. The **hydrolytic pathway** results in the production of glucose whereas the **phosphorolytic pathway** results in the accumulation of hexose phosphates. These pathways will now be discussed separately in more detail.

10.2.1 THE HYDROLYTIC DEGRADATION OF STARCH PRODUCES GLUCOSE

Because most plants store their carbohydrate as starch or sucrose (Chapter 9), the breakdown of these carbohydrates is an appropriate point at which to begin the path of respiratory carbon. Starch normally consists of a mixture of two polysaccharides: amylose and amylopectin. **Amylose**, which probably represents no more than one-third of the starch present in most

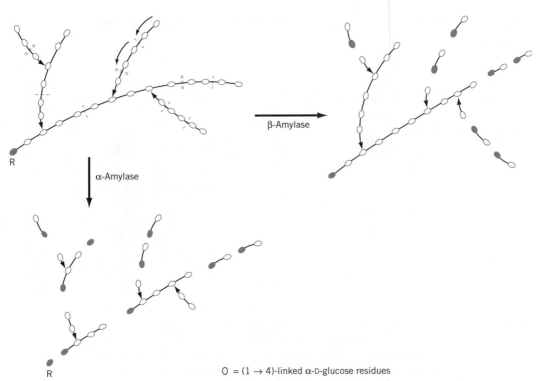

O = (1 → 4)-linked α-D-glucose residues

FIGURE 10.2 **A schematic representation of starch (amylopectin) degradation by α-
and β-amylases. Circles indicate (1→4)-linked α-D-glucose residues. Filled circles
indicate the reducing end of the chain.**

higher plants, consists of very long, straight chains of
(1→4)-linked α-D-glucose units. **Amylopectin**, on the
other hand, is a highly branched molecule in which
relatively short (1→4)-linked α-D-glucose chains are
connected by (1→6) links (Figure 10.2). Starch is normally deposited in plastids as water-insoluble granules
or grains. The complete breakdown of starch to its
component glucose residues requires the participation
of several hydrolytic enzymes.

10.2.2 α-AMYLASE PRODUCES MALTOSE AND LIMIT DEXTRINS

α-**Amylase** randomly cleaves α-(1→4) glucosyl bonds
in both amylose and amylopectin (Figure 10.2).
α-Amylase, however, does not readily attack terminal
α-(1→4) bonds. In the case of amylopectin, α-amylase
will not cleave the α-(1→6) glucosyl bonds, nor
those α-(1→4) bonds in the immediate vicinity of
the branch points. Consequently, about 90 percent
of the sugar released on hydrolysis of amylose and
amylopectin by α-amylase consists of the disaccharide
maltose ((1→4)-α-D-glucosylglucose). The balance
consists of a small amount of glucose and, in the
case of amylopectin, **limit dextrins**. Limit dextrins
are comprised of a small number of glucose residues,
perhaps 4 to 10, and contain the original branch
points. α-Amylase is not restricted to plants but
can be found widely in nature, including bacteria

and mammals (including human saliva). Indeed, this
enzyme can be expected in any tissue that rapidly
metabolizes starch. A unique and important property
of α-amylase is its ability to use starch grains as a
substrate. α-Amylase plays an important role in the
early stages of seed germination, where it is regulated
by the plant hormones gibberellin and abscisic acid
(Chapter 19, 21).

10.2.3 β-AMYLASE PRODUCES MALTOSE

β-**Amylase** degrades amylose by selectively hydrolyzing
every second bond, beginning at the nonreducing end of
the chain. β-Amylase thus produces exclusively maltose.
β-Amylase will degrade the short chains in amylopectin
molecules as well. However, because the enzyme can
work only from the nonreducing end and cannot cleave
the (1→6) branch points, β-amylase will degrade only
the short, outer chains and will leave the interior of the
branched molecule intact (Figure 10.2).

10.2.4 LIMIT DEXTRINASE IS A DEBRANCHING ENZYME

Limit dextrinase acts on limit dextrins and cleaves
the (1→6) branching bond. This allows both α- and
β-amylase to continue degrading the starch to maltose.

10.2.5 α-GLUCOSIDASE HYDROLYZES MALTOSE

The final step is the hydrolysis of maltose to two molecules of glucose by the enzyme α-glucosidase. All of the above enzymes mediate the **hydrolytic** breakdown of starch to free sugar; that is, the molecule is cleaved essentially by the addition of water across the bond.

10.2.6 STARCH PHOSPHORYLASE CATALYZES THE PHOSPHOROLYTIC DEGRADATION OF STARCH

When the inorganic phosphate level is high (greater than 1 mM), the breakdown of starch is accompanied by an accumulation of *phosphorylated* sugars. This is due to the action of the enzyme **starch phosphorylase**, which catalyzes the **phosphorolytic** degradation of starch:

$$\text{starch} + n\text{P}_i \rightarrow n(\text{glucose} - 1 - \text{phosphate}) \quad (10.3)$$

We will return to this point later, but because the end product is glucose-1-phosphate rather than free glucose, the action of phosphorylase offers a slight energetic advantage. Phosphorylase cannot operate alone—it is unable to degrade starch grains and, like β-amylase, its action is confined to the outer chains of amylopectin molecules. Phosphorylase thus can work only in conjunction with α-amylase, which initiates degradation of the insoluble grains, and debranching enzymes,

which render the interior glucose chains accessible to the phosphorylase enzyme. The relative importance of phosphorylase *in vivo* is not known, but in laboratory experiments phosphorylase accounts for less than half of the degradation of potato starch. The balance is degraded via α- and β-amylase. Starch is stored and degraded inside plastids (either chloroplasts or amyloplasts—see Chapter 5, Box 5.1), but the initial stages of cellular respiration occur in the cytosol. The products of starch degradation must therefore make their way across the plastid envelope in order to gain access to the respiratory machinery. This is accomplished by two transporter systems located in the membranes of the plastid envelope (Figure 10.3). The product of phosphorolytic breakdown, glucose-1-phosphate, is a component of the hexose phosphate pool and subsequently is converted to triose-P in the chloroplast. Triose-P exits the plastid via the P$_i$-triose phosphate transporter described earlier in Chapter 8. Free glucose is able to exit the plastid via a separate hexose transporter, a protein complex present in the inner envelope membrane of the chloroplast, which specifically moves glucose from the stroma to the cytosol. For a more detailed discussion of membrane transport review Chapter 3.

Sucrose synthesis has been described in Chapter 9. Two enzymes are responsible for its breakdown: sucrose synthase and invertase. **Invertase** occurs in two forms, alkaline invertase, with a pH optimum near 7.5, and acid invertase, with a pH optimum near 5.

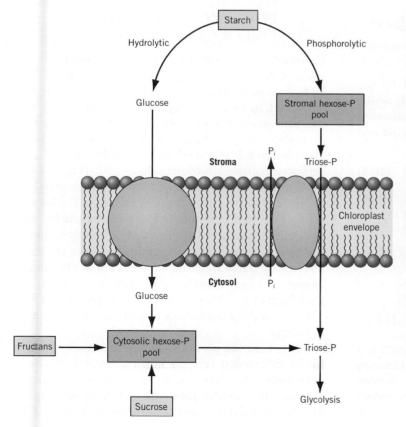

FIGURE 10.3 Mobilization of starch in the stroma of the chloroplast and sucrose in the cytosol. Starch is broken down either by hydrolytic or phosphorolytic enzymes. Glucose is transported through the chloroplast envelope membrane by a specific glucose transporter to the cytosol. In the cytosol, the exported glucose enters the cytosolic hexose phosphate (hexose-P) pool. The breakdown of sucrose also feeds hexose phosphates into the cytosolic hexose-P pool. The degradation of fructans in the vacuole also feeds hexose phosphates into the cytosolic hexose-P pool. Components of the cytosolic hexose-P pool are converted to cytosolic triose-P. The phosphorolytic pathway feeds hexose phosphates into the stromal hexose-P pool. Hexose-P is converted to stromal triose-P which is transported out of the chloroplast to the cytosol by the specific P$_i$-translocator, which imports one P$_i$ molecule for each triose-P exported to the cytosol. Cytosolic triose-P is then oxidized through glycolysis and respiration. Note that the two hexose-P pools are present in different compartments that are connected metabolically.

Sucrose synthase and alkaline invertase appear to be localized in the cytosol, while acid invertase is found associated with cell walls and vacuoles. Clearly the relative contributions of these three enzymes will depend to some extent on the cellular location of the sucrose being metabolized. Acid invertase, for example, would be important to the mobilization of sucrose in sugarcane (*Saccharum spontaneum*), which stores excess carbohydrate primarily as sucrose in the vacuoles of stem cells.

10.3 FRUCTAN MOBILIZATION IS CONSTITUTIVE

Fructans are soluble polymeric forms of fructose biosynthesized in the vacuole as major storage carbohydrates of certain plants species such as grasses (see Chapter 9). Although the enzymes involved on the biosynthesis of fructans are inducible through an increase in cytosolic sucrose concentrations, it appears that enzymes involved in the hydrolysis of fructans are consitutively expressed irrespective of sucrose concentrations. Thus, the net accumulation of fructans in the vacuole is the result of the differential, regulated rate of biosynthesis versus the unregulated rate of hydrolysis. The major enzyme involved in fructan hydrolysis in the vacuole is fructan exohydrolase (FEH). This enzyme is an exohydrolase which hydrolyzes one terminal fructosyl unit at a time from the fructan polymer (Equation 10.4).

$$\text{Glucosyl-1,2-fructosyl-1,2-fructosyl-(fructosyl)}_N \rightarrow$$

$$\text{glucosyl-1,2-fructosyl-1,2-(fructosyl)}_{N-1} + \text{fructose} \tag{10.4}$$

The hydrolysis of this polymer is completed by the action of the vacuolar invertase (Equation 10.5) which breaks down the initial sucrose acceptor molecule used to synthesize the fructan (Chapter 9) into glucose and fructose.

$$\text{Sucrose} \rightarrow \text{glucose} + \text{fructose} \tag{10.5}$$

The free hexoses are then transported from the vacuole to the cytosol where they are phosphorylated by cytosolic hexokinase and enter the cytosolic hexose phosphate pool (Figure 10.4).

10.4 GLYCOLYSIS CONVERTS SUGARS TO PYRUVIC ACID

The first stage of respiratory carbon metabolism is a group of reactions by which hexose sugars undergo a partial oxidation to the three-carbon acid pyruvic acid or **pyruvate**. These reactions, collectively known as **glycolysis**, are catalyzed by enzymes located in the cytosol of the cell. Parallel reactions occur independently in plastids, in particular amyloplasts and some chloroplasts. Thus, unlike animal cells, glycolysis in plants is not restricted to the cytosol. The reaction of glycolysis, which literally means the lysis or breakdown of sugar, was originally worked out by Meyerhof and others in Germany during the early part of the twentieth century, in order to explain fermentation in yeasts and the breakdown of glycogen in animal muscle tissue. Like much of respiratory metabolism, glycolysis is now known to occur universally in all organisms. It is also believed to represent the most primitive form of carbon catabolism since it can lead to fermentation products such as alcohol and lactic acid in the absence of molecular oxygen. Although the energy yield of glycolysis is low, it can be used to support growth in anaerobic organisms or in some aerobic organisms or tissues under anaerobic conditions. Under normal aerobic conditions, however, the pyruvate formed by glycolysis will be further metabolized by the mitochondria to extract yet more energy.

Glycolysis is conveniently considered in two parts. The first is a set of reactions by which the several forms of glucose and fructose derived from storage carbohydrate are converted to the common intermediate triose-phosphate (triose-P) via the hexose-phosphate pool (Figure 10.4). Triose-P is then converted to pyruvate, the end product of glycolysis (Figure 10.5).

10.4.1 HEXOSES MUST BE PHOSPHORYLATED TO ENTER GLYCOLYSIS

In order for carbon from storage carbohydrate to enter glycolysis, the glucose and fructose derived from hydrolysis of starch, sucrose, or fructans must first be converted to hexose phosphates. The cytosolic **hexose phosphate pool** consists of glucose-1-phosphate, glucose-6-phosphate, and fructose-6-phosphate. These phosphorylated intermediates are subsequently converted to fructose-1,6-bisphosphate (FBP) (Figure 10.4). Note that the conversion of glucose and fructose to FBP requires an initial expenditure of energy in the form of ATP. Two molecules of ATP are consumed for each molecule of sucrose that enters glycolysis (reactions 5 and 6). This is analogous to priming a pump; glucose is a relatively stable molecule and the initial phosphorylations, first to glucose-6-P and then to fructose-1,6-P, are a form of activation energy (see Chapter 8, Box 8.1). These two ATP molecules will be recovered during glycolysis.

It is here the phosphorolytic breakdown of starch in the chloroplast offers a slight energetic advantage over hydrolytic degradation. Because the product is glucose-1-P, for each molecule of hexose entering via the phosphorolytic route the initial expenditure of ATP

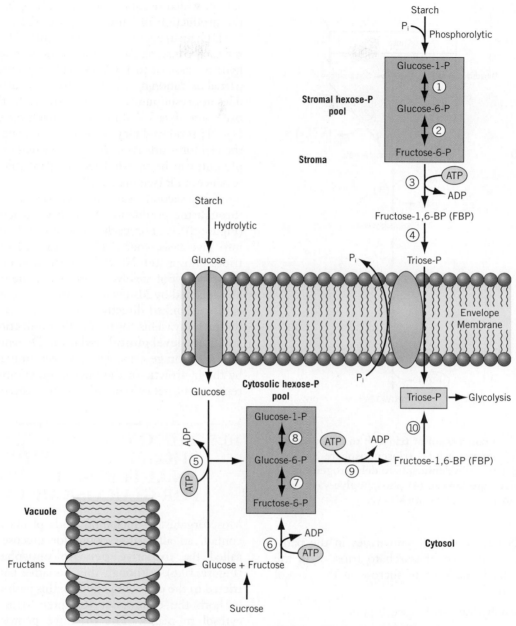

FIGURE 10.4 **The conversion of storage carbohydrate to triose phosphate in the chloroplast and the cytosol. In the chloroplast, starch is either broken down hydrolytically to glucose, which is exported to the cytosol, or broken down phosphorolytically to intermediates of the stromal hexose-P pool (glucose-1-P, glucose-6-P, and fructose-6-P). In the cytosol, sucrose is hydrolyzed to glucose plus fructose. The hydrolysis of fructans in the vacuole also supply glucose and fructose to the cytosol. Glucose and fructose are converted to intermediates of the cytosolic hexose-P pool by the enzymes hexokinase (5) and fructokinase (6), respectively. The intermediates of the stromal and cytosolic hexose-P pools are interconverted by chloroplastic and cytosolic isoforms of the enzymes phosphoglucomutase (1 and 8, respectively) and hexosephosphate isomerase (2 and 7, respectively). Carbon exits the stromal and cytosolic hexose-P pools through the conversion of fructose-6-P to fructose-1,6-BP (FBP) by the ATP-dependent phosphofructokinase present in the stroma (3) and in the cytosol (9). Fructose-1,6-BP is converted to triose phosphate (triose-P) by chloroplastic and cytosolic isoforms of the enzyme aldolase (4 and 10, respectively). Triose-P is exported from the stroma to cytosol by the P_i-transporter.**

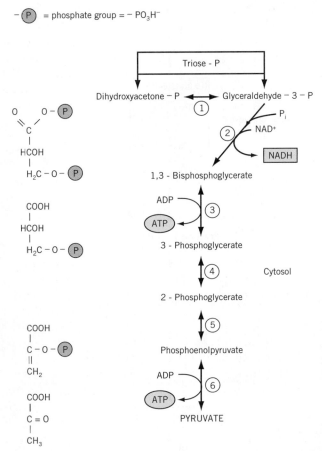

−(P) = phosphate group = − PO$_3$H$^-$

FIGURE 10.5 **The conversion of triose-P to pyruvate via glycolysis.** Enzymes are (1) triosephosphate isomerase, (2) glyceraldehydephosphate dehydrogenase, (3) phosphoglycerate kinase, (4) phosphoglycerate mutase, (5) enolase, (6) pyruvate kinase.

A principal function of glycolysis is energy conservation, which occurs in two ways. The first is through the production of **reducing potential** in the form of NADH. In reaction 2, two molecules of NADH (one for each triose phosphate) are produced as glyceraldehyde is oxidized to 1,3-bisphosphoglycerate. Since this partial oxidation does not require molecular oxygen and does not result in the release of any CO_2, the glycolytic oxidation of carbohydrate is an anaerobic process. The NADH produced may be used as reducing potential by the cell for synthesis of other molecules or, if oxygen is present, can be metabolized by plant mitochondria to produce ATP (see Figure 10.9).

The second way that energy is conserved is through the production of ATP via reactions 3 and 6 (Figure 10.5). For each molecule of hexose entering into glycolysis, four ATP are formed (two for each triose phosphate). Note that formation of ATP at this point does not involve a proton gradient and cannot be explained by Mitchell's chemiosmotic hypothesis. It is instead linked directly to conversion of substrate in the pathway. This form of ATP production is called a **substrate-level phosphorylation**. Depending on whether the storage carbohydrates were initially degraded by the hydrolytic or the phosphorolytic pathways, this represents a net gain of either two or three ATP.

10.5 THE OXIDATIVE PENTOSE PHOSPHATE PATHWAY IS AN ALTERNATIVE ROUTE FOR GLUCOSE METABOLISM

Most organisms, including both plants and animals, contain an alternative route for glucose metabolism called the **oxidative pentose phosphate pathway** (Figure 10.6). Although this oxidative pathway is restricted to the cytosol in animals, this pathway is present in both the chloroplast (Chapter 8) as well as the cytosol in plants. The oxidative pentose phosphate pathway shares several intermediates with glycolysis and is closely integrated with it. The first step in the oxidative pentose phosphate pathway is the oxidation of glucose-6-P to 6-phosphogluconate (Figure 10.6). This initial step, which is sensitive to the level of NADP$^+$, is apparently the rate-determining step for the oxidative pentose phosphate pathway. This is the reaction that determines the balance between glycolysis and the oxidative pentose phosphate pathway. The second step is another oxidation accompanied by the removal of a CO_2 group to form ribulose-5-P. The electron acceptor in both reactions is NADP$^+$, rather than NAD$^+$. Subsequent reactions in the pathway result in the formation of glyceraldehyde-3-P and fructose-6-P, both of which are then further metabolized via glycolysis.

is reduced. Note the overall similarities in the pathways for the breakdown of starch to triose-P in the stroma with the breakdown of sucrose in the cytosol (Equation 10.6):

$$\text{storage carbohydrate} \rightarrow \text{hexose-P pool}$$
$$\rightarrow \text{FBP} \rightarrow \text{triose-P} \quad (10.6)$$

10.4.2 TRIOSE PHOSPHATES ARE OXIDIZED TO PYRUVATE

The reactions for the further conversion of triose-P to pyruvate are summarized in Figure 10.5. The triose phosphates, dihydroxyacetone phosphate and glyceraldehyde-3-phosphate, are readily interconvertible (reaction 1), which means that all of the carbon in the original hexose molecule will eventually be converted to pyruvate. In other words, one molecule of hexose phosphate will yield two molecules of pyruvate. Thus, in order to account for the hexose molecule originally entering the pathway, everything from this point on must be multiplied by 2.

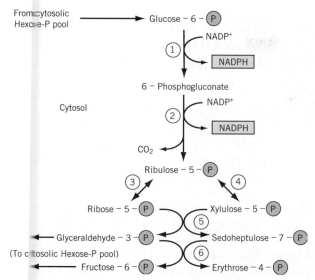

FIGURE 10.6 **The oxidative pentose phosphate pathway. A principal function of this alternative pathway is to generate reducing potential in the form of NADPH and pentose sugars for nucleic acid biosynthesis. The origin of glucose-6-P is the cytosolic hexose-P pool (see Figure 10.4). Glyceraldehyde-3-P and fructose-6-P may be returned to the glycolytic pathway for further metabolism. Enzymes are: (1) glucose-6-phosphate dehydrogenase, (2) 6-phosphogluconate dehydrogenase, (3) phosphoriboisomerase, (4) phosphopentoepimerase, (5) transketolase, (6) transaldolase.**

The role of the oxidative pentose phosphate pathway and its contribution to carbon metabolism overall is difficult to assess because the pathway is not easily studied in green plants. This is largely because many of the intermediates and enzymes of this respiratory cycle are shared by the more dominant **reductive pentose phosphate pathway**, or PCR cycle, in the chloroplasts (Chapter 8). From studies of animal metabolism, however, it can be concluded that the oxidative pentose phosphate pathway has two significant functions. The first is to generate reducing potential in the form of NADPH. $NADP^+$ is distinguished from NAD^+ by an extra phosphoryl group. NADPH serves primarily as an electron donor when required to drive normally reductive biosynthetic reactions, whereas NADH is used predominantly to generate ATP through oxidative phosphorylation (see below). This distinction allows the cell to maintain separate pools of NADPH and NAD^+ in the same compartment: a high $NADPH/NADP^+$ ratio to support reductive biosynthesis and a high $NAD^+/NADH$ ratio to support glycolysis. The oxidative pentose phosphate pathway is therefore thought to be a means to generate NADPH required to drive biosynthetic reactions in the cytosol. In animals, for example, the oxidative pentose phosphate pathway is extremely active in fatty tissues where NADPH is required for active fatty acid synthesis. The second

function for the oxidative pentose phosphate pathway is the production of pentose phosphate, which serves as a precursor for the ribose and deoxyribose required in the synthesis of nucleic acids. Another intermediate of the oxidative pentose phosphate pathway with potential significance to plants is the 4-carbon erythrose-4-P, a precursor for the biosynthesis of aromatic amino acids, lignin, and flavonoids.

10.6 THE FATE OF PYRUVATE DEPENDS ON THE AVAILABILITY OF MOLECULAR OXYGEN

The fate of pyruvate produced by glycolysis depends primarily on whether oxygen is present (Figure 10.7). Under normal aerobic conditions, pyruvate is transported into the mitochondrion, where it is further oxidized to CO_2 and water, transferring its electrons ultimately to molecular oxygen. We will address mitochondrial respiration further in the following section.

Although higher plants are obligate aerobes and are able to tolerate anoxia for only short periods, tissues or organs are occasionally subjected to anaerobic conditions. A typical situation is that of roots when the soil is saturated with water. When there is no oxygen to serve as the terminal electron acceptor, mitochondrial respiration will shut down and metabolism will shift over to **fermentation**. Fermentation converts pyruvate either to ethanol through the action of the enzyme **alcohol dehydrogenase (ADH)** or to lactate via **lactate dehydrogenase (LDH)**. In most plants, the principal products of fermentation are CO_2 and ethanol (Figure 10.7,

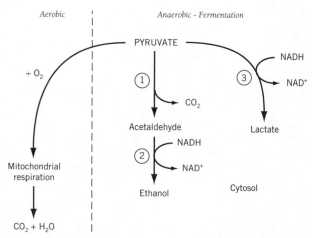

FIGURE 10.7 **The fate of pyruvate depends largely on available oxygen. Enzymes are: (1) pyruvate decarboxylase, (2) alcohol dehydrogenase (ADH), (3) lactate dehydrogenase (LDH).**

reactions 1, 2). Some lactate may be formed, primarily in the early stages of anoxia. However, lactate lowers the pH of the cytosol, which in turn activates pyruvate decarboxylase and initiates the production of ethanol. Note that either one of the fermentation reactions (Figure 10.7, reactions 2 and 3) consumes the NADH produced earlier in glycolysis by the oxidation of glyceraldehyde-3-P (Figure 10.5, reaction 2). Although this means there is no net gain of reducing potential in fermentation, this recycling of NADH is still important to the cell. The pool of NADH plus NAD^+ in the cell is relatively small and if the NADH is not recycled, there will be no supply of NAD^+ to support the continued oxidation of glyceraldehyde-3-P. If this were the case, glycolysis and the production of even the small quantities of ATP necessary to maintain the cells under anaerobic conditions would then grind to a halt.

10.7 OXIDATIVE RESPIRATION IS CARRIED OUT BY THE MITOCHONDRION

10.7.1 IN THE PRESENCE OF MOLECULAR OXYGEN, PYRUVATE IS COMPLETELY OXIDIZED TO CO_2 AND WATER BY THE CITRIC ACID CYCLE

The second stage of respiration is the complete oxidation of pyruvate to CO_2 and water through a series of reactions known as the **citric acid cycle (CAC)** (Figure 10.8). The citric acid cycle is also known as the tricarboxylic acid (TCA) cycle or the Krebs cycle, in honor of Hans Krebs, whose research in the 193Cs was

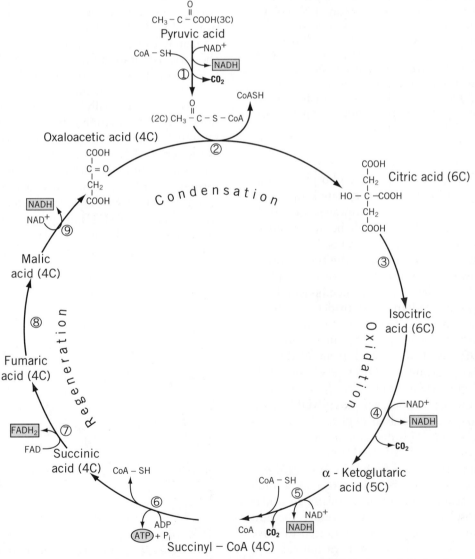

FIGURE 10.8 **The reactions of the citric acid cycle (CAC). The citric acid cycle completes the oxidation of pyruvate to carbon dioxide. Reducing potential is stored as NADH and FADH$_2$.**

responsible for elucidating this central metabolic process. Krebs was awarded the Nobel Prize in medicine in 1954 for his outstanding contribution.

Schemes for the citric acid cycle such as that shown in Figure 10.8 invariably begin with pyruvate. Although pyruvate is technically not a part of the cycle, it does provide the major link between glycolysis and subsequent carbon metabolism. Note that pyruvate is produced in the cytosol while the enzymes of the citric acid cycle are located in the matrix space of the mitochondrion (Chapter 5). Thus in order for pyruvate to be metabolized by the citric acid cycle, it must first be translocated through the inner membrane. This is accomplished by a pyruvate-OH^- antiport carrier—that is, pyruvate is taken up by the mitochondrion in exchange for a hydroxyl ion carried into the intermembrane space.

Once inside the matrix, pyruvate is oxidized and decarboxylated by a large multienzyme complex **pyruvate dehydrogenase**. Pyruvate dehydrogenase catalyzes a series of five linked reactions, the overall effect of which is to oxidize one molecule of pyruvate to a two-carbon acetate group:

$$\text{pyruvate} + NAD^+ + CoA \rightarrow$$
$$\text{acetyl-CoA} + NADH + H^+ + CO_2 \quad (10.7)$$

The resulting two-carbon acetyl group is finally linked via a thioester bond to a sulphur-protein **coenzyme A(CoA)**. In the process, NAD^+ is reduced to NADH. The CO_2 given off represents the first of three carbon atoms in the degradation of pyruvate.

The citric acid cycle proper begins with the enzyme **citrate synthase**, which condenses the acetyl group from acetyl-CoA with the four-carbon **oxaloacetate** to form the six-carbon, tricarboxylic **citric acid** (hence the designation **citric acid cycle, CAC**). The next step is an isomerization of citrate to isocitrate (Figure 10.8, reaction 3), followed by two successive oxidative decarboxylations (Figure 10.8, reactions 4, 5). The two CO_2 molecules given off effectively completes the oxidation of pyruvate which adds two more molecules of NADH to the pool of reductant in the mitochondrial matrix.

The balance of the citric acid cycle serves two functions. First, additional energy is conserved at three more locations. One molecule of ATP is formed from ADP and inorganic phosphate when succinate is formed from succinyl-CoA (Figure 10.8, reaction 6). Because the ATP formation is linked directly to conversion of substrate, this is another example of **substrate-level phosphorylation**. Additional energy is conserved with the oxidation of succinate to fumarate (Figure 10.8, reaction 7) and the oxidation of malate to oxaloacetate (reaction 9). Second, the cycle serves to regenerate a molecule of oxaloacetate and so prepare the cycle to accept another molecule of acetyl-CoA. Regeneration of oxaloacetate is critical to the catalytic nature of the cycle in that it allows a single oxalacetate molecule to

mediate the oxidation of an endless number of acetyl groups.

In summary, the citric acid cycle consists of eight enzyme-catalyzed steps, beginning with the condensation of a two-carbon acetyl group with the four-carbon oxaloacetate to form a molecule of the six-carbon citrate. The acetyl group is then degraded to two molecules of CO_2. The cycle includes four oxidations, which yield NADH at three steps and $FADH_2$ at one step. One molecule of ATP is formed by substrate-level phosphorylation. Finally, the oxaloacetate is regenerated, which allows the cycle to continue.

Before leaving the citric acid cycle for the moment, it is useful to point out that the cycle must turn twice to metabolize the equivalent of one hexose sugar.

10.7.2 ELECTRONS REMOVED FROM SUBSTRATE IN THE CITRIC ACID CYCLE ARE PASSED TO MOLECULAR OXYGEN THROUGH THE MITOCHONDRIAL ELECTRON TRANSPORT CHAIN

We noted earlier that one of the principal functions of the respiration is to retrieve, in useful form, some of the energy initially stored in assimilates. Our traditional measure of useful energy in most processes is the number of ATP molecules gained or consumed. By this measure alone, the yield from both glycolysis and the citric acid cycle is quite low. After two complete turns of the cycle, one molecule of glucose has been completely oxidized to six molecules of CO_2, but only four molecules of ATP have been produced (a net of two ATP from glycolysis plus one for each turn of the cycle). At this point, most of the energy associated with the glucose molecule has been conserved in the form of electron pairs generated by the oxidation of glycolytic and citric acid cycle intermediates. For each molecule of glucose, a total of 12 electron pairs were generated; 10 as NADH ($\Delta G^{\circ\prime} = -222 \text{ kJ mol}^{-1}$) and two as $FADH_2$ ($\Delta G^{\circ\prime} = -180 \text{ kJ mol}^{-1}$). Thus, the total energy that has been trapped as reducing power through the action of glycolysis and CAC is about 2580 kJ mol^{-1} [$(10 \times 222 \text{ kJ}) + (2 \times 180 \text{ kJ})$]. In addition, the net production of four ATP by substrate phosphorylation in glycolysis and the CAC traps a total of about 125 kJ energy. Therefore, the aerobic oxidation of one glucose molecule traps a total of about 2705 kJ, which represents an efficiency of about 94 percent ($2709/2869 \times 100\%$) given that the $\Delta G^{\circ\prime}$ for the oxidation of glucose is about $-2869 \text{ kJ mol}^{-1}$.

In this section we will discuss the third stage of cellular respiration—the transfer of electrons from NADH and $FADH_2$ to oxygen and the accompanying conversion of redox energy to ATP. The transfer of electrons from NADH and $FADH_2$ to oxygen involves a sequence of electron carriers arranged in an **electron transport chain**. Membrane fractionation studies have shown that the enzymes and electron carriers making up the electron transport chain are organized predominantly into four large multimolecular complexes (complexes I—IV) and two mobile carriers located in the inner mitochondrial membrane (Figure 10.9). In this sense, there are a great number of similarities between the mitochondrial inner membrane and the thylakoid membranes of the chloroplast (compare Figure 10.9 with Figure 7.6). This is not unexpected, since the principal function of each membrane is energy transformation and many of the same or similar components are involved. The path of electrons from NADH to oxygen can be summarized as follows. Electrons from NADH enter the electron transport chain through Complex I, known as **NADH-ubiquinone oxidoreductase**. In addition to several proteins, this complex also contains a tightly bound molecule of **flavin mononucleotide (FMN)** and several nonheme iron-sulphur centers. Complex I conveys the electrons from NADH to ubiquinone. **Ubiquinone** is a benzoquinone—its structure and function are similar to the plastoquinone found in the thylakoid membranes of chloroplasts (see Chapter 5). Like plastoquinone, ubiquinone is highly lipid soluble and diffuses freely in the plane of the membrane. It is not permanently associated with Complex I, but forms a pool of mobile electron acceptors that conveys electrons between Complex I and Complex III. Ubiquinol (the fully reduced form of ubiquinone) is oxidized by Complex III, or **cytochrome c reductase**. Complex III contains cytochromes b and c_1 and an iron-sulphur center. Complex III in turn reduces a molecule of cytochrome c. Cytochrome c is a peripheral

protein, located on the side of the membrane facing the intermembrane space. Like plastocyanin in chloroplasts, cytochrome c is a mobile carrier and conveys electrons between Complex III and the terminal complex in the chain, Complex IV. Also known as **cytochrome c oxidase**, Complex IV contains cytochromes a and a_3 and copper. Electrons are passed first from cytochrome c to cytochrome a, then to cytochrome a_3, and finally to molecular oxygen.

All of the oxidative enzymes of the citric acid cycle, with one exception, are located in the matrix. The one exception is succinic dehydrogenase (Figure 10.8, reaction 7). This enzyme is an integral protein complex (Complex II) that is tightly bound to the inner mitochondrial membrane (Figure 10.9). In fact, succinic dehydrogenase is the preferred marker enzyme for inner membranes when doing mitochondrial fractionations. Complex II, known as **succinate-ubiquinone oxidoreductase**, contains flavin adenine di-nucleotide (FAD), several nonheme iron proteins, and iron-sulphur centers. Like Complex I, succinic dehydrogenase transfers electrons from succinate to a molecule of ubiquinone from the membrane pool. From there the electrons pass through Complexes III and IV to molecular oxygen.

It is important to note that Figure 10.9 presents a static, essentially linear representation of electron flow in mitochondria. *In vivo*, the organization is far more dynamic. In Chapter 7 we showed that the electron transport complexes of the photosynthetic membranes were independently distributed, rather than organized in one supermolecular complex. Similar arguments apply here. The several complexes are not found in equal stoichiometry and are free to diffuse independently within the plane of the inner membrane. The large complexes

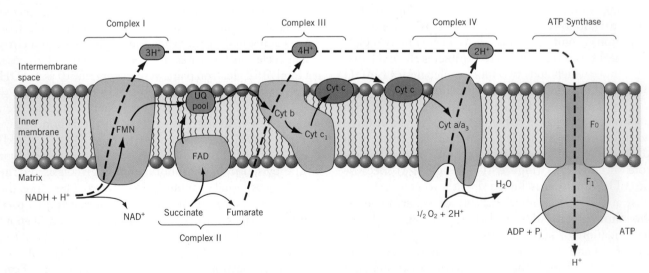

FIGURE 10.9 **A schematic representation of the electron transport chain and proton "pumping" sites in the inner membrane of a plant mitochondrion. Solid arrow indicates the path of electrons from NADH or succinate to molecular oxygen. Energy conserved in the proton gradient is used to drive ATP synthesis through the F_0—F_1-ATPase coupling factor elsewhere in the membrane.**

are functionally linked through the pools of ubiquinone and cytochrome *c*, which function as mobile carriers and convey electrons from one complex to the other largely on the basis of random collision.

10.8 ENERGY IS CONSERVED IN THE FORM OF ATP IN ACCORDANCE WITH CHEMIOSMOSIS

As electrons are passed from NADH (or FADH$_2$) to oxygen through the electron transport chain, there is a substantial drop in free energy. The actual free energy change is quantitatively the same, but of opposite sign, to the amount consumed when electrons are moved from water to NADPH in photosynthesis (Chapter 7). This energy is conserved first in the form of a proton gradient and ultimately as ATP. The energetics for ATP synthesis is explained by Mitchell's chemiosmotic hypothesis, described earlier in Chapter 5. Here the focus is on specific, biochemical mechanistic details of the proton gradient and ATP synthesis as it applies to mitochondria.

Studies of **P/O ratios** (the atoms of phosphorous esterified as ATP relative to the atoms of oxygen reduced) and various inhibitors have established that there are three transitions in the electron transport chain that are associated with ATP synthesis. Put another way, when internal or matrix NADH is oxidized, the P/O ratio is approximately 3. According to Mitchell's hypothesis, then, these transitions represent locations, generally described as proton pumps, where contributions are made to a proton gradient across the mitochondrial inner membrane. The three locations, associated with Complexes I, III, and IV, respectively, are identified in Figure 10.9. The resulting proton gradient then drives ATP synthesis via a F$_0$-F$_1$-ATP synthase complex located in the same membrane (see below). Because mitochondrial ATP synthesis is closely tied to oxygen consumption, it is referred to as **oxidative phosphorylation**.

In the course of mitochondrial electron transport, protons are extruded from the matrix into the intermembrane space. Proton extrusion associated with Complex I (site 1) can be explained by the vectorial arrangement of the complex across the membrane. When a pair of electrons is donated to the complex by NADH, a pair of protons are picked up from the matrix. When the electrons are subsequently passed on to ubiquinone, the protons are released into the intermembrane space. Proton extrusion associated with Complex III (site 2) is probably due to the operation of a "Q-cycle," described in Chapter 7 (see Chapter 7, Figure 7.10). The contribution of cytochrome *c* oxidase (site 3) to the proton gradient has been the subject of some discussion for many years. Experiments with isolated enzyme incor-

porated into lipid vesicles indicate that cytochrome *c* oxidase was capable of transferring protons across membranes. These results are difficult to explain, because cytochromes exchange only electrons, not protons, when reduced and oxidized. However, the H$^+$/electron pair ratio for site 3 is about 2, which can readily be explained by the two protons consumed from the matrix when oxygen is reduced to water. This is similar in principle to the production of protons in the intrathylakoid space, as water is oxidized early in photosynthetic electron transport (see Chapter 7, Figure 7.6). The stoichiometry of proton "extrusion" (the term is applied whether or not the protons are physically carried across the membrane) has been studied extensively. It appears that approximately nine protons are extruded for each pair of electrons conveyed from internal (or matrix) NADH to oxygen.

The link between a proton gradient and ATP synthesis in the mitochondrion embodies the same principles previously described for ATP synthesis in the chloroplast: (1) the inner membrane is virtually impermeable to protons; (2) a proton motive force (pmf) is established across the membrane by a combination of membrane potential and proton disequilibrium; and (3) ATP synthesis is driven by the return of protons to the matrix through an integral membrane protein complex known variously as **ATP synthase**, coupling factor, or F$_0$-F$_1$-ATPase. According to equation 10.8, pmf consists of two principal components: a chemical component, that is, ΔpH, which reflects the difference in H$^+$ concentration, and an electrical component ($\Delta\Psi$), which reflects a difference in charge between the matrix and the intermembrane space.

$$\text{pmf} = -59\text{V}\Delta\text{pH} + \Delta\Psi \qquad (10.8)$$

Due to the capacity of the cytosol to buffer changes in pH in the intermembrane space, ΔpH contributes minimally to the overall pmf in mitochondria. Thus, in contrast to the chloroplast, the difference in electrical charge ($\Delta\Psi$) across the inner membrane is the major factor contributing to the proton motive force generated by mitochondria.

Mitochondrial ATP synthase is structurally and functionally similar to the chloroplast enzyme. It consists of a hydrophobic, channel-forming portion (F$_0$) that spans the membrane plus a multimeric, matrix-facing peripheral protein (F$_1$) that couples proton translocation to ATP synthesis. As in the case of chloroplasts, the H$^+$/ATP ratio is approximately 3. Because nine protons are extruded for each pair of electrons moving through the entire chain, this means that a total of three ATP molecules could be formed from each NADH produced in the matrix. For electrons entering the chain from extramitochondrial NADH, succinate, or via the rotenone-insensitive dehydrogenase (see below), all three of which bypass site 1, a maximum of two ATP could be formed.

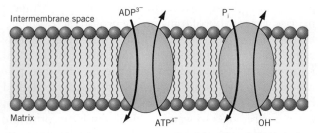

FIGURE 10.10 The adenine nucleotide transporter. The one-for-one exchange of mitochondrial ATP and cytosolic ADP across the inner membrane is driven by the membrane potential. Inorganic phosphate is returned to the matrix in exchange for hydroxyl ions.

Unlike the chloroplast, most of the ATP synthesized in the mitochondrion is utilized elsewhere in the cell. This requires that the ATP be readily transported out of the organelle. As well, a supply of ADP and inorganic phosphate is required in order to maintain maximum rates of electron transport and ATP synthesis. This is accomplished by two separate translocator proteins located in the inner membrane. An **adenine nucleotide transporter** located in the inner membrane (Figure 10.10) exchanges ATP and ADP on a one-for-one basis. An inorganic phosphate translocator exchanges P_i for hydroxyl ions.

10.9 PLANTS CONTAIN SEVERAL ALTERNATIVE ELECTRON TRANSPORT PATHWAYS

The electron transport chain described above is shared in essentially the same form by virtually all organisms:

plants, animals, and microorganisms. Plant mitochondria contain, in addition, several other redox enzymes, at least two of which are unique to plants (Figure 10.11). These enzymes have been discovered largely by virtue of their insensitivity to certain classic inhibitors of electron transport.

10.9.1 PLANT MITOCHONDRIA CONTAIN EXTERNAL DEHYDROGENASES

Unlike animal mitochondria, plant mitochondria contain "external" dehydrogenases that face the intermembrane space and are capable of oxidizing cytosolic NADH and NADPH respectively (Figure 10.11). As a consequence, electrons from the oxidation of either cytosolic NADH or NADPH are donated directly to the ubiquinone pool. Because the external dehydrogenase enzymes do not span the membrane, they will not translocate protons as Complex I does. Consequently, only two ATP can be formed from the transfer of each pair of electrons to oxygen.

10.9.2 PLANTS HAVE A ROTENONE-INSENSITIVE NADH DEHYDROGENASE

The reduction of ubiquinone by Complex I is sensitive to inhibition by **rotenone** and **amytal**. Plants, however, appear to have another NADH dehydrogenase that is insensitive to both of these electron transport inhibitors. Called the **rotenone-insensitive dehydrogenase**, this enzyme will oxidize only internal, or matrix, NADH (Figure 10.11). The enzyme must therefore be located on the inner surface of the membrane, facing the matrix.

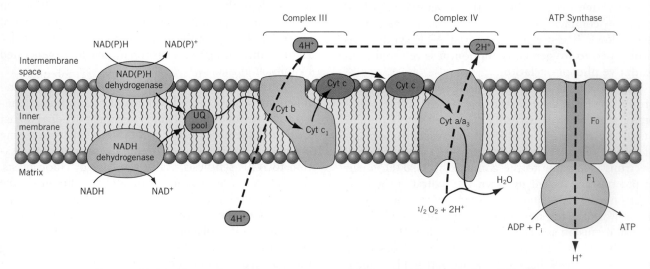

FIGURE 10.11 Alternative electron transport pathways in plant mitochondria. Electrons entering the chain through the alternative dehydrogenases will pass through two phosphorylating sites rather than three.

As with the external NADH and NADPH dehydrogenase enzymes, electrons entering the chain via the rotenone-insensitive dehydrogenase can generate only two ATP per electron pair.

Thus, the inner membrane of plant mitochondria contain four distinct NAD(P)H dehydrogenases exhibiting different P/O ratios: (1) an internal NADH dehydrogenase; (2) a rotenone-insensitive NADH dehydrogenase; (3) an external NADH dehydrogenase; and (4) an external NADPH dehydrogenase.

10.9.3 PLANTS EXHIBIT CYANIDE-RESISTANT RESPIRATION

Cytochrome *c* oxidase (Complex IV) is inhibited by cyanide (CN^-), carbon monoxide (CO), and azide (N_3^-). In many animals, all three of these inhibitors completely inhibit respiratory O_2 uptake. By contrast, most plants or plant tissues show considerable resistance to these inhibitors. In tissues such as roots and leaves of spinach (*Spinacea oleraceae*) or pea (*Pisum sativum*), for example, cyanide-resistant respiration may account for as much as 40 percent of total respiration. Cyanide-resistant respiration is, however, sensitive to inhibition by hydroxamic acid derivatives such as **salicylhydroxamic acid (SHAM)**. This cyanide-resistant, SHAM-sensitive respiration is attributed to a so-called **alternative oxidase** (Figure 10.12). The pathway is commonly referred to as the **alternative respiratory pathway**, or, simply, the alternative pathway. Although the existence of a cyanide-resistant alternative respiratory pathway has been widely accepted for more than a decade, the nature of the oxidase enzyme itself proved difficult to unravel. The enzyme has been difficult to study by conventional biochemical techniques; the protein appears to be relatively unstable and loses its activity rapidly upon isolation from the membrane. However, through a molecular biological approach, a gene encoding the oxidase protein has been cloned, first from *Sauromatum guttatum*, the voodoo lily, and since from tobacco, soybean, and other plants. This has led to significant advances in our understanding of the regulation of the enzyme.

The alternative oxidase is composed of two identical subunits (a homodimer) that span the inner mitochondrial membrane, with the active site facing the matrix side of the membrane. It functions as a ubiquinone O_2 oxidoreductase; that is, it accepts electrons from the ubiquinone pool and transfers them directly to oxygen. This is an important characteristic of the alternative oxidase because it means that electrons processed by this enzyme bypass at least two sites for proton extrusion. Consequently, energy that would otherwise be conserved as ATP is, in the case of the alternative oxidase, converted to heat instead. Depending on whether electrons are initially donated to Complex I, the rotenone-insensitive NADH-dehydrogenase, or succinic dehydrogenase, electrons passing through the alternative oxidase will contribute to the synthesis of one ATP or none (Figure 10.12).

The physiological role of alternative pathway respiration is still uncertain. One possible role is **thermogenesis**, a hypothesis based largely on events in the floral development in certain members of the family Araceae. Just prior to pollination in species such as skunk cabbage (*Symplocarpus foetidus*), the tissues of the spadix (the structure that bears both male and female flowers) undergoes a surge in oxygen consumption, called a **respiratory crisis**. The respiratory crisis is attributed almost entirely to an increase in alternative pathway respiration and can elevate the temperature of the spadix by as much as 10°C above ambient. The high temperature volatilizes certain odoriferous amines (hence, *skunk* cabbage) that attract insect pollinators. Thermogenesis does not, however, appear to be the function of the alternative pathway in roots and leaves. In one study of an arctic herb, for example, the alternative pathway

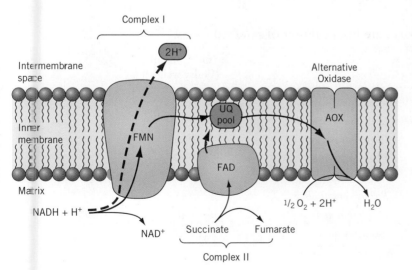

FIGURE 10.12 **The alternative respiratory pathway. Electrons intercepted by the alternative oxidase (AOX) pass through one or no phosphorylating sites.**

accounts for up to 75 percent of total respiration but, in part because the heat is rapidly dissipated, accounts for no more than a $0.02°C$ rise in leaf temperature.

A second hypothesis to explain the alternative pathway is referred to as the **energy overflow hypothesis**. This hypothesis is based on two general observations. First, in most tissues the alternative pathway is inoperative until the normal cytochrome pathway has become saturated. Second, the rate of the alternative pathway can be increased by increasing the supply of carbohydrate to cells. In spinach (*Spinacea oleraceae*), for example, the alternative pathway is engaged only after photosynthesis has been in operation for several hours and has built up a supply of carbohydrate. In other words, the alternative pathway is generally engaged when there is an excess supply of carbohydrate, over and above what is required for metabolism or processed for storage. The function of the alternative pathway, according to this hypothesis, would be to burn off temporary accumulations of excess carbon that might otherwise interfere with source-sink relationships and inhibit translocation.

In addition, the induction of the alternative oxidase represents a mechanism to prevent the overreduction of the respiratory electron transport chain, which would diminish the probability of superoxide formation and **oxidative stress** under conditions where ATP consumption has been slowed by either low temperature or other stresses.

10.10 MANY SEEDS STORE CARBON AS OILS THAT ARE CONVERTED TO SUGAR

Although lipids are a principal constituent of membranes and are stored by many tissues, they are not frequently used as a source of respiratory carbon. A major exception to this rule is found in germinating seeds, many of which store large quantities of lipids, principally triglycerides, as reserve carbon (Table 10.1). Storage lipids are deposited as **oil droplets** (also called oil bodies, oleosomes, or spherosomes), which are normally found in storage cells of cotyledons or endosperm.

Since fats and oils are not water soluble, plants are unable to translocate fats and oils through the phloem by pressure flow from seed storage tissues to the elongating roots and shoots where the energy and carbon are required to support growth. The fatty acids must first be converted to a form that is more readily translocated by the aqueous phloem. Usually this is sucrose (or sometimes stachyose), which is readily translocated from the storage cells containing the oil droplets to the embryo where the sucrose is metabolized. Complete conversion of triglycerides to sucrose is a complex process, involving the interaction of the oil bodies, glyoxysomes, mitochondria, and the cytosol (Figure 10.13).

We can summarize the conversion of triglycerides to sucrose as follows. The first step is the hydrolysis of triglycerides to free fatty acids and glycerol. This is accomplished through the action of **lipase** enzymes, which probably act at the surface of the oil droplet. The fatty acid then enters the **glyoxysome**, an organelle similar in structure to the peroxisome found in leaves but with many different enzymes. In the glyoxysome, the fatty acid undergoes β-oxidation; the fatty acid chain is cleaved at every second carbon, resulting in the formation of acetyl-CoA.

Some of the acetyl-CoA combines with oxaloacetate (originating in the mitochondrion) to form citrate (6 carbons) in what is known as the **glyoxylate cycle**. The citrate in turn is converted to isocitrate, which then breaks down into one molecule of succinate (4 carbons) and one molecule of glyoxylate (2 carbons). The succinate returns to the mitochondrion where it enters the citric acid cycle, regenerating oxaloacetate, which is necessary to keep the glyoxylate cycle turning. Glyoxylate combines with another acetyl-CoA to produce malate. The malate then enters the cytosol where it is first oxidized to oxaloacetate and

TABLE **10.1**   **Approximate lipid content of selected seeds.**

Species		Oil Content (% dry weight)
Macadamia nut	*Macadamia ternifolia*	75
Hazel nut	*Coryllus avellana*	65
Safflower	*Carthmus tinctoris*	50
Oil palm	*Elaeis guineensis*	50
Canola	*Brassica napus*	45
Castor bean	*Ricinus communis*	45
Sunflower	*Helianthus annum*	40
Maize	*Zea mays*	5

Hopkins & Huner.

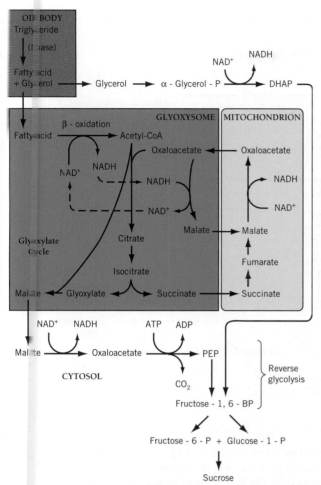

FIGURE 10.13 Lipid catabolism, the glyoxylate cycle, and gluconeogenesis.

energy changes are highly unfavorable in the direction of glucose synthesis. During gluconeogenesis, these reactions are replaced by reactions that make glucose synthesis more thermodynamically favorable. The conversion of fructose-1,6-bisphosphate to fructose-6-P is catalyzed by cytosolic fructose-1,6-bisphosphatase and the conversion of glucose-6-P to glucose is catalyzed by glucose-6-phosphatase. These differences are significant because they allow both directions to be thermodynamically favorable, yet be independently regulated. One direction can be activated while the other is inhibited, thus avoiding what might otherwise end up as a futile cycle. The glycerol resulting from lipase action in the oil droplet also enters the cytosol, where it is first phosphorylated with ATP to form α-glycerolphosphate and then oxidized to dihydroxyacetone phosphate (DHAP). The DHAP can also be converted to sucrose by reversal of glycolysis. Some of the energy stored in triglycerides is conserved in the sucrose formed by gluconeogenesis, but β-oxidation of fatty acids in the glyoxysome also produces a large amount of NADH. The glyoxysome is unable to reoxidize NADH directly, but it can be used to reduce oxaloacetate to malate (Figure 10.13). The malate then moves into the mitochondrion where it is reoxidized by malate dehydrogenase. Malate thus serves as a shuttle, carrying reducing equivalents between the glyoxysome and the mitochondrion. Reoxidation of malate inside the mitochondrion yields NADH, which can then enter the electron transport chain and drive ATP synthesis.

10.11 RESPIRATION PROVIDES CARBON SKELETONS FOR BIOSYNTHESIS

Before leaving the subject of cellular respiration, it is important to note that production of reducing potential and ATP is not the sole purpose of the respiratory pathways. In addition to energy, the synthesis of nucleic acids, protein, cellulose, and all other cellular molecules requires carbon skeletons as well. As noted at the beginning of this chapter, respiration also serves to modify the carbon skeletons of storage compounds to form these basic building blocks of cell structure. A few of the more important building blocks that can be formed from intermediates in glycolysis and the citric acid cycle are represented in Figure 10.14.

The withdrawal of glycolytic and citric acid cycle intermediates for the synthesis of other molecules means, of course, that not all of the respiratory substrate will be fully oxidized to CO_2 and water. The flow of carbon through respiration no doubt represents a balance between the metabolic demands of the cell for ATP to drive various energy-consuming functions on the one hand and demands for the reducing equivalents

decarboxylated to phosphoenolpyruvate (PEP). The glyoxylate cycle thus involves enzymes of both the glyoxysome and the mitochondrion. Two enzymes of the cycle are unique to plants: **isocitrate lyase**, which converts isocitrate to succinate plus glyoxylate, and **malate synthase**, which condenses an acetyl group with glyoxylate to form malate. The malate is then translocated from the glyoxysome into the cytosol where it is quickly oxidized to oxaloacetate by the enzyme malate dehydrogenase. The overall effect of the glyoxylate cycle is to catalyze the formation of oxaloacetate from two molecules of acetyl-CoA.

In the cytosol, oxaloacetate derived from the glyoxylate cycle is decarboxylated via the enzyme **phosphoenolpyruvate carboxykinase (PEPCK)** to form phosphoenolpyruvate (PEP). Through a sequence of reactions that is essentially a reversal of glycolysis, PEP is converted to glucose. The conversion of PEP to glucose by a reversal of glycolysis is known as **gluconeogenesis**. Gluconeogenesis utilizes the enzymes of glycolysis, with significant differences. The glycolytic phosphofructokinase and hexokinase reactions (Figure 10.4) are effectively irreversible—their free

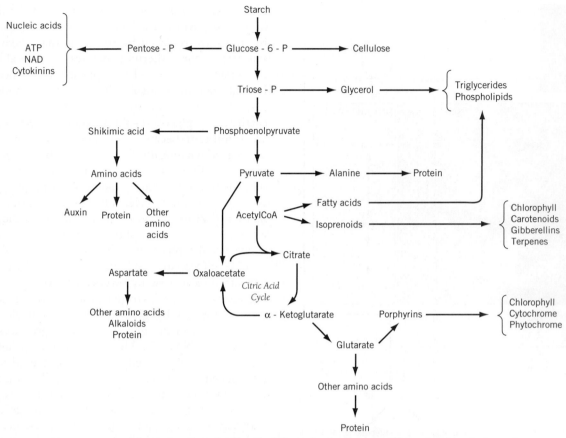

FIGURE 10.14 The role of respiration in biosynthesis. Intermediates in glycolysis and the citric acid cycle are drawn off to serve as building blocks for the synthesis of cellular molecules. Carbon in the cycle is maintained by the synthesis of oxaloacetate through anaplerotic reactions. This scheme is incomplete and is intended only to give some indication of the importance of these two schemes in biosynthesis.

and carbon skeletons required to build cell structure on the other.

It is also important to note that during periods of active synthesis, diversion of carbon from the citric acid cycle for synthetic reactions will lead to a significant reduction in the level of oxaloacetate. These synthetic reactions require not only carbon, but energy in the form of reducing potential and ATP as well. Without some means of compensating for this loss of oxaloacetate, the cycle will slow down or, in the extreme case, come to a complete halt and energy production will be impaired. This eventuality is precluded by the action of two cytosolic enzymes: phosphoenolpyruvate (PEP) carboxylase (see Chapter 15) and malate dehydrogenase. All plants, not just those with C_4 photosynthetic activity (see Chapter 15), have some level of PEP carboxylase activity that converts phosphoenolpyruvate (PEP) into oxaloacetate:

$$PEP + HCO_3^- \rightarrow oxaloacetate \qquad (10.9)$$

In this case the PEP is derived from glycolysis.

Although there is some evidence that oxaloacetate may be translocated directly into the mitochondrion, it

is more likely that oxaloacetate is quickly reduced to malate by the action of cytosolic malate dehydrogenase:

$$oxaloacetate + NADH \rightarrow malate + NAD^+ \qquad (10.10)$$

The malate would then pass into the mitochondrion, via a malate (or dicarboxylate) translocator, where it is reoxidized to oxaloacetate by the action of a mitochondrial malate dehydrogenase:

$$malate + NAD^+ \rightarrow oxaloacetate + NADH \qquad (10.11)$$

The replenishment of oxaloacetate in this way is an example of a "filling-up" mechanism or **anaplerotic** pathway. Thus carbon from glycolysis is delivered to the citric acid cycle through two separate but equally important streams: (1) to citrate via pyruvate and acetyl-CoA and (2) from PEP via oxaloacetate and malate to compensate for carbon "lost" to synthesis. Anaplerotic reactions such as the latter help to ensure that diversion of carbon for synthesis does not adversely influence the overall carbon balance between energy-generating catabolic reactions and biosynthetic anabolic reactions.

In addition to the normal enzymes of the citric acid cycle, plant mitochondria tend to have significant levels of **NAD$^+$ -malic enzyme**, which catalyzes the oxidative decarboxylation of malate:

$$malate + NAD^+ \rightarrow pyruvate + CO_2 + NADH \quad (10.12)$$

The pyruvate may be further metabolized by pyruvate dehydrogenase to acetyl-CoA and from there enter the citric acid cycle. Thus the mitochondrial pool of malate may replenish the citric acid intermediates through either oxaloacetate or pyruvate.

In addition to serving an anaplerotic role, the uptake and oxidation of malate by mitochondria via either malic enzyme or malate dehydrogenase also provides an alternative pathway for metabolizing malate. This alternative pathway may be particularly significant in plants such as those of the family Crassulaceae (see Chapter 15) and others that store significant levels of malate in their vacuoles. Finally, it should be noted that diversion of pyruvate through oxaloacetate and malate bypasses the pyruvate kinase step in glycolysis (Figure 10.5, reaction 6) and thus reduces the yield of ATP by one. This reduction is, however, offset by gains achieved by reduction of malate in the cytosol and its subsequent reoxidation in the mitochondrion. This sequence of reactions effectively shuttles extramitochondrial NADH (generated during glycolysis) into the mitochondrion, where it can be used to generate three molecules of ATP. This is a gain of one ATP over the two ATP generated via the NADH-reductase route described earlier for extramitochondrial NADH.

10.12 RESPIRATORY RATE VARIES WITH DEVELOPMENT AND METABOLIC STATE

The study of respiration at the level of individual organs or the whole plant becomes much more difficult than it is for the study of individual cells. Whole plant respiration is normally studied by measuring the uptake of oxygen or the evolution of CO_2, but respiration rates obtained this way are highly variable. The balance of O_2 and CO_2 exchange is dependent on the substrate being respired and the balance of fermentation, citric acid cycle, and alternative pathway activities at any point in time. In addition, respiration rates differ between organs, change with age and developmental state, and are markedly influenced by temperature, oxygen, salts, and other environmental factors. Nevertheless, the study of respiration at the organ and plant level is a field of active study. Understanding respiration at this level has important implications for the plant physiologist interested in growth and development, for the physiological ecologist interested in plant biomass production, and the agricultural scientist because of its impact on productivity and yield.

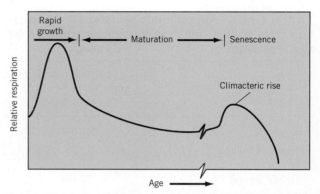

FIGURE 10.15 **Respiratory rate as a function of age. This type of curve applies generally to most herbaceous plants, tissues, and organs. The magnitude of the climacteric will vary—some organs exhibit little or no climacteric rise.**

As a general rule, respiratory rate is a reflection of metabolic demands. Younger plants, organs, or tissues respire more rapidly than older plants, organs, or tissues (Figure 10.15). The rapid rate of respiration during early stages of growth is presumably related to synthetic requirements of rapidly dividing and enlarging cells. As the plant or organ ages and approaches maturity, growth and its associated metabolic demands decline. Many organs, especially leaves and some fruits, experience a transient rise in respiration, called a **climacteric**, that marks the onset of senescence and the degenerative changes that precede death. Typically the climacteric rise in O_2 consumption is accompanied by a decline in oxidative phosphorylation, indicating that ATP production is no longer tightly coupled to electron transport. The respiration rate of woody stems and branches, expressed on a weight or mass basis, also declines as they grow. This is because as the diameter increases, the relative proportion of nonrespiring woody tissue also increases.

Carbon lost to the plant due to respiration can represent a significant proportion of the available carbon. Actual respiration rates for plant tissues range in the extreme between barely detectable (0.005 μmol CO_2 gW_d^{-1} h^{-1}) in dormant seeds to 1000 μmol CO_2 gW_d^{-1} h^{-1} or more in the spadix of skunk cabbage during the respiratory crisis. More typically, rates for vegetative tissues range from 10 to 200 (mol CO_2 gW_d^{-1} h^{-1} (Table 10.2). This may represent a considerable fraction of the carbon assimilated by photosynthesis during a 24-hour period. (Recall that photosynthesis occurs only during daylight hours, but respiration, especially of roots and similar tissues, is ongoing 24 hours a day.) On the average, 30 to 60 percent of daily photoassimilate is lost as respiratory CO_2. In tropical rainforest species, probably because of accelerated enzymic activities at higher temperatures, this loss may exceed 70 percent. Of the total daily

TABLE 10.2 **Approximate specific dark respiration rates at 20°C for crop species, deciduous foliage, and conifers.**

	Specific Respiration Rate μ mol CO_2 evolved g^{-1} dry mass h^{-1}
Crops	70–180
Deciduous foliage	
(sun leaves)	70–90
(shade leaves)	20–45
Conifers	4–25

Hopkins & Huner.

carbon loss in any given plant, some 30 to 70 percent is accounted for by respiration in the roots alone, that "hidden half" of the plant.

10.13 RESPIRATION RATES RESPOND TO ENVIRONMENTAL CONDITIONS

10.13.1 LIGHT

The effects of light on mitochondrial respiration have been the subject of considerable debate for some time. Traditionally, photosynthesis investigators and crop physiologists have tacitly assumed that respiration continues in the light at a rate comparable to that in the dark. The true rate of photosynthesis is therefore taken as equal to the apparent rate (measured as CO_2 uptake) plus the rate of respiration (CO_2 evolved) in the dark. However, attempts to study respiration in green leaves have led to alternative and conflicting conclusions. These range from complete inhibition of mitochondrial activities, to partial operation of the citric acid cycle, or to stimulation of respiration by light. The problem lies in the difficulty of measuring respiration during a period when gas exchange is dominated by the overwhelming flux of CO_2 and O_2 due to photosynthesis, the recycling of CO_2 within the leaf, and the exchange of metabolites by chloroplasts and mitochondria.

Light effects on respiration during a subsequent dark period have been demonstrated. For example, dark respiratory rates in leaves adapted to full sun (sun leaves) are generally higher than those of leaves of the same species adapted to shade (shade leaves) (Table 10.2). As well, the rate is consistently higher in mature leaves of shade-intolerant species than in shade-tolerant species. Indeed, a reduced respiratory rate appears to be a fairly consistent response to low irradiance. This is probably related to lower growth rates also observed in shade-grown plants, but it is not known which is cause

and which is effect. It is projected that low respiratory and growth rates may confer a survival advantage under conditions of deep shade. The basis for light regulation of respiratory rate is unknown, although some have suggested that low respiratory rates under conditions of low irradiance may reflect availability of substrate. For example, the respiration rate 1 to 2 hours following a period of active photosynthesis is higher than after a long dark period.

Other experiments, however, have shown that dark respiratory rates do not correlate with CO_2 supply during the previous light period. This seems to suggest a more direct effect on respiration. The pyruvate dehydrogenase complex (PDH) of the mitochondrion can exist either in an active, nonphosphorylated form or an inactive, phosphorylated form (Figure 10.16). It has been shown that the photorespiratory-generated NH_4^+ (Chapter 8) stimulates the phosphorylation of the mitochondrial pyruvate dehydrogenase complex, thereby inhibiting the rate of respiration in the light by decreasing the rate at which acetyl-CoA is generated for the CAC. Light regulation of respiration remains a controversial issue in plant physiology.

10.13.2 TEMPERATURE

One of the most commonly applied quantitative measures used to describe the effect of temperature on a process is the **temperature coefficient**, or Q_{10}, given by the expression:

$$Q_{10} = \frac{\text{rate at} (t + 10)°C}{\text{rate at } t°C} \qquad (10.13)$$

At temperatures between 5°C and about 25°C or 30°C, respiration rises exponentially with temperature and the

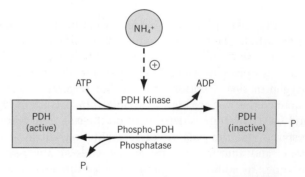

FIGURE 10.16 Regulation of mitochondrial pyruvate dehydrogenase complex. The pyruvate dehydrogenase complex (PDH) exists in two forms: an active, nonphosphorylated form and an inactive, phosphorylated form. The reversible interconversion of the active and inactive forms of this enzyme is the result of the activity of the PDH kinase, which consumes ATP to phosphorylate PDH, and the activity of the phospho-PDH phosphatase, which dephosphorylates PDH. Ammonium ion (NH_4^+) stimulates the PDH kinase and hence inactivates the PDH complex.

Q_{10} value is approximately 2.0 in many but not all plants (see Chapter 14). Within this temperature range, a doubling of rate for every $10°C$ rise in temperature is typical of enzymic reactions. At temperatures above $30°C$, the Q_{10} in most plants begins to fall off as substrate availability becomes limiting. In particular, the solubility of O_2 declines as temperatures increase and the diffusion rate (with a Q_{10} close to 1) does not increase sufficiently to compensate. As temperatures approach $50°C$ to $60°C$, thermal denaturation of respiratory enzymes and damage to membranes bring respiration to a halt.

Some investigators have observed differences in the rate of respiration in tropical, temperate, and arctic species at different temperatures. For example, the respiration rate of leaves of tropical plants at $30°C$ is about the same as that of arctic species at $10°C$. The temperature coefficient (Q_{10}) for respiration is the same in both cases and there is no evidence of intrinsic differences in the biochemistry of respiration. It is likely that the differences reflect differences in temperature optima for growth of arctic and tropical species—optima determined by factors other than respiration—and the consequent metabolic demand for ATP. A detailed discussion of the effects of temperature stress and acclimation on respiration will be focused in Chapters 13 and 14.

10.13.3 OXYGEN AVAILABILITY

As the terminal electron acceptor, oxygen availability is obviously an important factor in determining respiration rate. The oxygen content of the atmosphere is relatively stable at about 21 percent O_2. The equilibrium concentration of oxygen in air-saturated water, including the cytosol, is approximately $250\,\mu M$. However, cytochrome c oxidase has a very high affinity for oxygen with a K_m (see Chapter 8, Box 8.1) less than $1\,\mu M$. Under normal circumstances, oxygen is rarely a limiting factor.

There are some situations, however, where oxygen availability may become a significant factor. One is in bulky tissues with low surface-to-volume ratios, such as potato tubers and similar storage tissues, where the diffusion of oxygen may be slow enough to restrict respiration. This may not be a serious problem, however. A significant volume—as much as 40 percent—of roots and similar tissues may be occupied by intercellular air spaces that aid in the rapid distribution of O_2 absorbed from the soil or, in some cases, from the aerial portions of the plant. Plants are most likely to experience oxygen deficits during periods of flooding, when air in the large pore spaces of the soil is displaced by water, thereby decreasing the oxygen supply to the roots. For similar reasons, plants grown in hydroponic culture must be aerated to maintain adequate oxygen levels in the vicinity of the roots (Chapter 3).

SUMMARY

Cellular respiration consists of a series of interdependent pathways by which carbohydrate and other molecules are oxidized for the purpose of retrieving the energy stored in photosynthesis and to obtain the carbon skeletons that serve as precursors for other molecules used in the growth and maintenance of the cell. Plants store excess photosynthate either as starch, a long linear or branched polymer of glucose, in the chloroplast stroma or as fructans, a polymer of fructose, in the vacuole. Storage carbohydrates such as starch and fructans are enzymatically degraded to glucose or fructose, which then enter the cytosolic hexose phosphate pool as either glucose-1-phosphate, glucose-6-phosphate, or fructose-6-phosphate. Hexose phosphates exit the hexose-P pool by conversion to fructose-1,6-bisphosphate (FBP). FBP is subsequently converted to triose-P, which is the starting point for glycolysis, a series of reactions that ultimately produce pyruvate. In the process, a small amount of ATP and reducing potential is generated. The intermediates in glycolysis are three-carbon sugars, many of which are precursors to triglycerides and amino acids. Precursors with four and five carbons are produced by an alternative route for glucose metabolism called the oxidative pentose phosphate pathway. The oxidative pentose phosphate pathway also produces NADPH (as opposed to NADH), which provides reducing potential when required for biosynthetic reactions in plants.

The fate of pyruvate depends on the availability of oxygen. In an anaerobic environment, pyruvate is reduced (usually to ethanol), while in the presence of oxygen pyruvate is first oxidized to acetyl-CoA and carbon dioxide. The acetate group is then further oxidized to carbon dioxide and water through the citric acid cycle (CAC). The CAC enzymes are located predominantly in the matrix of the mitochondrion. Altogether, eight enzyme-catalyzed steps degrade the acetate group to carbon dioxide and water. The cycle includes four oxidations that yield NADH at three steps and $FADH_2$ at another. One molecule of ATP is generated in a substrate-level phosphorylation and the original acetate acceptor, oxaloacetate, is regenerated, which allows the cycle to continue. The NADH and $FADH_2$ produced in the CAC are oxidized via an electron transport chain found in the mitochondrial inner membrane. The chain consists of four multiprotein complexes linked by mobile electron carriers. The final complex, cytochrome oxidase, transfers the electrons to molecular oxygen, forming water. At three points in the chain, the free energy drop associated with electron transport is used to establish a proton gradient across the membrane. As in the chloroplasts, this proton gradient is used to drive ATP synthesis.

The CAC and electron transport chain are virtually identical in all organisms. Plants, however, have an alternative oxidase that intercepts electrons early in the chain, thus bypassing two of the three proton-pumping sites. When using this route, at least two-thirds less ATP is formed and much of the electron energy is converted to heat. The alternative oxidase, at least in some plants, has been associated with thermogenesis, particularly in certain members of the Araceae where the higher temperatures volatilize amines that appear to attract insect pollinators. The alternative oxidase may also serve to "burn off" excess carbohydrate as well as protect against oxidative stress by preventing the overreduction of the respiratory electron transport chain.

Many seeds store carbon as fats and oils, which must first be converted to sugar in order to be respired. After the fatty acids are broken down into acetyl-CoA units, a complex series of reactions involving enzymes of the mitochondrion, the glyoxysome, and the cytosol convert the acetate units to phospho*enol*pyruvate. The pyruvate is then converted to glucose by gluconeogenesis, a process that is essentially a reversal of glycolysis.

The respiratory rate of whole plants and organs varies widely with age, metabolic state, and environmental conditions.

CHAPTER REVIEW

1. Compare respiration and fermentation. Aerobic organisms are generally much larger than anaerobic organisms. Can you suggest how this may be related to respiration?

2. Phosphorous plays an important role in respiration, photosynthesis, and metabolism generally. What is this role?

3. Oils are a common storage form in seeds, particularly small seeds. What advantage does this offer the seed?

4. There are a number of similarities between chloroplasts and mitochondria. Compare these two organelles from the perspective of:

(a) ultrastructure and biochemical compartmentation;
(b) organization of the electron transport chain;
(c) proton motive force (pmf) and ATP synthesis;
(d) alternative oxidases.

5. How does cyanide-resistant respiration differ from normal respiration? Of what value might cyanide-resistant respiration be to plants?

6. Discussions of respiration most often emphasize energy retrieval and ATP production. What other very important metabolic role(s) does respiration fulfill?

7. Give an example of an anaplerotic pathway. What is the role of this pathway?

FURTHER READING

Buchanan, B. B., W. Gruissem, R. L. Jones. 2000. *Biochemistry and Molecular Biology of Plants*. Rockville, MD: American Society of Plant Physiologists.

Foyer, C. H., S. Ferrario-Méry, S. C. Huber. 2000. Regulation of carbon fluxes in the cytosol: Coordination of sucrose synthesis, nitrate reduction and organic and amino acid biosynthesis. In: *Photosynthesis: Physiology and Metabolism* R. C. Leegood, T. D. Sharkey, S. von Caemmerer, (eds.), *Advances in Photosynthesis and Respiration*, Vol. 9, pp. 177–203. Dordrecht: Kluwer.

Gardeström, P., A. U. Igamberdiev, A. S. Raghavendra. 2002. Mitochondrial functions in the light and significance to carbon-nitrogen interactions. In *Photosynthesis: Physiology and Metabolism* R. C. Leegood, T. D. Sharkey, S. von Caemmerer (eds.), *Advances in Photosynthesis and Respiration*, Vol. 9, pp. 151–172. Dordrecht: Kluwer.

Vanlerberghe, G. C., L. McIntosh. 1997. Alternative oxidase: From gene to function. *Annual Review of Plant Physiology and Plant Molecular Biology* 48: 703–734.

Vanlerberghe, G. C., S. H. Ordog. 2002. Alternative oxidase: Integrating carbon metabolism and electron transport in plant respiration. *Photosynthesis: Physiology and Metabolism* In: R. C. Leegood, T. D. Sharkey, S. von Caemmerer (eds.), *Advances in Photosynthesis and Respiration*, Vol. 9, pp. 173–191. Dordrecht: Kluwer.

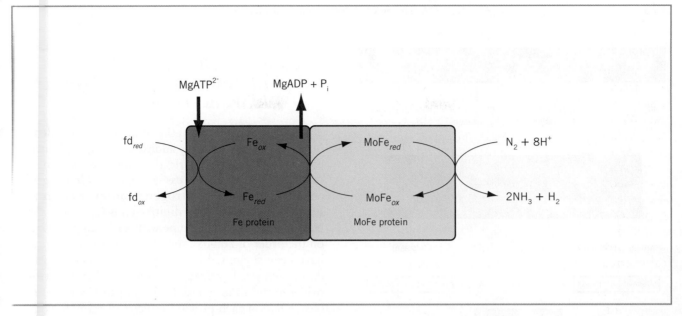

11

Nitrogen Assimilation

On a dry-weight basis, nitrogen is the fourth most abundant nutrient element in plants. It is an essential constituent of proteins, nucleic acids, hormones, chlorophyll, and a variety of other important primary and secondary plant constituents. Most plants obtain the bulk of their nitrogen from the soil in the form of either nitrate (NO_3^-) or ammonium (NH_4^+), but the supply of nitrogen in the soil pool is limited and plants must compete with a variety of soil microorganisms for what nitrogen is available. As a result, nitrogen is often a limiting nutrient for plants, in both natural and agricultural ecosystems.

The bulk of the atmosphere, 78 percent by volume, consists of molecular nitrogen (N_2, or **dinitrogen**), an odorless, colorless gas. In spite of its abundance, however, higher plants are unable to convert dinitrogen into a biologically useful form. The two nitrogen atoms in dinitrogen are joined by an exceptionally stable bond ($N \equiv N$) and plants do not have the enzyme that will reduce this triple covalent bond. Only certain prokaryote species are able to carry out this important reaction. This situation presents plants with a unique problem with respect to the uptake and assimilation of nitrogen; plants must depend on prokaryote organisms to convert atmospheric dinitrogen into a usable form. The nature of this problem and the solutions that have evolved are the subject of this chapter.

The principal topics discussed in this chapter include:

- a review of the nitrogen cycle: the flow of nitrogen between three major global nitrogen pools,
- the biology and biochemistry of biological nitrogen-fixing systems, and
- pathways for assimilation of ammonium and nitrate nitrogen by plants.

11.1 THE NITROGEN CYCLE: A COMPLEX PATTERN OF EXCHANGE

The global nitrogen supply is generally distributed between three major pools: the atmospheric pool, the soil (and associated groundwater) pool, and nitrogen contained within the biomass. Central to the idea of a nitrogen cycle (Figure 11.1) is the pool of nitrogen found in the soil. Nitrogen from the soil pool enters the biomass principally in the form of nitrate (NO_3^-) taken up by plants and microorganisms. Once assimilated, nitrate nitrogen is converted to organic nitrogen in the form of amino acids and other nitrogenous building blocks of proteins and other macromolecules. Nitrogen moves further up the food chain when animals consume

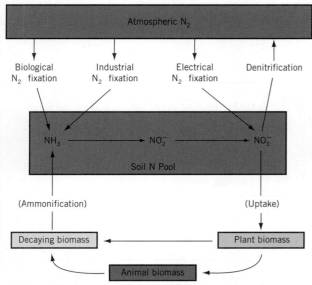

Figure 11.1 **The nitrogen cycle, illustrating relationships between the three principal nitrogen pools: atmospheric, soil, and biomass.**

plants. Nitrogen is returned to the soil through animal wastes or the death and subsequent decomposition of all organisms.

11.1.1 AMMONIFICATION, NITRIFICATION, AND DENITRIFICATION ARE ESSENTIAL PROCESSES IN THE NITROGEN CYCLE

In the process of decomposition, organic nitrogen is converted to ammonia by a variety of microorganisms including the fungi. This process is known as **ammonification** (Figure 11.1). Some of the ammonia may volatilize and reenter the atmosphere, but most of it is recycled to nitrate by soil bacteria. The first step in the formation of nitrate is the oxidization of ammonia to nitrite (NO_2^-) by bacteria of the genera *Nitrosomonas* or *Nitrococcus*. Nitrite is further oxidized to nitrate by members of the genus *Nitrobacter*. These two groups are known as nitrifying bacteria and the result of their activities is called **nitrification**. Nitrifying bacteria are **chemoautotrophs**; that is, the energy obtained by oxidizing inorganic substances such as ammonium or nitrite is used to convert carbon dioxide to organic carbon.

In taking up nitrate from the soil, plants must compete with bacteria known as **denitrifiers** (e.g., *Thiobacillus denitrificans*). By the process of **denitrification**, these bacteria reduce nitrate to dinitrogen, which is then returned to the atmosphere. Estimates for the amount of nitrogen lost to the atmosphere by denitrification range from 93 million to 190 million metric tons annually.

11.2 BIOLOGICAL NITROGEN FIXATION IS EXCLUSIVELY PROKARYOTIC

The loss of nitrogen from the soil pool through denitrification is largely offset by additions to the pool through conversion of atmospheric dinitrogen to a combined or *fixed* form. The process of reducing dinitrogen to ammonia is known as **nitrogen fixation** or **dinitrogen fixation**. Just how much dinitrogen is fixed on a global basis is difficult to determine with any accuracy. Figures on the order of 200 to 250 million metric tons annually have been suggested. Approximately 10 percent of the dinitrogen fixed annually is accounted for by nitrogen oxides in the atmosphere. Lightning strikes and ultraviolet radiation each provide sufficient energy to convert dinitrogen to atmospheric nitrogen oxides (NO, N_2O) (Figure 11.1). Another 30 percent of the total dinitrogen fixed, 80 million metric tons, is accounted for by industrial nitrogen fixation for the production of agricultural fertilizers through the Haber-Bosch process (Figure 11.1). However, by far, the bulk of the nitrogen fixed on a global scale, about 60 percent or 150 to 190 million metric tons annually, is accounted for by the reduction of dinitrogen to ammonia by living organisms (Figure 11.1).

Plants are **eukaryotic** organisms, distinguished by the presence of a membrane-limited nuclear compartment. Eukaryotic organisms are unable to fix dinitrogen because they do not have the appropriate biochemical machinery. Bacteria and cyanobacteria are **prokaryotic** organisms; the genetic material is not contained within a membrane-limited organelle. Nitrogen fixation is a prokaryote domain, because only prokaryote organisms have the enzyme complex, called **dinitrogenase**, that catalyzes the reduction of dinitrogen to ammonia. Simple as this may seem, biological nitrogen fixation turns out to be a complex biochemical and physiological process.

Prokaryotes that fix nitrogen, called **nitrogen-fixers**, include both free-living organisms and those that form symbiotic associations with other organisms.

11.2.1 SOME NITROGEN-FIXING BACTERIA ARE FREE-LIVING ORGANISMS

Free-living, nitrogen-fixing bacteria are widespread. Their habitats include marine and freshwater sediments, soils, leaf, and bark surfaces, and the intestinal tracts of various animals. Although some species are aerobic (e.g., *Azotobacter*, *Beijerinckia*), most will fix dinitrogen only under anaerobic conditions or in the presence of very-low-oxygen partial pressures (a condition known as **microaerobic**). These include both nonphotosynthetic genera (*Clostridium*, *Bacillus*, *Klebsiella*) and photosynthetic genera (*Chromatium*, *Rhodospirillum*) of bacteria. In addition to the bacteria, several genera of cyanobacteria (principally *Anabaena*, *Nostoc*, *Lyngbia*, and *Calothrix*) are represented by nitrogen-fixing species.

Although free-living nitrogen-fixing organisms are widespread, most grow slowly and, except for the photosynthetic species, tend to be confined to habitats rich in organic carbon. Because a high proportion of their respiratory energy is required to fix dinitrogen, less energy is therefore available for growth.

11.2.2 SYMBIOTIC NITROGEN FIXATION INVOLVES SPECIFIC ASSOCIATIONS BETWEEN BACTERIA AND PLANTS

Several types of symbiotic nitrogen-fixing associations are known, including the well-known association between various species of bacteria and leguminous plants. Some of the more important associations are listed in Table 11.1. In symbiotic associations the plant is identified as the **host** and the microbial partner is known as the **microsymbiont**. The most common form of symbiotic association results in the formation of enlarged, multicellular structures, called **nodules**, on the root (or occasionally the stem) of the host plant (Figure 11.2). In the case of legumes,[1] the microsymbiont is a bacterium of one of three genera: *Rhizobium*, *Bradyrhizobium*, or *Azorhizobium*. Collectively, these organisms

TABLE 11.1　Some examples of specificity in rhizobia-legume symbiosis.

Bacterium	Host
Azorhizobium	*Sesbania*
Bradyrhizobium japonicum	*Glycine* (soybean)
Rhizobium meliloti	*Medicago* (alfalfa),
	Melilotus (sweet clover)
Rhizobium leguminosarum	
biovar viciae	*Lathyrus* (sweet pea),
	Lens (lentil),
	Pisum (garden pea),
	Vicia (vetch, broad bean)
biovar trifolii	*Trifolium* (clover)
biovar phaseoli	*Phaseolus* (bean)
Rhizobium loti	*Lotus* (bird's-foot trefoil)

[1]The legumes are a heterogeneous group traditionally assigned to the family Leguminosae. Modern treatments split the group into three families: Mimosaceae, Caesalpiniaceae, and Fabaceae (S. B. Jones, A. E. Luchsinger, *Plant Systematics*, New York: McGraw-Hill, 1986). Most of the economically important, nitrogen-fixing legumes are assigned to the Fabaceae.

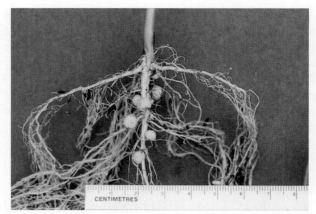

FIGURE 11.2　**Nitrogen-fixing nodules on roots of soybean (*Glycine max*).**

are referred to as **rhizobia**. Curiously, only one non-leguminous genus, *Parasponia* (of the family Ulmaceae), is known to form root nodules with a rhizobia symbiont.

The rhizobia are further divided into species and subgroups called **biovars** (a biological variety) according to their host range (Table 11.1). Most rhizobia are restricted to nodulation with a limited number of host plants while others are highly specific, infecting only one host species.

Nodules are also found in certain nonleguminous plants such as alder (*Alnus*), bayberry (*Myrica*), Australian pine (*Casuarina*), some members of the family Rosaceae, and certain tropical grasses. However, the microsymbiont in these nonleguminous nodules is a filamentous bacterium (*Frankia*) of the group actinomycetes. Both *Rhizobium* and *Frankia* live freely in the soil but fix dinitrogen only when in symbiotic association with an appropriate host plant.

A limited number of non-nodule-forming associations have been studied, such as that between *Azolla* and the cyanobacterium *Anabaena*. *Azolla* is a small aquatic fern that harbors *Anabaena* in pockets within its leaves. In southeast Asia, *Azolla* has proven useful as green manure in the rice paddy fields where it is either applied as a manure or co-cultivated along with the rice plants. Because more than 75 percent of the rice acreage consists of flooded fields, free-living cyanobacteria and anaerobic bacteria may also make a significant contribution. These practices have allowed Asian rice farmers to maintain high productivity for centuries without resorting to added chemical fertilizers.

11.3 LEGUMES EXHIBIT SYMBIOTIC NITROGEN FIXATION

It is generally agreed that symbiotic nitrogen fixers, particularly legumes, contribute substantially more nitrogen to the soil pool than do free-living bacteria.

Typically a hectare of legume-*Rhizobium* association will fix 25 to 60 kg of dinitrogen annually, while nonsymbiotic organisms fix less than 5 kg ha^{-1}. There are over 17,000 species of legumes. Even though only 20 percent have been examined for nodulation, 90 percent of those examined do form nodules. Given the obvious significance of the legume symbiosis to the nitrogen cycle, it is worth examining in some detail.

11.3.1 RHIZOBIA INFECT THE HOST ROOTS, WHICH INDUCES NODULE DEVELOPMENT

The sequence of events beginning with bacterial infection of the root and ending with formation of mature, nitrogen-fixing nodules has been studied extensively in the legumes, historically from the morphological perspective and more recently from the biochemical/molecular genetic perspective. Overall the process involves a sequence of multiple interactions between the bacteria and the host roots. In effect, the rhizobia and the roots of the prospective host plant establish a dialogue in the form of chemical messages passed between the two partners. Based on studies carried out primarily with *Glycine*, *Trifolium*, and *Pisum*, as many as nine or ten separate developmental stages have been recognized. In order to simplify our discussion, however, we will consider the events in four principal stages:

1. Multiplication of the rhizobia, colonization of the rhizosphere, and attachment to epidermal and root hair cells.

2. Characteristic curling of the root hairs and invasion of the bacteria to form an infection thread.

3. Nodule initiation and development in the root cortex. This stage is concurrent with stage 1.

4. Release of the bacteria from the infection thread and their differentiation as specialized nitrogen-fixing cells.

The four principal stages are illustrated in Figure 11.3. In this section, we will concentrate on the physiology and morphology of infection and nodule development. Genetic aspects will be addressed in a subsequent section.

11.3.1.1 Early stage involves colonization and nodule initiation. Rhizobia are free-living, saprophytic soil bacteria. Their numbers in the soil are highly variable, from as few as zero or 10 to as many as 10^7 gram^{-1} of soil, depending on the structure of the soil, water content, and a variety of other factors. In the presence of host roots, the bacteria are encouraged to multiply and colonize the rhizosphere. The initial attraction of rhizobia to host roots appears to involve **positive chemotaxis**, or movement toward a chemical stimulant. Chemotaxis is an important adaptive feature in microorganisms generally. It allows the organism

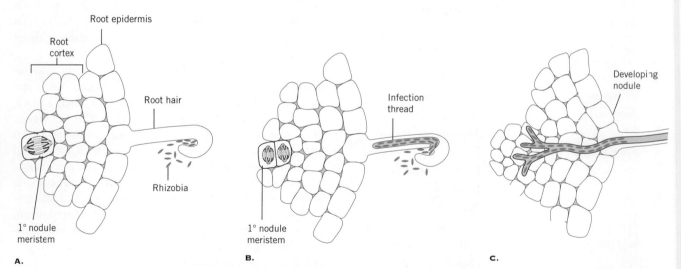

FIGURE 11.3 **Schematic diagram of the infection process leading to nodule formation. (*A*) Rhizobia colonize the soil in the vicinity of the root hair in response to signals sent out from the host root. The rhizobia in turn stimulate the root hair to curl while, at the same time, sending mitogenic signals that stimulate cell division in the root cortex. (*B*) Rhizobia invade the root by digesting the root hair cell wall and forming an infection thread. The rhizobia continue to multiply as the infection thread elongates toward the root cortex. (*C*) The infection thread branches to penetrate numerous cortical cells as a visibly evident nodule develops on the root. The final stage (not shown) is the release of rhizobia into the host cells and the activation of the nitrogen-fixing machinery.**

to detect nutrients and other chemicals that are either beneficial or required for their growth and reproduction. A group of chemicals that have been implicated in attraction of rhizobia are the **flavonoids** (Figure 11.4).

Once rhizobia have colonized the rhizosphere, they begin to synthesize morphogenic signal molecules called nodulation factors, or **nod factors**. Nod factors are derivatives of chitin, a β-$(1 \rightarrow 4)$-linked polymer of N-acetyl-D-glucosamine found in the cell walls of fungi and exoskeletons of insects. Nod factors are similar polymers except that a fatty acid replaces the acetyl group at one end of the molecule. Nod factors are consequently considered **lipo-chitooligosaccharides**. Nod factors secreted into the soil solution by the rhizobia induce several significant changes in the growth and metabolism of the host roots as a prelude to rhizobial invasion of the root hair and subsequent nodule development. These changes (Figure 11.3A) include increased root hair production and the development of shorter, thicker roots. Stimulated by the nod factors to renew their growth, the root hairs develop branching and curl at the tip.

Before actually invading the host, rhizobia also release mitogenic signals that stimulate localized cell divisions in the root cortex. These cell divisions form the **primary nodule meristem**, defining the region in which the nodule will eventually develop (Figure 11.3A). A second center of cell division arises in the pericycle. Eventually these two masses of dividing cells will fuse to form the complete nodule.

Rhizobia-host specificity is probably determined when the rhizobia attach to the root hairs and must involve some form of recognition between symbiont and host. As a general principle, recognition between cells involves chemical linkages that form between unique molecules on cell surfaces. In the case of rhizobia-host interactions, recognition appears to involve two classes of molecules: **lectins** and complex polysaccharides. Lectins are small, nonenzymatic proteins synthesized by the host and have the particular ability to recognize and bind to specific complex carbohydrates.

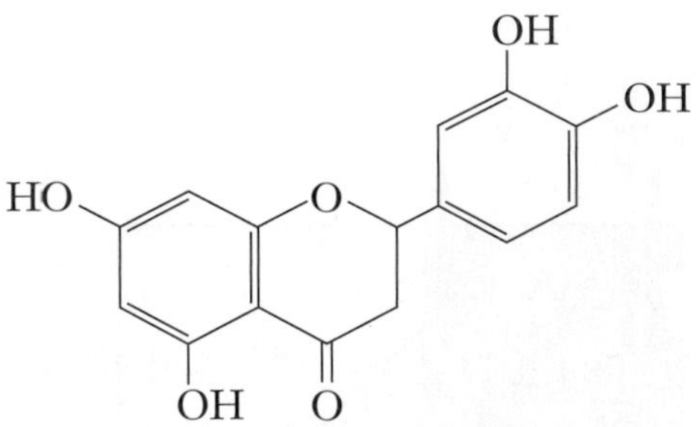

Individual legume species each produce different lectins with different sugar-binding specificities. Lectins appear to recognize complex polysaccharides found on the surface of the potential symbiont. Although bacterial surfaces normally contain an array of complex extracellular polysaccharides, the synthesis of additional nodulation-specific extracellular polysaccharides is directed by bacterial genes that are activated in the presence of flavonoids in the host root exudate. Host range specificity would thus result from attachment of the rhizobium to the host root hair because of specific lectin-surface polysaccharide interactions. Support for this hypothesis comes from experiments in which the gene for pea lectin was introduced into roots of white clover. The result was that clover roots could be nodulated by strains of *Rhizobium leguminosarum*, biovar *viciae*, which are usually specific for peas.

11.3.1.2 Second stage involves invasion of the root hair and the formation of an infection thread.
In the second stage of nodulation, the bacterium must penetrate the host cell wall in order to enter the space between the wall and the plasma membrane. In pea, the preferred attachment site is the tip of the growing root hair. The root hairs of pea grow by **tip growth**; that is, new wall material is laid down only at the tip of the elongating hair cell. Colonies of attached rhizobia become entrapped by the tip of the root hair as it curls around. How rhizobia actually breach the cell wall is not known, but the process almost certainly includes some degree of wall degradation. There is some evidence that rhizobia release enzymes such as pectinase, hemicellulase, and cellulase, which degrade cell wall materials. These enzymes could result in localized interference with the assembly of the growing wall at the root tip and allow the bacteria to breach the cell wall and gain access to the underlying plasma membrane.

Once the rhizobia reach the outer surface of the plasma membrane, tip growth of the root hair ceases and the cell membrane begins to invaginate. The result is a tubular intrusion into the cell called an **infection thread**, which contains the invading rhizobia (Figure 11.3B). The infection thread elongates by adding new membrane material by fusion with vesicles derived from the Golgi apparatus. As the thread moves through the root hair cell, a thin layer of cellulosic material is deposited on the inner surface of its membrane. Because this new wall material is continuous with the original cell wall, the invading bacteria never actually enter the host cell but remain technically outside the cell.

The infection thread continues to elongate until it reaches the base of the root hair cell. Here it must again breach the cell wall in order for the bacteria to gain access to the next cell in their path. This is apparently

Luteolin

FIGURE 11.4 **Structure of a common flavonoid implicated in rhizobia-host interactions. Leuteolin (flavone) is released by the host root. The flavonoid interacts with the product of the bacterial *nod*D gene, leading to the induction of other nodulation genes.**

accomplished by fusing the infection thread membrane with the plasma membrane. In the process, some bacteria are released into the apoplastic space. These bacteria apparently degrade the walls of the next cell in line, thus allowing the infection process to continue into successive cells in the cortex. As the infection thread moves through the root hair into the cortex, the bacteria continue to multiply. When the thread reaches the developing nodule, it branches so that many individual cells in the young nodule become infected (Figure 11.3C).

11.3.1.3 *Finally bacteria are released.*

The final step in the infection process occurs when the bacteria are "released" into the host cells. Actually the membrane of the infection thread buds off to form small vesicles, each containing one or more individual bacteria. Shortly after release, the bacteria cease dividing, enlarge, and differentiate into specialized nitrogen-fixing cells called **bacteroids**. The bacteroids remain surrounded by a membrane, now called the **peribacteroid membrane**. Differentiation into a bacteroid is marked by a number of metabolic changes, including the synthesis of the enzymes and other factors that the organism requires for the principal task of nitrogen fixation.

The infection process continues throughout the life of the nodule. As the nodule increases in size due to the activity of the nodule meristem, bacteria continue to invade the new cells. Also as the nodule enlarges and matures, vascular connections are established with the main vascular system of the root (Figure 11.5). These vascular connections serve to import photosynthetic carbon into the nodule and export fixed nitrogen from the nodule to the plant.

11.4 THE BIOCHEMISTRY OF NITROGEN FIXATION

Dinitrogen is not easily reduced because the interatomic nitrogen bond ($N \equiv N$) is very stable. In the industrial process, reduction of the dinitrogen triple bond with hydrogen can be achieved only at high temperature and pressure and at the cost of considerable energy. Biological reduction of dinitrogen is equally costly, consuming a large proportion of the photoassimilate provided by the host plant.

11.4.1 NITROGEN FIXATION IS CATALYZED BY THE ENZYME DINITROGENASE

Only prokaryote cells are able to fix dinitrogen principally because only they have the gene coding for this enzyme (see Chapter 8, Box 8.1). The enzyme **dinitrogenase** has been purified from virtually all known nitrogen-fixing prokaryotes. It is a multimeric protein complex made up of two proteins of different size (Figure 11.6). The smaller protein is a **dimer** consisting of two identical subunit polypeptides. The molecular mass of each subunit ranges from 24 to 36 kD, depending on the bacterial species. It is called the **Fe protein** because the dimer contains a single cluster of four iron atoms bound to four sulphur groups (Fe_4S_4). The larger protein in the dinitrogenase complex is called the **MoFe protein**. It is a **tetramer** consisting of two pairs of identical subunits with a total molecular mass of approximately 220 kD. Each MoFe protein contains two molybdenum atoms in the form of an iron-molybdenum-sulphur cofactor. The MoFe protein also contains Fe_4S_4 clusters, although the exact number is uncertain. It varies as a function of the species or its physiological condition.

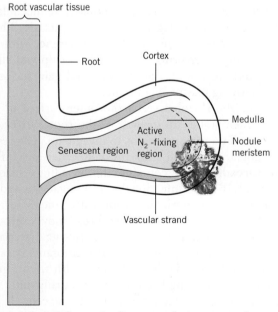

FIGURE 11.5 Schematic diagram of a cross-section through a mature nodule. Vascular connections with the host plant provide for the exchange of carbon and nitrogen between the host and the microsymbiont.

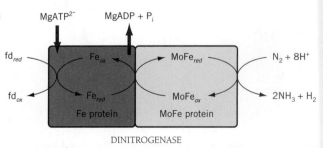

FIGURE 11.6 Schematic diagram of the dinitrogenase reaction in bacteroids. Electron flow is from left to right. The principal electron donor is ferredoxin (fd), which receives its electron from respiratory substrate.

The overall reaction for reduction of dinitrogen to ammonia by dinitrogenase is shown in the following equation:

$$8H^+ + 8e^- + N_2 + 16\,ATP \rightarrow 2NH_3$$
$$+ H_2 + 16\,ADP + 16P_i \qquad (11.1)$$

Note that the principal product of biological nitrogen fixation is ammonia, but that for every dinitrogen molecule reduced, one molecule of hydrogen is generated. We will return to the problem of hydrogen evolution later. Also note that reduction of dinitrogen is a two-step process. In the first step, the Fe protein is reduced by a primary electron donor, usually ferredoxin. Ferredoxin is a small (14 to 24 kD) protein containing an iron-sulphur group. Electrons are carried by the iron moiety, which can exist in either the reduced ferrous (Fe^{2+}) or the oxidized ferric (Fe^{3+}) states. It is of interest to note that ferredoxin not only participates in nitrogen fixation, but is an important electron carrier in photosynthesis as well (see Chapter 7).

In the second step, the reduced Fe protein passes electrons to the MoFe protein, which catalyzes the reduction of both dinitrogen gas and hydrogen. The precise role of ATP in the reaction is not yet clear, but it is known to react with reduced Fe protein and to cause a conformational change in this protein that alters its redox potential. This facilitates the transfer of electrons between the Fe protein and the MoFe protein.

11.4.2 NITROGEN FIXATION IS ENERGETICALLY COSTLY

Biological reduction of dinitrogen, as is industrial nitrogen fixation, is very costly in terms of energy. One measure of energy cost is the number of ATP required. At least *16 ATP* are required for each molecule of dinitrogen reduced—two for each electron transferred (Equation 11.1). By comparison, only 3 ATP are required to fix a molecule of carbon dioxide in photosynthesis (see Chapter 8). The total energy cost of biological nitrogen fixation, however, must take into account the requirement for reduced ferredoxin as well. Were this reducing potential not required for the reduction of dinitrogen, it could have been made available for the production of additional ATP or other uses by the plant. It has been estimated that the reducing potential used in nitrogen fixation is equivalent to at least a further 9 ATP, bringing the total investment to a minimum of 25 ATP for each molecule of dinitrogen fixed. A similar calculation for CO_2 fixation brings the total to 9 ATP, about one-third the cost for nitrogen.

Another and perhaps better way to assess the cost of nitrogen fixation is by measuring the amount of carbon utilized in the process. The ultimate source of energy for symbiotic nitrogen fixation is carbohydrate produced by photosynthesis in the host plant. A portion of that carbohydrate is diverted from the plant to the bacteroid, where it is metabolized to produce the required reducing potential and ATP. It has been calculated that, in soybean, approximately 12 grams of carbon are required to fix a gram of dinitrogen. It is clear that nitrogen fixation represents a considerable drain on the carbon resources of the host plant. A diagram summarizing the integration of photosynthesis, respiration, and nitrogen fixation is presented in Figure 11.7.

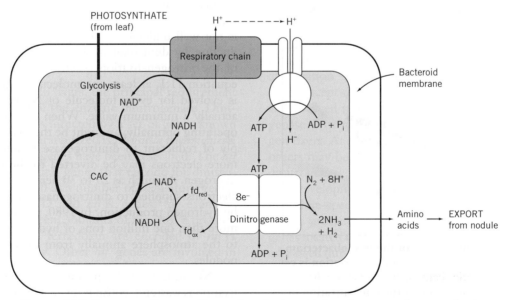

FIGURE 11.7 **Summary diagram illustrating the interactions between photosynthesis, respiration, and nitrogen fixation in bacteroids.**

of nodulation, prior to infection of the root, a set of rhizobial *nod* genes is switched on by flavonoids in the enzymes such as uricase and glutamine synthetase involved in the metabolic processing of fixed nitrogen.

11.5.3 WHAT IS THE SOURCE OF HEME FOR LEGHEMOGLOBIN?

Early experiments with a mutant of *Rhizobium melilotii* indicated that the heme was supplied by the bacteroid. The mutant, unable to synthesize the heme precursor δ-aminolevulinic acid (ALA), produced white nodules that were unable to fix dinitrogen. These results suggest the host plant is unable to provide sufficient heme to build adequate levels of leghemoglobin. However, in later experiments with soybean, plants infected with *Bradyrhizobium japonicum* carrying the same mutation produced fully competent nodules. Thus, the source of the heme component of leghemoglobin is not yet clear.

Symbiotic nitrogen fixation clearly requires the coordinated expression of many genes of both the host and microsymbiont. Understanding how these genes are regulated and how the complex processes of infection and nodule development are coordinated constitutes one of the more challenging problems facing plant physiologists today. Armed with sufficient understanding of the process and the tools of modern molecular genetics, plant scientists may one day be able to extend the range of biological nitrogen fixation to other important crop species—thus extending the benefits of nitrogen fixation to nitrogen-poor soils and reducing the economic and environmental costs of chemical nitrogen fertilizers.

11.6 NH₃ PRODUCED BY NITROGEN FIXATION IS CONVERTED TO ORGANIC NITROGEN

The first stable product of nitrogen fixation is ammonia (NH₃), although at physiological pH ammonia is almost certainly protonated to form ammonium ion:

$$NH_3 + H^+ \leftrightarrow NH_4^+ \qquad (11.2)$$

Plants that cannot fix dinitrogen meet their nutritional needs by taking in nitrogen from the soil. While there are exceptions, most plants are able to assimilate either NH_4^+ or NO_3^-, depending on their relative availability in the soil. In most soils, ammonia is rapidly converted to nitrate by the nitrifying bacteria described earlier in this chapter. Nitrifying bacteria do not grow well under anaerobic conditions and consequently ammonia will accumulate in soils that are poorly drained. Nitrification itself is also inhibited in strongly acidic soils. Some members of the family Ericaceae, typically found on acidic soils, have adapted by preferentially utilizing ammonium as their nitrogen source. One extreme example is the cranberry (*Vaccinium macrocarpon*), native to swamps and bogs of eastern North America, which cannot exploit NO_3^- as a nitrogen source and must take up nitrogen in the form of ammonium ion.

Regardless of the route taken, assimilation of mineral (inorganic) nitrogen into organic molecules is a complex process that can be very energy intensive. It has been estimated, for example, that assimilation of ammonium nitrogen consumes from 2 to 5 percent of the plant's total energy production. Nitrate, on the other hand, must first be reduced to ammonium before it can be assimilated, at a cost of nearly 15 percent of total energy production. In this section, we will review the assimilation first of ammonium nitrogen and then of nitrate nitrogen.

11.6.1 AMMONIUM IS ASSIMILATED BY GS/GOGAT

Although NH_4^+ is readily available to many plants, either as the product of nitrogen fixation or by uptake from the soil, it is also quite toxic to plants. In nitrogen-fixing systems, NH_4^+ will inhibit the action of dinitrogenase. Ammonium also interferes with the energy metabolism of cells, especially ATP production. Even at low concentrations, NH_4^+ has the potential to uncouple ATP formation from electron transport in both mitochondria and chloroplasts (see Chapters 7 and 10). Consequently, it appears that plants can ill afford to accumulate excess free NH_4^+. It is assumed that most plants avoid any toxicity problem by rapidly incorporating the NH_4^+ into amino acids.

The general pathway for NH_4^+ assimilation in nitrogen-fixing symbionts has been worked out largely by supplying nodules with labeled dinitrogen ($^{13}N_2$ or $^{15}N_2$). These studies have indicated that the initial organic product is the amino acid **glutamine**. Assimilation of NH_4^+ into glutamine by legume nodules is accomplished by the **glutamate synthase cycle**, a pathway involving the sequential action of two enzymes: **glutamine synthetase (GS)** and **glutamate synthase (GOGAT)**[2] (Figure 11.9). Both GS and GOGAT are nodulin proteins that are expressed at high levels in the host cytoplasm of infected cells, outside the peribacteroid membrane. The NH_4^+ formed in the bacteroid must therefore diffuse across the peribacteroid membrane before it can be assimilated.

In the first reaction of the glutamate synthase cycle, catalyzed by GS, the addition of an NH_4^+ group to **glutamate** forms the corresponding amide, **glutamine**:

$$\text{glutamate} + NH_4^+ + ATP \rightarrow \text{glutamine} + ADP + P_i \qquad (11.3)$$

---

[2]The acronym *GOGAT* refers to glutamine-2-oxoglutarate-amino-transferase. 2-Oxoglutarate is an alternative name for α-ketoglutarate.

Energy to drive the amination of glutamate is provided by ATP, yet an additional cost of nitrogen fixation. Glutamine is then converted back to glutamate by the transfer of the amide group to a molecule of **α-ketoglutarate**.

$$\text{glutamine} + \alpha\text{-ketoglutarate} + \text{NADH} \rightarrow$$
$$2\text{glutamate} + \text{NAD}^+ \qquad (11.4)$$

Reaction 11.4 is catalyzed by GOGAT and requires reducing potential in the form of NADH. The α-ketoglutarate is probably derived from photosynthetic carbon through respiration in the host cell. α-Ketoglutarate is an intermediate in the respiratory pathway for the oxidation of glucose. Note that reaction equation 11.4 gives rise to two molecules of glutamate. Since only one molecule of glutamate is required to keep the cycle going, the other is available for export to the host plant (Figure 11.9). Overall, then, carbon skeletons originating with photosynthesis and nitrogen fixed by the microsymbiont are combined to form organic nitrogen that is exported out of the nodule for use by the host.

There is a possible alternative pathway for nitrogen assimilation, involving the direct reductive amination of α-ketoglutarate by the enzyme **glutamate dehydrogenase (GDH)**. However, although GDH activity has been detected in nodules, there is no convincing evidence that it plays a significant role in NH_4^+ assimilation under normal circumstances. Both the quantities and activities of GS and GOGAT are much higher than GDH. GS alone may account for as much as 2 percent of the total soluble protein outside the bacteroid. In addition, GDH has a much lower affinity (see Chapter 8, Box 8.1) for NH_4^+ than does GS and could hardly be expected to compete with GS for available NH_4^+. The cost of the glutamate synthase cycle is one ATP for each NH_4^- assimilated, but the benefit is rapid assimilation. The high affinity of GS for NH_4^+, together with the high concentration of the enzyme, ensures that the free NH_4^+ concentration is kept below toxic levels.

Before leaving the glutamate synthase cycle, it is important to note that GS and GOGAT are not restricted to nodules. These enzymes are located in the roots and leaves of non-nitrogen-fixing plants where they also catalyze the assimilation of NH_4^+ nitrogen.

11.6.2 PII PROTEINS REGULATE GS/GOGAT

As illustrated in Figure 11.9, GS/GOGAT represents a critical metabolic point of coordination between nitrogen assimilation (NH_4^+) and carbon metabolism (α-ketoglutarate). Cellular carbon/nitrogen status is sensed by the signal sensing protein, PII. **PII proteins** are one of the most ubiquitous and highly conserved regulatory proteins found in nature. They are present in the ancient archeabacteria, eubacteria as well as eukaryotes such as plants. This relatively small chloroplastic protein of about 14 kDa is nuclear encoded and senses both the ATP status and α-ketoglutarate levels by allosteric means (see Chapter 8, Box 8.1). Cellular nitrogen status is sensed through changes in glutamine availability which results in post-translational modification of PII by **uridylylation** (Figure 11.10). Like the light-regulated enzymes of the Calvin Cycle (see Chapter 8), GS/GOGAT can exist in either an active or an inactive state. When glutamine levels in the GS/GOGAT cycle are low, PII proteins are uridylylated using uridine triphosphate (UTP) to form PII-UMP (Figure 11.10). This modified PII protein interacts with inactive GS/GOGAT to convert the latter from the inactive to the active form which stimulates glutamine synthesis and hence the stimulates assimilation of NH_4^+. Conversely, when glutamine levels are high, PII-UMP is converted back to PII which converts the active GS/GOGAT to the inactive form which slows the rate of NH_4^+ assimilation.

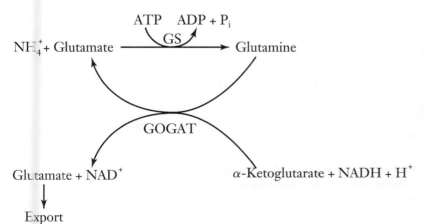

FIGURE 11.9 Assimilation of ammonium by the glutamate synthase cycle. The enzyme glutamate synthetase (GS) converts ammonium plus glutamate to glutamine. Glutamine undergoes a transamination reaction with α-ketoglutarate, which results in the production of two molecules of glutamate via the enzyme glutamate synthase (GOGAT). One glutamate is exported while the other is recycled through the reaction catalyzed by GS.

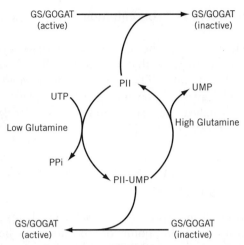

FIGURE 11.10 **Regulation of GS/GOGAT by Uridyly-lation of PII Proteins. Uridine triphosphate (UTP) is a phosphorylated nucleotide analogous to ATP (see Chapter 5) except that the nitrogenous base, uracil, replaces adenine. UTP, uridine triphopshate; UMP, uridine monophosphate; PPi, pyrophosphate.**

11.6.3 FIXED NITROGEN IS EXPORTED AS ASPARAGINE AND UREIDES

The final step in nitrogen fixation is the export of the fixed nitrogen from the nodule to other regions of the host plant. Export of the organic nitrogen products from nodules is primarily through the xylem. Consequently, the form in which the nitrogen is exported has been identified primarily by analysis of xylem sap. There are some pitfalls to such analyses, however, as there is no guarantee that all of the nitrogen present in sap represents current nodule production. Some of the better analyses have been conducted directly on detached nodules or by monitoring the flow of organic nitrogen following fixation of $^{15}N_2$. These studies have shown that although glutamine is the principal organic product of nitrogen fixation, it rarely accounts for a significant fraction of the nitrogen exported, at least in legumes. In some groups of legumes, largely those of temperate origins, such as pea and clover, the amino acid **asparagine** is the predominant form translocated. Legumes of tropical origins, for example, soybean and cowpea, appear to export predominantly derivatives of urea, known as **ureides** (Figure 11.11).

The biosynthetic pathway for asparagine in nodules involves two **transamination** reactions. A transamination reaction is the transfer of an amino group from an amino acid to the carboxyl group of a keto acid. Transamination reactions, catalyzed by a class of enzymes known as **aminotranferases**, enable nitrogen initially fixed in glutamate to be incorporated into other amino acids and, ultimately, into protein. Aminotransferases are found throughout the plant—in the cytosol, in chloroplasts, and in microbodies—wherever protein synthesis activity is high. The enzymes involved in asparagine biosynthesis in nodules appear to be similar to those found elsewhere in the plant. The first step is the transfer of an amino group from glutamate to **oxaloacetate**, catalyzed by the enzyme **aspartate aminotransferase**.

glutamate + oxaloacetate

$$\rightarrow \alpha\text{-ketoglutarate} + \text{aspartate} \qquad (11.5)$$

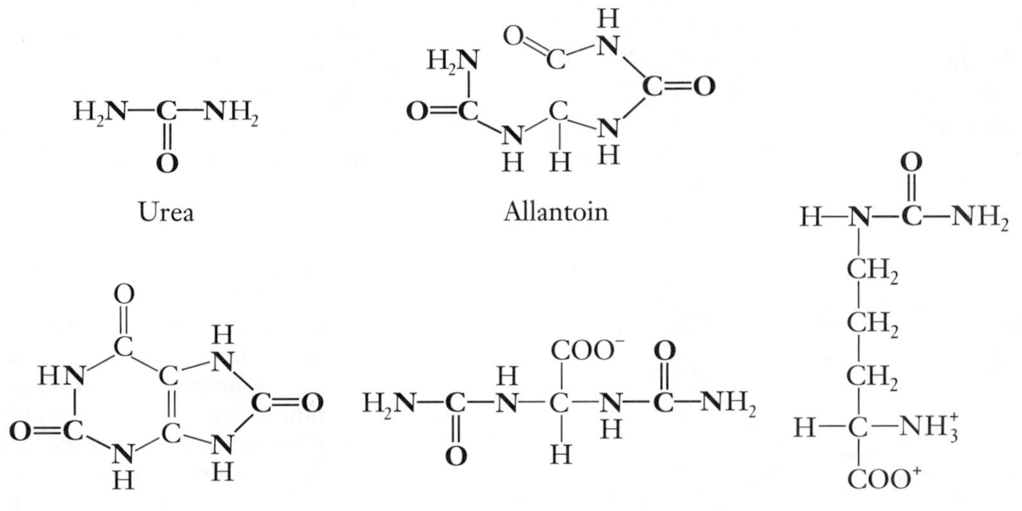

FIGURE 11.11 **Structures of the principal ureides used in the transport of assimilated nitrogen in some nitrogen-fixing species. Ureides are considered derivatives of urea and are formed principally from uric acid (urate). The N—C—N urea backbone is shown in bold print.**

The glutamate used in this reaction is derived from the GS-GOGAT reactions in the nodule. In order to continue the synthesis and export of asparagine, the nodule requires a continued supply of the 4-carbon acid oxaloacetate. This could be provided through the oxidation of carbon in the nodule; oxaloacetate is another intermediate in the respiratory oxidation of glucose. However, nodules from a number of species exhibit high activities of the enzyme *phosphoenolpyruvate carboxylase* (**PEP carboxylase**). PEP carboxylase catalyzes the addition of a carbon dioxide (a **carboxylation** reaction) to the 3-carbon **phosphoenolpyruvate (PEP)** to form **oxaloacetate (OAA)** (Equation 11.6). PEP carboxylase is involved in a number of important metabolic pathways in plants and animals, including respiration and photosynthesis.

$$PEP + CO_2 \rightarrow OAA \qquad (11.6)$$

In the second step of asparagine biosynthesis, the amide nitrogen is transferred from glutamine to aspartate.

$$\text{glutamine} + \text{aspartate} + ATP$$
$$\rightarrow \text{glutamate} + \text{asparagine} + ADP + P_i \qquad (11.7)$$

The enzyme for this reaction is **asparagine synthetase** and the reaction is driven by the energy of one molecule of ATP for each asparagine synthesized.

The synthesis of ureides is more complex, both biochemically and with respect to the division of labor between the microsymbiont and tissues of the host plant. **Allantoin** and **allantoic acid** (Figure 11.11) are formed by the oxidation of purine nucleotides, which apparently requires an active symbiosis. Ureides apparently serve specifically for the transport of nitrogen. They are translocated through the xylem to other regions of the plant, where they are rapidly metabolized. In the process, NH_4^+ is released, which is then reassimilated via GS and GOGAT in the target tissue.

Although the synthesis of asparagine, and especially the ureides, both appear to be complex processes, there are some advantages relating to the energy costs and efficiencies of nitrogen export. It has been estimated that the carbon metabolism associated with nitrogen export may consume as much as 20 percent of the photosynthate diverted to nitrogen fixation. One way to judge efficiency is to consider the amount of carbon required for each nitrogen exported. The ureides, with a carbon to nitrogen ratio of 1 (C:N=1), are the most economic in the use of carbon. Both asparagine and citrulline (C:N=2) require more carbon in their transport and glutamine (C:N=2.5) would be the least economic. The energy costs of ureides, asparagine, and citrulline, in terms of ATP consumed, are about the same, so the principal advantage to be gained by the ureide-formers appears to be a favorable carbon economy.

11.7 PLANTS GENERALLY TAKE UP NITROGEN IN THE FORM OF NITRATE

Except in extreme situations noted earlier, nitrate (NO_3^-) is the more abundant form of nitrogen in soils and is most available to plants that do not form nitrogen-fixing associations. However, in spite of numerous studies describing the physiology of NO_3^- uptake, there is a great amount of uncertainty surrounding the mechanism of NO_3^- transport into roots. It has been shown in various studies that uptake of NO_3^- is sensitive to (1) low temperature, (2) inhibitors of both respiration and protein synthesis, and (3) anaerobic conditions. All of these results support the hypothesis that NO_3^- transport across the root cell membrane is an energy-dependent process mediated by a carrier protein (see Chapter 3).

In root cells that have never been exposed to nitrate, there appears to be a limited capacity for NO_3^- uptake. This suggests a small amount of carrier is present in the membrane at all times (that is, a **constitutive** protein). On exposure to external nitrate, the rate of uptake increases from two- to fivefold, but addition of inhibitors of protein synthesis causes the rate to fall rapidly back to the constitutive level. This pronounced sensitivity of NO_3^- uptake to inhibitors of protein synthesis suggests that the bulk of the carrier protein is **inducible**, that is, the presence of NO_3^- in the soil stimulates the synthesis of new carrier protein. Once inside the root, NO_3^- may be stored in the vacuole, assimilated in the root cells, or translocated in the xylem to the leaves for assimilation.

Nitrate cannot be assimilated directly but must first be reduced to NH_4^+ in order to be assimilated into organic compounds. This is a two-step process, the first being the reduction of NO_3^- to nitrite (NO_2^-) by the enzyme **nitrate reductase (NR)**.

$$2H^+ + NO_3^- + 2e^- \rightarrow NO_2^- + H_2O \qquad (11.8)$$

NR is generally assumed to be a cytosolic enzyme. The product NO_2^- then moves into plastids (in roots) or chloroplasts (in leaves) where it is quickly reduced to NH_4^+ by the enzyme **nitrite reductase (NiR)**. In leaves, the electrons required for the reduction of NO_2^- to NH_4^+ are generated

$$8H^+ + NO_2^- + 6e^- \rightarrow NH_4^+ + 2H_2O \qquad (11.9)$$

by photosynthetic electron transport. Thus, the assimilation of NO_3^- competes with the assimilation of CO_2 (Chapter 8) for photosynthetic electrons. As a consequence, NO_3^- assimilation in leaves can also be considered a photosynthetic process similar to CO_2 assimilation. The interactions between carbon and nitrogen metabolism indicate the importance of photosynthetic electron transport in overall primary reductive metabolism.

Nitrite is toxic and is rarely found at high concentrations in plants. This is no doubt because the activity of NiR (per gram dry weight of tissue) is normally several times higher than the activity of NR. The resulting ammonia is then rapidly assimilated into organic compounds via the GS/GOGAT system already described. In non-nitrogen-fixing systems, both GS and GOGAT are commonly found in root and leaf cells. GS is found in the cytosol of root cells and in both the cytosol and chloroplasts of leaf cells. GOGAT is a plastid enzyme, localized in the chloroplasts of leaves and in plastids in roots. Depending on its location, GOGAT may use ferredoxin, NADH, or NADPH as electron donors.

Nitrate reductase is a ubiquitous enzyme found in both prokaryote and eukaryote cells. In prokaryotes, the principal electron donor is ferredoxin, while in higher plants electrons are donated by the reduced forms of one of the pyrimidine nucleotides, *n*icotinamide *a*denine *d*inucleotide (**NAD**) or *n*icotinamide *a*denine *d*inucleotide *p*hosphate (**NADP**) (Chapter 7). The enzyme isolated from a variety of higher plants is composed of two identical subunits with a molecular mass of approximately 115 kD. A key constituent of NR is molybdenum; NR is the principal Mo-protein in non-nitrogen-fixing plants. One of the results of Mo deficiency is markedly reduced levels of nitrate reductase activity and consequent nitrogen starvation (see Chapter 4).

NR is a highly regulated, inducible enzyme. It has long been recognized that both substrate (NO_3^-) and light are required for maximum activity and that induction involves an increase in the level of NR messenger RNA followed by *de novo* synthesis of NR protein. Treatment of cereal seedlings such as barley (*Hordeum vulgare*) or maize (*Zea mays*) with nitrate in the dark induces relatively low levels of NR activity, but activity is strongly promoted if seedlings are also exposed to light. Induction by light is eliminated by various treatments that interfere with chloroplast development or photosynthetic energy transformations, implying a requirement for photosynthetic energy. NR activity can also be reversibly regulated by red and far-red light, indicating control by the phytochrome system (Chapter 22).

More recent work has established that NR activity is also subject to post-translational regulation by a specific NR **protein kinase**. Protein kinases, first characterized by Edwin Krebs and Edmund Fischer in the 1950s, are a ubiquitous class of enzymes that phosphorylate other proteins by transferring a phosphate group from adenosine triphosphate (ATP). The phosphate group can then be removed by a second enzyme called **protein phosphatase**. It is increasingly evident that, by switching enzymes and other proteins on and off, reversible protein phosphorylation plays a central role in regulating metabolism. The fundamental role of protein kinases was recognized by the award of the Nobel Prize to Krebs and Fischer in 1992.

In the case of NR, the enzyme appears to be active in both the phosphorylated and nonphosphorylated states. When NR is phosphorylated and the leaf is transferred from the light to dark, however, the enzyme is rapidly inactivated by binding with a small inhibitor protein. NR activity is slowly restored on return to light by a release of the inhibitor protein and subsequent phosphatase action. The question of why there exists such a complex system for regulation of NR activity has yet to be answered. The overall effect, however, is to coordinate nitrate reduction with photosynthetic activity. It ensures that nitrate reduction is engaged only after photosynthesis is fully active and able to provide both the energy required and the carbon skeletons necessary for incorporation of ammonia.

As indicated earlier, nitrate assimilation can be carried out in either the root or shoot tissues in most plants. Several studies have shown that the proportion of NO_3^- reduced in the root or shoot depends to a large extent on the external NO_3^- concentration. At low concentrations, most of the NO_3^- can be reduced within the root tissues and translocated to the shoot as amino acids or amides. At higher concentrations of NO_3^-, assimilation in the roots becomes limiting and a higher proportion of the NO_3^- finds its way into the translocation stream. Thus, at higher concentrations, a higher proportion of the nitrogen is assimilated in the leaves.

Not all plants have the same capacity to metabolize NO_3^- in their roots. In the extreme, NO_3^- is virtually the sole nitrogen source in the xylem sap of cocklebur (*Xanthium strumarium*). This is because cocklebur has no detectable NR in its roots. On the other hand, plants such as barley (*Hordeum vulgare*) and sunflower (*Helianthus annus*) translocate roughly equal proportions of NO_3^- and amino acid/amide nitrogen, and radish (*Raphanus sativus*) translocates only about 15 percent of its nitrogen as NO_3^-.

11.8 NITROGEN CYCLING: SIMULTANEOUS IMPORT AND EXPORT

Nitrogen uptake by most plants is highest during its early rapid-growth phase and declines as reproductive growth begins and the plant ages. Cereals, for example, take up as much as 90 percent of their total nitrogen requirement before the onset of reproductive growth. Most of this nitrogen is directed toward young, expanding leaves, which reach their maximum nitrogen content just prior to full expansion. The leaf then begins to *export* nitrogen. Several studies have shown that mature leaves continue to import nitrogen, even though they have become net nitrogen exporters and the total nitrogen content of the leaf is in decline. This simultaneous import and export of nitrogen is known as **nitrogen cycling**.

The export of nitrogen from leaves becomes particularly significant as the seed begins to develop. The nitrogen requirement of developing seeds is sufficiently great that it cannot be met by uptake from the soil (in the case of cereals, for example) or by nitrogen fixed in nodules. The additional nitrogen must come from vegetative parts, principally leaves (Figure 11.12). This may have significant implications for the photosynthetic capacity of leaves. The major leaf protein is Rubisco, the enzyme that catalyzes photosynthetic incorporation of carbon dioxide (see Chapter 8). Rubisco may comprise from 40 to 80 percent of the total soluble protein in leaves of soybean and cereal grains. Perhaps because of its abundance, Rubisco also functions as a storage protein; it may be degraded when nitrogen is required elsewhere in the plant, such as developing seeds. With the loss of Rubisco there is a concomitant decline in photosynthetic carbon fixation. In the case of soybean and other symbiotic legumes, this means less energy available to the nodule to support nitrogen fixation. This competition between nitrogen and carbon supply may be a major factor limiting seed development in legumes. In the case of cereals or plants growing in

nitrogen-poor soils, mobilization of accumulated nitrogen from the leaves represents the principal source of nitrogen for developing fruits and seeds. In perennial plants, nitrogen from senescing leaves is mobilized and translocated to the roots for storage over the winter. In this way, the nitrogen is conserved and made available to support the first flush of renewed growth the following spring.

11.9 AGRICULTURAL AND ECOSYSTEM PRODUCTIVITY IS DEPENDENT ON NITROGEN SUPPLY

In terms of quantity, nitrogen is the fourth most abundant element in plants and is the most abundant mineral element (see Chapter 4). On a dry-weight basis, herbaceous plant material typically contains between 1 and 4 percent nitrogen, mostly in the form of protein. At the same time, the availability of nitrogen in the soil may be limited by a number of environmental factors, such as temperature, oxygen, water status, and pH, which influence the activity of microorganisms responsible for nitrogen fixation, nitrification, and ammonification. Moreover, a substantial quantity of nitrogen is removed each year with the harvested crop. It is not too surprising, then, that crop growth is most often limited by nitrogen supply.

In agricultural situations, the application of nitrogen fertilizers overcomes environmentally imposed nitrogen limitation. Most crops respond to applied nitrogen with increases in yield (Figure 11.13). At sufficiently high application rates, factors other than nitrogen become limiting and there is no further gain. At even higher application rates, yield may decline slightly, but this is probably due to excess salt in the soil rather than some form of nitrogen toxicity. Data such as those shown in Figure 11.13 are of considerable practical value to farmers, who want to maximize the ratio of yield to input costs. Throughout North America, corn is the leading consumer of nitrogen fertilizer and farmers typically apply 100 to 150 kg N ha^{-1} each growing season. During its early rapid growth phase, a well-irrigated stand of corn will take up as much as 4 kg of nitrogen ha^{-1} day^{-1}. The use of such large amounts of nitrogen fertilizers is also costly in terms of energy. It has been estimated, for example, that fully one-third of the energy cost of a corn crop is accounted for by the production and distribution of nitrogen fertilizers.

Without continued application of fertilizers, yields of nonleguminous crops traditionally decline over a period of years. In a few situations where records have been kept, yields on plots from which nitrogen fertilizers have been withheld will eventually stabilize at

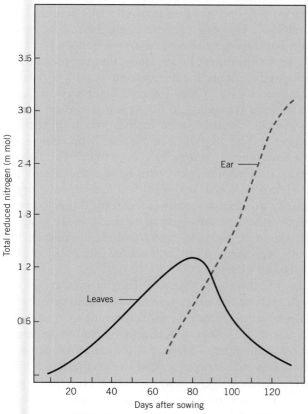

FIGURE 11.12 **Nitrogen redistribution in wheat. (From Abrol et al., in H. Lambers, J. J. Neeteson, I. Stuhlen (eds.), *Fundamental, Ecological and Agricultural Aspects of Nitrogen Metabolism in Higher Plants*, Dordrecht: Marinus Nijhoff, 1986. Reprinted by permission of Kluwer Academic Publishers).**

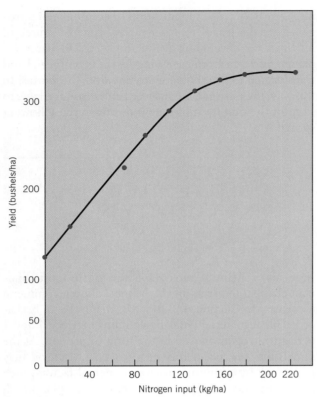

FIGURE 11.13 **The effect of applied nitrogen fertilizer on corn yield.**

a lower level that can be sustained indefinitely. Sustained low yields are possible because the extraction of nitrogen (and other nutrients) from the soil is balanced against replenishment from all sources, including rainfall, irrigation water, dust, and weathering of parent rock.

The role of nitrogen in natural ecosystems is much more difficult to define, in part because the level of inputs is very low relative to the total nitrogen pool and much of the nitrogen is recycled. Still, it is generally agreed that nitrogen is limiting in most natural ecosystems just as it is in agriculture. In forest ecosystems, approximately two-thirds of the annual nitrogen input is contributed by nitrogen fixation, while the other third is believed to be derived primarily from atmospheric sources: either through rainfall or dry deposition of nitrogen oxides. A recent study has shown that, on average, close to half of the incoming fixed nitrogen is retained in the canopy. This should not be surprising since the photosynthetic apparatus present in the leaves of the canopy contains the bulk of the assimilated organic nitrogen in form of proteins such as Rubisco (see Chapter 8), the most abundant protein in nature, the major light harvesting polypeptides present in thylakoid membranes, and of course, the pigments, chlorophyll a and chlorophyll b, which are also rich in nitrogen (see Chapter 4). This implies that foliar absorption of nitrogen could play a significant role in nitrogen uptake by forest species.

This seems particularly true of trees growing at high elevations, which are frequently bathed in cloud cover, or trees that grow near urban industrialized areas.

Except in mature, slowly growing forests, most of the nitrogen is taken up and either retained in the canopy or held in long-term storage in the litter on the forest floor where it is slowly recycled. The nitrogen content of the litter is slowly leached into the soil by rain and surface water or is broken down into simpler compounds by a variety of soil bacteria, fungi, earthworms, and other decomposing organisms. The final step in the breakdown is **mineralization**, or the formation of inorganic nitrogen from organic nitrogen. Mineralization is largely due to the process of ammonification described earlier. Mineralization is invariably accompanied by **immobilization**, or the retention and use of nitrogen by the decomposing organisms. Availability of litter nitrogen to plants depends above all on **net mineralization**, or the extent to which mineralization exceeds immobilization.

The balance between mineralization and immobilization and oxidation of the mineralized NH_4^+-nitrogen by nitrifying bacteria is regulated by environmental parameters. Principal among these are temperature, pH, soil moisture, and oxygen supply. The optimum temperature for nitrification generally falls between 25°C and 35°C, although climatic adaptations of indigenous nitrifying bacteria to more extreme temperatures have been demonstrated. In one study, for example, the optimum temperature for nitrification in soils of northern Australia was 35°C, in Iowa (U.S.) it was 30°C, and in Alberta (Canada), 20°C. As noted earlier, soil pH is a major limiting factor in the growth of nitrifying bacteria. The growth of *Nitrobacter* is probably inhibited by ammonium toxicity at high pH (7.5 and above) while aluminum toxicity is suspected as the cause of limited nitrification in acid soils. Soil moisture and oxygen supply go hand-in-hand—little nitrification occurs in water-saturated soils because of the limited O_2 supply. At the other extreme, the rate of nitrification declines with decreasing soil water potential ($\Psi/_{soil}$) below about −0.03 to −0.04 MPa.

The relative significance of nitrification in the nitrogen cycle of natural ecosystems is not altogether clear. Most studies indicate a relatively minor role, since little if any surplus NO_3^- is found in the soil or streams of most undisturbed ecosystems. Experimental deforestation, however, leads to a rapid rise (as much as 50-fold) in the levels of NO_3^- in stream water. NO_3^- levels gradually returned to normal only as the vegetation began to regrow. These results suggest that nitrification is a significant source of nitrogen, but rapid uptake of NO_3^- by plants is an important factor in maintaining low levels of NO_3^- in the soil solution.

Trees and other plants also tend to conserve a large proportion of their nitrogen, withdrawing nitrogen from the leaves and flowers before they are shed and placing it in storage in the roots and stem tissues. Between one-third and two-thirds of a plant's nitrogen may be conserved by such internal cycling. In the case of deciduous trees, for example, this stored nitrogen offers a degree of nutritional independence from the often nitrogen-poor soil during the flush of growth in early spring.

SUMMARY

Nitrogen is often a limiting nutrient for plants, even though molecular nitrogen is readily available in the atmosphere. Plants do not have the gene coding for dinitrogenase but must depend instead on the nitrogen-fixing activities of certain prokaryote organisms to produce nitrogen in a combined form.

Nitrogen-fixing organisms may be free-living or form symbiotic associations with plants. Symbiotic nitrogen fixation involves complex genetic and biochemical interactions between host plant roots and bacteria. The invading rhizobia induce the formation of root nodules, where the protein leghemoglobin helps to ensure a low-oxygen environment in which the enzyme dinitrogenase can function. The host plant provides energy in the form of photosynthate and, in turn, receives a supply of combined nitrogen for its own growth and development.

The product of nitrogen fixation is ammonium, which is rapidly incorporated into amino acids through GS/GOGAT before it is exported from the nodule. PII proteins sense cellular carbon/nitrogen balance and regulate GS/GOGAT activity. Plants that do not form nitrogen-fixing associations generally take up nitrogen in the form of nitrate. Nitrate must first be reduced to ammonium before it can be incorporated into organic molecules. In leaves, this reducing power is generated by photosynthetic electron transport. Consequently, the reduction of nitrate to NH_4^+ in leaves can be considered photosynthetic since it competes with the reduction of CO_2 for photosynthetically generated electrons.

CHAPTER REVIEW

1. What are ammonification, nitrification, and denitrification? What are their respective contributions to the nitrogen cycle?

2. What is meant by the statement that biological nitrogen fixation is exclusively a prokaryote domain?

3. Describe the process of rhizobial infection and nodule development in a legume root.

4. Review the biochemistry of nitrogen fixation. How does a bacteroid differ from a bacterium?

5. What is the function of leghemoglobin in symbiotic nitrogen fixation?

6. The product of nitrogen fixation is ammonia. Trace the path of nitrogen as the ammonia is converted to organic nitrogen and translocated to a leaf cell.

7. What is the role of PII proteins in the assimilation of NH_4^+?

8. While most plants take up nitrogen in the form of nitrate ion, there are some that seem to prefer ammonium. Can you suggest a possible biochemical basis for this difference?

9. Why can nitrate conversion to NH_4^+ in leaves be considered photosynthetic?

10. Heavy fertilization of agricultural crops with nitrogen is a costly process, both economically and energetically. Is it feasible to produce crops without nitrogen fertilizers? If so, what would be the consequences with respect to yields?

FURTHER READING

Buchanan, B. B., W. Gruissem, R. L. Jones. 2000. *Biochemistry and Molecular Biology of Plants*. Rockville MD: American Society of Plant Physiologists.

Foyer, C. H., G. Noctor. 2002. *Photosynthetic Nitrogen Assimilation and Associated Carbon and Respiratory Metabolism. Advances in Photosynthesis and Respiration, Vol. 12*. Dordrecht: Kluwer Academic Publishers.

Giraud, E., et al. 2007. Legumes symbioses: Absence of nod genes in photosynthetic Bradyrhizobia. *Science* 316:1307–1312.

Guerts, R., T. Bisseling. 2002. Rhizobium Nod factor perception and signalling. *The Plant Cell* (Supplement) 14:S239–S249.

Lam, H.-M., K. T. Coschigano, I. C. Oliveira, R. Melo-Oliveira, G. M. Coruzzi. 1996. The molecular genetics of nitrogen assimilation into amino acids in higher plants. *Annual Review of Plant Physiology and Plant Molecular Biology* 47:569–593.

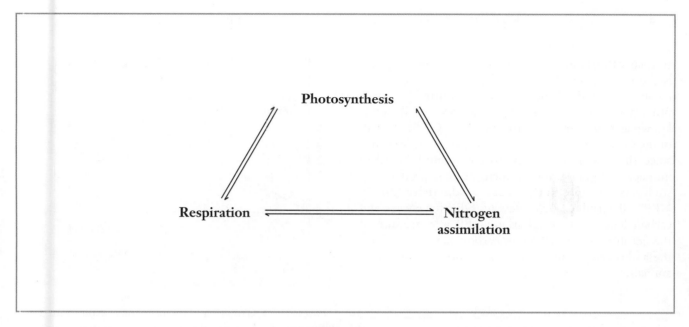

12

Carbon and Nitrogen Assimilation and Plant Productivity

In an earlier chapter, the theme was introduced that photosynthesis is the fundamental basis of competitive success in green plants (Chapters 7 and 8). The significance of carbon assimilation, however, extends well beyond the performance of green plants. Carbon assimilation creates plant biomass, or dry matter, which in turn supports humans and virtually all other heterotrophic organisms in the biosphere. Since the beginnings of the industrial revolution, growth of the human population and industries has been putting increasing pressures on the biosphere. These pressures have in turn stimulated interest in studies of plant productivity, which is directly related to carbon gain through photosynthesis. Productivity and its relationship to yield is of obvious concern to agriculture, but it also has relevance in broader ecological terms because it provides the energetic and material basis for other organisms. Knowledge of plant productivity provides a basis level for ecosystem research, helping to elucidate problems of energy and nutrient flow and their relationships to the structures of communities. Plant productivity also carries implications for the upper limit of the earth's sustainable human population. Knowing how plants function in a natural environment, their potential for harvest, and how they might respond to potentially stressful environmental change is essential to learning how to manage world resources in a time of burgeoning world population.

The physiology of photosynthesis—light capture, energy conversion, and partitioning of carbon—are at the root of productivity. This chapter will address

- the concepts of carbon gain and productivity,

- the interactions between respiration, photosynthesis and nitrogen assimilation and how these relationships determine the overall carbon gain of a plant,

- environmental factors such as light, available carbon dioxide, temperature, availability of soil water and nutrients, which influence photosynthesis and productivity, and various aspects of leaf and canopy structure and their impact on productivity.

The chapter will finish with a brief discussion of primary productivity on a global scale.

12.1 PRODUCTIVITY REFERS TO AN INCREASE IN BIOMASS

Although inorganic nutrients are a part of this dry matter, by far the bulk of dry matter for any organism consists of carbon. The basic input into the biosphere is the conversion of solar energy into organic matter by photosynthetic plants and microorganisms, known as primary productivity (PP). Total carbon assimilation is known as **gross primary productivity (GPP)**. Not all of the GPP is available for increased biomass, however, since there is a respiratory cost that must be taken into account. The principal focus of most productivity studies is therefore **net primary productivity (NPP)**. NPP is determined by correcting GPP for energy and carbon loss due to respiration. NPP is a measure of the net increase in carbon, or carbon gain, and reflects the additional biomass that is available for harvest by animals.

12.2 CARBON ECONOMY IS DEPENDENT ON THE BALANCE BETWEEN PHOTOSYNTHESIS AND RESPIRATION

Because photosynthesis occupies such a prominent position in the metabolism of higher plants, its rate is often regarded as the primary factor regulating biomass production and crop productivity. Yet it has often been observed that plants with similar photosynthetic rates may differ markedly with respect to growth rate and biomass accumulation. Clearly, other factors such as partitioning and allocation of carbon, translocation rates, and respiration rates must be considered when attempting to understand the overall carbon budget or carbon economy of a plant. Carbon economy is the term used to describe the balance between carbon acquisition and its utilization. Respiration is the principal counterbalance to photosynthesis. Respiration consumes assimilated carbon in order to obtain the energy required to increase and maintain biomass. Respiratory loss of carbon constitutes one of the most significant intrinsic limitations on plant productivity.

In an effort to better understand the impact of respiration on the carbon economy of plants, some physiologists have sought to distinguish experimentally between the carbon and energy costs of *growth* on the one hand and *maintenance* on the other. The term **growth respiration** has been coined to account for the carbon cost of growth. Growth respiration includes the carbon actually incorporated plus the carbon respired to produce the energy (in the form of reducing potential and ATP) required for biosynthesis and growth. **Maintenance respiration**, on the other hand, provides the energy for processes that do not result in a net increase in dry matter, such as turnover of organic molecules, maintenance of membrane structure, turgor, and solute exchange. These distinctions are not easily made, but

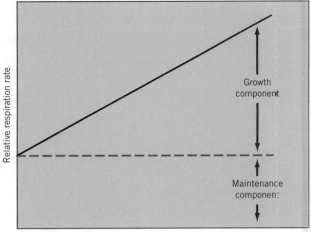

FIGURE 12.1 Growth and maintenance respiration. The proportion of respiration devoted to maintenance can be estimated by extrapolation to zero growth rate.

may be estimated by relating respiration rate to the relative growth rate (Figure 12.1). When respiration rate is extrapolated back to zero growth, it can be assumed that the residual respiration represents the carbon and energy requirements for maintenance of the nongrowing cells.

From Figure 12.1 it can be seen that maintenance respiration relative to total respiration will vary with the growth rate. The proportion devoted to maintenance will be least in a young, rapidly growing plant or organ while it will account for the bulk of respiration in a nongrowing organ such as a mature leaf. Indeed, measuring maintenance respiration by one commonly used experimental approach assumes that the respiration of a mature leaf is essentially 100 percent maintenance, although a small (but unknown) amount must be used for the translocation of solutes into and out of the leaf. Maintenance respiration also tends to be higher in roots than in shoots and other above-ground organs. This may be related to the expenditure of maintenance energy by roots for ion uptake to satisfy not only their own needs but the needs of the shoot as well. It may also reflect the observation that the cyanide-resistant alternative pathway tends to be higher in roots.

Respiration produces the metabolic energy that is required for various growth processes that increase biomass and agricultural yield, but it can also consume carbon with little or no apparent yield of useful energy. Since the latter situation represents a loss of carbon to the plant, it has been assumed that lower respiration rates would establish a more positive carbon economy that might, in turn, result in more rapid growth and increased productivity. Is it possible to manipulate respiration rates in favor of higher productivity? In a series of studies, genotypes of perennial rye grass (*Lolium perenne*) were selected for respiratory rates ranging from "slow" ($2.0 \, \text{mg CO}_2 \, \text{g}^{-1} \, \text{h}^{-1}$) to "rapid" ($3.5 \, \text{mg CO}_2 \, \text{g}^{-1} \, \text{h}^{-1}$).

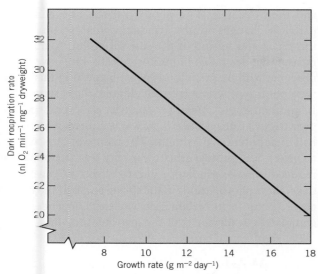

FIGURE 12.2 **The inverse correlation between respiratory rate and growth rate in genotypes of perennial rye grass (*Lolium perenne*). (From Wilson, D. *Annals of Botany* (London) 49:303–312, 1982. With permission of the Annals of Botany Company).**

Selection was based on the specific respiration rate of mature leaves at 25°C (i.e., maintenance respiration). The results (Figure 12.2) establish a negative correlation between respiration and growth rate; that is, the highest growth rates were recorded for genotypes having the lowest rates of respiration. It was concluded that the higher growth rates resulted from a more efficient use of carbon as a consequence of reduced respiratory evolution of CO_2 from fully grown tissues. In other words, with less of the carbon consumed in maintenance respiration, a higher proportion of the carbon was available for allocation to growth. Other investigators have found evidence of similar negative correlations between respiration and growth with wheat, barley, and oats.

Another method for improving respiratory efficiency might be to reduce the contribution, if any, of cyanide-resistant alternative pathway activity (Chapter 10). The alternative pathway oxidizes carbon with most of the energy released as heat. The overall impact of the alternative pathway remains to be firmly established, but there are some promising indications. Certain cultivars have been identified that differ with respect to the engagement of alternative pathway respiration at two different CO_2 levels. The cultivars were crossed to produce progeny that either strongly expressed the alternative pathway or showed no expression. The accumulation of biomass was greater in the progeny from which the alternative pathway was absent.

Studies such as these suggest that respiration has a significant impact on biomass accumulation and yield. They also indicate that improving yield by manipulating respiration might be feasible. Given that approximately half of the carbon assimilated in photosynthesis is eventually lost by respiration, reducing either the level of maintenance respiration or the engagement of the alternative pathway should shift the carbon balance in favor of growth respiration and increased biomass production. It is assumed, of course, that maintenance respiration and/or the alternative pathway can be reduced without detrimental effects on the plant, but this remains to be clearly established. Certainly some level of maintenance respiration is essential to the health of the plant—this level is yet to be determined.

One approach to improving the efficiency of net primary production is through breeding and selection programs. It is quite possible that this has already been achieved to some extent in existing breeding programs, without consciously evaluating the role of respiration. Another approach would be to manipulate respiration through genetic engineering. However, respiration is a complex process, both biochemically and physiologically. It is central to the metabolism of the cell, and involves many different enzymes and ultimately the coordinated activities of cellular organelles plus the cytosol. It is difficult to know which enzymes or reactions—that is, which genes—might be profitably manipulated or how an altered respiratory balance will affect other physiological processes. Although the prospects for improving productivity by manipulating respiration are encouraging, there is clearly a great deal of fundamental research yet to be conducted before significant progress can be expected.

12.3 PRODUCTIVITY IS INFLUENCED BY A VARIETY OF ENVIRONMENTAL FACTORS

The rate of photosynthesis may be limited by a host of variables. Which of these variables have an influence and the extent to which the influence is felt depend on whether one is concerned with a single leaf, a whole plant, or a population of plants that form a canopy. Included in these variables are both environmental factors and genetic factors. Environmental factors include light, availability of CO_2, temperature, soil water, nutrient supply, pathological conditions, and pollutants. Major factors include leaf age, and morphology, leaf area index, leaf angle, and leaf orientation. The influence of environmental stress, acclimation and adaptation on plant productivity will be discussed in Chapters 13, 14 and 15.

12.3.1 FLUENCE RATE

Typical responses of photosynthesis to fluence rate are illustrated in Figure 12.3. At very low fluence rates the

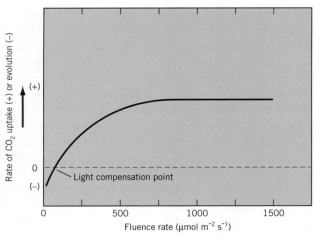

FIGURE 12.3 **A graph showing the typical light response of photosynthesis in C3 plants.**

rate of CO_2 evolution due to dark respiration exceeds the rate of photosynthetic CO_2 uptake. This results in a negative CO_2 uptake, or net CO_2 evolution. As fluence rate increases, photosynthesis also increases and so does CO_2 uptake until the rate of CO_2 exchange equals zero. This is the fluence rate, known as the **light compensation point**, at which the competing processes of photosynthesis and respiration are balanced. The light compensation point for most plants falls somewhere in the range of 10 to 40 $\mu mol\ m^{-2}\ s^{-1}$, roughly equivalent to the light level found in a well-lighted office, laboratory, or classroom.

At fluence rates above the compensation point, the rate of photosynthesis continues to increase until, at least in C3 plants, it reaches light saturation. In most C3 plants at normal atmospheric CO_2 levels, photosynthesis saturates with light levels of about 500 to 1000 μmol photons $m^{-2}\ s^{-1}$, that is, about one-quarter to one-half of full sunlight. Light saturation occurs because some other factor, usually CO_2 levels, becomes limiting. In most cases, both the saturation rate of photosynthesis and the fluence rate at which saturation occurs can be increased by increasing the CO_2 level above ambient.

A small number of C3 plants, such as peanut (*Arachis hypogea*), do not light saturate. It is not clear why this is the case, but these are exceptions to the rule. Individual leaves and plants will also acclimate to the light environment in which they are grown. The light-saturated rate of photosynthesis, for example, is lower in leaves that have acclimated to growth at low irradiance (shade leaves) than in those that have acclimated to higher irradiance (sun leaves). Acclimation and adaptation to light environments is discussed in more detail in Chapters 14 and 15.

In a natural environment, even C3 plants rarely light saturate and then only for relatively brief periods. Between dawn and dusk, the rate of photosynthesis gradually increases, reaching a maximum near midday,

and then declines. The photosynthetic rate generally parallels changes in the irradiance that accompanies the rising and setting of the sun. Even during midday, measurable decreases in photosynthetic rate have been observed with passing cloud cover, suggesting that even then photosynthesis was barely, if at all, light saturated. In another study, annual productivity of several species growing in a European hedgerow was limited to less than half their potential maximum. The failure of carbon gain to match leaf photosynthetic capacity was attributed to reduced average irradiance due to effects of dawn and dusk, short photoperiods in the spring and fall, and cloud cover. Long-term carbon gain is clearly dependent on cumulative irradiance over the growing season.

12.3.2 AVAILABLE CO_2

The carbon dioxide concentration of the atmosphere is relatively low, at least over the short term, at about 0.035 percent by volume or 350 $\mu l\ l^{-1}$. This is below the CO_2 saturation level for most C3 plants at normal fluence rates (Figure 12.4), which means that availability of CO_2 is often a limiting factor in photosynthesis. In C3 plants, increased photosynthetic rates with higher CO_2 levels results from two factors: increased substrate for the carboxylation reaction and, through competition with oxygen, reduced photorespiration. Note the interaction between ambient CO_2 levels and light. At higher fluence rates, both the maximum rate of photosynthesis and the CO_2 saturation level increase. Furthermore, under high light, C3 plants typically exhibit a lower CO_2 compensation point. This represents the CO_2 concentration where the rate of CO_2 uptake by photosynthesis equals the rate of CO_2 evolution due to respiration.

Assessing the impact of CO_2 levels on photosynthesis is not quite as straightforward as it might at first

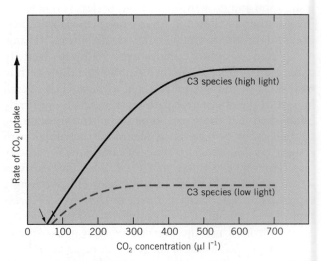

FIGURE 12.4 **A graph showing the typical response of C3 species to ambient CO_2 concentration. Arrow indicate CO_2 compensation concentration.**

appear. The rate of photosynthesis is actually determined not by the ambient CO_2 concentration, as much as by the **intracellular CO_2 concentration**, that is, the supply of CO_2 at the carboxylation site in the chloroplast. It is assumed that the intracellular CO_2 concentration is in equilibrium with the intercellular spaces. Since diffusion rates depend in part on concentration gradients, the primary effect of increasing ambient CO_2 levels would be to increase the intercellular CO_2 concentration by increasing the rate of diffusion into the leaf. Here it is assumed that water supply is adequate and, consequently, stomatal CO_2 conductance is not limiting.

Although it was once thought that stomatal CO_2 conductance was the principal factor limiting photosynthesis, more recent studies suggest it may be the other way around—stomatal conductance varies in response to photosynthetic capacity. **Photosynthetic capacity** is determined by the balance between carboxylation capacity and electron transport capacity (Figure 12.5). At low CO_2 concentrations, the rate of photosynthesis is limited by available CO_2 and, hence, the carboxylation capacity of the system, but is saturated with respect to availability of the acceptor molecule, ribulose-1,5-bisphosphate (RuBP) (see Chapter 8). However, any excess generation of RuBP, which is in turn dependent on the electron transport reactions, over that required to support carboxylation would represent an inefficient use of resources. Conversely, at high CO_2 concentrations or in low light, the limiting factor would be the energy-limited capacity to regenerate the acceptor molecule, ribulose-1,5-bisphosphate. In this case, an excess of carboxylating capacity—that is, an excess of Rubisco—would be an inefficient use of resources.

The most efficient use of resources for the plant would be to maintain intercellular CO_2 levels in the transition zone, where there is neither an excess of electron transport capacity nor an excess of carboxylating capacity. Because intercellular CO_2 levels are at least partly determined by stomatal conductance, it appears that the principal function of the stomata might be to regulate CO_2 uptake in order to keep intercellular CO_2 levels as much as possible within the transition range. Note that this is not the traditional view of stomatal function, which says that stomata operate principally to regulate water loss. Note also that CO_2 enrichment at high fluence rates leads to both higher photosynthetic maxima and higher CO_2 saturation levels (Figure 12.4). These observations suggest that plants are also able to compensate for higher light levels by increasing their carboxylating capacity. Such an increase could be achieved by regulating the amount of catalytic activities of photosynthetic enzymes, principally Rubisco.

CO_2 limitation is a particular problem in greenhouses, especially in winter when the greenhouses are closed and CO_2 levels are reduced due to photosynthesis. Even under more normal conditions, most plants will grow significantly faster and increase yields when the atmosphere is enriched with CO_2. For these reasons, CO_2 enrichment has become common practice for commercial growers of vegetable crops such as lettuce, tomato, and cucumbers. In practice, the CO_2 content of the greenhouse atmosphere is increased up to twice present atmospheric levels. Much beyond $700 \ \mu l \ l^{-1}$ there is an increasing risk of stomatal closure as well as the potential to induce feedback-limited photosynthesis (Chapter 9) that will cause a reduction in the rate of photosynthesis. Thus, there is an upper limit to which CO_2 concentrations can be increased with an increase in plant productivity. This limitation may occur when the photosynthetic capacity of source leaves far exceeds sink capacity of the plant (see Chapter 9). CO_2 enrichment might be expected to improve growth and productivity of field crops, but there are obvious technical and economic problems related to controlling the supply of gas in an open environment as well as potential physiological problems related to possible **feedback inhibition** of photosynthesis. The rate of photosynthesis is reduced when the rate of CO_2 assimilation exceeds the capacity for carbon utilization and export. The rate of photosynthesis is down-regulated through complex metabolic feedback loops.

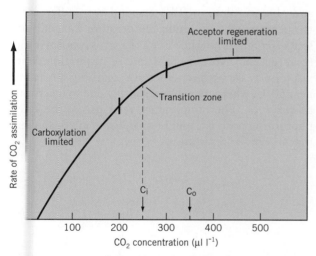

FIGURE 12.5 A model to describe limitation of photosynthetic rate as a function of CO_2 concentration. At low CO_2 concentrations, photosynthesis is limited by the carboxylation capacity of the enzyme Rubisco. At high concentrations of CO_2, the rate is limited by the rate of regeneration of the acceptor molecule, ribulose-1,5-bisphosphate. Stomata probably operate to keep the intercellular CO_2 concentration within the transition zone where there is neither an excess of carboxylating capacity nor an excess of electron transport. C_i and C_0 indicate the intercellular and ambient CO_2 concentrations, respectively. (Redrawn from Farquhar, G., T. Sharkey. 1982. *Annual Review of Plant Physiology* 33:317–345).

12.3.3 TEMPERATURE

Photosynthesis, like most other biological processes, is sensitive to temperature. The temperature response for most biological processes reflects the temperature dependence of the enzymic and other chemical reactions involved. The temperature response curve can be characterized by three **cardinal point**s: the **minimum** and **maximum** temperatures (T_{min} and T_{max}, respectively) at which the reaction can proceed and the **optimum** temperature (T_{opt}) (Figure 12.6). Thus there is a range of temperatures below the optimum over which the rate of the reaction or process is stimulated with increasing temperature, and a range beyond the optimum over which the rate declines. These points are largely determined by biochemical factors such as the binding of substrate with active sites (Chapter 8, Box 8.1) and protein (enzyme) stability.

The temperature response of chemical and biological reactions can generally be characterized by comparing the rate of the reaction at two temperatures $10°C$ apart, a value known as the Q_{10}:

$$Q_{10} = R_T + 10/R_T \qquad (12.1)$$

The value of Q_{10} for enzyme-catalyzed reactions is usually about 2, meaning that the rate of the reaction will approximately double for each $10°C$ rise in temperature. This value for Q_{10} applies primarily to stimulation of the reaction by temperatures between T_{min} and T_{opt}. Once the optimum is reached, the reaction rate may decline sharply due to enzyme inactivation (Figure 12.6). Because photosynthesis, respiration, and nitrogen assimilation are complex, multienzyme processes, it is generally assumed that temperature responses will tend to reflect the average temperature characteristics for all of the enzymes. As we will see in Chapter 14, plants can adjust their Q_{10} for respiration during acclimation to temperature.

A basic characteristic of photochemical reactions is that they occur largely independent of temperature in the biologically relevant range of $0°–50°C$. Thus, the Q_{10} for photochemical reactions is close to 1.0. Consequently, the short-term temperature response of photosynthesis largely reflects the effect of temperature on the reactions of carbon metabolism and intersystem electron transport but not the photochemical reactions within the reaction centers of photosystem II and photosystem I. However, prolonged exposure to high temperature can cause the destabilization of photosystem II, which may lead to a decreased capacity for photosynthetic photochemistry and hence an inhibition in the rates of photosynthetic electron transport. Photosystem I is generally much more stable to changes in environmental conditions than is photosystem II.

Measurement of photosynthetic activity in leaves and whole plants is normally based on net gas exchange, that is, it is based on **apparent rates of photosynthesis (AP)**—the difference between the actual rate of photosynthetic CO_2 uptake (gross photosynthesis, GP), CO_2 evolution due to respiration (R), and photorespiration (PR) (Figure 12.7; Equation 12.2).

$$AP = GP - (R + PR) \qquad (12.2)$$

Because gross photosynthesis (GP), respiration (R), and photorespiration (PR) respond very differently to temperature, the optimum temperature for net photosynthesis (AP) is not the same as for gross photosynthesis (Figure 12.8). Note that the rate of respiration continues to increase with temperature, reaching a maximum near $50°C$, where it drops off sharply due to inactivation of enzymes. The temperature response of the rate of photorespiration is thought to follow a curve similar to that shown for respiration.

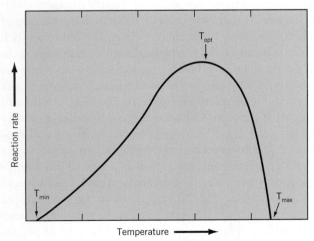

FIGURE 12.6 **Temperature dependence and cardinal points for a typical biological reaction.**

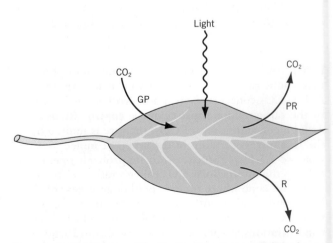

FIGURE 12.7 **Diagram illustrating processes giving rise to** CO_2 **exchange in the light in a C3 leaf.**

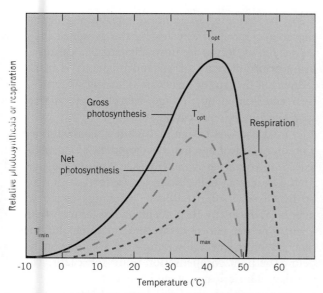

FIGURE 12.8 Diagram illustrating the temperature dependence of gross photosynthesis, respiration, and net photosynthesis. Gross photosynthesis increases with thermal activation of the participating enzymes until inhibitory factors (enzyme inactivation, stomatal closure) take effect. Respiration increases more slowly with temperature and has a higher temperature optimum, but declines more rapidly at high temperature. Net photosynthesis (dashed curve) is determined as the difference between gross photosynthesis and respiration. The resulting cardinal points for net photosynthesis are indicated.

12.3.4 SOIL WATER POTENTIAL

The importance of available water in determining productivity cannot be underestimated. The rate of photosynthesis declines under conditions of water stress, and in cases of severe water stress may cease completely. Stomatal closure and the resultant decrease in CO_2 supply due to water stress imposes a major limitation on photosynthesis. When this occurs in the presence of light for prolonged periods of time, this lack of CO_2 supply may lead to photoinhibition of photosynthesis (Chapters 13 and 14). Photorespiration may protect the photosynthetic apparatus from excess light under such conditions because the energy absorbed can be used to fix O_2 when the CO_2 supply is limiting due to stomatal closure.

Low water potentials reduce turgor pressure in leaf cells, which in turn reduces leaf expansion (see Chapter 17). Under prolonged water stress this results in a reduced photosynthetic surface area. There may be some compensation as stored reserves are mobilized to offset the loss of new assimilate, but overall even mild water stress causes a reduction in net productivity. C4 plants enjoy some advantage over C3 plants with respect to photosynthesis and water stress because of their higher water use efficiency (Chapter 15).

12.3.5 NITROGEN SUPPLY LIMITS PRODUCTIVITY

The maximum possible photosynthetic rate of a leaf, known as the leaf photosynthetic capacity, is determined as the rate of photosynthesis per unit leaf area under conditions of saturating incident light, normal CO_2 and O_2 concentrations, optimum temperature, and high relative humidity. Although leaf photosynthetic capacity may vary as much as a hundredfold, it is generally highest in plants acclimated to resource-rich environments; that is, where light, water, and nutrients are abundant. Reduced photosynthesis is a consequence of deficiencies of virtually all essential elements, but photosynthetic capacity is particularly sensitive to nitrogen supply. As a basic constituent of chlorophyll, redox carriers in the photosynthetic electron transport chain, and all of the enzymes involved in carbon metabolism, nitrogen plays a critical role in primary productivity.

In a C3 species, Rubisco alone will account for more than half of the total leaf nitrogen. In one study, net photosynthesis increased linearly with nitrogen content (Figure 12.9). In barley seedlings, a 5-fold increase in nitrate supply stimulated a 25-fold increase in net photosynthesis. One impact of nitrogen deficiency is to reduce the amount and activity of photosynthetic enzymes, but leaf expansion and other factors no doubt contribute to reductions in photosynthetic capacity as well. Thus, to maximize plant biomass requires the integration of complex metabolic pathways including photosynthesis, respiration, and nitrogen assimilation (Figure 12.10). This is due to the fact that photosynthesis provides energy in the form of fixed carbon as well as basic carbon skeletons for all the necessary cellular constituents. The energy stored as fixed carbon through photosynthesis (Chapters 7 and 8) is retrieved by the cell for all other

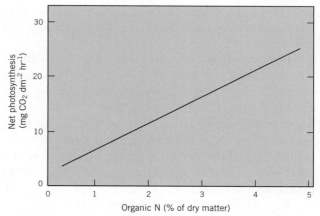

FIGURE 12.9 The relationship between leaf organic nitrogen content and net photosynthesis for the C3 species, Tall fescue. (Adapted from *Plant Physiology* 66:97–100. 1980. Copyright American Society of Plant Physiologists).

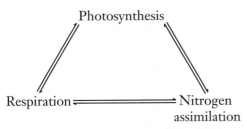

FIGURE 12.10 **Diagram illustrating the interdependence of photosynthesis, respiration, and nitrogen assimilation.**

processes such as growth and development through the process of respiration (Chapter 10). Intermediates of the TCA cycle combine with nitrate assimilation to provide amino acid biosynthesis (Chapters 10 and 11). Photosynthetic energy is used directly in nitrate reduction (Chapter 11) and nitrogen assimilation provides amino acids for the synthesis of enzymes and proteins involved in both photosynthesis and respiration. Thus, all three primary metabolic pathways are intimately connected to each other through GS/GOGAT and its regulation by PII proteins (Chapter 11).

12.3.6 LEAF FACTORS

The net carbon gain of an individual leaf depends on its photosynthetic capacity limited by environmental parameters and balanced against its construction and maintenance costs. For example, the net carbon gain of a leaf varies markedly during leaf development and aging. During initial development and the rapid growth phase, the photosynthetic capacity of a leaf also increases. However, the developing leaf functions as a sink, utilizing carbon assimilated locally as well as importing carbon to support its expansion. Leaf photosynthetic capacity then declines as the aging leaf undergoes senescence, a progressive deterioration of the leaf characterized in part by the loss of chlorophyll and photosynthetic enzymes. Only in the period between full expansion and the onset of senescence does a leaf produce a profit in terms of carbon gain.

Different types of leaves may also have different photosynthetic capacities. Evergreen leaves, for example, have a lower photosynthetic capacity than deciduous leaves and, because they take longer to develop, their construction costs are also higher. Still, the evergreen leaf may be favored if the cost of maintaining the leaf over winter together with the cost of a lower photosynthetic capacity are less than the cost of producing a new leaf.

Regardless of leaf photosynthetic capacity, unfavorable environmental conditions will cause a reduction in long-term carbon gain (Chapter13). In any natural environment, available water, the quantity of light, and temperature will all vary widely and to some extent independently, often keeping the photosynthetic rate well below full capacity. In addition, a prominent feature in the environment of any leaf is the presence of other leaves; that is, leaves are normally part of a **canopy**. Net primary productivity of a stand of plants is marked by influenced by canopy structure. Canopy structure is in turn determined by the age, morphology, angle, and spacing of individual leaves.

A herbaceous C3 annual plant is characterized by a gradient in leaf age and development along the stem axis. The young, growing leaves at the top are exposed to full sunlight while older leaves further down may be heavily shaded. Irradiance reaching shaded leaves may be reduced to 10 percent or less, thus producing a very low net photosynthesis. Very often the fluence rate reaching leaves lowermost in a canopy may fall below the light compensation point for a large part of the day. Those leaves would not only no longer contribute to net photosynthesis, but would incur a negative carbon gain through respiratory loss. Many herbaceous annuals avoid the costs of maintaining such nonproductive leaves by undergoing **sequential senescence**; that is, the older leaves lower in the canopy senesce as new leaves are being formed at the top of the canopy. Senescing leaves may lose as much as 50 percent of their dry weight, largely in the form of soluble organic nitrogen compounds, before dying and falling to the ground. These compounds are exported to developing leaves and other sinks where they are reused. In this way, limited resources are redistributed among leaves of varying ages in order to maximize whole plant carbon gain.

Canopy architecture is important when considering agricultural crops and natural ecosystems because it determines how efficiently light is absorbed. High productivity is in part dependent on the extent to which ground area is covered with photosynthetic surface. Because sunlight striking exposed soil does not contribute to productivity, most agricultural systems are designed so that the young plants fill in the canopy rapidly in order to maximize interception of available light. On the other hand, planting at too high a density will introduce mutual shading by leaves in the canopy. Shading reduces the overall efficiency of light interception and, consequently, reduces long-term carbon gain (Figure 12.11). The ratio of photosynthetic leaf area to covered ground area is known as the **leaf area index (LAI)**. Leaf area is usually taken as the area of a single surface (or projected area, where leaves are not planar). Because both leaf surface and the covered ground are measured as areas (m^2), LAI is dimensionless. Values of LAI in productive agricultural ecosystems typically fall in the range of 3 to 5.

The optimum LAI for a given stand of plants depends on the angle between the leaf and the stem. Horizontal leaves, typical of beans (*Phaseolus*) and similar crops, are efficient light absorbers because of the

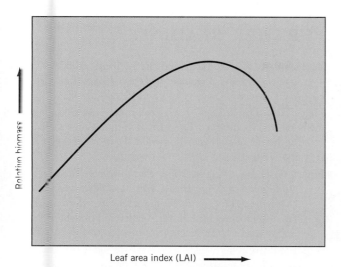

FIGURE 12.11 The relationship between biomass and leaf area index (LAI) in a crop. LAI is varied by varying the density of plants in the stand. Higher planting densities lead to a decline in biomass because of mutual shading of leaves and loss of carbon by respiration in the shaded, nonproductive leaves.

broad surface presented to the sun, but they also more effectively shade leaves lower down in the canopy. Erect leaves, typical of grasses like wheat (*Triticum*) and maize (*Zea mays*), produce less shading but, because of their steeper angle, are not as efficient at intercepting light. Experiments with closely spaced rice plants have shown that carbon gain dropped by a third when leaves were weighted in the horizontal position. With maize, tying upper leaves into a more erect posture increased yield by as much as 15 percent. In some crop plants, such as sugarbeet (*Beta vulgaris*), leaf angle varies from near vertical at the top of the plant to horizontal at the base. This arrangement reduces light interception by the uppermost leaves, but allows light to penetrate more deeply into the canopy. Overall, the more uniform distribution of light tends to improve the efficiency of light interception by the canopy and thus the efficiency of carbon gain.

The relationship between LAI, leaf angle, and photosynthetic rate has been tested by computer models. The results show that with low values of LAI leaf angle has little effect, but above a LAI of 3, more layers of vertical leaves are required to maximize photosynthesis. Field studies have confirmed that canopies with predominantly horizontal leaves have LAI values of 2 or less, while vertical leaf canopies support LAI values of 3 to 7.

In cases where the leaves are more or less fixed in space, the efficiency of light interception will change with the angle of the sun. Many desert and agricultural species, however, have the capacity to alter the orientation of their leaves, allowing them to track the sun as it moves across the sky through the day. Called

heliotropism, or **solar tracking**, the leaf blades move in such a way that their surfaces remain perpendicular to the sun's direct rays. Solar tracking would help to maximize daily carbon gain in those plants that must complete their life cycle in a brief period before the onset of unfavorable conditions such as drought or high temperatures.

SUMMARY

Carbon assimilation by plants creates the plant biomass that supports humans and virtually all other heterotrophic organisms. The study of carbon gain, or productivity, at both the organismal and population level is an important component of agricultural and ecosystem research. Overall carbon gain depends on net primary productivity—the balance between carbon uptake by photosynthesis and carbon loss to respiration. Carbon loss to respiration can be divided into the carbon cost of growth, or growth respiration, and the cost of simply maintaining structure and processes that do not result in a net increase in dry matter. Several studies have shown a negative correlation between respiration and growth rate. The implication is that respiration has a significant impact on biomass accumulation and yield, and that improving yield by manipulating respiration might be feasible.

Productivity is also influenced by a variety of genetic and environmental factors that influence photosynthesis. These include light, available carbon dioxide, temperature, soil water, nutrients, and canopy structure. The rate of photosynthesis increases between the light compensation point and saturation. Because irradiance changes constantly throughout the day, long-term carbon gain depends on the cumulative irradiance over the growing season.

Although the carbon dioxide content of the atmosphere is relatively constant in the short term, there is evidence that plants use the stomata to keep the internal carbon dioxide concentration in balance with electron transport capacity. On the other hand, with adequate light, plants will respond to carbon dioxide enrichment by increased productivity. However, ultimately productivity is also dependent on source-sink relationships.

Photosynthesis, respiration, and photorespiration respond differently to temperature. Thus the optimum temperature for net photosynthesis is not the same as the optimum temperature for gross photosynthesis. Plants also require an adequate water and nitrogen supply in order to maximize their leaf photosynthetic capacity. Productivity in a stand depends on the pattern of leaf senescence and the structure of the canopy. The ideal canopy maximizes the efficiency of light interception and carbon gain by balancing leaf area, leaf angle, leaf orientation, plant density, and senescence of older leaves.

CHAPTER REVIEW

1. Distinguish between growth respiration and maintenance respiration. Which might best be manipulated in order to improve productivity?

2. Describe how various environmental factors influence plant productivity.

3. Why do large values for leaf area index lead to an overall decline in productivity of a stand?

4. How do plants alter leaf structure and morphology to protect themselves from excess light?

FURTHER READING

Buchanan, B. B., W. Gruissem, R. L. Jones. 2000. *Biochemistry and Molecular Biology of Plants*. Rockville, MD: American Society of Plant Physiologists.

Field, T. S., D. W. Lee, N. M. Holbrook. 2001. Why leaves turn red in autumn: The role of anthocyanins in senescing leaves of red-osier dogwood. *Plant Physiology* 127: 566–574.

Foyer, C. H., G. Noctor. 2002. *Photosynthetic Nitrogen Assimilation and Associated Carbon and Respiratory Metabolism. Advances in Photosynthesis and Respiration*, Vol. 12. Dordrecht: Kluwer Academic Publishers.

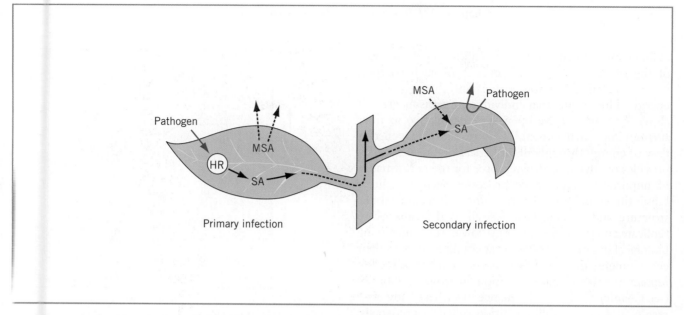

Pathogen

MSA

Pathogen

MSA

HR

SA

SA

Primary infection

Secondary infection

13

Responses of Plants to Environmental Stress

The previous chapters have focused on the underlying processes by which plant roots acquire water and nutrients from the soil (Chapters 1 to 4), leaves harvest light energy and convert atmospheric CO_2 and soil NO_3^- into stable chemical forms of sucrose, starch, and and amino acids (Chapters 5, 7, 8, 11) and subsequently unlock the stored chemical energy through the process of glycolysis and mitochondrial respiration for growth and development (Chapter 10). By necessity, to elucidate the biochemical and molecular mechanisms by which these processes occur, plants are studied under normal or ideal environmental conditions for growth and development. However, plants often encounter unusual or extreme conditions: trees and shrubs in the northern temperate latitudes experience the extreme low temperatures of winter; alpine plants experience cold and drying winds; and agricultural crops may experience periods of extended drought as well as high and low temperatures. Extremes in environmental parameters create stressful conditions for plants, which may have a significant impact on their physiology, development, and survival.

The study of plant responses to environmental stress has long been a central theme for plant environmental physiologists and physiological ecologists. How plants respond to stress helps to explain their geographic distribution and their performance along environmental gradients. Because stress invariably leads to reduced productivity, **stress responses** are also important to

agricultural scientists. Understanding stress responses is essential in attempts to breed stress-resistant cultivars that can withstand drought, and other yield-limiting conditions. Finally, because stressful conditions cause perturbations in the way a plant functions, they provide the plant physiologist with another very useful tool for the study of basic physiology and biochemistry.

This chapter will examine some of the stresses that plants encounter in their environment. The principal topics to be addressed include

- the basic concepts of plant stress, acclimation, and adaptation,
- the light-dependent inhibition of photosynthesis through a process called photoinhibition,
- the effects of water deficits on stomatal conductance,
- the effects of high- and low-temperature stress on plant survival,
- the challenge of freezing stress on plant survival, and
- the responses of plants to biotic stress due to infestations by insects and disease.

13.1 WHAT IS PLANT STRESS?

Because life is an endergonic process, that is $\Delta G > 0$ (Chapter 5), energy is an absolute requirement for the maintenance of structural organization over the lifetime of the organism. The maintenance of such complex order over time requires a constant through put of energy. This means that individual organisms are not closed systems but are open systems relative to their surrounding environment. This results in a constant flow of energy through all biological organisms, which provides the dynamic driving force for the performance of important maintenance processes such as cellular biosyntheses and transport to maintain its characteristic structure and organization as well as the capacity to replicate and grow. Such energy flow ensures that living biological organisms are never at equilibrium with their environment, that is, ΔG is never equal to zero, but remain in a steady-state condition far from equilibrium (see Chapter 5). The maintenance of such a steady-state results in a meta-stable condition called **homeostasis**. As a consequence, all life forms may be considered transient energy storage devices with finite but varying lifetimes (Figure 13.1).

Any change in the surrounding environment may disrupt homeostasis. Environmental modulation of homeostasis may be defined as **biological stress**. Thus, it follows that **plant stress** implies some adverse effect on the physiology of a plant induced upon a sudden transition from some optimal environmental condition where homeostasis is maintained to some suboptimal condition which disrupts this initial homeostatic state (Figure 13.2). Thus, plant stress is a relative term since the experimental design to assess the impact of a stress always involves the measurement of a physiological phenomenon in a plant species under a suboptimal, stress condition compared to the measurement of the same physiological phenomenon in the same plant species under optimal conditions. Since the extent of a stress can be quantified by assessing the difference in these measurements under optimal versus suboptimal conditions, the basis of stress physiology is **comparative**

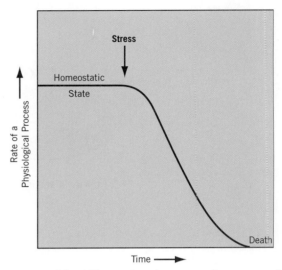

FIGURE 13.2 **The effects of environmental stress on plant homeostasis. Under some optimal environmental condition, a plant is in homeostasis as indicated by a constant rate of some important physiological process over time. Upon the imposition of an external stress, the rate of this physiological process decreases rapidly. It is fatal for some plants that are unable to adjust to an imposed stress and can not establish a new homeostatic state. Such plants are classified as susceptible to the stress.**

physiology. However, plant species are highly variable with respect to their optimum environments and their susceptibility to extremes of, for example, irradiance, temperature, and water potential. Is stress a function of the environment or the organism? For example, are the extreme environments encountered in deserts or arctic tundra stressful for plants that normally thrive there? Are these environments stressful only to some species but not to others?

13.2 PLANTS RESPOND TO STRESS IN SEVERAL DIFFERENT WAYS

Plant stress can be divided into two primary categories. **Abiotic stress** is a physical (e.g., light, tempera,-ture) or chemical insult that the environment may impose on a plant. **Biotic stress** is a biological insult, (e.g., insects, disease) to which a plant may be exposed during its lifetime (Figure 13.3). Some plants may be **injured** by a stress, which means that they exhibit one or more metabolic dysfunctions. If the stress is moderate and short term, the injury may be temporary and the plant may recover when the stress is removed. If the stress is severe enough, it may prevent flowering, seed formation, and induce senescence that leads to plant death. Such plants are considered to be **susceptible** (Figure 13.3). Some plants escape the stress altogether, such as ephemeral, or short-lived, desert plants.

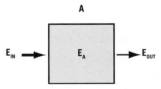

FIGURE 13.1 **Biological life forms as energy-storing devices. A is any biological life form. E_{IN} (thick arrow) is the energy flowing from the surrounding environment into the biological organism, A. E_{OUT} (thin arrow) is the energy flowing out of the biological organism back into the environment. The thicknesses of the arrows indicate the differences in the relative flux of energy flowing in and out of a living organism. E_A is the steady-state energy stored or trapped by a living organism. According to the First Law of thermodynamics (Chapter 5), E_{IN} + E_A + E_{OUT} = 1. Thus, $E_A = E_{IN} - E_{OUT}$.**

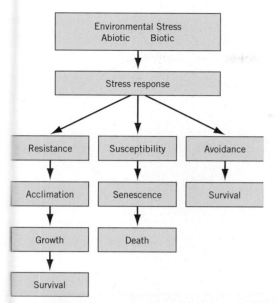

FIGURE 13.3 **The effect of environmental stress on plant survival.**

Ephemeral plants germinate, grow, and flower very quickly following seasonal rains. They thus complete their life cycle during a period of adequate moisture and form dormant seeds before the onset of the dry season. In a similar manner, many arctic annuals rapidly complete their life cycle during the short arctic summer and survive over winter in the form of seeds. Because ephemeral plants never really experience the stress of drought or low temperature, these plants survive the environmental stress by **stress avoidance** (Figure 13.3).

Avoidance mechanisms reduce the impact of a stress, even though the stress is present in the environment. Established plants of alfalfa (*Medicago sativa*), for example, survive dry habitats as adult plants by sending down deep root systems that penetrate the water table. Alfalfa is thereby ensured an adequate water supply under conditions in which more shallow-rooted plants would experience drought. Other plants develop fleshy leaves that store water, thick cuticles or pubescence (leaf hairs) to help reduce evaporation, or other modifications that help to either conserve water or reduce water loss. Cacti, with their fleshy photosynthetic stems and leaves reduced to simple thorns, are another example of drought avoiders. Most drought avoiders would be severely injured should they ever actually experience desiccation.

Many plants have the capacity to tolerate a particular stress and hence are considered to be **stress resistant** (Figure 13.3). Stress resistance requires that the organism exhibit the capacity to adjust or to **acclimate** to the stress.

Finally, a controversy over terminology is concerned with use of the term **strategy**. Strategy is often used to describe the manner in which a plant responds successfully to a particular stress. Some physiologists object to use of the term for the reason that strategy implies a conscious plan; that is, it is teleological.[1] However, strategy can validly describe a genetically programmed sequence of responses that enable an organism to survive in a particular environment.

13.3 TOO MUCH LIGHT INHIBITS PHOTOSYNTHESIS

In Chapters 7 and 8, we discussed the conversion of visible light energy into ATP and NADPH through photosynthetic electron transport. In addition to forming Triose-P that can be converted to sucrose or starch, the ATP and NADPH are required to regenerate RuBP by the Calvin Cycle (Figure 13.4A). The continuous regeneration of RuBP is an absolute requirement for the continuous assimilation of CO_2 by Rubisco. This requirement which is satisfied by the light-dependent biosynthesis of ATP and NADPH is what makes CO_2 assimilation light dependent. However, although light is required for the photosynthetic assimilation of CO_2, too much light can inhibit photosynthesis.

In all plants, the light response curve for photosynthesis exhibits saturation kinetics as illustrated in Figure 13.4B. At low irradiance, the rate of CO_2 assimilation increases linearly with an increase in irradiance. This is to be expected since more absorbed light means higher rates of electron transport which, in turn, means increasing levels of ATP and NADPH for the regeneration of RuBP (Figure 13.4A). Thus, under low, light-limiting conditions, the rate at which RuBP is regenerated through the consumption of ATP and NADPH by the Calvin Cycle limits the rate of photosynthesis measured either as CO_2 assimilation or O_2 evolution. The maximum initial slope of the photosynthetic light response curve under low, light-limiting conditions provides a measure of **photosynthetic efficiency** measured either as moles of CO_2 assimilated per photon absorbed, or alternatively, moles of O_2 evolved per photon absorbed if photosynthesis is measured as the rate of O_2 evolution (Figure 13.4B).

Upon further increases in irradiance, the rate of photosynthesis is no longer a linear function of irradiance but rather levels off. At these higher light intensities, the rate of photosynthesis is said to be **light saturated** (Figure 13.4B, red shaded area). This means that the Calvin Cycle is saturated with ATP and NADPH which, in turn, means that Rubisco is saturated with one of its substrates, RuBP. The maximum light saturated rate is a measure of **photosynthetic**

---

[1]Teleology is the doctrine of final causes, assigning purpose to natural processes. Teleological arguments are considered inappropriate in natural science.

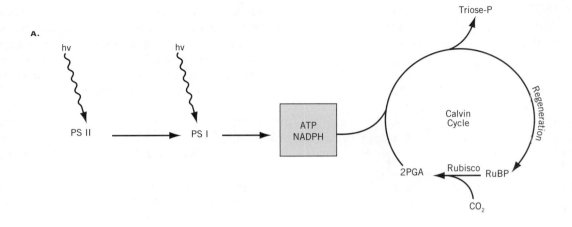

FIGURE 13.4 **A schematic illustration of the response of photosynthesis to increasing irradiance.** (*A*) **A model illustrating the interaction between photosynthetic linear electron transport and the Calvin Cycle.** (*B*) **A schematic light response curve for photosynthesis measured as either the rate of CO_2 assimilation or the rate of O_2 evolution. The area above the light response curve represents excess irradiance that is not used in photosynthesis.**

capacity and will have the units of either moles of CO_2 assimilated or moles of O_2 evolved per leaf area per unit time. Thus, under light-saturated conditions, the rate of regeneration of RuBP no longer limits the rate of CO_2 assimilation but rather it is the rate at which Rubisco can consume RuBP and CO_2 that limits the rate of photosynthesis. Consequently, under light-saturated conditions, the rate of photosynthesis becomes light-independent, that is, the exposure of a plant to higher irradiance no longer changes the rate of photosynthesis. Under light-saturated conditions, plants

become exposed to increasing levels of **excess light** (Figure 13.4B, yellow shaded area), that is, they become exposed to more light than the plant can use for photosynthesis. If the plant continues to be exposed to higher and higher levels of excess light, the rate of photosynthesis begins to decrease (Figure 13.4B). This is called **photoinhibition** of photosynthesis and is defined as the light-dependent decrease in photosynthetic rate that may occur whenever the irradiance is in excess of that required either for the photosynthetic evolution of O_2 or the photosynthetic assimilation of CO_2.

There is a consensus in the literature that in most plants, PSII is more sensitive to photoinhibition than PSI. The effects of exposure to photoinhibition by shifting plants from low light to high light can be assessed either by monitoring changes in photosynthetic efficiency and photosynthetic capacity (Figure 13.5A) or by monitoring chlorophyll fluorescence (Box 13.1) to assess changes in maximum PSII photochemical efficiency measured as Fv/Fm (Figure 13.5B). Photoinhibition that results in a decrease in photosynthetic efficiency as well as photosynthetic capacity (Figure 13.5A) usually reflects

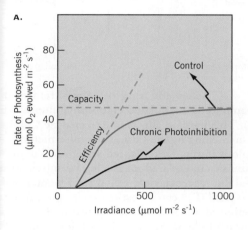

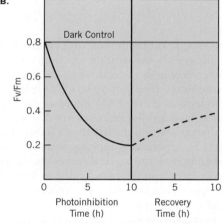

FIGURE 13.5 **The effects of chronic photoinhibition on photosynthesis.** (*A*) Schematic light response curves for control plants which have not been pre-exposed to high light and plants that have been pre-exposed to high light to induce chronic photoinhibition prior to measuring their light response curve. The maximum initial slope is a measure of photosynthetic efficiency. The maximum light saturated rate is a measure of photosynthetic capacity. (*B*) A schematic representation illustrating the effect of exposure time to high light to induce chronic photoinhibition on the maximum photochemical efficiency of PSII measured as Fv/Fm (see Box 13.1). The broken line illustrates the slow rate of recovery of Fv/Fm when the chronically photoinhibited plants are shifted back to low-light conditions. Control plants were kept in the dark over the course of the experiment.

chronic photoinhibition which is the result of **photodamage** to PSII. Specifically, the site of damage is the D1 reaction center polypeptide of PSII which causes a decrease in the efficiency of PSII charge separation (Chapter 7). This decrease in the PSII photochemical efficiency can be measured as Fv/Fm (Figure 13.5B) (Box 13.1). The decrease in PSII photochemical efficiency is usually paralleled by a decrease in photosynthetic efficiency of O_2 evolution. A characteristic of chronic photoinhibition is that it is only very slowly reversible after plants are shifted from excess light to low light (Figure 13.5B).

13.3.1 THE D1 REPAIR CYCLE OVERCOMES PHOTODAMAGE TO PSII

PSII reaction centers exhibit an inherent lifetime. This was first indicated by the fact that the D1 polypeptide of PSII reaction centers exhibits the fastest turnover rate of any plant protein. The D1 polypeptide is degraded and resynthesized in the time span of approximately 30 minutes. Recently, it has been proposed that PSII actually exhibits the properties consistent with a **photon counter**, that is, its lifetime is dependent upon the number of photons absorbed and not the absolute time. It has been calculated that, under normal growth conditions, each PSII reaction center is irreversibly damaged presumably due to photooxidative damage and spontaneously degraded after the absorption of 10^5 to 10^7 photons. Thus, exposure to excess light simply shortens the lifetime of PSII because these conditions would enhance the probability of photooxidative damage and shorten the time to absorb the necessary photons to cause the degradation of D1.

How do plants ensure a constant supply of functional PSII reaction centers? Plants and green algae exhibit a single chloroplastic gene that encodes the D1 polypeptide called *psbA*. These organisms have evolved a D1 repair cycle which repairs photodamage to PSII (Figure 13.8). When the D1 polypeptide is damaged, it is marked for degradation by protein phosphorylation. After phosphorylation of D1, PSII is partially disassembled and the D1 polypeptide is degraded by proteolysis. Subsequently, the *psbA* gene is transcribed and translated using the chloroplastic transcriptional and translational machinery with the subsequent accumulation of a new D1 polypeptide. This new D1 polypeptide is inserted into the nonappressed, stromal thylakoids and a new, functional PSII complex is reassembled. Given the lateral heterogeneity present in thylakoid membranes, the mechanism by which damaged PSII complexes migrate laterally from granal stacks to nonappressed, stromal thylakoids for disassembly and reassembly remains to be elucidated.

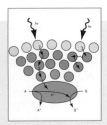

Box 13.1
MONITORING PLANT STRESS BY CHLOROPHYLL FLUORESCENCE

In Chapter 6 we defined fluorescence as the emission of a photon of light by an excited molecule as it returns to ground state from its lowest singlet excited state. Due to the initial thermal deactivation to the lowest singlet excited state, chlorophyll emits red light as it fluoresces to ground state. Thus, regardless of the wavelength used to excite chlorophyll, it always emits red light. Of the total energy absorbed by a leaf, less than 3 percent of that energy is lost due to chlorophyll fluorescence. Due to the differences in the structure and composition of PSII and PSI, chlorophyll fluorescence measured at room-temperature emanates primarily from PSII (Figure 13.6). Due to this property, room-temperature chlorophyll fluorescence is exploited as a sensitive, intrinsic probe not only of PSII function but also of the overall function of photosynthesis. As a consequence, chlorophyll fluorescence is undoubtedly the most widely used spectroscopic technique in photosynthesis and plant stress research.

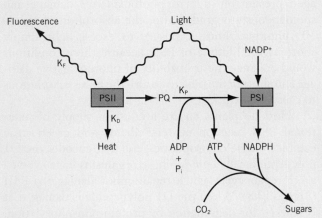

FIGURE 13.6 **A model illustrating the possible fates of absorbed light energy in the photosynthetic apparatus. The primary fates of photosynthetically absorbed light energy are thought to include useful photochemistry designated by the rate constant, K_P, nonphotochemical dissipation of excess absorbed light energy as heat designated by the rate constant, K_D, and chlorophyll fluorescence designated by the rate constant, K_F. Since less than 3 percent of the absorbed light energy is ever lost as chlorophyll fluorescence, this pathway is not considered a major pathway for the safe dissipation of excess light energy. However, chlorophyll fluorescence is a sensitive, intrinsic probe for the function of the photosynthetic apparatus and overall physiological status of the plant.**

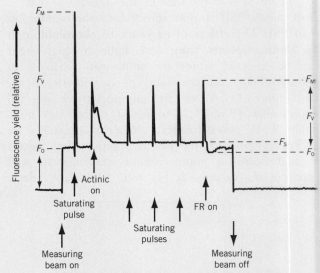

FIGURE 13.7 **Schematic trace of a measurement of pulse-amplitude-modulated chlorophyll fluorescence (PAM). This method allows one to analyze Chl fluorescence quenching by the saturation pulse method. When a sample is in the dark-adapted state, turning on the low-intensity modulated measuring beam results in a minimal fluorescence yield called F_o. The maximal fluorescence yield (F_m) can be estimated by the subsequent application of a high-intensity millisecond saturating pulse of white light. Since variable fluorescence (F_v) is defined as $F_m - F_o$, therefore,**

$$\frac{F_m - F_o}{F_m} = \frac{F_v}{F_m}.$$

When the sample is illuminated with actinic light sufficient to induce photosynthesis, the fluorescence yield (F_S) undergoes complex transitions until eventually F_S reaches a minimal, steady-state level. This is called the Kautsky induction curve. PAM fluorescence exploits the fact that maximal fluorescence yield (F'_m) can be assessed during active photosynthesis induced by the actinic light by repetitively applying saturating pulses of light as indicated by the vertical spikes emanating from the Kautsky induction curve indicated in Figure 13.7. To a first approximation, $F_m - F'_m$ represents nonphotochemical quenching of the fluorescence yield, whereas $F'_m - F_s$ represents photochemical quenching.

Kautsky and Hirsch were the first to report the variable nature of the chlorophyll fluorescence signal. Subsequent detailed studies of this complex fluorescence signal (Figure 13.7) showed it to be rich in information with respect to the properties of the PSII reaction center as well as its association with overall photosynthetic electron transport and CO_2 assimilation. When a leaf or an algal suspension is dark adapted for anywhere from 5 minutes to 60 minutes at room temperature depending on the species used, all electrons are drained from the photosynthetic electron transport chain by PSI causing all PSII reaction centers to be in the open configuration

[P680 Pheo Q$_A$] (Chapter 7). The minimal fluorescence yield with all PSII reaction centers in the open configuration is called the background or F_o fluorescence (Figure 13.7). If the samples are now exposed to a light intensity sufficiently strong to initiate photosynthesis, a fluorescence signal characterized by complex transients can be detected (Figure 13.7). This is called the **Kautsky effect**. The fast rise from F_o to an initial maximum fluorescence (F_m) (Figure 13.7) is due to the fact that the light absorbed closes the PSII reaction centers [P680$^+$ Pheo Q_A^-] (Chapter 7) much faster through energy transfer and photochemistry than they can be reopened by PSI, intersystem electron transport and ultimately CO_2 assimilation (Figure 13.7). The subsequent decrease in the fluorescence yield or fluorescence quenching reflects the reopening of PSII reaction centers. This is due to the induction of the much slower, enzyme-catalyzed reactions involved in ATP and NADPH biosynthesis, which are in turn ultimately consumed by biochemical reactions involved in CO_2 assimilation. Thus, chlorophyll fluorescence yield is inversely proportional to the rate of photosynthesis.

It is incumbent upon all photosynthetic organisms to maintain a balance in energy budget, that is, a balance between the energy absorbed through photochemistry versus energy either utilized through metabolism and growth and/or dissipated nonphotochemically as heat. There is a consensus that most photosynthetic organisms possess two primary mechanisms to maintain an energy balance. (1) **Photochemical quenching**, measured as qP (Equation 13.1), reflects the capacity to utilize excess absorbed energy through metabolism and growth.

$$qP = Fm' - Fs/Fm' - Fo \qquad (13.1)$$

$$NPQ = Fm - Fm'/Fm' \qquad (13.2)$$

This involves the upregulation of the expression and activity of specific enzymes involved in stromal CO_2 assimilation (Rubisco), cytosolic sucrose biosynthesis (SPS), as well as vacuolar fructan biosynthesis in certain plant species such as the cereals (Chapter 9). (2) **Nonphotochemical quenching**, measured as NPQ (Equation 13.2) reflects the capacity to dissipate excess absorbed energy as heat (Figure 13.6). This involves primarily the stimulation of the thylakoid, xanthophyll cycle enzymes involved in the reversible de-epoxidation of violaxanthin to antheraxanthin and zeaxanthin. Photochemical-quenching capacity (qP) and nonphotochemical-quenching capacity (qN, NPQ) can be estimated *in vivo* using **pulse-amplitude-modulated chlorophyll fluorescence (PAM)**. The extent of photoinhibition induced by any environmental stress can be rapidly assessed by measuring the maximum photochemical efficiency of PSII (Fv/Fm) before and after a photoinhibition treatment. Fv/Fm can be calculated from the following equation (Equation 13.3). As expected, the decrease in Fv/Fm follows a similar pattern as the decrease in photosynthetic efficiency of O_2 evolution in response to time under conditions of photoinhibition.

$$F_v/F_m = (F_m - F_o)/F_m \qquad (13.3)$$

REFERENCES

Adams III, W. W., B. Demmig-Adams. 2004. Chlorophyll fluorescence as a tool to monitor plant response to the environment. *Chlorophyll a Fluorescence—A Signature of Photosynthesis. Advances in Photosynthesis and Respiration*, Vol. 19, pp. 583–604. Dordrecht: Springer.

Schreiber, U. 2004. Pulse-amplitude-modulation (PAM) fluorometry and saturation pulse method: An overview. *Chlorophyll a Fluorescence—A Signature of Photosynthesis. Advances in Photosynthesis and Respiration*, Vol. 19, pp. 279–319. Dordrecht: Springer.

Baker, N. R. 2008. Chlorophyll fluorescence: a probe of photosynthesis in vivo. *Annual Review of Plant Biology* 59: 89–113.

13.4 WATER STRESS IS A PERSISTENT THREAT TO PLANT SURVIVAL

Water stress may arise through either an excess of water or a water deficit. An example of excess water is flooding. Flooding stress is most commonly an oxygen stress, due primarily to reduced oxygen supply to the roots. Reduced oxygen in turn limits respiration, nutrient uptake, and other critical root functions. Stress due to water deficit is far more common, so much so that the correct term **water deficit stress** is usually shortened to simply **water stress**. We will focus on water deficit stress in this chapter. Because water stress in natural environments usually arises due to lack of rainfall, a condition known as drought, this stress is often referred to as **drought stress**. In the laboratory, water stress can be simulated by allowing transpirational loss from leaves (Chapter 2), a condition commonly referred to as **desiccation stress**.

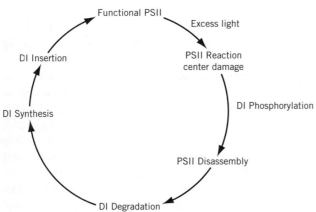

FIGURE 13.8 **The D1 repair cycle. Although light is the ultimate source of energy for photosynthesis, too much light can be dangerous because it may result in damage to the photosystems, especially the D1 reaction center polypeptide present in PSII. Plants and algae have evolved an elaborate mechanism to repair photodamage called the D1 repair cycle. Exposure to an irradiance that exceeds the capacity of the plant either to utilize that energy in photosynthesis or to dissipate it safely as heat results in damage to the D1 polypeptide. Functional PSII reaction centers are converted to damaged PSII reaction centers. When this happens, PSII is disassembled and D1 is degraded by thylakoid proteolytic enzymes. Subsequently, new D1 is synthesized de novo by the chloroplast translational machinery and inserted into the thylakoid membrane to form functional PSII reaction centers.**

13.4.1 WATER STRESS LEADS TO MEMBRANE DAMAGE

Damage resulting from water stress is related to the detrimental effects of desiccation on protoplasm. Removal of water, for example, leads to an increase in solute concentration as the protoplast volume shrinks, which may itself have serious structural and metabolic consequences. The integrity of membranes and proteins is also affected by desiccation, which in turn leads to metabolic dysfunctions. Stresses may alter the lipid bilayer and cause the displacement of membrane proteins, which, together with solute leakage, contributes to a loss of membrane selectivity, a general disruption of cellular compartmentation, and a loss of activity of membrane-based enzymes.

In addition to membrane damage, numerous studies have shown that cytosolic and organellar proteins may undergo substantial loss of activity or even complete denaturation when dehydrated. Loss of membrane integrity and protein stability may both be exacerbated by high concentrations of cellular electrolytes that accompany dehydration of protoplasm. The consequence of all these events is a general disruption of metabolism in the cell upon rehydration.

13.4.2 PHOTOSYNTHESIS IS PARTICULARLY SENSITIVE TO WATER STRESS

Photosynthesis can be affected by water stress in two ways. First, closure of the stomata normally cuts off access of the chloroplasts to the atmospheric supply of carbon dioxide. Second, there are direct effects of low cellular water potential on the structural integrity of the photosynthetic machinery. The role of water stress in stomatal closure will be discussed in the following section.

Direct effects of low water potential on photosynthesis have been studied extensively in chloroplasts isolated from sunflower (*Helianthus annuus*) leaves subjected to desiccation. Sunflower has proven useful for these studies because stomatal closure has only a minor effect on photosynthesis. This is because direct effects on the photosynthetic activity of chloroplasts decrease the demand for CO_2 and the CO_2 level inside the leaf remains relatively high. Both electron transport activity and photophosphorylation are reduced in chloroplasts isolated from sunflower leaves with leaf water potentials below about -1.0 MPa. These effects reflect damage to the thylakoid membranes and ATP synthase protein (CF_0-CF_1) complex (Chapters 5 and 7).

The direct effects of water stress on photosynthesis are exacerbated by the additional effects of light. Since water stress inhibits CO_2 assimilation, this means that water stress will expose plants to excess light. The light absorbed by the photosynthetic pigments of the leaf continue to absorb light but this absorbed light energy can not be processed because photosynthetic electron transport is inhibited. Thus, a concomitant effect of exposure of plants to a water deficit is chronic photoinhibition.

13.4.3 STOMATA RESPOND TO WATER DEFICIT

Plants are often subjected to acute water deficits due to a rapid drop in humidity or increase in temperature when a warm, dry air mass moves into their environment. The result can be a dramatic increase in the vapor pressure gradient between the leaf and the surrounding air. Consequently, the rate of transpiration increases (Chapter 2). An increase in the vapor pressure gradient will also enhance drying of the soil. Because evaporation occurs at the soil surface, the arrival of a dry air mass has particular consequences for the uptake of water by shallow-rooted plants.

Plants generally respond to water stress by closing their stomata in order to match transpirational water loss through the leaf surfaces with the rate at which water can be resupplied by the roots. It has been shown in virtually all plants studied thus far, including plants from desert, temperate, and tropical habitats, that stomatal opening and closure is responsive to ambient humidity. Unlike the surrounding epidermal cells, the surfaces of the guard cells are not protected with a heavy cuticle. Consequently, guard cells lose water directly to the atmosphere. If the rate of evaporative water loss from the guard cells exceeds the rate of water regain from underlying mesophyll cells, the guard cells will become flaccid and the stomatal aperture will close. The guard cells may thus respond directly to the vapor pressure gradient between the leaf and the atmosphere. Closure of the stomata by direct evaporation of water from the guard cells is sometimes referred to as **hydropassive closure**. Hydropassive closure requires no metabolic involvement on the part of the guard cells; guard cells respond to loss of water as a simple osmometer (Chapter 1).

Stomatal closure is also regulated by *hydroactive processes*. **Hydroactive closure** is metabolically dependent and involves essentially a reversal of the ion fluxes that cause opening (Chapter 1). Hydroactive closure is triggered by decreasing water potential in the leaf mesophyll cells and appears to involve abscisic acid (ABA) and other hormones. Since the discovery of ABA in the late 1960s, it has been known to have a prominent role in stomatal closure due to water stress. ABA accumulates in water-stressed (that is, wilted) leaves and external application of ABA is a powerful inhibitor of stomatal opening. Furthermore, two tomato mutants, known as *flacca* and *sitiens*, fail to accumulate normal levels of ABA and both will wilt very readily. The precise role of ABA in stomatal closure in water-stressed whole plants has, however, been difficult to decipher with certainty. This is because ABA is ubiquitous, often occurring in high concentrations in nonstressed tissue. Also, some early studies indicated that stomata would begin to close before increases in ABA content could be detected.

In most well-watered plants, ABA appears to be synthesized in the cytoplasm of leaf mesophyll cells but, because of intracellular pH gradients, ABA accumulates in the chloroplasts (Figure 13.9). At low pH, ABA exists in the protonated form ABAH, which freely permeates most cell membranes. The dissociated form ABA⁻ is impermeant; because it is a charged molecule it does not readily cross membranes. Thus, ABAH tends to diffuse from cellular compartments with a low pH into compartments with a higher pH. There, some of it dissociates to ABA⁻ and becomes trapped. It is well established that in actively photosynthesizing mesophyll cells the cytosol will be moderately acidic (pH 6.0 to 6.5) while the chloroplast stroma is alkaline (pH 7.5 to 8.0) (Chapter 7). It has been calculated that if the stroma pH is 7.5 and cytosolic pH is 6.5, the concentration of ABA in the chloroplasts will be about tenfold higher than in the cytosol.

According to the current model, the initial detection of water stress in leaves is related to its effects on photosynthesis, described earlier in this chapter. Inhibition of electron transport and photophosphorylation in the chloroplasts would disrupt proton accumulation in the thylakoid lumen and lower the stroma pH. At the same time, there is an increase in the pH of the apoplast surrounding the mesophyll cells. The resulting pH gradient stimulates a release of ABA from the mesophyll cells into the apoplast, where it can be carried in the transpiration stream to the guard cells (Figure 13.10).

Just how ABA controls turgor in the guard cells remains to be determined. Evidence indicates that ABA does not need to enter the guard cell, but acts instead on the outer surface of the plasma membrane. Presumably ABA interacts with the high-affinity binding sites on the plasma membrane (see Chapter 21), although the existence of such sites has yet to be confirmed. Nonetheless, there are strong indications that ABA interferes with plasma membrane proton pumps and, consequently, the uptake of K⁺, or that it stimulates K⁺ efflux from the guard cells. Either way, the guard cells will lose turgor, leading to closure of the stomata.

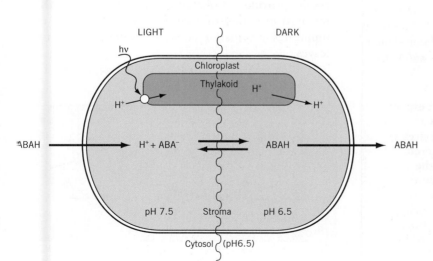

FIGURE 13.9 ABA storage in chloroplasts. In the light, photosynthesis drives protons into the interior of the thylakoid, creating a pH gradient between the stroma and the cytosol. The pH gradient favors movement of ABAH into the chloroplast, where it dissociates to ABA⁻. The membrane is less permeable to ABA⁻. In the dark, protons leak back into the stroma, the pH gradient collapses, and ABAH moves back into the cytosol.

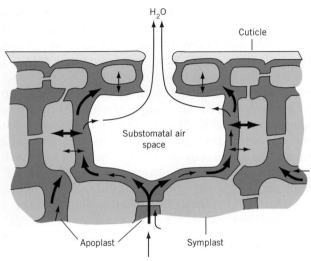

FIGURE 13.10 **ABA movement in the apoplast. ABA synthesized in the roots is carried to the leaf mesophyll cells (heavy arrows) in the transpiration stream (light arrows). ABA equilibrates with the chloroplasts of the photosynthetic mesophyll cells or is carried to the stomatal guard cells in the apoplast.**

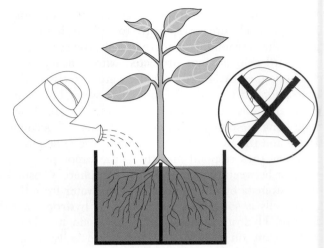

FIGURE 13.11 **An experimental setup for testing the effects of desiccated roots on ABA synthesis and stomatal closure. Roots of a single plant are divided equally between two containers. Water supplied to one container maintains the leaves in a fully turgid state while water is withheld from the second container. Withholding water from the roots leads to stomatal closure, even though the leaves are not stressed.**

As noted above, wilted leaves accumulate large quantities of ABA. In most cases, however, stomatal closure begins before there is any significant increase in the ABA concentration. This can be explained by the release of stored ABA into the apoplast, which occurs early enough and in sufficient quantity—the apoplast concentration will at least double—to account for initial closure. Increased ABA synthesis follows and serves to prolong the closing effect.

Stomatal closure does not always rely on the perception of water deficits and signals arising within the leaves. In some cases it appears that the stomata close in response to soil desiccation **before** there is any measurable reduction of turgor in the leaf mesophyll cells. Several studies have indicated a feed-forward control system that originates in the roots and transmits information to the stomata. In these experiments, plants are grown such that the roots are equally divided between two containers of soil (Figure 13.11). Water deficits can then be introduced by withholding water from

one container while the other is watered regularly. Control plants receive regular watering of both containers. Stomatal opening along with factors such as ABA levels, water potential, and turgor are compared between half-watered plants and fully watered controls. Typically, stomatal conductance, a measure of stomatal opening, declines within a few days of withholding water from the roots (Figure 13.12), yet there is no measurable change in water potential or loss of turgor in the leaves. In experiments with day flower (*Commelina communis*), there was a significant increase in ABA content of the roots in the dry container and in the leaf epidermis (Figure 13.13). Furthermore, ABA is readily translocated from roots to the leaves in the transpiration stream, even when roots are exposed to dry air. These results provide reasonably good evidence that ABA is involved in a kind of early warning system that communicates information about soil water potential to the leaves.

FIGURE 13.12 **Stomatal closure in a split-root experiment. Maize (*Zea mays*) plants were grown as shown in Figure 13.11. Control plants (open circles) had both halves of the root system well-watered. Water was withheld from half the roots of the experimental plants (closed circles) on day zero. Stomatal opening, measured as leaf conductance, declined in the plants with water-stressed roots. (From Blackman, P. G., W. J. Davies. 1985. *Journal of Experimental Botany* 36:39–48. Reprinted by permission of The Company of Biologists, Ltd).**

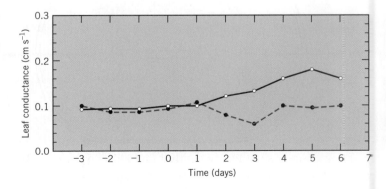

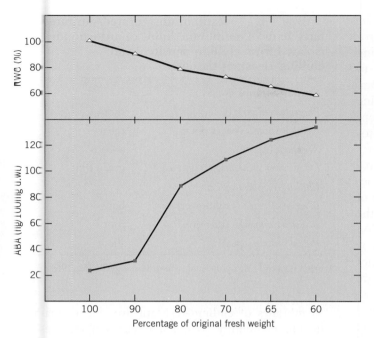

FIGURE 13.13 **Effect of air drying on the ABA content of *Commelina communis* root tips. Root tips were air dried to the relative water contents shown in the upper curve. Lower curve shows the dramatic increase in ABA content as the fresh weight decreases. (From J. Zhang, W. J. Davies, *Journal of Experimental Botany* 38:2015–2023, 1987. Reprinted by permission of The Company of Biologists, Ltd.)**

Hormones other than ABA may also be involved in communication between water-stressed roots and leaves. In an experiment with half-watered maize (*Zea mays*) plants, results similar to those with *Commelina* were obtained. One notable exception was that the ABA content of the leaves did not increase, and the application of cytokinins to the leaves prevented stomatal closure. At least in *Zea*, it appears that closure is brought about by decreased movement of cytokinins out of the drying roots.

Both hydropassive and hydroactive closure of stomata represent mechanisms that enable plants to anticipate potential problems of water availability through either excessive transpirational loss from the leaves or chronic, but nonlethal, soil water deficit. Although considerable progress has been achieved in this field over the last decade, there is clearly much yet to be learned about stomatal behavior and the response of plants to water stress.

13.5 PLANTS ARE SENSITIVE TO FLUCTUATIONS IN TEMPERATURE

Plants exhibit a wide range of sensitivities to extremes of temperature. Some are killed or injured by moderate chilling temperatures while others can survive. Each plant has its unique set of temperature requirements for growth and development (Chapter 27). There is an optimum temperature at which each plant grows and develops most efficiently, and upper and lower limits (Chapter 12). As the temperature approaches these limits, growth diminishes, and beyond those limits there is no growth at all. Except in the relatively stable climates of tropical forests, temperatures frequently exceed these limits on a daily or seasonal basis, depending on the environment. Deserts, for example, are characteristically hot and dry during the day but experience low night temperatures because, in the absence of a moist atmosphere, much of this heat is reradiated into space. Plants at high altitudes, where much of the daily heat gain is radiated into the thin atmosphere every night, experience similar temperature excursions. Plants native to the northern temperate and boreal forests must survive temperatures as low as −70°C every winter. How plants respond to temperature extremes has long captivated plant biologists. In this section, we will consider how chilling stress and high temperature stress effect the physiology of the plant.

13.5.1 MANY PLANTS ARE CHILLING SENSITIVE

Plants native to warm habitats are injured when exposed to low, nonfreezing temperatures and are considered to be **chilling sensitive**. Plants such as maize (*Zea mays*), tomato (*Lycopersicon esculentum*), cucumber (*Cucurbita* sp.), soybean (*Glycine max*), cotton (*Gossypium hirsutum*), and banana (*Musa* sp.) are particularly susceptible and will exhibit signs of injury when exposed to temperatures below 10 to 15°C. Even some temperate plants such as apple (*Malus* sp.), potato (*Solanum tuberosum*), and asparagus (*Asparagus* sp.) experience injury at temperatures above freezing (0 to 5°C).

Outward signs of chilling injury can take a variety of forms, depending on the species and age of the plant and the duration of the low-temperature stress. Young seedlings typically show signs of reduced leaf expansion, wilting, and chlorosis. In extreme cases, browning and the appearance of dead tissue (necrosis) and/or death of the plant will result. In some plants, reproductive development is especially sensitive to chilling temperature. Exposure of rice plants, for example, to chilling temperatures at the time of anthesis (floral opening) results in sterile flowers. Symptoms of chilling injury reflect a wide range of metabolic dysfunctions in chilling-sensitive tissues, including: impaired protoplasmic streaming, reduced respiration, reduced rates of protein synthesis as well as altered patterns of protein synthesis. One of the immediate plant responses to a chilling stress is the light-dependent inhibition of photosynthesis. Because low temperature inhibits the D1 repair cycle, this leads to chronic photoinhibition of PSII and PSI in cucumber. Indeed, there appear to be few aspects of cellular biochemistry that are not impaired in chilling-sensitive tissues following exposure to low temperature.

One explanation for this is that low temperature causes reversible changes in the physical state of cellular membranes. Membrane lipids consist primarily of diacylglycerides containing two fatty acids of either 16- or 18-carbon atoms. Some fatty acids are unsaturated, which means that they have one or more carbon-carbon double bonds ($-CH = CH-$), while others are fully saturated with hydrogen ($-CH_2-CH_2-$). Because saturated fatty acids—and lipids that contain them—solidify at higher temperatures than unsaturated fatty acids, the relative proportions of unsaturated and saturated fatty acids in membrane lipids have a strong influence on the fluidity of membranes. A change in the membrane from the fluid state to a gel (or semicrystalline) state is marked by an abrupt transition that can be monitored by a variety of physical methods. The temperature at which this transition occurs is known as the **transition temperature**.

Chilling-sensitive plants tend to have a higher proportion of saturated fatty acids (Table 13.1) and a correspondingly higher transition temperature. For mitochondrial membranes of the chilling-sensitive plant mung bean (*Vigna radiata*), for example, the transition temperature is 14°C. Mung bean seedlings grow poorly below 15°C. Chilling-resistant species, on the other hand, tend to have lower proportions of saturated fatty acids and, therefore, lower transition temperatures. During acclimation to low temperature, the proportion of unsaturated fatty acids increases and transition temperature decreases.

The net effect of the transition from a liquid membrane to a semicrystalline state at low temperature is similar to the effects of water stress described above. The integrity of membrane channels is disrupted, resulting in loss of compartmentation and solute leakage, and the operation of integral proteins that make up respiratory assemblies, photosystems, and other membrane-based metabolic processes is impaired.

TABLE 13.1   Ratio of unsaturated/saturated fatty acids of membrane lipids of mitochondria isolated from chilling-sensitive and chilling-resistant tissues.

Chilling-sensitive tissues		
Phaseolus vulgaris (bean)	shoot	2.8
Ipomoea batatas (sweet potato)	tuber	1.7
Zea mays (maize)	shoot	2.1
Lycopersicon esculentum (tomato)	green fruit	2.8
Chilling-resistant tissues		
Brassica oleracea (cauliflower)	buds	3.2
Brassica campestris (turnip)	root	3.9
Pisum sativum (pea)	shoot	3.8

From data of J. M. Lyons et al., *Plant Physiology* 39:262, 1964.

Membranes of chilling-resistant are able to maintain membrane fluidity to much lower temperatures and thereby protect these critical cellular functions against damage. However, it is important to appreciate that these differences in membrane lipid unsaturation, by themselves, cannot fully account for the differences in chilling sensitivity between chilling resistant and chilling sensitive plant genotypes.

13.5.2 HIGH-TEMPERATURE STRESS CAUSES PROTEIN DENATURATION

The traditional view is that the high-temperature limit for most C3 plants is determined by irreversible protein denaturation of enzymes. Reactions in the thylakoid membranes of higher plant chloroplasts are most sensitive to high-temperature damage, with consequent effects on the efficiency of photosynthesis. Photosystem II and its associated oxygen-evolving complex (Chapter 7) are particularly susceptible to injury. The oxygen-evolving complex is directly inactivated by heat, thereby disrupting electron donation to PSII resulting in the accumulation of P680$^+$, the strongest oxidizing agent in nature (Chapter 7). The result is that an increasing portion of the absorbed energy cannot be used photochemically by PSII which leads to chronic photoinhibition. This is easily monitored in intact leaves using chlorophyll fluorescence (Box 13.1). Other studies have indicated that the activities of Rubisco, Rubisco activase, and other carbon-fixation enzymes, may also be severely compromised at high temperatures.

Exposure of most organisms to supraoptimal temperatures for brief periods also suppresses the synthesis of most proteins including the PSII reaction center polypeptide, D1. This leads to an inhibition of the D1 repair cycle that is critically important to overcome the effects of chronic photoinhibition. In contrast, high-temperature stress also induces the synthesis of a new family of low molecular mass proteins known as **heat shock proteins (HSPs)**. This interesting class of proteins was originally discovered in *Drosophila melanogaster* (fruit fly) but they have since been discovered in a variety of animals, plants, and microorganisms. Exposures in the range of 15 minutes to a few hours at temperatures 5°C to 15°C above the normal growing temperature are usually sufficient to cause full induction of HSPs. HSPs are either not present or present at very low levels in nonstressed tissues. Initially, interest in HSPs centered on their potential for the study of gene regulation. There are, however, several aspects of HSPs that are of physiological interest.

There are three distinct classes of HSPs in higher plants, based on their approximate molecular mass: HSP90, HSP70, and a heterogeneous group with a molecular mass in the range of 17 to 28 kDa (Table 13.2). One in particular, HSP70, has a high degree of structural similarity—about 70 percent identical—in both plants and animals. Another protein, **ubiquitin**, is also found in all eukaryote organisms subjected to heat stress

and is considered a HSP. Ubiquitin has an important role in marking proteins for proteolytic degradation. HSPs are found throughout the cytoplasm as well as in nuclei, chloroplasts, and mitochondria. As well, induction of HSPs does not require a sudden temperature shift: they have been detected in field-grown plants following more gradual temperature rises of the sort that might be expected under normal growing conditions.

HSPs are synthesized very rapidly following an abrupt increase in temperature; new mRNA transcripts can be detected within 3 to 5 minutes and HSPs form the bulk of newly synthesized protein within 30 minutes. Within a few hours of return to normal temperature, HSPs are no longer produced and the pattern of protein synthesis returns to normal. The speed of their appearance suggests that HSPs might have a critical role in protecting the cell against deleterious effects of rapid temperature shifts. HSP70, for example, appears to function as a molecular chaperone, or **chaperonin**. Chaperonins are a class of proteins normally present in the cell that direct the assembly of multimeric protein aggregates. There is in the chloroplast, for example, a Rubisco-binding protein (HSP60) that helps to assemble the large and small subunits of Rubisco into a functional enzyme. It has been suggested that HSP70 functions to prevent the disassembly and denaturation of multimeric aggregates during heat stress. At the same time, increased ubiquitin levels reflect an increased demand for removal of proteins damaged by the heat shock.

In nature, high temperature stress is usually associated with water stress. Since both of these stresses inhibit photosynthesis, both of these stresses will concomitantly predispose the plant to photoinhibition. This illustrates the interactive effects of abiotic stresses to which plants are constantly exposed in a natural, fluctuating environment. Furthermore, exposure of plants to abiotic stresses such as drought and temperature extremes usually increases the susceptibility of these plants to attack by insects and plant diseases.

TABLE 13.2　Principal heat shock proteins (HSP) found in plants and their probable functions. Families are designated by their typical molecular mass. The number and exact molecular mass of proteins in each family vary depending on plant species.

HSP Family	Probable Function
HSP 110	Unknown.
HSP 90	Protecting receptor proteins.
HSP 70	ATP-dependent protein assembly or disassembly reactions; preventing protein from denaturation or aggregation (molecular chaperone). Found in cytoplasm, mitochondria, and chloroplasts.
HSP 60	Molecular chaperone, directing the proper assembly of multisubunit proteins. Found in cytoplasm, mitochondria, and chloroplasts.
LMW HSPs (17–28 kDa)	Function largely unknown. LMW (low-molecular-weight) HSPs reversibly form aggregates called "heat shock granules". Found in cytoplasm and chloroplasts.
Ubiquitin	An 8 kDa protein involved in targeting other proteins for proteolytic degradation.

Based on Vierling, 1990. E. Vierling. 1990. Heat shock protein function and expression in plants. In: R. G. Alscher, J. R. Cumming (eds.), *Stress Responses in Plants: Adaptation and Acclimation Mechanisms*. New York: Wiley-Liss, pp. 357–375.

13.6 INSECT PESTS AND DISEASE REPRESENT POTENTIAL BIOTIC STRESSES

Typically, a plant challenged by insects or potentially pathogenic microorganisms responds with changes in the composition and physical properties of cell walls, the biosynthesis of secondary metabolites that serve to isolate and limit the spread of the invading pathogen. These responses are collectively known as a **hypersensitive reaction**.

The hypersensitive reaction is commonly activated by viruses, bacteria, fungi, and nematodes and occurs principally in plants outside the pathogen's normal specificity range. Although the hypersensitive reaction is complex and can vary depending on the nature of the causal agent, there are common features that generally apply. An early event in this sensing/signaling pathway is the activation of defense-related genes and synthesis of their products, **pathogenesis-related (PR) proteins**. PR proteins include proteinase inhibitors that disarm proteolytic enzymes secreted by the pathogen and lytic enzymes such as β-1,3-glucanase and chitinase that degrade microbial cell walls. Also activated are genes that encode enzymes for the biosynthesis of isoflavonoids and other phytoalexins that limit the growth of pathogens. Lignin, callose, and suberin are accumulated in cell walls along with hydroxyproline-rich glycoproteins that are believed to provide structural support to the wall. These deposits strengthen the cell wall and render it less susceptible to attack by the invading pathogen. Finally, the invaded cells initiate **programmed cell death**, a process that results in the formation of necrotic lesions at the infection site. Cell necrosis isolates the pathogen, slowing both its development and its spread throughout the plant. It is not clear at this time to what extent these components of the hypersensitive reaction are sequential or parallel events.

13.6.1 SYSTEMIC ACQUIRED RESISTANCE REPRESENTS A PLANT IMMUNE RESPONSE

Some secondary metabolites associated with the hypersensitive reaction appear to constitute signal transduction pathways that prepare other cells and tissues to resist secondary infections. Initially the hypersensitive reaction is limited to the few cells at the point of invasion, but over a period of time, ranging from hours to days, the capacity to resist pathogens gradually becomes distributed throughout the entire plant. In effect, the plant reacts to the initial infection by slowly developing a general immune capacity. This phenomenon is known as **systemic acquired resistance (SAR)**.

The development of SAR is still not completely understood, but one component of the signaling pathway appears to be **salicylic acid** (Figure 13.14). Salicylic

acid (2-hydroxybenzoic acid) is a naturally occurring secondary metabolite with analgesic properties. Native North Americans and Eurasians have long used willow bark (*Salix* sps.), a source of the salicylic acid glycoside, **salicin**, to obtain generalized relief from aches and pains.

The relationship between salicylic acid and resistance to pathogens did not become apparent until the early 1990s, when it was observed that both salicylic acid and its acetyl derivative (aspirin), when applied to tobacco plants, induced PR gene expression and enhanced resistance to tobacco mosaic virus (TMV). Since then, it has been shown in a variety of plants that infection is followed by increased levels of salicylic acid both locally and in distal regions of the plant (Figure 13.15). For example, when tobacco plants are inoculated with TMV, the salicylic acid level rises as much as 20-fold in the inoculated leaves and 5-fold in the noninfected leaves. Furthermore, the appearance of PR proteins rises in parallel with salicylic acid. The rise in salicylic acid levels usually precedes the development of SAR. There are also a number of *Arabidopsis* mutants and transgenic plants that are characterized by constitutively high levels of both salicylic acid and SAR and, consequently, enhanced resistance to pathogens.

On the other side of the coin, plants with artificially low levels of salicylic acid generally fail to establish SAR. For example, bacteria have a gene designated *nahG* that encodes the enzyme salicylate hydroxylase. *Arabidopsis* plants transformed with the *nahG* gene thus contain little or no salicylic acid. Plants transformed with the *nahG* gene also fail to establish SAR and are compromised in their ability to ward off pathogen attack. Salicylic acid levels can also be reduced by direct inhibition of the enzyme phenylalanine-ammonia lyase (PAL), which catalyzes the first step in the biosynthesis of salicylic acid. PAL-limited *Arabidopsis* plants lose their

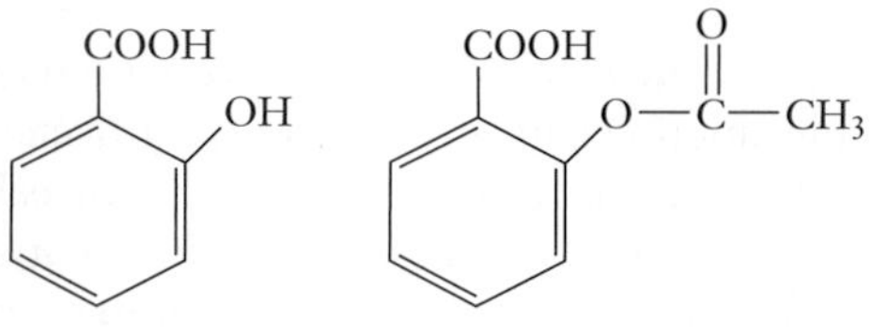

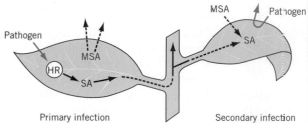

FIGURE 13.15 The possible role of salicylic acid in systemic acquired resistance (SAR). The first pathogens to infect the plant (primary infection) stimulate a localized hypersensitive reaction (HR) and the synthesis of salicylic acid (SA). Salicylic acid is translocated through the phloem to other regions of the plant where it prevents secondary infection by other pathogens. Alternatively, salicylic acid may be converted to methylsalicylic acid (MSA). MSA is moderately volatile and may function as an airborne signal.

COOH
OH

Salicylic
acid

COOH
O—C—CH₃
O

Aspirin
(acetylsalicylic acid)

FIGURE 13.14 The chemical structure of salicylic acid and its commercial derivative acetylsalicylic acid. Salicylic acid has been implicated in the immune strategies of plants.

resistance to disease, but resistance can be restored by applying salicylic acid. Based on results such as these, it is clear that salicylic acid has a significant role in plant defense responses. However, the mechanism whereby salicylic acid establishes and maintains SAR is yet to be determined.

13.6.2 JASMONATES MEDIATE INSECT AND DISEASE RESISTANCE

On the basis of recent experiments, it appears that **jasmonates**, especially **jasmonic acid** and its methyl ester (**methyljasmonate**) (Figure 13.16), also mediate insect and disease resistance. Jasmonates have been found to occur throughout plants, with highest concentrations in young, actively growing tissues. Methyljasmonate is the principal constituent of the essential oil of *Jasminium* and high concentrations of jasmonic acid have been isolated from fungal culture filtrates.

There are some similarities in the action of salicylic acid and jasmonates with respect to insect and disease resistance, but there are also some important distinctions. In a study of two fungal resistance genes in *Arabidopsis*, for example, it was found that expression of one gene was induced by salicylic acid, but not jasmonic acid, while the second gene was induced by jasmonic acid but not salicylic acid. Apparently there are at least two defensive pathways, one mediated by salicylic acid and one mediated by jasmonates. Jasmonic acid is synthesized from the unsaturated fatty acid, linolenic acid, which has led to the proposal that jasmonic acid functions as a type of second messenger.

Another very interesting but somewhat complicating aspect of jasmonates is that their action is not limited to insect and disease resistance. Through their effect on gene expression, jasmonates modulate a number of other physiological processes. These include seed and pollen germination, vegetative protein storage, root development, and tendril coiling. In most of these effects, the jasmonates appear to work in concert with ethylene.

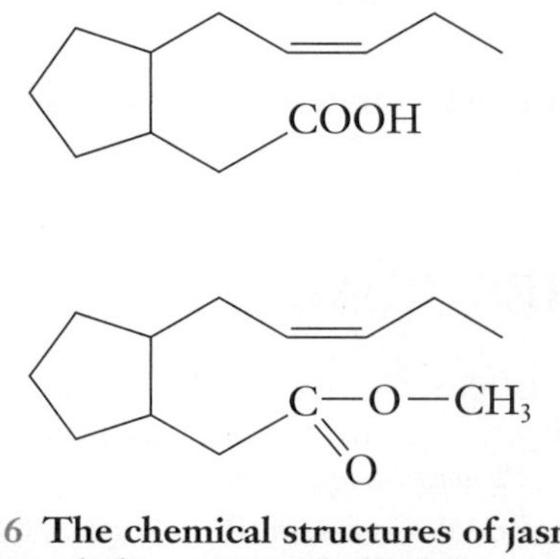

FIGURE 13.16 **The chemical structures of jasmonic acid (above) and methyljasmonate (below). Jasmonic acid is synthesized from linolenic acid (18:3).**

This breadth of jasmonate effects has led some to suggest that jasmonates should be elevated to the status of plant hormones.

13.7 THERE ARE FEATURES COMMON TO ALL STRESSES

The maintenance of cellular homeostasis is the result of a complex network of genetically regulated biochemical pathways. Thus, modulation of cellular homeostasis by abiotic and biotic stresses that we have discussed in this chapter can be detected at all levels of cellular organization—from genes to physiological function. Thus, any stress may induce or repress specific sets of genes or gene families through the regulation of transcription which will reflect changes in protein complement within the cell. Many of these proteins will be enzymes which catalyze specific reactions and consequently their presence or absence will alter cellular physiology. Because homeostasis is a complex network of interacting genes and biochemical pathways, different stresses may affect common sets of genes or gene families which is consistent with the observation that there appears to be overlap or **cross talk** between signal transduction pathways involved in the physiological response of plants to various stresses.

All stresses discussed in this chapter inhibit photosynthesis in one way or another, which may lead to chronic photoinhibition. Since chlorophyll fluorescence (Box 13.1) can be exploited as an intrinsic probe of the overall function of photosynthesis, chlorophyll fluorescence is the most widely used technique in plant stress research. Since photosynthesis is extremely sensitive to environmental stress, it should not surprise us that all abiotic and biotic stresses have a negative effect on plant productivity and survival (Table 13.3) (also see Chapter 12). If one assumes that the record yield represents the maximal yield under near-optimal, natural growth conditions, then the difference between the average yields and the record yields may be interpreted to indicate the average losses in yield due to suboptimal growth conditions as consequence of the combined effects of abiotic and biotic stresses. The data in Table 13.3 illustrate that environmental stresses have a staggering effect on the crop yields of corn, wheat, sorghum, and potato. The combined effects of abiotic and biotic stress reduces the yield of these crops by 70 to 87%! This should be cause for concern for regarding future world food production in the context of climate change predictions coupled with continued human population growth. However, plants also exhibit an astounding capacity to acclimate to myriad environmental conditions. The capacity of plants to sense and subsequently acclimate on a short-term and long-term basis to environmental stress is the subject of Chapter 14.

TABLE 13.3   **The effects of abiotic and biotic stress on average crop yields.**

Crop	Record Yield (kg/hectare)	Average Yield (kg/hectare)	Average Loss (kg/hectare)	(% loss)
Maize	19,300	4,600	14,700	76
Wheat	14,500	1,880	12,620	87
Sorghum	20,100	2,830	17,270	86
Potato	94,100	28,300	65,800	70

Adapted from Bray, E. A., J. Bailey-Serres, E. Weretilnyk. 2000. Responses to Abiotic Stresses. *Biochemistry and Molecular Biology of Plants*, pp. 1158–1203. Rockville, MD: American Society of Plant Physiologists.

SUMMARY

The maintenance of a cellular steady-state far from equilibrium results in an apparently stable condition called homeostasis. Environmental modulation of homeostasis may be defined as biological stress. Plant stress usually implies some adverse effect on the physiology of a plant. Plants may respond to stress in several ways.

Susceptible plants succumb to a stress, other plants avoid stress by completing their life cycle during periods of relatively low stress whereas stress resistant plants are able to tolerate a stress. Although light is required for the photosynthetic assimilation of CO_2, too much light can inhibit photosynthesis and results in photodamage from chronic photoinhibition. One role of the D1 repair cycle is to overcome the effects of photodamage to PSII reaction centers. Water stress in natural environments usually arises due to lack of rainfall, a condition known as drought. Thus, water deficit stress is often referred to as drought stress. This leads to desiccation of the protoplasm and cellular dysfunction. An immediate response of most plants to water stress is stomatal closure due to low turgor in the guard cells. Stomatal closure is triggered by decreasing water potential in the leaf mesophyll. The hormone abscisic acid (ABA) appears to have a significant role in stomatal closure. Chilling stress refers to exposure of plants to temperatures near but above the freezing point of water. The membranes of chilling-sensitive plants tend to have a higher proportion of unsaturated fatty acids and, consequently, change from a fluid to semicrystalline gel state at higher temperatures than chilling-resistant plants. The upper temperature limit for most plants is determined by a combination of irreversible denaturation of enzymes and problems with membrane fluidity. Plants subjected to heat stress respond by synthesizing a new family of low-molecular-weight heat shock proteins. Plants respond to insect damage and microbial pathogen infection with a hypersensitive reaction. The hypersensitive reaction includes changes in the composition and increased strength of the cell wall and the formation of necrotic lesions at the site of infection. These responses serve to isolate the potential pathogen and prevent its development and spread through the plant. Salicylic acid or its methyl ester may serve as a mobile signal, participating in systemic acquired resistance, a form of generalized immune response. Another possible signaling agent is jasmonic acid, a derivative of the fatty acid linolenic acid. A common feature of all stresses is that they induce or repress specific genes or gene families and they affect photosynthesis negatively. As a consequence, the combination of abiotic and biotic stresses reduces plant productivity and crop yield.

CHAPTER REVIEW

1. Define homeostasis. Define environmental stress.

2. If plants require light for photosynthesis, explain why plants can be exposed to too much light.

3. What is the role of the D1 repair cycle?

4. Describe how plants may be injured by water stress.

5. How does stomatal closure come about in response to water stress?

6. How do chilling-sensitive and a chilling-tolerant plants differ in their response to sudden exposures to low, nonfreezing temperatures?

7. What are heat shock proteins?

8. Define biotic stress. What are the roles of salicylic acid and jasmonates in a plant's response to biotic stress?

9. What are two common features of all stresses?

FURTHER READING

Bostock, R. M. 2005. Signal crosstalk and induced resistance: Straddling the line between cost and benefit. *Annual Review of Phytopathology* 43:545–580

Buchanan, B. B., W. Gruissem, R. L. Jones. 2000. *Biochemistry and Molecular Biology of Plants*. Rockville MD: American Society of Plant Physiologists.

Creelman, R. A., J. E. Mullet. 1997. Biosynthesis and action of jasmonates in plants. *Annual Review of Plant Physiology and Plant Molecular Biology* 48:355–381.

Demmig-Adams, B., W. W. Adams III, A. Mattoo. 2006. *Photoprotection, Photoinhibition, Gene Regulation and Environment. Advances in Photosynthesis and Respiration*, Vol. 21. Dordrecht: Springer.

Harwood, J. L. 1998. Involvement of chloroplast lipids in the reaction of plants submitted to stress. *Lipids in Photosynthesis: Structure, Function and Genetics. Advances in Photosynthesis*, Vol 6, pp. 287–302. Dordrecht: Kluwer Academic Publishers.

Howe, G. A., G. Jander. 2008. Plant immunity to insect herbivores. *Annual Review of Plant Biology* 59:41–66.

Iba, K. 2002. Acclimative response to temperature stress in higher plants: Approaches of gene engineering for temperature tolerance. *Annual Review of Plant Biology* 53:224–245.

Ingram, J., D. Bartels. 1996. The molecular basis of dehydration tolerance in plants. *Annual Review of Plant Physiology and Plant Molecular Biology* 47:377–403.

Kessler, A., I. T. Baldwin. 2002. Plant responses to insect herbivory: The emerging molecular analysis. *Annual Review of Plant Biology* 53:299–328.

Melis, A. 1999. Photosystem-II damage and repair cycle in chloroplasts: What modulates the rate of photodamage *in vivo*? *Trends in Plant Science* 4:130–135.

Nishida, I., N. Murata. 1996. Chilling sensitivity in plants and cyanobacteria: The crucial contribution of membrane lipids. *Annual Review of Plant Physiology and Plant Molecular Biology* 47:541–568.

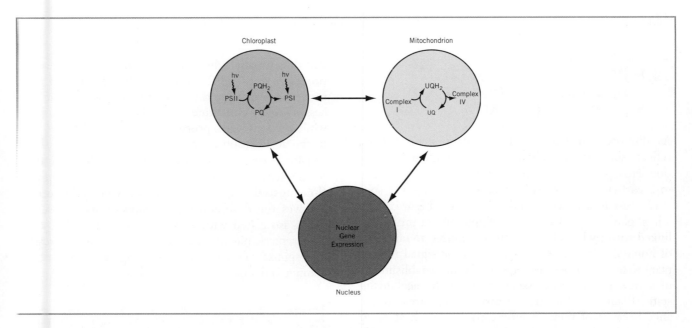

14

Acclimation to Environmental Stress

In Chapter 13 we defined stress as a negative effect on plant homeostasis which can be measured when a plant is exposed to a sudden change from an optimal condition for growth to some suboptimal abiotic condition such high light, low temperature, drought, or some biotic infestation. Many stress-sensitive plants may succumb to such a stress and die. However, in this chapter we focus on plants that exhibit the capacity to tolerate a particular stress over time. Such plants are called **stress resistant** or **stress tolerant** and reflects the ability of these plants to **acclimate** or adjust to the stress. The capacity of a plant to acclimate is, of course, a genetic trait. However, the specific changes brought about in response to stress are not themselves passed on to the next generation and thus are **nonheritable**. The capacity of plants to acclimate to changing environmental conditions reflects their remarkable **plasticity**. As a consequence, a plant's physiology and morphology are not static but are very dynamic and responsive to their environment. The ability of biennial plants and winter cultivars of cereal grains to survive over winter is an example of acclimation to low temperature. The process of acclimation to a stress is known as **hardening** and plants that have the capacity to acclimate are commonly referred to as **hardy** species. In contrast, those plants that exhibit a minimal capacity to acclimate to a specific stress are referred to as **nonhardy** species. Thus, frost-hardy plants are those that are able to acclimate to low temperature and are able to survive the freezing stress of winter, and drought-hardy plants are able to survive water stress. The physiological bases of acclimation to various environmental stresses will be the focus of this chapter and will include a discussion of

- the theoretical basis of plant acclimation and plasticity as a time-nested phenomenon,

- the role of state transitions in response to changes in light quality,

- the xanthophyll cycle and protection against photodamage,

- the mechanisms that allow plants to survive water limitations,

- low temperature acclimation and freezing tolerance,

- excitation pressure as a redox signal for retrograde regulation of nuclear genes,

- photosynthetic acclimation to high temperature, and

- the protective role of O_2 as an alternative photosynthetic electron acceptor during acclimation.

14.1 PLANT ACCLIMATION IS A TIME-DEPENDENT PHENOMENON

As discussed in Chapter 13, a plant stress usually reflects some sudden change in environmental condition. However, in stress-tolerant plant species, exposure to a particular stress leads to acclimation to that specific stress in a time-dependent manner (Figure 14.1). Thus, plant stress and plant acclimation are intimately linked with each other. The stress-induced modulation of homeostasis can be considered as the signal for the plant to initiate processes required for the establishment of a new homeostasis associated with the acclimated state. Plants exhibit stress resistance or stress tolerance because of their genetic capacity to adjust or to acclimate to the stress and establish a new homeostatic state over time. Furthermore, the acclimation process in stress-resistant species is usually reversible upon removal of the external stress (Figure 14.1).

The establishment of homeostasis associated with the new acclimated state is not the result of a single physiological process but rather the result of many physiological processes that the plant integrates over time, that is, integrates over the acclimation period. Plants usually integrate these physiological processes over a short-term as well as a long-term basis. The short-term processes involved in acclimation can be initiated within seconds or minutes upon exposure to a stress but may be transient in nature. That means that although these processes can be detected very soon after the onset of a stress, their activities also disappear rather rapidly. As a consequence, the lifetime of these processes is rather short (Figure 14.2, processes a, b and c). In contrast, long-term processes are less transient

and thus usually exhibit a longer lifetime (Figure 14.2, processes d, e and f). However, the lifetimes of these processes overlap in time such that the short-term processes usually constitute the initial responses to a stress while the long-term processes are usually detected later in the acclimation process. Such a hierarchy of short- and long-term responses indicates that the attainment of the acclimated state can be considered a complex, **time-nested response** to a stress. Acclimation usually involves the differential expression of specific sets of genes associated with exposure to a particular stress. The remarkable capacity to regulate gene expression in response to environmental change in a time-nested manner is the basis of plant plasticity.

14.2 ACCLIMATION IS INITIATED BY RAPID, SHORT-TERM RESPONSES

In this section, we will discuss examples of initial, rapid responses to changes in light, water availability and temperature that are part of the acclimation process.

14.2.1 STATE TRANSITIONS REGULATE ENERGY DISTRIBUTION IN RESPONSE TO CHANGES IN SPECTRAL DISTRIBUTION

Optimal CO_2 reduction requires an efficient supply of NADPH and ATP that would, in turn, require a steady, balanced electron flow through PSII and PSI (Chapter 7). But if light is not saturating and the delivery of excitation energy to PSII and PSI is not balanced, then the rate of electron transport and, consequently,

FIGURE 14.1 **A schematic relationship between stress and acclimation. Under some optimal environmental condition, a plant is in a homeostatic state A as indicated by a constant rate of some physiological process measured over time. Upon the imposition of an external stress, the rate of this physiological process in most cases decreases rapidly which indicates a disruption of the homeostatic state A. Plants that are able to adjust to this stress over time may establish a new physiological rate that is either lower (homeostatic state B) or higher (homeostatic state C) than the original rate (homeostatic state A). Such plants are capable of acclimation and are considered stress tolerant. The plant may remain in this new homeostatic state B or state C through completion of its life cycle. Alternatively, the plant may return to the original homeostatic state A when the stress is removed (broken lines). The black line indicates the most probable acclimation response.**

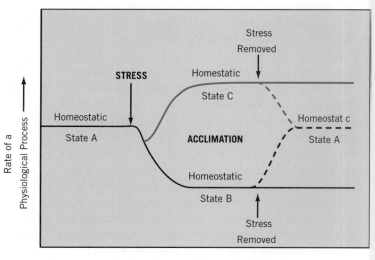

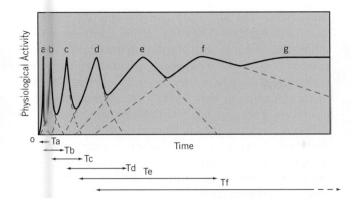

FIGURE 14.2 **A schematic graph illustrating plant acclimation as a time-nested response. Peaks labeled a, b, c, d, e, f, and g represent the appearance and disappearance of different physiological processes that are initiated by the onset of a stress at time 0. Arrows labeled Ta, Tb, Tc, Td, Te, Tf, and Tg approximates the lifetime of each process. The longer the arrow, the longer the lifetime. Thus, process a is the most transient but least stable response whereas process g is least transient but most stable physiological response to the stress. Processes a to g are said to be nested in a time with short-term processes occurring first, followed by the more long-term processes.**

photosynthesis will be limited by the photosystem receiving the least energy. Under natural conditions, the amount of light driving PSII and PSI is not necessarily balanced or consistent. For example, less energy is required to excite P700 (171 kJ) than P680 (175 kJ). Moreover, because of differences in the number of antenna molecules, their absorption coefficients, and a variety of other factors that influence absorption, the capacities of PSII and PSI to absorb light, often referred to as their **absorption cross-section (σ)**, are not equal. This might be expected to place unique constraints on the overall efficiency of photosynthesis. State transitions are a short-term mechanism to regulate excitation energy distribution between the two photosystems that is required to maintain an efficient flow of electrons to NADP$^+$.

In addition to the inherent inequities of spectral distribution, plants often face other situations requiring rapid adjustments in the amount of energy being fed into PSII and PSI. Plants growing in the shade of a canopy, for example, are frequently subject to

sudden transient fluctuations in fluence rate, known as **sunflecks**. A sunfleck is a spot of direct sunlight impinging on the leaf through an open gap in the canopy. Sunfleck lifetimes are variable, from a few seconds' duration due to wind flutter up to 20 minutes or longer under woodland canopies. Similar fluctuations occur when the sun suddenly reappears after having been blocked by extensive cloud cover. In situations such as these the leaf may be subject to as much as tenfold increases in energy that, particularly when directed to PSII, could have severe damaging effects.

State transitions are one of the best-understood mechanisms for short-term regulation of energy distribution and is based on reversible phosphorylation of LHCII protein (Figure 14.3). The phosphorylation of proteins is a ubiquitous mechanism for regulating many aspects of gene regulation and response to environmental stimuli in all eukaryote organisms. The phosphorylation of proteins is catalyzed by a class of enzymes known as **protein kinases**. Chloroplasts contain a thylakoid

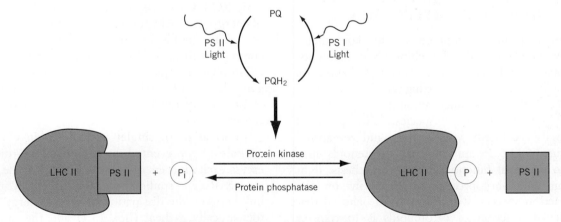

FIGURE 14.3 **Reversible phosphorylation of LHCII. When PSII is overexcited relative to PSI, plastoquinol (PQH$_2$) accumulates. A high level of plastoquinol activates a protein kinase that phosphorylates LHCII. Addition of a phosphate group weakens the interaction between LHCII and the PSII core antenna, causing LHCII to dissociate from PSII. The input of light to PSII is diminished and PSII slows down, thus allowing PSI to oxidize excess PQH$_2$ to plastoquinone (PQ), which, in turn, deactivates the protein kinase. LHCII is dephosphorylated by a protein phosphatase, allowing LCHII to reform in association with PSII.**

membrane-bound protein kinase capable of phosphorylating LHCII. The activity of this kinase is sensitive to the redox state of the thylakoid membrane and is activated when excess energy drives PSII, resulting in a buildup of reduced plastoquinone (PQH_2). Plants that are exposed to conditions that result in the preferential excitation of PSII are considered to be in **state 2**.

The resulting phosphorylation of LHCII increases the negative charge of the protein, causing LHCII to dissociate from PSII. The same negative charge also loosens the appression of the thylakoid membranes in the grana stacks, freeing a certain portion of LHCII to migrate into the PSI-rich stroma thylakoids. This shifts the balance of energy away from the PSII complexes, which remain behind in the appressed region, in favor of PSI. The preferential excitation of PSI is referred to as **state 1**. As the plastoquinone pool becomes reoxidized, due to increased PSI activity, the protein kinase is deactivated and a protein phosphatase enzyme dephosphorylates the LHCII, causing it to migrate back into the appressed region and recombine with PSII. Recently, it has been shown that the PSI-H subunit of *Arabidopsis thaliana* is required for reversible transitions between state 1 and state 2. In *Arabidopsis thaliana* lacking the PSI-H subunit due to antisense suppression of the gene which codes for PSI-H, LHCII is unable to transfer energy to PSI, thus impairing state transitions. The net result of these state transitions is a very dynamic, continuous adjustment of excitation energy distribution between PSII and PSI. This continual adjustment of energy input in turn maintains an optimal flow of electrons through the two photosystems.

14.2.2 CAROTENOIDS SERVE A DUAL FUNCTION: LIGHT HARVESTING AND PHOTOPROTECTION

The pigment-protein complexes of thylakoid membranes contain not only the chlorophyll pigments, but carotenes and xanthopylls as well (Chapter 6). The principal carotene in most higher plant species is β-carotene, although smaller amounts of α-carotene may be present in some species. The principal xanthophylls are lutein, violaxanthin, and zeaxanthin (see Figure 6.11). It appears that carotenoids may serve two principal functions in photosynthesis: light harvesting and photoprotection. Primarily on the basis of action spectra, it has long been believed that the principal function of carotenoids is to transfer absorbed light energy to chlorophyll. In this sense, the carotenoids serve a light-harvesting function and the energy they absorb is eventually transferred from the light-harvesting complex to reaction centers for use in photosynthetic electron transport for generation of ATP and NADPH. In contrast to the light-harvesting role of carotenoids, there is now substantial evidence that

carotenoids also play an important role in protecting the photosynthetic system from chronic photoinhibition. This is called **photoprotection**. During periods of peak irradiance, plants typically absorb more energy than they can utilize in the reduction of carbon dioxide, that is, they are exposed to excess light (see Figure 13.4). Rapidly growing crops, for example, may utilize no more than 50 percent of absorbed radiation, while other species, such as evergreens, may utilize as little as 10 percent. Any excess absorbed energy must be dissipated. If not, the excess absorbed energy may lead to photoinhibition of photosynthesis (see Chapter 13). Prolonged exposure to photoinhibitory conditions may lead subsequently to photodamage and uncontrolled destruction of the photosynthetic apparatus due to the accumulation of **reactive oxygen species (ROS)**. These are toxic forms of oxygen that may lead to cell damage and ultimately cell death.

A unique feature of O_2 is that, in its ground state, it is in a triplet state ($^3O_2$) rather than the usual singlet state that characterizes almost all other molecules. This prevents O_2 from chemically reacting with most organic molecules. Thus, this difference between ground state O_2 and almost all other molecules is a major deterrent to spontaneous combustion and allows life to persist even in an oxygen-rich environment! However, whenever the energy absorbed by the photosynthetic apparatus exceeds the capacity to utilize that energy, there is an increased probability that singlet excited chlorophyll (^1Chl) will be converted to triplet excited chlorophyll (^3Chl) (Chapter 6). As a consequence of such an intersystem crossing event, triplet chlorophyll may interact with ground state O_2 and convert ground state oxygen to **singlet excited O_2($^1O_2$)**. This is one example of a toxic ROS. Clearly light can be a very dangerous form of energy, especially for oxygenic photosynthetic organisms. Carotenoids present in thylakoid membranes can compete with ground state oxygen for triplet excited chlorophyll and prevent the formation of singlet oxygen. Thus, carotenoids such as β-carotene are important in preventing the formation of toxic reactive oxygen species.

Formation of singlet excited O_2 in chloroplasts can also be prevented by trapping and dissipating excess excitation energy before it reaches the reaction center. Recent studies have established an important link between the dissipation of excess energy nonphotochemically as heat (Box 13.1) and the presence of the xanthophyll, zeaxanthin. Zeaxanthin is formed from violaxanthin by a process known as the **xanthophyll cycle** (Figure 14.4A). Violaxanthin is a diepoxide; it contains two epoxy groups, one on each ring. Under conditions of excess light, violaxanthin (V) is enzymatically converted to zeaxanthin (Z) through the removal of those two oxygens (de-epoxidation).

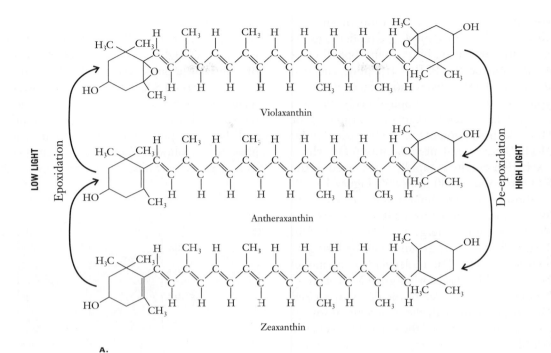

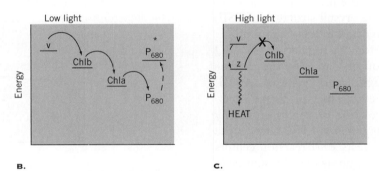

FIGURE 14.4 **The xanthophyll cycle and photoprotection.** (*A*) Under high light, de-epoxidation removes the two oxygen (epoxy) groups from the rings of violaxanthin. This is a two-step reaction with the intermediate antheraxanthin containing only one epoxy group. De-epoxidation is induced by high light, low lumenal pH, and high levels of reduced ascorbate. Note that in zeaxanthin, the number of carbon–carbon double bonds is increased by two relative to violaxanthin. The xanthophyll cycle is reversible since, under low-light conditions, zeaxanthin is converted back to violaxanthin. (*B*) Violaxanthin (V) acts as a light-harvesting pigment. Violaxanthin is able to transfer its absorbed excitation energy, via chlorophyll b (Chl b) and chlorophyll a (Chl a) to the PSII reaction center to excite P680. (*C*) Under excess light, most of the violaxanthin (V) is converted to zeaxanthin (Z), which is unable to transfer its absorbed excitation energy to the PSII reaction center via chlorophyll b. Zeaxanthin loses its absorbed energy as heat.

De-epoxidation is stepwise—removal of the first oxygen generates an intermediate monoepoxide (antheraxanthin)(A). De-epoxidation is also induced by a low pH in the lumen, which is a normal consequence of electron transport under high light conditions. The reaction is reversed in the dark as zeaxanthin is again enzymatically converted back to violaxanthin. Xanthophyll cycle activity can be assessed by measuring the de-epoxidation state (DEPS) of the cycle pool which is a measure of the concentration of zeaxanthin (Z) and antheraxanthin (A) relative to the total pool (V + A + Z) (Equation 14.1).

$$DEPS = Z + \tfrac{1}{2}A/V + A + Z \qquad (14.1)$$

Although it has been established that the xanthophyll cycle plays a key role in photoprotection of the chloroplast through **nonphotochemical quenching (NPQ;** Box 13.1), the precise molecular mechanism is still disputed. Through the generation of an *Arabidopsis thaliana* mutant named *npq4*, one model suggests that the zeaxanthin produced by the xanthophyll cycle may be bound to a specific PSII polypeptide encoded by the *PsbS* gene. These observations form the basis for the hypothesis that the xanthophyll cycle acts as a reversible, molecular switch to regulate the capacity for NPQ and the safe dissipation of excess absorbed. Under low light, LHCII has abundant violaxanthin (V) whose absorbed excitation energy is rapidly transferred to the reaction center to excite P680 to P680* (Figure 14.4B). However, exposure to high light rapidly converts to violaxanthin to zeaxanthin within LHCII. Since Z can not transfer its excitation energy to chlorophyll, Z in the excited state decays to its ground state by losing its excitation energy as heat. The molecular switch model assumes that zeaxanthin (Z) absorbs light energy directly and that NPQ is a consequence of the decay of excited state Z to its ground state with the concomitant loss of heat. Thus, zeaxanthin decreases the efficiency of energy transfer to PSII reaction centers by acting as a direct **quencher** of energy. Support for the molecular switch model is still equivocal since the presence of PsbS still has not been detected in the crystal structure of PSII, and furthermore, NPQ can occur in the absence of Z.

An alternative model for the regulation of NPQ suggests that the conversion of V to Z by the xanthophyll cycle regulates the aggregation state of LCHII. In the aggregated state, the efficiency of energy transfer from LHCII to PSII reaction centers is drastically decreased. In this model, aggregated LHCII acts as the energy quencher which protects PSII reaction centers from excess excitation. This suggests that Z is indirectly involved in quenching energy by affecting the physical structure of LHCII. Regardless of which model is correct, it appears that the xanthophyll cycle is a ubiquitous process for protecting the chloroplast against potentially damaging effects of excess light. However, it is interesting to note that the xanthophyll cycle is absent in cyanobacteria. The regulation of NPQ by the xanthophyll cycle continues to be an intensive and exciting area of photosynthesis research.

What are the functional consequences of a stimulation of NPQ by high light? Irrespective of the mechanism underlying photoprotection through NPQ, the functional consequence of NPQ is a decrease in the efficiency of energy transfer to PSII reaction centres. Because less energy is transferred to PSII reaction centers per photon absorbed by LCHII under such conditions, the efficiency of PSII photochemistry (P680 + energy $\rightarrow$ P680$^+$ + e), will decrease per photon absorbed. Thus, the light-inducible xanthophyll cycle protects PSII reaction centers by dissipating excess excitation nonphotochemically as heat.

When plants are subjected to excess light, photoinhibition will occur (Chapter 13) which can be measured as a decrease in Fv/Fm as a function of time under high light. However, concomitantly, exposure to light also stimulates the de-epoxidation of the xanthophyll cycle (DEPS) and NPQ over time (Figure 14.5A, left panel). However, when the plants are allowed to recover from the photoinhibition by removal from the high light condition, Fv/Fm and DEPS rapidly recover to their original values prior to the photoinhibition treatment (Figure 14.5A, right panel). Thus, the responses of Fv/Fm

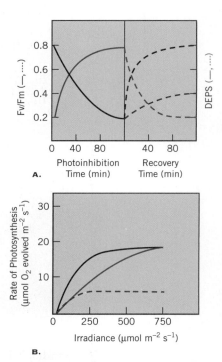

FIGURE 14.5 A schematic graph illustrating photoprotection and dynamic photoinhibition. (*A*) The effects of time of exposure to photoinhibition and time of recovery from this photoinhibition on Fv/Fm (black curve) and DEPS (red curve). The panel on the left illustrates that the time-dependent decrease in Fv/Fm is almost a mirror image of time-dependent increase in DEPS due to exposure to high light. The panel on the right illustrates that the time-dependent recovery of Fv/Fm (black broken line) is almost a mirror image of the time-dependent decrease in DEPS (red broken line). Such mirror image responses are characteristic of dynamic photoinhibition. In contrast to dynamic photoinhibition, recovery from chronic photoinhibition is much slower (blue broken line). (Adapted from Pocock, T., D. Koziak, D., Rosso, N. P. A. Huner. 2007. *Journal of Phycology* 43:924–936.) (*B*) The effects of exposure to either dynamic (red line) or chronic photoinhibition (broken red line) on the light response curves for O$_2$ evolution. The black line represents the light response curve for control, non-photoinhibited plants. (Adapted from Osmond, C. B. 1994. In: N. R. Baker, J. R. Bowyer (eds.), *Photoinhibition of Photosynthesis: From Molecular Mechanisms to the Field*, pp. 1–24. Oxford: Bios Scientific Publishers.)

and DEPS to photoinhibition and recovery time appear to be mirror images of one another. The *rapidly reversible* inhibition of PSII reaction centers that is usually a consequence of an increase in thermal energy dissipation through NPQ is defined as **dynamic photoinhibition**. This reflects **photoprotection**. In contrast, **photodamage** or **chronic photoinhibition** is defined as the *slowly reversible* inhibition of PSII reaction centers that is usually a consequence of damage to the D1 reaction center polypeptide. Thus, the time required for the recovery of Fv/Fm from chronic photoinhibition is much longer than that from dynamic photoinhibition. Photodamage is only slowly reversible because of its dependence on protein synthesis for the D1 repair cycle (Chapter 13). The difference between dynamic and chronic photoinhibition can also be seen at the level of O_2 evolution (Figure 14.5B). Dynamic photoinhibition usually results in a rapidly reversible decrease in photosynthetic efficiency (Chapter 13) measured as the maximum slope of the CO_2 light response curve but not a decrease in photosynthetic capacity measured as the maximum light saturated rate of photosynthesis. In contrast, chronic photoinhibition usually results in a decrease in both photosynthetic efficiency and photosynthetic capacity which recover very slowly (Figure 14.5B). Clearly, both photodamage and photoprotection may lead to photoinhibition of photosynthesis as reflected in a decrease in photosynthetic efficiency, albeit for different reasons. The former causes a reduction in photosynthetic efficiency due to an alteration in the antenna that reduces the efficiency of resonance energy transfer to the PSII reaction center. The latter is due to damage to the reaction center that reduces the efficiency of charge separation rather than energy transfer in the antenna.

14.2.3 OSMOTIC ADJUSTMENT IS A RESPONSE TO WATER STRESS

A pronounced response to water stress in many plants is a decrease in osmotic potential resulting from an accumulation of solutes (see Chapter 1 and 2). This process is known as **osmotic adjustment**. While some increase in solute concentration is expected as a result of dehydration and decreasing cell volume, osmotic adjustment refers specifically to a net increase in solute concentration due to metabolic processes triggered by stress. Osmotic adjustment generates a more negative leaf water potential, thereby helping to maintain water movement into the leaf and, consequently, leaf turgor.

Solutes accumulate during osmotic adjustment and the decreases in osmotic potential due to osmotic adjustment are relatively small, less than 1.0 MPa. Nevertheless, the role of solutes in maintaining turgor at relatively low water potentials represents a significant form of acclimation to water stress. Osmotic adjustment may also play an important role in helping partially wilted leaves to regain turgor once the water supply recovers. By helping to maintain leaf turgor, osmotic adjustment also enables plants to keep their stomata open and continue taking up CO_2 for photosynthesis under conditions of moderate water stress. Solutes implicated in osmotic adjustment include a range of inorganic ions (especially K^+), sugars, and amino acids (Figure 14.6). One amino acid that appears to be particularly sensitive to stress is **proline**. A large number of plants synthesize proline from glutamine in the leaves. The role of proline is demonstrated by experiments with tomato cells in culture. Cells subjected to water (osmotic) stress by exposure to hyperosmotic concentrations of polyethylene glycol (PEG) responded with an initial loss of turgor and rapid accumulation of proline. As proline accumulation continued, however, turgor gradually recovered. **Sorbitol**, a sugar alcohol, and **betaine** (N,N,N-trimethyl glycine) are other common accumulated solutes. Most chemicals associated with osmotic adjustment share the property that they do not significantly interfere with normal metabolic processes. Such chemicals are called **compatible solutes**.

Although osmotic adjustment appears to be a general response to water stress, not all species are capable of adjusting their solute concentrations. Sugarbeet (*Beta vulgaris*), on the one hand, synthesizes large quantities of betaine and is known as an **osmotic adjuster**. Osmotic adjustment in cowpea (*Vigna unguiculata*), on the other hand, is minimal and cowpea is known as an osmotic nonadjuster. Cowpea instead has very sensitive stomata and avoids desiccation by closing the stomata and maintaining a relatively high water potential. It is interesting

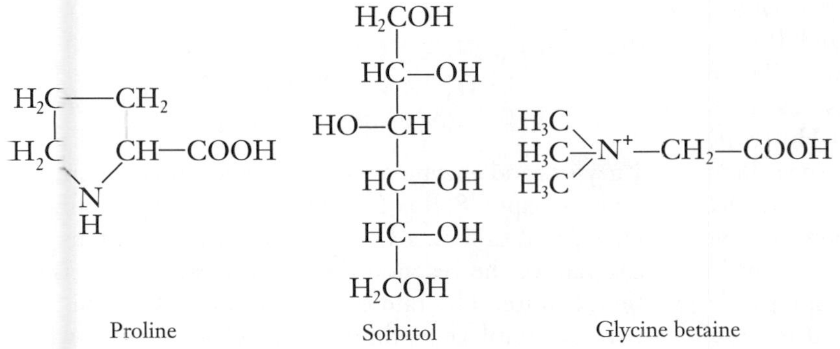

FIGURE 14.6 **Three solutes typically involved in osmotic adjustment.**

to note that there is no long-term advantage of osmotic adjustment over stomatal closure, at least with regard to net carbon gain. While sugarbeet is able to continue photosynthesis at lower water potentials, the advantage over cowpea is shortlived. After one or two days, excessive water loss overrides osmotic adjustment and over the long term carbon assimilation in sugarbeet declines.

14.2.4　LOW TEMPERATURES INDUCE LIPID UNSATURATION AND COLD REGULATED GENES IN COLD TOLERANT PLANTS

The study of cold tolerant, herbaceous plants such as wheat (*Triticum aestivum*), barley (*Hordeum vulgare*), alfalfa (*Medicago*), spinach (*Spinacea oleracea*), and the model plant species *Arabidopsis thaliana* has enhanced our understanding of the metabolic and molecular events before, during, and after acclimation. This has assisted greatly in the search for metabolic and genetic factors involved in cold tolerance. One of the immediate responses of cold tolerant plants to low temperature is an increase in the proportion of unsaturated fatty acids (Chapter 13) bound to lipids associated with the plasma membrane, mitochondrial membranes as well as thylakoid membranes. Various biophysical measurements indicate that this ensures that the membrane can remain in a more fluid and less gel-like state at lower temperature which enhances membrane stability and function at these low temperatures. A change in the membrane from the fluid state to a more solid state is marked by an abrupt change in the membrane activity. The temperature at which this transition occurs is known as the **transition temperature**. This means that at temperatures above the transition temperature, the membrane remains fluid but becomes more solid or gel-like at temperatures below the transition temperature. This allows higher activity of membrane process at lower temperatures.

Cold acclimation of herbaceous plants induces changes in gene expression. During acclimation, there are changes in mRNA transcription, increases in protein synthesis, and qualitative changes in the pattern of proteins synthesized. A major class of cold-induced genes encode homologs of *late embryogenesis active* proteins (**LEA-proteins**) that are synthesized late in embryogenesis and during dehydration stress. These polypeptides fall into a number of families based on amino acid sequence similarities (Table 14.1). However, these proteins encoded by cold-regulated genes share common physical properties. (1) They are unusually hydrophilic. (2) They remain soluble upon boiling in dilute aqueous buffer. (3) They exhibit relatively simple amino acid sequences that form amphipathic α-helices. It appears that cor15a and wsc120 interact

TABLE 14.1　Plant cold-regulated genes.

Plant Source	Gene	Polypeptide	Mol. Mass (kD)
Arabidopsis	*cor15a*	cor15a	15
Alfalfa	*cas15*	cas15	15
Wheat	*wcs120*	wcs120	39
Barley	*hva1*	hva1	22

with membranes, which enhances their stability to low temperature as well as against freezing.

The promoter regions of certain cold-regulated genes are activated in response to low temperature and dehydration stress. Analyses of these promoter regions of cold-regulated genes of *Arabidopsis thaliana* led to the identification of a DNA regulatory element called the *d*ehydration *r*esponsive *e*lement (**DRE**). The DRE has a conserved core C-repeat sequence of CCGAC that imparts responsiveness to low temperature and dehydration. Specfic proteins that bind to the DRE are called *C*-repeat *b*inding *f*actors (**CBFs**). Thus, CBFs are transcriptional activators that are involved in regulating the expression of cold-regulated genes. It is concluded that cold acclimation is regulated by a family of DRE-containing genes that are, in turn, induced by a family of CBF transcriptional factors.

There is good reason to believe that ABA might be involved in cold acclimation of herbaceous tissues. An increase in endogenous ABA levels has been observed during cold acclimation in several species and the amount of increase is greater in cold-tolerant varieties than in cold-sensitive varieties. In addition, significant levels of cold tolerance can be induced by the application of ABA to intact plants, callus, and suspension cultures. In both intact alfalfa seedlings and suspension-cultured cells of winter rapeseed (*Brassica napus*), exogenous ABA can induce up to 50 to 60 percent survival compared with a normal cold-acclimation treatment. ABA also induces the synthesis of new proteins. In alfalfa, some of the induced proteins are unique to ABA treatment, but some are common to both low-temperature and ABA treatment. Further research is required to elucidate the precise role of ABA in cold acclimation and the induction of genes associated with cold tolerance.

14.2.5　Q₁₀ FOR PLANT RESPIRATION VARIES AS A FUNCTION OF TEMPERATURE

Enzymes and enzyme reactions are sensitive to temperature (Chapter 8, Box 1). Enzyme reactions typically are considered to have a Q_{10} of about 2, which means that the rate of the reaction doubles for each $10°C$ rise in temperature. The rate of reaction increases with temperature until an optimum is reached, beyond which

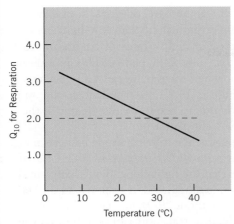

FIGURE 14.7 **The temperature dependence of the Q_{10} for plant respiration. The broken line indicates the expected relationship between Q_{10} and temperature whereas the solid line represents the actual relationship for plant respiration. (Adapted from Atkin, O. K., M. G. Tjoelker. 2003.** *Trends in Plant Science* **8:343–351.)**

the rate usually declines sharply. The decline in enzyme activity is normally caused by **thermal denaturation** as a result of protein unfolding. It is usually assumed that Q_{10} is independent of the temperature range over which it is measured. However, this is not true for plant respiration. In fact, the Q_{10} increases linearly upon short-term increases in temperature from $40°$ to $0°C$ (Figure 14.7). This dynamic temperature response of Q_{10} for respiration appears to be consistent across diverse plant taxa. Since Q_{10} measures the temperature-dependent *change* in respiration rate, an increase in ambient temperature will cause a greater change in rates of respiration in plants native to cold, Arctic climates than plants native to hotter climates. Further, other abiotic factors such as irradiance and water deficit can also influence the Q_{10} for plant respiration.

Why does the Q_{10} for plant respiration vary as a function of short-term exposure to temperature? This may be explained on the basis of two primary factors: (1) the effect of temperature on the Vmax (Chapter 8, Box 1) of enzymes involved in respiratory carbon metabolism as well as on the maximum rate of respiratory electron transport (Chapter 10); (2) the effect of temperature on substrate availability. Short-term exposures to low temperature reduces the flux of carbon through glycolysis and the TCA cycle because low temperature will reduce the activities of the various enzymes involved in these pathways. In addition, low temperature will decrease the fluidity of the inner membrane, which decrease the rate of respiratory electron transport. As a consequence, short-term exposure to low temperature will reduce the rates of CO_2 evolution and O_2 consumption. However, at moderate to high temperatures, it is not enzyme activity that limits the rate of reaction but rather the availability of substrates such as ADP and ATP. At high

temperatures, mitochondrial membranes may become leaky to protons, and therefore, reduce the capacity to synthesize ATP by chemiosmosis (Chapter 5 and 10).

14.3 LONG-TERM ACCLIMATION ALTERS PHENOTYPE

In this section, we will discuss specific examples of slower, long-term responses to changes in light, water availability and temperature that are part of the acclimation process that result in phenotypic alterations.

14.3.1 LIGHT REGULATES NUCLEAR GENE EXPRESSION AND PHOTOACCLIMATION

The process whereby adjustments are made to the structure and function of the photosynthetic apparatus in response changes in growth irradiance is called **photoacclimation**. One consequence of photoacclimation is a change in pigment composition which results in an altered visible phenotype. It is important to note that photoacclimation requires growth and development. For example, photoautrophs grown under high light typically exhibit as decrease in total chlorophyll per leaf area compared to the same plants grown at low irradiance. Thus, the leaves of high-light plants are usually a pale green or yellow-green compared to a dark green phenotype of the same species grown at low light. Functionally, high-light plants exhibit a photosynthetic light response curve for CO_2 assimilation that is distinct from that of plants grown under low light, whether measured as net photosynthesis (Figure 14.8A) or as gross photosynthesis (Figure 14.8 B). Typically, plants grown under high light have a higher photosynthetic capacity, that is, a higher light saturated rate of photosynthesis than low-light plants. In contrast, high-light plants may have a lower photosynthetic efficiency, that is, a lower initial slope, compared to the same plants grown at low light (Figure 14.8B).

Many green algae may exhibit an even more dramatic change in phenotype in response to growth at either high or low light than terrestrial plants (Figure 14.9A). The high light phenotype illustrates one mechanism of photoacclimation which involves the modulation of the size and composition of the light-harvesting complex (LHCII) of PSII coupled with a change in Rubisco content (see Chapter 7). There is now a consensus that the content of LHCII decreases on a leaf area basis as the growth irradiance increases (Figure 14.9B), which is coupled with increased xanthophyll cycle activity and increased photoprotection through NPQ. Since the bulk of the chlorophyll a and chlorophyll b is bound to LHCII (Chapter 7), a decrease in the amount of LHCII results in a decrease

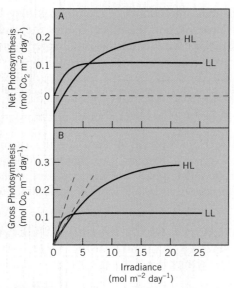

FIGURE 14.8 **Photosynthetic light response curves for** *Alocasia macrorrhiza*. **This C3 plant was grown under either high light (25 mol m^{-2} day^{-1}) or low light (1.7 mol m^{-2} day^{-1}).** (*A*) **Light response curves for net CO_2 gas exchange. Recall (Chapter 8) that the apparent or net rate of CO_2 uptake by a leaf (AP) = the gross rate or actual photosynthetic rate (GP)—[rate of respiration (R) + the rate of photorespiration (PR)]. In the dark, GP = 0 and PR = 0. In the dark gas exchange is due to respiration and measured as CO_2** *evolution*. **Thus, in (*A*) the value for net photosynthetic rate in the dark (irradiance = 0) is a negative number indicating net CO_2 evolution from the leaf. At higher irradiance, the rate of photosynthetic CO_2 uptake exceeds both respiration and photorespiration such that there is a net uptake of CO_2. Rates of light-dependent CO_2 uptake are arbitrarily given positive values whereas rates of CO_2 evolution are given negative values. Thus, gross photosynthesis (GP) represents photosynthetic CO_2 uptake rates that have been corrected for rates of respiratory CO_2 evolution, that is, GP = AP + R +PR. In (*B*), gross photosynthetic rates were calculated assuming minimal contributions from photorespiration. (Adapted from Pearcy, R. W. 1996.** *Photosynthesis: A Comprehensive Treatise*, **A. S. Raghavendra (ed.), pp. 250–263. Cambridge: Cambridge University Press.)**

in total chlorophyll content as well as an increase in the ratio of chlorophyll a/chlorophyll b. Functionally, this results in a decrease in light-harvesting efficiency under low, light-limiting irradiance which may be detected as a decrease in the initial slope of the photosynthetic light response curve in high-light plants compared to low-light plants (Figure 14.8B). Under light-saturated conditions, the activity of the Calvin Cycle limits the rate of photosynthesis (Chapter 13). Thus, high-light plants exhibit higher photosynthetic capacity because they exhibit a higher total Rubisco content per leaf area than the same plants grown under low light.

Furthemore, the rate of growth of plants is usually faster under high-light than low-light conditions. This is reflected in higher rates of respiration in high-light plants than low-light plants. This is indicated by the fact that when measuring net CO_2 assimilation in the dark (Figure 14.8A, 0 irradiance), the rate of CO_2 evolution due to respiration is greater in the high-light than low-light plants.

Experiments utilizing single cell green algal species such as *Dunaliella tertiolecta* and *Chlorella vulgaris* indicate that the light-dependent change in the content of LHCII is modulated in response to the redox state of the plastoquinone (PQ) pool (see Chapter 7). Photosynthetic electron transport can be inhibited specifically at the Cyt b_6/f complex with a compound called DBMIB (2,5-dibromo-6-isopropyl-3-methyl-1,4-benzoquinone) (Figure 14.9C). In the presence of this compound, there is a net accumulation of PQH_2 because, although PSII is able to convert PQ to PQH_2, PSI can not oxidize this pool because of the chemical block at the Cyt b_6/f complex (Figure 14.9C). Thus, under these conditions, the PQ pool remains largely reduced and transcription of the nuclear *Lhcb* genes coding for the major LHCII polypeptides is repressed. This results in an inhibition of the biosynthesis of LHCII polypeptides, which decreases the LHCII polypeptide content. As a consequence, this results in a yellow phenotype typical of high-light grown algal cells (Figure 14.9A, compare DBMIB and HL). Alternatively, photosynthetic electron transport can also be inhibited specifically at PSII (Figure 14.9C) in the presence of DCMU (3-(3,4-dichlorophenyl)-1,1-dimethylurea; see Figure 7.13). Under these conditions, PSII is unable to reduce PQ to PQH_2 and PQ accumulates because any PQH_2 pool is oxidized by PSI. This produces a green phenotype (Figure 14.9, DCMU) which mimics low-light-grown cells (Figure 14.9A, LL).

How can we rationalize this phenotypic response to growth under high light? Exposure to high light potentially exposes plants to an imbalance in energy budget. This energy imbalance is a consequence of the fact that more light is absorbed than can be utilized metabolically either through the reduction of CO_2 and NO_3^- or by the oxidation of carbon by respiration (Chapter 8, 10, and 11). Under such conditions, the PQ pool tends to be in a reduced state. Continued exposure to such conditions can lead to chronic photoinhibition (Chapter 13) which, if it persists over the long-term, may lead to cell death. Energy balance can be attained under high light by reducing the efficiency of light absorption under high light. This is accomplished by producing a smaller LHCII (Figure 14.9B) with high concentrations of zeaxanthin which decreases the absorption cross section (σ) of PSII and decreases the probability of light absorption by PSII. One consequence of

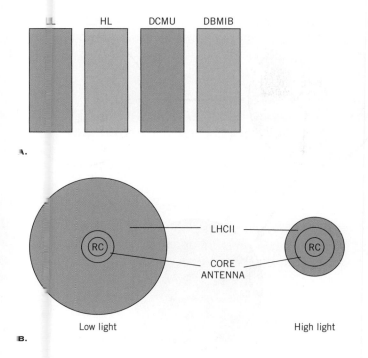

A.

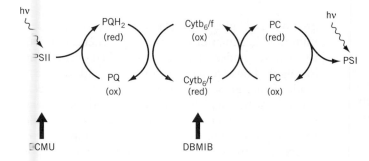

B.

C.

FIGURE 14.9 **Photoacclimation to high light. (*A*) An illustration of cell cultures of the green alga, *Dunaliella tertiolecta* grown under either low light (LL) or high light (HL) and either in the presence of DBMIB or DCMU. (*B*) A model to illustrate the difference in LHCII between an organism grown at low light and the same organism grown at high light. RC represents the reaction centre complex of PSII. Photoacclimation is thought to be a result of changes in LHCII content with little or no change in either the content of the core antenna (Chapter 7; CP47 and CP43) or the reaction center complex (D1 and D2, Chapter 7). (*C*) A simplified model of photosynthetic linear electron transport illustrating the coupled reduction and oxidation of PSII, the PQ pool, the Cyt *b₆/f* complex, plastocyanin (PC) and PSI. (See Chapter 7 for details.) DCMU inhibits PSII activity. DBMIB inhibits the Cyt *b₆/f* complex.**

this decreased efficiency for light absorption and energy transfer is enhanced photoprotection and increased tolerance to growth conditions that typically cause photoinhibition.

An energy imbalance due to high light appears to be sensed through changes in the redox state of the PQ pool. Such an energy imbalance due to overexcitation of PSII is called **excitation pressure**. Thus, the redox state of the PQ pool appears to induce a signal transduction pathway which regulates the expression of the *Lhcb* genes present in the nucleus. The regulation of nuclear genes by organelles such as the chloroplast is called **retrograde gene regulation** (Figure 14.10). However, retrograde regulation is not restricted to chloroplast-nucleus interactions. Like the chloroplast, the mitochondrion also regulates nuclear gene expression. This illustrates that important molecular communication pathways exist between the chloroplast, the mitochondrion, and the nucleus to regulate nuclear gene expression. The chloroplast-nucleus and the

mitochondrion-nucleus signal transduction pathway(s) remain elusive but an intensive area of research.

The maintenance of cellular energy balance is called **photostasis**, which is dependent upon chloroplast-mitochondrial interactions. For example, the mitochondrial Moc1 protein is thought to regulate the transcription of mitochondrial genes involved in the maintenance of the mitochondrial respiratory electron transport. However, under high light, the *moc1* mutant, which lacks this mitochondrial protein, is unable to up-regulate rates of respiration to match the production of fixed carbon by photosynthesis. The block in mitochondrial electron transport slows the rate of respiratory carbon metabolism which, in turn, causes a feedback inhibition in the rate of photosynthetic electron transport. This also results in the reduction of the PQ pool in the chloroplast. This is an excellent example of the link between chloroplast and mitochondrial metabolism and its importance in the regulation of cellular redox balance.

FIGURE 14.10 **A model illustrating retrograde regulation. The redox state of the plastoquinone pool (PQ) in the chloroplast can regulate nuclear gene expression. Similarly, the redox state of the mitochondrial ubiquinone pool (UQ) can regulate nuclear gene expression. The regulation of nuclear gene expression by the chloroplast and the mitochondrion is called retrograde regulation. The chloroplast and the mitochondrion communicate through carbon metabolic pathways.**

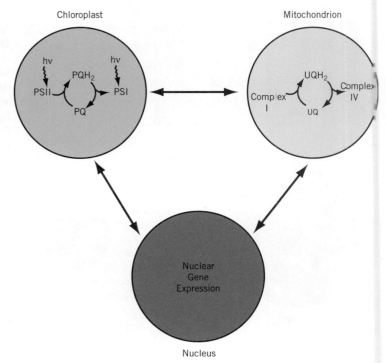

14.3.2 DOES THE PHOTOSYNTHETIC APPARATUS RESPOND TO CHANGES IN LIGHT QUALITY?

Terrestrial plants growing in extremely shaded habitats below the canopy of a tropical rainforest floor receive about 1 percent or less of the photosynthetically active radiation (PAR) (Chapter 6) incident at the canopy level. Furthermore, the light that reaches the forest floor is enriched in far-red and green light because the red and blue light are absorbed by the canopy leaves of the taller trees. Thus, acclimation to natural shade conditions would appear to be a complex interaction of responses to both light intensity and light quality. Acclimation in response to light quality would involve regulation by the plant photoreceptor, phytochrome (Chapter 22). A major problem in past experiments to test the differential effects of light quality versus light intensity on acclimation of the photosynthetic apparatus is the fact that differences in light quality were not applied at equal photon fluence rates (Chapter 6). Thus, this makes interpretations of many of these experiments equivocal. Experiments with pea and corn where equal photon fluence rates of red and far-red light were used, the ratio of chlorophyll a : chlorophyll b did not change significantly. This is consistent with the observations no major changes in thylakoid membrane structure and composition were observed in pea and corn plants exposed to differences in light quality of equal photon fluence rates. However, the ability to adjust the structure and composition of the photosynthetic apparatus in response to light quality appears to be strongly species dependent. When the shade fern, *Asplenium australasicum*, was grown in red light, changes in leaf pigment composition mimicked acclimation to high light whereas growth of this species under blue light mimicked acclimation to low light. These results are consistent with the fact that shade plants grow in habitats enriched in far-red and green light but depleted in the blue and red regions of the visible spectrum. Furthermore, very little is known regarding the contents of phytochrome and other photoreceptors in shade adapted plants. Thus, it appears that acclimation of the photosynthetic apparatus to light quality is highly species dependent.

A dramatic example of acclimation to light quality is exhibited by cyanobacteria (Figure 14.11). Recall that the photosynthetic apparatus of these prokaryotic photoautrophs exhibit similarities as well as important differences from that of their eukaryotic photoautotrophic counterparts (Chapter 7). Like plants and green algae, cyanobacteria contain both photosystem I (PSI) and photosystem II (PSII) and, as a consequence, are oxygenic. However, unlike these eukaryotes, cyanobacteria use extrinsic, thylakoid membrane, pigment-protein complexes called phycobilisomes to harvest light energy for photosynthesis. The major pigments of many cyanobacteria covalently bound to the phycobilisomes include phycocyanin (PC) and phycoerythrin (PE). Cyanobacteria that accumulate PC and PE such as *Fremyella diplosiphon* are capable of adjusting the pigment composition of phycobilisomes in response to changes in light quality. This ability to adjust to ambient light color is called **complementary chromatic adaptation (CCA)**. CCA is a consequence of a change in the proportion of PC and PE in response to light quality (Figure 14.11). When exposed to growth

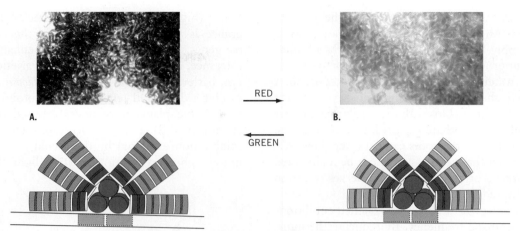

FIGURE 14.11 **The effect of light quality on the phenotypes of *Fremyella diplosiphon*** The filamentous cyanobacterium, *F. diplosiphon*, grown on agar plates and acclimated to either green light (*A*) or to red light (*B*). Below each culture is a model illustrating the change in pigment composition of the phycobilisome associated with PSII in response to growth under either green or red light. (Adapted from Kehoe, D. M., A. Gutu. 2006. Responding to color: The regulation of complementary chromatic adaptation. *Annual Review of Plant Biology* 57:127–150 (with permission)).

under red light, *Fremyella diplosiphon* exhibits the characteristic blue-green phenotype due to the fact that PC most effectively absorbs red light. In contrast, during growth under green light, *Fremyella diplosiphon* exhibits a distinctive red phenotype because PE most effectively absorbs green light. Since this acclimation process is photoreversible, it appeared to share features that were common to the red/far red photoreversible, phytochrome response characteristic of terrestrial plants. However, recently it has been shown that CCA is regulated by both the redox state of the PQ pool as well as by multiple, as yet unidentified, photoreceptors.

Clearly, CCA is due to changes in the expression of genes present in the genome of a cyanobacterium to produce the different phenotypes. The reversible changes in phenotype associated with CCA reflect a remarkable plasticity of cyanobacteria to respond to changes in ambient light color to maximize absorption of light energy for photosynthesis and growth. Thus, this phenomenon should be called complementary chromatic *acclimation* rather than complementary chromatic adaptation.

14.3.3 ACCLIMATION TO DROUGHT AFFECTS SHOOT–ROOT RATIO AND LEAF AREA

One of the long-term effects of water deficit is a reduction in vegetative growth. Shoot growth, and especially the growth of leaves, is generally more sensitive than root growth. In a study in which water was withheld from maize (*Zea mays*) plants, for example, there was a significant reduction of leaf expansion when tissue water potentials reached −0.45 MPa and growth was completely inhibited at −1.00 MPa. At the same time,

normal root growth was maintained until the water potential of the root tissues reached −0.85 MPa and was not completely inhibited until the water potential dropped to −1.4 MPa. Reduced leaf expansion is beneficial to a plant under conditions of water stress because it leads to a smaller leaf area and reduced transpiration. Traditionally, the effect of low water potential on cell enlargement has been attributed to a loss of turgor in the cells in the growing region. Plant cell enlargement occurs when water moves in to establish full turgor following stress relaxation in the cell wall (Chapter 16 and 17). It should not be too surprising, then, that an early consequence of limited water supply would be reduced growth. Thus, although the cells are able to maintain turgor, it does not appear to be sufficient to maintain a full rate of growth based on cell enlargement.

The preceding discussion applies primarily to shoots and leaves that are actively growing. Many mature plants, such as cotton (*Gossypium hirsutum*), subjected to prolonged water stress will respond by accelerated senescence and abscission of the older leaves. In the case of cotton, only the youngest leaves at the apex of the stem will remain in cases of severe water stress. This process, sometimes referred to as **leaf area adjustment**, is another mechanism for reducing leaf area and transpiration during times of limited water availability. So long as the buds remain viable, new leaves will be produced when the stress is relieved.

As noted above, roots are generally less sensitive than shoots to water stress. Apparently, osmotic adjustment in roots is sufficient to maintain water uptake and growth down to much lower water potentials than is possible in leaves. Relative root growth may actually be enhanced by low water potentials, such that the root–shoot ratio will change in favor of the proportion of roots. An increase in the root–shoot ratio as the water supply becomes depleted is clearly advantageous, as it improves the capacity of the root system to extract more water by exploring larger volumes of soil. A changing root–shoot ratio is accompanied by a change in source–sink relationships with the result that a larger proportion of photosynthate is partitioned to the roots. Delivery of carbon to the roots can continue, however, only to the extent that carbon supply can be maintained by photosynthesis or mobilization of reserves stored in the leaves. In addition, to changes in leaf area, some plants respond to growth under water deficit conditions by reducing stomatal frequency. By reducing the number of stomates per leaf area, a plant can reduce potential water loss due to transpiration when water is limiting growth.

An early response to water deficits is closure of stomates to conserve water. In some plants this may lead to low internal leaf CO_2 concentrations which will limit photosynthetic capacity. A limited photosynthetic capacity in the presence of light can result in exposure of plants to excess light and photoinhibition. Since leaf O_2 levels will remain higher than CO_2 when leaf stomates are closed, O_2 will be preferentially consumed through photorespiration and the action of the enzyme Rubisco (Equation 14.1). Just as for the reduction of CO_2 (Equation 14.1), the continuous consumption of O_2 by Rubisco (Equation 14.2)

$$RuBP + CO_2 \rightarrow 2PGA \tag{14.1}$$

$$RuBP + O_2 \rightarrow PGA + P\text{-Glycolate} \tag{14.2}$$

requires the continuous regeneration of RuBP by the Calvin Cycle (Chapter 8). This requires a constant supply of NADPH and ATP generated by photosynthetic electron transport (see Figure 13.4). The fixation of one mole of O_2 through photorespiration consumes more energy (5ATP + 3NADPH) than the fixation of one mole of CO_2 (3ATP + 2NADPH). Consequently, the photorespiratory pathway may play an important role in maintaining photostasis when CO_2 is limiting. Such a role for photorespiration is supported by the fact that photorespiratory mutants of *Arabidopsis thaliana* are more sensitive to high light under water stress.

14.3.4 COLD ACCLIMATION MIMICS PHOTOACCLIMATION

For more than a century it has been known that the growth of winter varieties of cold tolerant herbaceous plant species such as rye (*Secale cereale* L. cv Musketeer) and wheat (*Triticum aestivum* L. cv Kharkov) at low temperatures results in enhanced freezing tolerance measured as LT50, the freezing temperature at which 50 percent of a population of plants are

killed (Figure 14.12A). The enhanced freezing tolerance is strongly correlated with the expression of *cor* genes. However, in addition to enhanced freezing tolerance, cold acclimated winter varieties of wheat, rye, barley, spinach as well as *Arabidopsis thaliana* also exhibit a decreased sensitivity to photoinhibition even though the plants were never exposed to high light (Figure 14.12B). Thus, the decreased sensitivity to photoinhibition exhibited by these cold acclimated plants mimics photoacclimation to high light. However, in

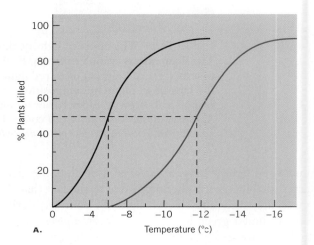

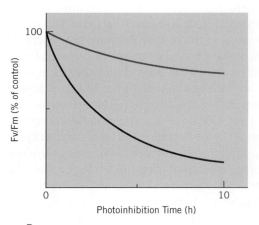

FIGURE 14.12 (*A*) The schematic graph illustrating the effect of growth temperature on freezing tolerance measured as LT50. Cold-tolerant plants grown at 25°C (black line) exhibit an LT50 of −7°C whereas the cold-tolerant plants that are grown at 5°C (red line) exhibit an LT50 of about −12°C. Thus, cold-acclimated plants exhibit an increased freezing tolerance. (*B*) A schematic graph illustrating the effect of growth temperature on sensitivity to photoinhibition. The maximal photochemical efficiency of PSII measured as Fv/Fm of the plants grown at 25°C (black line) decreases to a greater extent than that of plants grown at 5°C (red line). Thus, cold-acclimated plants exhibit a decreased sensitivity to photoinhibition (Adapted from Gray, G. R., L. V. Savitch, A. G. Ivanov, N. P. A. Huner. 1996. *Plant Physiology* 110:61–71).

these winter varieties, this is accomplished with minimal changes in the structure and composition of LHCII.

How is this possible? Growth of cold-tolerant winter wheat and winter rye at low temperature stimulates photosynthetic capacity with minimal changes in photosynthetic efficiency or in the ratios of chlorophyll a/chlorophyll b. This stimulation in photosynthetic capacity is the result of the following. First, cold acclimation enhances the transcription and translation of genes encoding major regulatory enzymes of stromal and cytosolic carbon metabolism such as Rubisco, chloroplastic FBPase, cytosolic FBPase, and sucrose-P synthase (SPS), as well as increased fructan biosynthesis in the vacuole (Chapter 9). This results in higher total enzyme activity and a higher flux of carbon through the sucrose biosynthetic pathway. Second, this is coupled to higher rates of carbon export from the leaves in the light due to enhanced sink activity (Chapter 9). In contrast to short-term exposure to low temperature, growth at low temperature stimulates rates of respiratory carbon metabolism. Third, cold acclimation suppresses photorespiration which also enhances net carbon gain. Thus, subsequent exposure of these cold-acclimated plants to increasing irradiance stimulates their photosynthetic capacity even further, which is translated into increased growth rates and biomass production at low temperature in addition to the stimulation of NPQ via the xanthophyll cycle. Thus, these cold-acclimated plants exhibit a resistance to photoinhibition because of their enhanced capacity to utilize the absorbed light for carbon metabolism, biomass production and growth coupled with the dissipation of any excess absorbed light through NPQ. It is important to appreciate that, to exhibit these characteristics, cold-tolerant plants must grow and develop at low temperatures.

Similar to cold-acclimated winter varieties of wheat, rye, barley, spinach, as well as *Arabidopsis thaliana*, cold-acclimated *Chlorella vulgaris* also exhibits a decreased sensitivity to photoinhibition. However, in contrast with these terrestrial plants, cold-acclimated *Chlorella vulgaris* exhibits the same yellow-pale-green phenotype as high-light grown cells (Figure 14.9, HL) even though these cold-acclimated cells have not grown under high light. It has been shown that cold acclimation of *Chlorella vulgaris* mimics photoacclimation because growth at low temperature induces a comparable energy imbalance or excitation pressure as growth at high light. Why does this occur? Unlike wheat, rye, and Arabidopsis, *Chlorella vulgaris* is unable to up-regulate photosynthetic capacity during cold acclimation when measured on a per-cell basis. Low growth temperature reduces the rate of intersystem electron transport as well as enzyme activity involved in carbon metabolism without affecting light absorption, and energy transfer from LHCII to P680 and its subsequent photooxidation to P680$^+$. Under these conditions, PQH$_2$ accumulates because PSII reduces the PQ pool faster than PSI and ultimately CO$_2$ assimilation can oxidize this pool. Concomitantly, this chloroplast redox signal represses *Lhcb* gene expression by retrograde regulation (Figure 14.10). The cyanobacterium, *Plectonema boryanum*, also shows a similar phenotypic response to growth at either low temperature or high light as *Chlorella vulgaris*. However, the cyanobacterium responds by decreasing the size of its phycobilisome associated with PSII (Chapter 7).

The discussion above illustrates the remarkable plasticity with which photoautrophs respond to energy imbalances as a consequence of growth at either high light or low temperature. Although low temperature and high light cause a similar imbalance in cellular energy budget, the response of photoautrophs to energy imbalance or excitation pressure is species dependent. It appears that many terrestrial cold-tolerant plants can attain photostasis by combining an up-regulation of photosynthetic capacity and increased growth rates to utilize absorbed light energy with photoprotection through xanthophyll cycle activity. In contrast, many green algae attain photostasis by reducing light-harvesting efficiency by making a smaller light-harvesting complex through retrograde regulation due to their inability to utilize the absorbed energy through carbon metabolism and growth.

14.4 FREEZING TOLERANCE IN HERBACEOUS SPECIES IS A COMPLEX INTERACTION BETWEEN LIGHT AND LOW TEMPERATURE

Cold acclimation and the development of maximum freezing tolerance in overwintering herbaceous plants such as winter wheat, winter rye, spinach, and *Arabidopsis thaliana* requires active growth and development at low temperature. As a result, leaves of these plant species developed at low temperature are anatomically, morphologically, physiologically, and biochemically distinct from the same plants developed at warm temperatures. For example, these herbaceous species grown at low temperature exhibit a short, compact growth habit, thicker leaves due to an increase in leaf mesophyll cell size and/or an increase in the number of palisade cell layers, and an increase in cell cytoplasm associated with a decrease in leaf water content. Photosynthesis continues during the cold acclimation period and provides the necessary energy required for this low-temperature growth, which leads to the cold-acclimated state and maximum freezing tolerance (Equation 14.3).

TABLE **14.2**   **Effects of growth temperature and growth irradiance on LT50 of winter rye (*Secale cereale* L. cv Musketeer). Growth irradiance is in units of μmol photons m^{-2} s^{-1}).**

Growth Temperature (°C)	Growth Irradiance		LT50 (°C)
20	50	low light	−4
20	250	moderate light	−6
20	800	high light	−8
5	50	low light	−8
5	250	moderate light	−16

Cold Stress → Growth/Development

→ Cold Acclimation → Freezing Tolerance   (14.3)

As indicated in Table 14.2, exposure to low temperature (5°C) is an absolute requirement for the attainment of *maximum* freezing tolerance in winter rye (LT50 = −16°C). However, low temperature by itself is not sufficient to induce maximum freezing tolerance since growth at low temperature but low light reduces LT50 by 50% (Table 14.2). Thus, light is also an absolute requirement for the attainment of maximum freezing tolerance in winter cereals (Table 14.2). This light-dependent increase in freezing tolerance at constant low-growth temperature occurs independently of either photoperiod or light quality. Thus, the light dependence of LT50 in herbaceous winter cereals does not appear to be phytochrome-dependent. Furthermore, increasing the growth irradiance at temperatures that normally do not induce freezing tolerance (20°C) results in a doubling of LT50 from −4 under low light to −8°C under high light, which is consistent with the light dependence of freezing tolerance. However, high light can not compensate for low temperature for the induction of maximum freezing tolerance in winter cereals. Thus, the attainment of maximum freezing tolerance appears to be the result of an additive effect of both low-growth temperature and growth irradiance.

14.4.1 COLD ACCLIMATED PLANTS SECRETE ANTIFREEZE PROTEINS

Overwintering plants can tolerate freezing because of their ability to control the freezing event itself. As long as freezing of water is confined to the apoplast, that is the cell wall and the extracellular space, the plant will survive. Alternatively, if freezing occurs intracellularly, the plant will die. Cold acclimation in many plants is associated with the secretion of **antifreeze proteins (AFPs)** from the cytoplasm into the apoplast. AFPs have been reported in ferns, gymnosperms, as

well as mono- and dicotyledonous angiosperms. AFPs inhibit ice crystal growth by binding to the surface of a growing ice crystal via hydrogen bonding between specific hydrophilic amino acids present in the AFP and water within the crystal lattice of ice. The presence of AFPs in cold-tolerant plants is not constitutive but requires exposure to low temperature and they accumulate in virtually all plant tissue including seeds, stems, leaves, flowers, and roots. When winter rye plants cold acclimate, the gaseous plant hormone, ethylene, is produced. Ethylene induces the transcription of the family of genes that encode AFPs. Upon translation, the AFPs are secreted via the endoplasmic reticulum, Golgi bodies, and vesicles that fuse with the plasma membrane and are deposited on the surface of the cell wall where they inhibit ice crystal formation.

Do AFPs alter LT50? The answer is an unequivocal no. If this is so, what is the role of AFPs? Although AFPs have a minimal effect on LT50, AFPs have a significant effect on the rate of ice crystal formation. Thus, AFPs most likely enhance winter survival by slowing the rate of extracellular freezing.

14.4.2 NORTH TEMPERATE WOODY PLANTS SURVIVE FREEZING STRESS

Boreal deciduous trees, conifers, and shrubs such as paper birch (*Betula papyrifera*), trembling aspen (*Populus tremuloides*), and willow (*Salix* sp.)—all found as far north as the arctic circle—survive because they are able to acclimate to the below-freezing winter temperatures. During their normal growing season, these plants will suffer injury or death if exposed to freezing temperatures. Even a light frost during the spring or summer may be lethal to plants that are actively growing. Yet cold-acclimated stems of these species may survive temperatures as low as −196°C (liquid nitrogen) without apparent injury.

Acclimation of woody species to freezing stress is a common phenomenon in nature, but the precise mechanism by which acclimation is achieved is not well understood. It is known that acclimation in woody tissues occurs in two distinct stages (Figure 14.13). It begins in the autumn when growth and photosynthesis ceases and the plant enters **dormancy**. This first stage of acclimation is induced by short days and is thought to be under the control of phytochrome. Acclimation at this stage can be inhibited by long days and early frost. Thus, in woody species, it is essential that the plant enter the dormant stage prior to the onset of frost to prevent freezing damage.

The second stage of acclimation is triggered by exposure of the overwintering tissue to low temperature, corresponding to the first frost (Figure 14.13). At this stage, respiratory activity is sufficient to provide the

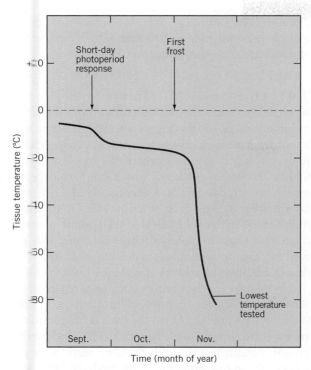

Figure 14.13 **Acclimation to low temperature in woody stems.** The curve depicts the lowest survival temperatures as a function of time of year. Note that significant decreases in survival temperature correspond to shortening daylength and the time of the first frost.

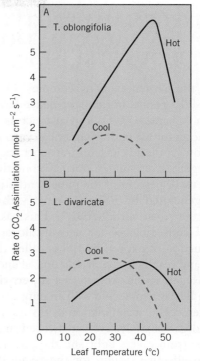

Figure 14.14 **A schematic graph illustrating the ability of thermotolerant C3 and C4 plants to adjust photosynthetic capacity. Temperature profiles for light-saturated photosynthetic rates are plotted as a function of leaf temperature for a C4 plant, *T. oblongifolia* (A), and a C3 plant, *L. divaricata* (B). (Adapted from Berry, J. A., O. Björkman. 1980. *Annual Review of Plant Physiology* 31:491–543.)**

energy necessary for the numerous metabolic changes required to attain the maximum cold-acclimated state. There are increases in the level of organic phosphates and the conversion of starch to sugars. Glycoproteins accumulate and the protoplasm becomes generally more resistant to dehydration. Fully acclimated cells can withstand temperatures far below those normally experienced in nature.

14.5 PLANTS ADJUST PHOTOSYNTHETIC CAPACITY IN RESPONSE TO HIGH TEMPERATURE

Plants that can acclimate to high temperatures are called **thermotolerant** plants. Photosynthetic capacity measured as the maximum light-saturated rate of CO_2 assimilation is sensitive to temperature (Figure 14.14). The C3 and the C4 plant illustrated in Figure 14.14 exhibit temperature optima for photosynthesis that are dependent upon the growth condition to which these plants were exposed. Plants exposed to cool temperatures generally exhibit a lower temperature optimum for photosynthesis than those exposed to high temperatures. Such a shift in temperature optima reflects photosynthetic acclimation to temperature which has been shown

to be reversible and not due to temperature-dependent stomatal limitations. Rather, such photosynthetic acclimation appears to be due to a combination of changes in the temperature stability of thylakoid membranes as well as the enzymes of the PCR cycle such as Rubisco. Although photosynthetic acclimation of C3 overwintering species such as wheat and rye requires growth and development at cold temperatures, this is not the case for all plants. For example, reversible photosynthetic acclimation in the C3 species, *Nerium oleander*, is observed in fully expanded, mature leaves exposed to either high or low temperatures. This may be due, in part, to the different absolute temperature ranges to which a particular species can acclimate.

The maximum rate of photosynthesis of the C4 plant, *T. oblongifolia*, appears to be more sensitive to leaf temperature than that of *L. divaricata*, a C3 plant (Figure 14.14). However, such differential responses do not reflect a general difference between C3 and C4 species. Rather, such differences have been shown to be very species dependent within plants that exhibit either C3 or C4 photosynthesis. Clearly, plants exhibit extraordinary plasticity to adjust photosynthetically to changes in temperature.

using stromal reductants (stromal pool). However, it has been proposed that subsequently the thylakoid-bound plastid terminal oxidase (PTOX) couples the oxidation of plastoquinol with the reduction of O_2 to water even in the light. Recently, the gene for a plastid terminal oxidase (PTOX) was identified in *Arabidopsis thaliana* as well as *Chlamydomonas reinhardtii*. The sequence of PTOX is very similar to the alternative oxidase (AOX) present in mitochondria (Chapter 10).

Since O_2 can act as an alternative electron acceptor for the photosynthetic electron transport chain either through the Mehler reaction or through the chlororespiratory pathway, both of these processes represent potential mechanisms to keep the PQ pool oxidized and decrease the probability of chronic photoinhibition during acclimation irrespective of the stress which initiates the short-term and long-term mechanisms of acclimation. However, the specific contribution of the Mehler reaction and chlororespiration to photoprotection during acclimation to stress appears to be both species dependent as well as dependent upon the specific stress to which a plant is exposed. Furthermore, irrespective of whether the short-term or long-term mechanism of acclimation is in response to high light, water deficits, or temperature, the maintenance of photostasis, that is a balance in energy budget, appears to be an important feature of the newly attained acclimated state.

SUMMARY

Stress-resistant or stress-tolerant plants exhibit the ability to acclimate or adjust to the environmental stress. Although specific changes in physiology brought

A unique characteristic of plant respiration is that its Q_{10} is not constant but varies as a function of short-term changes in temperature. In contrast to the short-term mechanisms that initiate acclimation responses, the long-term mechanisms of acclimation may result in phenotypic changes. Photoacclimation to high light leads to a reduction in chlorophyll content with concomitant decreases in the abundance of PSII light-harvesting complex polypeptides and increases in Rubisco content relative growth at low light. This response is a consequence of retrograde regulation of nuclear genes coding for PSII light-harvesting polypeptides by the redox status of the PQ pool. This response to high light is mimicked by acclimation to low temperature because both excess irradiance and low temperature increase the reduction state of the PQ pool in a similar way. This can be measured as excitation pressure which is a measure of the proportion of closed PSII reaction centers. Although photoautotrophs acclimate to light quality, the extent of such acclimation appears to be species dependent. The most dramatic example of acclimation to light quality is exhibited by many cyanobacteria through a process called complementary chromatic adaptation. Growth and development of winter cereals at low temperature stimulates photosynthetic and respiratory capacity which results in an increased tolerance to photoinhibition. In response to growth under water deficit conditions, plants reduce shoot–root ratios and total leaf area to reduce water loss due to transpiration. Cold acclimation of herbaceous as well as woody plants leads to increased freezing tolerance which is measured as LT50. Antifreeze proteins do not affect LT50 but control the rate of extracellular ice formation. Plants capable of acclimating to high temperature are considered thermotolerant and exhibit the ability to shift

14.6 OXYGEN MAY PROTECT DURING ACCLIMATION TO VARIOUS STRESSES

Although the oxygen evolving complex (OEC) associated with PSII results in the light-dependent evolution of oxygen (Chapter 7), O_2 can also act as alternative electron acceptor for photosynthetic electron transport. Thus, photosynthetic electron transport may also consume oxygen. Even under normal conditions, up to 5 percent to 10 percent of the photosynthetic electrons that are generated by PSI may react with molecular oxygen rather than with $NADP^+$. This has important functional consequences for active chloroplasts. The photoreduction of oxygen by PSI is called the **Mehler reaction** and results in the production of another toxic, reactive oxygen species known as a **superoxide radical** (O_2^-) (a radical is a molecule with an unpaired electron). To counteract the accumulation of this radical, photosynthetic organisms have evolved mechanisms to protect themselves from excess light and the potential ravages of O_2. An effective system for the removal of superoxide is the ubiquitous enzyme **superoxide dismutase (SOD)**. SOD is found in several cellular compartments including the chloroplast. It is able to scavenge and inactivate superoxide radicals by forming hydrogen peroxide and molecular oxygen (Equation 14.4):

$$2O_2^- + 2H^+ \rightarrow H_2O_2 + O_2 \qquad (14.4)$$

The H_2O_2, in turn, is reduced to water in the chloroplast by sequential reduction with ascorbate (vitamin C), glutathione, and NADPH (Figure 14.15A). It is interesting to note that plant chloroplasts normally exhibit relatively high concentrations of ascorbate (0.5 to 1.0 µmol mg^{-1} Chl in *Arabidopsis thaliana*), which can vary depending on the growth conditions. Reduction of H_2O_2 is necessary in order to prevent its reaction with O_2^- to form the highly toxic **hydroxyl radical (OH·)**, another example of an ROS that can rapidly damage proteins. This pathway for the protection against ROS is known as the **Asada-Halliwell Pathway** (Figure 14.15A).

In addition to the photoreduction of oxygen by PSI through the Mehler reaction, chloroplasts also exhibit the capacity to reduce O_2 in the dark through the **chlororespiratory pathway** (Figure 14.15B). Under normal growth conditions, this chloroplastic respiratory pathway results in the reduction of the thylakoid PQ pool in the dark. This pathway involves an NAD(P)H dehydrogenase (Ndh) that reduces PQ nonphotochemically by

FIGURE 14.15 **Oxygen as an alternative electron acceptor in chloroplasts.** (*A*) The Asada-Halliwell pathway. O_2 can be photoreduced by PSI directly to generate the superoxide free radical, O_2^- (Mehler reaction). Superoxide dismutase (SOD) then converts this radical to hydrogen peroxide (H_2O_2). Hydrogen peroxide is also toxic and is reduced via the chloroplastic enzyme, ascorbate peroxidase, to water and ascorbate is oxidized to monodehydroascorbate (MDHA). Ascorbate (vitamin C) is regenerated through the action of the enzyme, dehydroascorbate reductase, through the consumption of reduced glutathione (GSH). Oxidized glutathione (GSSH) is, in turn, reduced by the enzyme glutathione reductase, which uses NADPH as reductant. (*B*) Chlororespiratory pathway. NAD(P)H dehydrogenase (Ndh) present in thylakoid membranes consumes stromal NAD(P)H and passes the electrons (e) directly to plastoquinone (PQ). The plastid terminal oxidase (PTOX) present in thlylakoid membranes oxidizes plastoquinol and reduces O_2 to water. The stromal pool represents any metabolic pathway present in the stroma that generates reducing power (see Chapter 8). Ndh may also participate in cyclic electron transport around PSI.

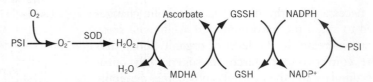

A. Asada-Halliwell pathway

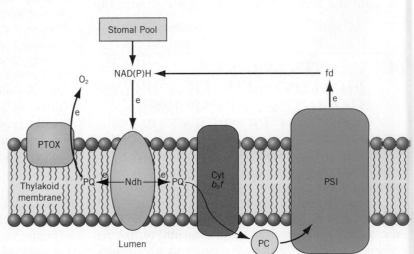

B. Chlororespiratory pathway

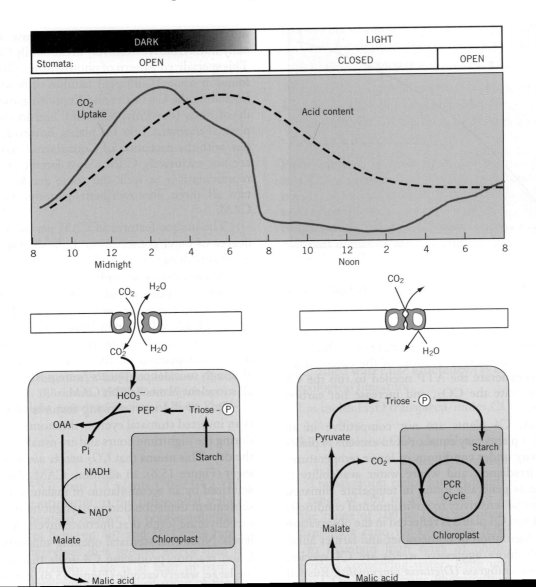

15.3.2 CAM PLANTS ARE PARTICULARLY SUITED TO DRY HABITATS

As mentioned above, CAM represents a particularly significant adaptation to exceptionally dry habitats. Many CAM plants are true desert plants, growing in shallow, sandy soils with little available water. Nocturnal opening of the stomata allows for CO_2 uptake during periods when conditions leading to evaporative water loss are at a minimum. Then, during the daylight hours when the stomata are closed to reduce water loss, photosynthesis can proceed by using the reservoir of stored CO_2. This interpretation is supported by the transpiration ratio for CAM plants, in the range of 50 to 100, which is substantially lower than that for either C3 or C4 plants. There is a price to be paid, however. Rates for daily carbon assimilation by CAM plants are only about one-half those of C3 plants and one-third those of C4. CAM plants can be expected to grow more slowly under conditions of adequate moisture. On the other hand, CO_2 uptake by CAM plants will continue under conditions of water stress that would cause complete cessation of photosynthesis in C3 plants and severely restrict carbon uptake by C4 plants. CAM plants enjoy the further advantage of being able to retain and reassimilate respired CO_2, thus preventing loss of carbon and helping to maintain a favorable dry weight through extended periods of severe drought.

While some species, in particular the cacti, are obligatory CAM plants, many other succulents exhibit a facultative approach to CAM. One well-studied example is the ice plant, *Mesembryanthemum crystallinum*, a fleshy annual of the family Aizoaceae. Under conditions of abundant water supply, *Mesembryanthemum* assimilates carbon as a typical C3 plant—there is no significant uptake of CO_2 at night and no diurnal variation in leaf cell acidity. Under conditions of limited water availability or high salt concentration in the soil, CAM metabolism is switched on. Although carbon assimilation by CAM is slower than with conventional C3 photosynthesis, its higher water use efficiency permits photosynthesis to continue in times of water stress and the plant is better able to complete its reproductive development.

15.4 C4 AND CAM PHOTOSYNTHESIS REQUIRE PRECISE REGULATION AND TEMPORAL INTEGRATION

In addition to regulation of PCR cycle enzymes discussed earlier, successful operation of C4 photosynthesis and CAM also requires regulation of starch-PEP interconversions; storage and retrieval of malate, PEPcase and Rubisco competition for CO_2; and the temporal operation of PEP carboxylation. PEPcase is a cytoplasmic enzyme found in virtually all cells of higher plants where it serves a variety of important metabolic functions. However, plants with the C4 and CAM modes of photosynthesis contain specific isozymes[1] with higher activity levels than associated with C3 or nonphotosynthetic cells. Based on a variety of physiological and biochemical considerations, it appears that PEPcase activity in C4 and CAM plants must be regulated by light–dark transitions. In C4 plants, for example, PEPcase activity should be high in the light in order to maximize availability of CO_2 for the PCR cycle in the bundle-sheath cells. Although the substrate PEP consumed in C4 photosynthesis is derived from the bundle-sheath chloroplast, PEP is also a critical intermediate in glycolysis. Glycolysis is also a cytoplasmic metabolic sequence that represents the first stage in respiration.

Continued high PEPcase activity at night could result in uncontrolled utilization of PEP and seriously impair respiratory metabolism. In the case of CAM it is clear that efficient operation requires the carboxylation and decarboxylation reactions—which occur within the same cell—not be allowed to compete with each other at the same time. In CAM, PEP carboxylase activity must be high at night when atmospheric CO_2 is available, but should be switched off during daylight hours in order to avoid a futile recycling of CO_2 derived from malate. Competition for CO_2 is not a problem at night since Rubisco and the PCR cycle are inoperative in the dark.

PEPcase regulation was initially studied in CAM plants but it is now evident that PEPcase activity in C4 plants as well as CAM is subject to reversible activation by light–dark conditions and inhibition by malate—a form of feedback inhibition in which accumulated product reduces the activity of the enzyme. In C4 plants, light induces an increase in the catalytic activity of the enzyme while at the same time reducing its sensitivity to inhibition by the product molecule, malate. This results in a fivefold increase in activity in the light when CO_2 assimilation and malate production is required to boost CO_2 levels in the bundle-sheath cells. In the case of CAM, the situation is reversed. Enzyme extracted during the night part of a diurnal cycle exhibits a high affinity for PEP and is relatively insensitive to inhibition by malate. Enzyme extracted during the day has a low affinity for PEP and is more sensitive to inhibition by malate.

---

[1]Isozymes are different species of enzyme that catalyze the same reaction. Isozymes are frequently coded by different genes and consequently have different protein structures. Although they act on the same substrate, isozymes may be tissue- or organ-specific and subject to regulation by different chemical or environmental factors.

Recent evidence from studies of *Bryophyllum* and *Kalanchoe*, both CAM plants, and the C4 plants *Zea mays* and *Sorghum* have shown that PEPcase exists in two states; the biochemically more active form of the enzyme is phosphorylated but the less active form is not. From this it can be concluded that PEPcase activity is regulated by a light-sensitive protein kinase. Evidence for reversible activation of a protein kinase has been obtained from both *in vivo* and *in vitro* experiments. The reaction requires ATP and phosphorylation occurs at a serine residue near the N-terminal end of the enzyme molecule. Just how light activates the protein kinase is not known at this stage, although there is some indication that it is related to photosynthetic electron transport. It may be that photophosphorylation supplies the ATP required for phosphorylation of the enzyme.

It is interesting to note that light appears to have the opposite effect in the two systems—the protein kinase is activated by light in C4 plants but inactivated during the day in CAM. The effect of light and darkness in CAM plants, however, may be indirect. Studies with *Bryophyllum* have indicated that CAM physiology and the sensitivity of PEPcase to inhibition by malate in CAM plants may be controlled by an endogenous circadian rhythm. Whatever the mechanism, spatial and temporal coordination of C4 and C3 metabolism in C4 and CAM plants is an important area where we can expect to see exciting advances in the future.

15.5 PLANT BIOMES REFLECT MYRIAD PHYSIOLOGICAL ADAPTATIONS

Physiology has an impact beyond the individual plant. The physiology of a plant community or ecosystem is defined as **ecophysiology**. At an even larger scale, **biomes** represent a collection of ecosystems that are characterized by distinctive vegetation that is related to a particular set of physical and environmental conditions. As examples of plant communities adapted to different environments, we will discuss two specific plant biomes: the tropical rainforest biome and the desert biome.

15.5.1 TROPICAL RAIN FOREST BIOMES EXHIBIT THE GREATEST PLANT BIODIVERSITY

A tropical rainforest is a diverse and complex ecosystem with a number of unique geographic and physical characteristics. All of the principal rainforest systems of the world are located between the latitudes of 23° 30′ N (Tropic of Cancer) and 23° 30′ S (Tropic of Capricorn). The equatorial location ensures that rainforest vegetation is subject to both high irradiance and high temperature. Moreover, at these latitudes seasonal

variations in both the quantity of light and photoperiod, as well as temperature, are very small. Rainfall is very high, with annual rainfalls of at least 1800 to 2000 mm per year and as high as 4000 mm per year. In this prevailing ever-wet tropical climate, with neither water nor temperature limiting, the rainforests rank among the most highly productive ecosystems in the world. Covering only about 8 percent of the global land mass, tropical rainforests contain more than 40 percent of the world's biomass.

There are three principal rainforest regions worldwide. The largest by far is located in the Amazon River basin of northern South America and extends, in small patches, through Central America into southern Mexico. The Amazon basin rainforest itself, stretching from the Atlantic Ocean in the east to the Andean mountains in the west, covers more than 4 million km². The second region is the African rainforest, located along the equatorial coast of eastern Africa and extending inland through the Congo River basin to the mountains of east-central Africa. Third, the Malaysian rainforest occurs in patches throughout the Malaysian Archipelago of southeastern Asia, primarily on the islands of Sumatra, Borneo, and New Guinea. While there are other areas, such as along the coast of northwestern United States and southwestern Canada, that also experience heavy rainfalls, these are not true rainforests. These coastal forests represent a southern extension of the northern coniferous, or boreal, forest, which is characterized by shorter growing seasons and persistent snow cover in the winter.

It is generally believed that rainforests are populated by a larger number of plant species than the rest of the world's ecosystems combined. At the same time, no one species accounts for more than 10–15 percent of the trees. There may be several hundred species per hectare and the individuals of each species may be widely scattered. Rainforest vegetation is dominated by tall, broad-leaved woody evergreens that form a canopy so dense that little light reaches the forest floor. As a result, available light on the forest floor is very low, limiting understory vegetation. The bulk of the canopy is located about 30 to 50 m above the ground, although many individuals, called **emergents**, may stand as high as 45 to 70 m. Some mature trees may reach diameters of 3 m, but most rainforest trees are relatively slender compared with their height.

Also abundant in the rainforests are **lianas**, large woody vines that entwine the trees, and **epiphytes**, plants that grow on the trunks and branches of other plants. In one report, it was found that as many as 60 percent of the trees in a Liberian rainforest carried epiphytes, with as many as 45 to 65 species on a single tree. Although epiphytes do grow on other plants, they are not parasites. Epiphytes are photoautotrophs that grow only in the illuminated portion of the canopy. It is believed that they obtain water either directly from rain and/or from the humid atmosphere of the canopy. Nutrients are obtained from atmospheric dust and the surfaces of the plants on which they grow. Common epiphytes include orchids, ferns, and bromeliads.

Soils of tropical rainforests typically retain very little in the way of plant nutrients and are effectively infertile. But how can a forest that grows on such a poor, nutrient-free soil achieve such extraordinarily high levels of productivity and biomass? The answer to this question is found in the rapid and thorough recycling of nutrients. In the moist, warm climate of the rainforest, litter is decomposed much more rapidly than it is in a comparable temperate forest. Large populations of termites, ants, fungi, and other **detritivores** ensure that the litter is rapidly broken down. The high average temperature, coupled with the absence of any cold season, allows decomposers to work with high efficiency year-round. Rainforest trees also have a high capacity for nutrient uptake. The shallow, spreading root systems of rainforest trees form extensive mycorrhizal associations that sweep the passing ground water clean of dissolved nutrient ions. There is some evidence that the mycorrhizal fungi (Chapter 3) may also be in contact with the decomposing litter, such that the transfer of nutrients is immediate and direct. This means that, unlike in a temperate forest where the soil is the principal nutrient reservoir, the trees themselves represent the largest reserve of nutrients in a rainforest. When vegetation is shed through storm damage or natural senescence, it is rapidly decomposed and the nutrients captured for recycling into new growth before they have much of a chance to enter the soil.

15.5.2 EVAPOTRANSPIRATION IS A MAJOR CONTRIBUTOR TO WEATHER

In any heavily vegetated region, there is to be expected a continual cycling of water between land surfaces (including vegetation) and the overlying atmosphere. The overall balance of land–atmosphere water exchange is known as the **hydrologic cycle**. Rainwater is returned to the atmosphere in the form of water vapor. Depending on the amount of vegetation cover, transpiration may be the principal source of water vapor (Chapter 2). In a rainforest, direct evaporation of soil water appears to be negligible, in part because of the normally heavy canopy cover and in part because the litter on the forest floor may act as a mulch to reduce evaporation of soil water. However, another significant source of water vapor is the direct evaporation of water intercepted by the canopy during a rainfall (Figure 15.7). Evaporation of intercepted rainwater occurs simultaneously with transpiration and it is not possible to distinguish

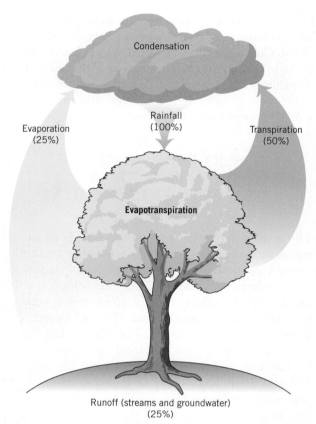

FIGURE 15.7 Rainfall and latent energy cycle in a tropical rainforest. Fifty percent of the rainfall is returned to the atmosphere through transpiration. Another 25 percent is returned by evaporation, primarily of water intercepted by the canopy, and 25 percent is accounted for as runoff.

between water vapors arising from the two sources. For this reason, the term **evapotranspiration** is used to identify the transfer of water vapor from vegetated land surfaces to the atmosphere, regardless of the source of the vapor.

Recycling of water through evapotranspiration is clearly a significant factor in the hydrologic cycle of rainforests. In the case of the Amazon basin, for example, water vapor enters the region from the Atlantic Ocean and is driven inland by the predominantly easterly trade winds. To this vapor is added at least an equal amount of vapor produced by evapotranspiration as the winds sweep across the dense forests. Vapors from these two sources mix to form the clouds that eventually produce rain across the basin.

The significance of evapotranspiration is not limited to the water economy of a group of plants or its contribution of water vapor to the atmosphere. The **latent heat flux** (LE) that accompanies the evaporation and condensation of water over tropical forests is an important factor in the redistribution of energy throughout that region of the atmosphere between ground level and an altitude of 5 to 10 miles (8 to 16 km), known as the **troposphere**. The troposphere contains most of the moisture in the atmosphere and is therefore the region where clouds and convection currents, or winds, are found. Consequently, evapotranspiration has the potential to influence both local climate and weather patterns throughout the atmosphere. The basis for this influence is the high positive heat exchange that occurs when water evaporates and condenses.

15.5.3 DESERT PERENNIALS ARE ADAPTED TO REDUCE TRANSPIRATION AND HEAT LOAD

The word *desert*[2] often conjures up images of an arid region with no plants, just a hot sun over barren, shifting sands. Most deserts do receive some precipitation and are populated by a diverse and interesting plant life that is uniquely adapted to extremely dry conditions. In contrast to our usual image of a hot desert, cold deserts also exist on Earth. The Taylor Dry Valley in Antarctica is the coldest, most arid, barren desert on Earth. Although no terrestrial plant life exists on this desert, the biodiversity of microorganisms in the perennially ice covered lakes in this area of Antarctica is astounding.

Most deserts have one or, at most, two seasons of predictable rainfall interspersed with extended periods of drought. Other deserts may receive a significant portion of the total annual precipitation as **episodic rainfall**, which comes as a single brief period of high-intensity rainfall. The result is that seasonality of precipitation, rather than mean annual rainfall values, has greater relevance to the distribution of desert vegetation and their life cycles. The most distinctive property of deserts is not so much low precipitation *per se*, but that potential evapotranspiration far exceeds rainfall on an annual basis. Because deserts typically have long periods of cloudless weather, the solar radiation load is high and most of what little rainfall is received is rapidly lost through evaporation. As a result, there is very little opportunity to store any significant amount of moisture in the soil.

The two principal constraints to plant survival in a desert environment are the availability of water and excessive heat load. Availability of nutrients, with the possible exception of nitrogen, is not normally a problem in desert soils. In order to cope with extended drought interspersed with only occasional periods of light rainfall and high summer temperatures, desert perennials

have developed numerous morphological and metabolic adaptations that are designed either to conserve water or to increase water-use efficiency and reduce heat load. *Encelia farinosa* (brittlebrush) is a C3 broadleaf, drought-deciduous shrub which withstands drought by exploiting seasonal **leaf polymorphism**. In response to winter rains, *Encelia* produces large, glabrous (smooth and hairless) leaves. As the dry season approaches and water supply becomes increasingly tenuous, *Encelia* produces progressively smaller, thicker leaves. In addition, both the adaxial and abaxial surfaces of the smaller leaves develop dense mats of dead, air-filled **trichomes** (epidermal hairs). This mat of trichomes gives the leaves a "white" appearance and increases the reflectance of the leaf. The early "green" leaves reflect approximately 15 percent of the incident solar radiation, while the late-season "white" leaves reflect up to 70 percent. This represents a significant reduction in the heat load on the leaf, enough to reduce leaf temperature by as much as 5 to 10°C at midday in the summer. A reduced heat load serves to reduce transpiration and thus improve water-use efficiency. It will also help to maintain leaf temperature in the optimum range for photosynthesis. These morphological changes thus enable *Encelia* and similar broadleaf desert species to resist drought stress and maintain a positive net photosynthesis well into the dry season. Sagebrush (*Artemisia tridentata*), like most cold-desert shrubs, has a relatively high proportion of its biomass invested in roots and probably occupies a greater volume of soil than most of its competitors. The high root–shoot ratio appears to be a means to ensure efficient extraction of water from the soil when it is recharged by precipitation during the winter/spring season.

Desert succulents such as members of the Cactaceae (the true cacti) and their relatives in the family Euphorbiaceae are often viewed as the quintessential desert plants. However, as succulents the cacti are not true xerophytes and are actually found more commonly in semideserts and coastal temperate deserts. The cacti are stem-succulents: leaves have been reduced to thorns and photosynthesis is taken over by the stem. The thick, fleshy stems are composed of very large cells that store copious amounts of water, and stomates are few and sunken. Another critical aspect is that they utilize crassulacean acid metabolism (CAM), a carbon-assimilation pathway that is inextricably tied to morphological succulence. Cacti and other succulents appear to reach their greatest development in the Sonoran, where they make use of the bimodal rainfall regime in order to maintain their tissue water charge. The Sonoran desert, for example, is where one finds the giant saguaro cactus (*Carnegiea gigantea*) and the organ-pipe cactus (*Stenocereus thurberi*). Also prominent in the warm deserts are leaf-succulent members of the family Agavaceae, such as *Agave deserti*. The succulent photosynthetic stems of

---

[2]The term *desert* is an anthropocentric term referring to dry places that are usually hot for at least a portion of the year. Arid and semiarid are classifications based on annual rainfall. Arid regions receive less than 250 mm of rain annually and semiarid regions between 250 and 450 mm.

most cacti are oriented vertically rather than horizontally. Such stems are thus designed not only to conserve water but also to minimize heat load. An elongated, or columnar, shape presents a minimum surface area to the sun when it is overhead during midday.

15.5.4 DESERT ANNUALS ARE EPHEMERAL

Annual plants are well represented in desert and semiarid regions. What perennials there are tend to be deep-rooted but widely spaced, so that they provide relatively little competition for annual species. Desert annuals are not xerophytic, but survive arid and semiarid conditions as dormant seed. Unlike perennials, which rely on physiological and morphological adaptations to help them survive periods of drought, desert annuals must germinate, establish their entire biomass, and set seed each season. Their entire life cycle must be completed during periods of adequate moisture and, because those wet period may be relatively short, these annuals are often referred to as **ephemerals**. Not surprisingly, the life cycle of desert annuals is keyed to seasonal distribution of precipitation. Where precipitation falls predominantly in the winter, the annual flora is active in the winter and early spring. In regions experiencing predominantly summer rains, the annual flora is predominantly summer-active. In those regions of the Sonoran Desert that experience bimodal precipitation, there are distinct winter- and summer-active annual flora. Desert winter annuals tend to exhibit C3 photosynthesis, reflecting the fact that winter/spring temperatures are lower and moisture is generally available over a longer period of time. Summer annuals, on the other hand, tend to be C4 species, reflecting the advantages offered by higher water-use efficiency in view of the short-lived water availability from summer rains.

The key to survival of desert annuals is a relatively rapid life cycle. Their seeds, able to survive in the soil through extended periods of drought, will quickly germinate when adequate water is available during the normal *wet* season. Desert annuals characteristically exhibit high growth rates, which allows them to reach maturity, flower, and set seed within the relatively brief period of time that moisture is available. High growth rates are supported by at least three factors: a high shoot–root ratio, high stomatal conductance (supporting rapid CO$_2$ uptake), and high photosynthesis rates. An example is *Cammisonia* claviformis, a Mojave Desert winter annual with relatively large leaves. Its roots are shallow, which allows them to absorb more efficiently light rainfall and drippings from nighttime condensation. Leaves tend to be amphistomatic (i.e., stomata are found on both surfaces), which, accompanied by a high stomatal conductance, supports a high rate of carbon dioxide uptake.

This chapter has indicated that the inherent plasticity of plants is reflected in the myriad adaptations to specific and, at times, extreme environments to which plant communities are exposed. The adaptations discussed not only have a significant impact on the biodiversity and geographical distribution of plant biomes but also influence local weather patterns.

SUMMARY

The physiological plasticity for acclimation to growth irradiance varies considerably between genotypes. Obligate shade adaptation appears to preclude acclimation to high light whereas obligate sun adaptation appears to preclude acclimation to extreme low light conditions. Plants with the C4 photosynthetic pathway are adapted to hot, dry climates. C4 species evolved a mechanism for avoiding the impact of photorespiratory carbon dioxide loss by concentrating carbon dioxide in the carbon-fixing cells. C4 plants exhibit a division of labor between mesophyll cells, which pick up carbon dioxide from the ambient air, and bundle-sheath cells, which contain the PCR cycle and actually fix carbon. Mesophyll cells contain the enzyme PEP carboxylase, which catalyzes the carboxylation of phospho*enol*pyruvate. The product C4 acid is transported into the bundle-sheath cell. There it is decarboxylated and the resulting carbon dioxide is fixed via the PCR cycle. C4 plants have a very low carbon dioxide compensation concentration and low transpiration ratio. This means C4 plants are able to maintain higher rates of photosynthesis at lower carbon dioxide levels, even when the stomata are partially closed to conserve water during periods of water stress. Plants that exhibit crassulacean acid metabolism (CAM) are adapted to hot, desert conditions. CAM plants maintain higher rates of photosynthesis in habitats with minimal access to water. CAM plants exhibit an inverted stomatal cycle, opening for carbon dioxide uptake at night (when water stress is lower) and closing during the day (when water stress is high). The carbon dioxide is stored as organic acids, again through the action of PEP carboxylase. Decarboxylation during the day furnishes the necessary carbon dioxide for photosynthesis. The water use efficiency of CAM is higher than that of C4 plants which, in turn, is higher than that of C3 plants. The difference in water use efficiency is an important factor that determines the geographical distribution of C3, C4 and CAM plants.

Plant biomes reflect myriad adaptations of plant communities to specific environments. Although the soils of tropical rainforests are extremely poor in nutrients, these biomes exhibit the greatest plant biodiversity. Since water is not limiting in rainforest biomes, the extent of evapotranspiration from tropical rainforest biomes is so significant that it affects weather patterns. Desert perennials are adapted to maximize photosynthesis but at the same time minimize heat load and water loss under arid conditions. They exhibit rather slow rates of vegetative growth. In contrast to desert perennials, desert annuals are ephemeral. They exploit short, episodic rainfall to germinate, grow rapidly, senescence, and produce seed to survive the next dry spell.

CHAPTER REVIEW

1. Define adaptation. List five traits that are associated with shade adaptation.

2. Trace the path of carbon in a typical C4 leaf, from entry into the leaf through the stomata to its export in the vascular tissue. How does this differ from the C3 pathway?

3. A distinctive feature of the bundle-sheath cells in a typical C4 leaf is a high density of chloroplasts. What advantage does this offer? Why is it advantageous to have the PCR cycle located in the bundle-sheath cells in a C4 leaf?

4. Review the ecological significance of the C4 cycle. In what situations can a C3 species be more productive than a C4 species?

5. In what significant way does crassulacean acid metabolism (CAM) differ from C4 metabolism?

6. Under which conditions would C3 plants outcompete C4 and CAM plants? Why is this so?

7. Define biome. How does the extent of evapotranspiration affect weather patterns?

8. Would you predict that a tropical rainforest has a greater proportion of C3 versus C4 species? Explain why this would be so.

9. How do adaptations of perennial desert plants to arid conditions differ from those of desert annuals?

FURTHER READING

Anderson, J. M. 1986. Photoregulation of the composition, function and structure of thylakoid membranes. *Annual Review of Plant Physiology* 37:93–136.

Buchanan, B. B., W. Gruissem, R. L. Jones. 2000. *Biochemistry and Molecular Biology of Plants*. Rockville MD: American Society of Plant Physiologists.

Bush, M. B., J. R. Flenley. 2007. *Tropical Rainforest Responses to Climate Change*. Berlin: Springer.

Cushman, J. C., H. J. Bohnert. 1999. Crassulacean acid metabolism: Molecular genetics. *Annual Review of Plant Biology* 50:305–332.

Edwards, G. E., V. R. Franceschi, E. V. Voznesenskaya. 2004. Single-cell C4 photosynthesis versus the dual-cell (Kranz) paradigm. *Annual Review of Plant Biology* 55:173–196.

Leegood, R. C., T. D. Sharkey, S. von Caemmerer. 2000. *Photosynthesis: Physiology and Metabolism. Advances in Photosynthesis*, Vol. 9. Dordrecht: Kluwer.

Matsuoka, M., R. T. Furbank, H. Fukayama, M. Miyao. 2001. Molecular engineering of C4 photosynthesis. *Annual Review of Plant Biology* 52:297–314.

Nobel, P. S. 2005. *Physicochemical and Environmental Plant Physiology*. Burlington: Elsevier Science & Technology.

Potters, G., T. P. Pasternak, Y. Guisez, K. J. Palme, M. A. K. Jansen. 2007. Stress-induced morphogenic responses: Growing out of trouble? *Trends in Plant Science* 12:98–105.

Robichaux, R. H. 1999. *Ecology of Sonoran Desert Plants and Plant Communities*. Tuscon: University of Arizona Press.

Sack, L., N. M. Holbrook. 2006. Leaf hydraulics. *Annual Review of Plant Biology* 57:361–381.

Smith, S. D., R. K. Monson, J. E. Anderson. 1997. *Physiological Ecology of North American Desert Plants*. Berlin: Springer Verlag.

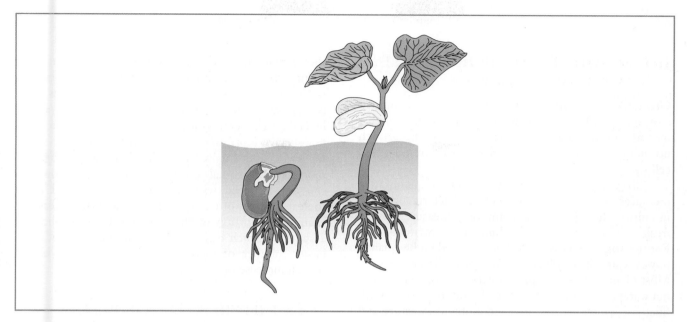

16

Development: An Overview

The development of a mature plant from a single fertilized egg follows a precise and highly ordered succession of events. The fertilized egg cell, or zygote, divides, grows, and differentiates into increasingly complex tissues and organs. In the end, these events give rise to the complex organization of a mature plant that flowers, bears fruit, senesces, and eventually dies. These events, along with their underlying genetic programs and biochemistry, and the many other factors that contribute to an orderly progression through the life cycle, constitute development. Understanding development is one of the major goals of plant biologists and looking for changes in the pattern of development is one way to enhance that understanding

In this chapter we highlight major stages in plant development, including

- the meaning of the terms growth, differentiation, and development,

- the nature of plant meristems,

- the development, maturation, and germination of seeds,

- the pattern of development from embryo to adult, and

- senescence and programmed cell death.

16.1 GROWTH, DIFFERENTIATION, AND DEVELOPMENT

Three terms routinely used to describe various changes that a plant undergoes during its life cycle are **growth, differentiation**, and **development**. The chapters that follow focus on factors that regulate growth, differentiation, and development, so it is necessary to clarify how these three terms are used.

16.1.1 DEVELOPMENT IS THE SUM OF GROWTH AND DIFFERENTIATION

Development is an umbrella term, referring to the sum of all of the changes that a cell, tissue, organ, or organism goes through in its life cycle. Development is most visibly manifested as changes in the form of an organ or organism, such as the transition from embryo to seedling, from a leaf primordium to a fully expanded leaf, or from the production of vegetative organs to the production of floral structures.

16.1.2 GROWTH IS AN IRREVERSIBLE INCREASE IN SIZE

Growth is a *quantitative* term, related only to changes in size and mass. For cells, growth is simply an irreversible increase in volume. For tissues and organs, growth normally reflects an increase in both cell number and cell size.

Growth can be assessed by a variety of quantitative measures. Growth of cells such as bacteria or algae in culture, for example, is commonly measured as the fresh weight or packed cell volume in a centrifuge tube. For the **angiosperms** and other so-called higher plants, however, **fresh weight** is not always a reliable measure. Most plant tissues are approximately 80 percent water, but water content is highly variable and fresh weight will fluctuate widely with changes in ambient moisture and the water status of the plant. **Dry weight**, determined after drying the material to a constant weight, is a measure of the amount of protoplasm or dry matter (i.e., everything but the water). Dry weight is used more often than fresh weight, but even dry weight can be misleading as a measure of growth in certain situations. Consider the example of a pea seed that is germinated *in darkness*. In darkness, the embryo in the seed will begin to grow and produce a shoot axis that may reach 10 to 12 inches in length. Although we intuitively sense that considerable growth has occurred, the total dry weight of the seedling plus the seed will actually decrease compared with the dry weight of the seed alone prior to germination. The dry weight decreases in this case because some of the carbon stored in the respiring seed is lost as carbon dioxide. In the light, this lost carbon would be replaced and even augmented by photosynthesis but photosynthesis doesn't operate in darkness. In a situation such as this, either fresh weight or the length of the seedling axis would be a better measure of growth. Length, and perhaps width, would also be suitable measures for an expanding leaf. Length and width would not only provide a measure of the amount of growth, but a length-to-width ratio would also provide information about the pattern of leaf growth.

It should be obvious that many parameters could be invoked to measure growth, dependent to some extent on the needs of the observer. Whatever the measure, however, all attempts to quantify growth reflect a fundamental understanding that *growth is an irreversible increase in volume or size*.

While cell division and cell enlargement normally go hand-in-hand, it is important to keep in mind that they are separate events; growth can occur without cell division and cell division can occur without growth. For example, cell division is normally completed very early in the development of grass coleoptiles and the substantial elongation of the organ that follows is due almost entirely to cell enlargement. Years ago, it was shown that wheat (*Triticum* sp.) seeds could be made to germinate even after having been irradiated with gamma radiation sufficient to block both DNA synthesis and cell division. The result was a small seedling produced by cell enlargement alone. Such seedlings generally did not survive more than two or three weeks but, except for having abnormally large cells, their morphology was more or less normal. On the other hand, cell division can also proceed without cell enlargement. During the early stages of embryo development in flowers, for example, a portion of the **embryo sac** goes through a stage in which cell division continues to produce a larger number of increasingly smaller cells, with no overall increase in the size of the embryo sac.

16.1.3 DIFFERENTIATION REFERS TO QUALITATIVE CHANGES THAT NORMALLY ACCOMPANY GROWTH

Differentiation refers to differences, other than size, that arise among cells, tissues, and organs. Differentiation occurs when cells assume different anatomical characteristics and functions, or form patterns. Differentiation begins in the earliest stages of development, such as, when division of the zygote gives rise to cells that are destined to become either root or shoot. Later, unspecialized parenchyma cells may differentiate into more specialized cells such as xylem vessels or phloem sieve tubes, each with a distinct morphology and unique function. Differentiation does not lend itself easily to quantitative interpretation but may be described as a series of *qualitative*, rather than quantitative, changes. Finally, although growth and differentiation are normally concurrent events, examples abound of growth without differentiation and differentiation without growth.

Differentiation is a two-way street and is not determined so much by cell lineage as by cell position with respect to neighboring cells. Thus, even though some plant cells may appear to be highly differentiated or specialized, they may often be stimulated to revert to a more embryonic form. For example, cells isolated from the center of a tobacco stem or a soybean **cotyledon** and cultured on an artificial medium may be stimulated to reinitiate cell division, to grow as undifferentiated **callus** tissue, and eventually to give rise to a new plant (Figure 16.1). It is as though the cells have been genetically reprogrammed, allowing them to reverse the differentiation process and to differentiate along a new and different path. This ability of differentiated cells to revert to the embryonic state and form new patterns without an intervening reproductive stage is called **totipotency**. Most living plant cells are totipotent—something akin to mammalian stem cells—and retain a complete genetic program even though not all of the information is used by the cell.

FIGURE 16.1 **Shoot regeneration in callus culture. A piece of pith tissue from the center of a tobacco stem was explanted onto a medium containing mineral salts, vitamins, sucrose, and hormones. The tissue proliferated as an undifferentiated callus (left) for several weeks before regenerating new plantlets (right). These plantlets can be eventually planted into soil and will produce a mature tobacco plant.**

at any given time. In this sense, development does not reflect a progressive loss of genetic information, only the selective use of that information in order to achieve particular developmental ends.

Not all cells are totipotent. Highly specialized cells whose development has been locked in, such as by exceptionally thick and rigid secondary cell walls or severely modified protoplasts, are not capable of renewed differentiation. On the other hand, it is probable that all *tissues* contain at least some potentially totipotent *cells*—cells that have the morphogenetic potential of a zygote. Plant development proceeds in an orderly fashion because that potential is carefully limited. When those limitations are removed, totipotent cells simply revert to the zygotic state and begin the developmental program anew.

There is increasing evidence that the genetic program being read in a particular cell may also depend on the position of that cell with respect to other cells and tissues and the inputs it receives from its neighbors. The position of a cell determines its interaction with its neighbors as well as its relationship to nutrient and hormone gradients. The importance of position has recently been demonstrated in a study of young *Arabidopsis* roots. *Arabidopsis* roots lend themselves well to this kind of study because of their relatively simple structure. The root tip is comprised of a single layer each of epidermis, cortex, endodermis, and pericycle surrounding a vascular bundle. This simplified organization makes it easier to track individual cells than is possible in more complex roots with multiple cell layers. Root cells were labeled with marker genes that were expressed differently in vascular and root cap cells and then a laser microscope was used to surgically remove cells in a region adjacent to the root cap called the quiescent center. The dead cells were displaced toward the root tip and replaced by

daughters of adjacent vascular cells. In their new position, the former vascular cells expressed the root cap marker. This is a strong indication that the reading of the genetic program in a cell is at least to some extent determined by positional information rather than cell lineage.

16.2 MERISTEMS ARE CENTERS OF PLANT GROWTH

Unlike animals, which are characterized by a generalized growth pattern, plant growth is limited to discrete regions where the cells retain the capacity for continued cell division. These regions are called **meristems** (Gk. *merizeim*, to divide). Two such regions are the **apical meristems** located at the tips of roots and stems. These regions of active cell division are responsible for **primary growth**, or the increase in the *length* of roots and stems.

The tip of the root is covered by a **root cap**, which provides mechanical protection for the meristem as the root grows through the abrasive soil medium. The root cap also secretes polysaccharides, which form a mucilaginous matrix called **mucigel**. Mucigel lubricates the root tip as it moves through the soil. The root cap along with its coating of mucigel is also involved in perception of gravity by roots (Chapter 23). The **root apical meristem (RAM)** is a cluster of dividing cells located at the tip of the root just behind the root cap (Figure 16.2). Each time a cell in the meristem divides, one of the two daughter cells will be retained to continue cell division while the second daughter cell proceeds to elongate, thus increasing the length of the root and pushing the root tip through the soil. In the center of the meristem is a region of slowly dividing cells called the **quiescent zone**. Cell divisions responsible for new tissues in the elongation root and regeneration of the root cap take place around the periphery of the quiescent zone.

The **shoot apical meristem (SAM)** is structurally more complex than the root apical meristem. This is understandable because in addition to producing new cells that elongate and extend the length of the axis of the shoot, the shoot apical meristem must also form primordia that give rise to lateral organs such as leaves, branches, and floral parts. At the same time it must perpetuate itself by maintaining a small population of undifferentiated, dividing cells. Similar to the root apical meristem, each time a cell divides in the SAM, one daughter cell is left behind to elongate and move the shoot apex forward while the other daughter cell remains within the meristem to continue dividing.

The meristem of a typical dicot shoot is a small, shiny dome that can be seen in detail with the aid of an electron microscope (Figure 16.3). A microscopic

FIGURE 16.2 **Schematic diagram of a young *Ara-bidopsis* root tip showing the principal regions of the root apical meristem.**

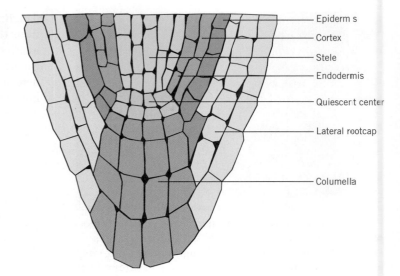

Epiderm s

Cortex

Stele

Endodermis

Quiescer t center

Lateral rootcap

Columella

examination of a typical SAM reveals that the cells of the meristem are usually organized into two distinct regions. Cells in the outermost two to four layers, called the **tunica**, undergo only anticlinal divisions, or divisions perpendicular to the surface of the meristem. The tunica thus contributes only to surface growth and its products contribute to the peripheral tissues of the stem. Underlying the tunica is the **corpus**, a body of cells that divide in various planes and contribute to the bulk of the shoot.

One of the ongoing mysteries of plant development is why and how leaves and branches form where they do. Part of the reason for this uncertainty is that a lot of things happen at the apex in rapid succession and it is difficult to separate these various events. Newer

methods for the study of gene expression in situ, however, are beginning to provide some insights as to which genes appear to be turned on and which genes are turned off when the lateral appendages begin to develop (see Chapter 20).

Leaves form as small swellings or **primordia** on the lateral flanks of the meristem. In some plants a pair of leaf primordia will arise on opposite sides of the meristem with successive pairs arising at 90 degrees to the last. In other plants, the leaf primordia arise in a spiral pattern. The points at which the leaf primordia form are referred to as nodes. In its early stages, the leaf primordium develops as a peg-like extension. As a leaf primordium elongates, however, marginal meristems develop on opposite sides of the primordium, leading

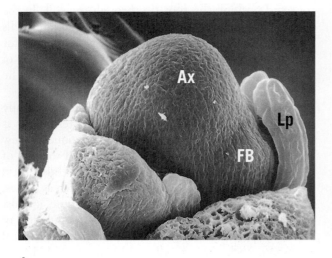

A.

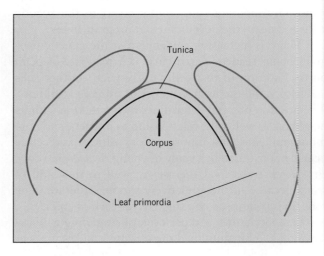

B.

FIGURE 16.3 **Shoot apical meristem. (*A*) Scanning electron micrograph of a flowering meristem of *Brassica napus*. Ax, apex; Lp, leaf primordium; FB floral bud (*B*). Schematic cross-section of a shoot apical meristem, showing position of the tunica, corpus, and leaf primordia. (A from Sawhney, V. K., P. L. Polowick. 1986. *American Journal of Botany* 73:254–263. With permission of the *American Journal of Botany*. Original photograph kindly provided by V. K. Sawhney and P. L. Polowick.)**

to lateral growth that gives rise to the typical flattened blade. As each leaf primordium forms, a bud primordium is also formed in the axil where the leaf joins the stem. These axillary buds will eventually give rise to branches. While all this has been going on, cells in the regions between successive leaves, or the internodes, continue to divide and elongate, thereby increasing the length of the stem.

Tissues that are derived directly from the root and shoot apical meristems are called **primary tissues**. The stems and roots of woody plants, however, grow in diameter as well. An increase in diameter results from the activity of a meristem called the **vascular cambium**. Tissues laid down by the vascular cambium are called **secondary tissues**, so the vascular cambium is responsible for **secondary growth**. The primary tissue of roots and shoots contains a central core of vascular, or conducting, elements. Characteristically, the xylem lies toward the center of the vascular core and the phloem lies at the outer edge of the core (Figure 16.4). The vascular cambium develops between the xylem and phloem and produces new xylem toward the inside and new phloem toward the outside. Because of its heavy cell walls and eventual lignification, xylem is a rigid and long-lasting tissue that eventually occupies the bulk of most woody stems or trunks. Phloem is a more fragile tissue and with each year's new growth the previous year's cells tend to be pushed outward and crushed. As a result, the xylem continues to transport water and minerals for several years, but a large tree seldom has more than one year's worth of functioning phloem.

16.3 SEED DEVELOPMENT AND GERMINATION

The life of an individual plant begins when an egg nucleus in the maternal organs of a flower is fertilized by a sperm nucleus to form a **zygote**. Growth and differentiation of the zygote produces an **embryo** contained within a protective structure called a **seed**. Under appropriate conditions, the embryo within the seed will renew its growth and will continue to develop into a mature plant.

16.3.1 SEEDS ARE FORMED IN THE FLOWER

Flowers appear to vary enormously in structure, yet all flowers follow the same basic plan. A generic flower consists of four whorls or circles. The two outermost whorls—the sepals and petals—are vegetative structures; and the two innermost—the stamens and pistil—are the male and female reproductive structures, respectively. At the base of the pistil, or female structure, is the ovary, which contains one or more ovules (Figure 16.5).

Within each ovule, a single large diploid cell, called the megaspore mother cell, undergoes mitosis to produce four megaspore cells. Only one megaspore cell survives and that cell undergoes meiotic division to produce an embryo sac with eight haploid nuclei. Subsequent cell division produces a mature embryo sac in which the eight nuclei are segregated into seven cells (Figure 16.6). One of those cells is the egg. Another is the large central cell containing two polar nuclei.

The male structures, or stamens, surround the pistil and consist of an anther perched on a stalk, or filament.

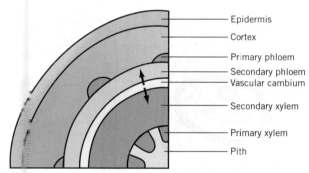

FIGURE 16.4 **A schematic transverse section through a one-year-old elderberry (*Sambucus*) stem showing the location of the secondary vascular cambium. The vascular cambium arises between the primary phloem and primary xylem and adds new, or secondary, xylem cells to the inside and new, or secondary, phloem cells to the outside (arrows). This is repeated annually to add girth to the stem. The secondary xylem develops heavy, lignified walls and becomes the woody tissue of the stem. The phloem is a soft tissue and each year's new growth crushes the previous year's phloem. When you peel the "bark" off a young stem, the vascular cambium is where the tissues separate.**

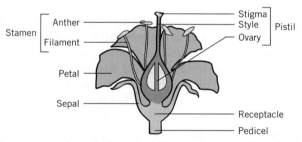

FIGURE 16.5 **A typical perfect flower consists of 4 whorls: sepals, petals, stamens, and the pistil. All four whorls are considered modified leaves. The sepals are often green or inconspicuous and the petals are brightly colored. In some flowers, the sepals and petals may both be colored. Pollen, containing the sperm nucleus, is produced in the anthers of the stamens. The female egg cells are produced in the ovary at the base of the pistil. Pollen is transferred to the stigma or stigmatic surface of the pistil, where it sends out a pollen tube that grows down the style and delivers the sperm nucleus to the egg.**

Box 16.1
DEVELOPMENT IN A MUTANT WEED

Many aspects of plant development have proven difficult to dissect using exclusively physiological and biochemical approaches. This is especially true of the complex metabolic sequences that connect a developmental event with the original signal that initiates that event. Metabolic events associated with a signal transduction pathway often represent a very small portion of all the biochemical processes within a plant and may not be discernible against this background of biochemical "noise."

Because all development can be traced back to the expression of genes, one way around this dilemma is to incorporate genetic mutations into a research program. The use of mutant genes to study metabolism and development has been a constructive approach since the pioneering work of G. Beadle and E. Tatum with *Neurospora* mutants in the 1940s. Over the years, mutants in maize, tomato, pea, and a host of other plants have provided important insights into normal development. More recently, however, recombinant DNA techniques have enabled us to ask questions about genes and developmental events in ways never before possible. The principal strategy is to identify and genetically map a mutation that modifies the physiological response of interest. It is then possible to physically isolate and clone the wildtype gene and, based on its nucleotide sequence, deduce the amino acid sequence of the encoded protein that normally operates in the pathway. This provides investigators with useful information about the function of the wildtype gene and a means to further probe the step that it controls. Furthermore, by studying interactions between mutant genes and multiple input signals, it is possible to dissect complex interactions between different pathways.

One organism that has come to dominate this new approach to the study of plant development is *Arabidopsis thaliana*, a member of the Brassicaceae, or mustard family. *A. thaliana* (mouse-ear cress) is a small, herbaceous weed that grows in dry fields and along roadsides throughout the temperate regions of the northern hemisphere. Most *Arabidopsis* are winter annuals. Their seeds germinate in the fall, forming an overwintering vegetative rosette. In the spring, as the days grow longer, the stem elongates and flowers.

There are many reasons why *Arabidopsis* has become such a popular experimental tool and model system for molecular genetic experiments on development. *Arabidopsis* is easily grown in the laboratory and its life cycle is complete in 5 to 6 weeks. It is also easily crossed or self-fertilized and produces prodigious numbers of seeds (up to 10,000 per plant). The *Arabidopsis* genome, which has recently been completely sequenced, is one of the smallest known plant genomes (approximately 10^8 nucleotides, or 25,498 genes). Finally, mutants are easily induced by treating the seeds with chemical mutagens. The surviving seeds are then germinated and mutant progeny are recovered for analysis. The small size of the plant together with its rapid growth and fecundity make it easy to screen for mutants with reasonable frequency.

One must always be cautious with so-called model systems. What is learned about *Arabidopsis*, for example, may well extrapolate to all brassicas, but not necessarily to all plants. Nonetheless, *Arabidopsis* is an ideal plant for doing molecular genetic and developmental experiments. The isolation and study of mutants in *Arabidopsis* has already made significant contributions to many areas of plant development. Moreover, what is learned about physiology and development in *Arabidopsis* gives investigators important clues about what to look for in other plants.

The initial step in germination of seeds is the uptake of water and rehydration of the seed tissues by the process of **imbibition**. Like osmosis, imbibition involves the movement of water down a water potential gradient. Imbibition differs from osmosis, however, in that it does not require the presence of a differentially permeable membrane and is driven primarily by surface-acting or **matric forces**. In other words, imbibition involves the chemical and electrostatic attraction of water to cell walls, proteins, and other hydrophilic cellular materials. Matric potential, like osmotic potential, is always negative. Hydration causes a swelling of the imbibing material, which may generate substantial forces (called **imbibition pressure**). Imbibition pressure developed by a germinating seed will cause the seed coat to rupture thus permitting the embryo to emerge.

One of the first events to follow the rapid influx of water is the leakage of low molecular weight solutes from the seed tissues. This apparently reflects an unstable configuration of cell membranes related to the severe desiccation. The leakage does not persist, however, as the cells quickly repair themselves and a more stable membrane configuration returns shortly after rehydration.

Imbibition of water is followed by a general activation of seed metabolism within minutes of water entering the cells, initially utilizing a few mitochondria and respiratory enzymes that had been conserved in the dehydrated state. Renewed protein synthesis is also an early event, utilizing preexisting RNA transcripts and ribosomes, as existing organelles are repaired and new organelles are formed. This is followed closely by (1) the release of hydrolytic enzymes that digest and mobilize the stored reserves, and (2) renewed cell division and cell enlargement in the embryonic axis. Detailed respiratory pathways have been studied thoroughly in only a few species of seeds, but it is believed that glycolysis and the citric acid cycle are active to varying degrees in most, if not all, seeds. These pathways produce the carbon skeletons and ATP that are required to support growth and development of the embryo. The pentose phosphate pathway is also important in seeds as it produces the reducing potential (in the form NADPH) required for the reductive synthesis of fatty acids and other essential cellular constituents. The pentose phosphate pathway also generates intermediates in the synthesis of aromatic compounds and perhaps nucleic acids. Seeds that store carbon reserves principally in the form of fats and oils will carry out the synthesis of hexose sugars via gluconeogenesis.

The mobilization of stored carbon in seeds has been most extensively studied in cereals (see Chapter 19). This is because the endosperm of cereals has long served as a principal source of nutrition for man and domesticated animals, as well as a basic feedstock in the brewing industries. These needs have provided a strong incentive for research into the mobilization of starch reserves in cereal grains. In nonendospermic dicot seed such as the legumes (peas, beans), the initial stages of radicle elongation appear to depend on reserves stored in the tissues of the radicle itself. Later, carbon reserves are mobilized from the cotyledons and transported to the elongating axis.

In most species, germination is considered complete when the radicle emerges from the seed coat. Radicle emergence occurs through a combination of cell enlargement within the radicle itself and imbibition pressures developed within the seed. Rupture of the seed coat and protrusion of the radicle allows it to make direct contact with water and nutrient salts required to support further growth of the young seedling.

16.3.4 THE LEVEL AND ACTIVITIES OF VARIOUS HORMONES CHANGE DRAMATICALLY DURING SEED DEVELOPMENT

Seed development is characterized by often dramatic changes in the levels of the principal plant hormones (Figure 16.10). In most seeds, cytokinin (CK) levels are highest during the very early stages of embryo development when the rate of cell division is also highest. As the

A.

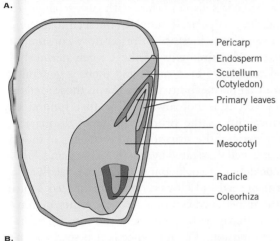

Pericarp
Endosperm
Scutellum (Cotyledon)
Primary leaves
Coleoptile
Mesocotyl
Radicle
Coleorhiza

B.

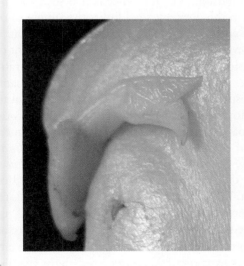

C.

FIGURE 16.9 (*A*) Maize (corn, (*Zea mays*)) kernel showing the location of the embryo underneath the pericarp. Each corn grain is actually an entire fruit and the seed coat has fused with the surrounding ovary wall to form a compound structure called the pericarp. (*B*) Cross section of a mature maize kernel (along the dotted line in A) showing the principal structures of the embryo. (*C*) A bean seed (*Phaseolus vulgaris*) showing the embryo and part of one cotyledon. The seed coat and one cotyledon have been removed in order to reveal the embryo.

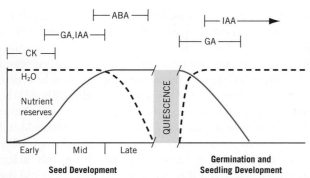

FIGURE 16.10 **The activity of hormones during seed development and germination. Cytokinins (CK) are present during the early stages of development, when cell division is most active. The concentration of cytokinins then declines while the concentrations of gibberellin (GA) and auxin (IAA) increase to support active cell enlargement. Abscisic acid (ABA) concentration increases in the later stages to prevent precocious germination. After a quiescent period, the release of gibberellins activates nutrient mobilization and IAA levels increase to stimulate cellular enlargement in the young seedling.**

cytokinin levels decline and the embryo enters a period of rapid cell enlargement and differentiation, both auxin and gibberellin (GA) levels increase. In the early stages of embryogenesis, there is little or no detectable abscisic acid (ABA). It is during the latter stages of embryo development, as GA and IAA levels begin to decline, that ABA levels begin to rise. ABA levels generally peak during the maturation stage when seed volume and dry weight also reach a maximum. The significance of these changes will become more apparent as we look at the specific effects of individual hormones in later chapters. Maturation of the embryo is characterized by cessation of embryo growth, accumulation of nutrient reserves, and the development of tolerance to desiccation.

16.3.5 MANY SEEDS HAVE ADDITIONAL REQUIREMENTS FOR GERMINATION

Many seeds will not germinate even though the minimal environmental conditions have been met. These seeds are said to be dormant and will not germinate until additional conditions have been met.

The most common causes of seed dormancy are the impermeability of the seed coat to water or oxygen or physiological immaturity of the embryo at the time the seed is shed from the mother plant. Immature seeds must undergo complex biochemical changes, collectively known as after-ripening, before they will germinate. After-ripening is usually triggered by low temperature, a mechanism that appears to ensure that the seed will not germinate precociously in the fall but will germinate when favorable weather returns in the spring.

Dormancy imposed by restricting the uptake of water and exchange of oxygen can be removed by mechanically disrupting or removing the seed coat, a process called **scarification**. In the laboratory, scarification may be accomplished with files or sandpaper. In nature, abrasion by sand, microbial action, or passage of the seed through the gut of an animal will accomplish the same end. Seed coats can be very tough. Uniformity and rate of germination of morning glory (*Pharbitis nil*), cotton, and some tropical legume seeds, for example, can be improved by soaking the dry seed in concentrated sulphuric acid for up to an hour. Scarification by passage through animal gut no doubt occurs as a result of the acidic conditions in the gut.

There is a considerable body of evidence to suggest that seed coats also interfere with gas exchange, oxygen uptake in particular. Removal of the seed coat often leads to a significant increase in respiratory consumption of oxygen. Measurements of the oxygen permeability of seed coats have been made and there is general agreement that permeability is very low in those seeds tested. However, it is not always clear that limited oxygen permeability is the primary cause of dormancy. The complexity of the situation and problems of interpretation are well illustrated by studies of the genus *Xanthium*, or cocklebur. A cocklebur contains two seeds: an upper, dormant seed and a lower, nondormant seed. Dormancy of the upper seed can be overcome either by removing the seed coat or by subjecting the intact seed to high oxygen tension. The inference is that seed coat permeability in the dormant seed limits the supply of oxygen to the embryo and thus prevents germination. However, several other observations have cast doubt on this conclusion. There are, for example, no measurable differences between the dormant and nondormant seed with respect to the permeability of the seed coat to oxygen. Moreover, the rate of oxygen diffusion through the seed coats is more than sufficient to support measured rates of oxygen consumption by the embryos inside. Clearly, dormancy of the upper seed in *Xanthium* cannot be due to limited permeability of the seed coat to oxygen. Why then, do the upper, dormant seeds require a higher oxygen level to elicit germination? It appears that the seed coat is a barrier, not to the uptake of oxygen but to the removal of an inhibitor from the embryo. Aqueous extracts of *Xanthium* seeds have revealed the presence of two unidentified inhibitors, based on tests of the extracts in a wheat coleoptile elongation assay. The same two inhibitors are found in diffusate collected from isolated embryos placed on a moist medium, but not in diffusates from seeds surrounded by an intact seed coat. Thus germination in the dormant seed appears to be prevented by the presence of these inhibitors and the seed coat serves as a barrier that prevents those inhibitors from being leached out. The apparent oxygen requirement can be explained by the observation that high oxygen tension reduces the quantity of an extractable inhibitor, presumably by some oxidative degradation.

The seed coat may contain inhibitors that prevent growth of the embryo but the role of inhibitors in seed dormancy is not clear. Along with hormones such as auxins and gibberellins, a large number of potential inhibitors have been identified in seeds, fruits, and other dispersal units. These include the hormone abscisic acid, unsaturated lactones (e.g., coumarin), phenolic compounds (e.g., ferulic acid), various amino acids, and cyanogenic compounds (i.e., compounds that release cyanide) characteristic of apple and other seeds in the family Rosaceae. The simple presence of an inhibitor does not, however, prove a role in dormancy. The inhibitors could be localized in tissues not directly involved in growth of the embryo or otherwise sequestered so as to preclude any role in preventing germination. Evidence in support of a role for inhibitors is generally limited to leaching experiments such as that described above for *Xanthium*. In some cases, dormancy can then be restored by exposing the leached seed to the inhibitor. In order to clearly establish whether an inhibitor has an active role in regulating germination, it is necessary to establish whether inhibitor levels in the seed correlate with the onset and termination of dormancy. In spite of the voluminous literature relating inhibitors to dormancy, there is very little critical support for a direct role. For the present, evidence for the imposition and maintenance of dormancy by inhibitors remains largely circumstantial.

The dormancy and germination of many seeds are also influenced by light and hormones. These factors will be discussed in later chapters.

16.4 FROM EMBRYO TO ADULT

The first structure to emerge when a seed germinates is the radicle. The radicle, which is the nascent primary root, anchors the seed in the soil and begins the process of mining the soil for water and nutrients. As the primary roots elongates, it gives rise to branch, or lateral, roots. Unlike the situation in the shoot apical meristem, lateral roots do not originate in the root apical meristem. Lateral root primordia originate in the **pericycle**, a ring of meristematic cells that surround the central vascular core, or **stele**, of the primary root. The growing lateral root works its way through the surrounding cortex, either by mechanically forcing its way through or by secreting enzymes that digest the cortical cell walls. Lateral root primordia arise in close proximity to the newly differentiated xylem tissue, which allows vascular elements developing behind the growing tip of the lateral root to maintain connections with the xylem and phloem of the primary root.

Emergence of the radicle is followed by elongation of the shoot axis. In some dicot seedlings, such as the bean (*Phaseous vulgaris*) the **hypocotyl** (*hypo*, below the cotyledons) is the first part of the axis to elongate. The hypocotyl is hooked so that it pulls rather than pushes the cotyledons and the enclosed first foliage leaves and shoot tip, called the **plumule**, up through the soil. This pattern is known as **epigeal** germination (Figure 16.11A). In other dicots, such as garden pea, the hypocotyl remains short and compact and the cotyledons remain underground. Instead, the **epicotyl** (*epi*, above the cotyledons) is the part of the shoot that elongates. The epicotyl in this case forms a hook so

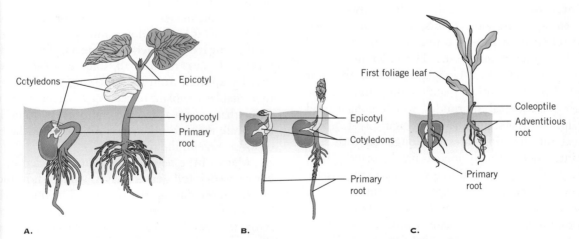

FIGURE 16.11 Germination and seedling development. (*A*) Stages in the germination of bean (*Phaseolus vulgaris*), an example of epigeal germination. (*B*) Hypogeal germination of pea (*Pisum sativum*). (*C*) Germination of corn (maize). See text for details.

that the clasping leaves of the plumule are pulled rather than pushed through the soil. This pattern of germination, where the cotyledons remain in the soil is known as **hypogeal** (Figure 16.11B). In either case, once the seedling axis breaks through the soil surface and emerges into the light, the hook straightens and the primary leaves unfold begin to expand. The principal objective of the hypocotyl or epicotyl hook is to bring the first leaves and meristem (the plumule) safely above the soil line in order to establish active photosynthesis. Without the hook, the plumule would no doubt be subject to considerable abrasion and damage as it moves through the soil.

A third example of seedling development is provided by corn (maize), oats, and other cereal grains (Figure 16.11C). In the seed, both the plumule, including the first leaves and meristem, and the radicle are surrounded by sheath-like structures—the **coleoptile** and the **coleorhiza**, respectively (see Figure 16.9B). The coleorhiza is the first structure to emerge from the pericarp. (In cereals, each grain is actually a fruit and the surrounding tissues are the mature ovary walls which function as a seed coat. In corn this is called the pericarp.) Growing rapidly, the radicle quickly penetrates the coleorhiza and assumes the function of the primary root. Meanwhile, the coleoptile is pushed upward by elongation of the mesocotyl or first internode. The coleoptile serves the same purpose as the plumular hook in dicots. It encases and protects the fragile first leaves until they emerge from the soil. Then the coleoptile splits open, allowing the leaves to grow into the sunlight.

Elongation of the shoot axis proceeds through a combination of cell division and enlargement of the cells laid down by the meristem. The rate and extent of elongation is subject to a variety of controls, including nutrition, hormones, and environmental factors such as light and temperature. The final height of a shoot is determined by the rate and extent to which **internodes**—the sections of stem between leaf nodes—elongate. In some plants, such as pea (*Pisum sativum*), elongation occurs primarily near the apical end of the youngest internode. The older internodes effectively complete their elongation before the next internode begins. In other plants, elongation may be spread through several internodes, which elongate and mature more or less simultaneously (Figure 16.12). Still others exhibit changing rates of elongation with successive internodes, usually increasing. In some plants, internodes fail to elongate, thus giving rise to the rosette habit in which all the leaves appear to originate from more or less the same point on the stem. This rosette habit is common in biennial plants (those that flower in the second year) such as cabbage and root crops such as carrot (*Daucus carota*) and radish (*Raphanus sativus*) before they reach the flowering stage. Failure of internode elongation is commonly related to low levels of the

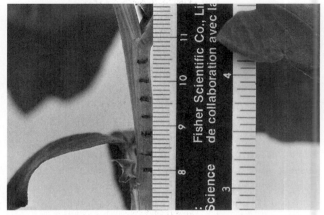

FIGURE 16.12 Internode elongation in broad bean (*Vicia faba*) over 48 hours. The initial spacing between marks was 2 mm. This internode has elongated more or less uniformly over its entire length.

plant hormone, gibberellin, since application of the hormone usually stimulates internode elongation in rosette plants.

16.5 SENESCENCE AND PROGRAMMED CELL DEATH ARE THE FINAL STAGES OF DEVELOPMENT

The final stage in the development of cells, tissues, and organs is **senescence**, an aging process characterized by increased respiration, declining photosynthesis, and an orderly disassembly of macromolecules. Senescing cells and tissues are metabolically very active—a number of metabolic pathways are turned off and new pathways, principally catabolic in character, are activated. Catabolism of proteins, for example, releases organic nitrogen and sulfur in the form of soluble amines, while nucleic acids release inorganic phosphate. Chlorophyll is broken down and lipids are converted to soluble sugars via **gluconeogenesis.** The products of these pathways are all small, soluble molecules that are readily exported from the senescing tissue. Senescence thus enables the plant to recover nutrients from cells or tissue that have reached the end of their useful life and reallocate them to other parts of the plant that survive or for storage in the roots.

Many of the new pathways are encoded by **senescence associated genes** (SAGs), which show increased transcription during senescence. The more than 50 SAGs that have been identified to date fall generally into one of two broad groups. One group includes genes that encode components of the system, such as the enzymes of the largely catabolic pathways that are newly activated during senescence. These include products such as ubiquitin, a range of proteolytic enzymes, and

metallothioneins—proteins that chelate metal ions. The second group of genes regulates the initiation of senescence or its rate of progress. Among this group are the so-called "stay-green" mutants characterized by delayed leaf senescence. An interesting consequence of the "stay-green" phenotype is the potential for higher crop yields. Because senescence of the leaf is delayed, the period of active photosynthetic output and, consequently, yield are extended.

Senescence is a normal consequence of the aging process and will occur even when the supply of water and minerals is maintained. For example, older leaves at the base of a plant commonly senesce in favor of younger leaves near the top. Flower petals senesce once the ovule has been fertilized and seed development is underway. The leaves of perennial plants senesce at the end of the growing season and export the nutrients to the roots for storage over the winter.

Senescence is one form of **programmed cell death (PCD)**, broadly defined as a process in which the organism exerts a measure of genetic control over the death of cells. PCD requires energy and is normally regulated by a distinct set of genes.

PCD is essential for normal vegetative and reproductive development. One example is the development of xylem tracheary elements. In order to function efficiently as a conduit for water transport, the protoplast of the developing tracheary element must die and be removed at maturity. PCD also operates in the formation of **aerenchyma**, a loose parenchymal tissue with large air spaces. Aerenchyma normally forms in the stems and roots of water lilies and other aquatic plants. These air spaces, created by a cell death program, provide channels for oxygen transport to the submerged portions of the plant. Even corn (*Zea mays*) and other terrestrial plants can be induced to form aerenchyma when subject to flooding.

In the development of unisexual flowers, primordia for both the male and female flowers are present in the early stages. One or the other then aborts via a cell death program, leaving only one type of organ to complete development.

PCD is also an important factor in plant responses to invading pathogens and abiotic stress. When a plant recognizes a pathogen, for example, host cells in the immediate area of the infection undergo PCD. This deprives the invading pathogen of living tissue and either slows or prevents it spread.

SUMMARY

Plants, like any other multicellular organism, build complexity by combining growth with differentiation. Growth is simply an irreversible increase in size that reflects an increase in cell number as well as the size of individual cells. Differentiation refers to the qualitative changes, or specialization in cell structure and function, that normally a accompany growth. The sum of growth and differentiation is development.

Growth in plants is limited to discrete regions called meristems. The principal meristems that give rise to the primary plant body are located at the apices of stems and roots. Secondary meristems in stems and roots are responsible for increases in girth.

The life of an individual plant begins with fertilization or the union of a haploid sperm nucleus (supplied by a pollen grain) with a haploid egg nucleus in the ovary, located at the base of the pistil. The resulting zygote goes through a series of stages to produce a mature embryo consisting of one or more cotyledons, a plumule consisting of primary leaves and a shoot apical meristem, and a radicle that is destined to become a primary root. The embryo develops within a nutritive tissue called the endosperm.

Seed germination is initiated when the seed takes up water to re-hydrate the dry tissues. Respiration is one of the first metabolic activities to be detected as nutrient reserves are mobilized from the endosperm or cotyledons and the embryo renews its growth.

Genetic control over the death of cells, called programmed cell death, is a normal component of development and gives rise to structures as diverse as mature xylem vessels and unisexual flowers.

CHAPTER REVIEW

1. Distinguish between growth, differentiation, and development. Can you give examples of each?

2. Describe the significance of meristems.

3. Where are the principle meristems located in plants? What is their contribution to development of the plant body?

4. If you had nailed an object to a young tree at a height of four feet off the ground and then returned after the tree had grown for several years, the object would still be four feet off the ground. Why has the object not moved further from the ground as the tree grew?

5. Where is the egg cell in a plant and how is the sperm delivered to effect fertilization?

6. Describe the process of seed formation from a fertilized egg cell?

7. Compare and contrast imbibition and osmosis.

8. What is a plumular hook and what function does it serve in young seedling development?

FURTHER READING

Bewley, J. D., M. Black. 1994. *Seeds: Physiology of Development and Germination*. New York: Plenum Press.

Bewley, J. D., M. Black, P. Halmer (eds.). 2006. *The Encyclopedia of Seeds: Science, Technology, and Uses*. CABI Publishing.

Buchanan B. B., W. Gruissem, R. L. Jones. 2000. *Biochemistry and Molecular Biology of Plants*. Rockville, MD: American Society of Plant Physiologists.

Esau, K. 1977. *Anatomy of Seed Plants*. New York: Wiley.

Raven, P. H., R. F. Evert, S. E. Eichhorn. 2005. *Biology of Plants*. 7th ed. New York: Bedford, Freeman & Worth.

Steeves, T. A., I. M. Sussex. 1989. *Patterns in Plant Development*. 2nd ed. Cambridge: Cambridge University Press.

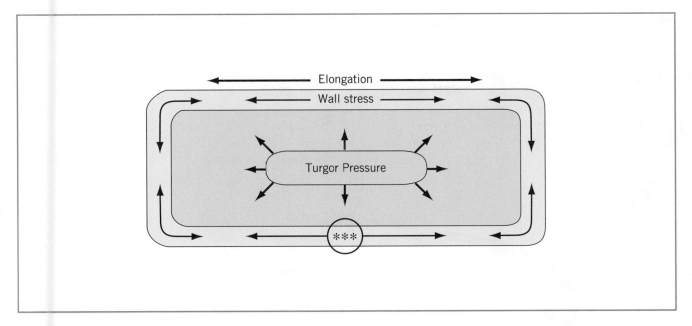

17

Growth and Development of Cells

Cells are the basic unit of life. The development of a whole plant, as intricate as it may seem, ultimately depends on the growth and coordinated development of individual cells and groups of cells. The growth and development of cells, both individually and collectively, is in turn directed by a variety of internal and external signals such as hormones, light, temperature, gravity, insect predation, disease, and even the position of a cell with respect to other cells. This chapter provides a general introduction to selected aspects of cellular development, including

- the structure of plant cell walls,
- the dynamics of cell division and cell growth, and
- an overview of signal perception and the signal chains that enable cells to respond to those signal.

17.1 GROWTH OF PLANT CELLS IS COMPLICATED BY THE PRESENCE OF A CELL WALL

Plant cells are characterized by a complex mixture of materials, called the **extracellular matrix (ECM),** that lie outside the plasma membrane. The ECM is dominated by the cell wall, which provides protection for the underlying protoplast and is ultimately responsible for maintaining cell shape and the structural integrity of the plant. Two types of cell walls are recognized: **primary walls,** which surround young, actively growing cells, and **secondary walls** that are laid down as the cells mature and are no longer growing.

17.1.1 THE PRIMARY CELL WALL IS A NETWORK OF CELLULOSE MICROFIBRILS AND CROSS-LINKING GLYCANS

The primary wall is very thin, measuring only a few micrometers in thickness. Its principal constituent is long, threadlike chains of $\alpha - 1 \rightarrow 4$-linked glucose units, called cellulose. The individual cellulose molecules are bundled in long, parallel arrays called **microfibrils** (Figure 17.1). Each microfibril is approximately 5 to 12 nm in diameter and contains, in higher plants, approximately 36 individual cellulose chains in cross-section. A single cellulose chain may contain as many as 3,000 or more glucose units but, because the chains begin and end at different places within the microfibril, an individual microfibril may contain several thousand cellulose chains and reach lengths of several hundred micrometers. Adjacent cellulose chains within a microfibril are held together by hydrogen bonds between hydroxyl (—OH) groups on adjacent glucose units. This bonding arrangement makes a microfibril

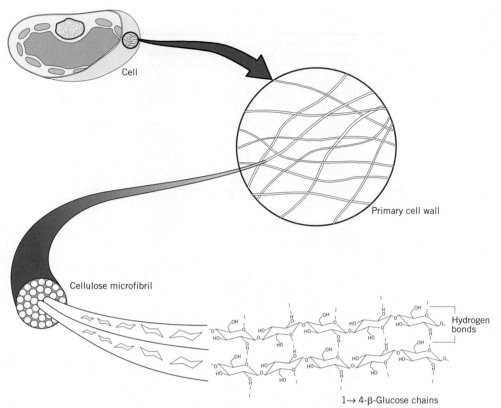

FIGURE 17.1 **The principal structural components of cell walls are cellulose microfibrils constructed of (1→4)-linked β-glucose chains. Adjacent cellulose chains within the microfibril are joined by intermolecular hydrogen bonds.**

very strong. In fact, the tensile strength of a microfibril, or its ability to withstand tension without breaking, is similar to that of a steel wire of the same size.

The orientation of microfibrils in the primary wall is more-or-less random (Figure 17.2), although in elongating cells they tend to orient parallel to the direction of growth. The cellulose microfibrils within the primary wall are held in position by cross-linking glycans.[1] Cross-linking glycans are noncellulosic polysaccharides that bind to the cellulose microfibrils, but are also long enough to bridge the distance between neighboring microfibrils and link them into a semi-rigid network (Figure 17.3). In the primary cell wall of all dicotyledonous (dicot) species and about one-half of the monocotyledons (monocot) species, the principal cross-linking glycans are xyloglycans (XyGs). Xyloglycans are linear (1→4)β-glucose chains (like cellulose), but with a 5-carbon sugar, xylose, linked to the oxygen at the carbon-6 position on many of the glucose units (Figure 17.4). In some species, some of the xylose units may be replaced by other sugars, such as arabinose, fructose, galactose, or glucuronic acid. Some side chains may be composed of two or three of these sugars instead of one.

17.1.2 THE CELLULOSE–GLYCAN LATTICE IS EMBEDDED IN A MATRIX OF PECTIN AND PROTEIN

As much as 35 percent of the primary wall consists of **pectins,** or pectic substances. Pectin is a complex, heterogeneous mixture of noncellulosic polysaccharides especially rich in galacturonic acid—the acid form of the sugar galactose. When the pectin chain is secreted into the wall space, most of the free carboxylic acid groups are esterified with methyl groups. Once in the wall, however, the enzyme pectin methylesterase cleaves some of the methyl groups, leaving the acid groups free to bind with calcium. The divalent calcium ions form cross-links between pectin chains and contribute to the stability of the cell wall. It is significant that calcium concentrations (and, consequently, calcium bridging) are kept low in the walls of actively growing cells.

Pectins are also the principal constituent of the **middle lamella.** The middle lamella lies between the primary walls of adjacent cells. It is the "cement" that

---

[1]Cross-linking glycans are the principal constituent of what was formerly referred to as "hemicellulose," a heterogeneous but generally uncharacterized mixture of cell wall polysaccharides characterized solely by their extractability into strong alkali solution.

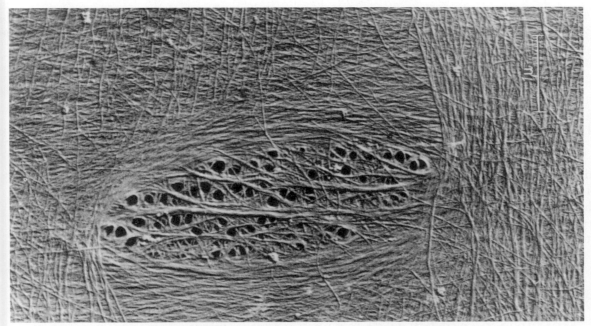

FIGURE 17.2 **The primary cell wall. An electron micrograph of the primary wall of a parenchyma cell from the coleoptile of an oat (*Avena*) seedling. Note the pores through which plasmodesmata pass. (From Böhmer, H. 1958. Untersuchungen über das Wachstum und den Feinbau der Zellwände in der *Avena*-Koleoptile. *Planta* 50:461–497, Figure 20. Copyright Springer-Verlag, Heidelberg.)**

holds the cells together. The softening of fruit as it ripens, for example, is due in part to the action of the enzyme polygalacturonase, which degrades the pectic substances in the middle lamella and loosens of the bonds between the cells.

Primary cell walls also contain approximately 10 percent glycoprotein which have an unusually high content of the amino acid hydroxyproline. These hydroxyproline-rich glycoproteins are called **extensin.** Although the precise function of extensin is unknown,

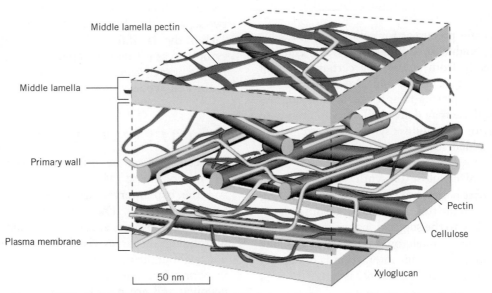

FIGURE 17.3 **A simplified model to illustrate how cellulose microfibrils, cross-linking xyloglucans, and pectic substances might be arranged in the cell wall. Not shown are extensin and other cell wall proteins. (From McCann, M. C., K. Roberts. 1991. Architecture of the primary cell wall. In: C. W. Lloyd (ed.), *The Cytoskeletal Basis of Plant Growth and Form.* Academic Press. Original figure courtesy of Dr. M. C. McCann.)**

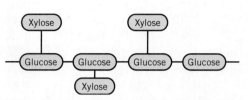

FIGURE 17.4 **A schematic diagram illustrating the structure of a simple cross-linking xyloglycan.**

it is thought to be one component of a structural network that adds strength to the wall and locks the wall into shape once the cell stops growing. Several other families of cell wall proteins rich in proline, glycine, or threonine have been described more recently, adding to the apparent complexity of cell wall structure.

17.1.3 CELLULOSE MICROFIBRILS ARE ASSEMBLED AT THE PLASMA MEMBRANE AS THEY ARE EXTRUDED INTO THE CELL WALL

Cellulose synthesis is catalyzed by cellulose synthase, a multimeric enzyme that is localized in the plasma membrane (Figure 17.5). Although active cellulose synthase has proven difficult to isolate, the enzyme complex can be visualized in electron micrographs of membranes prepared by a technique known as freeze-fracturing. At least in angiosperms, the enzyme appears to be organized in the form of a rosette composed of six subunits. The rosette not only synthesizes the cellulose chain, but assembles the chains into microfibrils and extrudes the microfibrils into the cell wall.

The immediate precursor for cellulose synthesis is uridine diphosphoglucose (UDP-Glu). UDP-Glu can be formed directly from sucrose by certain forms of the enzyme sucrose synthase, and also appears to be associated with the plasma membrane. It thus appears that sucrose synthase may deliver UDP-Glu directly to the catalytic site of the cellulose synthase rosette, where the glucose is added to the terminus of the growing microfibril.

The details of cellulose biosynthesis are not well understood, in part because of the difficulty in isolating active enzyme. Even the rosettes are absent from isolated membranes, leading to the suggestion that some form of interaction with the cytoskeleton is required to stabilize the enzyme in situ. When this interaction is disrupted, the complex dissociates and its activity is lost.

17.2 CELL DIVISION

The division of plant cells, like any eukaryotic cell, occurs in two steps: **mitosis** or **karyokinesis,** which

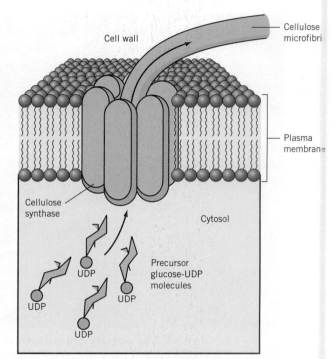

FIGURE 17.5 **A model for cellulose synthesis from precursor molecules, uridine diphosphoglucose (UDP-glucose). The enzyme, formally known as UDP-*glucose:(1→4)-β-glucan glucosyltransferase*, is believed to be located in the plasma membrane where it simultaneously synthesizes the cellulose chains, assembles them into microfilaments, and extrudes the microfilaments into the cell wall.**

results in the creation of two new nuclei, and **cytokinesis,** the subsequent division of the cytoplasm to create two separate daughter cells. Between divisions, however, each of the daughter cells must increase the amount of cytoplasm as it enlarges, replicate its DNA, and prepare for the next division. This sequence of events is known as the **cell cycle.**

17.2.1 THE CELL CYCLE

The cell cycle can be described as four distinct phases: the mitotic phase (M), the DNA synthesis phase (S), and the two intervening phases or "gaps" (G1 and G2) (Figure 17.6). The M phase is the *division phase* in which chromosomes are condensed and their distinctive morphology becomes visible with a light microscope. Throughout the other three phases, collectively referred to as *interphase*, the chromosomes are fully extended and uncoiled and are not visible with a light microscope. During S phase, DNA replication leads to the formation of two identical copies of the chromosomes. In the G2 phase, the chromosomes begin to condense and the cell assembles the rest of the machinery necessary to move the chromosomes apart during mitosis.

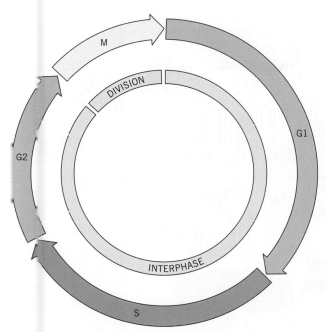

FIGURE 17.6 Phases of the cell cycle. M is the nuclear division phase, or mitosis. During mitosis the chromosomes condense and the sister chromatids are separated. S is the phase during which the chromosomes are diffuse, but DNA synthesis occurs. Mitosis and the synthesis phases are separated by two gaps in time. The progress of the cell cycle is normally controlled at either the G1 to S transition, or the G2 to M transition.

The time required to complete a cell cycle is highly variable, depending on the cell type and various developmental cues, but the length of the S phase and the M phase remain roughly constant for a given cell type. G1 is the most variable phase and may account for the major portion of a cell's life span.

In order for development to proceed in an orderly manner, the timing and rate of cell division and consequently entry into the cell cycle must be precisely controlled. The required control is achieved through a complex interplay of various kinases, phosphatases, and proteases that respond to both intrinsic and external signals. Central to cell cycle control are the activities of **cyclin-dependent kinases (CDKs)**. CDKs are composed of two subunits—one subunit functions as a catalyst and the other serves to activate the complex. The activating subunit is a small, unstable protein called **cyclin**, so named because when they were first discovered in *Xenopus* eggs, it was observed that during each cell cycle the concentration rose steadily from zero and then suddenly collapsed.

In plants, there are at least four classes of cyclins (A, B, D, and H) and the association of CDKs with specific cyclins is a key regulatory factor. Equally important is the availability of cyclins as the levels of the activating subunit oscillate during the cell cycle due to regulated synthesis and degradation. Activation of CDKs also

requires phosphorylation of a specific threonine residue (Thr160) on the catalytic subunit by a CDK-activating kinase (CAK), while phosphorylation of other threonine and tyrosine residues (Thr14, Tyr15) inhibits CDK activity. Phosphatase enzymes are available to remove the inhibitory phosphate groups at the appropriate time. The two principal sites of CDK control appear to be at the G1 to S or G2 to M transitions. It is assumed that the function of an active CDK complex is to phosphorylate and thus activate other proteins involved in the synthesis of DNA (S phase) or initiation of mitosis (M phase).

CDK inhibitors (CKI) also play a pivotal role by arresting cell division when necessary, such as at the end of a developmental program or in response to hormonal or environmental cues. Operating principally at the G1 to S transition, a CKI will bind with an active CDK complex in order to prevent it from phosphorylating substrate. When conditions are appropriate for the cell cycle to continue, the CKI can be removed by proteolysis. First, CKI is marked for degradation by attachment of ubiquitin (See Box 17.2). The ubiquitinated CKI is then degraded, the CDK complex reactivated, and the G1 to S transition is allowed to proceed.

17.2.2 CYTOKINESIS

Cytokinesis in animal cells is relatively simple, involving a constriction of the cell membrane which advances toward the center until the one cell becomes two cells. The existence of the cell wall prevents a similar pattern of cell division in plant cells. Plant cells must instead construct new membranes and a new *extracellular matrix*, including a middle lamella and two new primary walls, *inside* a living cell.

The formation of the new walls begins in the late mitotic or M phase of the cell cycle, after the two sets of chromosomes have separated and moved toward opposite poles in the cell and the new nuclear membranes have begun to form. At this point, the mitotic spindle disappears and the microtubules that had made up the spindle disassemble and then reassemble to form a cluster of interdigitating microtubules oriented perpendicular to the plane of the new crosswall (Figure 17.7) (Box 17.1). Called the **phragmoplast**, these microtubules serve to direct the movement of small secretory vesicles, derived from nearby **Golgi complexes**, into the equatorial region of the cell. The vesicles begin collecting in the center of the cell where they align along the equatorial plane and begin to fuse with one another. The resulting aggregate of vesicles is called the **cell plate.**

As newly arrived vesicles continue to add to the cell plate, the plate grows outward in all directions until its leading edge makes contact with the existing plasma membrane surrounding the parent cell. Fusion of the plate membranes with the parent cell membrane

FIGURE 17.7 **The phragmoplast and cell plate in a dividing plant cell.** Following nuclear division, tubulin subunits originating from remnants of the mitotic spindle reorganize as microtubules oriented perpendicular to the plane of cell division. These microtubules, collectively called the phragmoplast, serve to track small, Golgi-derived secretory vesicles into the equatorial plane of the cell. The vesicles fuse to form the cell plate, which gradually grows outward toward the lateral membranes of the dividing cell. The growing cell plate eventually fuses with the lateral membranes. The membranes of the cell plate form the new plasma membranes of the daughter cells and the contents of the original Golgi vesicles form the middle lamella between them. Once membrane fusion is complete, each daughter cell completes a new crosswall by depositing a new primary wall.

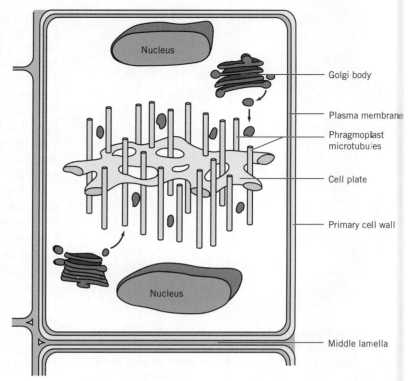

effectively completes the separation of the two daughter cells. The interior of the plate—between the two membranes—is filled with the pectic substances delivered by the Golgi vesicles and that now make up the middle lamella between the daughter cells. Cytokinesis is complete when cellulose synthase complexes are inserted into the new membranes and each daughter cell deposits a new primary wall adjacent to the middle lamella.

17.2.3 PLASMODESMATA ARE CYTOPLASMIC CHANNELS THAT EXTEND THROUGH THE WALL TO CONNECT THE PROTOPLASTS OF ADJACENT CELLS

When the cell plate forms, membrane fusion is not complete. This leaves some locations where cytoplasm continuity is maintained between daughter cells. As the cellulose is laid down and the new wall increases in thickness, these connections form membrane-encased channels called **plasmodesmata** (sing. *plasmodesma*) (Figure 17.9). The membrane that encases the channel is a continuation of the plasma membranes from adjacent cells. Running through the center of the plasmodesma is a second membranous tube—a tube within a tube—called the desmotubule. The desmotubule is formed as an extension of the endoplasmic reticulum that was entrapped during the formation of the cell plate. A sleeve of cytosol fills the space between the desmotubule and plasmodesmata itself.

Plasmodesmata are not large—approximately 60 nm in diameter—but there are often large numbers of them. Estimated frequencies are in the range of 0.1 to 10.0 μm^{-2} of cell wall, although there is a tendency for plasmodesmata to be grouped in roughly oval areas called primary pit fields. Plasmodesmata are small enough to preclude the exchange of organelles between cells but large enough to permit the diffusion of small solute molecules, including infectious viral RNA and plant transcription factors, through the cytosolic sleeve. Plasmodesmata thus provide a measure of membrane and cytosolic continuity between cells and allow for supracellular control over developmental programs.

The connection of neighboring protoplasts through plasmodesmata creates a continuous cytoplasmic network, referred to as the **symplast,** throughout the plant. In a similar manner, the apoplast consists of continuous extracellular or noncytoplasmic space. The apoplast consists of interconnected cell walls, intercellular spaces, and mature, nonliving vascular tissue. The concept of a symplast and an apoplast is especially useful when considering the movement of water and dissolved solutes throughout the plant.

17.3 CELL WALLS AND CELL GROWTH

Although the primary wall is not very thick, its interlocking network of cellulose microfibrils, cross-linking xyloglycans, and structural proteins confers remarkable

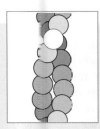

Box 17.1
CYTOSKELETON

Virtually all eukaryotic cells, both animal and plant, contain a three-dimensional, interconnected network of fibrous protein called the **cytoskeleton.** The cytoskeleton plays vital roles in determining the organization of cytoplasm and cell shape, and in cell division, growth, and differentiation.

The cytoskeleton of plant cells is composed of two different elements: **microtubules** and **microfilaments.** Microtubules are long rods approximately 24 nm in diameter and a hollow core about 12 nm in diameter. They are assembled from subunits of a globular protein called **tubulin,** which has a molecular mass of approximately 100,000 daltons (100 kD) and is made up of two globular polypeptides (α-tubulin and β-tubulin).[1] Microtubules are formed when tubulin subunits spontaneously self-assemble into long chains called **protofilaments.** Protofilaments then line up laterally to form the microtubule wall (see Figure 17.8). Microfilaments are solid threads composed of **actin,** also a globular protein. Two chains of actin subunits self-assemble in a helical fashion to form a microfilament approximately 6 nm in diameter.

The term *cytoskeleton* is an unfortunate choice because it implies a rigid structure with a static function. Instead, the cytoskeleton is very dynamic. Microtubules in particular are constantly being assembled, disassembled, and rearranged as the cell divides, enlarges, and differentiates. Microtubules form the mitotic spindle, which plays a significant role in the movement of chromosomes during cell division. Microtubules also determine the orientation and location of the new cell wall between daughter cells, and the deposition of cellulose in growing cell walls.

Microfilaments appear to control the direction of **cytoplasmic streaming,** the continuous flow of cytoplasmic particles and organelles around the periphery of the cell. The microfilaments form aggregates or bundles oriented parallel to the direction of cytoplasmic flow. Microfilaments are also involved in

the growth of pollen tubes. When a pollen grain germinates, it develops a tubular extension that grows down the stigma of the flower and serves to deliver the male nucleus to the egg. Growth of the tube is only at the tip, and vesicles that contain cell wall precursors are guided through the cytoplasm to the growing tip by a network of microtubules.

The importance of the cytoskeleton in organizing and coordinating the dynamic properties of growing cells is only beginning to be appreciated. Techniques for the study of plant cell cytoskeleton are rapidly improving, and we can expect exciting advances in the future.

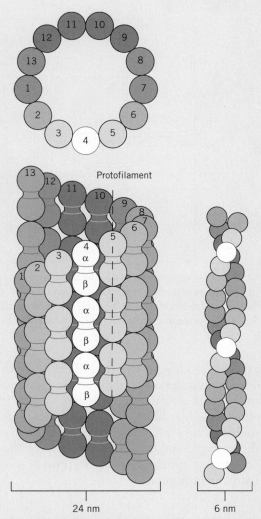

FIGURE 17.8 Microtubules and microfilaments. Left: Diagram of a microtubule in cross-section and longitudinal view. The number of vertical protofilaments varies from 11 to 15 but is usually 13, as shown here. Protofilaments are offset to form a helix. Right: A microfilament is composed of two parallel strands of globular subunits twisted to form a helix.

---

[1] The **molecular mass** of a molecule or particle is expressed in units of Daltons, defined as 1/12 the mass of a carbon atom. Molecular mass should not be confused with **molecular weight,** which is a dimensionless quantity expressing the ratio of particle mass to 1/12 the mass of a carbon atom. Molecular weight is symbolized by M, for *relative* molecular mass.

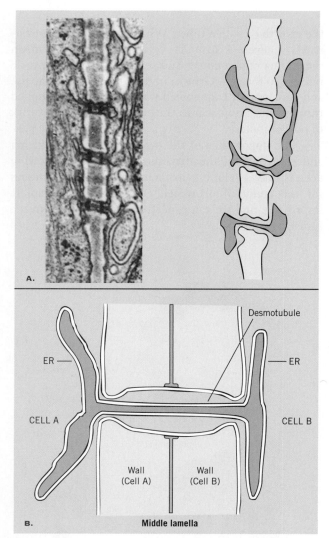

FIGURE 17.9 **Plasmodesmata. (*A*) Electron micrograph showing plasmodesmata connecting adjacent cells. (*B*) Diagram of a plasmodesmata showing the relationship between plasma membranes, ER, and desmotubules.**

mannitol or similar solute that cannot enter the cell, the cell will also not take up water and it will not grow. *The driving force for cell enlargement is the uptake of water.*

Recall that cells take up water by the process of osmosis. The high solute concentration of the protoplasm and vacuolar sap decreases the water potential of the cell to the point where water diffuses into the cell. With no means to compensate, a cell surrounded by pure water might continue to swell indefinitely or at least until internal pressures exceeded the tensile strength of the membrane. The consequences of such a situation are vividly demonstrated when mammalian red blood cells are placed in water. The cells quickly swell until the plasma membrane bursts, releasing their contents into the medium. Most animal cells avoid such osmotic disaster by using metabolic energy to excrete either solute or water and thus maintain a favorable pressure balance. Plant cells have found a different solution—they surround the plasma membrane with a strong, more or less rigid cell wall. Turgor pressure (a positive pressure) developed as the expanding protoplast pushes against the wall rises until it balances the negative osmotic pressure of the protoplast. At that point the water potential of the cell approaches zero and no further *net* water uptake—or cell growth—will occur. It is the strength and rigidity of the cell wall that imposes critical restrictions on the capacity of plant cells to grow. Thus, for a cell to increase in size, the strength and rigidity of the cell wall must be modified in order to reduce the water potential of the cell, permit water uptake, and, consequently, allow the cell to enlarge.

strength and rigidity to the wall. This strength and rigidity maintains the shape of the cell and its structural relationship with neighboring cells and, ultimately, the support of the entire plant. Thus a growing cell faces a delicate balancing act. It must, at the same time, maintain the strength and structural integrity of the wall while remaining sufficiently pliant to provide space for the expanding protoplast.

17.3.1 CELL GROWTH IS DRIVEN BY WATER UPTAKE AND LIMITED BY THE STRENGTH AND RIGIDITY OF THE CELL WALL

Since most of the volume of any cell is water, it follows that for a cell to increase its volume it must take up water. If, for example, a cell is bathed in an *isotonic* solution of

17.3.2 EXTENSION OF THE CELL WALL REQUIRES WALL-LOOSENING EVENTS THAT ENABLE LOAD-BEARING ELEMENTS IN THE WALL TO YIELD TO TURGOR PRESSURE

Any increase in volume of a cell requires a corresponding increase in the surface area of the surrounding wall, or wall extension. Investigators know that wall extension is related to turgor pressure—this has been demonstrated empirically. For example, when turgor pressure is experimentally reduced, the rate of cell expansion also declines. Furthermore, wall extension and growth do not occur in cells at very low or zero turgor pressure, even though the cells remain metabolically active and appropriate growth stimuli are present. This interdependence of wall extension and turgor pressure can be summarized by the following simple relationship:

$$dV/dt = m(P - Y) \tag{17.1}$$

The term dV/dt is the change (d) in volume (V) of a cell over time (dt). dV/dt is thus a simple way to express the growth rate of a cell. Y is the yield threshold, or the

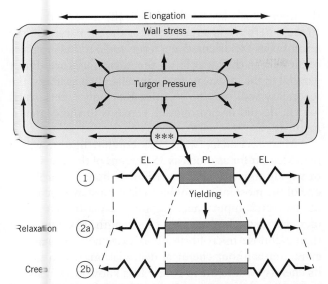

FIGURE 17.10 **A model for stress relaxation in the wall of a growing cell.** Both elastic (EL) and plastic (PL) components of the wall bear the stress of an expanding protoplast. In 1, turgor causes stress in the wall and extension of the elastic component, represented as springs. In 2a, yielding of the plastic components allows relaxation in the elastic component, illustrated by contraction of the springs. In 2b, turgor reestablishes wall stress and the wall expands to the same extent that the plastic component has lengthened. (From Cosgrove, D. I. 1987. *Plant Physiology* 84:561–564. Copyright American Society of Plant Physiologists.)

minimum turgor pressure necessary for cell expansion to occur. The term m is wall extensibility, a proportionality constant between growth rate and turgor pressure (P) in excess of the yield threshold. Wall extensibility is a quantitative measure of the capacity of the wall to irreversibly increase its surface area. According to this relationship, the growth of a cell depends principally on the amount by which turgor pressure exceeds the yield threshold.

It is apparent that a growing cell is faced with conflicting roles of turgor pressure. On the one hand, turgor pressure opposes the continued uptake of water, which is the driving force for cell expansion. At the same time, turgor pressure promotes irreversible wall extension and cell enlargement. How does the cell resolve this conflict? The answer to this question is provided by the concept of stress relaxation. Stress relaxation in the wall is central to the process of cell growth.

Turgor pressure develops because the cell wall—cellulose microfibrils cross-linked with xyloglucans—resists deformation as the protoplast attempts to expand. The force of the expanding protoplast pushing against the wall thus generates stress (defined as force/unit area) within the wall. Cell growth appears to be initiated when these stresses are relaxed by wall-loosening events that cause load-bearing elements in the wall to

yield (Figure 17.10). The most likely candidate for the load-bearing element is the xyloglucan linking two adjacent cellulose microfibrils. If the xyloglucan prevents displacement of the microfibrils, then a loosening of the xyloglucan cross link would constitute a stress relaxation and allow the cellulose microfibrils to move apart. The result would be a *simultaneous and proportionate reduction in turgor pressure*. A reduction in turgor pressure leads to a decrease in the water potential of the cell ($\Delta\Psi$ becomes more negative), followed by the passive uptake of water. The influx of water in turn increases cell volume, extends the cell wall, and tends to restore both wall stress and turgor pressure. The process of cell growth is thus seen as a continuous adjustment of turgor pressure through stress relaxation in order to balance its conflicting roles in water uptake and cell wall extension.

17.3.3 WALL LOOSENING AND CELL EXPANSION IS STIMULATED BY LOW PH AND EXPANSINS

It has been known since the 1930s that plant tissues elongate faster when bathed in a medium with a low pH (Figure 17.11). In the 1970s, a large body of work established that this "acid-growth" phenomenon applied not only to living tissues, but heat-killed tissues and isolated cell walls as well. These experiments gave rise to the concept of a wall-loosening enzyme with a low pH optimum.

In the early 1990s, two cell wall proteins were isolated and found to stimulate the expansion of cucumber hypocotyl sections that had been heat-treated in

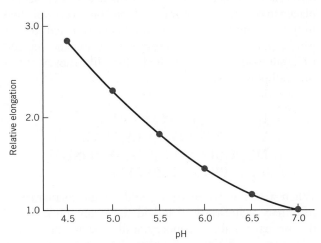

FIGURE 17.11 **An experiment demonstrating the acid-growth response. Apical segments (5 mm) from dark-grown oat coleoptiles (*Avena sativa*) were floated on buffers at the indicated pH. The increase in length of the segments was measured after 18 hours. (From data of Hopkins & Hillman. 1965. *Planta* 65:157–166.)**

order to inactivate endogenous wall-loosening activity. These proteins, called **expansins,** characteristically induce stress relaxation and extension of isolated cell walls at low pH. Expansins are small proteins, with a relative molecular mass of 26,000. In terms of dry mass—one part protein per 5,000 parts cell wall—they might be considered minor proteins. However, expansins are very active proteins and will induce cell wall extension when added in amounts as low as 1 part expansin per 10,000 parts of cell wall. The response is also very rapid—extension can be detected within seconds of adding expansins to the cell wall test system. In addition to stimulating extension of isolated cell walls, expansins enhance the rate of growth when applied to living cells. Moreover, expression of expansin genes is highest in growing and differentiating tissues, adding further weight to the hypothesis that expansins are significant wall-loosening agents.

Just how expansins achieve their unique effects on the physical properties of cell walls is not yet known, although two possibilities present themselves. The first possibility is that expansin hydrolyzes the cross-linking glycans or other elements of the wall matrix that hold the cellulose microfibrils in place. However, no hydrolytic activity on the part of expansins has been demonstrated to date. There are known enzymes whose activity is to hydrolyze components of the wall matrix, but these enzymes do not induce stress relaxation and extension. The second possibility is that expansins attack and weaken the noncovalent bonds by which the cross-linking glycans attach to the cellulose microfibrils. According to this proposal, expansin would migrate along the surface of the cellulose microfibril and weaken the attachment of the xyloglucan. The partial detachment of a xyloglucan that was under tension would allow it to relax. The cellulose microfibrils would then be displaced in response to turgor until tension is reestablished in the cross-link and new cross-links are formed. In this way, expansin would not progressively weaken the cell wall, but would catalyze an inchworm-like movement of the wall polymers.

17.3.4 IN MATURING CELLS, A SECONDARY CELL WALL IS DEPOSITED ON THE INSIDE OF THE PRIMARY WALL

When a cell stops enlarging and begins to mature, a secondary cellulose wall is laid down on the *inside* of the primary wall. Secondary walls are much thicker and more rigid than primary walls. They contain up to 45 percent cellulose, correspondingly lower amounts of cross-linking glycans, and relatively little pectic substance. In thick-walled woody cells, the secondary wall frequently consists of two distinct zones, characterized by differing orientation of the microfibrils. In both zones

the microfibrils are oriented helically around the cell. In the outermost layer, adjacent to the primary wall, the microfibrils are oriented at a large angle to the long axis of the cell. In the inner layer, the microfibrils are almost parallel to the long axis. In some cells, such as fibers, the secondary wall may be so thick that only a very small lumen, devoid of protoplasm, remains in the center at maturity.

Most secondary walls also contain **lignin,** which may account for as much as 35 percent of the dry weight of woody tissues. Next to cellulose, lignin is probably one of the most important biological substances in terms of structural importance. Lignin is a plastic-like polymer that has a high degree of strength, stronger even than cellulose microfibrils. It is extremely resistant to extraction without chemical degradation, which makes its chemistry difficult to study. It is known to consist of a complex system of interlocking bonds between several relatively simple phenolic alcohols. The combination of cross-linked cellulose microfibrils embedded in a matrix of pectic substances and lignin is responsible for the exceptional strength of wood. In engineering terms, secondary cell walls are a composite material, similar to modern fiber-reinforced plastics and with many of the same properties. The composite structure of cell walls is what enables tall trees to withstand the stresses of high winds and that makes wood such a useful and important building material.

17.4 A CONTINUOUS STREAM OF SIGNALS PROVIDES INFORMATION THAT PLANT CELLS USE TO MODIFY DEVELOPMENT

Plant cell development is highly regulated by a variety of signals that operate at several levels. **Hormones** are chemical messengers that enable cells to communicate with one another. Several classes of plant hormones, including auxin, gibberellins, cytokinins, abscisic acid, ethylene, and brassinosteroids, are known to promote or inhibit various developmental responses, either singly or in combination. Other intracellular factors such as turgor, mineral status, and internal clocks may also cause a cell to modify its metabolism and development.

Plant cells are also bombarded with external signals that provide information about their environment and that are used to modulate development accordingly. Light, temperature, gravity, and insect predation have the most obvious and dramatic impact. Factors such as magnetic field, sound, and wind (a mechanical stimulus) may have more subtle effects, but these have been difficult to establish experimentally. Other environmental factors such as soil moisture, humidity, nutrition, and pathogens may also influence development in some cases. More recently it has become evident that air and water pollutants represent not only an important environmental challenge to plants but may modify developmental patterns as well.

17.4.1 SIGNAL PERCEPTION AND TRANSDUCTION

A signal, regardless of whether it originates inside or outside the plant, can have no effect unless there is a receptor that allows it to be detected or perceived by the cell. The presence or absence of the appropriate receptor determines which cells are able to respond to a particular signal. There must then be a mechanism that converts, or transduces, the signal to some change in the chemistry or metabolism of the cell, or gene expression, that will ultimately give rise to the intended response. All developmental signals, regardless of their nature, share this sequence of *signal perception, transduction*, and *response*.

There are two principal mechanisms for signal perception and transduction. The first involves a receptor protein associated with the plasma membrane. The receptor is both specific to the signal (e.g., hormone molecule) and characteristic of the target cell. The formation of a signal-receptor complex then sets into motion a cascade of biochemical events that alters some aspect of cellular metabolism. In the second mechanism, the signal (e.g., a hormone molecule) is taken up by the cell and migrates into the nucleus where it reacts with a nuclear-based receptor to either activate or repress gene expression.

The two mechanisms are not mutually exclusive. The hormone auxin, for example, controls cell enlargement through a plasma membrane-based receptor while it controls more complex developmental responses through a nuclear-based receptor.

17.4.2 THE G-PROTEIN SYSTEM IS A UBIQUITOUS RECEPTOR SYSTEM

One receptor system that is proving to be ubiquitous in plants and animals is the G protein-coupled receptor. **G-proteins** are a large family of guanosine triphosphate (GTP) binding proteins that have long been known for their role in the response of animal cells to hormones, neurotransmitters, and a variety of other signals that operate across the cell membrane. However, recent genetic evidence from *Arabidopsis* and rice (*Oryza sativa*) has linked the presence of G proteins to signal perception in plants as well (Figure 17.12).

The two major players in the G protein system are the G protein itself, which consists of three distinct subunits (designated Gα, Gβ, and Gγ) and is located on the cytoplasmic side of the plasma membrane, and a transmembrane protein known as the G-protein-coupled receptor, or GPCR. GPCR is commonly referred to either as a 7-trans-membrane (7TM) receptor or heptahelical protein because the protein consists of seven α-helix domains that span the membrane. The seven helices are in turn connected by three hydrophilic loops that extend into the aqueous environment on each of the outer and inner surfaces of the membrane.

In effect, the G protein system functions as a biochemical on/off switch. In the absence of a hormone or other regulatory signal, the three subunits of the G protein combine to form an inactive complex that is located on the cytoplasmic surface of the membrane. In the inactive form (i.e., when the switch is "off"), the Gα subunit carries a molecule of guanosine diphosphate

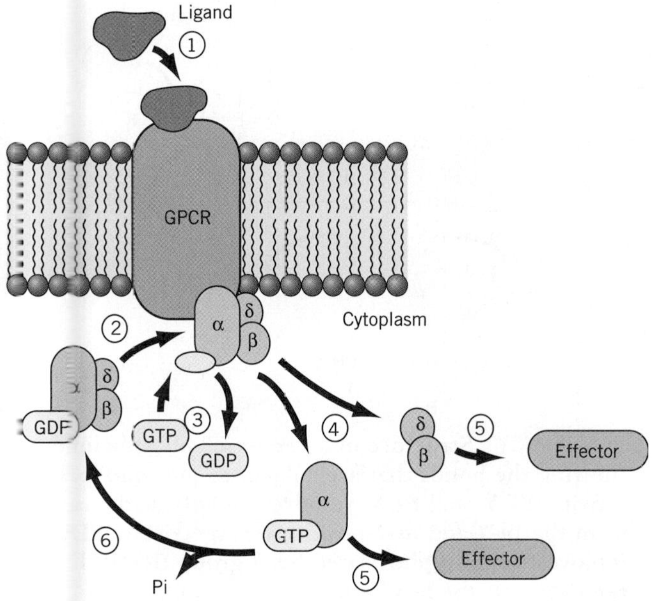

FIGURE 17.12 **A model for signal transduction by the G protein receptor system. In step 1, a signal molecule (the ligand) binds to the transmembrane G protein coupled receptor (GPCR), increasing its affinity for the G protein (step 2). In step 3, the G protein exchanges its GDP for GTP, which causes the dissociation of the Gα subunit from the G$\beta\gamma$ dimmer and the release of both subunits from the receptor (step 4). Both the Gα subunit and the G$\beta\gamma$ dimer may bind to an effector, thus activating or deactivating the effector (step 5). In step 6, the inherent GTPase activity of Gα hydrolyses the GTP, which deactivates Gα and allows the two subunits to recombine.**

(GDP). Signal transduction begins when a **ligand** (a hormone or other small regulatory molecule) binds to GPCR on the outer cell surface. Ligand binding causes the GPCR molecule to alter its conformation, or shape, such that it now has a higher affinity for the G protein. The G protein binds to the cytoplasmic surface of the receptor and the Gα subunit exchanges its GDP for a molecule of guanosine triphosphate (GTP). The G protein is now active, or "on."

In the next step, the activated G protein dissociates from the receptor and the Gα subunit dissociates from the combined G$\beta\gamma$ subunit. Each Gα and G$\beta\gamma$ subunit is now free to activate an effector protein. Effectors are enzymes that in turn control the amount of a secondary messenger produced or that regulate the flow of ions across membranes. Examples of effector molecules include calmodulin, phospholipases, and protein and lipid kinases discussed in the following sections. Gα subunits also possess GTPase activity, so when the Gα subunit has completed its job, the GTP is hydrolyzed to GDP and inorganic phosphate. Following hydrolysis of the GTP, the Gα subunit dissociates from the effector protein and is free to recombine with a G$\beta\gamma$ subunit and thus reform the inactive heterotrimeric complex. As long as the GPCR itself remains activated, it can turn on multiple G proteins, thereby amplifying the stream of signals to the effector molecules. The study of G proteins in plants is still in its infancy, but they have thus far been implicated in signaling responses to gibberellins, brassinosteroids, abscisic acid, and auxin.

17.5 SIGNAL TRANSDUCTION INCLUDES A DIVERSE ARRAY OF SECOND MESSENGERS

Signal perception is followed by a diverse array of biochemical events, referred to as **signal transduction** or **signaling,** that ultimately determines the cell's response to that signal. This often complex web of interacting pathways usually involves a variety of small, mobile molecules known as **second messengers.** The function of a second messenger is to relay information from the primary receptor to the biochemical machinery inside the cell. Second messengers commonly amplify the original signal by initiating a cascade of biochemical events. The principal species of second messengers in plants include protein kinase enzymes, calcium ions, and phospholipid derivatives.

17.5.1 PROTEIN KINASE-BASED SIGNALING

Protein kinases are enzymes that activate other proteins by catalyzing their phosphorylation. The action of protein kinases is balanced by the action of phosphatase enzymes that deactivate the protein by removing the phosphate group. Protein kinases are able to amplify weak signals because one active kinase is able to phosphorylate hundreds of target proteins. Often specific protein kinases act in series, creating a **protein kinase cascade.** Protein kinases and kinase cascades have been implicated in a wide array of plant signal transduction pathways.

17.5.2 PHOSPHOLIPID-BASED SIGNALING

Lipids are emerging as an important class of second messengers. These second messengers are generated by the action of enzymes known as **phospholipases** that hydrolyze phospholipids. Recall that phospholipids are a major constituent of cellular membranes. Four different phospholipases are known: phospholipase A$_1$ (PLA$_1$), phospholipase A$_2$ (PLA$_2$), phospholipase C (PLC), and phospholipase D (PLD). Each of these enzymes catalyzes the hydrolysis of a specific bond in the phospholipid molecule as shown in Figure 17.13. Virtually all of the products of phospholipase activity, including free fatty acids, appear to be involved in further signaling chains.

Lipid-based signaling in plants can be illustrated by the **inositol triphosphate** system. In this system, the signal receptor-complex activates PLC, possibly involving a G protein (Figure 17.14). PLC catalyzes the release of inositol triphosphate (IP$_3$) and diacylglycerol (DAG) from the membrane phospholipid phosphatidylinositol bisphosphate (PIP$_2$). Both IP$_3$ and DAG are second messengers. IP$_3$ diffuses into the cytoplasm where it activates calcium channels and stimulates the release of calcium from intracellular stores, most probably from the vacuole. In plants, DAG remains within the membrane where it is immediately phosphorylated to phosphatidic

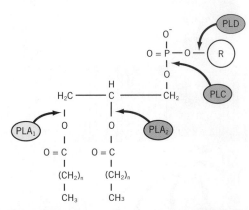

FIGURE 17.13 Structure of a "generic" phospholipid, showing the bonds that are subject to phospholipase activity. PLA$_1$ and PLA$_2$ remove the fatty acid chains from the *sn*-1 and *sn*-2 positions, respectively. PLC removes the phosphorylated head group (R). PLD removes only the head group.

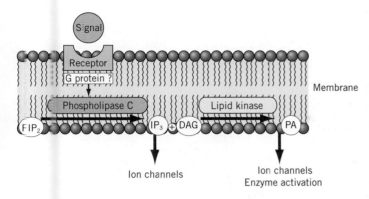

FIGURE 17.14 (*A*) The generation of second messengers by phospholipase C (PLC). PLC hydrolyzes phosphotidylinositol bisphosphate (PIP$_2$), producing a molecule each of inositol triphosphate (IP$_3$), and diacylglycerol (DAG). R^1 and R^2 are fatty acid–based acyl groups. (*B*) A model for signaling by the inositol triphosphate system. The hormone or other external signal activates a plasma membrane enzyme phospholipase C (PLC), which catalyzes the breakdown of PIP$_2$ to IP$_3$ and DAG. IP$_3$ diffuses into the cytoplasm where it stimulates the release of calcium. DAG is immediately phosphorylated to phosphatidic acid (PA) by the plasma membrane lipid kinase DAG kinase. PA then diffuses into the cytoplasm where it may also activate ion channels or various effector proteins.

acid (PA). PA then diffuses into the cytoplasm where it regulates ion channels or activates various enzymes.

17.5.3 CALCIUM-BASED SIGNALING

Calcium ions (Ca^2) are involved in the regulation of numerous physiological processes in plants, including cell elongation and division, protoplasmic streaming, the secretion and activity of various enzymes, hormone action, and tactic and tropic responses. In order for calcium to function effectively as a second messenger, the cytosolic Ca^{2+} concentration must be low and under metabolic control. Large amounts of calcium are stored in the endoplasmic reticulum, the mitochondria, and the large central vacuole but the cytosolic Ca^{2+} concentration is kept low largely through the action of membrane-bound, calcium-dependent ATPases. Activity of the ATPase and, consequently, the cytoplasmic Ca^{2+} concentration, is presumably under control of various stimuli such as light and hormones (Figure 17.15).

Calcium concentration throughout the cell is in part regulated by calcium channels located in the plasma membrane, endoplasmic reticulum, and vacuolar membrane (tonoplast). Some channels, which control the flow of Ca^{2+} between compartments, are voltage-gated, which means that their opening is determined by a particular value of membrane potential. Others are regulated either directly by signal receptors or by second messengers such as inositol triphosphates or cyclic nucleotides. Still others are able to sense tension in the membrane and open in response to turgor or mechanical stimuli such as touch and wind. Free Ca^{2+} also diffuses very slowly in the cytosol—much more slowly than in free solution. Because Ca^{2+} does not quickly disperse, Ca^{2+} gradients are easily established and maintained.

The principal calcium receptor in plant and animal cells is **calmodulin,** a highly conserved, ubiquitous protein that can be isolated from a variety of higher plants, yeasts, fungi, and green algae. Calmodulin from several plant sources, including spinach, peanut, barley,

corn, and zucchini, has been well characterized, and many of its properties are similar to calmodulin isolated from bovine brain tissue. Plant and bovine calmodulin have similar molecular mass (17 to 19 kDa), amino acid composition, and calcium-binding properties. When it binds with calcium, calmodulin undergoes a change in conformation that allows it to recognize and activate target proteins.

Several classes of enzymes, including NAD$^+$ kinases, protein kinases, and Ca^{2+}-ATPases, are known to be stimulated by calmodulin. NAD kinase catalyzes the phosphorylation of NAD to NADP in the presence of ATP. Because many redox enzymes are specific for one of these cofactors, regulating the balance between NAD and NADP is an effective way to regulate metabolism. Similarly, as noted above, many other enzymes are activated by protein kinase–catalyzed phosphorylation. Several calcium-dependent and calcium/

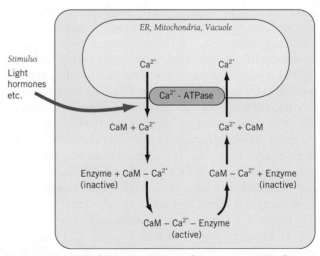

FIGURE 17.15 **Calcium as a second messenger. Exchange of calcium between the vacuole and the cytosol may be regulated by hormones or other factors such as light. Cytosolic Ca^{2+} forms an active complex with calmodulin (CaM) or other calcium-binding protein.**

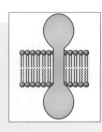

Box 17.2
UBIQUITIN AND PROTEASOMES— CLEANING UP UNWANTED PROTEINS

At any given time, a plant cell may contain upward of 10,000 or more individual proteins. Many of these proteins may no longer be required because the time of their function has passed. Some may contain errors introduced during their synthesis or may have been damaged due to excessive heat or some other stress. Others, in particular certain enzymes, may be short-lived by design because they catalyze the first or perhaps rate-limiting step in a multistep metabolic sequence. Still others may be key regulatory proteins, such as effectors of the cell cycle, which depend on rapid turnover for maximum effectiveness. Finally, ubiquitin-mediated degradation of repressor proteins is now known to be a key step in the regulation of gene transcription by hormones and other developmental signals.

For all of these reasons, proteins are continually subject to degradation by proteolytic enzymes called proteases. Protein degradation is a form of cellular housekeeping, in which unneeded or damaged proteins are broken down and their component amino acids recycled. Indeed, typically about half the protein complement of a cell is replaced every 4 to 7 days. However, proteases cannot be allowed uncontrolled access to cellular protein or the result would be chaos. There must be some mechanism both for controlling the access of proteases to protein and for marking or identifying the right proteins for degradation at the appropriate time.

Although protein is degraded in several cellular compartments, including chloroplasts, nuclei, mitochondria, and vacuoles, the process as it occurs in the cytosol is best understood. In the cytosol, protein degradation is accomplished by two principal components—a small, highly conserved protein, **ubiquitin,** and a large, oligomeric enzyme complex, the **proteasome** (Figure 17.16). The role of ubiquitin is to mark a protein for degradation by forming a conjugate with the target protein and delivering it to the proteosome where the protein is degraded. This is accomplished by the activity of three enzymes: E1, E2, and E3. E1, the ubiquitin-activating enzyme, activates the ubiquitin molecule with ATP. E2, the ubiquitin-conjugating enzyme, recruits the ubiquitin to E3, the ubiquitin-protein ligase.

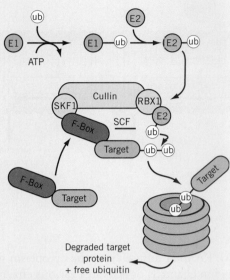

FIGURE 17.16 **Ubiquitin-mediated protein degradation. Through the action of a large, multienzyme complex, a targeted protein is marked for degradation by tagging the protein with ubiquitin. The complex consists of three enzymes: (1) a ubiquitin-activating enzyme (E1) that forms a covalent bond with ubiquitin through a cysteine residue; (2) a ubiquitin-conjugating enzyme (E2) that transfers the covalently bound ubiquitin; and (3) a ubiquitin-ligating enzyme (E3) that recognizes both the target protein and the E2-ubiquitin complex. E3, also known as the SCF complex, is composed of four subunits: SKP1, cullin, RBX1, and an F-box protein. The F-box protein recruits the target protein to the SCF complex, which then catalyzes the transfer of ubiquitin to the target protein. The action of E3 is repeated several times, marking the target protein with a chain of four or more ubiquitin molecules. The ubiquitin-protein complex is then delivered to a 26S proteasome for degradation. The proteasome consists of four stacked disks that form a hollow cylinder. Not shown are two 19S regulatory proteins that bind to the ends of the stack and direct the ubiquitin-protein complex into the hollow, where the protein-degrading active site is located.**

There are several different kinds of ubiquitin-protein ligases. In plants the most common kind appears to be a multimeric protein called the **SCF complex.** The complex is named for the first three subunits that were discovered: Skp1, cullin, and F-box protein. Later a fourth subunit, RBX1, was identified. Skp1, culllin, and Rbx 1 are common to all SCF complexes. They, in effect, form a scaffold onto which different F-box proteins can be assembled, each with a unique substrate-specificity. The *Arabidopsis* genome, for example contains nearly 700 putative F-box proteins. An F-box protein typically contains a highly conserved recognition site for the scaffold (the "F-box") and a recognition site for a specific target protein which it recruits to the SCF complex for ubiquitination. At least some F-box proteins also contain a recognition site for hormones or other developmental control factors.

Once the F-box protein has delivered the target protein to the scaffold, the now complete SCF complex then facilitates the transfer of the ubiquitin from E2 to the target protein. Additional ubiquitin molecules are added to form a ubiquitin chain. Normally, four ubiqitins are sufficient, at which point the ubiquitinated target protein is delivered to the proteasome.

The actual degradation of the protein is carried out by the proteasome, a complex with a molecular mass of more than 1.5 megadaltons (Mda). The proteasome

consists of two parts: a 20S core proteasome and two19S regulatory complexes.[1] Together the three parts make up the active 26S proteasome. The core proteosome is made up of four stacked rings with a central channel that contains the active sites for proteolysis. The regulatory complex governs access to the channel, probably unfolding the target protein and injecting it into the channel for degradation. When a ubiquitinated protein is delivered to the proteasome, the ubiquitin is released to be reused as the target protein is inserted into the proteasome for degradation.

[1]S values refer to the behavior of a complex with respect to sedimentation in a centrifugal field. It is dependent on the shape and density of the complex.

calmodulin-dependent NAD and protein kinases have been isolated from both soluble and membrane fractions from a large number of plants.

17.5.4 TRANSCRIPTIONAL-BASED SIGNALING

There is increasing evidence that some plant signals, such as the light-sensitive pigment phytochrome and some hormones, may by-pass plasma membrane receptors in favor of nuclear receptors and direct intervention in gene expression. They do this by diffusing into the nucleus where they interact with specific **transcription factors.**

Transcription factors are small, DNA-binding proteins that control the transcription of messenger RNA by binding to specific regulatory (i.e., noncoding) DNA sequences in particular genes. Transcription factors may either activate (or, up-regulate) the transcription of that particular gene, or it may act to repress (or, down-regulate) transcription. Some transcription factors have two binding sites: a DNA-binding site and a binding site for a regulatory molecule. Binding of the signal molecule, such as a hormone, to the transcription factor induces changes in the expression of the target gene. Exactly how these changes are brought about has not yet been determined with any certainty. However, in the case of at least three signal molecules—phytochrome and the two hormones auxin and gibberellins—the evidence indicates the presence of a transcription factor that, in the absence of the signal, represses expression of the target gene. When this is the case, it appears that binding of the signal molecule to the transcription factor flags the transcription factor for degradation by the ubiquitin-26S proteasome system

(see Box 17.2). Degrading the repressor allows RNA polymerase to bind with the promoter, thus enabling full expression of the gene. These pathways will be discussed more fully in later chapters.

17.6 THERE IS EXTENSIVE CROSSTALK AMONG SIGNAL PATHWAYS

Traditionally, signaling pathways were considered as though each operated as an independent chain of events. For example, the inositol triphosphate pathway was thought to represent one signal pathway while another used protein kinases and yet others used the G-protein or calcium pathways. As more becomes known about signal transduction pathways, it is increasingly apparent that this view of segregated pathways is far too simplistic. Signaling pathways more closely resemble an interconnected web and established pathways are linked by many connections, a situation commonly referred to as **crosstalk.** The *Oxford English Dictionary* defines crosstalk as "unwanted transfer of signals between communication channels" but to the biologist crosstalk refers to interactions between and within various classes of signals, the branching and merging of transduction pathways, and common use of second messengers—all of which help to coordinate developmental signals. Thus three classes of hormones (auxin, gibberellin, and brassinosteroids) stimulate hypocotyl elongation, while two hormones (cytokinin and ethylene) and light have an inhibitory effect. The extent of these connections is not surprising when one recognizes that there is a large number of rapidly changing environmental signals that must be integrated with various intrinsic control systems in order to execute a finely tuned developmental program.

SUMMARY

The cell cycle describes in effect describes the history of DNA and chromosomes during cell division. The actual division of the cell into two daughter cells, or cytokinesis, begins during the final stages of mitosis when the microtubules from the mitotic spindle reorganize as the phragmoplast. The phragmoplast directs Golgi-derived secretory vesicles to the equatorial plane of the cell where new plasma membranes form and the new crosswalls are laid down.

The driving force for cell enlargement is water uptake. However, in order to prevent excessive water uptake and to avoid rupturing the plasma membrane due to high turgor pressure, plant cells are surrounded by a very strong and relatively rigid wall. In order for a cell to enlarge, the strength and rigidity of the wall must be modified. In a turgid cell, the force of water pressing against the wall generates stress within the extensively cross-linked wall components. Growth is initiated when these stresses are relieved by wall-loosening events, which causes the load-bearing cross-links between wall polymers to yield. Relieved of stress, the wall expands, turgor is reduced, and more water moves in until both turgor and wall stresses are restored.

The orderly development of a complex multicellular organism is coordinated by a combination of intrinsic and extrinsic controls. Intrinsic controls are expressed at both the intracellular and intercellular levels. Intracellular controls are primarily genetic, requiring a programmed sequence of gene expression. Intercellular controls are primarily hormonal, chemical messengers that allow cells to communicate with one another. Extrinsic controls are environmental cues such as light, temperature, and gravity. Most environmental cues appear to operate at least in part by modifying gene expression or hormonal activities.

All developmental stimuli are characterized by a sequence of signal perception, signal transduction, and response. Signal perception requires a receptor molecule, which is normally a protein. Receptors are now known for red and blue light, and most hormones. Signal perception involves a diverse array of interacting metabolic pathways. Prominent in these pathways are small, mobile second messengers, such as the G-proteins, lipid kinases, protein kinases, cyclic nucleotides, calcium ion, and lipids. There is a high degree of interaction between second messengers, forming a complex web of signal transduction pathways.

CHAPTER REVIEW

1. Describe the process of cell division in plants.
2. What structural characteristics contribute strength and rigidity to a cell wall?
3. What limitations does the cell wall place on the growth of plant cells?
4. Describe the conflicting roles of turgor in the growth of plant cells.
5. Describe the importance of signal perception? What kinds of signals contribute to the development of plant cells?
6. What is a receptor protein and what is its role in development?
7. Describe second messengers. What is their role in signal transduction?
8. What is a "signal cascade"? How does a signal cascade amplify a signal?
9. How does ubiquitin assist in the removal of unwanted proteins?
10. What is meant by crosstalk (with respect to signal transduction) and what function might it serve in plant development?

FURTHER READING

Buchanan, B. B., W. Gruissem, R. L. Jones. 2000. *Biochemistry and Molecular Biology of Plants*. Rockville, MD: American Society of Plant Physiologists

Chory, J., D. Wu. 2001. Weaving the complex web of signal transduction. *Plant Physiology* 125:77–80.

Cosgrove, D. J. 2000. Loosening of plant cell walls by expansins. *Nature* 407:321–326.

Cosgrove, D. J. 2001. Wall structure and wall loosening: A look backwards and forwards. *Plant Physiology* 125:131–134.

Karp, G. 2008. *Cell and Molecular Biology*. 5th ed. New York: John Wiley & Sons.

Moon, J. G. Parry, M. Estelle. 2004. The Ubiqitin-Proteasome Pathway and plant development. *The Plant Cell* 16:3181–3195.

Pandy S., J. G. Chen, A. M. Jones, S. M. Assman. 2006. G-protein complex mutants are hypersensitive to abscisic acid regulation of germination and postgermination development. *Plant Physiology* 141:243–256.

Raven, P. H., R. F. Evert, S. E. Eichhorn. 2005. *Biology of Plants*. 7th ed. New York: Bedford, Freeman & Worth

Ryu, S. B. 2004. Phospholipid-derived signaling mediated by phospholipase A in plants. *Trends in Plant Science* 9:229–235.

Van Leeuwen, W., L. Ökrész, L. Bögre, T. Munnik. 2004. Learning the language of plant signaling. *Trends in Plant Science* 9:378–384.

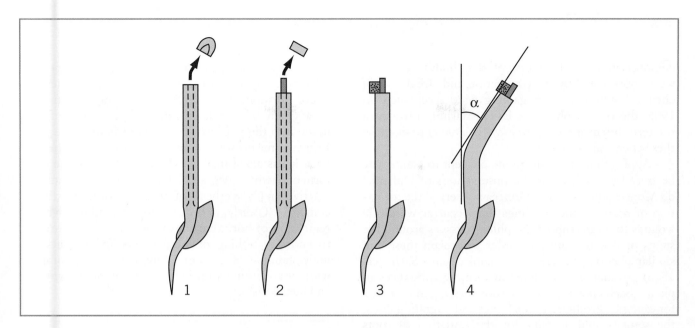

18

Hormones I: Auxins

Multicellular plants are complex organisms and their orderly development requires an extraordinary measure of coordination between cells. In order to coordinate their activities, cells must be able to communicate with each other. The principal means of intercellular communication within plants are the **hormones**. Hormones are signal molecules that individually or cooperatively direct the development of individual cells or carry information between cells and thus coordinate growth and development. Plant hormones have been the subject of intensive investigation since auxin was first discovered almost a century ago.

The discussion of each hormone in this and subsequent chapters will begin with a review of biosynthesis and metabolism. An understanding of hormone biochemistry makes it easier to understand what kinds of molecules they are and how they may function. In addition, a lot of what is known about what these molecules do and how they do it is based on studies of mutants that interfere with their biosynthesis or metabolism. The metabolic turnover of hormone molecules is also a significant factor in the regulation of cellular activities.

This first of four chapters on plant hormones is devoted to auxin. The following chapters will cover gibberellins, cytokinins, abscisic acid, ethylene, and brassinosteroids. In the case of each hormone, we will address the same three basic questions: what is it, what does it do, and how does it do it?

Because this is the first chapter on hormones, we will begin with an introduction to the hormone concept in plants. The balance of the chapter includes

- the biochemistry and metabolism of auxins,
- a review of auxin's principal effects on growth and development,
- how auxin controls cell enlargement,
- auxin transport in the plant, and
- auxin control of genetic expression.

18.1 THE HORMONE CONCEPT IN PLANTS

The concept of hormones, the chemical messengers that enable cells to communicate with one another, arose in the study of mammalian physiology. The latter half of the nineteenth century witnessed exciting advances in physiology and medicine. By 1850, it was known that blood-borne substances originating in the testis conditioned sexual characteristics. At the same time, physicians pursuing clinical studies had become interested in the effect of glandular extracts and secretions on the course of various diseases. By the turn of the century, a number of substances that elicited specific effects on the growth and physiology of mammals had been demonstrated and the concept that bodily functions were coordinated by the production and circulation of chemical substances was gaining wide acceptance. In 1905, the British physician E. H. Starling introduced the term **hormone** (Gr., to *excite* or *arouse*) to describe these chemical messengers.

Application of the hormone concept to plants may be traced as far back as the observations of Duhamel du Monceau in 1758. Du Monceau observed the formation of roots on the swellings that occur above girdle wounds that interrupted the phloem tissues around the stems of woody plants. In order to explain these and similar phenomena, German botanist Julius Sachs (ca. 1860) postulated specific organ-forming substances in plants. Sachs postulated that root-forming substances, for example, produced in the leaves and migrating down the stem, would account for the initiation of roots above the wound. The real beginning of plant hormone research, however, is found in a series of simple but elegant experiments conducted by Charles Darwin (see Box 18.1). It was Darwin's observations and experiments that ultimately led F. W. Went, almost half a century later, to describe a hormonal-like substance as the causative agent when plants grew toward the light. At about the same time, H. Fitting introduced the term *hormone* into the plant physiology literature.

What are hormones? Hormones are naturally occurring, organic molecules that, at low concentration, exert a profound influence on physiological processes. In addition, hormones, as defined by animal physiologists, are (1) *synthesized in a discrete organ or tissue*, and (2) *transported in the bloodstream to a specific target tissue* where they (3) control a physiological response *in a concentration-dependent manner*. While there are many parallels between animal and plant hormones, there are also some significant differences. Like animal hormones, plant hormones are naturally occurring organic substances that profoundly influence physiological processes at low concentration. The site of synthesis and mode of transport for plant hormones, however, is not always so clearly localized. Although some tissues or parts of tissues may be characterized by higher hormone levels than others, synthesis of plant hormones appears to be much more diffuse and cannot always be localized to discrete organs.

A hormone can serve effectively as a regulatory signal only if the molecule has a limited lifetime within the target cell. Any molecule sufficiently long-lived to be used repeatedly would sacrifice its dynamic, regulatory function. This means that the amount of a hormone in a cellular pool must be closely regulated and exhibit a rate of metabolic turnover that is rapid relative to the response that it controls.

The amount of hormone available to a target cell will be governed primarily by the rates at which active hormone molecules enter (input) and exit (output)

the hormone pool. Hormones may enter the pool by (1) *de novo* synthesis of the hormone, (2) retrieval of active hormone from an inactive storage form, such as a chemical conjugate, and (3) transport of hormone into the pool from a site elsewhere in the plant. Principal means for removing hormone from the pool once it has acted include: (1) oxidation or some other form of chemical degradation that renders the molecule inactive or (2) synthesis of an irreversibly deactivated conjugate. Clearly, in order to understand the dynamic regulation of hormone activity in plants, it is essential to know something of these inputs and outputs. No understanding of hormone function can be complete without a working knowledge of hormone biosynthesis and metabolism.

18.2 AUXIN IS DISTRIBUTED THROUGHOUT THE PLANT

Auxin (fr. G. *auxein*, to increase) is the quintessential plant hormone. Auxin was the first plant hormone to be discovered and it has a principal role in the most fundamental of plant responses—the enlargement of plant cells. Auxin is synthesized in meristematic regions and other actively growing organs such as coleoptile

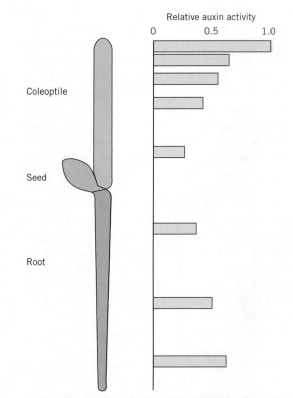

FIGURE 18.1 **Auxin distribution in an oat seedling (*Avena sativa*), showing higher concentrations of hormone in the actively growing coleoptile and root apices. (Based on data from Thimann, K. V. 1934. *Journal of General Physiology* 18:23–34.)**

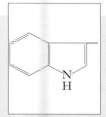

Box 18.1
DISCOVERING AUXIN

The experimental beginnings of plant hormone research in general and auxins in particular can be traced to the work of Charles Darwin. Although Darwin is best known for his work on evolution, later in his career he developed an interest in certain aspects of plant physiology. Some of these studies were summarized in the book *The Power of Movement in Plants*, co-authored by his son, Francis. One of several "movements" studied by the Darwins was the tendency of canary grass (*Phalaris canariensis*) seedlings to bend toward the light coming from a window, a phenomenon we now know as **phototropism**. The primary leaves of grass seedlings are enclosed in a hollow, sheath-like structure, called the **coleoptile**, which encloses and protects the leaves as they grow up through the soil. Darwin observed that coleoptiles, like stems, respond to unilateral illumination by growing toward the light source. However, curvature would not occur if the tip of the coleoptile were either removed or covered in order to exclude light. Since the bending response was observed over the entire coleoptile, Darwin concluded that the phototropic signal was perceived by the tip and "that when the seedlings are freely exposed to lateral light, some influence is transmitted from the upper to the lower part, causing the latter to bend." It was the implications of Darwin's "transmissible influence" that captured the imagination of plant physiologists and set into motion a series of experiments that culminated in the discovery of the plant hormone, auxin—the first plant hormone to be discovered.

Following the publication of Darwin's book, a number of scientists confirmed and extended their observations. In 1910, Boysen-Jensen demonstrated that the stimulus would pass through an agar block and was therefore chemical in nature. In 1918, Paal showed that if the apex were removed and replaced asymmetrically, curvature would occur even in darkness. In the climate of the time—Baylis and Starling's characterization of animal hormones had appeared only a few years earlier—plant physiologists were quick to interpret these observations as strong support for a plant hormone.

The active substance was first successfully isolated in 1928 by F. W. Went, then a graduate student working in his father's laboratory in Holland. Following up on the earlier work of Boysen-Jensen and Paal, Went removed the apex of oat (*Avena sativa*) coleoptiles and stood the apical pieces on small blocks of agar. Allowing a period of time for the substance to diffuse from the tissue into the agar block, he then placed each agar block asymmetrically on a freshly decapitated coleoptile. The substance then diffused from the block into the coleoptile, preferentially stimulating elongation of the cells on the side of the coleoptile below the agar block. Curvature of the coleoptile was due to differential cell elongation on the two sides. Moreover, the curvature proved to be proportional to the amount of active substance in the agar. Went's work was particularly significant in two respects: first, he confirmed the existence of regulatory substances in the coleoptile apex, and second, he developed a means for isolation and quantitative analysis of the active substance. Because Went used coleoptiles from *Avena* seedlings, his quantitative test became known as the **Avena curvature test**. Substances active in this test were called auxin, from the Greek *auxein* (to increase).

The results of Went's studies naturally stimulated intensive efforts to isolate and identify the active substance. One particularly active compound, indole-3-acetic acid (IAA), was isolated from human urine in 1934. This peculiar source was selected because it was suspected that female sex hormones, secreted in urine, might have some plant growth activity. In a beautiful piece of scientific serendipity, the impure urine preparation initially assayed was highly active, while subsequently purified hormone preparations were inactive. This led the investigators back to the material from which the female sex hormones were initially extracted—the urine of pregnant women—and the identification of IAA. At the same time, IAA was isolated from yeast extracts, and the following year, from cultures of *Rhizopus suinus*. IAA was isolated from immature corn kernels in 1946 and since then has been found to be ubiquitous in higher plants.

apices, root tips, germinating seeds, and the apical buds of growing stems (Figure 18.1). Young, rapidly growing leaves, developing inflorescences, and embryos following pollination and fertilization are also significant sites of auxin synthesis. Auxin, more than any other growth substance, appears to be actively distributed throughout the entire plant.

18.3 THE PRINCIPAL AUXIN IN PLANTS IS INDOLE-3-ACETIC ACID (IAA)

Although a large number of compounds have been discovered with auxin activity, **indole-3-acetic acid**

(IAA) is the most widely distributed natural auxin (Figure 18.2). In addition to IAA, several other naturally occurring indole derivatives are known to express auxin activity, including indole-3-ethanol, indole-3-acetaldehyde, and indole-3-acetonitrile. However, these compounds all serve as precursors to IAA and their activity is due to conversion to IAA in the tissue.

The initial discovery of IAA in plants and recognition of its role in growth and development stimulated the search for other chemicals with similar activity. The result has been an array of synthetic chemicals that express auxin-like activity. One of these chemicals was **indole-3-butyric acid (IBA)** (IV, Figure 18.2). More recently, IBA has been isolated from seeds and leaves of maize and several other species. A chlorinated analog of IAA (4-chloroindoleacetic acid, or 4-chloroIAA; II, Figure 18.2) has also been reported in extracts of legume seeds and a closely related, naturally occurring aromatic acid, phenyl acetic acid (PAA) (III, Figure 18.2) has recently been reported to have auxin activity. Because IBA, 4-chloroIAA, and PAA have now been isolated from plants, are structurally similar to IAA, and elicit many of the same responses as IAA, there is a strong argument for considering them natural hormones. However, it is not yet clear whether they are active on their own or whether they are first converted to IAA. Chemically, the single unifying character of molecules that express auxin activity appears to be an acidic side chain on an aromatic ring.

FIGURE 18.2 **The chemical structures of some naturally occurring and synthetic auxins. Indole-3-acetic acid (I) is believed to be the active auxin in all plants. Phenylacetic acid (III) is widespread and two others, 4-chlorindole-3-acetic acid and indole-3-butyric acid, have been identified in plant extracts. The latter three induce auxin responses when applied exogenously, but probably act via conversion to IAA. Structures VI, VII, and VIII are active herbicides.**

Naturally Occurring Auxins

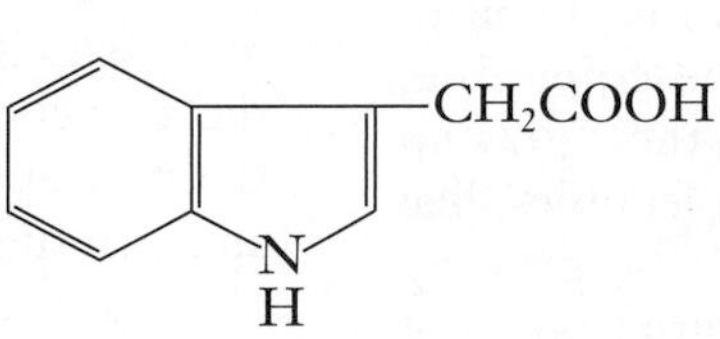

I. Indole-3-acetic acid (IAA)

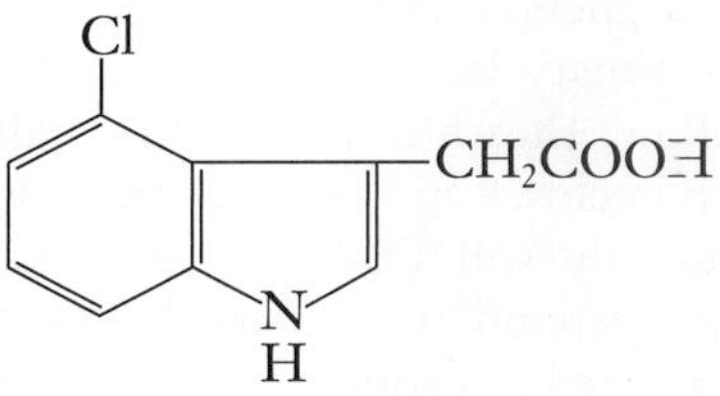

II. 4-Chloroindole-3-acetic acid

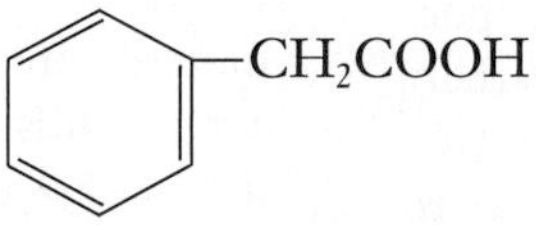

III. Phenylacetic acid

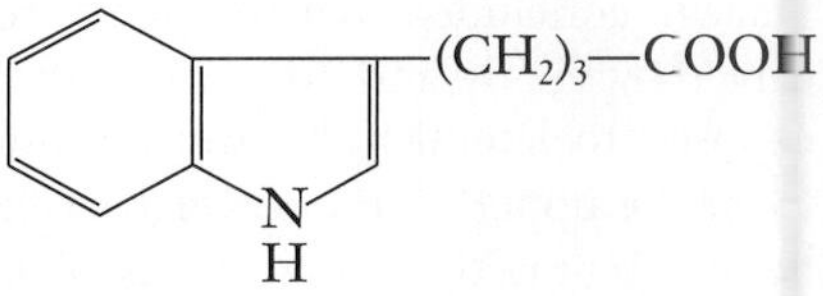

IV. Indole-3-butyric acid (IBA)

Synthetic Auxins

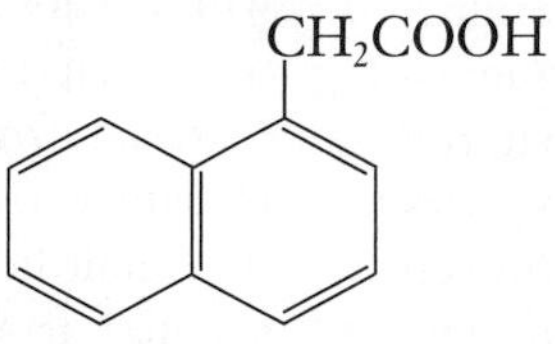

V. Naphthalene acetic acid (NAA)

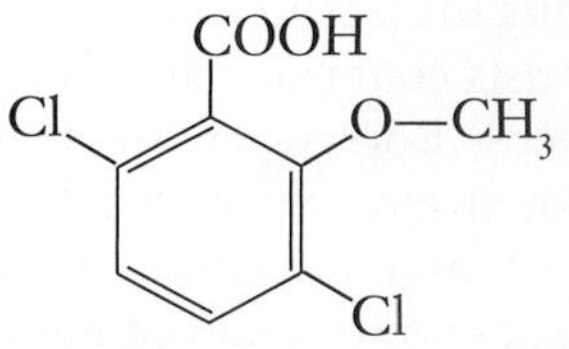

VI. 2-Methoxy-3,6-dichloro-benzoic acid (dicamba)

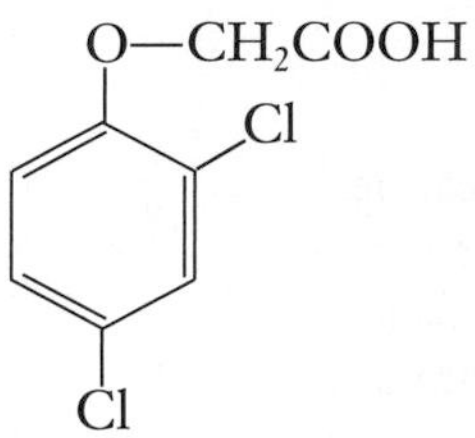

VII. 2,4-Dichlorophenoxyacetic acid (2,4-D)

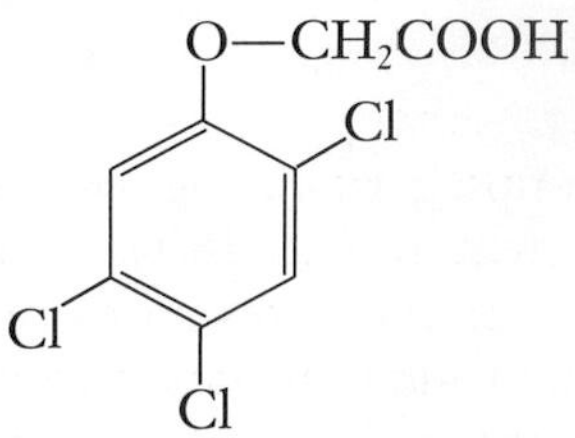

VIII. 2,4,5-Trichlorophenoxy-acetic acid (2,4,5-T)

The amount of IAA present will depend on a number of factors, such as the type and age of tissue and its state of growth. In vegetative tissues, for example, the amount of IAA generally falls in the range between 1 μg and 100 μg (5.7 to 570 nanomoles) kg^{-1} fresh weight, but in seeds it appears to be much higher. In one study, it was estimated that the endosperm of a single maize seed four days after germination contains 308 picomoles (pmole = 10^{-12} mole) of IAA. At the same time, the maize shoot contained 27 pmoles of IAA and required an estimated input of approximately 10 pmoles of IAA h^{-1} in order to support its growth. The high level of IAA in the seed apparently serves to support the rapid growth of the young seedling when the seed germinates.

18.4 IAA IS SYNTHESIZED FROM THE AMINO ACID l-TRYPTOPHAN

Since the 1930s, when K. V. Thimann first observed the synthesis of IAA in the mold *Rhizopus suinus*, which had been fed the amino acid tryptophan, the conversion of tryptophan to IAA has been studied *in vivo* in more than 20 different plant species and *in vitro* with at least 10 different cell-free enzyme preparations. The synthesis of IAA is normally studied by feeding plants tryptophan carrying a radioactive label, usually carbon (^{14}C) or tritium (^{3}H), and examining the radioactivity of subsequently isolated IAA or its intermediates.

Feeding experiments are complicated by several factors and the results must always be approached with caution. For example, radiolabeled tryptophan can apparently undergo radiochemical decomposition, thus giving rise to IAA by nonenzymatic reactions. In addition, the pool size of tryptophan (also a precursor for protein synthesis) is very large relative to that of IAA and there is little data on the actual quantity of IAA synthesized. Finally, care must be taken to ensure that experiments are conducted under sterile conditions, since many microorganisms readily convert tryptophan to IAA. While these complications make it difficult to ascertain the exact pathway that functions *in vivo*, the available evidence clearly establishes that plants are able to synthesize IAA from tryptophan.

In most plants, synthesis of IAA occurs in three steps, beginning with the removal of amino group on the tryptophan side chain. The product is indole-3-pyruvic acid (IPA) (Figure 18.3). This reaction is catalyzed by **tryptophanamino transferase**, a widely distributed multispecific enzyme that appears to act as well to remove amino groups from structural analogs of tryptophan such as phenylalanine and tyrosine. The second step is the decarboxylation of IPA to form indole-3-acetaldehyde (IAAld). The enzyme that catalyzes this step, **indole-3-pyruvate decarboxylase**,

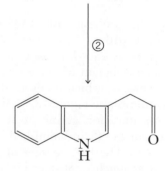

L-Tryptophan

①

Indole-3-pyruvic acid

②

Indole-3-acetaldehyde

③

Indole-3-acetic acid (IAA)

FIGURE 18.3 **Pathway for tryptophan-dependent biosynthesis of indole-3-acetic acid. The enzymes involved are (1) tryptophan aminotransferase; (2) indole-3-pyruvate decarboxylase; (3) indole-3-acetaldehyde oxidase.**

has been described in several plant tissues and cell-free extracts. Finally, IAAld is oxidized to IAA by a NAD-dependent **indole-3-acetaldehyde oxidase**. The presence of this enzyme has been demonstrated in a number of tissues, including oat coleoptile. IAAld may also be reversibly reduced to indole-3-ethanol.

Indole-3-ethanol is active in bioassays using stem sections, but this is probably due to its conversion to IAA in the tissue. Finally, IAA can be reversibly converted to IBA by the enzyme indole-3-butyric acid synthase.

There is some evidence for alternate biosynthetic pathways involving intermediates other than IPA, but the burden of biochemical evidence indicates that the IPA pathway is the principal pathway for the synthesis of IAA from tryptophan in higher plants. Although IAA-deficient mutants might be expected to provide further useful information, none have been identified to date. This is perhaps because an IAA deficiency would probably be lethal.

18.5 SOME PLANTS DO NOT REQUIRE TRYPTOPHAN FOR IAA BIOSYNTHESIS

Evidence for the biosynthesis of IAA via a tryptophan-independent pathway has been obtained from mutants of both maize and *Arabidopsis*. Seedlings of the *orange pericarp (orp)* mutant of *Zea mays* lack the enzyme tryptophan synthase, which catalyzes the final step in tryptophan synthesis (see Figure 18.3). Although seeds carrying the *orp* mutation germinate normally, they do not survive because of their diminished capacity for tryptophan synthesis. The IAA content of mutant seedlings, however, is as much as 50-fold higher than that of wildtype seedlings. Several tryptophan-requiring mutants have also been isolated from *Arabidopsis*. Two of these mutants, *trp2* and *trp3*, also lack tryptophan synthase and are unable to convert indole-3-glycerol phosphate to tryptophan. The *trp2* and *trp3* seedlings, unlike *orp*, do not accumulate free IAA but they do contain elevated levels of conjugated IAA (see below). Apparently, *trp2* and *trp3* store excess IAA in the conjugated form. Radioisotope-labeling experiments in both maize and *Arabidopsis* have confirmed that the IAA is synthesized from some precursor other than tryptophan.

The precise pathway for tryptophan-independent IAA synthesis is not known. However, the *trp2* and *trp3* *Arabidopsis* mutants do accumulate indole-3-acetonitrile. *Arabidopsis* also contains the nitrilase enzymes necessary for converting indole-3-acetonitrile to IAA, thus implicating indole-3-acetonitrile as an intermediate. The source of indole-3-acetonitrile is not known, although its accumulation in tryptophan mutants suggests a tryptophan-independent pathway for the biosynthesis of indole-3-acetonitrile as well. It is known that indole-3-acetonitrile can be derived from glucobrassicin, the principal **glucosinolate** present in members of the family Cruciferae. Details of the tryptophan-independent indole-3-acetonitrile pathway

for auxin biosynthesis and whether it is limited to *Arabidopsis* or the brassicas, or is more widespread, remain to be determined.

18.6 IAA MAY BE STORED AS INACTIVE CONJUGATES

Very early in the study of auxins, two populations of the hormone were recognized—one was free-moving and could be obtained by diffusion into agar; the other appeared to be bound in the cell and could be isolated only by extraction with solvents or by hydrolysis under alkaline conditions. This latter population, referred to as "bound auxin," is now recognized as IAA that has

A. Reversible Deactivation

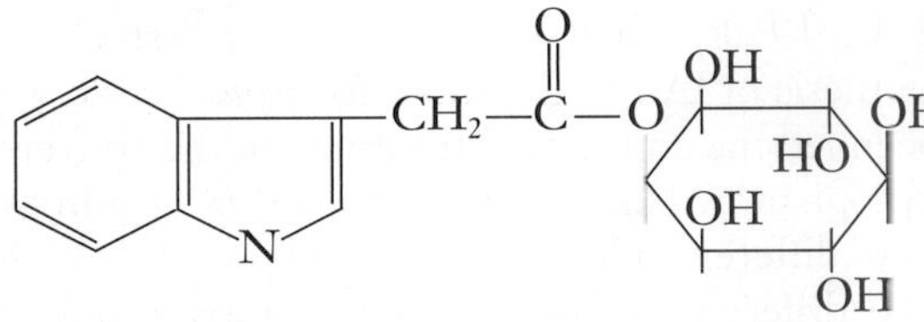

Indole-3-acetyl-*myo*-inositol

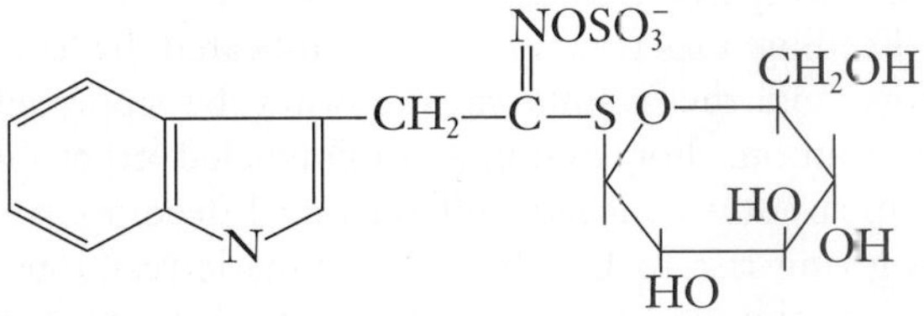

Glucobrassicin

B. Irreversible Deactivation

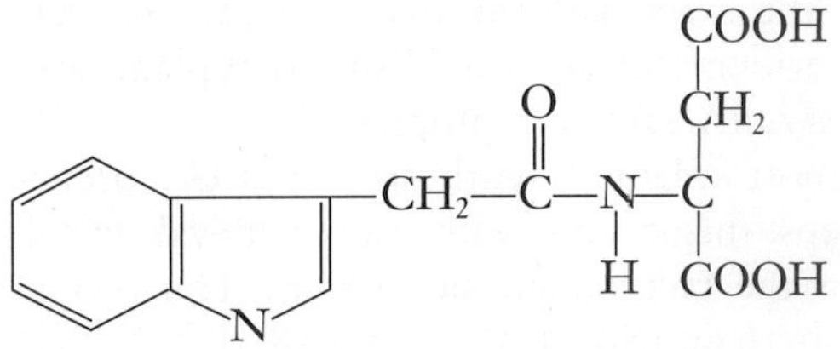

Indole-3-acetyl-L-aspartate

FIGURE 18.4 **Example of IAA conjugates. Conjugation ties up the side-chain carboxyl group, which is essential for auxin activity. Normally, conjugation with a sugar reversibly deactivates the auxin molecule while deactivation by conjugation with an amino acid is irreversible.**

formed chemical conjugates with sugars to form **glycosyl esters**. Conjugates are formed by esterification of a glucose or inositol molecule to the acid group of the side chain (Figure 18.4). IAA-glycosyl conjugates are themselves inactive but they do release free, biologically active IAA upon solvent extraction, alkaline hydrolysis, or enzymatic hydrolysis *in vivo*.

Although quantitative data are lacking for most plants, large pools of IAA glycosyl esters have been demonstrated in seeds of *Zea mays*. These pools of IAA conjugates are formed in the milky endosperm as the seed develops and appear to be an important source of active hormone for the embryo during the first few days of germination. It has been estimated, for example, that as much as 60 percent of the IAA requirement of a germinating maize shoot may be met by hydrolysis of IAA conjugates initially supplied by the endosperm. Since most of our knowledge of IAA release by hydrolysis of conjugates comes from studies with germinating seeds, it is not yet known whether conjugate hydrolysis is equally important in the growth of mature plants.

18.7 IAA IS DEACTIVATED BY OXIDATION AND CONJUGATION WITH AMINO ACIDS

IAA in aqueous solution is relatively unstable and is readily degraded by a variety of agents, including acids, ultraviolet and ionizing radiation, and visible light, the latter especially in the presence of sensitizing pigments such as riboflavin. IAA degradation in situ, however, appears primarily due to oxygen and peroxide, either separately or in combination, in the presence of a suitable redox system.

Inactivation of the *Avena* growth-promoting substance by aqueous extracts of leaves was first reported in the 1930s, even before the active principle was identified as IAA. An enzyme responsible for inactivating IAA was first isolated from plant extracts in the 1940s and was called **IAA oxidase**. Later, the enzyme **peroxidase**, in concert with a flavoprotein, was shown to catalyze the oxidation of IAA while at the same time releasing CO_2. The oxidative decarboxylation of IAA by peroxidase is now known to be synonymous with IAA oxidase. *In vitro* oxidative decarboxylation of IAA has been studied most extensively with purified horseradish peroxidase. Because the end products of IAA oxidation are physiologically inactive, IAA oxidation is an effective way of removing the hormone molecule once it has accomplished its purpose. More recent studies with green tomato fruits, *Vicia faba*, and other species have shown that conjugation of IAA with amino

acids such as alanine or aspartic acid also leads to irreversible deactivation (Figure 18.4).

18.8 AUXIN IS INVOLVED IN VIRTUALLY EVERY STAGE OF PLANT DEVELOPMENT

Auxins are characterized principally by their capacity to stimulate cell elongation in excised stem and coleoptile sections, but they are also involved in a host of other developmental responses, including secondary root initiation, vascular differentiation, and the development of axillary buds, flowers, and fruits. Auxins are also an important component in the signal chain that enables roots and shoots to respond to gravity and unilateral light. In fact, auxin is involved in virtually every stage of plant growth and development from the organization of the early embryo to flowering and fruit development.

18.8.1 THE PRINCIPAL TEST FOR AUXINS IS THE STIMULATION OF CELL ENLARGEMENT IN EXCISED TISSUES

Regulation of cell enlargement in *Avena* coleoptiles was the basis for its discovery and this action has been demonstrated repeatedly with excised plant tissues such as subapical coleoptile tissues and stem segments cut from dark-grown pea seedlings.

Auxin concentration-response curves typically show an increasing response with increasing concentrations of auxin until an optimum concentration is reached (Figure 18.5). Concentrations exceeding the optimum characteristically result in reduced growth. If the auxin concentration is high enough, growth may be inhibited compared with controls.

Another characteristic feature of auxin physiology is that intact stems and coleoptiles do not show a significant response to exogenous application of the hormone. Apparently the endogenous auxin content of intact tissues is high enough to support maximum elongation and added auxin has little or no additional effect. Thus, it is a general rule that the effect of exogenously supplied auxin on cell enlargement can be demonstrated only in tissues that have been removed from the normal auxin supply. These include excised segments of stems and coleoptiles or tissues cultured on artificial media.

18.8.2 AUXIN REGULATES VASCULAR DIFFERENTIATION

In addition to stimulating cell enlargement, auxin also has a role in regulating cellular differentiation. The most extensively studied system is the induction of vascular differentiation in shoots, which is under control of auxin

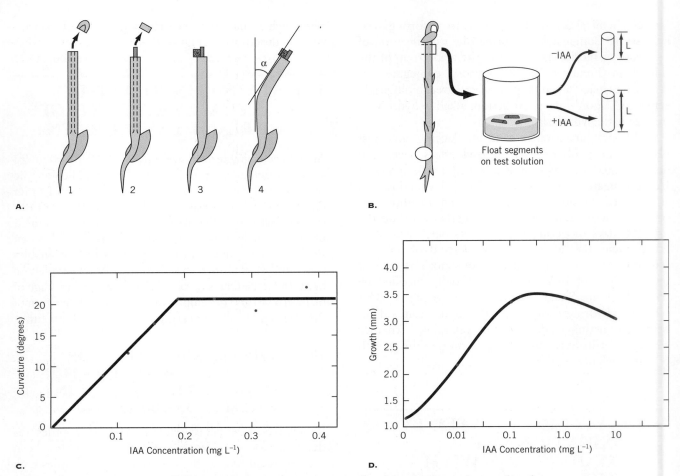

FIGURE 18.5 **Concentration response curves for two classic auxin-regulated responses.** (*A*) Went's *Avena* curvature test. A small cube of agar containing auxin is placed on the cut surface of a decapitated oat coleoptile. The auxin diffuses into the coleoptile, stimulating growth of the cells below the agar cube. The differential growth causes the coleoptile to curve away from the block. (*B*) Curvature in the *Avena* test is linearly related to auxin concentration. (Redrawn from the data of Went, F. W., K. V. Thimann. 1937. *Phytohormones*. By permission of K. V. Thimann.) (*C*) Pea stem segment test. Stem sections from dark-grown pea seedlings are floated on a medium with or without auxin. (*D*) Typical concentration-response in a pea stem section test. Note auxin concentration is expressed on a logarithmic scale. (Redrawn from the data of Galston, A. W., M. E. Hand. 1949. *American Journal of Botany* 36:85–94. With permission of the *American Journal of Botany*.)

produced in the young, rapidly developing leaves. The production of xylem strands at the base of a *Coleus* petiole, for example, is directly proportional to the stream of diffusible IAA moving through the petiole. Defoliation of *Coleus* epicotyls strongly reduces xylem differentiation in the petiole, but this effect can be reversed by applying equivalent amounts of IAA in lanolin paste.

A favorite system for the study of vascular differentiation is the regeneration of vessels and phloem sieve tubes around wounds in *Coleus* stems, which is also under the control of auxin (Figure 18.6). *Coleus*, like other members of the mint family (Lamiaceae), has characteristic square stems with a vascular bundle at each

corner. If a wedge-shaped incision is made that interrupts one of these vascular bundles, parenchyma cells in the region of the wound will differentiate into new vascular elements. These vascular elements will eventually reestablish continuity with the original bundle.

The differentiation of both xylem elements and phloem sieve tubes around the wound is limited and controlled by auxin supply. This can be shown by removal of leaves (a source of auxin) above the wound, for example, which reduces vascular regeneration. On the other hand, because auxin moves preferentially down the stem, removal of leaves below the wound has little or no effect. Furthermore, the extent of vascular regeneration is directly proportional to the auxin supply

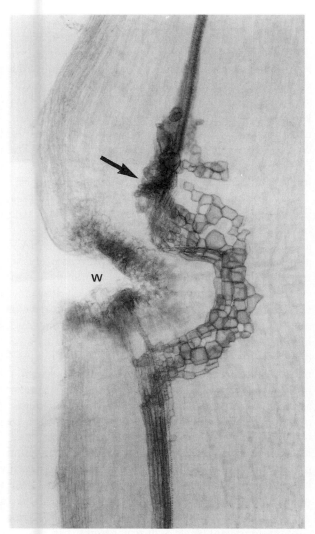

FIGURE 18.6 IAA-induced xylem regeneration. A longitudinal view of regenerated xylem vessel elements around a wound (W) in a decapitated internode of cucumber (*Cucumis sativus*). Lanolin containing 0.1 percent IAA was applied to the upper side of the internode immediately after wounding. Polar regeneration is indicated by the dense appearance of many xylem tracheary elements (arrow) in the region of the damaged vascular bundle *above* the wound. This is the region where the basipetally flowing IAA would initially accumulate because it was interrupted by the wound and forced to find a new pathway around the obstacle. (Magnification: × 60) (Photograph courtesy of Prof. R. Aloni, Tel Aviv University).

when exogenous auxin is substituted for the leaves. In general, differentiation of phloem sieve tubes is favored by low auxin concentrations (0.1% IAA w/w in lanolin) while xylem differentiation is favored by higher auxin concentrations (1.0% IAA w/w in lanolin).

Auxin is also required for vascular differentiation in plant tissue culture. When buds, which are a source of auxin, are implanted into clumps of undifferentiated callus tissue in culture, differentiation of callus parenchyma into vascular tissue occurs in regions adjacent to the implant. The same effect is achieved when agar wedges containing IAA and sugars are substituted for the implanted bud.

18.8.3 AUXIN CONTROLS THE GROWTH OF AXILLARY BUDS

As a shoot continues to grow and the apical meristem lays down new leaf primordia, small groups of cells in the **axil** (the angle between the stem and the leaf primordium) of the primordia become isolated from the apical meristem and produce an **axillary bud**. In some cases, such as the bean (*Phaseolus*), the bud continues to grow, although at a much slower rate than the apical bud. In many plants, however, mitosis and cell expansion in the axillary bud is arrested at an early stage and the bud fails to grow. It has been known for some time that removal of the shoot apex, a common horticultural technique for producing bushy plants, stimulates the axillary bud to resume growth (Figure 18.7). Apparently the apical bud is able to exert a dominant influence that suppresses cell division and enlargement in the axillary bud. For this reason, the phenomenon of coordinated bud development is known as **apical dominance**.

Shortly after auxin was first discovered, K. V. Thimann and F. Skoog questioned whether there might be a relationship between the capacity of the shoot tip to release auxin and its capacity to suppress axillary bud development—in other words, is apical dominance controlled by auxin? Thimann and Skoog tested this idea by decapitating broad bean (*Vicia faba*) plants and applying auxin to the cut stump. Axillary bud development remained suppressed in the presence of auxin. Since this initial demonstration, the capacity of auxin to substitute

FIGURE 18.7 Apical dominance in broadbean (*Vicia faba*). (*Left*) Control plants. (*Center*) Removal of the stem apex, a source of auxin, promotes axillary bud growth at the base of the young stem. (*Right*) Dominance can be restored by applying auxin (in lanolin paste) to the cut stem surface.

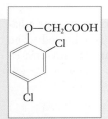

Box 18.2
COMMERCIAL APPLICATIONS OF AUXINS

Hormones and other regulatory chemicals are now used in a variety of applications where it is desirable for commercial reasons to control some aspect of plant development.

The synthetic auxins are used in commercial applications largely because they are resistant to oxidation by enzymes that degrade IAA. In addition to their greater stability, the synthetic auxins are often more effective than IAA in specific applications. One of the most widespread uses of auxin encountered by the consumer is the use of 2,4-D in weed control. 2,4-D and other synthetic compounds, such as 2,4,5-T and dicamba, express auxin activity at low concentrations, but at higher concentrations are effective herbicides.

The introduction of 2,4-D and 4-chlorophenoxyacetic acid (4-CPA) as herbicides in 1946 revolutionized our approach to agriculture. For reasons that are not clear, chlorinated phenoxyacetic acids are selectively toxic to broadleaf species. 2,4-D remains the principal component of "weed-and-feed" mixtures for home lawn care as well as for control of broadleaf weeds in cereal crops. The synthetic auxins are favored in commercial applications because of their low cost and greater chemical stability.

Indolebutyric acid and naphthaleneacetic acid are both widely used in vegetative propagation—the propagation of plants from stem and leaf cuttings. This application can be traced to the propensity for auxin to stimulate adventitious root formation. Generally marketed as "rooting hormone" preparations, the auxins, usually a synthetic auxin such as NAA or IBA, are mixed with an inert ingredient such as talcum powder. Stem cuttings are dipped in the powder prior to planting in a moist sand bed in order to encourage root formation.

4-CPA may be sprayed on tomatoes to increase flowering and fruit set while NAA is commonly used to induce flowering in pineapples. This latter effect is actually due to auxin-induced ethylene production. NAA is also used both to thin fruit set and prevent preharvest fruit drop in apples and pears. These seemingly opposite effects are dependent on timing the auxin application with the appropriate stage of flower and fruit development. Spraying in early fruit set, shortly after the flowers bloom, enhances abscission of the young fruits (again, due to auxin-induced ethylene production). Thinning is necessary in order to reduce the number of fruits and prevent too many small fruits from developing. Spraying as the fruit matures has the opposite effect, preventing premature fruit drop and keeping the fruit on the tree until it is fully mature and ready for harvest.

The use of synthetic auxins, especially the chlorinated forms, as herbicides has come under close scrutiny by environmental groups because of potential health hazards. 2,4,5-T, for example, has been banned in many jurisdictions because commercial preparations contain significant levels of dioxin, a highly carcinogenic chemical.

---

for the shoot tip in maintaining apical dominance has been confirmed repeatedly.

How does auxin from the shoot apex suppress axillary bud development? The most widely accepted theory holds that the optimum auxin concentration for axillary bud growth is much lower than it is for the elongation of stems. The stream of auxin flowing out of the shoot apex toward the base of the plant is thought to maintain an inhibitory concentration of auxin at the axillary bud. Removal of this auxin supply by decapitation reduces the supply of auxin in the region of the axillary bud and thereby relieves the bud of inhibition. More direct evidence for the role of auxin transport is offered by the observation that inhibitors of auxin transport (TIBA and NPA) stimulate release of buds from dominance when applied to the stem between the shoot apex and the bud. In addition, lines of tomato that exhibit prolific branching (that is, the absence of apical dominance) also fail to export radioactively labeled IAA from the shoot apex.

18.9 THE ACID-GROWTH HYPOTHESIS EXPLAINS AUXIN CONTROL OF CELL ENLARGEMENT

Whatever its primary action, auxin can alter the rate of cell expansion only by ultimately influencing one or more of the parameters previously identified in equation 17.1 (Chapter 17). An increase in growth rate, for example, would require an increase in wall extensibility (m), an increase in turgor pressure (P), or a decrease in yield threshold (Y). (Hydraulic conductance, L, of the plasma membrane depends on the presence of aquaporins and is not normally a limiting parameter.) Direct measurements of P, using a micropressure probe, have indicated that turgor pressure does not change significantly during auxin-stimulated increase in the growth rate of pea stem sections. Although Y cannot be measured directly, the results of indirect tests indicate that yield threshold does not change either. That leaves extensibility, *m*. Extensibility is difficult to assess. It is on the one hand a rate coefficient, but it is also a measure of the capacity of cell walls to undergo irreversible (plastic) deformation. A number of tests to measure extensibility have been devised. Whichever the method, however, the answer is invariably the same—the induction of rapid cell enlargement by auxin is accompanied by a large and rapid increase in wall extensibility.

The role of low pH in cell enlargement was introduced in Chapter 17. At the same time that this relationship between acid pH and cell enlargement was becoming clear, it was also discovered that auxin would cause growing cells to excrete protons. Several lines of evidence indicate that proton secretion is central to auxin-enhanced cell enlargement. (1) With *Avena* coleoptiles the pH of the **apoplastic**, or cell wall, solution drops from 5.7 to 4.7 within 8 to 10 minutes of auxin application. This lag period is consistent with the lag period observed between auxin addition and the beginning of the growth response. (2) Auxin-stimulated proton secretion is an energy-dependent process inhibited by both metabolic inhibitors and inhibitors of auxin-induced growth. (3) If the wall space of coleoptile sections is infiltrated with neutral buffers to prevent pH change, auxin-induced growth is almost completely

prevented. (4) Agents other than auxin that cause proton excretion have an effect similar to auxin on the promotion of growth. One such agent is **fusicoccin**, a phytotoxin from the fungus *Fusicoccum amygdali*, which causes cells to excrete protons at a great rate.

In 1970, R. Cleland and D. Rayle proposed a simple but rather provocative theory to explain auxin-stimulated increases in cell wall extensibility. They suggested that auxin causes acidification of the cell wall environment by stimulating cells to excrete protons. There the lower pH activates one or more wall-loosening enzymes, which have an acidic pH optimum. At about the same time, A. Hager, working in Germany, published a similar proposal but went further to suggest that auxin stimulated proton excretion by activating a plasma membrane–bound ATPase proton pump. The combined Cleland-Hager proposals are known as the **acid-growth hypothesis**. Although the acid-growth hypothesis has been tested in relatively few tissues (it has been tested thoroughly only in *Avena* coleoptiles), the evidence is generally supportive. In its present form, the acid-growth hypothesis proposes that auxin activates ATP-proton pumps located in the plasma membrane (Figure 18.8A). The resulting acidification of the cell wall space lowers the pH toward the optimum range for expansin activity. Increased expansin activity, in turn,

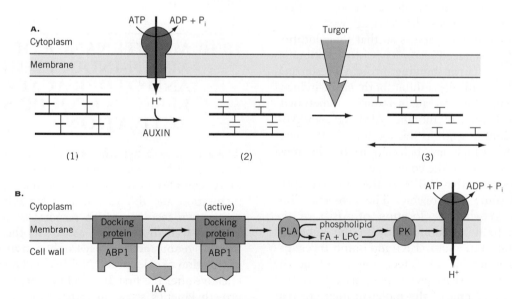

FIGURE 18.8 A schematic demonstrating the role of auxin in the acid-growth hypothesis for cell enlargement. (*A*) Cell wall polymers (cellulose microfibrils) are extensively cross-linked with load-bearing xyloglycans (1), which limits the capacity of the cell to expand. An auxin-activated ATPase-proton pump located in the plasma membrane acidifies the cell wall space by pumping protons from the cytoplasm. The lower pH activates wall-loosening enzymes, such as extensins, that loosen the load-bearing bonds (2). The forces of turgor acting on the membrane and cell wall cause the polymers to displace (3) and allow the cell to enlarge. (*B*) A hypothetical signal transduction chain linking auxin with activation of the ATPase-proton pump. See text for details. Abbreviations: ABP1, auxin-binding protein 1; PLA, phospholipase A$_2$; FA, fatty acids; LPC, lysophospholipid; PK, protein kinase.

increases wall extensibility and allows for turgor-induced cell expansion as described earlier in Chapter 17.

Although auxin does enhance the activity of ATPase-proton pumps in the plasma membrane, auxin itself does not bind to the ATPase. Therefore, there must be an auxin receptor that initiates a signal transduction chain linking the presence of auxin with increased ATPase activity. A putative auxin receptor has been isolated from maize (*Zea mays*), but details of the signal transduction chain itself remain obscure.

The maize auxin receptor is membrane-associated protein designated ABP1 (*A*uxin-*B*inding *P*rotein 1). ABP1 is a 43 kDa glycoprotein dimer of 22 kDa subunits that has a high affinity for IAA. ABP1 has been localized primarily in the endoplasmic reticulum, but small populations are also found associated with the plasma membrane and in the cell wall. ABP1 is a prime candidate for the auxin receptor that mediates cell elongation, although the evidence for this role is indirect. Perhaps the most compelling evidence comes from experiments with **antibodies**. Antibodies are proteins produced by the immune system of an animal in response to the presence of **antigens**. Antibodies will bind with the antigen, usually a "foreign" protein, to render that protein inactive. Antibodies (designated IgG) can be raised against plant proteins by injecting purified protein into an animal such as a mouse or rabbit. Antibodies are a useful tool because of the specificity of the antibody–antigen reaction. Antibodies can also be "tagged" with fluorescent chemicals or other markers so that their location can be readily visualized by microscopy. Antibodies raised against the auxin-binding protein (designated IgG-antiABP) specifically inhibit both auxin-induced coleoptile elongation and auxin-induced hyperpolarization of the plasma membrane. Also, IgG-antiABP applied to coleoptile sections was localized in the outer epidermal cells, which are believed to be the most auxin-responsive cells in the coleoptile.

The suggestion that ABP1 is the auxin-receptor has attracted some controversy. The principal difficulty has to do with the location of ABP1 in the cell. ABP1 is found predominantly in the lumen of the endoplasmic reticulum (ER) and some investigators have been unable to detect any ABP1 at the plasma membrane. ABP1 even contains amino acid sequences at either end of the molecule that are typical of proteins normally retained within the lumen of the ER. However, more sensitive immunolocalization techniques have now confirmed a small population (perhaps 1000 molecules) on the plasma membrane of maize protoplasts. A second problem is that, based on amino acid sequence, the ABP1 protein appears to have no lipophilic membrane-spanning domain. To reconcile these observations, it has been proposed that ABP1 forms a complex with a transmembrane docking protein. According to this model, the docking protein provides the necessary lipid solubility to anchor ABP1 to the membrane. The ABP1-docking protein complex is then exported from the ER to the plasma membrane where it is inserted with ABP1 facing the outside (Figure 18.8B). It has been proposed that the ABP1-docking protein complex is itself inactive, but attachment of an auxin molecule activates the complex and initiates the signal transduction pathway. The proposed docking protein has yet to be identified, but there is some suggestion that it might be a GCPR receptor in the family of G-proteins (Chapter 17).

Auxin also activates the enzyme phospholipase A_2 (PLA_2) and several experiments have implicated PLA_2 in the signal transduction chain. For example, activation of PLA_2 can be blocked by IgG-antiABP. Also, both **lysophospholipids** and fatty acids (the products of PLA_2) stimulate proton secretion and elongation. These effects are inhibited by vanadate, which specifically blocks the plasma membrane proton-ATPase. These data suggest that PLA_2 follows ABP1 in the chain and that lysophospholipids and fatty acids appear further along. Finally, both the IAA and lysophospholipids effects on proton secretion and elongation can be blocked by protein kinase inhibitors, suggesting that the lipids activate the proton-ATPase with the involvement of a protein kinase cascade. A model illustrating how these components might interact is presented in Figure 18.8B.

18.10 MAINTENANCE OF AUXIN-INDUCED GROWTH AND OTHER AUXIN EFFECTS REQUIRES GENE ACTIVATION

The acid-growth hypothesis does not alone resolve the question of how auxin regulates cell growth, let alone more complex developmental problems such as cell maturation and differentiation. One difficulty is that green stem sections, which respond to auxins, do not respond well (if at all) to acids. Another difficulty is that exogenous acid induces only a transitory growth stimulation of coleoptiles. Neither acid nor fusicoccin is effective after the first 30 to 60 minutes. Auxin-induced growth kinetics show an initial rapid increase in the growth rate that reaches a maximum within 30 to 60 minutes. This initial burst is followed by a steady or gradually declining rate over the next 16 hours (Figure 18.9). The most plausible explanation for such a two-phase response curve is that the acid-growth response is limited primarily to the rapid initial growth response. Additional auxin-regulated factors must then be required for the maintenance of growth over the longer term, including the well-defined progression of cells through the sequence division → expansion

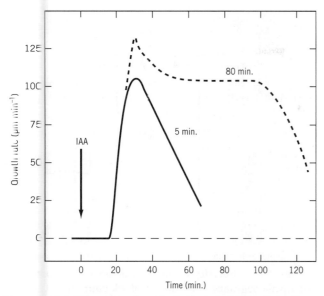

FIGURE 18.9 **Kinetics of auxin-induced elongation of maize (*Zea mays*) coleoptiles. The two curves differ in the duration of auxin action. In each case, auxin (10^{-5} M IAA) was added at time = 0 and removed after the indicated period (5 or 80 min). (From Dela, Fuente, R. K., A. C. Leopold. 1970. *Plant Physiology* 46:186. Copyright American Society of Plant Physiologists.)**

→ maturation → differentiation. These additional factors involve the transcription of genes and synthesis of growth-promoting proteins.

Auxin rapidly and specifically stimulates the transcription of a set of genes known as primary auxin responsive genes. These include both **SAUR** (small auxin upregulated RNAs) and *AUX/IAA. SAUR* genes encode short, relatively unstable RNA transcripts. In soybean hypocotyls, the expression of *SAUR* genes appears to be localized in tissues that normally respond to auxin and the RNA transcripts can be detected within 2 to 3 minutes of auxin application—even before auxin-induced elongation can be observed. Furthermore, an asymmetrical distribution of SAUR transcripts has been detected in gravity-stimulated seedlings. The asymmetry correlates with the differential cell elongation in responding seedlings, but can be detected even before any visible signs of curvature. Finally, several auxin-resistant mutants in *Arabidopsis* show low levels of SAUR expression in response to auxin treatment.

AUX/IAA genes are induced over a period of 4 to 30 minutes following application of auxin. This is a large family of genes—there are at least 29 different *AUX/IAA* genes in the *Arabidopsis* genome—that function as transcriptional regulators. The AUX/IAA proteins do not bind directly with DNA, but exert their regulatory effect by interacting with other proteins called the auxin response factor (ARF). ARFs bind to the promoter region of auxin-responsive genes and may act either to activate or to repress gene expression. Since

AUX/IAA proteins repress the activity of ARFs, they may act as either positive or negative regulators.

Early studies indicated that many of these responsive genes could also be induced by the protein synthesis inhibitor cycloheximide. This observation suggests that these genes may be controlled by short-lived repressor proteins that normally prevent transcription. According to one model, auxin was thought to initiate the ubiquitin-mediated degradation of these repressor proteins. This model was confirmed with the discovery of the *TIR1* gene in *Arabidopsis*. Originally identified in a genetic screen for auxin transport inhibitors (hence its name *Transport Inhibitor Response 1*), it was soon shown that TIR1 is a soluble, nuclear-located auxin-receptor protein that works in conjunction with auxin to derepress the transcription of auxin-responsive genes.

TIR1 is an F-box protein (see Chapter 17, Box 17.3, for the role of F-box proteins). However, in addition to having a recognition site that allows it to bind with the SCF scaffold, TIR1 also has a recognition site for auxin. A recent study of the crystal structure of TIR1 has shown that on the surface of the protein there is a pocket that accommodates the AUX/IAA peptide. However, the affinity of TIR1 for AUX/IAA is very low unless an auxin molecule is also present. The auxin molecule sits at the bottom of the pocket where it simultaneously interacts with both proteins. Auxin thus serves as a "molecular glue" that enhances TIR1-AUX/IAA binding. Once both the auxin and AUX/IAA proteins are in place, TIR1 is then able to link with the SCF complex for subsequent ubiquitination and degradation of the repressor by the ubiquitin-26S proteasome pathway (Figure 18.10). Removal of the AUX/IAA repressor protein derepresses the auxin-responsive gene, allowing the gene to proceed with the transcription of messenger RNA and, as a consequence, translation of auxin-induced proteins. Auxin appears to modulate development through depression of auxin-responsive genes, not through a simple activation.

As a side note, it is interesting that the crystallographic studies now make it easier to answer the long-standing question—"What makes an auxin?" Essentially any molecule that fits into the TIR1 binding pocket to enhance TIR1-AUX/IAA interaction will qualify as an auxin. The relative effectiveness of different auxin molecules depends on how well they fit in the pocket.

18.11 MANY ASPECTS OF PLANT DEVELOPMENT ARE LINKED TO THE POLAR TRANSPORT OF AUXIN

Auxin transport has naturally been studied almost exclusively in young seedlings, where synthesis takes place in the actively proliferating tissues. From these regions,

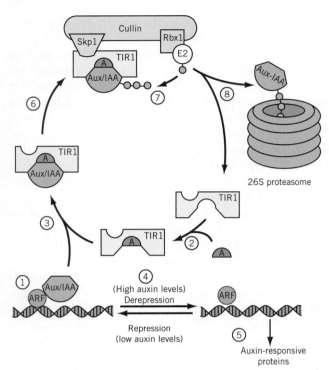

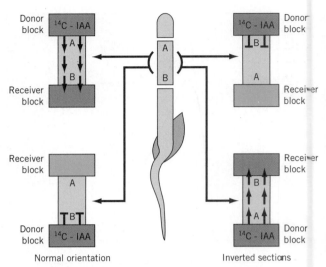

FIGURE 18.11 **Polarity in auxin transport in an oat coleoptile segment. The donor block contains ^{14}C-IAA. Regardless of the orientation of the segment, translocation of the radio-labeled IAA is always from the morphologically apical end (*A*) to the morphologically basal end (*B*) of the segment.**

FIGURE 18.10 **A model for auxin-induced gene derepression. (1) Auxin response factor protein (ARF) binds to the DNA in the promoter region of an auxin-responsive gene, but gene transcription is prevented by the presence of AUX/IAA repressor protein. When auxin levels are elevated, the auxin (*A*) combines with a nuclear-located auxin receptor, TRI1, to form an auxin-TRI1 complex (2). Auxin increases the affinity of TRI1 for AUX/IAA and facilitates the dissociation of AUX/IAA from the ARF (3). Removal of the AUX/IAA proteins from the ARF derepresses the gene (4), allowing transcription of mRNA and the translation of auxin-induced proteins, including AUX/IAA (5). Meanwhile, TRI1 recruits AUX/IAA to the E3 ubiquitin-ligating enzyme, or SCF complex (6), where (7) AUX/IAA is polyubiquitinated. The ubiqitinated protein is then recruited to the 26S proteasome (8), where it is degraded. The result is that when auxin levels are high, TRI1 facilitates active transcription of mRNA by continuously removing repressor protein. When auxin levels are low, TRI1 is unable to bind with the repressor, the repressor protein accumulates, and transcription is shut down.**

there appears to be a steady stream of auxin flowing down the shoot into the root. At least in *Arabidopsis* seedlings, some of this stream apparently moves down a concentration gradient in the phloem. A significant portion, however, moves through a complex, highly regulated polar transport mechanism.

Polar transport was originally described based on preferential movement either up or down in grass coleoptiles, stems, and roots (Figure 18.11). When movement is away from the morphological apex toward the morphological base of the transporting tissue, the

direction of movement is described as **basipetal**. Movement in the opposite direction, toward the morphological apex, is referred to as **acropetal**. When a stem or coleoptile section is inverted, as shown in Figure 18.11, the original direction of movement is maintained. However, as more is learned about auxin transport, the more evident it becomes that *directed* auxin transport may be lateral as well as up and down.

Polar transport of auxin in shoots tends to be predominantly basipetal at a velocity somewhere between 5 and 20 mm hr^{-1}. Acropetal transport in shoots is minimal. In roots, on the other hand, there appear to be two transport streams. An acropetal stream, arriving from the shoot, flows through xylem parenchyma cells in the central cylinder of the root and directs auxin toward the root tip. A basipetal stream then reverses the direction of flow, moving auxin away from the root tip, or basipetally, through the outer epidermal and cortical cell files.

The phenomenon of polar auxin transport has attracted widespread attention because of the assumption that auxin concentration is an important variable in several developmental responses. Auxin gradients due to polar transport have been invoked to explain, at least in part, developmental phenomena such as apical dominance, adventitious and secondary root formation, and differential growth responses to light and gravity. The flow of auxin in the root, for example, is intimately involved in the root response to gravity and will be covered in a later chapter.

Several observations indicate that polar transport involves a carrier-mediated, active transport mechanism in both shoots and roots. First, it can be shown that polar transport is inhibited by anaerobiosis or by respiratory poisons such as cyanide and 2,4-dinitrophenol. This is considered evidence that polar transport is an energy-requiring process dependent on oxidative metabolism in the mitochondrion. Second, certain chemicals, called **phytotropins**, have been known for some time to be specific, noncompetitive inhibitors of polar transport. These include **TIBA** (2,3,5-triiodobenzoic acid), **morphactin** (9-hydroxyfluorine-9-carboxylic acid), and **NPA** (N-1-naphthylphthalamic acid) (Figure 18.12). It is presumed that such inhibitors block auxin transport by binding to discrete carrier molecules that are involved in polar transport system. Third, the uptake of radioactive IAA is at least partially inhibited by nonradioactive IAA. This latter observation indicates that the labeled IAA and unlabeled IAA compete with each other for a finite number of carrier sites.

These observations formed the basis for the **chemiosmotic model** for auxin transport, proposed by P. H. Rubery, A. R. Sheldrake, and J. A. Raven in the mid-1970s. In its present form, the chemiosmotic model contains three essential features: (1) a pH gradient or proton motive force across the plasma membrane that provides the driving force for IAA uptake, (2) an IAA influx carrier, and (3) an IAA efflux carrier that is preferentially located at the base of auxin-transporting cells (Figure 18.13). The principles of the chemiosmotic model may be summarized as follows. IAA is a weakly acidic, lipophilic molecule. Depending on the pH, IAA may exist in either the protonated form (IAAH) or the unprotonated, anionic form (IAA⁻). The cell wall space is moderately acidic with a pH of about 5.5. At that pH, approximately 20 percent of the IAA will be protonated (IAAH). Consequently, the cell wall space will contain both anionic and protonated IAA. Based on its lipid solubility, a small proportion of uncharged IAAH molecule would be expected to diffuse slowly across the plasma membrane from the cell wall space

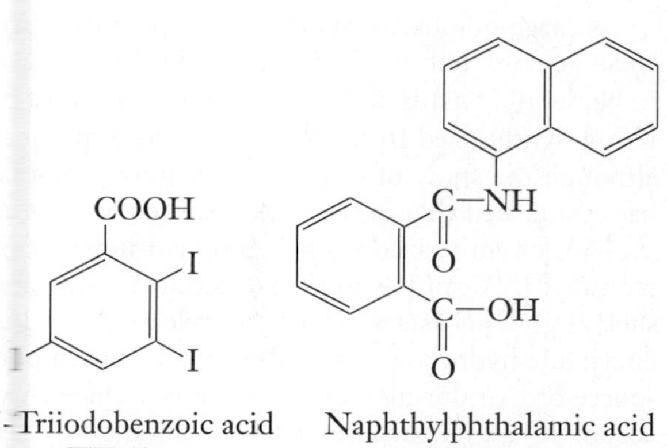

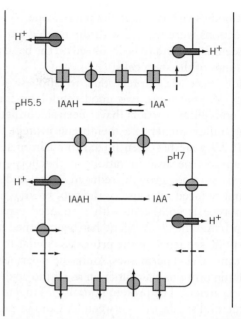

FIGURE 18.13 The chemiosmotic-polar diffusion model for polar transport of IAA. In the acidic cell wall space (pH 5.5) approximately 20% of the IAA is protonated. Protonated IAA (IAAH) may enter cells by diffusion across the cell membrane (dashed arrows) while the anionic form (IAA⁻) may be taken up through AUX1 (circles), a proton/IAA symport carrier located randomly in the plasma membrane. Inside the cell (pH 7.0) the deprotonated form IAA⁻ will dominate. IAA⁻ can exit the cell only through efflux carriers of the PIN family (squares) which are located preferentially at the base of the cell. Membrane-bound ATPase-proton pumps help to maintain the appropriate pH differential across the membrane and provide protons for IAA/H⁺ symport. The uniquely basal location of the efflux carriers is the key to polar transport.

into the cell. The bulk of the IAA, however, will enter the cell as IAA⁻ through an H⁺/auxin symport carrier (the influx carrier) that is uniformly distributed around the cell.

Once in the cytoplasm, where the pH is closer to 7.0, IAAH will dissociate to IAA⁻ and H⁺. The auxin is now trapped inside the cell because IAA⁻ can not readily diffuse across the membrane. The key to the chemiosmotic model, however, is the existence of a carrier, located only in the basal membranes of the cell, which mediates the efflux of IAA⁻ from the cell. It is the unique location of this efflux carrier, more than any other single factor, which establishes polarity in auxin transport.

The first direct evidence for the existence of a basal efflux carrier took advantage of the fact that the putative IAA-transport protein binds the phytotropin NPA. The NPA-binding protein was isolated, antibodies were raised against it, and the antibodies subsequently labeled with the fluorescent dye fluorescein in order to make

COOH

I

I

I

2,3,5-Triiodobenzoic acid
(TIBA)

C—NH
‖
O

C—OH
‖
O

Naphthylphthalamic acid
(NPA)

FIGURE 18.12 Phytotropins. Two examples of inhibitors of polar IAA transport.

the antibodies visible under the microscope. When pea stem sections were treated with the labeled antibodies, fluorescein was found to be localized on the basal plasma membranes of the stem cells.

More recently, largely due to studies of auxin mutants of *Arabidopsis*, two good candidates for auxin influx and efflux carriers have been identified. The putative influx carrier is a membrane protein, AUX1. The *AUX1* gene has been linked to auxin metabolism and transport because mutations at that locus exhibit IAA-resistant root growth, reduced lateral root initiation, and reduced response of the root to gravity. Such a phenotype is consistent with a reduced capacity to take up IAA. The *AUX1* gene has been cloned and the polypeptide sequence of the protein is similar to that of known amino acid permeases. Amino acid permeases are membrane proteins that function as amino acid/proton symport carriers. The protein homologies together with the structural similarities between IAA and its precursor amino acid, tryptophan, have led to the suggestion that the AUX1 protein functions as an auxin/proton symporter. Further support for this model is offered by the observation that the synthetic auxin NAA restores the gravitropic response to mutant (*aux1*) seedlings. NAA uptake by cells is not carrier-mediated, so the loss of AUX1 does not interfere with the response.

A family of genes, the *PIN* genes, that encode putative auxin efflux carriers have also been identified. (A total of eight *PIN* genes have now been identified in *Arabidopsis*.) One of the first to be discovered is the *PIN1* gene that controls flower development in *Arabidopsis*. The *pin1* mutant is characterized by influorescences that terminate in pinlike structures and show little or no evidence of floral bud development. Polar auxin transport is significantly reduced in mutant *pin1* influorescences and the characteristics of the mutant can be mimicked by blocking polar transport with phytotropins. As predicted by the chemiosmotic model, fluorescent-labeled antibodies have demonstrated that PIN1 protein is localized in the basal membranes of xylem parenchyma cells. Moreover, in keeping with the chemiosmotic model, AUX1 and PIN1 proteins are located at opposite ends of young root phloem cells (protophloem).

A second gene is variously called *PIN2*, *EIR1*, *WAV6*, or *AGR1*. The several names are due to the fact that the gene was identified independently in different laboratories, all studying mutants with impaired root response to gravity. *PIN2* was so named because it encodes a protein that is very similar to the protein encoded by *PIN1*. As with the PIN1 protein, immunolocalization experiments have shown that the PIN2 protein is localized in the basal membranes (i.e., furthest from the root tip) of cell files in the root cortex and epidermis. Moreover, like AUX1, the structure of the PIN proteins resembles the bacterial amino acid permeases and hence is a likely candidate for an IAA transporter.

Polarity in auxin transport is fundamental to plant development and PIN proteins direct this transport by moving from one cell surface to the other, in keeping with changing demands for auxin asymmetry. One problem, for example, that has long puzzled developmental biologists is how the apical-basal axis is established in young embryos. PIN proteins appear to part of the key. Immediately after the first division of the zygote, PIN proteins located acropetally in the basal cell direct the flow of auxin into the apical cell, specifying that cell as the founder of the proembryo. As the apical cells proliferate to form the globular embryo stage, they begin to synthesize auxin themselves. The PIN proteins then shift to a basipetal location and the direction of auxin flow reverses, thereby establishing the position of the developing root pole. Similar changing patterns of PIN distribution and the resulting changes in the polarity of auxin transport are equally important in other responses such as secondary root initiation and the responses of shoot and roots to gravity and unilateral illumination, which will be covered in detail in a later chapter.

SUMMARY

Hormones are numerous naturally occurring chemical substances that profoundly influence, at micromolar concentrations, the growth and differentiation of plant cells and organs. The effectiveness of a hormone depends on maintaining a closely regulated pool size, which is accomplished by a balance of biosynthesis, storage as inactive conjugates, and catabolic degradation of the molecule.

Auxins are characterized by their capacity to stimulate elongation in coleoptile and stem segment but are involved in virtually every aspect of plant development, including seed germination, vascular differentiation, lateral bud development, secondary root initiation, the response of roots and shoots to gravity, and flower and fruit development.

A large number of synthetic compounds express auxin activity, but indole-3-acetic acid (IAA) is thought to be the only naturally occurring auxin. In most plants IAA is synthesized from the amino acid tryptophan although the study of tryptophan-requiring mutants has established that in some plants, such as *Arabidopsis*, IAA is synthesized via a tryptophan-independent pathway. IAA can be stored as chemical conjugates such as glycosyl esters, which will release active IAA on enzymatic hydrolysis. Glycosyl esters are an important source of IAA during seed germination. Once IAA has accomplished its purpose, it can be removed by peroxidation to inactive products or conversion to amino acid conjugates. Auxins may be transported in the phloem or cell-to-cell by polar transport. The key to polar transport is the location of efflux carriers on specific cell walls.

The role of auxin in cell enlargement is best described by the acid-growth hypothesis. Central to this hypothesis is the activity of expansins; enzymes that weaken cross-links between cellulose molecules, increase wall extensibility, and allow for turgor-induced cell expansion. Auxin also acts to derepress genes by targeting repressor proteins for degradation by the 26S proteasome pathway, a process that accounts for auxin-induced developmental responses.

CHAPTER REVIEW

1. Why is it necessary for a hormone to be rapidly turned over? Describe how the size of the active hormone pool is regulated for auxins.

2. Some seeds appear to accumulate auxin conjugates. Can you suggest a physiological advantage for this?

3. Review the synthesis of IAA from tryptophan. Do all plants synthesize IAA from tryptophan? What is the evidence for alternative pathways?

4. The auxin needs of a germinating seed are initially met by stored auxin conjugates. Are all hormone-conjugates a form of hormone "storage"?

5. Describe the auxin-signaling pathway for cell enlargement.

6. Describe the auxin-signaling pathway for controlling gene expression.

7. Auxin transport is uniquely polar in character. How is this directional transport accomplished?

FURTHER READING

Badescu, G. O., R. M. Napier. 2006. Receptors for auxin: Will it end in TIRs? *Trends in Plant Science* 11:217–223.

Benjamins, R., N. Malencia, C. Luschnig. 2005. Regulating the regulator: The control of auxin transport. *BioEssays* 27:1246–1255.

Blakeslee, J. J., W. A. Peer, A. S. Murphy. 2005. Auxin transport. *Current Opinion in Plant Biology* 8:494–500.

Buchanan, B. B., W. Gruissem, R. L. Jones. 2000. *Biochemistry and Molecular Biology of Plants*. Rockville, MD: American Society of Plant Physiologists.

Davies, P. J. 2004. *Plant Hormones: Biosynthesis, Signal Transduction, Action*. Dordrecht: Kluwer Academic Publishers.

Kramer, E. M., M. J. Bennett. 2006. Auxin transport: A field in flux. *Trends in Plant Science* 11:382–386.

Napier, R. M. 2005. TIRs of joy: New receptors for auxin. *BioEssays* 27:1213–1217.

Tan, X. et al. 2007. Mechanism of auxin perception by the TIR1 ubiquitin ligase. *Nature* 446:640–645.

Woodward, A. W., B. Bartel. 2005. Auxin: Regulation, action, and interaction. *Annals of Botany* 95:707–735.

19

Hormones II: Gibberellins

Gibberellins are members of a large and varied family of plant constituents known as the terpenoids. Terpenes are normally recognized on the basis of their chemical structure, which may be dissected into one or more 5-carbon isoprene units. Gibberellins are thus biosynthetically related to the carotenoid pigments, sterols, latex, and other isoprene derivatives that will be discussed later. Gibberellins are noted for their capacity to stimulate hyper-elongation of intact stems, especially in dwarf plants and rosettes. They are also prominently involved in seed germination and mobilization of endosperm reserves during early embryo growth, as well as leaf expansion, pollen maturation, flowering, and fruit development. This chapter describes

- the biochemistry and metabolism of gibberellins,
- a review of gibberellins' principal effects on growth and development, and
- gibberellin receptors and the gibberellin signaling chain.

19.1 THERE ARE A LARGE NUMBER OF GIBBERELLINS

Gibberellins are a large class of molecules. In fact, more than 135 have now been identified in higher plants and fungi and additional members are added almost every year. Only a few of these are biologically active in their own right. The others are either intermediates in the biosynthetic pathway or products of inactivation. It is worth noting, however, that the number of gibberellins found in any one species or organ may be very small and the number of active gibberellins smaller yet. It is believed, for example, that GA_1 and GA_4 are the principal naturally occurring, active gibberellins in higher plants.

All gibberellins are diterpenes based on the 20-carbon **ent-gibberellane** structure (Figure 19.1). A little more than one-third of the gibberellins characterized to date have retained the full complement of 20 carbon atoms and are known as C_{20}-gibberellins. The others have lost carbon atom number 20 and are consequently known as C_{19}-gibberellins. With a complex ring structure and the number of possible substitutions on 19 or 20 carbon atoms, it is not difficult to see how there could be such a large number of gibberellins.

Gibberellins that are demonstrated to be naturally occurring and that have been chemically characterized are assigned an "A" number. This number does not imply chemical relationships; it is assigned roughly in order of discovery. A C_{20}-gibberellin commonly known as **gibberellic acid** was one of the first to be isolated and characterized. Because GA_3 is readily extracted from fungal cultures, it is also the most common

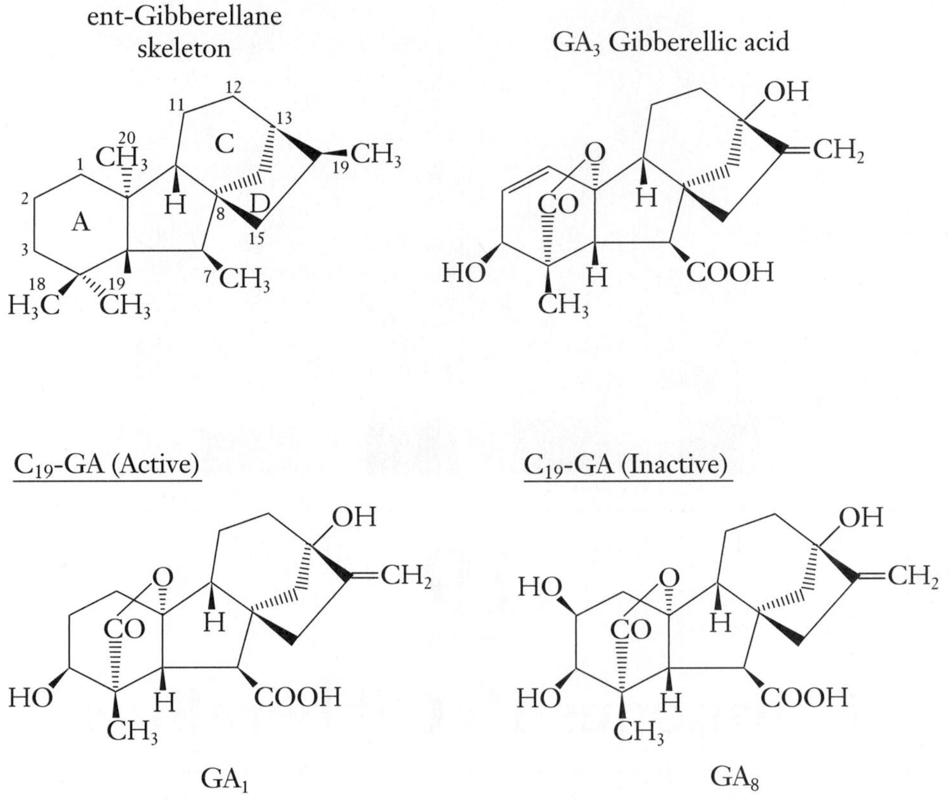

FIGURE 19.1 **The ent-gibberellane skeleton and chemical structures of selected active and inactive gibberellins. GA$_8$ is inactive because of the addition of the hydroxyl group in the 2 position.**

commercially available form. GA$_1$, GA$_3$, and GA$_4$, all of which promote vegetative growth, are the most active gibberellins and, consequently, the most widely used in gibberellin research.

There are certain structural requirements for GA activity. A carboxyl group at carbon-7 is a feature of all GAs and is required for biological activity, and C$_{19}$-GAs are more biologically active than C$_{20}$-GAs. In addition, those GAs with 3-β-hydroxylation, 3-β,13-dihydroxylation, or 1,2-unsaturation are generally more active; those with both 3-β-OH and 1,2-unsaturation exhibit the highest activity.

19.2 THERE ARE THREE PRINCIPAL SITES FOR GIBBERELLIN BIOSYNTHESIS

It is generally accepted that there are three principal sites of gibberellin biosynthesis: (1) developing seeds and fruits, (2) the young leaves of developing apical buds and elongating shoots, and (3) the apical regions of roots. Immature seeds and fruits are prominent sites of gibberellin biosynthesis. This is based on the observation that young fruits, seeds, and seed parts contain large amounts of gibberellins, particularly during stages of rapid increase in size. In addition, cell-free preparations from many seeds, such as wild cucumber (*Marah macrocarpus*) and pea (*Pisum sativum*), are able to actively synthesize gibberellins. The site of gibberellin biosynthesis may be the developing endosperm (as it is in the cucurbits), the young cotyledons of legumes, or the scutellum of cereal grains. As the seed matures, metabolism appears to shift in favor of gibberellin-sugar conjugates.

It is not as easy to obtain clear evidence that gibberellin biosynthesis occurs in shoots and roots. This is partly because gibberellin levels are much lower in vegetative tissues. Vegetative tissues also yield cell-free preparations that are less active, suggesting that enzyme levels for gibberellin metabolism are also lower than for reproductive tissues. Gibberellin synthesis in vegetative tissues is generally supported by the occurrence of gibberellins in tissue exudates and the effects of inhibitors of gibberellin biosynthesis. Gibberellins are apparently derived from fundamental precursors that are synthesized in chloroplasts (Section 19.3), which is consistent with the synthesis of gibberellins in green leaves. On the other hand, application of the inhibitor (2-chloroethyl) trimethylammonium chloride (CCC, an "**antigibberellin**") to roots rapidly decreases the

BOX 19.1
DISCOVERY OF GIBBERELLINS

During the late nineteenth and early twentieth centuries Japanese rice farmers grew concerned about a disease that seriously reduced the yield of their crops. Plants infected with the *bakanae* ("foolish seedling") disease exhibited weak, elongated stems and produced little or no grain. Japanese plant pathologists, interested in developing means for controlling the disease, soon established a connection with the presence of a fungus, *Gibberella fujikuroi*. In 1926, E. Kurosawa reported the appearance of symptoms of the disease in uninfected rice plants that had been treated with sterile filtrates from cultures of this fungus. By 1938, Japanese investi-

gators had isolated and crystallized the active material, which they called *gibberellin* after the genus name for the fungus.

Gibberellin did not come to the attention of Western plant physiologists until after the 1939–1945 war, when two groups—one headed by Cross in England and one by Stodola in the United States—isolated and chemically characterized *gibberellic acid* from fungal culture filtrates. At the same time, Japanese workers isolated three gibberellins, which they named gibberellin A_1, gibberellin A_2, and gibberellin A_3. Gibberellin A_3 proved to be identical with gibberellic acid. The known effect of gibberellins on rice and several other plant systems indicated that similar substances might be present in higher plants as well. The first higher-plant gibberellin to be characterized was isolated from immature seeds of runner bean (*Phaseolus coccineus*) and found to be identical with gibberellin A_1. Since then, gibberellins have been shown to be ubiquitous in higher plants.

---

amount of gibberellin appearing in exudates. Furthermore, removal of the root apices from seedlings of scarlet runner bean (*Phaseolus coccineus*) results in decreased shoot growth and the disappearance of GA_1. It remains possible that the synthesis of gibberellin precursors is not limited to chloroplasts, but occurs in other forms of plastids as well.

19.3 GIBBERELLINS ARE TERPENES, SHARING A CORE PATHWAY WITH SEVERAL OTHER HORMONES AND A WIDE RANGE OF SECONDARY PRODUCTS

Terpenes are a functionally and chemically diverse group of molecules. With nearly 15,000 structures known, terpenes are probably the largest and most diverse class of organic compounds found in plants. This large diversity arises from the number of basic units in the chain and the various ways in which they are assembled. Formation of cyclic structures, addition of oxygen-containing functions, and conjugation with sugars or other molecules all add to the possible complexity.

The terpene family includes, in addition to the gibberellins, the hormones abscisic acid and brassinosteroids, the carotenoid pigments (carotene and xanthophyll), sterols (e.g., ergosterol, sitosterol, cholesterol) and sterol derivatives (e.g., cardiac

glycosides), latex (the basis for natural rubber), and many of the essential oils that give plants their distinctive odors and flavors. Cytokinin hormones and chlorophyll, although not terpenes per se, do contain terpenoid side chains.

The distinguishing feature of terpenes and terpene derivatives is that they may be viewed as polymers of the simple 5-carbon unit 2-methyl-1,3-butadiene, or **isoprene** (Figure 19.2). Consequently, terpenes are often referred to as **isoprenoid** compounds. The actual building blocks for terpenoids, however, are not isoprene itself, but two phosphorylated derivatives known as **iso-pentenyl pyrophosphate** (**IPP**) and its isomer **dimethylallyl pyrophosphate** (**DMAP**). IPP is a product of two different biosynthetic pathways, one located in the cytoplasm and one located in chloroplasts (Figure 19.3).

The cytoplasmic pathway for IPP synthesis begins with acetyl-coenzyme A (acetyl-CoA), an intermediate in the respiratory breakdown of carbohydrate and fatty acid metabolism. Three molecules of acetyl-CoA condense in a two-step reaction to form hydroxymethylglutaryl-CoA, which is then reduced to **mevalonic acid** (**MVA**). Because the reduction of hydroxymethylglutaryl-CoA to mevalonic acid is virtually irreversible (at least *in vitro*), mevalonic acid

$$H_2C{=}\underset{\underset{\displaystyle CH_3}{|}}{C}{-}CH{=}CH_2$$

FIGURE 19.2 **The 5-carbon isoprene unit is the basic building block of terpenes and terpene derivatives.**

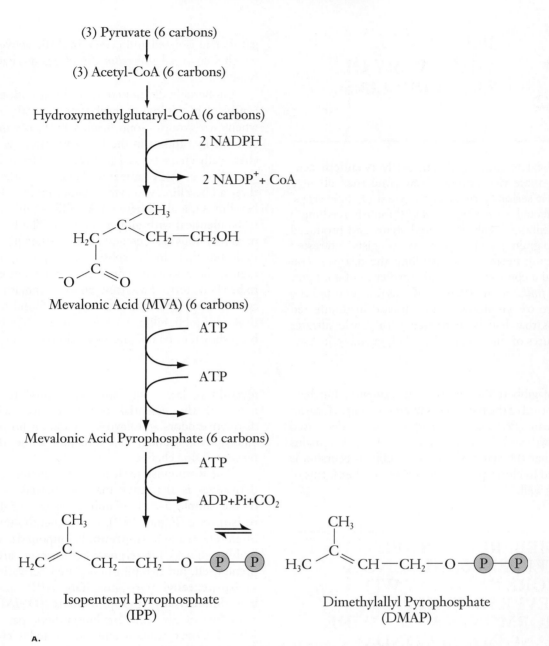

A.

Figure 19.3 **There are two pathways for the synthesis of isopentenylpyrophos-phate (IPP) and dimethylalylpyrophosphate (DMAP). (*A*) The mevalonic acid (MVA) pathway is located in the cytoplasm and is the source of cytoplasmic IPP. (*B*) The methylerythritol-4-phosphate (MEP) pathway is located in the chloroplast. The pathways are named after the first committed, or irreversible, intermediate.**

is the first committed precursor for terpenes and this pathway is therefore known as the **mevalonic acid pathway**. Mevalonic acid then undergoes a two-step phosphorylation, at the expense of two molecules of ATP, to form C_6 mevalonic acid pyrophosphate. Removal of a carboxyl group converts mevalonic acid pyrophosphate to the C_5 compound IPP, which is reversibly isomerized to DMAP.

The chloroplast pathway begins with the condensation of glyceraldehyde-3-phosphate (G3P) and pyru-

vic acid to form deoxyxylulose phosphate, a 5-carbon molecule. Deoxyxylulose phosphate is then reduced to **methylerythrltol-4-phosphate (MEP)**. MEP is subsequently phosphorylated and oxidized to form IPP.

Like the synthesis of IPP, the subsequent synthesis of terpenes is highly compartmentalized. In general, sesquiterpenes (C_{15}), triterpenes (C_{30}), and polyterpenes such as latex and other large molecules are produced in the cytosolic compartment while isoprene (a 5-carbon volatile organic hydrocarbon), monoterpenes

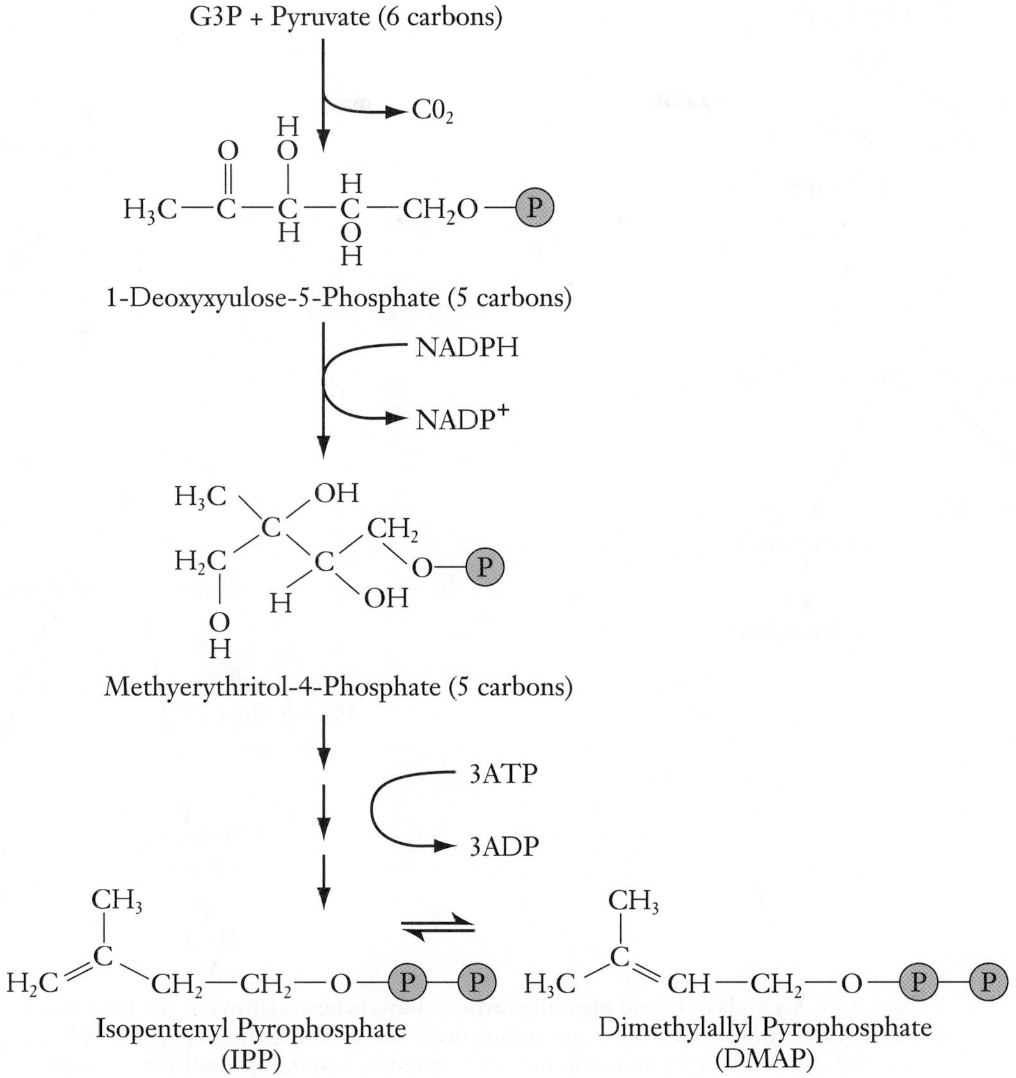

G3P + Pyruvate (6 carbons)

$\longrightarrow CO_2$

1-Deoxyxyulose-5-Phosphate (5 carbons)

NADPH

NADP$^+$

Methyerythritol-4-Phosphate (5 carbons)

3ATP

3ADP

Isopentenyl Pyrophosphate
(IPP)

Dimethylallyl Pyrophosphate
(DMAP)

B.

FIGURE 19.3 (Continued)

(C_{10}) diterpenes (C_{20}), and tetraterpenes (C_{40}) originate in plastids (Figure 19.4). In either case, the synthesis begins with the isomerization of IPP to DMAP. DMAP is, in effect, a reactive primer molecule that initiates chain formation by condensation with a molecule of IPP. Elongation of the chain then continues with the repetitive addition of IPP catalyzed by prenyltransferase enzymes. The distinction between cytosolic and plastid products is largely dependent on the compartmentalization of specific prenyltransferase enzymes. FPP is a branch point that can give rise to other C_{15} **sesquiterpenes** and, by head-to-head condensation of two C_{15} units, the C_{30} **triterpenes** (e.g., sterols). GGPP is a second branch point, giving rise to linear C_{20} **diterpenes, cyclic diterpenes** (including the gibberellins), and, by head-to-head condensation of two C_{20} units, the C_{40} **tetraterpenes** (e.g., carotenoid pigments).

19.4 GIBBERELLINS ARE SYNTHESIZED FROM GERANYLGERANYL PYROPHOSPHATE (GGPP)

The synthesis of gibberellins begins with the C_{10} isoprenoid geranylgeranyl phyrophosphate GGPP. The first two reactions involve the cyclization of GGPP to form copalylpyrophosphate and then kaurene (Figure 19.5). These two cyclization steps are inhibited by the antigibberellin or dwarfing agents, AMO-1618, CCC, and phophon-D, thus leading to a deficiency of gibberellin in the plant and reduced growth. Following cyclization, the carbon at position 19 on the kaurene molecule undergoes three successive oxidations in the sequence

CYTOPLASM

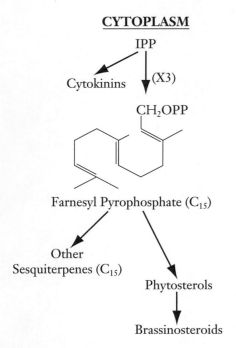

CHLOROPLAST

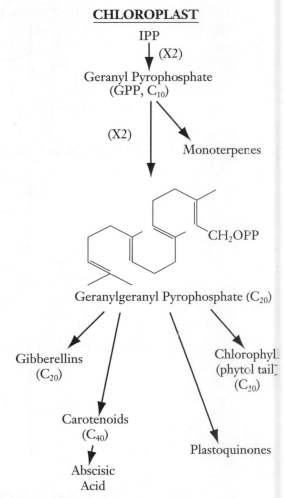

FIGURE 19.4 **Cytoplasmic and plastidic terpene biosynthesis, showing the structures of the key intermediates farnesyl pyrophosphate and geranylgeranyl pyrophosphate. Terpenes of increasing carbon number are formed by sequential addition of isopentenyl pyrophosphate units. Odd numbered terpenes (C_{15}, C_{30}) are synthesized in the cytoplasmic compartment while even numbered terpenes (C_{10}, C_{20}, C_{40}) are synthesized in the chloroplast.**

$CH_3 \rightarrow CH_2OH \rightarrow CHO \rightarrow COOH$ to form kaurenoic acid. The oxidation of kaurene to kaurenoic acid is inhibited by ancymidol, another dwarfing agent. The final two steps involve a hydroxylation at carbon-7 and contraction of the B ring with extrusion of carbon-7 to form GA_{12}-7-aldehyde.

The enzymes that convert *ent*-kaurene to GA_{12}-7-aldehyde are all NADPH-dependent, membrane-bound cytochrome P450 monooxygenases. The cytochrome P450 family is an interesting and important group of enzymes. They are generally noted for their lack of substrate specificity and are involved in numerous biosynthetic as well as degradative pathways in plants and animals. In mammals they are used to detoxify the liver by converting hydrophobic compounds to more hydrophilic derivatives that are more readily excreted.

As noted above, GA_{12}-7-aldehyde is the first compound with the true gibberellane skeleton and is the precursor to all other gibberellins. Oxidation of the aldehyde group on carbon-7 to a carboxyl group gives GA_{12}. This carboxyl group is a feature of all GAs and is required for biological activity. C_{19}-GAs arise by subsequent oxidative elimination of carbon-20. While the biosynthetic pathway up to GA_{12}-7-aldehyde is the same in all plants, subsequent pathways can vary substantially from genus to genus or even in different tissues in the same plant. A brief summary of demonstrated interconversions among gibberellins in pea seed and seedlings is presented in Figure 19.6 simply as one example. Similar pathways for biosynthesis of gibberellins have been demonstrated in cell-free endosperm and embryo preparations from pumpkin (*Cucurbita maxima*) or inferred from the knowledge of native gibberellins in *Arabidopsis*. The 13-hydroxylation pathway (bold arrows) leading to GA_{20} and GA_1 is probably of widespread occurrence in higher plants.

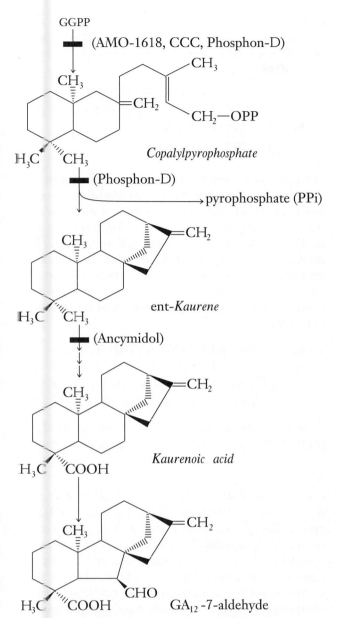

FIGURE 19.5 Gibberellin biosynthesis from geranyl-geranyl pyrophosphate (GGPP) to GA$_{12}$-7-aldehyde. GA$_{12}$-7-aldehyde is inactive, but serves as the precursor to all other gibberellins. The positions at which some antigibberellins (growth retardants) block gibberellin biosynthesis are indicated.

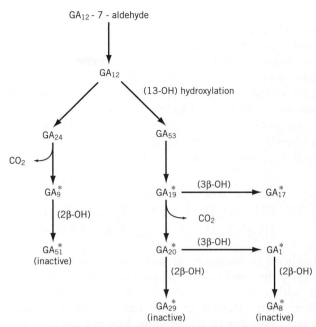

FIGURE 19.6 Proposed pathway for gibberellin biosynthesis in pea (*Pisum sativum*). The major pathway (bold arrows) occurs in seeds and shoots. The pathway shown in light arrows occurs only in shoots. The asterisk (*) indicates known endogenous forms.

19.5 GIBBERELLINS ARE DEACTIVATED BY 2β-HYDROXYLATION

There are several mechanisms for deactivating gibberellins and removing them from the active hormone pool. The principal mechanism, however, appears to be the introduction of a hydroxyl group at the 2 position, which renders the gibberellin inactive. Other inactivation products are possible, depending on the species. In

immature, actively developing seed of *Phaseolus vulgaris*, the principal free gibberellins are the active GA$_1$ and its inactive 2-β-hydroxyl analog GA$_8$, although small amounts of GA$_4$, GA$_5$, GA$_6$ (all C$_{19}$ GAs), and GA$_{37}$ and GA$_{38}$ (C$_{20}$ GAs) are also found. Mature seeds, however, contain mainly GA$_8$-glucoside, with smaller amounts of glucosyl esters of GA$_1$, GA$_4$, GA$_{37}$, and GA$_{38}$. In maize (*Zea mays*) the biologically inactive GA$_8$ is further converted to GA$_8$-glucoside, which is also inactive. In pea (*Pisum sativum*), however, GA$_8$ is oxidized to a 2-keto derivative referred to as a GA$_8$ catabolite.

19.6 GROWTH RETARDANTS BLOCK THE SYNTHESIS OF GIBBERELLINS

Since the 1950s, a number of synthetic compounds known as growth retardants have been developed. These compounds, also known as **antigibberellins**, have found commercial use, particularly in the production of ornamental plants. Growth retardants may be applied to potted plants either as a foliar spray or soil drench. The principal effect of growth retardants is to reduce stem elongation, resulting in plants that are shorter and more compact, with darker green foliage. Flower size, however, is unaffected. Because individual growth retardants block specific steps in gibberellin biosynthesis, they have also found use as tools in physiological research.

AMO-1618 and phosphon are antigibberellins that inhibit enzymes involved in the synthesis of kaurene while ancymidol blocks the subsequent oxidation of kaurene to kaurenoic acid. The effects of these inhibitors can be reversed by the application of gibberellins, such as GA_{20} or GA_3. Another inhibitor, known as BX-112, blocks the 3β-hydroxylation of GA_{20} to GA_1. In those plants where GA_1 is the active gibberellin, the effects of BX-112 can be reversed only by the application of GA_1 itself.

19.7 GIBBERELLIN TRANSPORT IS POORLY UNDERSTOOD

Gibberellin transport studies have been conducted largely by application of radioactively labeled GAs to either stem or coleoptile sections. Gibberellins have been detected in both the phloem and xylem saps. Transport of gibberellins does not appear to be polar, as it is with auxin, but moves along with other phloem-translocated organic materials according to a source-sink relationship. Whether gibberellins are actually transported in the xylem is not clear; they could end up there simply by lateral translocation from the phloem. On the other hand, it is likely that any gibberellins synthesized in the root tip are distributed to the aerial portions of the plant through the xylem stream. It is not known whether gibberellins are transported as free hormones or in conjugated form.

19.8 GIBBERELLINS AFFECT MANY ASPECTS OF PLANT GROWTH AND DEVELOPMENT

Since the discovery of gibberellins, it has been clear that this hormone regulates stem elongation and the mobilization of endosperm reserves during the early stages of seed germination. In addition, largely through studies of GA-deficient mutants, gibberellins have been implicated in a wide range of other developmental responses such as flowering and flower development, root and fruit growth, the development of seeds in the fruit, de-etiolation, and the initiation of leaf primordia in meristems.

19.8.1 GIBBERELLINS STIMULATE HYPER-ELONGATION OF INTACT STEMS, ESPECIALLY IN DWARF AND ROSETTE PLANTS

It was excessive stem elongation in infected rice plants that led to the discovery of gibberellins, and

Box 19.2
COMMERCIAL APPLICATIONS OF GIBBERELLINS

The principal commercial use of gibberellins is in the production of table grapes, such as the "Thompson Seedless." A gibberellin spray in the early stages of flowering thins the cluster by stimulating elongation of the floral stems. This spreads the flowers out, allowing for the development of larger fruit. Larger fruit size is encouraged by a second spray at time of pollination and fruit set. Gibberellins have also been used to enhance germination and stimulate early seedling emergence and growth in species such as grape, citrus, apples, peach, and cherry. Treatment of cucumber plants with gibberellin will promote formation of male flowers, which is useful in the production of hybrid seed. The GA-induced α-amylase in barley aleurone has led to widespread use of GA in the malting industry where it is used to speed up malt production.

The inhibition of gibberellin biosynthesis also has commercial applications. The growth of many stems can be reduced or inhibited by synthetic growth retardants or antigibberellins. These include AMO-1618, cycocel (or, CCC), Phosphon-D, ancymidol (known commercially as A-REST), and alar (or, B-nine). Growth retardants mimic the dwarfing genes by blocking specific steps in gibberellin biosynthesis, thus reducing endogenous gibberellin levels and suppressing internode elongation. These compounds have found significant commercial use, particularly in the production of ornamental plants. Growth retardants may be applied to potted plants either as a foliar spray or soil drench. Their principal effect is to reduce stem elongation, resulting in plants that are shorter and more compact, with darker green foliage. Flower size, however, is unaffected. Commercial flower growers have found these inhibitors useful in producing shorter, more compact poinsettias, lilies, and chrysanthemums, and other horticultural species.

In some areas of the world, wheat tends to "lodge" near harvest time, that is, the plants become top-heavy with grain and fall over. Spraying the plants with antigibberellins produces a shorter, stiffer stem and thus prevents lodging. Antigibberellins also have been used to reduce the need for pruning of vegetation under power lines.

hyper-elongation of stem tissue remains one of the more dramatic effects of gibberellins on higher plants. Unlike auxins, gibberellins promote elongation almost exclusively in intact plants rather than excised tissues. Nowhere is this more evident than in the control of internode elongation in genetic dwarfs. The relationship between dwarfing or internode-length genes and gibberellins was pioneered by the work of B. O. Phinney on maize (*Zea mays*) and P. W. Brian and coworkers on garden pea (*Pisum sativum*). Since these pioneering studies, experiments have been conducted with dwarf mutants of rice (*Oryza sativa*), bean (*Phaseolus vulgaris*), *Arabidopsis thaliana*, and several others. In all cases, application of exogenous gibberellin to the dwarf mutant restores a normal, tall phenotype (Figure 19.7). Exogenous gibberellin has no appreciable effect on the genetically normal plant.

In maize, more than 30 mutants that influence plant height have been described. Maize plants expressing these mutations have shortened internodes, due to reduced cell division and cell elongation, and at maturity reach only 20 to 25 percent of the height of normal plants. At least five of these mutants (*d1, d2, d3, d5, an1*) exhibit the normal phenotype when treated with GA$_3$, but show no response to other hormones or growth regulators. Activity assays such as these have been supported by other biochemical and radiotracer experiments, demonstrating conclusively that internode elongation in maize is under control of gibberellins. Specifically, each mutation blocks a different step in the biosynthetic pathway toward GA$_1$, which is the active form of gibberellin in maize. Similar experiments with the *Le* allele in garden pea (the same allele studied by Gregor Mendel in his pioneering genetic studies) have demonstrated that this dwarf genotype (the homozygous recessive *le/le*) also blocks the synthesis of GA$_1$. In both maize and pea, it has been shown that the dwarf genotype leads to a significant reduction in the gibberellin levels. While studies with dwarf plants have been instrumental in linking gibberellins with stem elongation, there are other dwarf mutants known that do not respond to application of gibberellin. These mutants may be unrelated to gibberellin-controlled growth and subject to other, as yet unknown, regulating factors.

Additional support for the role of gibberellins in stem elongation comes from the study of rosette plants. A rosette is essentially an extreme case of dwarfism in which the absence of any significant internode elongation results in a compact growth habit characterized by closely spaced leaves. The failure of internode to elongate may result from a genetic mutation, or may be environmentally induced. Regardless of the cause, hyper-elongation of stems in rosette plants is invariably brought about by the application of small amounts of gibberellin (Figure 19.8).

Environmentally limited rosette plants such as spinach (*Spinacea oleraceae*) and cabbage (*Brassica* sp.) generally do not flower in the rosette form. Just before flowering, these plants will undergo extensive internode elongation, a phenomenon known as **bolting**. Bolting is normally triggered by an environmental signal, either photoperiod (as in spinach) or a combination of low temperature and photoperiod (as in cabbage). We will return to the phenomena of photoperiod and cold requirement in later chapters. It is sufficient to note here that, under conditions normally conducive to the rosette habit, spinach, cabbage, and many other

FIGURE 19.7 The effect of gibberellic acid on dwarf pea seedlings. Left: Control, showing reduced internode elongation characteristic of the dwarf growth habit. Right: Gibberellin treated with a 5 × 10^{-4} M foliar-drench of GA$_3$. Note that gibberellin treatment increased stem elongation by simulating elongation of the internodes.

FIGURE 19.8 **Gibberellin-stimulated stem growth in a rosette genotype of *Brassica napus*. Treatments were (from left): 0, 0.5, 1.0, 10.0 ng GA$_3$ per plant, applied to the meristem.**

rosette plants can be induced to bolt by an exogenous application of gibberellic acid.

The above results suggest that (1) gibberellins are a limiting factor in the stem growth of rosette plants and (2) the effect of long days or cold treatment is to remove that limitation. These possibilities have been confirmed in spinach and *Silene armeria*, both photoperiodic plants requiring long days to flower, by the extensive investigations of J. A. D. Zeevaart and coworkers. Spinach contains six gibberellins, including GA$_{19}$ and GA$_{20}$. GA$_{20}$ will cause bolting in spinach under short day conditions while GA$_{19}$ is biologically inactive. Zeevaart and coworkers found that rosette plants of spinach contain high levels of the inactive form GA$_{19}$ and low levels of the active GA$_{20}$. Upon transfer to long day conditions, however, the level of GA$_{19}$ declined while the level of GA$_{20}$ increased (Figure 19.9). The reciprocal changes in GA$_{19}$ and GA$_{20}$ levels suggests a precursor-product relationship, which was confirmed in whole plants by feeding deuterium (^{2}H)-labeled precursors. In other experiments, ^{14}C-labeled GA$_{19}$ was converted to GA$_{20}$ by cell-free extracts from spinach plants maintained under long days, but not in extracts from plants maintained under short days. On the basis of these studies, it may be concluded that gibberellins have a significant role in the control of stem elongation in rosette plants.

The relationship between gibberellins and stem elongation in cold-requiring plants has not been studied as thoroughly as it has for plants that are sensitive to photoperiod. As mentioned above, exogenous application of GA$_3$ will substitute for the cold requirement in many plants and there is some evidence, based on bioassays,

that gibberellin-like activity increases in plants following cold treatment. It is reasonable to expect on the basis of these studies that changes in gibberellin biosynthesis or metabolism are involved, but a more thorough study is required.

19.8.2 GIBBERELLINS STIMULATE MOBILIZATION OF NUTRIENT RESERVES DURING GERMINATION OF CEREAL GRAINS

Gibberellins initiate the mobilization of nutrient reserves stored in the endosperm, while auxins promote elongation of the embryonic axis. The auxins that support early embryo growth are largely derived from the breakdown of stored conjugates to free, active IAA while the gibberellins, at least in cereal grains, appear to be released by the hydrated embryo from a preformed GA pool. In *Arabidopsis*, on the other hand, seeds carrying mutations such as *ga1*, *ga2*, and *ga3*, that act early in gibberellin biosynthesis, fail to germinate but germination can be rescued by applying exogenous gibberellin. More recently, it has been suggested that brassinosteroids might also have a role in germination of *Arabidopsis* seed (Chapter 21).

A role for gibberellins in mobilization of reserves during seed germination was first suggested by experiments on germinating cereal grains in the late 1950s. Cereal grains such as rye, barley, and wheat have a protein-rich layer of cells called the **aleurone** which surrounds the starchy endosperm tissue. During germination, cells in the aluerone secrete a range of hydrolytic enzymes, including α-amylase and proteases, which are involved in the hydrolysis of carbohydrate and protein stored in the endosperm.

The involvement of gibberellins in enzyme secretion can be shown by a relatively simple experiment. Seeds of cereals such as barley are transected to produce two half-seeds (Figure 19.10). One half-seed contains the embryo and the other half-seed does not. When imbibed, the embryo-containing half-seed will proceed to secrete α-amylase and other hydrolytic enzymes in order to digest the starchy endosperm, mobilize the resulting nutrients, and initiate germination. The half-seed without the embryo cannot, of course, germinate but neither does it produce elevated levels of α-amylase or any of the other hydrolytic enzymes required for germination. Treatment of the embryo-less half-seed with gibberellic acid, however, will stimulate the half-seed to produce high levels of α-amylase. Experiments of this general nature have shown that the germinating embryo sends a signal, probably gibberellin, to the aleurone cells. There the gibberellin either activates or derepresses transcription of genes encoding the necessary hydrolytic enzymes. These enzymes are then

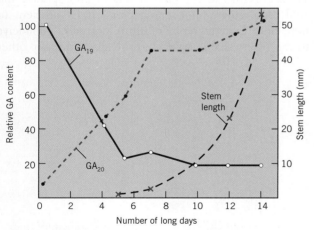

FIGURE 19.9 **Spinach plants (*Spinacea oleraceae*) exhibit extensive stem elongation when transferred from short days to long days. Stem elongation is accompanied by changes in gibberellin content as an inactive form is converted to an active form. A decrease in the level of the inactive GA$_{19}$ is matched by a corresponding increase in the active form G$_{20}$. (Redrawn from Metzger and Zeevaart 1980. *Plant Physiology* 66:844–846. Copyright American Society of Plant Physiologists.)**

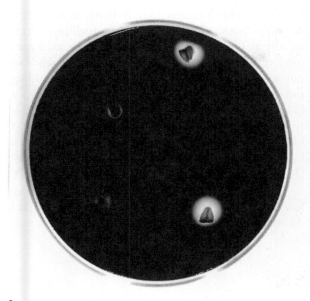
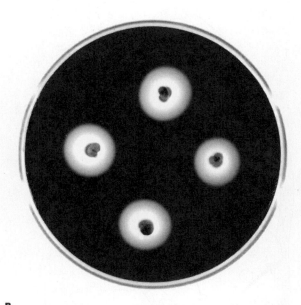

A.

B.

FIGURE 19.10 **Gibberellin-stimulated secretion of α-amylase from barley half-seeds. Embryo-less half-seeds were incubated on the surface of a starch-agar gel. After 48 hours, the gel was washed with iodine-potassium iodide (IKI), a reagent that reacts with starch to form a blue-black color. A clear circle, or halo, surrounding the half-seed indicates the digestion of starch by α-amylase. The control plate (left) contains four half-seeds but no added gibberellin. Two half-seeds produced no α-amylase while the other two exhibited low activity. The plate on the right contained 10 nanomoles gibberellic acid. Each of the gibberellin-treated half-seeds is surrounded by a large halo, indicating active α-amylase secretion. (From a student experiment.)**

released into the endosperm where they break down the starches and proteins to provide nutrients for the growing embryo (Figure 19.11).

It should be noted that much of this work has been conducted on a single cultivar (cv. "Himalaya") of barley and the responsiveness of seeds to gibberellin can be significantly affected by environment during seed development and maturation. Seed from barley grown at high temperatures, for example, produce high levels of α-amylase in the absence of added GA. Nonetheless, the principles that have emerged from the barley system appear to be widely applicable to wheat, rye, oats, and other cereal grains.

19.9 GIBBERELLINS ACT BY REGULATING GENE EXPRESSION

Early physiological studies found that GA-stimulated α-amylase secretion could be blocked by inhibitors such as actinomycin D and cycloheximide, which inhibit RNA and protein synthesis, respectively. Results like this indicated that gibberellin stimulated *de novo* synthesis of α-amylase by the aleurone layer.

Does gibberellin regulate transcription of α-amylase mRNA? An unequivocal answer to this question was possible only after techniques for isolation of protoplasts from barley aleurone cells were perfected. Aleurone protoplasts provide a useful experimental system. They respond normally to GA by exhibiting the same ultrastructural changes and producing the same isozymes with the same efficiency as intact aleurone cells. Most importantly, high yields of transcriptionally active nuclei can be readily isolated from protoplasts.

Based on evidence from several lines of investigation, it is clear that gibberellin dose regulates transcription of α-amylase mRNA. Both *in vivo* pulse labeling of protein and *in vitro* translation of protein from total aleurone RNA, followed by electrophoretic and autoradiographic analysis, show significant increases in the amount of α-amylase translated following the application of gibberellin (Figure 19.12). α-Amylase may constitute as much as 50 to 60 percent of the total translated aleurone protein.

The gibberellin response is apparently not limited to α-amylase alone, as a small number of other peptides either increase or decrease following gibberellin treatment. Furthermore, Northern blot hybridization of α-amylase cDNA clones with RNA from aleurone layers has confirmed that gibberellin causes an increase

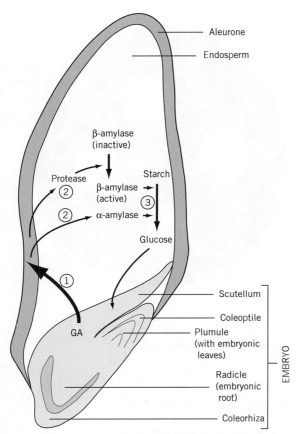

FIGURE 19.11 **A schematic illustrating gibberellin-induced release of enzymes and carbohydrate mobilization during germination of barley (*Hordeum vulgare*) seed. Gibberellin moves from the embryo (1) to the aleurone where it stimulates the synthesis of α-amylase and protease enzymes. (2) The protease converts an inactive β-amylase to the active form. α- and β-amylase together digest starch to glucose, (3) which is mobilized to meet the metabolic demands of the growing embryo.**

in the physical abundance of α-amylase mRNA. Following gibberellin treatment, α-amylase mRNA may comprise as much as 20 percent of the total translatable mRNA. Finally, time course studies have shown that the rate of α-amylase synthesis following gibberellin treatment closely correlates with the rate of mRNA accumulation. These studies pretty well established that a primary action of gibberellin, at least with respect to seed germination, is to regulate in some way gene transcription.

The challenge over the past ten years or so has been to identify the genes and proteins that are involved in the gibberellin signaling pathway. This has been accomplished primarily by screening for mutations that either enhance or lower gibberellin sensitivity. One such mutant is the rice mutant *slender rice1* (*slr1*). The *slr1* mutation results in what is called a *constitutive gibberellin*

phenotype, which means that it behaves as if it were saturated with gibberellin. Mutant seedlings elongate rapidly and excessively compared with the wildtype and α-amylase is produced by embryo-less half-seeds without added gibberellin. Excessive gibberellin production can be ruled out as the cause because the endogenous GA content of the mutant is lower than that of the wild-type and the response is not sensitive to inhibitors that block gibberellin synthesis. Similar "slender" mutants have been described in barley (*Hordeum vulgare*) and pea (*Pisum sativum*).

Another example is the gibberellic acid insensitive (*gai*) mutant in *Arabidopsis*. *gai* mutants are dwarfs that do not respond to applied gibberellin. Similar mutations

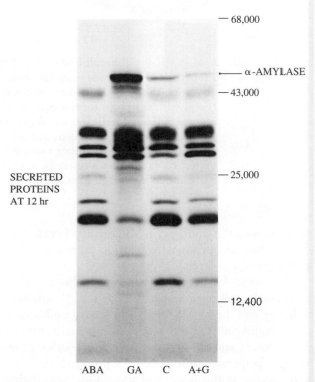

FIGURE 19.12 **Hormonal control of α-amylase biosynthesis by barley aleurone layers. Polypeptides synthesized by isolated aleurone layers were labeled with the radioactive amino acid [^{35}S]methionine incubated in the presence of ABA, GA$_3$, or ABA + GA$_3$ (A + G). Polypeptides secreted into the incubation medium were separated by electrophoresis on polyacrylamide gels. Polypeptides incorporating the radioactive label were detected by exposing the gel to X-ray film. Note the pronounced stimulation of α-amylase biosynthesis in the presence of GA$_3$, a stimulation almost completely abolished in the presence of ABA. C represents untreated controls. Numbers indicate approximate molecular mass. (From Higgins et al. 1982. *Plant Molecular Biology* 1:191. Reprinted by permission of Kluwer Academic Publishers. Original photograph kindly provided by Dr. J. V. Jacobsen.)**

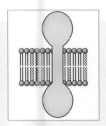

BOX 19.3
DELLA PROTEINS AND THE GREEN REVOLUTION

Throughout the 1950s and 1960s, the world witnessed advances in plant breeding and agronomy that led to significant increases in agricultural production worldwide. These transformations, collectively referred to as the "green revolution," were supported in part by the Rockefeller Foundation, other philanthropic organizations, and universities in an effort to help food production keep pace with worldwide population growth.

Some of the most significant advances were seen in wheat and maize breeding. Wheat yields, for example, were substantially increased by breeding new varieties that were shorter and thus more resistant to lodging damage by wind and rain. The new, shorter varieties also increased grain production at the expense of straw biomass. The significance of these efforts was recognized when Norman Borlaug of the International Maize and Wheat Improvement Center (CIMMYT) in Mexico was awarded the 1970 Nobel Peace Prize for his work in wheat breeding.

The new varieties were shorter because the breeders had introduced mutant dwarfing alleles, *Reduced height-1* or *Rht-B1*in wheat and *dwarf-8* or *d8* in maize, that conferred a reduced response to gibberellins. In 1999, J. Peng et al. (*Nature* 400:256) showed that the *Rht* and *d8* loci encode Della transcription factors that are analogous to the gibberellin insensitive (GAI) protein in *Arabidopsis*. Like the mutant *gai* protein in *Arabidopsis*, the rht-b1 and d8 proteins have an altered amino acid sequence at the N-terminal end that interferes with gibberellin signaling.

at the same genetic locus are responsible for dwarfing in wheat (*Triticum aestivum*) and maize (*Zea mays*). *slr1* and *gai* mutant plants respond the way they do because the wild-type *SLR1* and *GAI* genes encode transcription regulators referred to as **DELLA proteins**. DELLA proteins are a class of nuclear proteins that appear to function as repressors in gibberellin signaling. This conclusion is based on earlier observations that degradation of DELLA proteins *in planta* triggers gibberellin-type responses. Although *SLR1* is the only DELLA gene described for rice, a total of five, including *GAI*, have been described for *Arabidopsis*.

The discovery of DELLA proteins has been followed by the discovery of a soluble GA receptor protein, GA INSENSITIVE DWARF 1 (GID1), in rice. *GID1* encodes a nuclear protein that binds with gibberellic acid both *in vitro* and *in vivo*. Subsequently it was found that *Arabidopsis* carries three homologous genes, designated *AtGID1a*, *AtGID1b*, and *AtGID1c*, which encode similar proteins. Each of the *Arabidopsis* genes apparently has a specific, but overlapping, role in growth and development. For example, individually the mutant genes *atgid1a*, *atgid1b*, and *atgid1c* show no clear phenotype and even when any two mutant genes were present, germination was normal. The double mutant *atgid1a-atgid1c* exhibited a dwarf phenotype while other double mutant combinations were of normal height. However, the triple mutant *atgid1a-atgid1b-atgid1c* failed to germinate without peeling off the seed coat and the seedlings were severe dwarfs, achieving a height of only a few millimeters after a month's growth. On the other hand, the expression of any one of the *Arabidopsis* wildtype genes in *gid1* mutant rice plants produced normal rice plants with normal GA sensitivity.

All of this has been brought together recently by the demonstration that the GID1 receptor binds not only with gibberellin, but also with the DELLA protein SLR1 to form a GA-GID1-SLR1 complex. According to the current model, this hormone-receptor-repressor complex then results in the degradation of the DELLA repressor protein by the 26S proteasome pathway, thus relieving a DELLA-imposed repression and allowing expression of the gibberellin-responsive gene. In other words, the function of gibberellin—to derepress genes by facilitating the removal and ubiquitin-mediated degradation of a repressor protein—is similar to that of auxin described earlier in this chapter. This model would also explain why the *gai* mutants are dwarfs that do not respond to GA. Cloning of the *Arabidopsis GAI* gene has shown that the mutant gai protein is missing a group of amino acids at its N-terminal end. These missing amino acids presumably render the protein insensitive to the gibberellin-receptor complex without interfering with its role as a repressor (Figure 19.13).

GA elicits multiple responses in plants and it may yet be demonstrated that some of these responses involve membrane-bound receptors and a cascade of secondary messengers. However, it is interesting to note that, in 1957, P. W. Brian suggested that plants contained an endogenous inhibitor and gibberellin acted not by direct stimulation of a response but by relieving this inhibition. Fifty years later, we have come full circle and Brian's proposal now seems remarkably prescient.

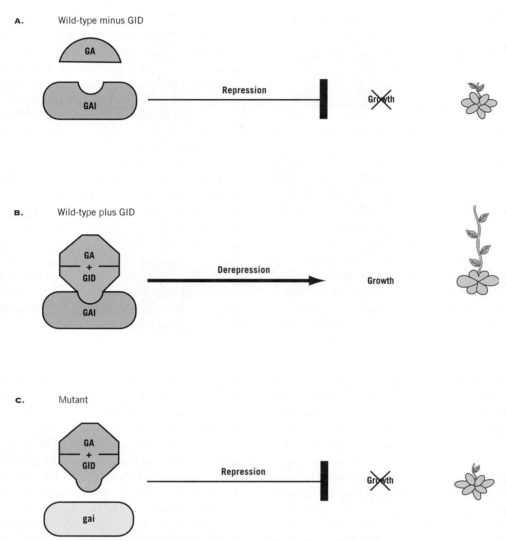

FIGURE 19.13 **A model for the roles of GA receptors and DELLA proteins in the gibberellin signaling pathway. (***A***) GAI represses growth-dependent gene expression.
GA alone cannot bind with GAI to reverse the repression. (***B***) The GA/GID-receptor
complex is able to bind with GAI, targeting the repressor for degradation and dere-
pressing the gene. (***C***) The mutant DELLA protein gai is unable to bind with the
GA/GID-receptor complex, but continues to repress gene expression. (After Harberd, N. P. 1998.** *BioEssays* **20:1001–1008.)**

SUMMARY

Gibberellins are diterpenes, noted for their capacity to
stimulate hyper-elongation of intact stems, especially
in dwarf and rosette plants and stimulate mobilization
of endosperm reserves during seed germination.

Gibberellins are related biosynthetically to
carotenes and other isoprene derivatives. Principal
intermediates are geranyl-geranyl-pyrophosphate and
GA_{12}-7-aldehyde. Gibberellins can be inactivated by
hydroxylation at the C-2 position or by conversion
to inactive conjugates. Gibberellins appear to be
transported primarily in the phloem in response
to source-sink relationships. Chemicals that block

gibberellin biosynthesis are used commercially to
produce shorter, more compact plants.

Gibberellin appears to act primarily by targeting the repressor proteins for degradation by the 26S
proteasome pathway.

CHAPTER REVIEW

1. Describe how the size of the active hormone pool is regulated for gibberellins.

2. What is the significance of the "G" number
   assigned to each gibberellin?

3. Identify structural characteristics that determine whether a gibberellin is active or inactive.

4. Describe hormonal involvement in nutrient mobilization in germinating cereal grains.

5. Compare and contrast the auxin and gibberellin signal chains for gene derepression.

FURTHER READING

Brian, P. W. 1957. The effects of some microbial metabolic products on plant growth. *Symposium of the Society for Experimental Biology* 11:168–182.

Buchanan, B. B., W. Gruissem, R. L. Jones. 2000. *Biochemistry and Molecular Biology of Plants*. Rockville, MD: American Society of Plant Physiologists.

Davies, P. J. 2004. *Plant Hormones: Biosynthesis, Signal Transduction, Action*. Dordrecht: Kluwer Academic Publishers.

Eckardt, N. A. 2007. GA perception and signal transduction: Molecular interactions of the GA receptor GID1 with GA and the DELLA protein SLR1 in Rice. *The Plant Cell*. 19:2095–2097.

Harberd, N. P. et. al. 1998. Gibberellin: Inhibitor of an inhibitor of . . . ? *BioEssays* 20:1001–1008.

Olszewski, N. T-p. Sun, F. Gubler 2002. Gibberellin signaling: Biosynthesis, catabolism, and response pathways. *The Plant Cell*. Supplement S61–S80.

Razem, F. A., K. Baron, R. D. Hill. 2006. Turning on gibberellin and abscisic acid signaling. *Current Opinion in Plant Biology* 9:454–459.

Schechheimer, C. 2008. Understanding gibberellic acid signaling—are we there yet? *Current Opinion in Plant Biology* 11:9–15.

Swain, S. M., D. P. Singh. 2005. Tall tales from sly dwarfs: Novel functions of gibberellins in plant development. *Trends in Plant Science* 10:123–129.

Weiss, D., N. Ori. 2007. Mechanisms of cross talk between gibberellins and other hormones. *Plant Physiology* 144:1240–1246.

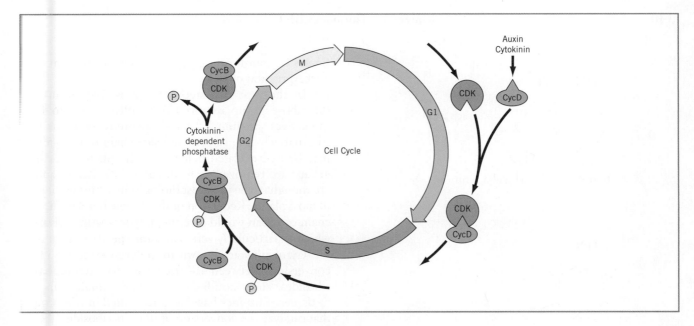

20

Hormones III: Cytokinins

Cytokinins (CK) are derivatives of the nitrogenous base adenine and are noted primarily for their capacity to stimulate cell division in tissue culture. Cytokinins also influence a number of other developmental responses, including shoot and root differentiation in tissue culture, the growth of lateral buds and leaf expansion, chloroplast development, and delay of senescence. Most recently it has been learned that cytokinins play a fundamental role in maintaining the indeterminate property of shoot apical meristems.

This chapter includes

* the structure, synthesis, and metabolism of cytokinins,

* the role of cytokinins in regulating cell division,

* the control of shoot meristem function by cytokinins,

* the role of cytokinins in apical dominance, senescence, and other developmental events, and

* cytokinin receptors and signal transduction.

20.1 CYTOKININS ARE ADENINE DERIVATIVES

Naturally occurring cytokinins are all adenine derivatives with either an isoprene-related side chain or an aromatic (cyclic) side chain. The former are called isoprenoid cytokinins and the latter are called aromatic cytokinins. Although there is some variation depending on species and developmental stage, the most common isoprenoid cytokinins are N^6-(Δ^2-isopentenyl)- adenine (iP), *trans*-zeatin (tZ), and dihydrozeatin (DZ) (Figure 20.1). The aromatic cytokinins, such as benzyladenine (BA) are less common and are found in only a few species.

The original cytokinin, **kinetin**, is a synthetic derivative that has not yet been identified in plants (Box 20.1). Curiously, however, there is a recent report that kinetin has been identified in human urine.

20.1.1 CYTOKININ BIOSYNTHESIS BEGINS WITH THE CONDENSATION OF AN ISOPENTENYL GROUP WITH THE AMINO GROUP OF ADENOSINE MONOPHOSPHATE

Enzymes that direct the synthesis of cytokinins have been isolated from the slime mold *Dictyostelium discoideum*, tobacco callus tissue, and crown gall tissue (see Figure 20.7). The key reaction is the addition of a dimethylallyl diphosphate (DMAP) group to the nitrogen at the 6-position of adenosine$-5'$-monophosphate

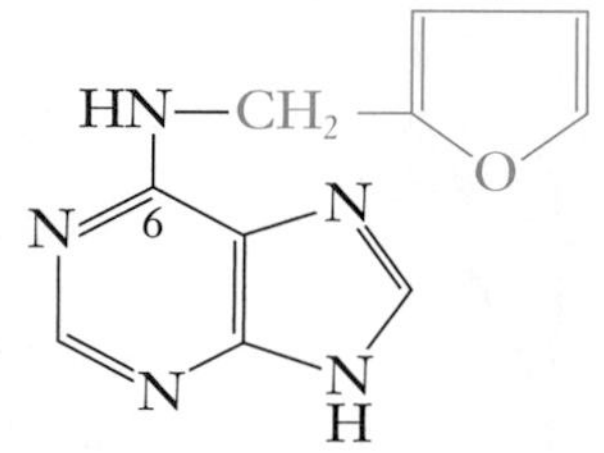

Kinetin (N⁶-furfuryl adenine)

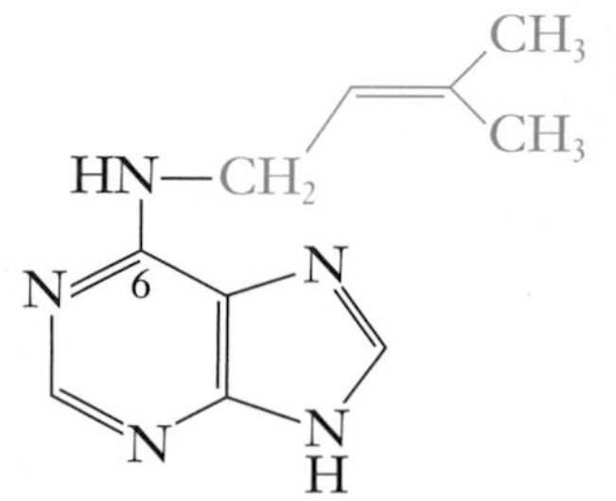

iP (N⁶-(Δ²-isopentenyl) adenine)

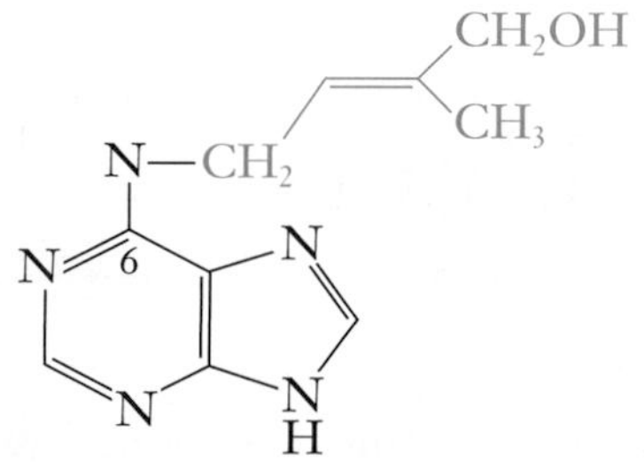

Z (*trans*-Zeatin)

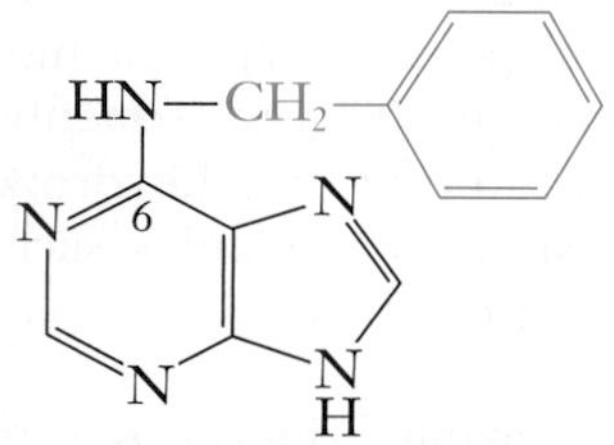

BAP (N⁶-(benzyl) adenine)

FIGURE 20.1 **The chemical structures of four representative cytokinins. Kinetin, the first compound found with cytokinin activity, is a synthetic cytokinin prepared by heating DNA. Isopentenyl adenine (iP) and *trans*-Zeatin, both isoprenoid-type cytokinins, are the most common naturally occurring cytokinins. Benzyladenine (BAP) is an aromatic cytokinin. The N⁶ position of adenine is indicated and the side chains are highlighted.**

(AMP) (Figure 20.2). DMAP, introduced previously in Chapter 19, is also known as Δ²-isopentenyl diphosphate or Δ²iPP. This addition reaction is catalyzed by the enzyme **adenosine phosphate-isopentenyl transferase (IPT)**. The product is N⁶-(Δ²-isopentenyl)-adenosine−5′-monophosphate or iPRMP. The IPT-catalyzed reaction is also the rate limiting reaction in cytokinin biosynthesis, a factor that has enabled many investigators to manipulate the cytokinin content of

tissues by transforming plants with genes that cause an overexpression of IPT.

In the next two steps, the phosphate group and the ribose group removed from [9R−5′P]iP to form the active cytokinin N⁶-(Δ²-isopentenyl)-adenine (iP). Alternatively, the isopentenyl side chain of [9R−5′P]iP may be hydroxylated before the phosphate and ribose groups are removed to form zeatin (Z). Zeatin and iP are thought to be the most biologically active cytokinins in most plants. Reduction of the double bond in the side chain of zeatin would give the dihydrozeatin derivative which is particularly active in some species of legumes.

Cytokinins are known to undergo extensive interconversions between the free base (or, **nucleobase**) ribosides, and ribotides when experimentally supplied to tissues. Enzymes have been identified in wheat germ that catalyze the conversion of iP to its riboside ([9R]iP) or to its ribotide ([9R−5′P]iP) as well as enzymes that catalyze the hydrolysis of the ribotides and ribosides to the free base (iP). Naturally, these rapid interconversions make it very difficult to ascertain which form is the truly "active" form of the hormone. However, this confusion has been at least partially resolved with the identification of cytokinin receptors, discussed later in this chapter. Individual receptors from different species appear to have different affinities for cytokinin nucleobases, ribosides and ribotides. This may be simply one way of conferring specificity through the cytokinin-receptor interaction.

20.1.2 CYTOKININS MAY BE DEACTIVATED BY CONJUGATION OR OXIDATION

There are two principal routes for regulating cytokinin activity levels by removal of cytokinins from the active pool: (1) conjugation with either glucose or amino acids, and (2) oxidation. Examples of glucosylation of the side chain hydroxyl group (**O-glucosylation**) of zeatin or dihydrozeatin, for example, are abundant in plants (Figure 20.3). O-glucosides are not themselves active, but are readily hydrolyzed to active cytokinin by glucosidase enzymes. O-glucosylation appears to be a mechanism for storing excess cytokinin for retrieval when physiological conditions warrant. O-glucosides are also resistant to oxidation and thus may serve to protect the hormone from oxidation while being transported to a target tissue. In some plants, such as radish, maize, or tobacco, cytokinins may also be glucosylated at one of the nitrogen positions on the purine ring. Both the 7- and 9-glucosides are biologically inactive. N-glucosides are also very stable and do not appear to be hydrolyzed readily to give the active free base. Their formation thus appears to be more a means for permanent inactivation of cytokinins rather than storage.

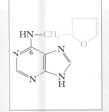

BOX 20.1
THE DISCOVERY OF CYTOKININS

The discovery of cytokinins came about because plant cells in culture would not divide. The first experimental evidence for chemical control of plant cell division was provided by Haberlandt in 1913, when he demonstrated that phloem sap could cause nondividing, parenchymatous potato tuber tissue to revert to an actively dividing meristematic state. Other cell-division factors were later demonstrated in wounded bean pod tissue, extracts of *Datura* ovules, and the liquid (milky) endosperm of coconut.

In the 1940s and 1950s, plant tissue culture was attracting the attention of physiologists as a tool for study of cell division and development. One group, under the direction of F. Skoog at the University of Wisconsin, was studying the nutritional requirements of tissue cultures derived from tobacco stem segments. Skoog and coworkers found that stem tissue explants containing vascular tissue would proliferate on a defined medium containing auxin. On the same auxin-containing medium, however, tissue explants freed of vascular tissue would exhibit cell enlargement, but failed to divide. They soon found that extracts of vascular tissue, coconut milk, and yeast would all stimulate cell division in the presence of auxin.

C. O. Miller, then working as a postdoctoral student in Skoog's laboratory, took on the task of isolating the active material. Miller was able to provisionally identify the active material as a adenine, one of the nitrogenous bases found in nucleic acids. This led to a search for active material in nucleic acid preparations, a source high in adenine. In a beautiful piece of serendipity, Miller sampled a bottle of herring sperm DNA which had been sitting on the laboratory shelf. The sample proved to be highly active, so a fresh supply of herring sperm DNA was ordered. Unfortunately, the fresh sample proved to be completely inactive. It turns out that the active principle in the original DNA sample had slowly accumulated as the DNA aged on the laboratory shelf. Activity could be generated in fresh DNA samples simply by artificially "aging" the sample with heat and acid.

In 1956, Miller and his colleagues reported the isolation and crystallization of a highly active substance, identified as the adenine derivative N^6-furfurylaminopurine, from autoclaved herring sperm DNA. Because the compound elicited cell division, or *cytokinesis*, in tissue culture, Miller and his colleagues named the substance kinetin. In 1965, Skoog and his colleagues proposed the term cytokinin.

Even though kinetin remains one of the most biologically active cytokinins, it is an artifact of isolation from DNA and has not been found in plants. However, the discovery of kinetin and its dramatic effect on cell division stimulated physiologists to look for naturally occurring cytokinins. In the early 1960s, Miller, then at Indiana University, and D. S. Letham, working in Australia, independently reported the isolation of a purine with kinetinlike properties from young, developing maize seed and plum fruitlets. This substance was characterized as 6-(4-hydroxy-3-methyl-*trans*-2-butenylamino) purine, which was given the trivial name *zeatin*. Since the discovery of zeatin, a number of other naturally occurring cytokinins have been isolated and characterized.

Cytokinins also form conjugates with the amino acid alanine (Figure 20.3). 9-Alanyl conjugates of zeatin and dihydrozeatin have been identified in tissues of lupine (*Lupinus*) fruit and root nodules, immature apple seeds, and bean (*Phaseolus*) seedlings. These are also very stable conjugates that probably serve to inactivate the cytokinin in the same manner as N-glucosides.

Another mechanism—possibly the major mechanism—for removing cytokinins from the hormone pool is their irreversible degradation by the enzyme **cytokinin oxidase/dehydrogenase (CKX)**. CKX cleaves the isopentenyl side chain from either zeatin or iP or their ribosyl derivatives. In order to function, however, CKX must recognize the double bond of the isoprenoid side chain. Consequently, dihydrozeatin and aromatic cytokinins are resistant to degradation by CKX. In the same way that the endogenous cytokinin content can be increased by overexpression of the gene Isopentenyl transferase (*IPT*) in transgenic plants, cytokinin content can be reduced in transgenic plants that overexpress cytokinin oxidase/dehydrogenase.

20.2 CYTOKININS ARE SYNTHESIZED PRIMARILY IN THE ROOT AND TRANSLOCATED IN THE XYLEM

A major site of cytokinin biosynthesis in higher plants is the root. High cytokinin levels have been found in roots, especially the mitotically active root tip, and in the xylem

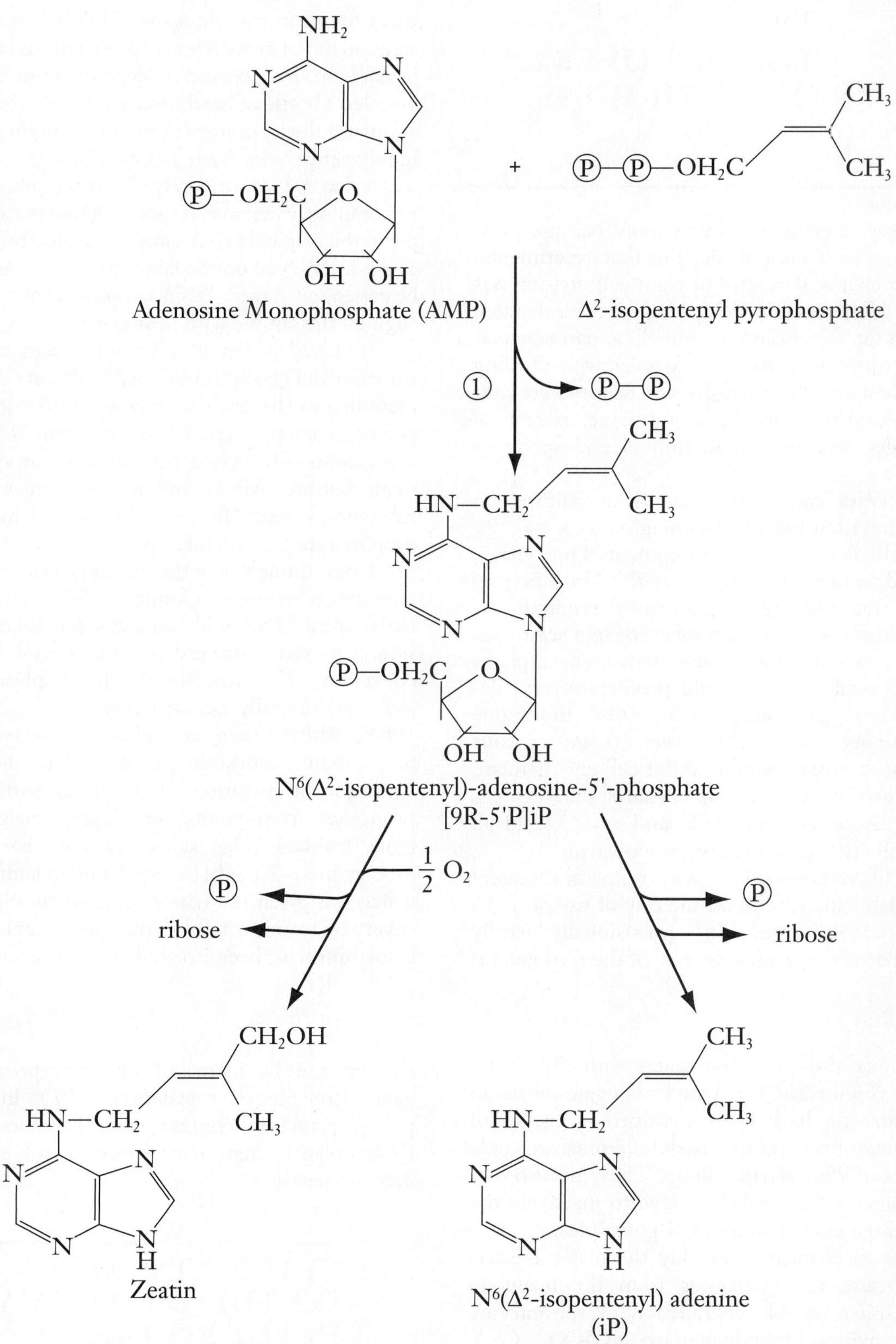

FIGURE 20.2 **A general outline for the biosynthesis of isopentenyl adenine (iP) and** *trans*-**zeatin (tZ) from adenosine monophosphate and isopentenyl pyrophosphate. Adenosine diphosphate and adenosine triphosphate may also be used as precursors. Reaction 1 is the rate limiting step and is catalyzed by the enzyme adenine phosphate-isopentenyl transferase (IPT).**

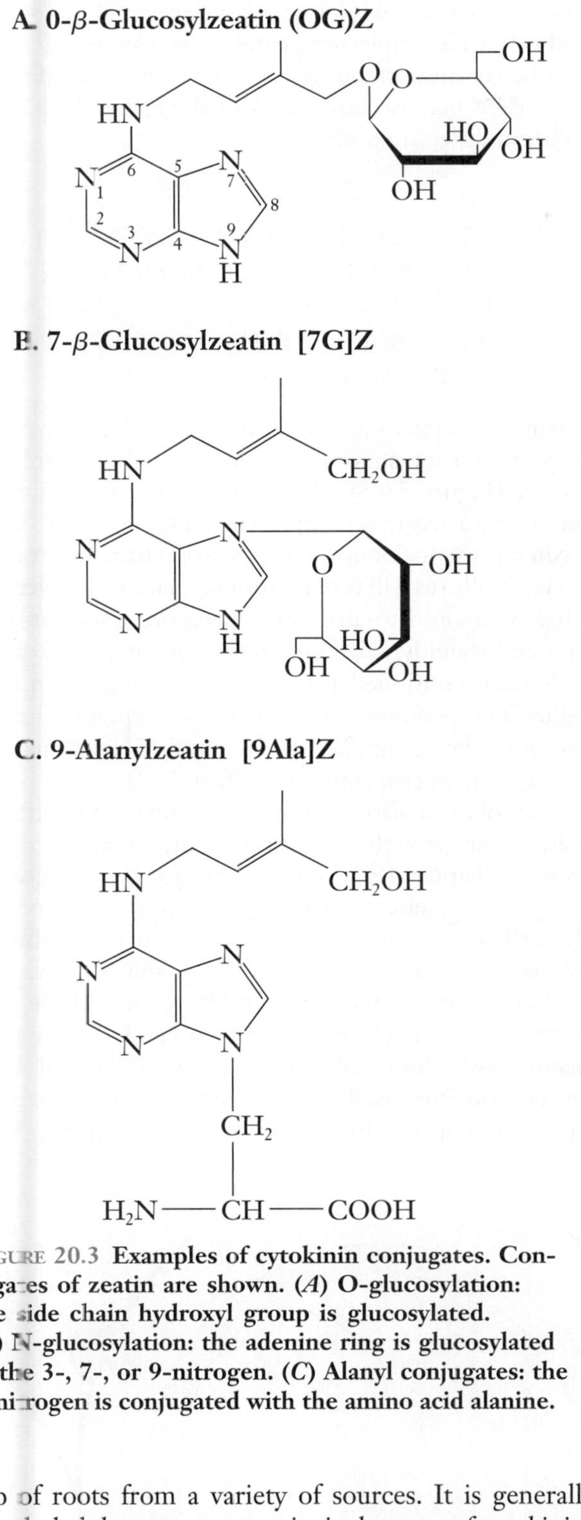

A. 0-β-Glucosylzeatin (OG)Z

B. 7-β-Glucosylzeatin [7G]Z

C. 9-Alanylzeatin [9Ala]Z

FIGURE 20.3 **Examples of cytokinin conjugates. Conjugates of zeatin are shown. (*A*) O-glucosylation: the side chain hydroxyl group is glucosylated. (*B*) N-glucosylation: the adenine ring is glucosylated at the 3-, 7-, or 9-nitrogen. (*C*) Alanyl conjugates: the 9-nitrogen is conjugated with the amino acid alanine.**

sap of roots from a variety of sources. It is generally concluded that roots are a principal source of cytokinins in most, if not all, plants and that they are transported to the aerial portion of the plant through the xylem. Indirect support for this conclusion is provided by the observation that excised leaves from many species can be maintained in a moist sand bed only if adventitious roots are permitted to form at the base of the petiole. If these roots do not form or are removed as they form, the leaves will quickly senesce. The delayed senescence

when roots are allowed to form is apparently due to the presence of cytokinins, which are synthesized in the root and transported to the leaf through the vascular tissue.

Immature seeds and developing fruits also contain high levels of cytokinins; the first naturally occurring cytokinins were isolated from milky endosperm of maize and developing plum fruits. While there is some evidence that seeds and fruits are capable of synthesizing cytokinins, there is also evidence to the contrary. Thus it remains equally possible that developing seeds, because of their high metabolic activity and rapid growth, may simply function as a sink for cytokinins transported from the roots. On the other hand, there is now evidence that cytokinins are not always a long-distance messenger. As we will see later, meristematic cells in the shoot apical meristem and floral meristems in particular are under the control of locally produced cytokinins.

20.3 CYTOKININS ARE REQUIRED FOR CELL PROLIFERATION

The role of cytokinins in regulating cell division first became apparent as a result of attempts to culture isolated carrot and tobacco tissues on defined media (see Box 20.1). Cell proliferation occurred only when auxin plus some "cell-division factor" was present in the medium. The "cell division factor" turned out, of course, to be kinetin. It was soon learned that kinetin and other cytokinins, *always in the presence of auxin*, stimulated cell division in a wide variety of tissues. The significance of these studies was that no cambium or other meristematic tissue was present in the cultures, therefore demonstrating that cytokinins have the capacity to initiate division in quiescent, or nondividing, cells.

20.3.1 CYTOKININS REGULATE PROGRESSION THROUGH THE CELL CYCLE

Although much remains to be learned about how cytokinins regulate cell division, studies of tobacco suspension-cultured cells and *Arabidopsis* indicate a direct role for cytokinins in regulating progression through the cell cycle.

Freshly established tobacco cell cultures require both auxin and cytokinin for continued cell division. The absence of either hormone causes the cells to be arrested in either the G1 or G2 phase of the cell cycle. Following addition of the missing hormone, the onset of cell division can be detected within 12 to 24 hours. In 1996, K. Zhang and coworkers reported that cultured cells arrested in G2 by the absence of cytokinin contained cyclin-dependent kinases (CDK) with reduced activity, due to a high level of phosphorylation on a tyrosine residue. When such cultures were re-supplied with cytokinin, the tyrosine was dephosphorylated, the enzyme reactivated, and cell division resumed.

The current evidence suggests that a specific tyrosine residue in the CDK catalytic unit is phosphorylated during the S phase by another kinase (Figure 20.4). Although the phosphorylated CDK catalytic unit is able to combine with cyclin (in this case, a B-type cyclin or CycB), the phosphorylated complex remains inactive until the inhibitory phosphate group is removed by a cytokinin-dependent phosphatase. Thus the principal role of cytokinin in cultured tobacco cells appears to be that of generating an active CDK complex that initiates cell division by catalyzing the transition from the G2 phase to mitosis, or M phase.

A second type of cytokinin action has been described for *Arabidopsis*. A mutant of *Arabidopsis* with a high level of cytokinin was found to contain as well a high level of a D-type cyclin (CycD3). The accumulation of CycD3 was rapidly induced when cytokinin was added to wildtype cell cultures or applied to whole plants. Moreover, cells of cultured callus tissues from transgenic plants that were constructed to overproduce CycD3 continue to divide in the absence of cytokinin.

While it is well established that the B-type cyclins act at the transition from G2 to mitosis, the level of D-type cyclins in plants remains fairly constant throughout the cell cycle and no specific activity for D-type cyclins has yet been described. However, based on analogy with similar cyclins in animals, it has been proposed that they control the transition from G1 to S phase. If this is true, the *Arabidopsis* results would indicate that cytokinin initiates cell division at the G1 to S transition through the induction of CycD3. Interestingly, two other *Arabidopsis* D-type cyclins (CycD2 and CycD4) are induced by sucrose. As more is learned about the

links between cytokinins or other developmental signals and cell cycle regulatory proteins, we can look forward to a better understanding not only of how cell division is initiated but also how cells exit the cell cycle in order to begin differentiation.

20.3.2 THE RATIO OF CYTOKININ TO AUXIN CONTROLS ROOT AND SHOOT INITIATION IN CALLUS TISSUES AND THE GROWTH OF AXILLARY BUDS

Auxin and cytokinins have antagonistic actions with respect to root shoot formation in cultured tobacco tissues (Figure 20.5). Both auxin and cytokinins are required to maintain callus cultures. However, when auxin is present alone, or if the ratio of auxin to cytokinin is high, cultures will initiate root formation. Conversely, a high cytokinin-to-auxin ratio promotes shoot production and roughly equal amounts of auxin and cytokinin will cause continued proliferation of undifferentiated callus. This phenomenon has been put to practical application in the technique of regenerating large numbers of plants by micropropagation (Box 20.2).

Cytokinins also antagonize the auxin effect in regulating the growth of axillary buds, or apical dominance (Chapter 18). In many species the application of cytokinins either to the shoot apex or directly to the axillary bud will release the bud from inhibition. Tomato mutants expressing strong apical dominance contain lower amounts of cytokinins than those with normal dominance. More recent studies with transgenic plants have confirmed that plants with elevated auxin levels show increased apical dominance and that overproduction of cytokinin reduces apical dominance. Most

FIGURE 20.4 **A simplified model for hormonal control of the cell cycle in plants. Cytokinin promotes the onset of mitosis (the G2 to M transition) by activating a phosphatase that removes an inhibitory phosphate group from the cyclin-dependent kinase (CDK)/cyclin B complex. Auxin and cytokinin also promote the accumulation of G1 cyclins (shown here as cyclin D), necessary for the onset of the S (synthesis) phase.**

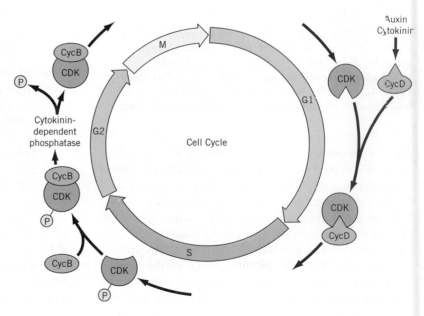

Box 20.2

TISSUE CULTURE HAS MADE POSSIBLE LARGE-SCALE CLONING OF PLANTS BY MICRO-PROPAGATION

With a relatively small investment in space, technical support, and materials, tissue culture has made it possible to produce literally millions of high-quality, genetically uniform plants. The process is known as **micropropagation**. The most common technique is to place excised meristematic tissue on an artificial medium containing a cytokinin/auxin ratio that reduces apical dominance and encourages axillary bud development (See Figure 20.5). The new shoots can be separated and sub-cultured to produce more axillary shoots, or placed on a medium that encourages rooting. Once roots appear, the plantlets can be planted out and allowed to develop into mature plants. Alternatively, excised tissues can be used to establish callus cultures, which may then be induced to form roots and shoots by manipulating the cytokinin/auxin ratio.

Micropropagation can also be an effective way to eliminate viruses and other pathogens and produce commercial quantities of pathogen-free propagules. The first plants to be mass-produced by tissue culture were

virus-free orchids of the genus *Cymbidium*, but the technique has also been found useful for potato, lilies, tulips, and other species that are normally propagated vegetatively. Potato, for example, is vegetatively propagated through buds on the tubers, a system that readily transmits viruses to the next generation. Micropropagation of potato from meristem cultures has proven to be an effective way to isolate virus-free lines.

Micropropagation is also used extensively in the production of forest tree species. Here the propagules are generated primarily from cultures of axillary and adventitious buds; callusing and differentiation of new buds is rarely used. A similar approach has been applied successfully to cultivars of apple (*Malus*), peach (*Pyrus*), and pear (*Prunus*). Because most temperate fruits are highly heterozygous, they do not breed true from seed but are propagated by vegetative cuttings. Rooting of microcuttings in culture is now a routine procedure in many commercial laboratories. By the early 1980s, growers in the Netherlands were producing more than 21 million plants by micropropagation.

In spite of the fact that plantlets derived from tissue culture are cloned from presumably identical somatic (nonsexual) cells, the regenerated plants can exhibit significant variation in their morphology and physiology. This is known as **somaclonal variation**. The cause of somaclonal variation is not clear, but it appears to involve spontaneous genetic mutation as a result of the culture conditions. The value of somaclonal variation is that occasionally the variants exhibit some agronomically useful trait, such as disease resistance or variations in floral color patterns.

interesting, however, is the observation that an exogenous application of cytokinin to the suppressed apical buds of auxin-overproducing plants will release the buds from dominance. Thus, it appears that it is the ratio of auxin to cytokinin, not the absolute level of auxin, which suppresses axillary bud growth.

The cytokinin–auxin antagonism is believed to account for the phenomenon of "witch's broom," an example of extreme axillary bud release (Figure 20.6). The witch's broom syndrome appears on a wide variety of plants and most often results from parasitism by fungi, bacteria, or mistletoes (*Arceuthobium* sp.), a dwarf flowering shrub that parasitizes predominantly spruce (*Picea*), larch (*Larix*), and pine (*Pinus*). Although the exact nature of the hormonal imbalance is not known, it is believed that the parasitism stimulates an overproduction of cytokinin. The resulting release of apical dominance produces a dense mass of short branches.

20.3.3 CROWN GALL TUMORS ARE GENETICALLY ENGINEERED TO OVERPRODUCE CYTOKININ AND AUXIN

Agrobacterium tumifaciens is a bacterium that causes neoplastic or tumorous growth, known as **crown gall**, on stems (Figure 20.7). Crown gall tissues can be excised and maintained on a simple medium containing mineral salts, some vitamins, and sucrose as a carbon source. No hormones need be added to the medium because the infected tissues have been naturally engineered with the capacity to synthesize both cytokinins and auxin.

A. tumefaciens contains a typical bacterial plasmid— a small circular strand of DNA separate from the rest of the bacterial genome. The *A. tumefaciens* plasmid is called a **tumor-inducing**, or **Ti**, plasmid because it carries the genes responsible for the neoplastic growth that occurs when the bacterium infects a wounded plant. A

FIGURE 20.5 **Shoot regeneration in callus culture. A piece of pith tissue from the center of a tobacco stem was explanted onto a medium containing mineral salts, vitamins, sucrose, auxin (left), or auxin plus cytokinin (right). The tissue grown on auxin alone proliferated as an undifferentiated callus. The tissue grown on auxin plus cytokinin regenerated shoots and roots. These plantlets can be eventually planted into soil and will produce a fully competent, mature tobacco plant.**

FIGURE 20.7 **Crown gall is a neoplastic, or tumorous, growth shown here on the stem of a *Bryophyllum* plant. Crown galls are the result of an infection with the bacterium *Agrobacterium tumefaciens*. The bacterium transforms the host plant cells with bacterial genes that cause an overproduction of auxin and cytokinin.**

portion of the plasmid DNA, known as T-DNA (for *transferred* DNA), contains the genes for three classes of proteins. Two of these are the enzymes necessary for the synthesis of cytokinins and auxins. The third is the enzyme that causes the plant to produce **opines**, unusual amino acids that serve as nutrients for the bacterium. When *A. tumefaciens* invades a host plant, the T-DNA portion of the plasmid is inserted into the host cell nuclear DNA. These genes are then replicated along with the host cell genome. *A. tumefaciens* is in fact a natural genetic engineer. It transforms the host cell, which is then programmed to overproduce cytokinins, auxin, and opines. The cytokinins and auxin together encourage

cell proliferation and neoplastic growth, while opines feed the invading bacterium.

20.3.4 CYTOKININS DELAY SENESCENCE

At present, there are three lines of evidence indicating a role for cytokinins as inhibitors of senescence. First is the observation that exogenous application of cytokinin to detached leaves will delay the onset of senescence, maintain protein levels, and prevent chlorophyll breakdown. Application of cytokinins will also delay the natural senescence of leaves on intact plants.

The second line of evidence consists of correlations between endogenous cytokinin content and senescence. For example, detached leaves that have been treated with auxin to induce root formation at the base of the petiole will remain healthy for weeks. In this case, the growing root is a site of cytokinin synthesis and the hormone is transported through the xylem to the leaf blade. If the roots are continually removed as they form, senescence of the leaf will be accelerated. It has also been observed that when a mature plant begins its natural senescence, there is a sharp decrease in the level of cytokinins exported from the root.

A third and particularly convincing line of evidence comes from recent studies employing recombinant DNA techniques. Tobacco plants (*Nicotiana tobacum*) have been transformed with the *Agrobacterium* gene for cytokinin biosynthesis, designated *TMR* (Figure 20.8). The *Agrobacterium TMR* gene encodes for the enzyme **isopentenyl transferase**, which catalyzes the rate-limiting step in cytokinin biosynthesis. In this case, the

FIGURE 20.6 **Witch's broom on white pine (*Pinus strobus*). Witch's broom is the result of a fungal infection that stimulates an overproduction of cytokinin. The result is an uncontrolled release of axillary bud development.**

A. B. C. D.

FIGURE 20.8 **Cytokinin control of senescence and bud growth in tobacco. Tobacco callus cells, genetically transformed such that cytokinin production could be stimulated by heat shock, were allowed to regenerate plantlets. (*A*) Transformed heat-shocked plantlets. (*B*) Untransformed heat-shocked plantlets. (*C*) Transformed controls (no heat shock). (*D*) Untransformed controls. Note especially the proliferation of lateral buds and absence of senescence in the transformed, heat-shocked plantlets. The large, white areas in *B* and *D* are senesced leaves. The transformed controls (*C*) do not, as expected, exhibit the cytokinin effect on lateral bud growth but do not exhibit senescence. This probably indicates the transformed gene is "leaky" and a small amount of cytokinin is produced in the absence of heat shock. (From Smart, C. et al. 1991. *The Plant Cell* 3:647. Copyright American Society of Plant Physiologists. Photo courtesy of C. Smart.)**

TMR gene was linked to a **heat shock promoter**. A **promoter** is a sequence of DNA that signals where the transcription of messenger RNA (mRNA) should begin. The heat shock promoter is normally involved in the heat shock response of plants, which is induced by a brief period of high temperature. Normally, the heat shock response involves the synthesis of a new set of proteins called **heat shock proteins**. The heat shock promoter is thus active only when subjected to a high temperature treatment. By linking the *TMR* gene to the heat shock promoter, cytokinin biosynthesis can be turned on in the transformed plants simply by subjecting the plants to a brief period of high temperature. A heat shock of 42°C for 2 hours caused a 17-fold increase in zeatin levels in transformed plants compared with untransformed control plants. When subjected to heat shock on a weekly basis over a 12-week period, transformed plants exhibited a marked release of lateral buds from apical dominance as well as delayed senescence. Transformed but non-heat-shocked plants also remained green longer than normal plants but did not exhibit release from apical dominance. This is probably due to "leakiness" on the part of the promoter, allowing production of a very small but effective amount of cytokinin even at normal temperature.

The mechanism by which cytokinins are able to delay senescence is not clear, but there is some evidence that cytokinins exert a role in mobilizing nutrients. The classic experiment of K. Mothes and coworkers, summarized in Figure 20.9, illustrates this point. In this experiment, a nutrient labeled with radioactive carbon (e.g., ^{14}C-glycine) is applied to a leaf after a portion of the leaf has been treated with cytokinin. Invariably the radioactivity is transported to and accumulates in the region of cytokinin treatment. A variety of similar experiments have led to the hypothesis that cytokinins direct nutrient mobilization and retention by stimulating metabolism in the area of cytokinin application. This creates a new sink—an area that preferentially attracts metabolites from the region of application (the source). It is unlikely that cytokinins act directly through stimulating protein synthesis since the mobilization of nonmetabolites such as α-aminoisobutyric acid is directed by cytokinins equally well.

20.3.5 CYTOKININS HAVE AN IMPORTANT ROLE IN MAINTAINING THE SHOOT MERISTEM

The fundamental characteristic of a plant meristem is the capacity to maintain the spatial distinction between dividing cells and differentiating cells. The entire course of plant development depends on maintaining that small population of perpetually dividing cells. Ever since Skoog and Miller demonstrated that cytokinins induced cell division and shoot regeneration fifty years ago,

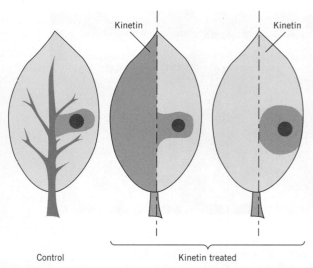

Control   Kinetin treated

FIGURE 20.9 **Diagram of an experiment demonstrating the role of cytokinin in nutrient mobilization. Radioactive ^{14}C-aminobutyric acid was applied to the area indicated by the dark spot. Left: Control, no cytokinin treatment. Radioactivity spreads into the vascular tissue for export through the petiole. Center: Radioactivity accumulates in the left half of the leaf, which has been treated with kinetin. Right: Radioactivity is retained near the point of application when the right half of the leaf is treated with kinetin. (Based on experiments of Mothes, K. 1963. *Régulateurs Naturels de la Croissance Végetale*, 123. Centre National de la Recherche Scientifique.)**

cytokinins have been routinely used to induce shoot formation in tissue culture for plant propagation and for the production of transgenic plants. It has always been assumed that cytokinins had a significant role in maintaining the meristem *in planta*, but direct evidence for such a role has been difficult to obtain. The traditional method involving exogenous application of cytokinins is something of a shotgun approach that provides relatively little solid information. When you simply spray a plant with hormone solution, for example, it is virtually impossible to know how much hormone actually gets into the plant or where it goes. Moreover, excessive hormone levels may cause artifactual, nonphysiologic effects.

Several new lines of evidence, however, now point to a positive role for cytokinins in the shoot apical meristem. Most of these experiments involve *reducing* the *in planta* concentration of cytokinins, either by overexpressing appropriate genes in transgenic plants or through loss-of-function mutants. For example, cytokinin levels can be reduced *in planta* by overexpressing the genes for cytokinin oxidase/dehydrogenase (CKX), which degrades active cytokinins. *Arabidopsis* has seven *CKX* genes and, depending on which of these genes is overexpressed, the cytokinin content can be reduced to 30 to 45 percent of wildtype plants. The result in all cases was retarded shoot development; dwarfed, late flowering plants; and reduced size of the

shoot meristem. The formation of leaf primordia was slower in cytokinin-deficient plants and the number of leaf cells was significantly reduced. Where CKX was most strongly expressed, growth of the shoot was arrested completely shortly after germination. Similar results were obtained in experiments where the cytokinin content was reduced through loss-of-function mutants of the gene for IPT or of genes for known cytokinin receptors (Section 20.4).

Further evidence for the maintenance of the shoot apical meristem by cytokinins is offered by the discovery of a cytokinin-deficient rice mutant that was given the intriguing name of *lonely guy* (*log*). Rice flowers are borne in a typical, highly branched grass inflorescence called a panicle (Figure 20.10). The flowers, or spikelets, normally contain a single pistil surrounded by several stamens. In *log* mutant plants, the size of the panicle was severely reduced. There were fewer branches and the branches bore abnormal flowers. Flowers were often reduced to no pistil and but a single stamen (hence the name lonely guy). Microscopic studies revealed that after the transition from vegetative to reproductive stage the normally dome-shaped floral meristem flattens and the differentiation of floral organs is prematurely shut down.

Clearly the *log* mutant is characterized by a deficient meristem, but why? When the *LOG* gene was isolated and cloned, it was found that expression of the *LOG* gene is localized in regions of active cell division in the meristem such as the apex of the meristem and branch primordia. It was also found that *LOG* encodes a phosphoribohydrolase enzyme. This means that the enzyme activates cytokinins by removing the ribose phosphate group from an inactive cytokinin nucleotide to leave the active free base. The absence of this enzyme in the *log* mutant thus reduces the level of active cytokinin in the critical region of cell division and the result is an improperly maintained meristem.

20.3.6 CYTOKININ LEVELS IN THE SHOOT APICAL MERISTEM ARE REGULATED BY MASTER CONTROL GENES

If cytokinins maintain the meristem by controlling cell division, then what triggers differentiation? Plants, fungi, and animals all contain genes that are involved in patterning and organ initiation. Most of these genes share a sequence of about 180 base pairs called the **homeobox** or **homeodomain**. The resulting proteins are therefore referred to as homeobox or homeodomain proteins. In plants, the homeobox proteins are identified as KNOX proteins, based on the KNOTTED1 protein in maize—the first homeobox protein to be identified in plants. Homeobox genes are often referred to as master control genes because of their fundamental role in development and their capacity to specify the fate of tissues

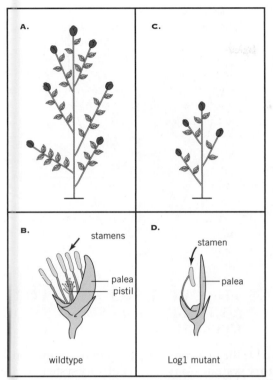

FIGURE 20.10 **Diagrammatic representations of the phenotypes of wildtype and the cytokinin-deficient *log-1* mutants in rice (*Oryza sativa*). The wildtype rice inflorescence (*A*) is a typical grass panicle with numerous branches covered with lateral flowers, or spikelets shown here in green). Each branch ends with a terminal spikelet (shown in red). The number of branches and spikelets is determined by the timing of when the shoot apical meristem of the main stem or branches is converted to a terminal spikelet meristem. The wildtype spikelet contains a pistil and numerous stamens (*B*). With less cytokinin in the mutant, the branch meristems are not properly maintained. The result is fewer, shorter branches with fewer spikelets (*C*). Floral meristems are also affected and the spikelet may be reduced to a single stamen and no pistil (*D*).** (Based on Kurakawa, T. et al. 2007. *Nature* 445:652; and Kyozuka, J. 2007. *Current Opinion in Plant Biology* 10:442.)

and organs. In the case of plants, one fundamental role of *KNOX* genes is to specify meristems by regulating the biosynthesis of cytokinins and gibberellins.

The impact of *KNOX* genes on development is illustrated by two types of mutations. Expression of wildtype *KNOX* genes is localized to the nuclei of shoot apical meristems. They are not normally expressed in leaf primordia or developed leaves of wildtype plants. There are several dominant mutations, however, that alter this pattern. These dominant mutations are *gain-of-function* mutations, meaning that the genes are overexpressed in the mutant seedlings. The result is that some leaf tissues fail to differentiate and continue to divide, forming sporadic outgrowths, or "knots," on the leaf blade. In extreme cases, **ectopic** shoots can actually be seen

developing on the leaf surface, indicating the presence of active meristem-like cells. In other words, when KNOX proteins accumulate due to overexpression of the *KNOX* genes, cells in the leaf which should have switched to differentiation fail to do so and continue dividing.

The *Arabidopsis* gene *SHOOTMEREISTEMLESS* (*STM*) also encodes a KNOX protein and the effect of the recessive loss-of-function mutant, *stm*, is just the opposite of the gain of function dominant mutatnts described above. Plants carrying the mutant *stm* fail to develop any shoot apical meristem during embryogenesis. The pattern of *STM* expression is very similar to the pattern of *KNOTTED1* expression in maize. Using techniques to visualize mRNA in situ, the pattern of STM mRNA was followed during embryogenesis. STM mRNA initially appears in one or two cells in the earlier globular stage embryos (Figure 20.11). As the embryo enters the heart-shaped and torpedo stages, *STM* expression is restricted to the notch between the embryonic cotyledons—i.e., expression of *STM* is limited to the cells that will eventually organize as the SAM. As the plant grows into the seedling stage and adult plant, *STM* expression persists into the vegetative, axillary, inflorescence, and floral meristems. But it is not expressed in leaves or leaf primordia.

It thus appears that the Knox proteins are transcription factors based almost exclusively in the meristem and that regulate the switch between indeterminant (i.e., meristematic) growth and differentiation. If so, what genes are targeted by these transcription factors? The primary candidates appear to be genes that control the synthesis of gibberellins and cytokinins. We've already seen that LOG maintains the meristem by catalyzing the formation of active cytokinins. In early studies of KNOX proteins, it was noted that overexpression of these transcription factors was accompanied by a significant increase in the cytokinin content, especially trans-zeatin (tZ) and isopentenyl adenine (iP). Recent studies in rice (*Oryza sativa*) have confirmed that the higher cytokinin levels found in transgenic plants overproducing KNOX proteins is a direct result of increased transcription of the rate-limiting enzyme isopentenyl transferase (IPT).

At the same time, it has been reported that KNOX proteins suppress the transcription of GA20-oxidase genes in at least four different species: *Arabidopsis*, tobacco (*Nicotiana tabaccum*), rice (*Oryza sativa*), and potato (*Solanum tuberosum*). GA$_{20}$-oxidase catalyzes the conversion of inactive GA$_{20}$ to the active gibberellin, GA$_1$. Thus it appears that cytokinins are primarily responsible for initiating and maintaining populations of dividing cells. Gibberellins, on the other hand, are more involved in the subsequent differentiation of cells. The balance between division and differentiation is maintained by KNOX proteins; transcription factors whose job it is to maintain a high CK/GA ratio in the shoot

FIGURE 20.11 **Schematic representation of *STM* expression during embryogenesis in *Arabidopsis*. STM mRNA, shown in blue, is first detected in a single cell in the early globular stage embryo (32–64 cell stage) (Left). In the later, torpedo-stage embryo, STM mRNA is confined to the nascent meristem (Right). C = cotyledons. (Based on Long, J. A. et al. 1996. *Nature* 379:66–69.)**

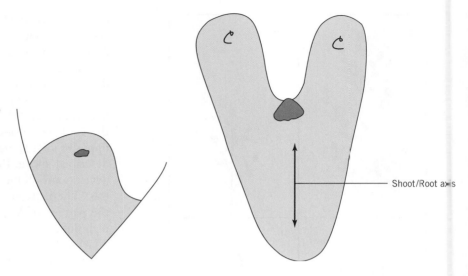

apical meristem. This is accomplished by simultaneously stimulating cytokinin biosynthesis and suppressing gibberellin biosynthesis. The resulting high-CK/low GA condition is required for both the formation and maintenance of the meristem. Outside the meristem, KNOX genes are apparently turned off, cytokinin concentrations decline, cell division effectively ceases, and gibberellins take over to encourage differentiation.

20.4 CYTOKININ RECEPTOR AND SIGNALING

In spite of the fundamental role played by cytokinins in cell division, the multiple other effects that cytokinins have on plant development have made it difficult to identify cytokinin receptors and signal chains. It has only been within the last decade, more than fifty years after Skoog and Miller purified the first cytokinin, that the first genes involved in cytokinin signaling have been identified.

The cytokinin receptor was finally discovered by T. Kakimoto and his colleagues who developed an *Arabidopsis* hypocotyl test to screen for mutants. Hypocotyl sections, or explants, respond to added cytokinins by typical cytokinin responses: rapid cell proliferation, greening, and shoot formation. The *cytokinin response 1(cre1)* mutant shows none of these responses, even with a tenfold increase in cytokinin concentration. This would be expected if the cytokinin receptor were either missing or nonfunctional in the mutant. Subsequent experiments confirmed that the wildtype protein CRE1 was in fact a cytokinin receptor. The same gene has also been identified as *WOODENLEG* (*WOL*), so named because its mutation retarded root growth, and *Arabidopsis HISTIDINE KINASE 4* (*AHK4*).

20.4.1 THE CYTOKININ RECEPTOR IS A MEMBRANE-BASED HISTIDINE KINASE

CRE1 is the first component of a **two-component regulatory system**—a type of regulatory system previously known to operate in bacteria and other prokaryotes. The name comes from the bacterial configuration where the **receptor** (or sensor)—the first component—activates a **response regulator (RR)**—the second component. Response regulators in turn either regulate the transcription of target genes or modulate other metabolic reactions. In addition to serving as hormone receptors, two-component regulatory systems also function in osmosensing (Chapter 1), light sensing, and other forms of sensory perception.

CRE1 is an intracellular **histidine kinase (HK)** with three domains (Figure 20.12). The **sensor domain**, at the N-terminal end of the protein, includes two small hydrophobic membrane-spanning regions that anchor the receptor in the plasma membrane. Between them is a hydrophilic loop that extends into the extracellular space. This loop includes the cytokinin-binding site since a mutation at this site interferes with cytokinin binding and renders the receptor inoperative. The **histidine kinase domain** is located on the cytoplasmic side of the membrane. The term *kinase* identifies HK as an enzyme involved in a phosphorylation reaction and the reference to *histidine* kinase means that the phosphoryl group is added to a histidine residue. The phosphate attaches to a specific histidine residue (His_{459}) in the histidine kinase domain. The histidine kinase also includes two receiver domains, Da and Db.

Once it became evident that the acceptor was a histidine kinase, the fully sequenced *Arabidopsis* genome could be searched for other potential components of the signaling system. *Arabidopsis* is now known to

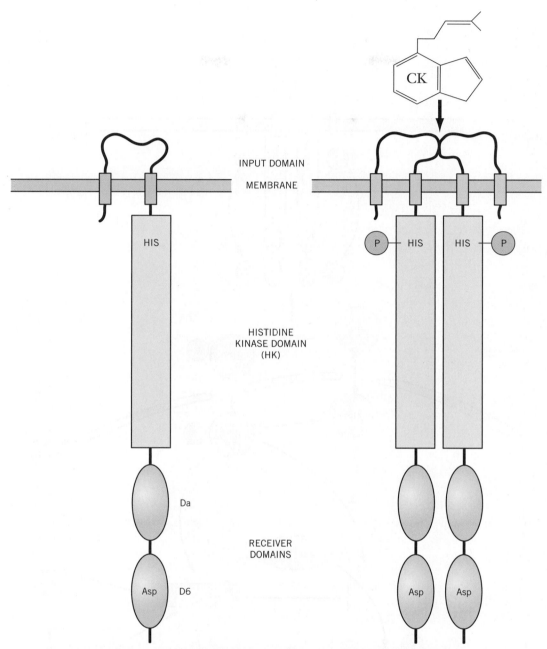

FIGURE 20.12 Predicted structure of the cytokinin receptor CRE1. The monomer (left) has three domains and two small membrane-spanning hydrophobic regions. Binding with cytokinin (right) induces dimerization and autophosphorylation. The locations of the histidine and aspartic acid phosphate-binding residues are indicated.

have the genes for eight different histidine kinases, six histidine-phosphotransfer proteins, and 23 response regulators. Only three of the HK genes (*CRE1*, *AHK2*, *AHK3*) encode cytokinin receptors. At least two are ethylene receptors and one is believed to be an osmosensor. The reason for such a large number of response regulators appears to be that many are expressed in a tissue-specific manner, which allows for a more finely tuned, tissue-specific signaling output.

20.4.2 THE CYTOKININ SIGNALING CHAIN INVOLVES A MULTISTEP TRANSFER OF PHOSPHORYL GROUPS TO RESPONSE REGULATORS

A general scheme for cytokinin signaling is shown in Figure 20.13. Binding of a cytokinin molecule to the sensor domain induces dimerization and subsequent

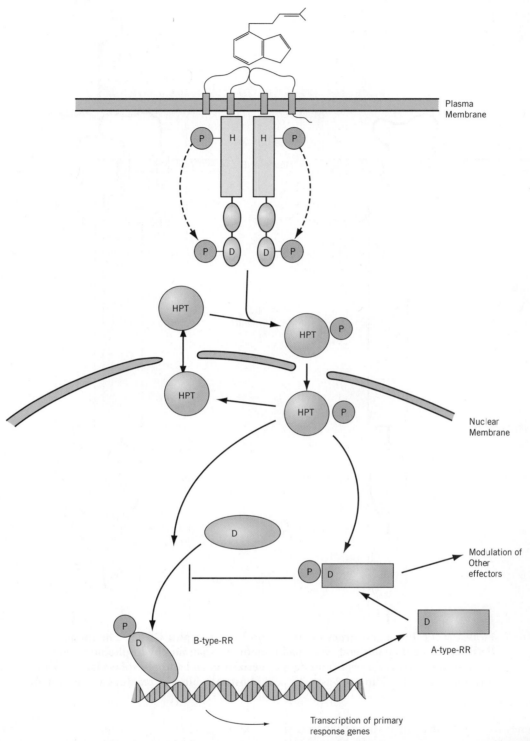

Figure 20.13 **A model for cytokinin signal transduction via a multistep phosphorelay system. Cytokinin sensing occurs when cytokinin binds with the input domain in the extracellular space. Cytokinin binding induces dimerization and autophosphorylation of the acceptor histidine kinase. The phosphorelay system begins with the transfer of the phosphoryl group first to an aspartic acid residue (D) in the receiver domain and then to a histidine residue in a separate histidine phosphotransfer protein (HPT). The phosphorylated HPT migrates into the nucleus where the phosphoryl group is transferred to either a B-type or A-type response regulator (RR). The activated B-type response regulator then activates transcription of cytokinin primary response genes, including the A-type response regulator. The A-type response regulators may down-regulate cytokinin responses by suppressing the activation of B-type response regulators. Alternatively, A-type response regulators may positively or negatively modulate other cytokinin.**

phosphorylation of a histidine residue in each of the two receptor molecules. Whereas most kinase enzymes catalyze the addition of a phosphoryl group to a second molecule, the histidine kinase receptor autophosphorylates, which means that it phosphorylates itself. The phosphoryl group is then spontaneously transferred to an aspartic acid residue on the Db receiver domain.

The classic prokaryote two-component system is comprised of only a receptor kinase and a response regulator and the response regulator is activated by receiving a phosphoryl group directly from the histidine kinase. In plants, the phosphoryl group is passed bucket-brigade fashion through one or more **histidine-phosphotransfer proteins (HPTs)**. The phosphoryl group is transferred from the receiver domain of the histidine kinase to a histidine residue on the HP protein. The phosphorylated HP protein then migrates into the nucleus where the phosphoryl group is transferred to an Asp residue in a response regulator. Note that the transfer is alternately from histidine to aspartic acid to histidine to aspartic acid. The system that transfers phosphoryl groups between the various HKs, HPTs, and RRs is referred to as a **phosphorelay network**.

There are two classes of response regulators—the A-type and the B-type. The role of response regulators is still being worked out, but in general, it appears that B-type response regulators are transcription factors. When activated by phosphorylation, B-type response regulators induce the expression of genes that are responsible for some cytokinin-regulated responses. Among the target genes for B-type response regulators are the genes for A-type response regulators. A-type response regulators, however, are not transcription factors and do not regulate gene expression. They apparently modulate cytokinin responses by influencing other aspects of metabolism.

The cytokinin system also has a built-in capacity for shutting down the phosphorelay network when no cytokinin is present. This conclusion is based on the finding that CRE1, aside from its kinase function, also exhibits phosphatase activity. The activity of a phosphatase enzyme is the opposite of a kinase—a phosphatase removes phosphoryl groups. Thus, in the absence of cytokinin, CRE1 reverses the process, unloads phosphoryl groups from HPTs, and quickly inactivates the cytokinin response pathway

SUMMARY

Cytokinins are N^6 adenine derivatives with either an isoprenoid-related side chain or an aromatic side chain.

The most common naturally occurring cytokinins are isopentenyl adenine and trans-zeatin.

Synthesis of cytokinins begins with the addition of dimethylallyl pyrophosphate to an adenine nucleotide (AMP, ADP, or ATP). The nucleotide-phosphate group is cleaved off to generate the active form. Although it has long been known that cytokinins are synthesized in the roots and translocated to the shoot through the xylem sap, it is now clear the some tissues, meristems in particular, are under the control of locally produced cytokinins. Cytokinins are deactivated by conjugation with sugars and amino acids, or by oxidative degradation.

Cytokinins, commonly in concert with auxin, are involved in numerous developmental responses. These include regulation of cell division, shoot and root initiation, release of axillary bud growth, delay of senescence, and maintenance of an actively dividing shoot apical meristem.

Cytokinin receptors are of a class known as histidine kinases. The signal chain involves a multistep transfer of phosphoryl group transfer referred to as a phsophorelay network. The final targets are A- and B-type response regulators. When activated by phosphorylation, the responses regulators either activate transcription of cytokinin primary response genes or modulate other aspects of cytokinin related metabolism.

CHAPTER REVIEW

1. Auxins are identified by their control of cell enlargement in excised tissues and gibberellins are identified largely on the basis of chemical structure. How are cytokinins identified?

2. Describe how cytokinins regulate the cell cycle.

3. The bacterium *Agrobacterium* is commonly used to produce transgenic plants. What characteristic(s) make it useful in this regard?

4. In controlling various developmental responses, cytokinins frequently require the presence of auxin. Can you think of a reason why cytokinins would require the presence of auxin?

5. What is the evidence that cytokinins are required for the maintenance of the shoot apical meristem?

6. What role do KNOX proteins have in maintaining the apical meristem?

7. What is the apparent role of gibberellins in the shoot apical meristem?

8. What is a two-component regulatory system? How does a two-component regulatory system relate to cytokinin activity?

FURTHER READING

Davies, P. J. 2004. *Plant Hormones: Biosynthesis, Signal Transduction, Action*. Dordrecht: Kluwer Academic Publishers.

Ferreira, F., J. J. Kleiber. 2005. Cytokinin signaling. *Current Opinion in Biology* 8:518–525.

Heyl, A., T. Schmülling. 2003. Cytokinin signal perception and transduction. *Current Opinion in Plant Biology* 6:480–488.

Kurakawa, T. et al. 2007. Direct control of shoot meristem activity by a cytokinin-activating enzyme. *Nature* 445:652–655.

Kyozuka, J. 2007. Control of shoot and root meristem function by cytokinin. *Current Opinion in Plant Biology* 10:442–446.

Mok, D. W., M. C. Mok. 2001. Cytokinin metabolism and action. *Annual Review of Plant Physiology and Plant Molecular Biology* 52:89–118. (This series of reviews is accessible in most data bases under the current title: *Annual Review of Plant Biology*.)

Müller, B., J. Sheen. 2007. Advances in cytokinin signaling. *Science* 318:68–69.

Sakakibara, H. 2006. Cytokinins: Activity, biosynthesis, and activity. *Annual Review of Plant Biology* 57:431–449.

Shani, E., O. Yanai, N. Ori. 2006. The role of hormones in shoot apical meristem function. *Current Opinion in Plant Biology* 9:484–489.

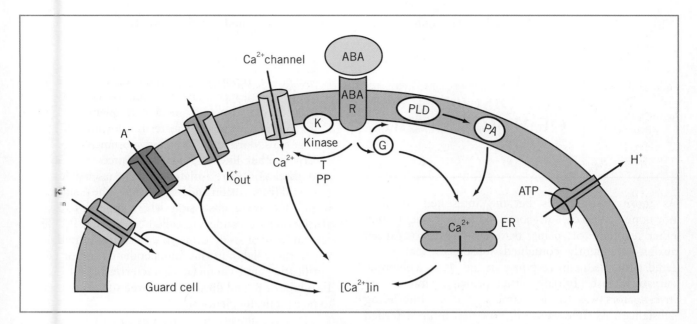

21

Hormones IV: Abscisic Acid, Ethylene, and Brassinosteroids

Abscisic acid (ABA), **ethylene**, and **brassinosteroids** (BR) are three hormone classes that are noted for their involvement in a wide variety of plant responses as well as extensive interactions with auxins and gibberellins and, to a lesser extent, with cytokinins. Most of the current work on each of these three hormones focuses on working out their physiology, metabolism, and mode of action by screening for mutations.

In this chapter, we will take each of these three hormones in turn and look at

- their biosynthesis and metabolism,
- the range of physiological effects, and
- our current understanding of signal perception and transduction.

21.1 ABSCISIC ACID

Unlike auxins, gibberellins, and cytokinins, the hormone **abscisic acid (ABA)** is represented by a single 15-carbon sesquiterpene (Figure 21.1). ABA also appears to have a more limited range of specific effects than auxins, gibberellins, and cytokinins. The name is based on the once held belief that it was involved in the abscission of leaves and other organs. It now appears to have nothing to do with abscision, but the name has stuck (Box 21.1).

The primary functions of ABA are (1) prohibiting precocious germination and promoting dormancy in seeds and (2) inducing stomatal closure and the production of molecules that protect cells against desiccation in times of water stress. ABA has also been implicated in other developmental responses, including the induction of storage protein synthesis in seeds, heterophylly (leaves of different shape on the same plant), initiation of secondary roots, flowering, and senescence.

21.1.1 ABSCISIC ACID IS SYNTHESIZED FROM A CAROTENOID PRECURSOR

Once the structure of ABA had been determined, two possible pathways for the synthesis of ABA were proposed. In the "direct pathway," ABA would be synthesized from a 15-carbon terpenoid precursor such as farnesyl diphosphate (see Figure 19.4). By the late 1970s it had been clearly established that this pathway was operative in certain fungal plant pathogens that actively synthesized ABA, but not in plants themselves. According to the second, or "indirect pathway," ABA

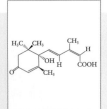

BOX 21.1
THE DISCOVERY OF ABSCISIC ACID

As more investigators became interested in plant hormone research, it soon became evident that ether extracts of plant material—used to extract auxins—frequently contained substances that interfered with the auxin response in the *Avena* coleoptile curvature test. Initially, the principal interest of investigators was to rid extracts of these interfering substances. As time went on, however, interest turned toward the possibility that these inhibitors might themselves be growth regulators in their own right. The advent of paper chromatography as an analytical tool made it possible to achieve better separation of the various substances in crude extracts. In 1953, Bennet-Clark and Kefford reported that plant extracts contained, in addition to IAA, a substance that inhibited growth of coleoptile sections, which they called inhibitor β. The observation that large amounts of inhibitor β could be isolated from axillary buds and the outer layer of dormant potato tuber led Kefford to suggest that it was involved in apical dominance and maintaining dormancy in potatoes. Meanwhile, other

investigators reported the occurrence of inhibitors in buds and leaves that appeared to correlate with the onset of dormancy in woody plants. In 1964, P. F. Waring proposed the term "dormin" for these endogenous, dormancy-inducing substances.

In another line of study, substances that accelerated abscission were isolated from senescing leaves of bean and from cotton and lupin fruits. These substances would accelerate abscission when applied to excised abscission zones and were called "abscission II." These several lines of study came to a head in the mid-1960s when three laboratories independently reported the purification and chemical characterization of abscisin II, inhibitor β, and dormin. All three substances proved to be chemically identical.

It is not unusual in such cases that there was some disagreement over what this substance should be called. Although abscisin II had priority (it was the first to be crystallized and chemically characterized), some felt the term awkward and argued it did not adequately describe its range of effects. Finally, a panel of scientists active in research on abscisin II and dormin was charged with proposing an acceptable name. The name *abscisic acid* and abbreviation *ABA* were recommended by this panel to the 1967 International Conference on Plant Growth Substances, which met in Ottawa. The recommendation was accepted by the Conference and the term abscisic acid is now in universal use.

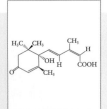

Abscisic acid

FIGURE 21.1 **Abscisic acid is a class of hormones represented by a single compound.**

was produced from the cleavage of a carotenoid such as β-carotene. Originally proposed in the late 1960s, the indirect pathway was based on structural similarities between carotenoid pigments and ABA and has since received support from a variety of biochemical studies, $^{18}O_2$-labeling experiments, and, most recently, the characterization of ABA biosynthetic mutants. The cleavage of carotenoids, especially β-carotene, to produce useful biochemicals is not without precedent. The cyanobacterium *Microcystis*, for example, produces a C_{10} metabolite by cleavage of β-carotene. Mammals produce vitamin A by cleavage of β-carotene and cleavage of β-carotene to produce 2 molecules of the photoreceptor retinal (C_{20}) has been reported.

There is now a growing body of evidence supporting the indirect synthesis of ABA from β-carotene via the 40-carbon terpene **violaxanthin** (Figure 21.2). First, a series of *viviparous* mutants in maize (described further below) were found to have reduced levels of both carotenoids and ABA. These mutants, shown to be affected in the early steps of carotenoid biosynthesis, establish a strong correlation between carotenoid and ABA biosynthesis. Second, the carbon skeleton of ABA and the position of the oxygen-containing substituents are very similar to that of violaxanthin. J. A. D. Zeevaart and his colleagues compared the incorporation of $^{18}O_2$, a stable isotope of oxygen, into ABA in water-stressed leaves and turgid leaves of several species. The pattern of $^{18}O_2$-enrichment in the carboxyl group of ABA was consistent with the cleavage of a xanthophyll and its rapid conversion to ABA in water-stressed leaves. Third, it is known that violaxanthin can be degraded in the light *in vitro* to a 15-carbon derivative, **xanthoxin**, a natural constituent of plants. If radio-labeled xanthoxin is fed to bean or tomato plants, some of the radioactivity appears in ABA. In ABA-deficient tomato mutants, however, conversion of radio-labeled xanthoxin into ABA is reduced relative to wildtype plants. Finally, at least

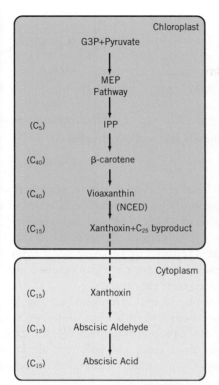

FIGURE 21.2 A flow sheet for the biosynthesis of abscisic acid. ABA biosynthesis begins in the chloroplast with the synthesis of isopentenylpyrophosphate (IPP) from glyceraldehydes-3-phosphate (G3P) and pyruviate via the methylerythritol-4-phosphate (MEP) pathway. IPP in the chloroplast gives rise to a variety of C_{10}, C_{20}, and C_{40} terpenoids, including β-carotene. β-Carotene is converted to violaxanthin, which is cleaved by the enzyme nine-cis-epoxycarotenoid dioxygenase (NCED) to yield xanthoxin, a C15 precursor to ABA, and a 25-carbon "byproduct."

two groups have reported a stoichiometric relationship between losses of violaxanthin and increases in ABA in stressed etiolated bean leaves.

Although ABA is synthesized in the cytosol, its biosynthetic pathway begins in the chloroplast (and possibly other plastids in nongreen cells), which is where carotenoid pigments are produced (Figure 21.2. See also Figure 19.4). The critical enzyme is nine-cis-expoycarotenoid dioxygenase (NCED). This enzyme cleaves the 40-carbon carotenoid violaxanthin to produce a 15-carbon product, xanthoxin, and a 25-carbon "by-product." Xanthoxin is then converted to abscisic aldehyde by an alcohol dehydrogenase. Abscisic aldehyde is in turn oxidized to abscisic acid by abscisic aldehyde oxidase. The enzyme NCED and, consequently xanthoxin production, is known to be targeted in the chloroplast while the alcohol dehydrogenase and abscisic aldehyde oxidase are located in the cytosol. This means that xanthoxin must migrate from the chloroplast into the cytosol, although the mechanism of migration is not yet known.

21.1.2 ABSCISIC ACID IS DEGRADED TO PHASEIC ACID BY OXIDATION

Abscisic acid is rapidly metabolized when it is applied exogenously to plant tissues. In wilted bean leaves, for example, the half-time for turnover (the time for one-half of the labeled ABA to be destroyed) was estimated to be about three hours. A glucose ester of ABA has been found in low concentration in a variety of plants, but the principal metabolic route seems to be oxidation to phaseic acid (PA) and subsequent reduction of the ketone group on the ring to form dihydrophaseic acid (DPA) (Figure 21.3). At least some tissues appear to carry the metabolism further to form the 4′-glucoside of DPA. DPA and its glucoside are both metabolically inactive.

21.1.3 ABSCISIC ACID IS SYNTHESIZED IN MESOPHYLL CELLS, GUARD CELLS, AND VASCULAR TISSUE

There are a lot of open questions about the sites of ABA synthesis in the plant. Earlier physiological studies indicated that abscisic acid was found predominantly in mature, green leaves, especially in water-stressed plants. This would fit with the more recent biochemical and genomic studies described above showing that ABA precursors originate in chloroplasts but ABA itself is formed in the cytoplasm. There is also evidence that ABA may be stored in the chloroplasts (Chapter 13). At low pH, ABA exists in the protonated form ABAH, which freely permeates most cell membranes. The dissociated form ABA⁻ is impermeant because it is a charged molecule that does not readily cross membranes. In actively photosynthesizing mesophyll cells the cytosol will be moderately acidic (pH 6.0 to 6.5) while the chloroplast stroma is alkaline (pH 7.5 to 8.0). Thus, ABAH diffuses readily from the cytosol into the chloroplast stroma, where it dissociates and beomes trapped. This stored ABA can later be released when photosynthesis shuts down and the stroma pH declines.

Abscisic aldehyde oxidase (*AAO*) expression is induced in guard cells under conditions of water stress and *NCED* expression has been detected in guard cells of senescing leaves and cotyledons. Thus it appears that ABA is also synthesized directly in the guard cells. Furthermore, expression of ABA biosynthetic genes (*NCED* and others) has been localized in phloem companion cells and xylem parenchyma cells of fully turgid plants. This indicates that vascular tissues are also a site of ABA synthesis in unstressed plants.

Abscisic acid is highly mobile and moves quickly out of the leaves to other parts of the plant, especially sink tissues. For example, radioactively labeled abscisic acid applied to soybean leaves can be detected in the roots within 15 minutes. Developing seeds also import

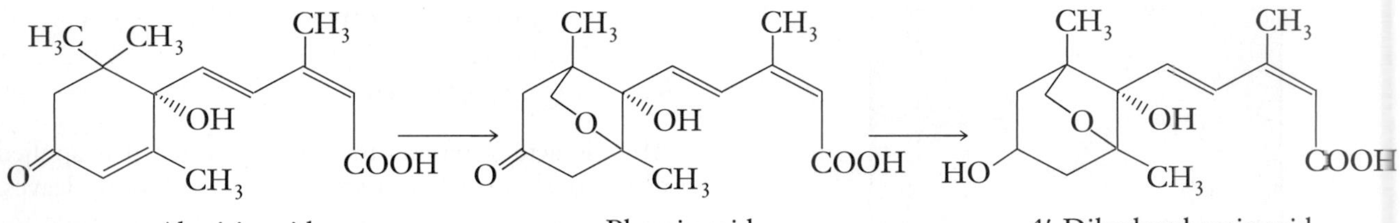

FIGURE 21.3 **Oxidative degradation of abscisic acid to phaseic acid and dihydrophaseic acid.**

large amounts of abscisic acid from the leaves. There is also some evidence that under conditions of water stress, ABA either stored or synthesized in the roots is rapidly exported to the leaves (See Section 21.1.5).

21.1.4 ABSCISIC ACID REGULATES EMBRYO MATURATION AND SEED GERMINATION

The development of embryos and subsequent germination of the seed is characterized by often dramatic changes in hormone levels (refer to Figure 16.10). In most seeds, cytokinin levels are highest during the very early stages of embryo development when the rate of cell division is also highest. As the cytokinin level declines and the seed enters a period of rapid cell enlargement, both GA and IAA levels increase. In the early stages of embryogenesis, there is little or no detectable ABA. It is only during the latter stages of embryo development, as GA and IAA levels begin to decline, that ABA levels begin to rise. ABA levels generally peak during the maturation stage, when seed volume and dry weight also reach a maximum, and then return to lower levels in the dry seed. Maturation of the embryo is characterized by cessation of embryo growth, accumulation of nutrient reserves in the endosperm, and the development of tolerance to desiccation.

The timing of ABA accumulation to coincide with embryo maturation reflects the critical role that ABA plays in the maturation process. One of the functions of a seed, of course, is to disperse the population and ensure survival of the species through unfavorable conditions. A seed would be of little value if the embryo did not enter dormancy but continued to grow and establish a new plant before dispersal could occur. One function of ABA is to prevent such precocious germination, or **vivipary**, while the seed is still on the mother plant.

The relationship between ABA and precocious germination is clear. Vivipary can be chemically induced in maize by treatment of the developing ear at the appropriate time with fluridone, a chemical inhibitor of carotenoid biosynthesis. Since carotenoids and ABA share early biosynthetic steps, fluridone inhibits biosynthesis of ABA as well. Fluridone-induced vivipary can be at least partially alleviated by application of exogenous ABA. Soybean embryos can be encouraged to germinate precociously by treatments such as washing or slow drying, both of which lower the endogenous ABA level. Precocious germination will occur when the ABA concentration is reduced to 3 to 4 µg per g fresh weight of seed, a level that is not normally reached until the late stages of seed maturation.

The strongest indication of a role for ABA in preventing precocious germination, however, comes from the study of viviparous mutants. At least four viviparous mutants in maize (*vp2*, *vp5*, *vp7*, *vp9*) are known to be ABA-biosynthetic mutants with reduced levels of ABA in the seeds. One maize mutant, *vp1*, appears to have normal ABA levels but is missing what is believed to be an ABA-specific transcription factor. All of these mutants germinate prematurely on the cob before the seeds have entered dormancy. Viviparous mutants are also known for *Arabidopsis*. ABA also stimulates protein accumulation in the latter stages of soybean embryo development and is known to prevent GA-induced α-amylase biosynthesis in cereal grains. All of these results establish a strong connection between ABA and seed maturation and/or prevention of precocious germination.

ABA also initiates desiccation of the seed, although the mechanisms are unknown. This may involve ABA regulation of genes which encode proteins that are involved in desiccation tolerance.

21.1.5 ABSCISIC ACID MEDIATES RESPONSE TO WATER STRESS

Plants generally respond to acute water deficits by closing their stomata in order to match transpirational water loss from the leaf surface with the rate at which water can be resupplied by the roots. Since the discovery of ABA in the late 1960s, it has been known to have a prominent role in stomatal closure during water stress. In fact, ABA has long been recognized as antitranspirant because of its capacity to induce stomatal closure and thus reduce water loss through transpiration.

ABA accumulates in water-stressed (that is, wilted) leaves and exogenous application of ABA is a powerful inhibitor of stomatal opening. Furthermore, two tomato mutants, known as *flacca* and *sitiens*, fail to accumulate normal levels of ABA and both wilt very readily. The precise role of ABA in stomatal closure in water-stressed

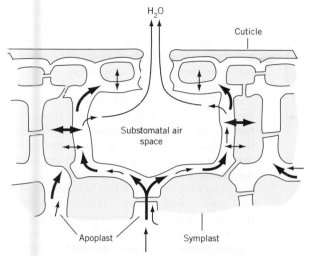

FIGURE 21.4 ABA movements in the apoplast. ABA synthesized in the roots is carried to the leaf mesophyll cells (heavy arrows) in the transpiration stream (light arrows). ABA equilibrates with the chloroplasts of the photosynthetic mesophyll cells or is carried to the stomatal guard cells in the apoplast.

whole plants has, however, been difficult to decipher with certainty. This is because ABA is ubiquitous, often occurring in high concentrations in nonstressed tissue. Also, some early studies indicated that stomata would begin to close before increases in ABA content could be detected.

According to current thinking, the initial detection of water stress in leaves is related to its effects on photosynthesis. Inhibition of electron transport and photophosphorylation in the chloroplasts would disrupt proton accumulation in the thylakoid lumen and lower the stroma pH. At the same time, there is an increase in the pH of the apoplast surrounding the mesophyll cells. The resulting pH gradient stimulates a release of ABA from the mesophyll cells into the apoplast, where it can be carried in the transpiration stream to the guard cells (Figure 21.4).

As noted above, wilted leaves accumulate large quantities of ABA. In most cases, however, stomatal closure begins before there is any significant increase in the ABA concentration. This could be explained by the release of stored ABA into the apoplast, which occurs early enough and in sufficient quantity—the apoplast concentration will at least double—to account for initial closure. Increased ABA synthesis follows and serves to prolong the closing effect.

Stomatal closure does not always rely on the perception of water deficits and signals arising within the leaves. In some cases it appears that the stomata close in response to soil desiccation well *before* there is any measurable reduction of turgor in the leaf mesophyll cells. Several studies have indicated a feed-forward control system that originates in the roots and transmits information to the stomata. In these experiments, plants are

grown such that the roots are equally divided between two containers of soil (Figure 21.5A). Water deficits can then be introduced by withholding water from one container while the other is watered regularly. Control plants receive regular watering of both containers. Stomatal opening along with factors such as ABA levels, water potential, and turgor are compared between half-watered plants and fully watered controls. Typically, stomatal conductance, a measure of stomatal opening, declines within a few days of withholding water from the roots (Figure 21.5B), yet there is no measurable change in water potential or loss of turgor in the leaves. In experiments with day flower (*Commelina communis*), there was a significant increase in ABA content of the roots in the dry container and in the leaf epidermis (Figure 21.6). Furthermore, ABA is readily translocated from roots to the leaves in the transpiration stream, even when roots are exposed to dry air. These results suggest that ABA is involved in some kind of early warning system that communicates information about soil water potential to the leaves.

21.1.6 OTHER ABSCISIC ACID RESPONSES

There is recent evidence that ABA may also have a role in lateral or secondary root development. The initiation and development of lateral roots is known to be primarily under the control of auxin, but lateral root development can be inhibited by ABA if the hormone is applied during early stages of lateral root development, before the lateral root meristem becomes organized.

Earlier studies also indicated an impact of exogenous ABA on flower formation under certain conditions, but the data was equivocal. In particular, no causal relationship could be established between endogenous ABA levels and flowering behavior. However, the prospect of a role for ABA in flowering has been revived recently with the discovery that, under conditions that would normally delay flowering, ABA-deficient mutants of *Arabidopsis* produce flowers somewhat earlier than wild-type plants. This observation suggests that endogenous ABA may normally inhibit or delay flowering in *Arabidopsis*. Further support comes from the discovery that a gene (*FCA*) previously known to be involved in controlling the time of flowering also has the properties of an abscisic acid receptor. We will take a closer look at this receptor in the next section and the role of the *FCA* gene in flowering in Chapter 25.

21.1.7 ABA PERCEPTION AND SIGNAL TRANSDUCTION

ABA perception and signaling appears to be particularly complex and, although its metabolism and physiology

A.

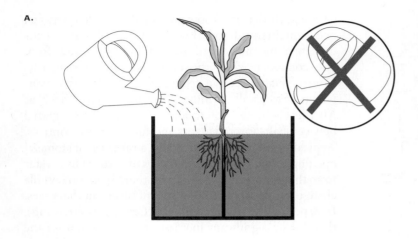

B.

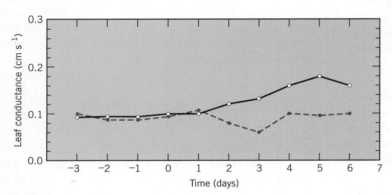

FIGURE 21.5 (*A*) An experimental setup for testing the effects of desiccated roots on ABA synthesis and stomatal closure. Roots of a single plant are divided equally between two containers. Water supplied to one container maintains the leaves in a fully turgid state while water is withheld from the second container. Withholding water from the roots leads to stomatal closure, even though the leaves are not stressed. (*B*) Stomatal closure in a split-root experiment. Maize (*Zea mays*) plants were grown as shown in (*A*). Control plants (open circles) had both halves of the root system well-watered. Water was withheld from half the roots of the experimental plants (closed circles) on day zero. Stomatal opening, measured as leaf conductance, declined in the plants with water-stressed roots. (B from Blackman, P. G., W. J. Davies. 1985. *Journal of Experimental Botany* 36:39–48. Reprinted by permission of The Company of Biologists, Ltd.)

have been studied for decades, the mechanism of ABA perception and its subsequent signal chain have remained elusive. As noted earlier, ABA is a weak acid. As such it is likely to exist in both the protonated and unprotonated forms in the relatively acidic environment of the apoplast. In the protonated state it may diffuse across the plasma membrane and react with an intracellular receptor or, in the unprotonated form, it may remain outside the cell to be sensed by a site on the plasma membrane. Indeed, experiments employing impermeable ABA derivatives and/or microinjection of ABA into cells have indicated multiple ABA receptors at multiple locations.

Over the last 20 years, methods that have normally been employed to identify hormone receptors have proven relatively unsuccessful in the search for ABA receptors. A more recent approach has made use of antigen-antibody reactions with what are called *anti-idiotypic antibodies*. In this method, antibodies raised against ABA are themselves used as antigens to raise a second group of antibodies—the anti-idiotypic antibodies—that have binding characteristics similar to ABA. Thus, any protein that binds with the anti-idiotypic antibodies could be a putative ABA receptor. The anti-idiotypic antibodies were then used to screen the proteins encoded by a **complimentary**

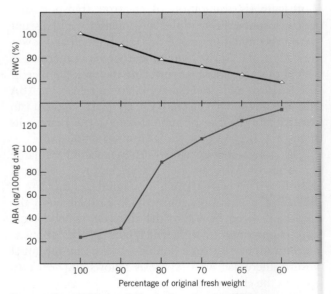

FIGURE 21.6 **Effect of air drying on the ABA content of *Commelina communis* root tips. Root tips were air dried to the relative water contents shown in the upper curve. Lower curve shows the dramatic increase in ABA content as the fresh weight decreases. (From Zhang, J., W. J. Davies. 1987. *Journal of Experimental Botany* 38:2015–2023. Reprinted by permission of The Company of Biologists, Ltd.)**

DNA (cDNA) library for barley aleurone cells. This approach led to the identification of ABAP1, a protein that is located in the plasma membrane of barley aleurone cells and that specifically and reversibly binds ABA *in vitro*.

Since the discovery of ABAP1, at least three other putative ABA receptors have been identified. One is a chloroplast protein Magnesium Protoporphyrin-IX Chetalase H subunit (CHLH, also known as ABAR). The second is a soluble, flowering-time control protein FCA isolated from *Arabidopsis*. Based on similarity of a ninc acid sequence, FCA is homologous to the barley protein ABAP1. FCA interacts with another protein (FY) to regulate the processing of functional mRNA (see Chapter 25 for the role of FCA in flowering). The third putative receptor is a membrane-localized G-protein coupled receptor (GPCR) identified as GCR2. The simple fact that these proteins bind ABA *in vitro*, however, does not prove they are true receptors. It still needs to be demonstrated that loss-of-function or gain-of-function mutants alter ABA functions in a predictable manner.

The signal chain for ABA effects, both upstream and downstream from the hormone, is a subject of intensive study. The apparently complex interactions between abiotic signals, receptors, second messengers, and ABA-induced gene transcription—let alone crosstalk with other signals—make it difficult to assemble a definitive scheme. Still, a number of components are beginning to fall into place. Most of the recent progress has been made through newly discovered ABA-insensitive gene mutations and can be summarized in the following points.

1. There appears to be rapid turnover of ABA in both stressed and unstressed plants, but the events that sense abiotic stress and initiate ABA accumulation remain unknown.

2. Ca^{2+} appears to be an important part of the ABA signal chain, especially in stomatal guard cells. Ca^{2+} mediates ABA-induced turgor adjustments by activating plasma membrane anion channels (Figure 21.7).

3. The promoter region of some genes contains a sequence called the ABA response element (ABRE). Transcription factors known as ABA response element binding factors (ABFs) bind to this promoter region to regulate the activity of many ABA-induced genes. These genes include putative protective proteins such as enzymes required for the synthesis of **osmolytes** or **compatible solutes** that help the plant adapt to water stress (Chapter 13), and transcription factors that in turn regulate other changes in gene expression

4. A number of ABA-insensitive (*abi*) mutants have been identified. At least three insensitive mutants, *abi3*, *abi 4*, and *abi 5*, impair only seed germination and early seedling development. All three wildtype genes (*ABI3, 4, 5*) encode transcription factors that are expressed mainly in seeds, suggesting that the role of ABA in seeds requires gene transcription.

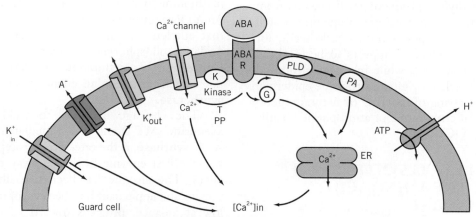

FIGURE 21.7 **A simplified schematic illustrating the coordination of ion pumps by ABA and Ca^{2+} during closure of stomatal guard cells. ABA is perceived by an unknown receptor (ABA R) that transmits the ABA signal to inward-rectifying calcium channels via a membrane-associated protein kinase. The kinase is antagonized by a protein phosphatase (PP). ABA also stimulates the release of Ca^{2+} from internal stores such as the endoplasmic reticulum (ER), possibly mediated by phospholipid signaling and/or G protein. The increased cytosolic Ca^{2+} concentration inhibits K_{in}^+ channels and opens both K_{out}^+ and anion channels (A^-). The result is a net loss of ions from the guard cell, followed by a loss of water and turgor, and closure of the stomatal pore.**

5. A number of ABA-activated protein kinases that positively regulate ABA responses have been identified. In addition, ABI1 and ABI2 are protein phosphatases that negatively regulate ABA responses. So, protein phosphorylation events are clearly important in ABA signaling.

It will no doubt take some time to sort out all of these components and those yet to be discovered and construct a clear model of the signaling chains for various ABA-mediated responses.

21.2 ETHYLENE

Ethylene is another class of hormones with a single representative. Ethylene is a simple gaseous hydrocarbon with the chemical structure $H_2C{=}CH_2$. Ethylene is apparently not required for normal vegetative growth, although it can have a significant impact on the development of roots and shoots. Ethylene appears to be synthesized primarily in response to stress and may be produced in large amounts by tissues undergoing senescence or ripening. Ethylene is commonly used to enhance ripening in bananas and other fruits that are picked green for shipment (Box 21.2). Ethylene is frequently produced when high concentrations of auxins are supplied to plant tissues and many of the inhibitory responses to exogenously applied auxin appear to be due to auxin-stimulated ethylene release rather than auxin itself.

Ethylene occurs in all plant organs—roots, stems, leaves, bulbs, tubers, fruits, seeds, and so on—although the rate of production may vary depending on the stage of development. Ethylene production will also vary from tissue to tissue within the organ, but is frequently located in peripheral tissues. In peach and avocado seeds, for example, ethylene production appears to be localized primarily in the seed coats, while in tomato fruit and mung bean hypocotyls it originates from the epidermal regions. The off-gassing of ethylene by natural vegetation is also a significant source of atmospheric volatile organic carbon (VOC).

21.2.1 ETHYLENE IS SYNTHESIZED FROM THE AMINO ACID METHIONINE

Despite the early discovery of ethylene, its known importance in plant development, and its relatively uncomplicated chemistry, the pathway for ethylene biosynthesis initially proved difficult to unravel. This is partly because there were a large number of potential precursors (sugars, organic acids, or peptides) known to be present in plant tissues. In addition, until recently, the enzymes involved have proven too labile to isolate and study *in vitro*. Consequently, most of the work has

been carried out *in vivo*, with all the pitfalls inherent in such experiments. Moreover, ethylene is a volatile gas and available analytical methods for its measurement were simply too insensitive. It wasn't until the early 1960s that developments in gas chromatography made it possible to analyze ethylene at physiologically active concentrations. With the advent of gas chromatography, the study of ethylene began to advance rapidly.

M. Lieberman and L. W. Mapson first demonstrated in 1964 that methionine was rapidly converted to ethylene in a cell-free, nonenzymatic model system. In subsequent studies, Lieberman and coworkers confirmed that plant tissues such as apple fruit converted [^{14}C]-methionine to [^{14}C]-ethylene and that the ethylene was derived from the third and fourth carbons of methionine. Little progress was made until 1977 when D. Adams and F. Yang demonstrated that **S-adenosylmethionine (SAM)** was an intermediate in the conversion of methionine to ethylene by apple tissue. In 1979, Adams and Yang further demonstrated the accumulation of **1-aminocyclopropane-1-carboxylic acid (ACC)** in apple tissue fed [^{13}C]-methionine under anaerobic conditions—conditions that inhibit the production of ethylene. However, upon reintroduction of oxygen, the labeled ACC was rapidly converted to ethylene. ACC is a nonprotein amino acid that had been isolated from ripe apples in 1957, but its relationship to ethylene was not obvious at that time. These results established that ACC is an intermediate in the biosynthesis of ethylene.

The three-step pathway for ethylene biosynthesis in higher plants is shown in Figure 21.8. In the first step, an adenosine group (i.e., adenine plus ribose) is donated to methionine by a molecule of ATP, thus forming SAM. An ATP requirement is consistent with earlier evidence that ethylene production is blocked by inhibitors of oxidative phosphorylation, such as 2,4-dinitrophenol. Conversion of methionine to SAM is catalyzed by the enzyme methionine adenosyltransferase or **SAM synthetase**.

The cleavage of SAM to yield 5'-methylthioadenosine (MTA) and ACC, mediated by the enzyme **ACC synthase**, is the rate-limiting step. ACC synthase was the first enzyme in the pathway to be studied in detail. The enzyme has been partially purified from tomato and apple fruit but, because of its instability and low abundance, progress toward its purification and characterization has been slow. More recently, genes for ACC synthase have been isolated from zucchini (*Cucurbita*) fruit and tomato pericarp tissue. The cloned genes direct the synthesis of active ACC synthase in the bacterium *E. coli* and yeast, making it possible to produce the enzyme in sufficient quantity for further study. The enzyme that catalyzes the oxidation of ACC to ethylene, previously referred to as the ethylene-forming enzyme but now known as **ACC oxidase**, proved

Box 21.2
THE DISCOVERY OF ETHYLENE

The effect of ethylene on plants was originally described by Dimitry Nikolayevich Neljubow, a graduate student in Russia in 1886, who found that abnormal growth of dark-grown pea seedlings could be traced to ethylene emanating from illuminating gas. Interest in ethylene as a plant growth factor, however, did not gain real momentum until it was found to have commercial implications.

Those whose business involves the shipping and storing of fruit have long been aware that ripe and rotting fruit could accelerate the ripening of other fruit stored nearby. For example, bananas picked in Cuba and shipped by boat often arrived in New York in an overripe and unmarketable condition. One of the earliest reports that these effects were due to a volatile substance given off by plant tissue was published in 1910 by H. H. Cousins in an annual report of the Jamaican Department of Agriculture. He discovered that ripe oranges released a volatile product that would accelerate ripening of bananas stored with them. A number of similar reports appeared in the early 1930s, showing that volatile emanations from apples caused epinasty in tomato seedlings and respiratory changes associated with the ripening process. In 1934, R. Gane provided indisputable evidence that the volatile substance was ethylene.

especially difficult to isolate in the active form. ACC oxidase was finally identified when a gene cloned from ripening tomato fruit (pTOM13) was linked to ethylene production.

Another important aspect of ethylene biosynthesis is the limited amount of free methionine available in plants. In order to sustain normal rates of ethylene production, the sulfur released during ethylene formation must be recycled back to methionine. This is accomplished by what is commonly referred to as the methionine cycle (Figure 21.8). This cycle is also known as the Yang cycle, after S. F. Yang, who carried out much of the pioneering work on ethylene biosynthesis. Double-labeling experiments have shown that the CH3S-group is salvaged and recycled as a unit. The remaining four carbon atoms of methionine are supplied by the ribose moiety of the ATP used originally to form SAM. The amino group is provided by a transamination reaction.

Ethylene production is promoted by a number of factors including IAA, wounding, and water stress, principally by the induction of the synthesis of ACC synthase. Induction of this enzyme in plant tissues is blocked by inhibitors of both protein and RNA synthesis, suggesting that induction probably occurs at the transcriptional level. In *E. coli* carrying the cloned ACC synthase gene, the physical abundance of ACC synthase messenger RNA also increases in response to IAA

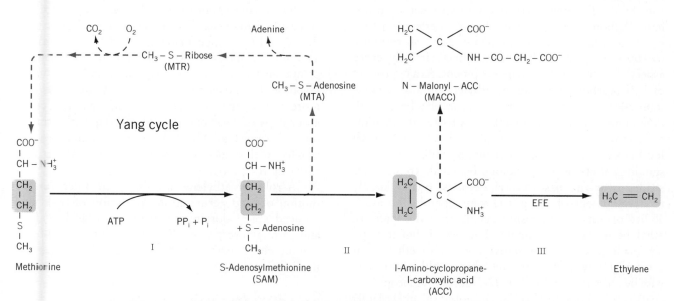

FIGURE 21.8 A scheme for ethylene biosynthesis in higher plants. The enzymes are I: SAM synthetase; II: ACC synthase; and III: ACC oxidase. The ethylene group is highlighted in yellow. The Yang cycle for sulfur recovery is highlighted in orange.

and wounding. Control of ethylene production thus appears to be exercised primarily through transcriptional regulation of the ACC synthase gene. Ethylene production is also stimulated by ethylene itself, a form of autocatalysis. This is commonly seen in ripening fruits (see Chapter 25) where ethylene apparently stimulates an increase in both ACC synthesis and its subsequent conversion to ethylene.

21.2.2 EXCESS ETHYLENE IS SUBJECT TO OXIDATION

Unlike other hormones, ethylene is a volatile gas that is readily given off to the atmosphere. Ethylene can, however, be metabolized by oxidation to carbon dioxide or by conversion to either ethylene oxide or ethylene glycol. It has not yet been established whether ethylene metabolism has any active role in the physiological action of the hormone. In fact, kinetic studies have indicated that ethylene metabolism is a straightforward chemical reaction not subject to normal physiological controls. It may thus be only a nonessential consequence of high ethylene levels in the tissue. It is therefore likely that most tissues lose excess ethylene by simple diffusion into the surrounding atmosphere.

21.2.3 THE STUDY OF ETHYLENE PRESENTS A UNIQUE SET OF PROBLEMS

Because ethylene is a simple gaseous hydrocarbon that readily diffuses from its site of synthesis, study of its role in development presents a unique set of problems. Although known primarily for its effects on fruit ripening and its synthesis by many tissues in response to stress, ethylene is known to affect virtually every aspect of plant growth and development. As a byproduct of hydrocarbon combustion, ethylene is also a common environmental pollutant that can play havoc with greenhouse cultures or laboratory experiments. In addition, ethylene biosynthesis is also stimulated by high levels of auxin and other hormones. Still, our understanding of ethylene physiology has made tremendous strides over the past several decades, owing largely to development of the gas chromatograph, the availability of ethylene-releasing agents, and the study of ethylene-insensitive mutants. The gas chromatograph has made possible quantitative analysis of ethylene at extremely low concentrations that could not otherwise be measured. **Ethephon** (2-chloroethylphosphonic acid) is a compound that, at physiological pH, readily decomposes to produce ethylene. Use of ethephon is advantageous in the laboratory as its application and concentration is often more easily controlled compared with gaseous ethylene.

21.2.4 ETHYLENE AFFECTS MANY ASPECTS OF VEGETATIVE DEVELOPMENT

Ethylene is known primarily for its promotion of fruit ripening and senescence. Ethylene control of fruit development has been studied extensively in tomato, which is a **climacteric** fruit. During the development of climacteric fruits there is a characteristic developmentally regulated burst in respiration called the climacteric rise. The climacteric rise is normally accompanied by ethylene production and is followed by the expression of a set of genes that enhance ripening-related activities such as development of fruit color, flavor, and texture. The tomato *never ripe* mutant is insensitive to ethylene because it has a defective ethylene receptor protein. Consequently, the "ripening genes" are not expressed and, although the fruit reaches full size, it never develops the red color, flavor, and texture characteristic of a ripe tomato.

Ethylene has been shown to stimulate elongation of stems, petioles, roots, and floral structures of aquatic and semiaquatic plants. The effect is particularly noted in aquatic plants because submergence reduces gas dispersion and thus maintains higher internal ethylene levels. In rice, ethylene is ineffective in the presence of saturating levels of gibberellins, which also promotes stem elongation. Moreover, ethylene promotes gibberellin synthesis in rice and the elongation effect can be blocked with antigibberellins, which suggests that gibberellin mediates this ethylene effect. By contrast, root and shoot elongation in nonaquatic plants such as peas (*Pisum sativum*) is inhibited by ethylene.

Ethylene stimulates many inhibitory and abnormal growth responses such as the swelling of stem tissues and the downward curvature of leaves, or **epinasty**. Leaf epinasty occurs because of excessive cell elongation on the adaxial (i.e., upper) side of the petiole. Epinasty is a common response to water logging of flood-sensitive plants such as tomato (*Lycopersicum*) and is actually a response to anoxia in the region of the roots. The more vertical orientation of epinastic leaves reduces the absorption of solar energy and, consequently, transpirational water loss. This helps to bring water loss more into line with reduced capacity for water uptake in plants suffering root anoxia.

A role for ethylene has also been noted for promotion of seed germination, inhibition of bud break, reduced apical dominance, fruit ripening, cell death, and pathogen responses. Ethylene can be a problem in commercial greenhouse that are heated with gas-fired heating systems.

21.2.5 ETHYLENE RECEPTORS AND SIGNALING

One of the best-known effects of ethylene is referred to as the "triple response" of etiolated (dark-grown) dico-

seedlings (Figure 21.9). This response is characterized by inhibition of hypocotyl and root cell elongation, a pronounced radial swelling of the hypocotyl, and exaggerated curvature of the plumular hook. The response is rapid (3 days post-germination) and allows large populations of seedlings to be screened for ethylene mutations. An absence of the triple response in the presence of exogenous ethylene has been used successfully to identify ethylene-resistant mutants and other ethylene response defects, especially in experiments conducted with *Arabidopsis*. These mutants generally fall into one of three distinct categories: (1) constitutive triple-response mutants that exhibit the triple response in the absence of ethylene, (2) ethylene-insensitive mutants, and (3) mutants in which ethylene-insensitivity is limited to specific tissues, such as the plumular hook or root elongation. As a result of these experiments, several ethylene receptors and downstream elements in the ethylene signal chain have been identified.

Ethylene is perceived by a family of five membrane-associated, two-component histidine kinase receptors. Unlike most other two-component receptors, which are localized in the plasma membrane, it has been shown convincingly that the *Arabidopsis* receptor ETR1 is associated with the membrane of the endoplasmic reticulum (ER). The specific advantage(s) to localization of the receptor in the ER rather than the plasma membrane is not clear, but since ethylene diffuses readily in both aqueous and lipid environments, ethylene would have ready access to any subcellular location.

The sensor domain of the ethylene receptor has three membrane-spanning regions and is assumed to function as a dimer. The sensor domain also contains a copper cofactor that is necessary for ethylene binding. A unique characteristic of the ethylene response system is that the receptors are believed to be constitutively active. This means that the receptors and subsequent signal chain are functional *in the absence of ethylene* and that ethylene, in effect, turns the system off.

Although there are some variations, ethylene signaling appears to follow the general model shown in Figure 21.10. In the absence of ethylene, the signal chain begins with a protein called Constitutive Triple

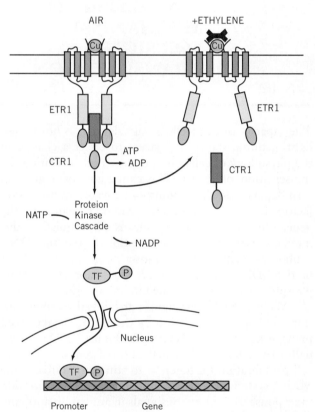

FIGURE 21.10 **A model for gene regulation by the ethylene perception and response pathway. Ethylene is sensed by a family of two-component histidine kinase receptors (ETR) that are located in the membrane of the endoplasmic reticulum. In the absence of ethylene the receptors are functionally active and activate a serine-threonine kinase (CTR1). CTR1 is the first component in a protein kinase cascade that ultimately targets one or more transcription factors. Phosphorylation activates the transcription factors that are then able to bind to the promoter regions of ethylene-sensitive genes and enable transcription of the genes. Ethylene binding inhibits receptor function and blocks the activation of CTR1, thus shutting down the protein kinase cascade, preventing phosphorylation of the transcription factor, and turning off the gene.**

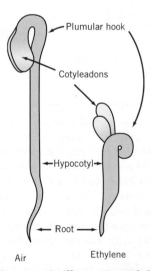

FIGURE 21.9 **Diagramatic illustration of the ethylene triple response in dark-grown dicot seedlings. Note the inhibition of hypocotyl and root cell elongation, a pronounced radial swelling of the hypocotyl, and exaggerated curvature of the plumular hook in the seedlings exposed to ethylene. (Based on Guzman, Ecker. 1990.** *Plant Cell* **2:513–523.)**

Response 1 (CTR1). CTR1 physically interacts with the histidine kinase domain of the receptor ETR1. This interaction leads to the phosphorylation of CTR1 and initiates the signal transduction stream. CTR1 is a serine/threonine protein kinase. According to this model, CTR1 initiates a protein kinase cascade that ultimately results in the phosphorylation of one or more transcription factors and the constitutive expression of certain genes. The protein kinase cascade is very similar to a widely known group of mitogen-activated protein kinases that serve a critical role in the transduction of many signals in animals, plants, and fungi (Box 21.3). CTR1 is equivalent to a MAPKKK and several potential candidates for the other kinases in the MAPK cascade

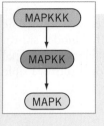

Box 21.3

MITOGEN-ACTIVATED PROTEIN KINASE: A WIDESPREAD MECHANISM FOR SIGNAL TRANSDUCTION

The capacity of extracellular signals such as hormones, light, osmotic status, and stress to effect a change in the physiology of cells often depends on regulating the transcription of genes. The expression of a gene in turn depends on the binding of activated transcription factors to the DNA in the promoter region of the gene. One of the primary means for regulating this interaction between a transcription factor and DNA is phosphorylation. Some transcription factors will not bind to DNA unless they are phosphorylated, while phosphorylation inhibits the binding of others.

Many extracellular signals are linked to transcription factor phosphorylation by the **mitogen-activated protein kinase (MAPK)** system (Figure 21.11). Originally named for its role in activation of genes involved in cell proliferation (mitogen = to stimulate mitosis), the MAPK system is known to mediate gene transcription in response to a variety of signals in animals, plants, and fungi.

The core of the MAPK system is a sequence of three kinase enzymes. Each enzyme acts to effect the phosphorylation of the next enzyme in the sequence. Starting at the bottom end, the third enzyme in the sequence, MAP kinase (MAPK), is responsible for phosphorylating a transcription factor. MAPK is in turn phosphorylated by an MAP kinase kinase (MAPKK or MAP2K), which is phosphorylated by an MAP kinase kinase kinase (MAPKKK, or MAP3K). MAPKKK can be activated directly by a signal-receptor complex, a G-protein, or some other secondary signal.

There are two advantages to this system. The first is that each kinase is able to phosphorylate multiple copies

of the next component—hence the name "cascade." The second advantage is that each of the components is represented by a small family of proteins. By utilizing different members of each family in various combinations, the cell is able to assemble a large number of different pathways that can interpret different extracellular signals and activate a variety of different transcription factors.

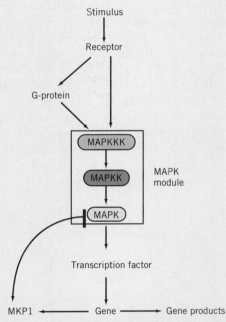

FIGURE 21.11 **The principal components of a generalized MAP kinase cascade. The activation of a receptor by an extracellular stimulus initiates the kinase cascade involving a sequence of serine-threonine kinases. The initial activation may be mediated by a G-protein or other second messenger system. The first kinase in the cascade, mitogen-activated protein kinase kinase kinase (MAPKKK) phosphorylates the second kinase, MAP kinase kinase (MAPKK), which in turn phosphorylates the third component, MAP kinase (MAPK). The ultimate target is the phosphorylation of a transcription factor. The activated transcription factor binds to the promoter region of the target gene and up-regulates transcription of the gene. One of the genes whose expression is stimulated is MKP1, a protein phosphatase that can remove the phosphate group from MAPK and block gene activation.**

stream have recently been identified in *Arabidopsis*. When ethylene binds with the receptor, it prevents the interaction of CTR1 with ETR1. This blocks the initiation of the protein kinase cascade and subsequent gene activation. The result is that in the absence of ethylene the expression of ethylene-controlled genes is always "on." The effect of ethylene is to turn these genes "off" by preventing the activation of the required transcription factors.

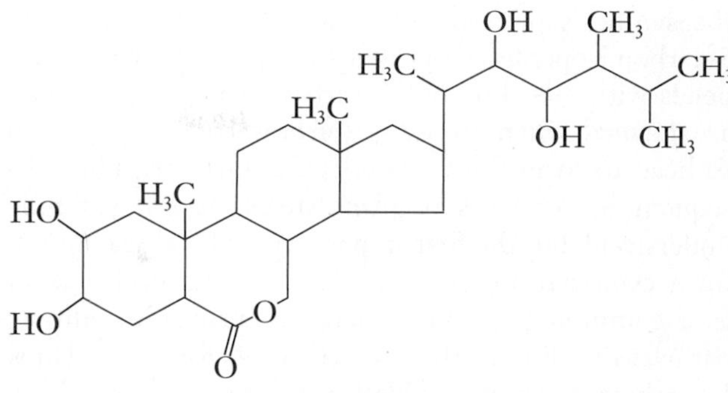

21.3 BRASSINOSTEROIDS

Brassinosteroids are steroid hormones with a chemical structure similar to the steroid hormones in animals. Brassinosteroids elicit an impressive array of developmental responses, including an increased rate of stem and pollen tube elongation, increased rates of cell division (in the presence of auxin and cytokinin), seed germination, leaf morphogenesis, apical dominance, inhibition of root elongation, vascular differentiation, accelerated senescence, and cell death. Brassinosteroids are also implicated in mediating responses to both abiotic and biotic stress, including salt, drought, temperature extremes, and pathogens.

The study of brassinosteroids as plant hormones dates back to the early 1970s, when a group of agricultural researchers began screening pollen, already known as a rich source of growth-promoting substances. The result was a complex mixture of lipids that stimulated elongation of bean second internodes. Because the most active preparations were isolated from pollen of the rape plant (*Brassica napus*), the active substances were referred to collectively as **brassins**.

Many of the effects of the brassins were similar to those of GA, leading many to believe the extracts were simply crude extracts of gibberellins, rather than a new class of hormones as originally proposed. However, in 1979, M. D. Grove and his coworkers identified the active component as **brassinolide (BL)** (Figure 21.12). The very low concentration of brassinosteroids in plant tissues is suggested by the fact that approximately 40 kilograms of bee-collected rape pollen were required in order to obtain only 4 milligrams of pure, crystalline BL! Fortunately, BL was chemically synthesized only two years later, eliminating the need for such massive and laborious extraction procedures.

For many years, the classification of brassinosteroids as hormones was not widely accepted. Effects could be demonstrated only by exogenous application and it was difficult to assess their function *in vivo*. This all changed with the discovery of mutants of *Arabidopsis*, pea, and tomato that are blocked in brassinosteroid biosynthesis. In all cases, the normal phenotype could be rescued by the application of brassinosteroids. These

Brassinolide

FIGURE 21.12 **The phytosterol brassinolide (BL) is an example of brassinosteroid hormones.**

studies clearly established that brassinosteroids have some well-defined functions in normal plant development and qualify as a distinct class of endogenous plant hormones. More than 40 analogs of BL have now been isolated from more than 60 plant species and virtually all types of tissues, including pollen, seeds, leaves, stems, roots, and flowers. However, brassinolide remains the most biologically active brassinosteroid and is widely distributed throughout the plant kingdom. Like other hormones, brassinolide is active in micromolar concentrations. Brassinolide at a concentration of 10^{-7} M is able to stimulate a fourfold increase in the length of soybean epicotyl sections.

Interestingly, most of the responses controlled by brassinosteroids are also controlled by auxin. Although the two hormones can act independently, it should not be too surprising that there is considerable crosstalk between brassinosteroid and auxin signaling pathways. Relatively little is known about the mechanisms for BR regulation of the many responses it appears to be involved in. With respect to cell elongation, however, the evidence indicated that BR increases plastic extensibility of the cell wall by regulating genes encoding wall-modifying enzymes such as expansins and cellulose synthase.

21.3.1 BRASSINOSTEROIDS ARE POLYHYDROXYLATED STEROLS DERIVED FROM THE TRITERPENE SQUALENE

Brassinosteroids are polyhydroxylated plant **sterols**—lipoidal substances related biosynthetically to the gibberellins and abscisic acid (refer to Figure 19.4). Plants synthesize a large number and variety of sterols, including sitosterol, stigmasterol, cholesterol, and campesterol. Sterols and sterol derivatives are discussed more extensively in Chapter 27. Sterols are triterpenoids, C_{30} molecules that are derived from acetate through the mevalonic acid pathway (see Figure 19.3A, 19.4). In the synthesis of terpenes, sequential additions of the 5-carbon isopentenyl pyrophosphate (IPP) produce terpenes with 10-, 15-, or 20-carbon atoms. Triterpenes are formed when two C_{15} (farnesyl) units join head to head to form the C_{30} molecule squalene. The subsequent biosynthesis of plant sterols is not yet fully understood, but the first step is a cyclization reaction to form **cycloartenol** (Figure 21.13). Using cycloartenol as a common precursor, there are probably multiple pathways leading to the several sterols found in plants. Decarboxylation and oxidation reactions are involved, as most common sterols have from 26 to 29 carbons and a single hydroxyl (—OH) group.

It is thought that most sterols, with the exception of stigmasterol, may serve as precursors for various brassinosteroids. However, the pathway for the biosynthesis of brassinolide is best understood. This pathway was established largely through studies using cultured cells of *Catharanthus roseus*. The precursor to brassinolide is **campesterol**, a C_{28} sterol. Through a series of largely oxidative steps, additional hydroxyl (—OH) groups are added and oxygen is introduced into the B ring. There are two parallel pathways for the conversion of campesterol to brassinolide, depending on whether the oxidation at carbon-6 occurs early or late in the pathway. In either case, approximately 12 steps are

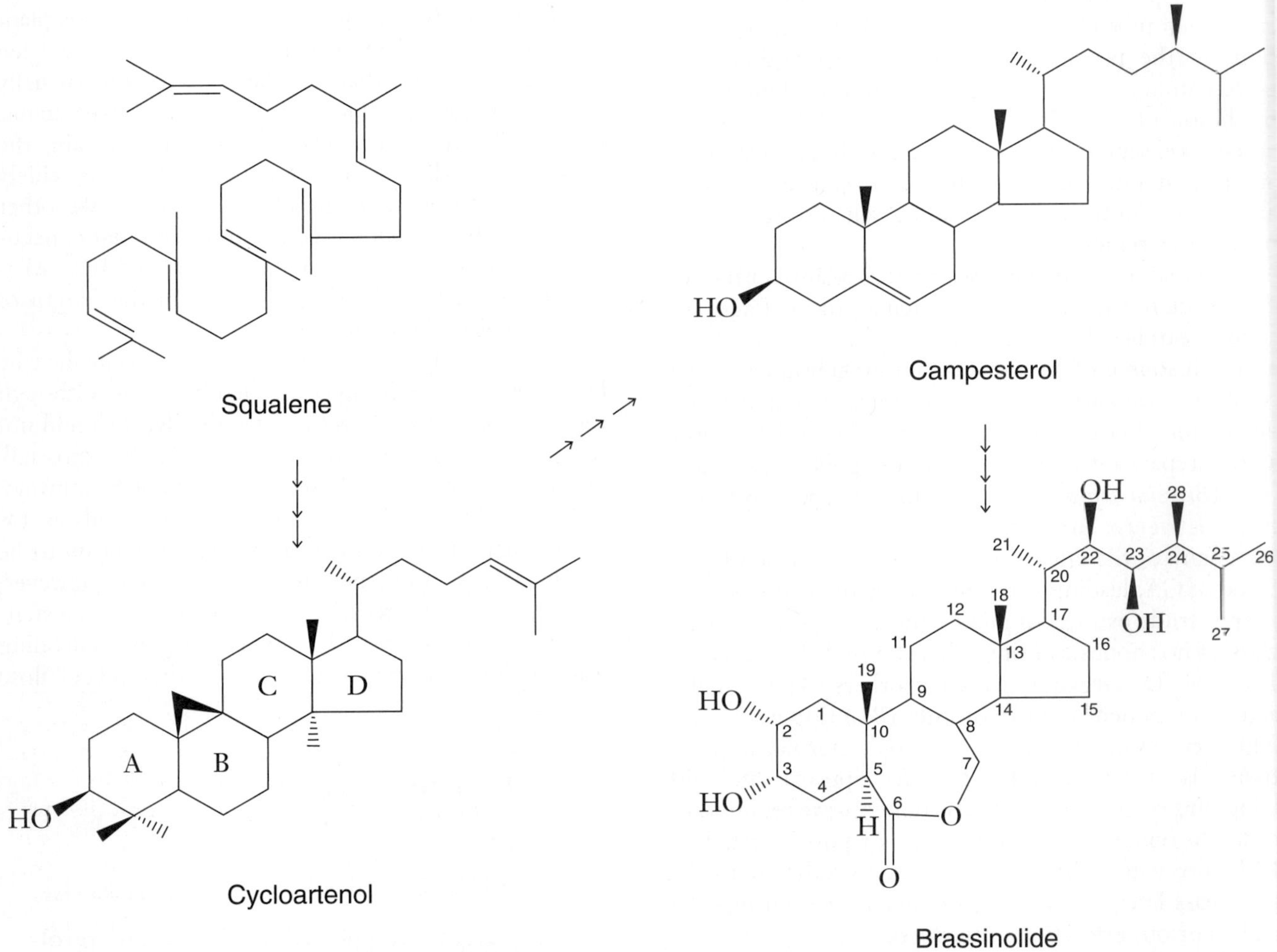

FIGURE 21.13 **Principal steps in the biosynthesis of brassinolide from the triterpenoid squalene. The biosynthesis of sterols in plants is poorly understood. Squalene undergoes a cyclization reaction to form cycloartenol. Cycloartenol is then subject to various oxidations and methylations to form campesterol and other sterols. The pathway for synthesis of brassinolide from campesterol has been established in cultured cells of *Catharanthus roseus*. Two alternate pathways are known, each involving at least 12 steps.**

required to complete the conversion of campesterol to brassinolide (Figure 21.13).

21.3.2 SEVERAL ROUTES FOR DEACTIVATION OF BRASSINOSTEROIDS HAVE BEEN IDENTIFIED

Studies into the metabolism of brassinosteroids in tomato and *Ornithopus* cells, cucumber, and mung bean have revealed several deactivation mechanisms. The α-hydroxyl groups on the A ring may be epimerized to form a β-hydroxyl, which is then followed by esterification with fatty acids or by glucosylation. Other possibilities include cleavage of the side chain or conjugation at other hydroxyl positions. The precise method for deactivation clearly depends on the species and possibly the tissue involved.

21.3.3 BRASSINOLIDE RECEPTORS AND SIGNALING

With the discovery of brassinosteroids in plants, it is now clear that steroid hormones are not restricted to animals. Unlike animals, however, which rely mainly on nuclear receptors for steroid hormones, plants appear to use membrane-based receptors to initiate a phosphorylation cascade that carries the signal into the nucleus. Although genetic studies are beginning to unmask some of the many components in brassinosteroid perception and signaling, there remains much to be learned about how they interact with each other and with signaling for other biotic and abiotic factors. Figure 21.14 depicts a general model for brassinosteroid signaling.

The principal receptor for brassinosteroids requires the interaction of two proteins that form a plasma membrane-associated **heterodimer**. The first is a serine/threonine kinase known as BRASSINOS-TEROID INSENSITIVE 1 (BRI1). (Keep in mind that the name of the gene and protein are based on observed mutations—hence the name "insensitive.") The second is the protein BRI1-ASSOCIATED RECEPTOR KINASE (BAK1). Brassinosteroids bind to the extracellular domain of BRI1, which first induces the dissociation of an inhibitory protein (BKI1) that inhibits the association of BAK1 with BRI1 and then promotes dimerization with BAK1 and autophosphorylation of BRI1. The phosphorylated complex initiates BR signaling. One target of brassinosteroid signaling is the protein BZR1 (BRASSINAZOLE RESISTANT 1). BZR1 is a transcription factor whose location is dependent on its phosphorylation status. In the phosphorylated state, BZR1 is trapped in the cytoplasm while dephosphorylation allows it to move into the nucleus. The phosphorylation

status of BZR1 is mediated by two competing factors; the BR signaling chain and a separate protein, BIN2. BIN2 mediates phosphorylation of BZR1 and thus keeps the protein in the cytoplasm. The brassinosteroid signal, on the other hand, mediates dephosphorylation of BZR1, which both activates the transcription factor and encourages its migration into the nucleus. Once in the nucleus, BZR1-P binds to target sites in the promoter region of BR-sensitive genes and initiates transcription. BR signaling may also inhibit the phosphorylating ability of BIN2, thus further ensuring activation and nuclear localization of BZR1.

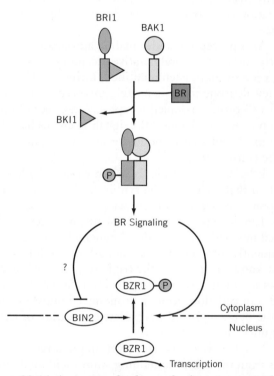

FIGURE 21.14 **A proposed scheme for brassinosteroid (BR) signaling. In the absence of BR, the two receptor kinases BRI1 and BAK1 exist independently in the membrane as monomers. Dimerization, necessary for activation, is inhibited by the presence of BKI1. BR induces the removal of BKI1, followed by the dimerization of the two receptors and auto phosphorylation of kinase region of BRI1. The targets of the BR signaling chain are transcription factors such as BZR1. The cellular location and activaty of BR-sensitive transcription factors is determined by their phosphorylation status. The BR signaling chain dephosphorylates cytoplasmic BZR1-P, which then migrates into the nucleus where it enables transcription of BR-sensitive genes. The protein BIN2 phosphorylates BZR1, causing it to migrate back into the cytoplasm, thus inhibiting transcription. BR signaling may also enhance transcription by blocking BIN2.**

SUMMARY

Abscisic acid (ABA) is a 15-carbon sesquiterpene that is synthesized by cleavage of the 40-carbon carotene violaxanthin. It is synthesized primarily in green leaves and can be stored in the chloroplast, although there is some evidence that ABA may also be either synthesized or stored in roots and exported to the leaves in times of water stress.

The primary functions of ABA are (1) prohibiting precocious germination and promoting dormancy in seeds and (2) inducing stomatal closure and the production of molecules that protect cells against desiccation in times of water stress. ABA has also been implicated in other developmental responses, including heterophylly (leaves of different shape on the same plant), initiation of secondary roots, flowering, and senescence.

ABA perception and signaling appears to be particularly complex. Several putative receptors or ABA binding proteins have been identified, including ABAP1 in barley aleurone tissue, a protein involved in flowering, and a G-protein coupled receptor. A number of other components of various ABA signaling chains have been identified, but a clear model for any one chain has yet to be constructed.

Ethylene is a simple 2-carbon gaseous hydrocarbon that appears to be synthesized primarily in response to stress or in senescing and ripening tissue. Ethylene is ubiquitous in plants, has been implicated in a wide range of developmental responses, and frequently interacts with auxin and gibberellin. It's best-known response is the triple response of etiolated dicot seedling, characterized by inhibition of hypocotyl and root cell elongation, a pronounced radial swelling of the hypocotyl, and exaggerated curvature of the plumular or epicotyl hook.

Ethylene is synthesized by a simple three-step process from the sulfur-containing amino acid methionine. In order to sustain normal rates of ethylene production, the sulphur released during ethylene formation is recycled back to methionine via the Yang Cycle.

Ethylene is sensed by a membrane-located histidine kinase receptor that initiates a mitogen-activated protein (MAP) kinase cascade that results in the activation of transcription factors. Uniquely in the case of ethylene, however, the receptor kinase system is constitutively active (i.e., "on") and is blocked, or turned "off," by ethylene.

Brassinosteroids (BR) are steroidal hormones that elicit a wide range of effects when applied to plants including an increased rate of stem and pollen tube elongation, increased rates of cell division (in the presence of auxin and cytokinin), seed germination, leaf morphogenesis, apical dominance, inhibition of root elongation, vascular differentiation, accelerated senescence, and cell death. Brassinosteroid sensing involves a serine/threonine kinase that regulates the phosphorylation and dephosphorylation of transcription factors.

CHAPTER REVIEW

1. Auxins, gibberellins, and brassinosteroids all influence stem elongation. In what ways do the responses of stems to these three hormones differ?

2. What is the evidence that ABA mediates responses to water stress?

3. What unique problems are related to the study of ethylene as a plant hormone?

4. Describe the hormonal changes that occur during seed development, maturation, and germination. Offer a rationale for the observed pattern.

5. What is a "MAPK module"? How is it involved in signal perception and transduction?

6. How are gibberellins and brassinosteroids related biosynthetically?

7. What is the evidence that ABA is synthesized in leaf mesophyll cells? In vascular tissue?

8. What is vivipary? How do we know that ABA is involved in regulating vivipary?

9. Compare and contrast the receptor systems for ABA, ethylene, and brassinosteroids.

FURTHER READING

Buchanan, B. B., W. Gruissem, R. L. Jones. 2000. *Biochemistry and Molecular Biology of Plants*. Rockville, MD: American Society of Plant Physiologists.

Chen, Y-F., M. D. Randlett, J. L. Findell, G. E. Schaller. 2002. Localization of the ethylene receptor ETR1 to the endoplasmic reticulum. *Journal of Biological Chemistry* 277:19861–19866.

Clouse, S. D. 2002. Brassinosteroids. In: C. R. Somerville, E. M. Meyerowitz (eds.), *The Arabidopsis Book*. American Society of Plant Biologists, Rockville, MD, doi: 10.1199/tab.0009, www.aspb.org/publications/arabidopsis/.

Davies, P. J. 2004. *Plant Hormones: Biosynthesis, Signal Transduction, Action*. Dordrecht: Kluwer Academic Publishers.

Gendron, J. M., Z-Y Wang. 2007. Multiple mechanisms modulate brassinosteroid signaling. *Current Opinion in Plant Biology* 10:436–441.

Guo, H., J. R. Ecker. 2004. The ethylene signaling pathway: New insights. *Current Opinion in Plant Biology* 7:40–49.

Guzman, P., J. R. Ecker. 1990. Exploiting the triple response of Arabidopsis to identify ethylene-related mutants. *Plant Cell* 2:513–523.

Hardtke, C. S. 2007. Transcriptional auxin-brassinosteroid crosstalk: Who's talking? *BioEssays* 29:1115–1123.

Hirayama, T., K. Shinozaki. 2007. Perception and transduction of abscisic acid signals: Keys to the function of the versatile plant hormone, ABA. *Trends in Plant Science* 12:343–351.

Karp, G. 2008. *Cell and Molecular Biology*. New York: John Wiley & Sons. (Includes a good review of MAP kinases.)

Li, J. 2005. Brassinosteroid signaling: From receptor kinases to transcription factors. *Current Opinion in Biology* 8:526–531.

Nambra, E., A. Marion-Poll. 2005. Abscisic acid biosynthesis and catabolism. *Annual Review of Plant Biology* 56:165–185.

Razem, F. A. et al. 2004. Purification and characterization of a barley aleurone abscisic acid-binding protein. *Journal of Biological Chemistry* 279:9922–9929.

Ryu, H. et al. 2007. Nucleocytoplasmic shuttling of BZR1 mediated by phosphorylation is essential in *Arabidopsis* brassinosteroid signaling. *The Plant Cell* 19:2749–2762.

Schwartz, S. H., X. Qin, J. A. D. Zeevaart. 2003. Elucidation of the indirect pathway of abscisic acid biosynthesis by mutants, genes, enzymes. *Plant Physiology*. 2003. 131:1591–1601.

Verslues, P., J-K. Zhu. 2007. New developments in abscisic acid perception and transduction. *Current Opinion in Plant Biology* 10:447–452.

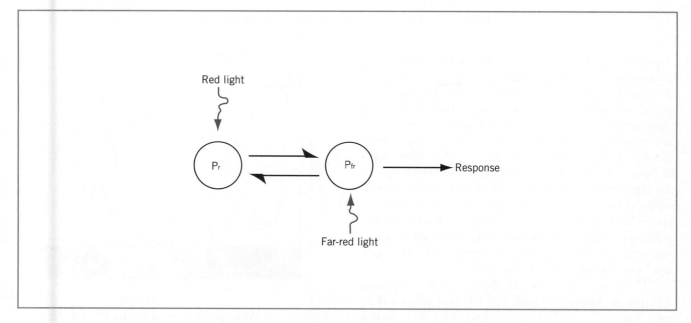

22

Photomorphogenesis: Responding to Light

Plants do not enjoy the luxury of being able to change their environment or to seek shelter from adverse conditions by changing their location. Yet species survival dictates that plants must be able to avoid adverse conditions. The germination of seeds and survival of the seedling that emerges are largely dictated by conditions in their immediate environment. Many seeds will not germinate if buried too deeply or if they lay in the shade of a forest canopy. Seedlings that emerge beneath a canopy tend to have elongated stems, as if reaching out for the light. Weeds growing in full sun at the edge of a wheat field will be shorter and more compact than plants of the same species competing with the crop in the center of the field. Plants flower at different times, spreading flowering through the season as though each species is awaiting some environmental cue. These and similar patterns of plant behavior have a significant survival advantage. They enable plants to make the best use of available resources, compete effectively with other species, or anticipate unfavorable environmental change.

How can seeds and seedlings know where they are? How can plants ensure they are in a position to maximize photosynthesis? How do plants measure the passing of the season? The answers to these and many other questions directly related to plant survival may be found in their capacity to detect and interpret a variety of environmental signals.

One particularly important environmental signal is light. The quantity and quality of light are constantly changing, often in predictable ways. Plants are able to monitor these changes and make use of this information to direct their growth, form, and reproduction. In this chapter we

- introduce the concept of light-regulated plant development or photomorphogenesis,

- describe the phytochromes; pigments that mediate plant responses to red and far-red light,

- describe the cryptochromes; pigments that mediate plant responses to blue light,

- review the principal developmental responses regulated by the phytochromes and cryptochromes,

- review the basic aspects of phytochrome and cryptochrome signal transduction, and

- present a case study of the role of phytochrome and cryptochrome in de-etiolation in *Arabidopsis*.

22.1 PHOTOMORPHOGENESIS IS INITIATED BY PHOTORECEPTORS

The regulation of plant development by light, or **photomorphogenesis**, is a central theme in plant development. In order to acquire and interpret the information that is provided by light, plants have developed sophisticated photosensory systems comprised of light-sensitive **photoreceptors** and signal transduction pathways. A photoreceptor "reads" the information contained in the light by selectively absorbing different wavelengths of light. Absorption of light normally induces a conformational change in the pigment or an associated protein, a photochemical oxidation-reduction reaction, or some other form of photochemical change. Whatever the nature of the primary event, absorption of light by the photoreceptor sets into motion a cascade of events that ultimately results in a developmental response.

There are four classes of photoreceptors in plants. The **phytochromes** absorb red (R) and far-red (FR) light (ca. 660 and 735 nm, respectively) and have a role in virtually every stage of development from seed germination to flowering. **Cryptochromes** and **phototropin** detect both blue (400–450 nm) and UV-A light (320–400 nm). The cryptochromes appear to play major roles during seedling development, flowering, and resetting the biological clock. Phototropin mediates phototropic responses, or differential growth in a light gradient, and will be discussed in the context of these responses in Chapter 23. A fourth class of photoreceptors that mediate responses to low levels of UV-B (280–320 nm) light have not yet been characterized.

Phytochrome, cryptochrome, and phototropin are all **chromoproteins** (Chapter 6). Chromoproteins contain a light-absorbing group, or **chromophore**, attached to a protein with catalytic properties, called the **apoprotein**. The chromophore plus the apoprotein is referred to as the holoprotein.

22.2 PHYTOCHROMES: RESPONDING TO RED AND FAR-RED LIGHT

Phytochrome is a unique pigment that can exist in two states—one with an absorption maximum in the red (R, or 665 nm) region of the spectrum and one with an absorption maximum in the far-red (FR, 730 nm) (Figure 22.1). The pigment is ubiquitous in plants and its discovery solely on the basis of simple but elegant physiological experiments ranks among the major achievements of twentieth-century plant biology (Box 22.1).

Early phytochrome studies employed almost exclusively a combination of physiological experiments and physical techniques such as *in vivo* spectroscopy. It is interesting to note that even these early physiological experiments predicted multiple forms of phytochrome.

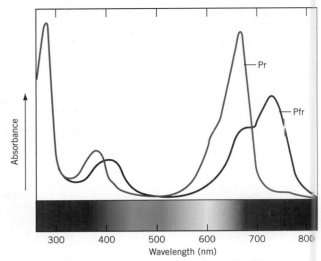

FIGURE 22.1 **Absorption spectra of the Pr and Pfr forms of purified phytochrome. The spectrum for Pfr is actually the spectrum for an equilibrium mixture of Pfr and Pr (see text for details). Note the differential absorption in the blue region of the spectrum as well as the red, far-red region. Some blue light effects are mediated by phytochrome, but photoconversion by red light is 50 to 100 times more effective than in the blue. Because both forms absorb equally in the green region (500 to 550 nm), green light does not appreciably change the state of the pigment and can, in most cases, be used as a safe light when setting up phytochrome experiments.**

With the advent of molecular genetics, the existence of multiple phytochromes in higher plants was confirmed. In most angiosperms, for example, there are at least three distinct phytochromes: phyA, phyB, and phyC, encoded by the genes *PHYA*, *PHYB*, and *PHYC*, respectively.[1] *Arabidopsis*, which has been studied most extensively, has five phytochromes (phyA–phyE). The differences between the several phytochromes are in the protein—the chromophore is common to all members of the phytochrome family. The different phytochromes are expressed in different tissues at different times in development and mediate different light responses, although it appears that phyA and phyB are the principal species that mediate red and far-red responses. Phytochromes also interact with each other as well as with other photoreceptors and developmental stimuli, rendering an understanding of their complex signal transduction pathways one of the more challenging areas of current study.

---

[1]The notation convention adopted for plant photoreceptors follows the recommendations of Quail et al. 1994. *Plant Cell* 6:468: e.g., phyA for the holoprotein, PHYA for the apoprotein, *PHYA* for the wildtype gene, and *phyA* for the mutant gene.

Box 22.1
HISTORICAL PERSPECTIVES— THE DISCOVERY OF PHYTOCHROME

It is now well established that phytochrome plays a critical role in almost every stage of plant development. Its existence was predicated on the basis of a simple physiological observation: seed germination and growth of etiolated seedlings exhibited photoreversible responses to red and far-red light. Because of its uniquely photoreversible character, however, the newly proposed pigment was initially greeted with skepticism in the scientific community.

It had long been recognized that light influenced plant development under conditions that excluded significant levels of photosynthesis. Indeed, dramatic differences in the growth and form of plants in darkness and light had fascinated botanists and physiologists for centuries. However, little real progress toward understanding these phenomena was achieved until the early 1950s. At that time, H. A. Borthwick, a botanist, and S. B. Hendricks, a physical chemist, began a study of action spectra for such diverse phenomena as seed germination, stem elongation, and photoperiodic control of flowering. It soon became apparent that all these phenomena shared similar action spectra, with peaks in the red and far-red. More interesting, however, was the discovery that a response potentiated by red light could be negated if the red light treatment were followed immediately with far-red light. Such clear photoreversibility had never before been described in biology. This distinctive characteristic led Borthwick and Hendricks to propose the existence of a novel pigment system, later called phytochrome.

This new, but hypothetical, pigment would exist in two forms: a red-absorbing form called Pr and a far-red–absorbing form called Pfr. The pigment would also be photochromic, which means that absorption of light would alter its absorbance properties. Absorption of red light by Pr would convert the pigment to the far-red–absorbing form while subsequent absorption of far-red light by Pfr would drive the pigment back to the red-absorbing form. Solely on the basis of simple physiological experiments, they were able to predict several other features of this hypothetical pigment system. First, because seeds and dark-grown seedling tissues responded initially to red light, not far-red, the pigment was probably synthesized as the Pr form. Moreover, Pr was stable and probably physiologically inactive. Second, because treatment with red light initiated germination and other developmental events, Pfr was probably the active form. On the other hand,

Pfr was apparently unstable and was either destroyed or could revert to Pr in darkness by a nonphotochemical, temperature-dependent reaction. Third, because the pigment could not be seen in dark-grown, chlorophyll-free tissue, it was no doubt present at very low concentration. Borthwick and Hendricks further surmised that the pigment must be acting catalytically and was therefore possibly a protein. It is a tribute to the scientific acumen of these investigators and their coworkers that every one of these predictions was later proven true. Yet, at the time, the existence of phytochrome was met with some skepticism within the scientific community, largely because there was no precedent for such a photoreversible pigment in the plant or animal research literature.

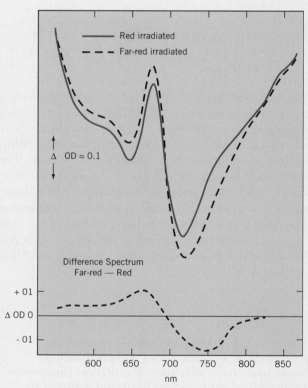

FIGURE 22.2 **Absorbance curves for maize shoots following red or far-red irradiations. Note that these curves represent the absorbance of whole tissue, not just the pigment. Note also that far-red irradiation of the tissue, which converts the pigment from Pfr (solid curve) to Pr (dashed curve), causes the expected increase in absorbance in the red and a decrease in the far-red regions of the spectrum. The difference spectrum effectively represents the absorption spectrum of the Pr form. These data represent the first physical demonstration of the existence of phytochrome. (From Butler, W. et al. 1959.** *Proceedings of the National Academy of Sciences* **USA 45:1703–1708. Reprinted by permission.)**

It was clearly necessary to obtain physical evidence for the existence of phytochrome and, ultimately, to isolate the pigment and characterize it *in vitro*. The strategy that led to a satisfactory resolution of this problem turned on the same unique photoreversible character that generated skepticism in the first place. Since phytochrome was the only known photochromic pigment present in plants, it should be possible to detect absorbance changes related to photoconversion of the pigment from one form to the other. Hendricks and his colleagues thus predicted that the conversion of Pr to Pfr in red light would be accompanied by a decrease in absorbance in the red (the maximum absorbance of Pr) and a corresponding increase in absorbance in the far-red. Subsequent irradiation with far-red light should cause an increase in absorbance in the red and a decrease in the far-red. These experiments would require a special kind of spectrophotometer, one capable of measuring very small absorbance changes in dense, light-scattering tissue samples. Fortunately, such an instrument was under development in another laboratory at Beltsville and relatively straightforward modifications were required to adapt its use for phytochrome detection.

The predicted photoreversible absorbance changes were first demonstrated in dark-grown maize shoots in 1959 (Figure 22.2). This spectral analysis was the first physical evidence that phytochrome actually existed. A short time later, the pigment was successfully isolated and purified from dark-grown cereal seedlings. In the years that followed, phytochrome was found to be ubiquitous in the plant kingdom. It is found in algae, bryophytes (mosses and liverworts), and probably all higher plants where it plays a significant role in biochemistry, growth, and development.

22.2.1 PHOTOREVERSIBILITY IS THE HALLMARK OF PHYTOCHROME ACTION

The key to the discovery of phytochrome was the finding that plant responses to weak red light could be nullified if the red treatment were followed immediately with far-red light (Table 22.1). Such clear **photoreversibility** had never before been described in biology. The data in Table 22.1 are for groups of seeds that were allowed to imbibe water in darkness for three hours before being subjected to various brief light treatments. The light treatments were either 1 minute of red light (R; λ ca. 660 nm) or 3 minutes of far-red light (FR; λ > 700 nm) at low fluence rates, or alternating R and FR in succession.

Following irradiation, the seeds were returned to darkness for 48 hours, and the number of germinated seeds in each lot was then counted. Note that red light promotes a high germination rate but that a R,FR treatment (red light followed immediately by far-red light) maintains germination at the dark level (22%). When the R and FR treatments are alternated, the germination rate depends solely on whether R or FR was presented last. It is as though germination were dependent on a switch that could be turned on by red light and turned off by far-red light. Similar results were observed with responses as diverse as stem elongation in dark-grown pea seedlings and photoperiodic control of flowering in cocklebur (*Xanthium strumarium*).

Photoreversible behavior can be explained by a simple, two-state photoequilibrium model for phytochrome (Figure 22.3). The red light-absorbing form of the pigment is designated **Pr** and the far-red-absorbing form, **Pfr**. Phytochrome is synthesized as the Pr form, which

TABLE 22.1 **Photoreversible control of germination. Lettuce seeds were imbibed for 3 hours prior to irradiations. Irradiation times were 1 min of low intensity red light and 3 min of Fr. Germination was scored after 48 h in subsequent darkness at 20°C.**

Irradiations	Germination (%)
R	88
R, Fr	22
R, Fr, R	84
R, Fr, R, Fr	18
R, FR, R, FR, R	72
R, Fr, R, Fr, R, Fr	22

Data from a student experiment.

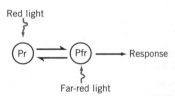

FIGURE 22.3 **The simple, two-state photoequilibrium model for phytochrome as originally postulated on the basis of seed germination and other plant responses to alternating red and far-red light. Absorption of red light by the red-absorbing form of the pigment (Pr) converts it to a form that absorbs far-red light (Pfr). Conversely, absorption of far-red light by Pfr returns the pigment to the Pr form. Pfr is considered the active form and initiates a signal transduction chain that leads to germination.**

accumulates in dark-grown tissue and is generally considered to be physiologically inactive. When Pr absorbs red light, it is converted to the Pfr form, which is the physiologically active form of the pigment for most known responses. Exposure of Pfr to far-red light returns the pigment to the Pr form. Both physiological and spectrophotometric experiments have also indicated that some Pfr may revert to Pr by a temperature-dependent process called dark reversion.

22.2.2 CONVERSION OF PR TO PFR IN ETIOLATED SEEDLINGS LEADS TO A LOSS OF BOTH PFR AND TOTAL PHYTOCHROME

Most of the early work on phytochrome was conducted with dark-grown, or etiolated, seedlings. Dark-grown seedlings grow quickly, they accumulate relatively large amounts of phytochrome, and the absence of chlorophyll makes it possible to measure phytochrome directly in tissue with spectrophotometers adapted for use with optically dense, light-scattering materials.[2] With the appropriate instrument, changes in the total amount of phytochrome and the relative proportions of Pr and Pfr could be monitored following controlled irradiations. These *in vivo* spectrophotometric studies confirmed many of the original predictions about the dynamic properties of phytochrome.

Typical phytochrome transformations in etiolated seedling tissue are shown in Figure 22.4. The most distinctive feature is that Pfr is relatively unstable. Note that when the tissue is returned to darkness after Pr is converted to Pfr with a brief pulse of low fluence red light, the concentration of Pfr declines with a half-life of 1 to 1.5 hours. This loss of Pfr is accompanied by a corresponding decline in the total amount of phytochrome. These kinetics can be explained by the fact that both forms of the pigment are subject to irreversible chemical degradation (Figure 22.5). In darkness, Pr accumulates until its rate of synthesis is matched by the rate of Pr degradation, which is relatively low. The loss of phytochrome following a red irradiation can be explained by two factors. First, the rate of Pfr degradation is approximately 100 times greater than the rate of Pr degradation. Immunochemical studies have demonstrated the conjugation of Pfr with ubiquitin, indicating that Pfr is subject to degradation by the ubiquitin/26S proteasome system. Second, it has been demonstrated that Pfr suppresses the transcription of the phytochrome gene by feedback inhibition. As little as 5 seconds of red light causes a

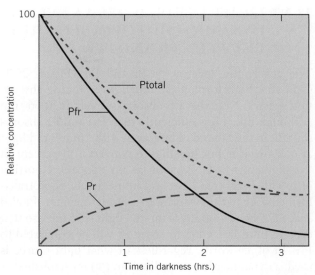

FIGURE 22.4 **Typical phytochrome transformations in etiolated seedling tissue. Dark-grown oat coleoptile tissue was given a short exposure to low fluence red light at time 0. Absorbance changes were then monitored for total pigment and Pfr in the ensuing dark period. Pr was calculated as the difference between total phytochrome and Pfr.**

rapid decline in translatable phytochrome mRNA in etiolated seedlings.

Although it was not known at the time of the original studies, it is now clear that only phyA accumulates to high levels in etiolated tissue and is rapidly degraded in the light. All other phytochromes (phyB–E) are stable when irradiated and are present in constant, although much lower, amounts regardless of the light conditions.

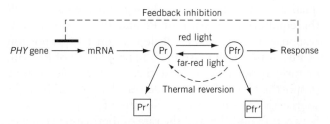

FIGURE 22.5 **The phytochrome system, based on the behavior of phytochrome in dark-grown seedling tissue. Phytochrome is synthesized as the physiologically inactive red-absorbing form (Pr), which accumulates in darkness. Red light (660 nm) drives a phototransformation to the far-red-absorbing form (Pfr). Absorption of far-red light (735 nm) returns the pigment to the Pr form. Pfr, the active form, induces a response. Pr′ and Pfr′ represent inactive degradation products of Pr and Pfr, respectively. Pfr is known to revert to Pfr in darkness by a temperature dependent process. Pfr also suppresses phytochrome mRNA transcription by feedback inhibition.**

---

[2] A spectrophotometer is an instrument for measuring the absorption of light by pigments in solution. Conventional spectrophotometers are by design limited to optically clear solutions, free of light-scattering particles.

22.2.3 LIGHT ESTABLISHES A STATE OF DYNAMIC PHOTOEQUILIBRIUM BETWEEN PR AND PFR

The absorption spectra for Pr and Pfr (Figure 22.1) show that both forms have broad, overlapping absorption spectra. Note that Pfr absorbs some light at 660 nm (although much less efficiently than Pr) and Pr absorbs slightly in the far-red. Thus, even with "pure" red light at 660 nm, it is not possible to convert 100 percent of the pigment to Pfr. As soon as Pfr appears, a portion of it will absorb red light and immediately phototransformed back to the Pr form. In a similar manner, Pr also absorbs a small amount of far-red light (735 nm), so that even in pure far-red light some Pr will be converted to Pfr. In other words, regardless of what light source is used, a dynamic **photoequilibrium** (Φ) is established as phytochrome cycles between Pr and Pfr. This photoequilibrium is conveniently defined as $\Phi = \text{Pfr}/\text{P}_{\text{TOT}}$, in which P_{TOT} is the total phytochrome or the sum of Pr and Pfr. The photoequilibrium established by red light (660 nm) in etiolated tissues, for example, is 0.8 while the value for far-red light at 720 nm is 0.03. In other words, pure red light will maintain about 80 percent Pfr and 20 percent Pr while far-red light will establish about 3 percent Pfr. For this reason, the absorption spectrum of Pfr shown in Figure 22.1 is actually the spectrum of an equilibrium mixture of Pr and Pfr following a saturating red light treatment.

Except in the laboratory, of course, plants do not grow in dark boxes with occasional flashes of red and far-red light. Moreover, sunlight contains a mixture of red and far-red wavelengths and, depending upon time of day and environmental conditions, the relative proportions of red and far-red wavelengths in sunlight will change. The result is that sunlight will also produce an equilibrium mixture of Pr and Pfr and, because the quality of sunlight changes throughout the day, so will the equilibrium mixture of Pr and Pf. The biological response in most cases will depend on the proportion of Pfr, or (Φ) in the system. The proportion of Pfr will, in turn, depend on at least three factors: the relative proportions of red and far-red wavelengths in the light source, the forward and reverse rates of photoconversion between Pr and Pfr, and the rate of thermal reversion of Pfr to Pr.

22.2.4 PHYTOCHROME RESPONSES CAN BE GROUPED ACCORDING TO THEIR FLUENCE REQUIREMENTS

Phytochrome-mediated responses are conveniently grouped into three categories on the basis of their energy requirements. The classical red, far-red photoreversible responses discovered by Hendricks and Borthwick and their colleagues are known as **low fluence responses (LFRs)**. LFRs are stimulated by light doses in the range of 1 µmole m^{-2} to 1000 µmole m^{-2}. This is equivalent to about a 0.1 second exposure under a dense plant canopy at the lower end and about one second of full daylight at the upper end. In addition, LFRs are FR-reversible.

Phytochrome responses stimulated by light levels in the range of 10^{-6} to 10^{-3} µmole^{-2} are called **very low fluence responses (VLFRs)**. Typically, such low levels of light (comparable to the light emitted from a firefly) convert only about 0.01 percent of the phytochrome. Several studies have indicated that dark-grown seedlings are capable of responding to such very low levels of light. Red light, for example, promotes an increase in sensitivity of cereal grain seedlings to a subsequent phototropic stimulus. But the red light fluence required to saturate the response was found to be at least 100 times less than that required to induce a *measurable* conversion of Pr to Pfr! A low far-red fluence also promotes phototropic sensitivity just as red light does, indicating that less than 1 percent of the pigment need be converted to Pfr in order to saturate the response. Exposure to even the traditional dim green safelights is sufficient to elicit or even saturate elongation responses in dark-grown *Avena* seedlings. For example, as little as 0.01 percent Pfr is required to elicit inhibition of mesocotyl elongation. This extreme sensitivity to light obviously makes the study of VLFRs technically difficult. VLFRs, for example, are not photoreversible. The principal evidence that a VLF response is mediated by phytochrome is the similarity of its action spectrum to the absorption spectrum of Pr. The phenomenon, however, raises perplexing yet intriguing questions about experimental photocontrol of plant development.

In the natural environment, plants are exposed to long periods of sunlight at relatively high fluence rates. Under such conditions, characterized by relatively high energy over long periods of time, the photomorphogenic program achieves maximum expression, and responses such as leaf expansion and stem elongation are far more striking. Such light-dependent responses are known as **high irradiance reactions (HIRs)**. High irradiance reactions generally share the following characteristics: (1) full expression of the response requires prolonged exposure to high irradiance, primarily with a high proportion of far-red light; (2) the magnitude of the response is a function of the fluence rate and duration; (3) like VLFRs, HIRs are not red, far-red photoreversible.

The HIR has been implicated in a wide range of responses that also qualify as LFRs, including stem growth, leaf expansion, and seed germination. However, HIRs may exhibit strikingly different action spectra depending on the species or growth conditions. Etiolated seedlings, for example, respond to blue, red, and far-red light. As de-etiolation progresses, there is a shift from a far-red-sensitive HIR to a red-sensitive HIR. Not surprisingly then, light-grown, green tissues are more responsive to red light rather than far-red. Some systems, such as anthocyanin synthesis in *Sorghum* seedlings, respond only to blue-UV-A light.

Genetic studies involving phytochrome mutants in *Arabidopsis* have identified that phyA is responsible for the VLFR and FR-sensitive HIR responses and phyB is responsible for the LFR and red-sensitive HIR.

22.3 CRYPTOCHROME: RESPONDING TO BLUE AND UV-A LIGHT

Charles Darwin was one of the first to note that plants respond to blue light when he observed that heliotropic movements were diminished in light passed through a solution of potassium dichromate, an effective absorber of blue light. It is now known that those portions of the spectrum that constitute blue and UV-A light regulate many aspects of growth and development in plants, fungi, and animals. Plant responses to blue and UV-A light include aspects of de-etiolation such as the inhibition of hypocotyl elongation and stimulation of cotyledon expansion, the opening and closure of stomata, gene expression, flowering time, the "setting" of endogenous clocks (Chapter 24), and the growth of shoots in response to light gradients, or phototropism (Chapter 23).

Most action spectra for blue-light responses had peaks in the blue and UV-A regions of the spectrum and closely resembled the absorption spectra of flavin molecules, such as riboflavin. This prompted A. W. Galston to postulate as early as 1950 that a flavoprotein was involved in blue light responses. Others, however, mounted strong arguments in favor of a carotenoid-based photoreceptor, and for many years the flavin–carotenoid controversy was hotly debated. The controversy was difficult to resolve because plants contain a bewildering array of flavoproteins and carotenoproteins, a factor that seriously complicated any efforts to identify the one that might be involved specifically in blue-light responses. Because of this elusive, or "cryptic," nature of the pigment and the pervasive blue-light responses in cryptograms, or nonflowering plants, the pigment was referred to as cryptochrome.

The first protein with the characteristics of a blue-light photoreceptor was isolated from *Arabidopsis* in 1993. Elongation of hypocotyls in wildtype *Arabidopsis* can be inhibited by red, far-red, or blue light. Several mutants (the *hy* mutants) were characterized by elongated hypocotyls because they had lost the capacity to respond to one or more regions of the spectrum. One of these, *hy4* (later named *cry1*), had lost the capacity to respond specifically to blue light. The protein encoded by the wildtype allele was subsequently isolated and, on the basis of its photobiological and genetic properties, identified as cryptochrome 1 (cry1). The argument between the flavin camp and the carotenoid camp was finally resolved after more than 40 years when it was shown that cry1 binds two flavin-related chromophores, one of which is flavin adenine dinucleotide (FAD) (Section 22.5.3).

A second cryptochrome, cry2, has also been identified. Cry2 mediates blue-light suppression of hypocotyl elongation, cotyledon expansion, and anthocyanin production in *Arabidopsis*. In addition, cry2 has a role in determining flowering time and is synonymous with FHA, the product of a flowering-time gene. The roles of cry2 and FHA in flowering are examined in Chapter 25.

22.4 PHYTOCHROME AND CRYPTOCHROME MEDIATE NUMEROUS DEVELOPMENTAL RESPONSES

Phytochrome and cryptochrome act both jointly and independently to regulate a wide range of developmental responses. Some of the better understood responses are described in the following sections.

22.4.1 SEED GERMINATION

The germination of many seeds is influenced by light as evident in the flush of germination in areas of cultivation or natural disturbance. Some seeds, known as **positively photoblastic seeds**, are stimulated to germinate by light. The germination of others, known as **negatively photoblastic seeds**, is inhibited by light. Some seeds, mostly agriculturally important species that have been selected for high germinability, are not affected by light. Many seeds, such as lettuce, may require only brief exposure to light, measured in seconds or minutes, while others may require as much as several hours or even days of constant or intermittent light (e.g., *Lythrum salicaria*, *Epilobium cephalostigma*). In all cases, the responsible pigment appears to be phytochrome.

Most seeds that require light for germination tend to be very small seeds that have comparably small embryos and limited endosperm. They need to be close to the surface when they germinate so that the young seedling can reach the sunlight before the reserves are exhausted. Most soils attenuate light very quickly. A 1 mm thickness of fine soil, for example, passes less than 1 percent of the incident light and then only at wavelengths longer than 700 nm. As a result, most light-requiring seeds need not be buried very deeply for germination to be held in check. However, some seeds (e.g., *Sinapis arvensis*) require very little Pfr ($\Phi = 0.05$) to stimulate germination and may exhibit germination when covered with up to 8 mm of soil. Thus, the role of phytochrome in both seed germination and seedling development appears to be one of conveying information to the seed or seedling about its position relative to the soil surface.

Interestingly, most common agricultural weeds such as *Amaranthus* (pigweed), *Ambrosia* (ragweed), and *Chenopodium* (lambs quarters) produce prodigious numbers of very small light-sensitive seeds. These seeds accumulate in the "seed bank" just below the surface of the soil where they will not germinate. Every time the soil is disturbed, however, a new batch of seeds germinates because they are exposed to light. This is a major factor in the competitive success of these species.

Suppression of germination in negatively photo-blastic seeds, such as wild oats (*Avena fatua*), generally requires long-term exposures at high fluence rates. Far-red and blue light are most effective, although in some cases (e.g., *Phacelia tanacetifolia*) red light is also effective. Photoinhibition of seed germination appears to be an example of a high irradiance reaction. In *Arabidopsis*, seed germination is controlled solely by phytochromes.

22.4.2 DE-ETIOLATION

Plants grown in darkness exhibit a distinct morphology. The details may vary from one species to another, but in a dicot such as bean (Figure 22.6) the hypocotyl is elongated and spindly, with a pronounced plumular hook, or recurve, just below the first leaves. The leaves themselves undergo limited development and remain small and clasping, as they were in the embryo. Chlorophyll is absent and the seedlings appear white or yellow in color. In monocot cereal grains the first internode, or mesocotyl, elongates excessively in the dark and the coleoptile, which is a modified leaf, grows slowly. The primary leaves remain within the coleoptile and stay tightly rolled around their midvein. This general condition exhibited by dark-grown seedlings is called **etiolation**. Other characteristics of the etiolated condition include arrested chloroplast development and low activities of many enzymes. When exposed to light, etiolated seedlings undergo **de-etiolation**, a process under control of both phytochromes and cryptochromes. Hypocotyl growth is arrested, the plumular hook gradually straightens, and elongation of the epicotyl accelerates. Light also stimulates the leaves to unfold, enlarge, and complete their development. Chloroplast development also proceeds and the leaves green up as chlorophyll accumulates and the leaves become photosynthetically competent. In

FIGURE 22.6 **The de-etiolation response in 7-day-old seedlings of bean (*Phaseolus vulgaris*). The seedling on the right was grown in darkness. Note the elongated hypocotyl, recurved plumular hook, an absence of primary leaf expansion, and absence of chlorophyll. The seedling at left was grown under normal white light conditions. Note the shortened hypocotyl, unfolding of the plumular hook, expanded primary leaves, and accumulation of chlorophyll. The center seedling was exposed to 5 minutes of weak red light daily for three days, which is sufficient to initiate the de-etiolation response.**

particular, light stimulates the synthesis of the chlorophyll a/b light harvesting pigment-protein.

The developmental significance of etiolation and de-etiolation is not difficult to construct. Remember that a plant is fundamentally a photosynthetic organism. A seed carries a limited amount of nutritive tissue that must support development of the seedling until such time as its leaves are established in the light and photosynthesis can take over the supply of energy and carbon. In the darkness experienced beneath the soil, the limited reserves of a seed are committed to extension growth of the hypocotyl in order to maximize the possibility that the plumule, composed of the young leaves, will reach the light and be able to carry out photosynthesis before the reserves are exhausted. Once established in the light, the remaining reserves may be invested in development of photosynthetic tissue, such as leaf expansion and chloroplast development.

22.4.3 SHADE AVOIDANCE

Plants need light for photosynthesis and those that find themselves growing in the shade of neighbors or under a canopy can adjust to the reduced availability of light in two ways. They can (1) adjust their light-harvesting capability by increasing the specific leaf area and the amount of chlorophyll a/b light-harvesting complex or they can (2) adjust their morphology in order to position their leaves out of the shade. Plants typically respond to shadelight with increased elongation of stem-like organs (including hypocotyls and leaf petioles, a more upward orientation of leaves (hyponasty), reduced branching, and reduced tillering (in grasses). In the end, shading leads to early flowering and seed set in an effort to "escape" shading by shortening generation time. These and other effects are collectively called the **shade avoidance syndrome**.

Radiation within and below a canopy is markedly deficient of red and blue light because these wavelengths are largely absorbed by the chlorophyll in the overlying leaves. By contrast, chlorophyll is transparent to far-red light; any attenuation of far-red is limited almost solely to reflection. Plants therefore use phytochrome (specifically phyB, phyD, and phyE) and cryptochrome to detect these characteristic differences in composition between shadelight and unfiltered daylight.

The effect of canopy shading can be described in terms of the ratio of red to far-red fluence rates (R/FR, or ζ; Gr. zeta). The value of ζ in unfiltered daylight is typically in the range of 1.05 to 1.25. The value in shadelight beneath a canopy will, of course, vary with the nature of the vegetation and the density of the canopy. Some representative values are listed in Table 22.2. These values fall well within the range where a small change in ζ would cause a relatively large change in the proportion of Pfr (Figure 22.7).

Shadelight can be mimicked in the laboratory or growth chamber by supplementing white fluorescent light ($\zeta = 2.28$) with various amounts of far-red light through the entire photoperiod. This can be done in such a way that the fluence rate of photosynthetically active radiation (PAR) remains constant from one

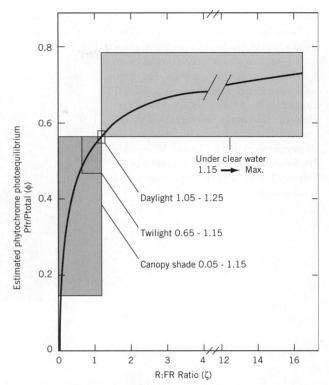

FIGURE 22.7 Phytochrome photoequilibrium (Φ) is related to the ratio of red to far-red light (ζ). Blocked areas indicate the range of values for ζ observed under indicated ecological conditions. (Reproduced with permission from the *Annual Review of Plant Physiology*, Vol. 33. Copyright 1982 by Annual Reviews, Inc.)

treatment regime to the next. This removes the impact of photosynthetic output and any differences in growth; morphogenesis can thus be attributed to the phytochrome photoequilibrium value (Φ), which can be estimated from the measured R/FR ratio for each regimen. When young plants of *Chenopodium album* were grown this way, the logarithmic stem extension rate was found to be linearly related to Φ (Figure 22.8). The response to light quality may be quite rapid. In *Chenopodium*, for example, an increase in the stem extension rate can be observed within seven minutes of adding FR light to the background fluorescent source. Even when not directly shaded, plants are able to anticipate shading. The reduction in red to far-red ratio reflected from the leaves of nonshading neighbors and propagated horizontally is often sufficient to initiate shade avoidance responses.

22.4.4 DETECTING END-OF-DAY SIGNALS

There are also substantial changes in the spectral energy distribution of natural daylight on a daily basis. At both dawn and dusk, when the sun sits low on the horizon, there are significant relative decreases in the value of ζ

TABLE 22.2 Approximate values of R/FR (ζ) for canopy filtered light.

Canopy	R/FR
Wheat	0.5
Maize	0.20
Oak woodland	0.12–0.17
Maple woodland	0.14–0.28
Spruce forest	0.15–0.33
Tropical rainforest	0.22–0.30

FIGURE 22.8 **The growth of *Chenopodium album* seedlings in simulated shadelight. Seedlings were grown for 14 days in shadelight simulated by providing supplementary far-red light sufficient to provide the indicated R:FR ratios (ζ). Note that internode elongation is inversely proportional to the amount of Pfr in the tissue. (Reproduced by permission of Dr. David Morgan and Professor Harry Smith, University of Leicester, UK. Photograph courtesy of Prof. Smith.)**

compared with the main part of the day. In one study, for example, ζ values at dusk were reduced by 14 to 44 percent of those at midday. A detailed examination of the response of pumpkin (*Cucurbita pepo*) to end-of-day red or far-red light reveals that a reduction in the proportion of phytochrome maintained as Pfr at the end of the photoperiod is associated with drastic changes in the developmental pattern. Lowering of Φ from a high value (ca. 0.65 to 0.75) to a very low value (ca. 0.03) accentuates stem and petiole extension, reduces leaf expansion and branching, and lowers the chlorophyll content. These experiments do not, of course, prove a causal link between phytochrome photoequilibrium and morphogenic responses to changes in the radiation environment. They do, however, demonstrate that plants have the capacity to respond to changes in spectral energy distribution similar to those that occur naturally in the environment. It also seems highly likely that phytochrome is the photoreceptor that detects end-of-day signals, which could reflect an important role in time measurement (Chapters 24, 25)

22.4.5 CONTROL OF ANTHOCYANIN BIOSYNTHESIS

Anthocyanins are the water-soluble red and blue pigments responsible for the color of many vegetables, fruits, and flowers. The biosynthesis of anthocyanins is a classical high irradiance reaction, first revealed in studies of red cabbage seedlings. Like other responses of etiolated seedlings, the *initiation* of anthocyanin accumulation is a classic phytochrome-dependent LFR. The red, far-red photoreversibility, however, is limited to brief irradiations. When longer-term irradiations are applied, the action peak for continued anthocyanin accumulation is shifted to the far-red, with reduced effectiveness in the red. This effect of prolonged far-red irradiation has been interpreted as a requirement for maintaining a low level of Pfr over time—long enough to avoid rapid depletion of the Pfr pool by degradation.

22.4.6 RAPID PHYTOCHROME RESPONSES

The response time for most phytochrome-mediated developmental effects is measured in hours or even days, but there are some responses with response times measured in minutes or seconds. Most, but not all, of these rapid responses appear to be related to membrane-based activities such as bioelectric potential or ion flux. One of the earliest indications that phytochrome influenced the electrical properties of tissues was a curious effect on the surface charge of root tips reported by T. Tanada in 1968. Tanada observed that dark-grown barley root tips would float freely in a glass beaker with a specially prepared negatively charged surface. Within 30 seconds following a brief red irradiation the root tips would adhere to the surface. A subsequent far-red treatment would release the root tips from the glass. It was later found that adhesion and release was correlated with phytochrome-induced changes in the surface potential of the root tips. A brief red treatment generated a positive surface potential, attracting the tips to the negatively charged surface. A far-red treatment generated a negative surface potential, thereby causing the tips to detach. Similar effects of red and far-red light on surface potential of *Avena* coleoptiles have also been demonstrated.

Phytochrome-modulated transmembrane potentials have since been reported for a variety of tissues from several laboratories. The results are not completely consistent, but in most cases red light induces a depolarization of the membrane within 5 to 10 seconds following a red light treatment. A subsequent far-red treatment causes a slow return to normal polarity or small hyperpolarization. At this point it is not known whether such effects are due to a direct action of phytochrome on the membrane or whether a second messenger system is involved.

22.4.7 PhyA MAY FUNCTION TO DETECT THE PRESENCE OF LIGHT

What purpose does it serve the plant to have multiple forms of phytochrome? More specifically, of what benefit is it for etiolated plants to accumulate an apparent excess of labile phyA, which is so quickly degraded in the Pfr form? One possibility is that phyA functions only to detect the presence of light, rather than to distinguish subtle differences in light quality. Note that phyA accumulates in two particular situations: (1) in seeds that require red light to germinate and consequently do not germinate when buried deep in the soil, and (2) in germinated seedlings in which phytochrome is used to detect light as the seedling approaches the soil surface. The large amount of phyA that accumulates under both of these conditions appears to function as a sensitive antenna or photon-counter that detects only the presence of light. Once the seed or seedling is exposed to adequate light, the excessive quantity of labile phytochrome disappears. This allows the more stable phyB to monitor the R-FR ratio over time

and direct development accordingly. It may be difficult to obtain direct evidence in support of such a scenario, but it is an important first step in taking phytochrome studies out of the laboratory and into the real world.

22.5 CHEMISTRY AND MODE OF ACTION OF PHYTOCHROME AND CRYPTOCHROME

22.5.1 PHYTOCHROME IS A PHYCOBILIPROTEIN

The phytochrome chromophore (Figure 22.9) is called **phytochromobilin** because it is a *linear* tetrapyrrole that is similar in structure to mammalian **bile** pigments. Phytochromobilin is virtually identical to phycocyanobilin, the chromophore of the pigment phyconcyanin found in cyanobacteria and the red algae. Phytochromobilin is also related to the chromophores of chlorophyll, the respiratory cytochrome pigments, and hemoglobin, except that these pigments are all *cyclic* tetrapyrroles.

The difference between the Pr chromophore and the Pfr chromophore is a rotation (a *cis–trans* isomerization) of the double bond between rings C and D. The absorption of red light provides the energy required to overcome the high activation energy for rotation around the double bond, a transition that is not normally possible at ambient temperature. The pigment can be returned to the more stable Pr configuration by either FR light or thermal-dependent reversion in darkness. There is also evidence that the change in the chromophore induces substantial conformational changes in the protein, which would account for its activation when converted to Pfr.

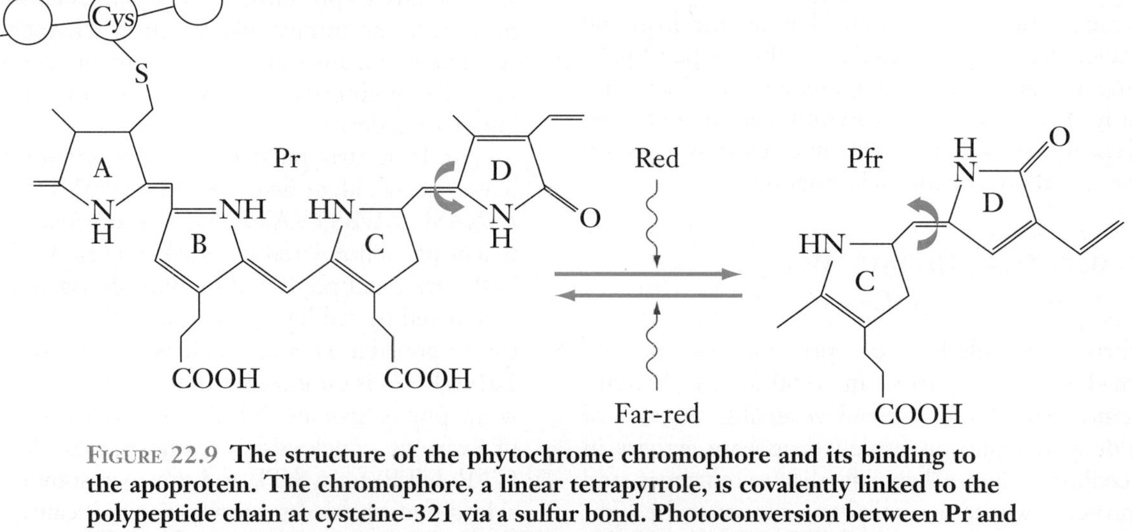

FIGURE 22.9 **The structure of the phytochrome chromophore and its binding to the apoprotein. The chromophore, a linear tetrapyrrole, is covalently linked to the polypeptide chain at cysteine-321 via a sulfur bond. Photoconversion between Pr and Pfr involves a rotation around the double bond linking the C and D rings.**

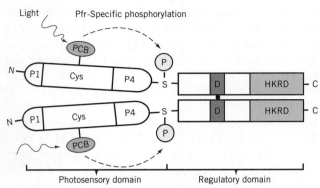

FIGURE 22.10 **The structure of phytochrome. This block diagram of two phytochrome molecules (a dimer) shows the principal domains and subdomains. The N-terminal photosensory domain binds the phytochromobilin chromophore (PCB) and is linked by a hinge region to the C-terminal regulatory domain. The regulatory subunit contains the dimerization domain (DD) and the histidine kinase-related domain (HKRD). One possible phosphorylation site is shown (dashed line); a serine residue (S) located in the hinge region that is phosphorylated specifically in the Pfr form. Other possible phosphorylation sites are serine residues located in the P1 subdomain at the N-terminus of the photosensory domain. The P4 domain is believed to interact with the D-ring of the chromophore to stabilize the energetically unfavorable Pfr form. The regulatory domain presumably initiates transduction of the light signal.**

Phytochrome apoprotein is a relatively small protein with a molecular mass of 125 kDa. It consists of two structural domains of approximately equal size; a globular N-terminal domain and an open, or extended, C-terminal domain (Figure 22.10). The A ring of the chromophore is covalently linked to the protein through a cysteine residue in the N-terminal domain. The N-terminal end of the molecule is thus called the **photosensory domain**. The C-terminal domain, or **regulatory domain**, contains a subdomain that has the characteristics of a histidine kinase (the histidine kinase-related domain, HKRD). In this respect, phytochrome resembles a two-component system like those previously encountered in osmoregulation and hormone sensing. *In vivo*, phytochromes exist as a dimer, with one chromophore on each monomer.

22.5.2 PHYTOCHROME SIGNAL TRANSDUCTION

Phytochrome regulation of proteins or protein function was first reported in 1960 by A. Marcus, who demonstrated red, far-red reversible control of glyceraldehyde-3-phosphate dehydrogenase activity in bean seedlings. Since then, the list of enzymes and other proteins whose activities are known to be light regulated, in most cases by phytochrome, has grown to

more than 60. These observations led to the obvious conclusion that phytochrome acted by controlling gene expression. Only recently, however, have the results of genetic studies suggested mechanisms for the control of gene expression by phytochrome. Although the detailed mechanism of phytochrome signaling is still not clear, at least three important factors have been identified:

1. the phytochromes are serine/threonine kinases that both autophosporylate and phosphorylate a number of other proteins,

2. phytochromes are imported into the nucleus in the active Pfr form, and

3. phytochrome responses are associated with significant changes in gene expression.

Early studies of the amino acid sequence of the phytochrome molecule revealed that its carboxy-terminal domain had amino acid sequences homologous to bacterial histidine kinase enzymes. Because of this similarity it was suggested in the 1980s that phytochromes might be protein kinases. There was, however, no direct evidence of such activity. This all changed with the discovery of phytochrome in the cyanobacteria. The cyanobacterial phytochrome (cph1) has an N-terminal domain similar to the chromophore-binding domain of plant phytochrome, has spectral properties similar to plant phytochromes, and, most importantly, exhibits light-mediated histidine kinase activity. Shortly afterward, it was shown that purified oat phytochrome A exhibited light-regulated autophosphorylation ability. However, plant phytochrome autophosphorylates serine residues, rather than histidine residues as its cyanobacterial counterpart does. Phytochrome is preferentially phosphorylated in the Pfr form (Figure 22.10) which may then initiate signaling by transferring the phosphate group to an appropriate substrate molecule. It is well known that the addition of a phosphate group or its removal has profound effects on the structure and stability of proteins, thereby influencing functional properties and intracellular location. Phytochrome may regulate various aspects of development simply by phosphorylating different substrates or initiating different kinase cascades.

At least two putative phosphorylation substrates have been identified. PKS1 (*PHYTOCHROME KINASE SUBSTRATE* 1) is a cytoplasmic protein that is phosphorylated by phyA *in vitro*. PKS1 binds to both phyA and phyB and its phosphorylation *in vivo* is stimulated by red light, suggesting that phytochrome is the responsible kinase. Nucleoside diphosphate kinase 2 (NDPK2) is an enzyme that also appears to interact with phytochrome. NDPK catalyzes the synthesis of various nucleoside triphosphates (e.g., CTP, GTP, UTP) from ATP and the corresponding nucleoside diphosphate. Its activity is significantly increased in the presence of the Pfr form of phyA *in vitro*, phosphorylation of NDPK2 is stimulated by red light *in vivo*, and *ndpk2* mutants exhibit altered responses to both red and far-red light. Purified phyA can also phosphorylate both cryptochromes *in vivo* and phosphorylation of cry1 is also stimulated by red light *in vivo*. This is particularly interesting in view of the known physiological interactions between red and blue light responses and raises the possibility that phyA might modulate blue light responses via phosphorylation of cryptochrome (see Section 22.5.3). Unfortunately, a direct link between kinase activity and phytochrome signaling has yet to be established.

In order to control gene expression, the phytochrome signal chain must extend into the nucleus where the genes are located. In fact the signal chain in this case is rather short, because phytochrome itself can move into the nucleus. For a long time it was assumed that phytochrome was strictly a cytosolic protein. This view changed when phytochrome was tagged by fusing it with a **green fluorescent protein** (**GFP**) and the fusion product was then expressed in transgenic plants. The GFP-fused phytochrome remains biologically active and the location of the tagged phytochrome can be visually confirmed by microscopic examination. These studies made it clear that phytochrome in its inactive Pr form does indeed accumulate in the cytoplasm, but conversion to the Pfr form unmasks a nuclear localization sequence that allows phytochrome to be recognized by the nuclear import machinery (Figure 22.11). The mechanism of nuclear import is not clear, but it has been shown that the recognition sequence is located in the carboxy-terminal domain of the phytochrome molecule and that a protein identified as FHY1 (from the mutant *fhy1*, or *far-red elongated hypocotyl 1*) is specifically required. The rate of nuclear import varies. PhyA,

for example, is transported into the nucleus within 15 minutes of the onset of light, but phyB is not detected for at least two hours.

As multiple phytochromes arrive inside the nucleus, they appear to form aggregates referred to as nuclear bodies or speckles. Phytochrome also physically interacts, at least *in vitro*, with a family of transcription factors called phytochrome-interacting factors (PIF), an interaction that establishes a direct link between phytochrome and gene expression. Genetic analysis has established that PIF and PIF-like (PIL) proteins are predominantly negative regulators of phytochrome-dependent pathways. For example, members of the PIF family inhibit seed germination, control protochlorophyllide accumulation and repress phytochrome-regulated gibberellin biosynthetic genes. Most PIF proteins are stable in the dark, but are rapidly degraded in the light. The degradation of at least two PIFs (PIF1 and PIF3) is mediated by the 26S-proteosome system in a phytochrome-dependent manner. Once the light is switched off, PIF protein degradation stops and the PIF proteins rapidly accumulates.

Based on the above observations, a general model for phytochrome action has begun to emerge

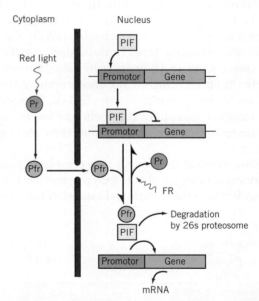

FIGURE 22.12 A model for the control of gene activation by phy. Phytochrome interacting factor (PIF) in the nucleus is a negative regulator and probably represses transcription by binding with the promoter region of a photoresponsive gene. Red light converts cytosolic phy to its active Pfr form, which then is imported into the nucleus. The Pfr recruits the promoter-bound PIF for degradation by the 26S-proteosome system, thus relieving the repression of the gene. Far-red light will convert the Pfr back to Pr, which immediately dissociates from PIF, thus allowing PIF to reassociate with the promoter and reestablish gene repression.

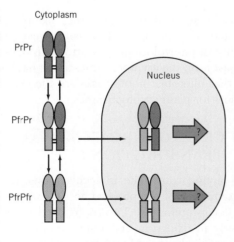

FIGURE 22.11 Import of phytochrome from the cytoplasm into the nucleus is an important step in phytochrome signal transduction. Nuclear import requires that at least one of the two molecules in the dimer be in the Pfr form.

(Figure 22.12). In the dark, phytochrome accumulates and remains in the cytoplasm as Pr, which is inactive. Meanwhile, nuclear PIF proteins inhibit the expression of phytochrome-dependent genes. Upon irradiation, Pr is converted to Pfr, which is then imported into the nucleus where it targets transcriptional regulators for degradation, thereby activating transcription of phytochrome-responsive genes. This model, however, does not explain all phytochrome-dependent responses. The rapid phytochrome responses described above (22.4.5) would seem to require a cytoplasmic, perhaps even membrane-located, phytochrome system. There is some evidence that these responses may involve cyclic GMP and/or calcium second messenger systems.

22.5.3 CRYPTOCHROME STRUCTURE IS SIMILAR TO DNA REPAIR ENZYMES

Most plant cryptochromes are 70 to 80 kDa proteins with two recognizable domains (Figure 22.13). The N-terminal domain shares amino acid sequence homologies with microbial DNA **photolyase**, a unique class of flavoenzymes that use blue light to catalyze repair of UV-induced damage to microbial DNA. The N-terminal domain of cryptochrome is therefore called the photolyase-related (PHR) domain. The PHR domain of cryptochrome binds two chromophores. The first is flavin adenine dinucleotide (**FAD**), the same redox cofactor encountered in respiratory metabolism. The second is 5,10-methenyltetrahydrofolate, or MTHF. MTHF is a member of another family of redox cofactors called **pterins** (Figure 22.14). The second, or C-terminal, domain of cryptochrome has no known homologies with other proteins and its function is not clear.

The principal differences between photolyase and the cryptochromes are that the cryptochromes have the distinguishing carboxy-terminal extension not found in

Flavin

Pterin

5-10-Methenyltetrahydrofolate (MTHF)

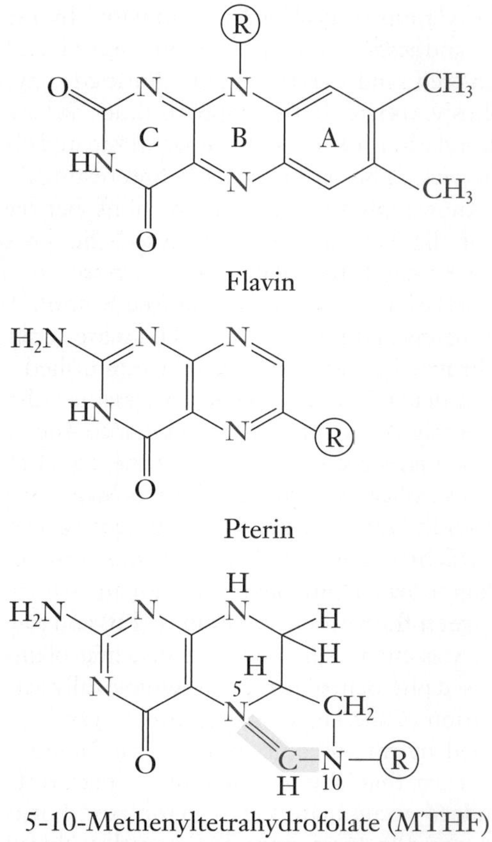

FIGURE 22.14 The core ring structures of cryptochrome chromatophores. Depending on the nature of the R group, the flavin nucleus forms either flavin-adenine dinucleotide (FAD), flavin mononucleotide (FMN) or riboflavin. 5,10-Methenyltetrahydrofollate (MTHF) is a pterin derivative. Note the similarity in the structures of pterin and the B and C rings of flavin. The R group of MTHF consists of 3 to 6 glutamate molecules. Both flavins and pterins absorb strongly at the blue end of the spectrum.

the photolyases and cryptochromes exhibit no photolyase activity. The similarity does, however, raise the possibility of an evolutionary relationship between the photolyases and the cryptochromes and it has been suggested that microbial photolyases are the evolutionary precursors for plant cryptochromes.

22.5.4 CRYPTOCHROME SIGNAL TRANSDUCTION

Unlike phytochrome, the molecular consequences of photoexcitation of cryptochrome remain unknown. Other flavoproteins are known to participate almost exclusively in oxidation–reduction reactions and photolyases repair damaged DNA by transferring electrons to pyrimidine dimers. It seems likely that the primary photochemical event when cryptochromes absorb blue light would involve a similar electron-transfer mechanism.

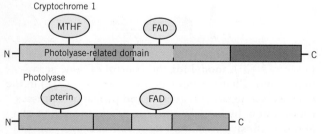

FIGURE 22.13 Cry1 and cry2 share a similar photolyase-related domain (the amino acid sequence is about 58 percent identical) but differ primarily in the length of the C-terminal domain. The two chromophores are attached to the photolyase-related domain. The C-terminal domain contains the amino acid sequence that determines nuclear localization of the pigment.

The signal transduction chain for the crypto-chromes appears to be relatively short since the cryptochromes are located mainly in the nucleus. Cry1 is located in the cytoplasm in the light but moves into the nucleus in the dark while cry2 seems to reside permanently in the nucleus. In the nucleus, cryptochrome interacts directly with the COP1 (CONSTITUTIVELY PHOTOMORPHOGENIC 1), an E3 ubiquitin ligase. COP1 is a major repressor of photomorphogenic responses by constantly degrading a number of transcription factors. The *cop1* mutant therefore displays all of the characteristics of a light-grown seedling (a **constitutive photomorphogenic** phenotype) in the dark. When irradiated with blue light, cryptochrome undergoes a conformational change which leads to the deactivation of COP1 and accumulation of transcription factors necessary for proper development in the light.

A number of experiments have shown that cryptochromes are subject to light-dependent phosphorylation. In one study, for example, it was shown that the C-terminal domain of cry1 was phosphorylated by phyA *in vitro*. Phosphorylation of cry1 by phyA occurred with both red and blue light. In other experiments, both cry1 and cry2 exhibited blue light-dependent phosphorylation *in vivo*. The *in vivo* blue light-dependent phosphorylation of cry2 was also detected in a range of phytochrome mutants—that is, in the absence of functional phytochrome. Thus, while phytochrome can phosphorylate cryptochrome, blue light-dependent phosphorylation of cryptochrome is independent of phytochrome. The implication is that cryptochrome in the dark is not phosphorylated and, consequently, inactive. The absorption of blue light by cryptochrome enables its phosphorylation by some unknown kinase. The phosphorylated form of cryptochrome is active and initiates signal transduction.

22.6 SOME PLANT RESPONSES ARE REGULATED BY UV-B LIGHT

A number of plant responses are attributed to radiation in the UV-B region of the spectrum. A positive effect of ultraviolet light on anthocyanin accumulation has been known since the mid-1930s. Later it was recognized that sunlight filtered through window glass, which absorbs ultraviolet rays, was less effective than unfiltered sunlight. This effect was finally characterized when it was demonstrated that flavonoid biosynthesis in parsley (*Petroselinum crispum*) cell suspension cultures and seedlings was induced by UV-B radiation (280–320 nm). Maximum effectiveness was at 290 to 300 nm, with little or no activity beyond 320 nm. By 1986, 11 species of higher plants were listed for which UV-B induced anthocyanin

and flavonoid biosynthesis in coleoptiles, hypocotyls, seedling roots, and cell culture.

In *Sorghum bicolor* the action spectrum for flavonoid biosynthesis shows three peaks: 290 nm, 385 nm, and 650 nm. Action at 385 nm and 650 nm could be reversed by a subsequent exposure to far-red, but the peak at 290 nm could not. The 385 nm and 650 nm peaks have been attributed to phytochrome, leaving the 290 nm peak due to a UV-B receptor. In parsley it appears that flavonoid biosynthesis results from the coaction of three pigments: phytochrome, a separate blue receptor (probably cryptochrome), and a UV-B receptor. The UV-B system is a necessary prerequisite to flavonoid biosynthesis since neither the blue receptor nor phytochrome is effective unless preceded by a UV-B light treatment.

The UV-B receptor has yet to be isolated and its identity remains unknown. Phytochrome has been suggested—the protein moiety does absorb UV-B light—but results such as those described above would argue against it. In members of the Leguminoseae family, ultraviolet-induced flavonoid biosynthesis can be reversed with blue light in a manner reminiscent of photoreactivation of UV damage in microorganisms. This could implicate DNA itself as a UV photoreceptor, but the action peak is shifted to wavelengths somewhat shorter than those normally characteristic of UV-B action.

22.7 DE-ETIOLATION IN ARABIDOPSIS: A CASE STUDY IN PHOTORECEPTOR INTERACTIONS

Arabidopsis is a typical dicot seedling in that growth in white light is accompanied by (1) arrested hypocotyl elongation, (2) a straightening of the hypocotyl or plumular hook, (3) unfolding of the cotyledons, and (4) expansion of the cotyledons. Studies with phytochrome- and cryptochrome-deficient mutants have confirmed that de-etiolation responses in *Arabidopsis* seedlings involve complex interactions between three different photoreceptors: phyA, phyB, and cry1. An experiment demonstrating some of these interactions is illustrated in Figure 22.15. In the experiment illustrated here, elongation of the hypocotyl in wildtype seedlings was suppressed by approximately 68 percent in white light. The phyA-deficient single mutant (*phyA*) had little effect on hypocotyl elongation, while in the absence of phyB, hypocotyl elongation was suppressed by only 20 percent. These results indicate that phyB is the principal photoreceptor for control of hypocotyl elongation. phyA still has a role, however, a fact which is uncovered in the double mutant, *phyAphyB*. In the absence of both phyA and phyB, not only is hypocotyl elongation not suppressed by white light,

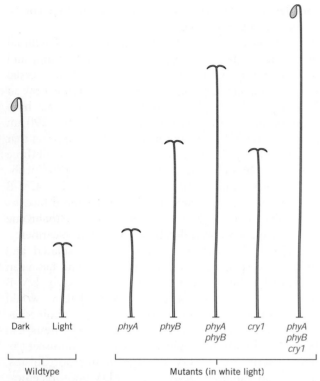

FIGURE 22.15 **The effect of phytochrome- and cryptochrome-deficient mutations on** *Arabidopsis* **hypocotyl elongation in white light. (Based on the data of Neff, M. M., J. Chory. 1998.** *Plant Physiology* **118:27–36.)**

the hypocotyls are even longer than in dark-grown controls. In other words, the *phyA* mutation appears to have enhanced the effect of the *phyB* mutation.

Both straightening of the hypocotyl hook and unfolding of the cotyledons in white light were relatively unaffected by any one of the three single mutants, *phyA*, *phyB*, and *cry1*. In the *phyAphyBcry1* triple mutant, however, the hook failed to straighten out and the cotyledons did not unfold. These results indicate that hook straightening and cotyledon unfolding are controlled in a completely redundant manner by all three photoreceptors. Only when all three photoreceptors are missing are these two aspects of de-etiolation significantly compromised.

Although not shown in Figure 22.15, cotyledon expansion in *Arabidopsis* appears to be controlled principally by phytochrome B and cryptochrome. In one experiment, cotyledon area in the *phyB* and *cry1* mutants was reduced by 50 percent and 64 percent, respectively. However, in the *phyBcry1* double mutant, cotyledon area was not significantly greater than in the dark-grown controls. This additive effect of the *phyB* and *cry1* mutants suggests that each photoreceptor independently controls a distinct aspect of cotyledon expansion.

The results described here represent only a brief sampling of the many interactions that occur between the several photoreceptors, but one thing is clear.

Redundancy, or the possibility of evoking similar effects through different photoreceptors, is common to light-induced phenomena. As well, numerous examples of synergism and antagonism between photoreceptors have been demonstrated. Whether such interrelationships are the results of parallel, independent signal pathways or of extensive crosstalk between pathways remains to be seen. Nevertheless, plants have clearly evolved a versatile system of multiple photoreceptors that allows them to respond efficiently and flexibly to dynamic changes in fluence rate and composition of light in their environment.

SUMMARY

The light in a plant's environment contains a significant amount of information and the use of that information by plants, called photomorphogenesis, is a central theme in plant development. In order to acquire the information provided by light, plants have developed a sophisticated array of photoreceptors and signal transduction pathways. There are three major classes of photoreceptors: the phytochromes that respond principally to red and far-red light, the cryptochromes that respond to blue light, and phototropin, which also responds to blue light.

Phytochrome is a family of photoreceptors; five are known in *Arabidopsis* (phyA–phyE). They are chromoproteins with a chromophore that consists of an linear tetrapyrrole similar to phycocyanin. Phytochrome *in vivo* is a dimer of a 125 kDa polypeptide chain. The hallmark of phytochrome action is photoreversibility: irradiation with red light converts the red-absorbing form of the pigment (Pr) to the far-red-absorbing form (Pfr). The physical presence of phytochrome was established by demonstrating the predicted photoreversible absorbency changes *in vivo*. PhyA accumulates in dark-grown seedlings in the Pr form, which is stable. PhyA$_{fr}$ is unstable and is destroyed with a half-life of 1 to 1.5 hours. PhyB is expressed at low levels in both light and dark. PhyB$_{fr}$ is light stable. A mixture of red and far-red light (or white light) will establish a photoequilibrium mixture of Pr and Pfr. Pfr is the physiologically active form.

Phytochrome-mediated effects are conveniently grouped into three categories on the basis of their energy requirements: very low fluence responses (VLFR), low fluence responses (LFR), and high irradiance reactions (HIR). LFRs include the classically photoreversible phytochrome responses such as seed germination and de-etiolation. LFRs convey information to the seed about its position relative to the soil surface and maximize the potential for a seedling to become established in light and initiate photosynthesis before the nutrient reserves of the seedling are exhausted. VLFRs are not photoreversible and are difficult to study because they saturate at light levels below those that cause a measurable conversion of Pr to Pfr. HIRs require prolonged exposure to high irradiance, are time dependent, and are not photoreversible.

Under natural conditions, the phytochrome photoequilibrium value (Pfr/P) is related to the red to far-red fluence rates. It is likely that phyB is the sensor that detects changes in red, far-red fluence ratio that occur under canopies and as end-of-day signal. In this way, phytochrome mediates the shade avoidance syndrome, provides a plant with information about the proximity of its neighbors, and contributes to the time-sensing mechanism.

Cryptochrome is also a chromoprotein. Cryptochrome has two chromophores—FAD and a pterin—and bears similarity to the microbial DNA repair enzyme, photolyase. Cryptochrome mediates hypocotyl elongation, cotyledon expansion, and setting the biological clock. Cryptochrome also has a role in determining time of flowering.

The study of photoreceptor signal transduction is in its infancy, although some significant advances have been made. Phytochrome appears to operate in two modes. Phytochrome has serine-threonine kinase activity and at least two signaling partners are known PKS1 and NDPK2 are both phosphorylated by phyA *in vitro*. phyA can also phosphorylate cry1, which may help to explain the extensive interactions between phytochrome and cryptochrome in regulating de-etiolation. Phytochrome also regulates gene action through the translocation of phyB into the nucleus where it has a role in activating or suppressing transcription. The mode of action of cryptochrome is unknown, but its similarity to photolyase suggests that cryptochrome may have a redox function. Cryptochrome is also readily phosphorylated, which may be significant in the signal transduction chain.

CHAPTER REVIEW

1. What unique character distinguishes phytochrome from all other plant pigments?

2. How is phytochrome uniquely suited to monitor the natural light environment?

3. Compare and contrast the structures of phytochrome and cryptochrome.

4. Distinguish between low fluence responses and high irradiance responses.

5. Irradiation of phytochrome establishes a photoequilibrium between Pr and Pfr. Can you think of a situation in which only one form of the pigment would be present?

6. What is a chromoprotein?

7. What is meant by the statement that cryptochrome is "homologous" with photolyase? What is the significance, if any, of this homology?

8. Two plant physiologists, H. Mohr and W. Shropshire, once wrote that "*Normal development in higher plants is photomorphogenesis.*" What do you think they meant by this statement?

9. How does the experiment described in section 22.7 demonstrate an interaction between phytochrome and cryptochrome in control of de-etiolation?

FURTHER READING

Banerjee, R., A. Batschauer. 2005. Plant blue-light receptors. *Planta* 220:498–502.

Briggs, W. R., M. Olney. 2001. Photoreceptors in plant photomorphogenesis to date. Five phytochromes, two cryptochromes, one phototropin and one superchrome. *Plant Phytisology* 125:85–88.

Chen, M., J. Chorey, C. Frankhauser. 2004 Light signal transduction in higher plants. *Annual Review of Genetics* 38:87–117.

Kim, J-I., J-E. Park, X. Zarate, P-S. Song. 2005. Phytochrome phosphorylation in plant light signaling. *Photochemical & Photobiological Sciences* 4:681–687.

Lin, C. 2002. Blue light receptors and signal transduction. *The Plant Cell*. Supplement 2002:S207–S225.

Lin, C., D. Shalitin. 2003. Cryptochrome structure and signal transduction. *Annual Review of Plant Biology* 54:469–496.

Lorrain, S., T. Genoud, C. Frankhauser. 2006. Let there be light in the nucleus! *Current Opinion in Plant Biology* 9:509–514.

Sage, L. C. 1992. *Pigment of the Imagination: A History of Phytochrome Research*. New York: Academic Press.

Shen, H., J. Moon, E. Huq. 2005. PIF1 is regulated by light-mediated degradation through the ubiquitin-26S proteosome pathway to optimize photomorphogenesis of seedlings in Arabidopsis. *The Plant Journal* 44:1023–1035.

Vandenbussche, F. et al. 2005. Reaching out of the shade. *Current Opinion in Plant Biology* 8:462–468.

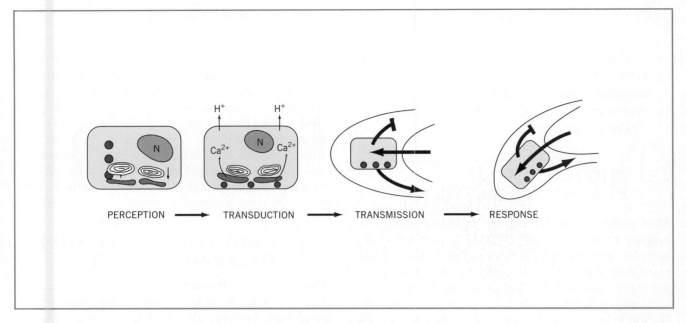

PERCEPTION   ⟶   TRANSDUCTION   ⟶   TRANSMISSION   ⟶   RESPONSE

23

Tropisms and Nastic Movements: Orienting Plants in Space

The power of movement is not normally associated with plants. Yet movement pervades the life of the green plant. Movement in higher plants does not involve locomotion as it does in animals, nor is it so obvious. Plant movement is mostly slow and deliberate, but it is a key factor in determining the orientation of plants in space. Plants that have been inadvertently placed in the horizontal position will reorient their root and shoot to the vertical. House plants will bend, appearing to seek light coming through a window. Leaves may periodically rise and fall throughout the day and night, while others track the sun as it moves across the sky. Leaves of the Venus flytrap snap closed on a hapless insect. While most plant movements are relatively slow, they nonetheless serve important functions by positioning organs for the uptake of nutrients and water and optimal interception of sunlight, or (in the case of the flytrap) obtaining nutrients such as nitrogen through the leaves.

There are two principal categories of movement in plants based on the distinctiveness of their mechanisms. **Growth movements** are *irreversible*. They arise as the result of differential growth within an organ or between two different organs. **Turgor movements** are *reversible*, resulting from simple volume changes in certain cells—most often in a special organ called the *pulvinus*. Within each group, we can further distinguish between **nutation**, **tropism**, and **nastic movement**. The term nutation (or circumnutations) denotes a regular rotary or helical movement of plant organs, most typically the stem apex, in space. Nutations are best demonstrated by time-lapse photography. Tropic responses are directionally related to the stimulus such as light (*phototropism*), gravity (*gravitropism*), water (*hydrotropism*), or touch (*thigmotropism*). Nastic responses are not obviously related to any vector in the stimulus. Directionality of nastic responses is inherent in the tissue and includes *epinasty* (bending down), *hyponasty* (bending up), *nyctinasty* (the rhythmic sleep movements of leaves), *seismonasty* (response to mechanical shock), *thermonasty* (temperature), and *thigmonasty* (touch).

This chapter will focus on the three plant movements that have been explored most thoroughly. These are

- phototropism, particularly the nature of the photoreceptor and the role of auxin in the signal transduction chain,

- gravitropism, including a brief discussion of the nature of the gravitational stimulus and the mechanism of gravity perception, the particular character of gravitropism in shoots and roots, and the role of auxin and calcium in the differential growth response, and

- nastic movements, including the structure of motor organs and the role of potassium flux in nyctinastic and seismonastic movements.

23.1 PHOTOTROPISM: REACHING FOR THE SUN

Most people are familiar with the sight of house plants bending toward the light from an open window, an everyday example of the phenomenon called **phototropism** (Figure 23.1). Phototropism is a classic plant physiology problem that has attracted the interest of botanists since the middle of the nineteenth century. Darwin's study of phototropism, published in his book *The Power of Movement in Plants* in 1881, is credited with overcoming the preoccupation of English-speaking botanists with descriptive and taxonomic biology and stimulating an interest in the more dynamic aspects of plant function. Cell elongation in phototropically stimulated grass coleoptiles also led to Went's discovery of plant hormones (Chapter 18).

Tropic responses may be either positive or negative. If a plant responds in the direction of the stimulus (e.g., toward a light source) it is said to be positive. If it grows away from the stimulus it is said to be negative. Whether the phototropic response is positive or negative depends largely on the nature of the organ or its age. For example, coleoptiles, hypocotyls, and the elongating portions of stems and other aerial organs are for the most part positively phototropic while the tendrils of most climbing plants are negatively phototropic. Leaves are normally **plagiotropic**, which means they orient at angles intermediate to the light. Roots, on the other hand, are largely nonphototropic, although some may exhibit a weakly negative response. The stems of ivy (*Hedera helix*) are negatively phototropic during the shade-loving juvenile stage, but older branches become positively phototropic. The stems of ivy-leafed toad flax (*Cymbalaria muralis*) become negatively phototropic following fertilization. This interesting behavior helps to place ripening seed pods into crevices in the walls on which the plant is normally found.

23.1.1 PHOTOTROPISM IS A RESPONSE TO A LIGHT GRADIENT

Phototropism is often defined as a response to unilateral light, and so it is in the laboratory. Under normal growth conditions, however, the bending response will occur even in plants that are receiving light from all sides. All that is required is that the fluence rate be unequally distributed. In experiments with bilaterally illuminated grass coleoptiles, for example, as little as 20 percent difference in fluence rate on the two sides of the organ will induce a bending response (Figure 23.2). Thus light can be presented unilaterally (as it is in most laboratory experiments), bilaterally, from all sides, and even from above, providing only that a gradient is created across the organ. Phototropism is thus a growth response to a light gradient.

The magnitude of a light gradient across an organ such as a coleoptile is dependent on optical properties of the tissue as well as differences in incident light. A light gradient across an organ, for example, can be

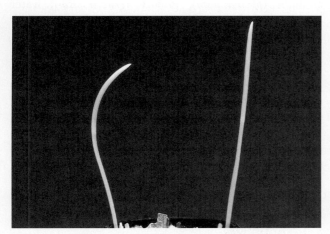

FIGURE 23.1 **The phototropic response in oat (*Avena sativa*) coleoptiles. Left: Dark-grown seedling placed in unilateral blue light, from the right, for 90 minutes. Right: Unlighted control. Note that the overall length of the coleoptile is approximately the same in the two seedlings.**

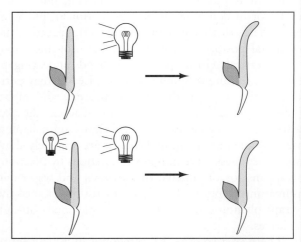

FIGURE 23.2 **A phototropic response will occur whenever there is a light gradient established across the stem or coleoptile axis. The upper part of the figure represents unilateral illumination, which is normal in experimental situations. The lower part of the figure illustrates bilateral illumination, with the highest fluence rate from the right. Phototropic curvature is the same in either case.**

intensified by screening within the organ. Pigments, including but not limited to the photoreceptor itself, will attenuate the light as it passes through the organ. Light can also be attenuated by scattering, reflection, or diffraction within the cells or as it passes between cells. Thus gradients across individual cells, measured by using microfiberoptic probes, may vary from 5:1 to 50:1. To further complicate matters, organs such as coleoptiles appear to function as light pipes. This means that light applied to the tip, for example, will be transmitted through the coleoptile to cells further down the organ. Thus the phototropic stimulus is far from being a simple matter. These complex interactions between light and the optical properties of tissue have led to significant difficulties in experimental design as well as in interpretation of the resulting data.

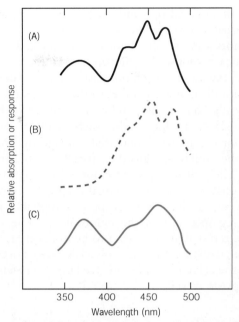

FIGURE 23.3 (*A*) The action spectrum for *Avena* coleoptile phototropism. The action spectrum shows peak activity in the blue and UV-A regions of the spectrum. For comparison, the absorption spectra for riboflavin, a common flavonoid (*B*), and β-carotene (*C*) are shown. Although β-carotene "fits" in the blue region (around 450 nm), riboflavin has the required absorption in both the blue and the UV-A regions (320 nm–400 nm) of the spectrum.

23.1.2 PHOTOTROPISM IS A BLUE-LIGHT RESPONSE

Since the 1930s, action spectra for phototropism have been repeatedly determined for a number of organisms, but have been most thoroughly documented for coleoptiles of oats (*Avena sativa*) and maize (*Zea mays*) and for sporangiophores of the fungus *Phycomyces*. The action spectra for oat coleoptiles and *Phycomyces* are virtually identical, indicating that they share a homologous, if not common, photoreceptor. All other phototropic action spectra are similar and consistently show two peaks in the blue region of the spectrum near 475 nm and 450 nm and a small peak or shoulder at 420 nm (Figure 23.3). In addition there is a broad action peak in the UV-A region near 370 nm. The action spectra for both oat and *Phycomyces* phototropism show an additional peak in the region of 280 nm, indicating that the photoreceptor is probably a chromoprotein.

As early as the 1940s, it was suggested that the photoreceptor could be a flavin molecule such as riboflavin. However, because of its action spectrum the photoreceptor for phototropism was subject to the same flavin-carotenoid controversy described in the previous chapter for other blue-light responses now known to be regulated by the cryptochromes. On the other hand, there were several physiological results that ruled out carotenoids as the photoreceptor for phototropism long before the responsible pigment, **phototropin**, was finally discovered and shown to be a flavo-protein. Carotenoid biosynthesis, for example, can be blocked, either by mutation or by treatment of seedlings with the herbicide norflurazon, which inhibits the enzyme phytoene desaturase. Yet carotenoid-deficient maize mutants, albino barley seedlings, and norflurazon-treated seedlings all exhibit a normal phototropic response to blue light.

23.1.3 PHOTOTROPISM ORIENTS A PLANT FOR OPTIMAL PHOTOSYNTHESIS

The phototropic blue-light response is distinct from the blue-light responses mediated by phytochrome and cryptochrome that were discussed in the previous chapter. Phytochrome and cryptochrome responses are morphogenetic responses—they alter the *pattern* of growth and development. The singular impact of phototropism, on the other hand, is that it orients growth and leaf angle toward incident light in order to maximize light interception for photosynthesis. The bending of coleoptiles and hypocotyls is only the most visible part of a larger blue-light syndrome that plants use to optimize photosynthesis. Plants also use blue light to control stomatal opening and facilitate gas exchange as well as to relocate chloroplasts within the cell.

It has long been known that stomatal opening is under the control of light. On the one hand, light absorbed by chlorophyll (i.e., red light) stimulates stomatal opening and obviously depends on photosynthetic reactions in the guard cell chloroplasts. However, there is a second, much more sensitive, system that is driven by low levels of blue light. Most of the evidence points to a dominant role of the blue-light response in the early phases of stomatal opening, such as when the stomata open at dawn, prior to the beginning of photosynthesis.

Plants also use blue light to control the high-light avoidance response of chloroplasts in the mesophyll cells. In low light, the chloroplasts always gather along the cell walls that are parallel to the surface, (i.e., periclinal walls) that are perpendicular to the incident light (Figure 23.4). In high light, such as direct sunlight, the chloroplasts avoid potential damage by lining up along the anticlinal walls (i.e., parallel to the incident light). The redistribution of chloroplasts appears to be in response to a light gradient through the cytoplasm, so the responsible photoreceptor is probably located in the cytoplasm, not the chloroplasts. The mechanism of redistribution has yet to be discovered, but the cytoskeleton is commonly involved in moving organelles within the cell and may be involved in the movement of chloroplasts as well.

23.1.4 FLUENCE RESPONSE CURVES ILLUSTRATE THE COMPLEXITY OF PHOTOTROPIC RESPONSES

Perhaps no aspect of phototropism has indicated the complexity of the process so much as attempts to define relationships between fluence and response. Phototro-

pism is characterized by a rather curious fluence-response curve, quite unlike most photobiological responses.

Fluence response curves are generally obtained by monitoring the response of the organ to different total amounts of light (fluence), usually by using a single fluence rate but varying the presentation time. Figure 23.5 shows a fluence-response curve determined for *Avena* coleoptile phototropism that illustrates the classic response to increasing fluence. There is an initial rise to a first peak, which is called **first positive curvature**. With increasing fluence, curvature declines, to the point that this may even result in a bending *away* from the light source. This decline and negative response is called **first negative curvature**. Note that first negative curvature is not necessarily "negative" in the sense of bending away from the light. It may be simply a reduced positive response. Following the region of first negative curvature, the response curve again rises into what is called **second positive curvature**. In some cases, a second negative and even a third positive curvature have been reported. First positive curvature is also known as tip curvature, because it is restricted to the apex of coleoptiles. Second positive curvature is also called basal curvature because the curvature extends more toward the basal region of the coleoptile.

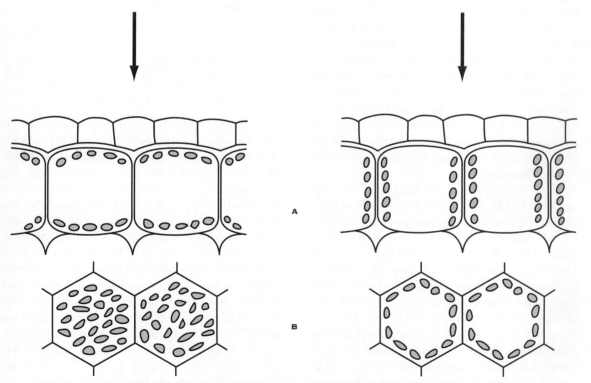

FIGURE 23.4 **The high-light avoidance response of chloroplasts in a typical mesophyll cell. In low light conditions (left), the chloroplasts gather perpendicular to the incident light along the upper and lower (periclinal) walls in order to maximize light interception. In high-light conditions (right) the chloroplasts gather along the side (anticlinal) walls, or parallel to the incident light, in order to avoid damage due to excessive light. Mesophyll cells are shown in cross-section (A) and surface view (B).**

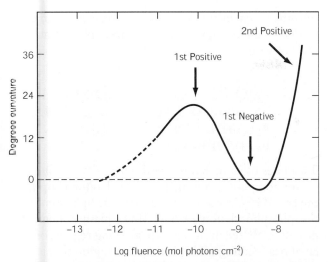

FIGURE 23.5 A phototropic fluence-response curve for *Avena* **coleoptiles. First positive, first negative, and second positive curvatures are indicated.**

There is a fundamental law of photochemistry, called the **Bunsen-Roscoe reciprocity law**, which says the product of a photochemical reaction is determined by the total amount of energy presented, regardless of fluence rate or presentation time. In other words, the same result is obtained with either a brief exposure to a high fluence rate or a longer exposure to a low fluence rate. Numerous experiments have established that the reciprocity law applies to first positive curvature but does not apply to first negative or second positive curvatures. Second positive curvature is instead very time-dependent. In other words, second positive curvature is more dependent on presentation time than on fluence rate. Failure of reciprocity for second positive curvature suggests the possibility that more than one photoreceptor might be involved. However, action spectra have been determined for both first and second positive curvature and they are identical. Apparently both first and second positive curvature are mediated by the same photoreceptor and the complexities of second positive curvature are due to subsequent events in the signal transduction chain.

23.1.5 THE PHOTOTROPIC RESPONSE IS ATTRIBUTED TO A LATERAL REDISTRIBUTION OF DIFFUSIBLE AUXIN

At the same time F. W. Went and his contemporaries had chosen to study the influence of the apex on coleoptile elongation, parallel studies on the role of the root apex were being conducted in Germany by N. Cholodny. The result was independent proposals by Cholodny, in 1924, and Went, in 1926, that the apex was able to influence cell elongation in the more basipetal extension region of the organ. These ideas of these 2 investigators

were drawn together in the late 1920s in an attempt to explain phototropism. The **Cholodny-Went hypothesis** states that unilateral illumination induces a *lateral redistribution* of endogenous auxin near the apex of the organ. This asymmetry in auxin distribution is maintained as the auxin is transported longitudinally toward the base of the organ. The higher concentration of auxin on the shaded side of the organ stimulates those cells to elongate more than those on the lighted side. It is this differential growth that causes curvature toward the light source.

The experimental basis for the Cholodny-Went hypothesis is derived largely from agar-diffusion experiments originally conducted by Went and described earlier in Chapter 18. In Went's experiments, oat coleoptiles were first stimulated with unilateral light. The coleoptile apices were then excised, split longitudinally, and the two halves placed on agar blocks in order to collect the auxin that diffused out of the base. The amount of auxin collected in the agar blocks was then assayed by the *Avena* curvature test (Chapter 18). Went reported that a significantly higher quantity of auxin was collected from the shaded half of the coleoptile apex than from the lighted half, indicating that unilateral lighting caused a greater proportion of the auxin to be transported down the shaded side of the coleoptile.

Doubts as to the validity of the Cholodny-Went hypothesis arose from numerous unsuccessful attempts to verify asymmetric auxin distribution by applying ^{14}C-labeled IAA to tropically stimulated coleoptiles. These problems, however, may be largely attributed to poor experimental technique. It is now evident that a large proportion of the radioactive auxin taken up by the tissue in those experiments did not enter the auxin transport stream. When care is taken to discount this nondiffusible auxin, a clear differential in auxin transport can be detected. For example, when maize coleoptile tips were supplied with ^{14}C-IAA, approximately 65 percent of the radioactivity was recovered from the shaded side. There was no significant asymmetry when subapical sections were used, further evidence that the lateral redistribution of auxin occurs at the very apex of the coleoptile.

In the 1960s, the Cholodny-Went hypothesis was systematically reevaluated by W. R. Briggs and his colleagues. Briggs repeated Went's original split-tip experiments but, unlike Went, he excised the tips and placed them on agar blocks *before* presenting the phototropic stimulus. The results (Figure 23.6) clearly demonstrate that when the tip is partially split, leaving tissue continuity only at the very apex of the coleoptile, exposure to unilateral light causes an increase in the amount of diffusible auxin on the shaded side and a decrease on the lighted side. The total amount of auxin recovered, however, remains effectively constant. When lateral diffusion of auxin is prevented throughout the entire

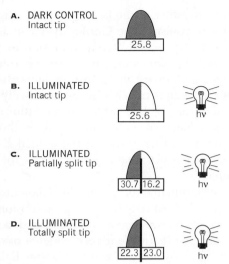

A. DARK CONTROL
Intact tip
25.8

B. ILLUMINATED
Intact tip
25.6
hv

C. ILLUMINATED
Partially split tip
30.7 16.2
hv

D. ILLUMINATED
Totally split tip
22.3 23.0
hv

FIGURE 23.6 **Phototropic stimulation establishes an asymmetric distribution of diffusible auxin in excised *Zea mays* coleoptile apices. (*A, B*) Intact control apices. *A* was maintained in darkness and *B* was provided light unilaterally from the right. (*C*) Tips were partially split, leaving tissue continuity only at the very apex. A microscope cover slip was inserted to provide a barrier to lateral diffusion. The tips were then presented with unilateral light from the right. (*D*) Tips were totally split and the diffusion barrier passed through the apex before being presented with unilateral light from the right. Numbers indicate the amount of auxin collected in the agar blocks over a 3-hour period, based on degrees of curvature in the *Avena* curvature bioassay. Values are for auxin collected from 3 tips (*A, B*) or 6 half-tips (*C, D*). (Data from Briggs, W. R. 1963. *Plant Physiology* 38:237.)**

length of the tip, no such asymmetric auxin distribution is observed. These results clearly support the principal tenet of the Cholodny-Went hypothesis, namely, that unilateral light induces a preferential migration of auxin down the shaded side of the coleoptile. Their experiments also confirmed that auxin production in coleoptiles of *Zea mays* is confined to the apical 1 to 2 mm and that lateral redistribution during phototropic stimulation probably occurs within the most apical one-half mm.

Compelling support for the Cholodny-Went theory has been provided by a more recent study of auxin redistribution in Brassica oleraceae hypocotyls. The free IAA concentration found on the shaded side of the hypocotyl was found to be at least 20 percent higher than on the lighted side following phototropic stimulation. Moreover, the differential auxin concentration was accompanied by a several-fold increase in the expression of auxin-regulated genes on the shaded side of the hypocotyls, including two members of the α-expansin family of genes that are necessary for cell wall extension (see Chapter 17). Finally, both the auxin differential and

the differential in auxin-regulated gene expression could be detected *before* there was any noticeable curvature of the hypocotyls.

23.1.6 PHOTOTROPISM AND RELATED RESPONSES ARE REGULATED BY A FAMILY OF BLUE-SENSITIVE FLAVOPROTEINS

Two lines of study led to the discovery of the photoreceptor for phototropism, now called **phototropin**. In the late 1980s, it was reported that blue light stimulated the phosphorylation of a 120 kDa plasma membrane protein localized in the actively growing regions of etiolated pea seedlings. This is the same region that is most responsive to the phototropic stimulus. After extensive biochemical and physiological characterization, the protein was found to be a kinase that autophosphorylates in blue light. There was also a strong suggestion that this kinase was the photoreceptor for phototropism.

A short time later, a mutant characterized by a failure to respond to the phototropic stimulus (*nonphototropic hypocotyl 1*, or *nph1*) was isolated from *Arabidopsis*. Plants carrying the *nph1* mutant not only failed to exhibit phototropism, but coincidentally lacked the 120 kDa membrane protein. When the *NPH1* gene was cloned, it was found, as expected, to encode the 120 kDa protein. The NPH1 holoprotein was subsequently renamed phototropin 1 (phot1) because of its functional role in phototropism.

Phototropin 1 is a flavoprotein with two flavin mononucleotide (FMN) chromophores (Figure 23.7). It has a carboxy-terminal domain with the characteristics of a serine-threonine kinase. The photosensory domain at the N-terminus has two distinctive domains called LOV domains, so named because they share characteristics with microbial proteins that regulate responses to light, oxygen, or voltage. Not surprisingly, the two LOV domains are the two sites that bind FMN and make phototropin responsive to light. A second phototropin, phototropin 2 (phot2), has since been discovered. On the basis of amino acid sequence, the two phototropins are about 60 % similar, but the two LOV domains are virtually identical and each phototropin binds two FMN chromophores.[1]

The two phototropins found in *Arabidopsis*, phot1 and phot2, exhibit some overlapping functions. Each also appears to have unique physiological roles. First positive curvature appears to be mediated solely by

---

[1]In 2001, the gene for photo2 was originally described as *NPL1* (NPH-like1), a homolog of the gene *NPH1* (nonphototropic hypocotyl 1) or *PHOT1*. *NPL1* has since been renamed *PHOT2* to bring the nomenclature in line with phototropin 1.

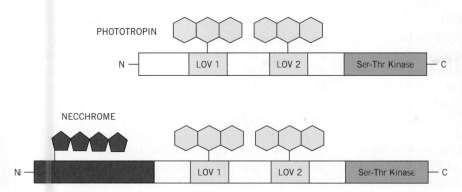

FIGURE 23.7 **The domain structures of phototropin and neochrome. Phototropin contains two LOV domains that are characteristic of proteins activated by light, oxygen, or voltage. Each LOV domain carries one flavin mononucleotide chromophore that absorbs in the blue region of the spectrum. At the carboxy-terminal end of the protein (–C) is a serine-threonine kinase domain. Note that neochrome is virtually identical to phototropin except that it has a phytochrome photosensory domain at the N-terminal end. Neochrome responds to both blue and red/far-red light.**

phot1, while second positive curvature is mediated by both phot1 and phot2. Phot1 and photo2 contribute equally to stomatal opening, while the avoidance movement of chloroplasts under high light intensities is mediated only by phot2.

Recent evidence has also indicated a role for phototropins in the promotion of cotyledon and leaf expansion and the rhythmic sleep movements of kidney bean leaves (see Chapter 24).

23.1.7 A HYBRID RED/BLUE LIGHT PHOTORECEPTOR HAS BEEN ISOLATED FROM A FERN

A particularly interesting photoreceptor has recently been isolated from the fern *Adiantum capillus-veneris*. Designated **neochrome** (formerly known as phy3), this photoreceptor has properties of both phytochrome and phototropin (Figure 23.7). The amino acid sequence of the amino-terminal domain shows a significant homology to the chromophore-binding domain of phytochrome. Furthermore, when the gene was expressed in yeast and the purified protein was reconstituted with a phycocyanobilin chromophore, it showed typical phytochrome photoreversible behavior. But neochrome also has the two LOV domains, which bind FMN, and a serine/threonine kinase domain at the C-terminus that are virtually identical to phototropin. Neochrome is required for phototropism in *Adiantum*, which is regulated by red as well as blue light. Neochrome is clearly a hybrid photoreceptor that mediates red, far-red, and blue-light responses. What is curious, however, is that *Adiantum* also has two fully functional phototropins like their higher plant counterparts and, again like their higher plant counterparts, phot2 is solely responsible for mediating high-light chloroplast avoidance movements.

Several other ferns, mosses, and algae have both phototropins and neochromes, which raises interesting

questions regarding the evolution of photoreceptors as well as their physiological action.

23.1.8 PHOTOTROPIN ACTIVITY AND SIGNAL CHAIN

Participants in the phototropin signal chain are only now just beginning to "come to light." Some of the more important factors can be summarized as follows:

1. *Autophosphorylation of phototropin plays a significant role in the phototropic response, probably by initiating a phosphorylation cascade.* Studies employing mutants of the *PHOT1* gene have shown that the protein is apparently folded in such a way that the phosphorylation site is blocked by the LOV2 domain in the dark. Absorption of blue light by the chromophore induces a change in the conformation of the protein so that the phosphorylation site is available and active. The role of the LOV1 domain is not clear. Mutants lacking the LOV1 domain have shown that LOV1 is not necessary for phosphorylation but its presence does increase kinase activity.

2. *Phototropins may be involved in gene regulation.* No substrates directly phosphorylated by phototropin *in planta* have yet been identified, but there are several proteins that interact with the photoreceptor and are necessary for a proper response. For example, the proteins NONPHOTOTROPIC HYPOCOTYL 3 (NPH3) from *Arabidopsis* and a homologous protein (called an *ortholog*) from rice, COLEOPTILE PHOTOTROPISM 1 (CPT1), include domains that are characteristic of transcriptional regulators or proteins that are involved in protein degradation. The mutants *nph3* and *cpt1*, in which these proteins are missing, show no phototropic response.

3. *Phototropin disrupts polar auxin transport.* One of the challenges presented by phototropism is to establish whether or not a link exists between the absorption

of blue light by phototropin and the asymmetrical auxin distribution proposed by the Cholodny-Went hypothesis. As described previously, asymmetric auxin distribution has been demonstrated experimentally. Recent experiments have focused on the relationship between phototropin and auxin efflux facilitator PIN1. PIN1 is normally localized at the basal ends of xylem-associated cells where it serves to facilitate the polar vertical flow of auxin (Chapter 18). When the location of PIN1 in Arabidopsis hypocotyls was monitored by immunofluorescence microscopy following phototropic stimulus, the basal location of PIN1 in wildtype plants was disrupted in the cortical cells *on the shaded side* of the hypocotyl. A similar disruption was not observed in *phot1* mutants. These results suggest that phototropic bending is initiated by a phototropin-mediated decrease in the vertical transport of auxin. This would lead to a retention or sequestering of auxin, and consequent increased growth, in those cells that are directly involved in phototropic bending.

23.1.9 PHOTOTROPISM IN GREEN PLANTS IS NOT WELL UNDERSTOOD

A final area of concern is phototropism in light-grown plants, where relatively little is known about the phototropic process. As with phytochrome, discussed in Chapter 22, most of what we know about phototropism is derived from laboratory studies with etiolated seedlings. However, in light-grown cucumber (*Cucumis sativus*) and sunflower (*Helianthus annuus*) seedlings subjected to uniform lighting, curvature of the stem can be induced by simply shading one of the

cotyledons (Figure 23.8), and the phototropic response of sunflower seedlings is markedly decreased if the leaves are removed. The cucumber response, at least, differs from the classical phototropic response in that it is induced by red light rather than blue light. This appears to be a phytochrome-mediated response and is related to inhibition of hypocotyl elongation below the irradiated cotyledon. Both the cucumber and sunflower responses may be attributed to the fact that the leaves are a prime source of auxin required for the growth response. Both cucumber and white mustard (*Sinapis alba*) also exhibit a classical phototropic response induced by irradiating the hypocotyls directly with blue light. Clearly the control of stem growth in green plants is an area where there is still much to learn.

23.2 GRAVITROPISM

Gravitropism is probably one of the most unfailingly obvious and familiar plant phenomena to most people (Figure 23.9). Everyone is aware that shoots always grow "up" and roots always grow "down." Or do they? A casual walk through the woods or garden should reveal how overly simplified this view is. The lateral branches of most trees and shrubs do not grow up; they grow outward in a more or less horizontal position. Stolons (or runners) of strawberry (*Fragaria*) plants and buttercups (*Ranunculus*) also grow horizontally along the soil surface. Dig into the soil and you will find rhizomes (underground stems) and many roots growing horizontally. Many pendulous inflorescences show no directional preference for growth, but hang down simply of their own weight.

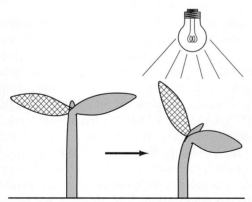

FIGURE 23.8 **Curvature in cucumber (*Cucumis sativus*) seedlings induced by shading cotyledons. The left-hand cotyledon was covered with aluminum foil and the seedling uniformly irradiated with white light for 8 hours. (After Shuttleworth, J. E., M. Black. 1977. *Planta* 135:51.)**

FIGURE 23.9 **Gravitropism in maize (*Zea mays*) seedlings. Four-day-old dark-grown seedlings were placed in the horizontal position for 3 hours. Note the shoot exhibits negative gravitropism and the root exhibits positive gravitropism.**

Unlike most other environmental stimuli, the force of gravity is omnipresent and nonvarying. It does not vary in magnitude as temperature does, for example. Gravity cannot be turned on and off, such as light at dawn and dusk. Moreover, gravity is not a unilateral stimulus—there is no gradient component in gravity. Cells on the lower side of a stem or root are subjected to the same gravitational force as those on the upper side. Consequently, it is likely that gravity can be detected only by the movement of some structure or structures within the cells—a movement that establishes an initial asymmetry in the cell and is translated in terms of pressure. The mass and movement of whatever structure is involved must be consistent with the sensitivity and speed of the gravitational response and there must be a mechanism for transducing the pressure signal into a biochemical signal that can lead to a differential growth response.

23.2.1 GRAVITROPISM IS MORE THAN SIMPLY UP AND DOWN

It is true that the root and shoot of the primary plant axis do align themselves parallel with the direction of gravitational pull. Such an alignment is said to be **orthogravitropic**. The primary root, which grows toward the center of the earth, exhibits **positive gravitropism**. The shoot, which grows away from the center of the earth, exhibits **negative gravitropism**. Organs such as stolons, rhizomes, and some lateral branches, which grow at right angles to the pull of gravity, are said to be **diagravitropic**. Organs oriented at some intermediate angle (between 0° and 90° to the vertical) are said to be **plagiogravitropic**. Lateral stems and lateral roots are commonly plagiogravitropic. Organs that exhibit little or no sensitivity to gravity are said to be **agravitropic**.

The advantages to the plant of positive and negative gravitropic growth responses are fairly obvious. Seeds may assume a random orientation in the soil, but in order to ensure survival, the shoot, with its photosynthetic structures, must be above ground in order to take advantage of sunlight. The root system must remain in the soil in order to secure anchorage and a reliable supply of nutrients and water. The primary root most often exhibits a strongly positive orthogravitropic response. Secondary roots (i.e., first-level branch roots) however, tend to grow more horizontally while tertiary roots are generally agravitropic. This hierarchy of gravitational responses ensures that the root system more effectively fills the available space and thus more efficiently mines the soil of water and nutrients (Figure 23.10). In a similar fashion, a hierarchy of negative orthogravitropic, diagravitropic, and plagiogravitropic responses in the shoot system

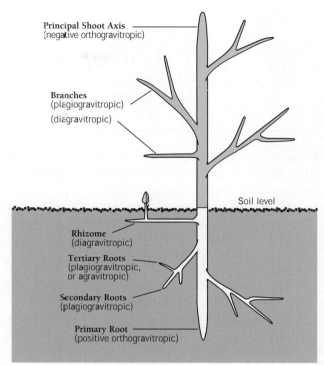

FIGURE 23.10 Diagram illustrating the range of gravitropic responses in shoots and roots. Note that differences in the gravitropic behavior among different levels of both shoot and root branches ensures the plant fills space.

helps to reduce mutual shading and ensures a more efficient capture of sunlight to drive photosynthesis.

23.2.2 THE GRAVITATIONAL STIMULUS IS THE PRODUCT OF INTENSITY AND TIME

Gravitational stimulation (stimulus quantity or dose) is the product of the intensity of the stimulus and the time over which the stimulus is applied:

$$d = t \cdot a$$

where a is the acceleration of mass due to gravity (in g), t is the time (in seconds) over which the stimulus is applied, and d is the dose in g seconds (g z s) (see Box 23.1: Methods in the Study of Gravitropism). The minimum dose required to induce gravitropic curvature is called the **threshold dose**. Threshold dose will vary depending on the organism or experimental conditions. Values of d in the range of 240 $g \cdot$ s (at 22.5°C) to 120 g z s (at 27.5°C) have been reported for *Avena* coleoptiles, but more careful mathematical analyses suggest that less than 30 $g \cdot$ s (i.e., an acceleration of 1 g for less than 30 seconds) is sufficient to induce gravitropic curvature in roots. Three other parameters are of interest when defining gravistimulation: presentation time, reaction time, and threshold intensity.

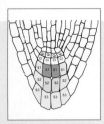

BOX 23.1
METHODS IN THE STUDY OF GRAVITROPISM

A fundamental requirement for any form of scientific experimentation is that of controlling the application of the stimulus in the form of intensity (or concentration) and duration. Since the gravitational field of earth cannot be extinguished (except in the microgravity conditions of space), experimentation on gravitational effects on plants and other organisms has required some unique approaches.

Most experiments require a mass acceleration in the range of 1 *g* or less, which can easily be achieved by simply orienting the organ (e.g., a coleoptile or primary root) away from the vertical. The force (at least for short-term stimulation) is generally proportional to the sine of the angular deviation from the vertical. Thus the force is greatest in the horizontal position since sine 90° = 1 (the sine of angles less than 90° is less than 1). Seedling shoots generally must be oriented between 0.5° to 10° from the vertical in order to induce curvature.

Forces greater than 1 *g* can be achieved by the use of specially designed centrifuges. Similar centrifuges have been used earlier in the century for studying the properties of animal (egg) membranes, and so forth. Centrifugation has not been used extensively in the study of gravitropisn, but those experiments in which it has been used have provided some useful insights.

The problem of extinguishing gravitational forces has been approached in two ways: **clinostats** and **space flight**. A clinostat is a device that holds the plant axis in a horizontal position while continuously rotating it about the horizontal axis. This does not actually extinguish the gravitational field, of course, but the summated effect is a nondirected constant stimulation. With the clinostat, plants can first be subjected to a brief stimulus and then rotated to, in effect, remove any further stimulus. As might be expected, continuous multilateral stimulation has been found to influence a variety of physiological parameters. *Avena* seedlings, for example, respond with increased growth rate and increased respiration. Some of these changes may be incidental, but others may influence the gravitropic response. Results must always be interpreted with caution.

SPACE—THE FINAL FRONTIER

The advent of space flight in the 1950s has provided plant scientists with unique opportunities to study responses to microgravity conditions. Since 1960, when the first wheat and maize seeds were carried aloft on Sputnik 4, experiments with plants have been conducted on manned and unmanned spacecraft from both the United States and the Soviet Union (now Russia). Perhaps not unexpectedly, physiological effects of microgravity are not limited to gravitropism but embrace a variety of other cellular events. Many of the effects are deleterious, including reduced growth, chromosomal aberrations, and other cytological abnormalities. Death at the flowering stage was common until, in 1982, *Arabidopsis thaliana* were successfully carried through a complete life cycle and produced viable seed. Many of the difficulties could be attributed simply to the logistics of trying to maintain plants in space, but even with improved methods, difficulties are still being encountered. As yet the returns may be modest, but the use of microgravity as an experimental tool is ripe for exploitation.

The minimum duration of stimulation required to induce a curvature that is just detectable is known as the **presentation time**. The intensity of stimulation should also be defined, although a stimulus of 1 *g* at 90° is more or less standard. A force of 1 *g* is easily obtained by simply placing the stem or root in a horizontal position. Presentation times of 12 seconds for cress roots and 30 seconds for *Avena* coleoptiles have been determined, but a brief 1-second stimulus will induce curvature in *Avena* coleoptiles if the stimulus is repeated every 5 seconds. This suggests that some cumulative receptive process begins the instant the plant assumes a horizontal position.

Presentation time should not be confused with **reaction time**, which is the interval between the presentation of the stimulus and the actual development of curvature. Reaction times involve the complete signal transduction sequence that leads to the asymmetric growth response. Typically, 10 minutes is required before curvature can be visually detected, although reaction times may vary from a few minutes to hours, depending on the species and conditions. In experiments employing sensitive electronic position-sensing transducers, curvature of maize coleoptiles could be detected within about 1.5 minutes of horizontal placement, while bending of the mesocotyl could not be detected before 3.5 minutes.

The minimum stimulus intensity required to induce a response is known as the **threshold intensity**. Threshold intensities have been determined for a variety of plant organs under different experimental conditions. The results are remarkably consistent and indicate that roots are perhaps an order of magnitude more sensitive than shoots. In land-based clinostat experiments

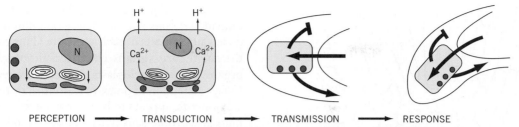

PERCEPTION ⟶ TRANSDUCTION ⟶ TRANSMISSION ⟶ RESPONSE

FIGURE 23.11 The four phases of root gravitropism. When the orientation of a root changes in a gravitational field, the change is perceived in the root cap by the settling of amyloplasts against intracellular membranes such as the endoplasmic reticulum. The biophysical signal is then converted to a biochemical signal through second messengers such as hydrogen ions, calcium ions, and the relocation of auxin transport facilitators (red circles). The signal is then transmitted to the elongation zone of the root via an altered flow of auxin (arrows) which results in the curvature response. N = nucleus.

(see Box 23.1), values for threshold intensity for *Avena* coleoptiles and roots were found to be $1.4 \times 10^{-3} g$ and $1.4 \times 10^{-4} g$. Values calculated for lettuce seedling hypocotyl and roots in experiments aboard the Salyut 7 spacecraft were $2.9 \times 10^{-3} g$ and $1.5 \times 10^{-4} g$, respectively. It is apparent that many plants are very sensitive to gravitational stimulus.

23.2.3 ROOT GRAVITROPISM OCCURS IN FOUR PHASES

Virtually all of the studies on root gravitropism have focused on primary roots and have identified four successive phases: perception, transduction, transmission, and growth response (Figure 23.11). Although the actual timing may vary depending on the conditions of the experiment, the initial perception phase occurs within perhaps one second of orienting a root off the vertical and involves biophysical mechanisms (e.g., pressure) for sensing the direction of gravitational pull. The transduction phase, occurring between 1 and 10 seconds following reorientation, involves the conversion of the biophysical single to a biochemical signal. The transmission phase occurs between 10 seconds and 10 minutes of reorientation and involves a redistribution of auxin within the root tip. The growth response, due to the unequal distribution of auxin, causes curvature of the root toward a more vertical orientation.

23.2.3.1 Gravity is perceived by the columella cells in the root cap. Gravitropic perception in the root is localized in the root cap, a thimble-like mass of cells that covers the tip of the root. The root cap consists of a central core of cells (the **columella**) arranged in regular tiers and one or more outer layers of **peripheral cells**. Traditionally, the function of the root cap was thought to be twofold; it provides physical protection for the root apical meristem and its peripheral cells secrete a mucilaginous polysaccharide that lubricates the path of the growing root.

A third function, that of gravity perception, has been established by experiments in which the root cap is wholly or partially surgically removed (Figure 23.12). Complete removal of the root cap does not interfere with the elongation of the root but completely abolishes any gravitropic response. Decapped roots will recover sensitivity to gravity after about 24 hours, which correlates with the regeneration of a new cap. Surgical experiments have also indicated that removal of the central core, or columella, cells caused the strongest inhibition of the response to gravity. Individual cells or pairs of columella cells can be selectively removed or ablated (L. *ablatus*, to take away) with a nitrogen laser, in conjunction with an optical microscope. Ablation of the innermost columella cells has the greatest impact on root curvature, without affecting overall growth rate of the root (Figure 23.13). Laser ablation of the root cap peripheral cells, on the other hand, has no effect on the gravitropic response.

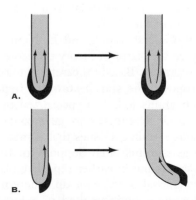

FIGURE 23.12 The role of the root cap in curvature of vertically oriented roots. (*A*) Control root. Growth is uniform when the root cap is left intact. (*B*) When the root cap is surgically removed from one-half of the root, the root grows toward the side with the cap remaining.

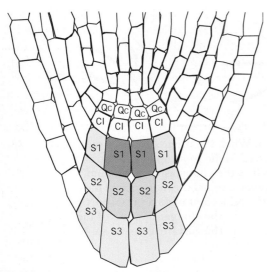

FIGURE 23.13 **Diagram of an *Arabidopsis* root cap showing the quiescent center (QC), columella initials (CI), and three ranks of columella cells. Colored cells indicate the cells that are sensitive to gravistimulation, based on laser ablation experiments. Relative sensitivity is indicated by the intensity of color. (Based on Perbal and Driss-Ecole, 2003).**

23.2.3.2 Gravity perception involves displacement of starch-filled amyloplasts.

A response to gravity must almost certainly involve sedimentation of some physical structure within the cell. F. Noll was the first to suggest, in 1892, that plants might sense gravity in a manner similar to some animals. Crustaceans, molluscs, and many other invertebrates have gravity-sensing organs called **statocysts**, small innervated cavities lined with sensory hairs. Within the cavity are one or more **statoliths**, tiny granules of sand or calcium carbonate that are pulled downward by gravity. When the statocyst changes position, the statoliths also shift position, bending the sensory hairs and sending an action potential to inform the central nervous system of the change.

In 1900, G. Haberlandt and E. Nemec independently adapted the statolith theory to account for plant responses to gravity. Based on careful cytological studies, they proposed the **starch-statolith hypothesis** in which starch grains found in specialized tissues function as statoliths. **Statocytes** are cells containing sedimentable starch grains. Tissues that contain statocytes are known as **statenchyma**. Support for the statolith hypothesis was found in earlier reports by Darwin and others that removal of the root tip, where most of the starch grains are found, resulted in a loss of gravitropic response. Nonetheless, the hypothesis was not universally accepted and over the decades a number of investigators have attempted to prove or disprove it.

A statolith is not a naked starch grain, but a group of starch grains contained within a membrane, called an

amyloplast (Figure 23.14). There may be 1 to several individual grains within an amyloplast and as many as a dozen amyloplasts in each statocyte. This compares with the single large grains characteristic of starch storage organs. Not all amyloplasts in all cells are readily mobile. In fact, detection of putative statoliths, or readily mobile amyloplasts, appears to be largely confined to regions of high gravitropic sensitivity. These include the mass of columella cells in the central core of the root cap and, in hypocotyls, a zone of endodermal cells that sheath the vascular tissues (also referred to as the starch sheath). Mobile amyloplasts may also be found in the inner cortical cells of aerial organs and the pulvini, or motor organs in the nodes of grass stems that are responsive to gravity.

Any gravity-sensing mechanism involving particle sedimentation would have to operate with a speed and sensitivity consistent with the known speed and sensitivity of the response. In the 1960s, L. J. Audus undertook a careful examination of various subcellular particles. Audus concluded that, of all the cellular organelles, only starch grains have the mass and density to move through the viscous cytoplasm within known presentation times. Ultrastructural examination has shown that other cellular organelles, such as the endoplasmic reticulum, may become shifted in cells subjected to gravitational

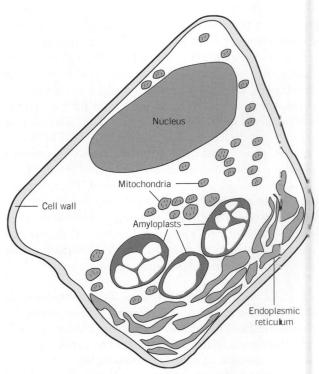

FIGURE 23.14 **A statocyte (columella cell) containing three statoliths (amyloplasts). (Based on an electron micrograph of *Lepidium* root, Volkmann and Sievers, 1979. In W. Haupt, M. E. Feinleib (eds.), *Encyclopedia of Plant Physiology*, NS, Vol. 7, pp. 573–600. Berlin: Springer-Verlag.)**

stimulus, but these movements are thought to be a consequence of starch grain sedimentation. Although there is still no direct proof for the starch-statolith hypothesis, there is a large body of evidence that is *more consistent with that idea* than any other that has been put forward to date. This evidence is summarized below:

1. *Gravitropism is generally absent in plant species that have no starch grains or amyloplasts.* In lower plants such as algae and fungi, excess carbohydrate may not be stored as starch. In these cases, some other substance may function as statoliths. In the alga *Chara*, for example, the role of starch grains is replaced by granules of barium sulphate.

2. *There is a strong correlation between the rate of starch sedimentation and presentation time.* In sweet pea (*Lathyrus odoratus*), for example, there is a parallel increase in both sedimentation time and presentation time as the temperature is lowered from 30°C to 10°C. The decline in both is presumably related to an increase in protoplasmic viscosity.

3. *Loss of starch by hormone treatment or mutation is accompanied by a loss of graviresponse.* For example, roots of cress seedlings (*Lepidium sativum*) treated with cytokinin or gibberellin at 35°C become starch-free in 29 hours. The growth rate of treated roots is reduced only slightly ($0.48 \, mm \, h^{-1}$ vs. $0.54 \, mm \, h^{-1}$) but any response to gravity is completely eliminated. Transfer of the roots to water in the light results in a parallel recovery of both amyloplasts and gravitropic responsiveness after 20 to 24 hours. In maize (*Zea mays*) the *amylomaize* mutant produces smaller amyloplasts than the wildtype. In studies of the percentage and speed of amyloplast sedimentation the degree of coleoptile curvature was strictly correlated with the size of the amyloplast. Another mutant of maize, *hcf-3* (*high chlorophyll fluorescence-3*) is unable to carry out photosynthesis and thus can form no starch in the leaf base statocytes when the endosperm reserves have been exhausted. Such seedlings do not respond to gravity unless fed sucrose, in which case recovery of both amyloplasts and sensitivity to gravity was noted.

4. *Amyloplasts can be displaced by a high-gradient magnetic field in place of gravity.* An intracellular magnetic field has been used to displace statoliths laterally (called magnetophoresis) in both roots and hypocotyls. The magnetic field also induces a curvature similar to the gravitropic response.

It is not known how the sedimentation of statoliths creates physiological asymmetry in the cell or tissue, although a number of models have been proposed. Most models agree that it is not the change in position of the statolith or the process of movement per se that is important. The preferred view is that the statolith exerts pressure on one or more membranes or other cellular components. Although there is no direct evidence for pressure-sensitive membranes in plants, both the plasma membrane and the endoplasmic reticulum (ER) have been suggested as likely targets. Electron micrographs of gravistimulated root statocytes, for example, commonly show amyloplasts sedimented on the endoplasmic reticulum against the lower side of the cell (Figure 23.14).

23.2.3.3 The transmission and response phases involve a lateral redistribution of auxin in the elongation zone.

Like phototropism, development of curvature in response to gravity ultimately involves a differential growth response. It is not surprising, then, that the Cholodny-Went hypothesis of asymmetric auxin distribution has dominated thinking and research into gravitropism for more than 60 years. Accordingly, the hypothesis states that horizontal orientation of the shoot or roots induces a lateral translocation of auxin toward the *lower* side of the organ. Auxin redistribution would bias the growth rate in favor of the lower side such that negatively gravitropic organs (e.g., coleoptiles and shoots) would turn upward. In positively gravitropic organs such as roots, the higher concentration of auxin is thought to *inhibit* elongation on the lower side relative to the upper, causing the organ to grow downward. Exogenous auxin, for example, very effectively inhibits root growth when applied at concentrations of 10^{-6} M or greater, concentrations that normally stimulate elongation of coleoptiles and shoots. In addition, gravitropic curvature is prevented by inhibitors of auxin transport (TIBA, NPA).

Auxin flow in the root is described by the "auxin fountain model" (Figure 23.15). Auxin synthesized in the shoot is transported basipetally through the shoot and into the root. There it continues to move acropetally toward the root tip through cells associated with the central vascular tissues, or stele. In the columella region of the root cap, the auxin flow is reversed and the auxin moves basipetally into the cortical cells of the elongation zone (Figure 23.15A). In a vertical root the flow of auxin into the cortical region is distributed uniformly around the root. When the root is displaced horizontally (Figure 23.15B), the flow of auxin from the columella is redistributed laterally (i.e., downward). In other words, auxin flow in a horizontal root is biased toward the lower side of the root. The implication is that the higher auxin content on the lower side of the root inhibits elongation relative to the upper side and the root curves downward.

23.2.3.4 Gravitropic induction in roots involves several second messengers and redistribution of auxin transporters.

An early event in gravity-stimulated roots is a change in the membrane

FIGURE 23.15 **The path of auxin flow in roots.** Auxin produced in the shoot flows into the root through the central vascular tissue. In the columella (yellow square), the auxin stream is diverted into the epidermal and cortical cell files, where it flows up toward the elongation zone. When the root is displaced from the vertical, events in the columella divert the main auxin flow toward the lower side of the root.

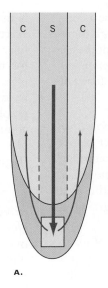

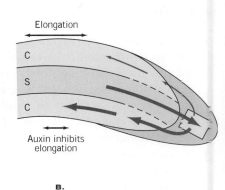

potential of the columella cells. Within seconds of reorienting *Lepidium sativum* roots to a horizontal position, the columella cells on the lower side of the root *depolarize* (the potential becomes less negative) and those on the upper side *hyperpolarize* (the potential becomes more negative). These changes in membrane potential appear to be in some way related to the cytoskeleton because they are inhibited by **cytochalasin D**, a drug that disrupts the cytoskeleton by binding to rapidly elongating actin filaments. There is also evidence that amyloplasts are connected to the plasma membrane by actin filaments. These observations have given rise to the hypothesis that displacement of amyloplasts stimulates stretch-activated ion channels. Stretch-activated channels are so named because they are activated in patch-clamp experiments when the membrane is stretched by applied suction. According to this hypothesis, stretch-activated ion channels would be responsible for the observed changes in membrane potential in the columella cells, which in turn would lead to the asymmetric distribution of auxin.

Changes in the pH of columella cells have also been observed in gravity-stimulated roots. By using pH-sensitive fluorescent proteins to monitor changes in pH during gravistimilation of *Arabidopsis* root, it has been shown that pH of the root cap apoplast decreases from pH 5.5 to 4.5 within 2 minutes of gravistimulation. Conversely, the cytoplasmic pH of columella cells increases slightly, from pH 7.2 to pH 7.6. These pH changes in the root cap precede auxin-related pH changes in the elongation zone by about 10 minutes.

Several investigators have highlighted a role for calcium in the gravity response of both coleoptiles and roots. Radio-labeled calcium (^{45}Ca) accumulated in the upper half of sunflower hypocotyls and maize coleoptiles within one hour of stimulation by gravity. Calcium redistribution also occurred following asymmetric application of exogenous IAA and could be

prevented in horizontal organs treated with the auxin transport inhibitor NPA. Histochemical techniques have demonstrated calcium localization in cells of the upper epidermis and underlying parenchyma of *Avena* coleoptiles within 10 minutes of gravistimulation. In addition, both calcium redistribution and gravitropism are prevented by prior treatment with EGTA (ethyleneglycol-bis-(β-aminoethyl ether)-N,N′-tetraacetic acid), a chelator that ties up free calcium, and the graviresponse of coleoptiles is prevented by treatment with chlorpromazine, an inhibitor of the calcium-binding protein calmodulin. This suggests that the response to gravity might at one stage involve a Ca^{2+}/calmodulin complex.

Experiments with exogenously applied calcium have provided equally convincing evidence of a role for calcium in root gravitropism. For example, asymmetrically applied agar blocks containing 10 mM $CaCl_2$ will induce curvature of maize roots, but only if the block is applied to the root tip (Figure 23.16). Migration of calcium in roots appears to be restricted to the root cap—migration is prevented if the cap is removed—and is directed toward the lower side of the horizontal root. Moreover, the calcium appears to move not through root cap cells but through the thick mucilaginous layer that coats the root cap. The importance of this mucilaginous coating is reflected in the observation that its continual removal by washing renders the root insensitive to gravity. Note that the direction of calcium asymmetry relative to auxin is opposite in gravistimulated coleoptiles and roots. In both organs, calcium migrates toward the potentially concave side. Thus in a horizontally oriented root, calcium moves downward to accumulate on the lower side of the root cap but it moves toward the upper side of a coleoptile. The source of this calcium is unknown, but it could be released from the ER as the result of an interaction between the amyloplasts and the ER.

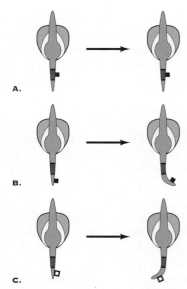

FIGURE 23.16 **Calcium-induced curvature of the primary root of maize (*Zea mays*). (*A*) An agar block containing calcium placed on the side of the root in the elongation zone has no effect. (*B*) When the agar block containing calcium is applied to the root tip, the root grows toward the source of calcium. (*C*) An agar block containing EGTA, a calcium chelator, causes the root to grow in the opposite direction.**

Finally, and perhaps not unexpectedly, recent experiments in which the PIN proteins were linked to a green fluorescent protein and examined microscopically have demonstrated that the auxin fountain and root gravitropism depend on the coordinated distribution and activities of the auxin efflux facilitators PIN1, PIN2, and PIN3 (Figure 23.17). PIN1 is localized at the apical end of cells in the stele and is responsible for delivering the auxin stream to the root apex. PIN3 is located along the lateral wall of the columella cells in a vertically oriented primary root, where it diverts the flow of auxin laterally, or centrifugally, toward the peripheral root tissues. PIN2 is located primarily on the basal walls of the peripheral root cap, epidermal, and cortical cells where it mediates the basipetal stream of auxin toward the cell elongation zone. The importance of PIN2 in gravitropism is indicated by the observation that *pin2* mutants fail to establish the required lateral auxin distribution following gravistimulation and do not exhibit a normal gravitropic response.

When a vertical root is rotated to the horizontal position, two significant events occur. First, PIN3 becomes redistributed, accumulating along the lower sidewalls of the columella cells. This redistribution of PIN3 presumably diverts the incoming auxin toward the lower side of the root. Second, PIN2 becomes asymmetrically distributed between the upper and lower sides of the root. The evidence suggests that the PIN2 protein is rapidly turned over due to ubiquitination and

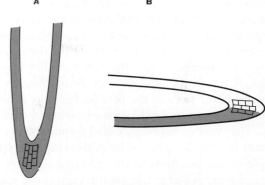

FIGURE 23.17 **The expression and distribution of the auxin transport facilitator PIN2 in vertical (*A*) and gravistimulated (*B*) roots. In vertical roots PIN2 is located more or less symmetrically throughout the peripheral root cap, epidermis, and cortex, mediating the basipetal flow of auxin toward the cell elongation zone uniformly around the root. In gravistimulated roots, enhanced degradation of PIN2 at the upper side of the root biases PIN2 location to the lower side.**

proteasome-dependent degradation. Gravistimulation apparently leads to an increased rate of PIN2 degradation in the upper side of the root. At the same time, higher auxin levels were found to promote increased retention of PIN2 in the lower side of the root. This unequal distribution of PIN2 would thus help to maintain the auxin gradient originally established by PIN3 in the columella cells.

But how does a root straighten out when it once again approaches a vertical orientation? It turns out that auxin can also promote the degradation of PIN2. It has been proposed that as auxin levels continue to build up in the lower side of the root, concentrations eventually reach a threshold where auxin-mediated PIN2 degradation begins to reduce auxin transport toward the elongation zone, thus reducing further curvature.

The big unresolved question at this time, of course, is how the settling of amyloplasts in response to gravity induces changes in pH, calcium flux, and other second messengers and how these changes are all linked to changes in the distribution and activity of the auxin transport proteins.

23.3 NASTIC MOVEMENTS

In addition to the directed movements of tropisms, many plants and plant parts, especially leaves, exhibit *nastic movements*, in which the direction of movement is not related to any vectorial component of the stimulus. Nastic responses may involve differential growth, in which case the movement is permanent. Alternatively the movement may be reversible, caused by turgor changes in a specialized motor organ.

Epinasty and thermonasty are examples of nastic responses involving differential growth. **Epinasty** is the downward bending of an organ, commonly petioles and leaves whose tips are inclined toward the ground. It is not a response to gravity, however, but appears to depend on an unequal flow of auxin through the upper and lower sides of the petiole. Epinasty is also a common response to ethylene or excessive amounts of auxin. The reverse response, called **hyponasty**, is less common but can be induced by gibberellins. A typical example of **thermonasty** is the repeated opening and closure of some flower petals, such as tulip and crocuses. In spite of their repeated nature, however, thermonastic movements are permanent and result from alternating differential growth on the two surfaces of the petals.

The most dramatic nastic movements are all turgor movements, which may be broadly separated into three categories: (1) leisurely rhythmic leaf movements in **nyctinastic** plants, (2) very rapid **seismonastic** movements in a limited number of species, and (3) **thigmonastic** or **thigmotropic** curling of threadlike appendages in climbing plants and vines. Nyctinastic and seismonastic responses depend on differential turgor movements in specialized motor organs, called the **pulvinus** (pl. *pulvini*). The pulvinus is a bulbous structure most often encountered in plant families characterized by compound leaves, such as the Leguminoseae and Oxalidaceae (Figures 23.18 and 23.19). It

occurs at the base of the petiole (*primary pulvinus*), the pinna (*secondary pulvinus*), or the pinnule (*tertiary pulvinus*). The pulvinus contains a number of large, thin-walled **motor cells**, which alter the position of the leaf by undergoing reversible changes in turgor.

23.3.1 NYCTINASTIC MOVEMENTS ARE RHYTHMIC MOVEMENTS INVOLVING REVERSIBLE TURGOR CHANGES

Nyctinastic movements (Gr. *nyctos*, night + *nastos* = closure) are most evident in leaves that take up a different position in the night from that taken during the day. Typically leaves or leaflets are in the horizontal, or open, position during the day and assume a more vertical, or closed, orientation at night. The primary leaf of common bean plants exhibits particularly strong nyctinastic movements but this can also be seen in *Coleus*, prayer plants, and other common garden and house plants. Observations of nyctinastic movements can be traced back as far as the writings of Pliny in ancient Greece. The Swedish botanist C. Linnaeus (in 1775) coined the term "plant sleep" to describe nyctinastic movements and they are commonly referred to as sleep movements today. The sleep movements of bean were prominent in the discovery of endogenous biological clocks and are described further in Chapter 24.

Sleep movements have been studied by several eighteenth- and nineteenth-century botanists, including Darwin. The process, however, has been studied most extensively by Ruth Satter and her colleagues in *Samanea samanan*, a member of the Leguminoseae with doubly compound leaves (Figure 23.18). In *Samanea* the paired pinnae and pinnules are normally separated and spread apart, while in closing they fold toward each other. In *Samanea*, the paired pinnules fold basipetally (i.e., downward), but in other species, such as *Mimosa pudica* and *Albizzia julibrissin*, closure of paired pinnules is upward, or acropetal. The doubly compound leaves of *Mimosa*, *Albizzia*, and *Samanea* all have three pulvini, but the simple leaf of *Phaseolus* (bean) has only two. It is the secondary pulvinus that generally exhibits the more rapid or dramatic change and has consequently been studied most extensively (Figure 23.19). They are also relatively large (2 to 3 mm diameter, 4 to 7 mm long in *Samanea*) and the changes in curvature are readily visible to the naked eye.

All nyctinastic responses depend on reversible turgor changes in the pulvinus. The pulvinus is typically cylindrical in shape, with prominent transverse furrows which facilitate bending, on the adaxial and abaxial sides (Figure 23.19). It contains a central vascular core with both xylem and phloem surrounded by sclerenchyma tissue. The vascular tissue assumes a linear arrangement

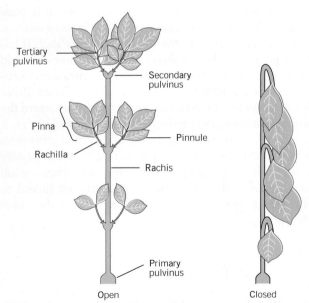

FIGURE 23.18 A leaf of *Samanea samanan*, illustrating the location of primary and secondary pulvini. Activation of the primary pulvinus causes leaflets to fold upward, parallel to the rachis. Activation of the secondary pulvinus causes the rachilla to fold downward. (Reproduced from the *Journal of General Physiology* 40:413–430, 1974, by copyright permission of the Rockefeller University Press.)

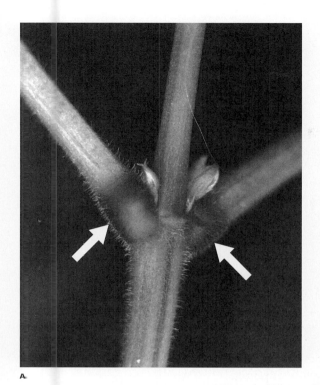

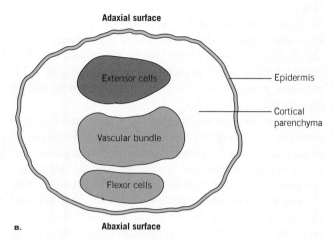

FIGURE 23.19 **The pulvinus. (*A*) Secondary pulvini of bean (*Phaseolus vulgaris*). The secondary pulvinus is the swollen area (arrows) at the juncture of the petiole with the stem, just below the axillary bud. In bean, the primary pulvinus is located at the distal end of the petiole, where it attaches to the leaf. During closure in bean, the petioles fold toward the stem axis while the leaf blade drops from the horizontal to the vertical position. (*B*) Schematic diagram of the bean secondary pulvinus in cross-section, showing the enlarged cortical parenchyma region. During closure, the motor cells of the extensor region lose turgor, while the motor cells of the flexor region gain turgor and the petiole moves toward the stem axis. During opening, the extensor cells gain turgor and the flexor cells loose turgor, thus driving the petiole away from the stem axis.**

as it passes through the pulvinus, apparently enhancing the flexibility of the pulvinar region. Outside the vascular core is a cortex comprised of 10 to 20 layers of parenchyma cells. The cells of the outer cortex have thin, elastic walls and exhibit large changes in size and shape during movement. These are called **motor cells**. Changes in the size and shape of the motor cells are responsible for leaf movement.

The opposite sides of the pulvinus are known as the **extensor** and **flexor** regions (Figure 23.19B). The extensor region is formed by motor cells that lose turgor during the bending movement, or "closure." Motor cells in the flexor region gain turgor during closure and lose turgor during opening. Thus, the swelling of extensor motor cells and shrinkage of flexor motor cells straightens the pulvinus and opens or spreads apart leaves or leaflets. The relative positions of extensor and flexor regions in the pulvinus (whether adaxial or abaxial, for example) will be reversed, depending on whether closure is basipetal or acropetal.

Nyctinastic movements are sensitive to blue light, the physiological status of phytochrome, and endogenous rhythms. Although the mechanism of signal perception and how these three stimuli interact is not known, it is clear that both the receptors and the responding system (the motor cells) are located in the pulvinus—at most a few cells apart. It is known that phytochrome can "reset" the endogenous clock that regulates nyctinastic leaf oscillations. The relationship between phytochrome and endogenous rhythms will be discussed in the next chapter.

23.3.2 NYCTINASTIC MOVEMENTS ARE DUE TO ION FLUXES AND RESULTING OSMOTIC RESPONSES IN SPECIALIZED MOTOR CELLS

Regardless of the nature of the stimulus, motor cell volume changes are due to osmotic water uptake (or loss) as a result of ion accumulation (or loss) across the cell membrane. What ions are exchanged, how are they transported, and what is the driving force for ion movement? These questions have been approached with a variety of techniques, including histochemical and radiochemical methods, and scanning electron microscopy coupled with X-ray analysis. The results show that leaf movement in all nyctinastic plants studied thus far is associated with a massive redistribution of potassium ion (K^+) between the symplast and apoplast in both the extensor and flexor regions of the pulvinus (Figure 23.20). Swollen extensor cells are characterized by high protoplasmic K^+ and low apoplastic K^+. In *Phaseolus vulgaris*, fully 30 percent of the osmotic potential change can be accounted for by K^+ movement. The charge carried by K^+ is compensated primarily by chloride and possibly small organic anions such as malate and citrate.

Beyond the central role of K^+ flux in nastic movements, it is difficult to pin down the specific sequence of molecular and biophysical events in the signal chain. Patch-clamp experiments with isolated *Samanea* motor cell protoplasts have established that K^+ exchange across the plasma membrane occurs through K^+ channels and that these channels can be regulated by changing the membrane polarity. Depolarization of the membrane opens the channels and allows K^+ to move out of the cell down its electrochemical gradient. It has also been established that there are pH gradients across the plasma membranes of motor cells. In the case of open *Samanea* pulvina, the pH of the apoplast is about 5.5 in the flexor region and 6.2 in the extensor region. The cytoplasmic pH is approximately neutral in both regions. Upon a light/dark transition (leading to closure), the pH gradient in the extensor region dissipates while in the flexor region the gradient increases. Although quantitative relationships between H^+ flux and K^+ flux have yet to be tested, H^+ extrusion could contribute to the electrochemical gradient necessary to drive K^+ uptake. Any observed changes in membrane potential are undoubtedly the consequence, not the cause, of the cross-membrane traffic in osmotically active ions.

The prominent roles of K^+ channels and H^+ pumps have been incorporated into the current model for motor cell movement. A simplified version of the model is shown in Figure 23.21. In this model, the light signal activates phytochrome (or cryptochrome), which accelerates **inositol phospholipid** turnover. Recent experiments have shown that light that stimulates opening of *Samanea* pulvini also decreases the level of phosphotidylinositol 4,5-bisphosphate (PIP_2) and increases the level of the second messenger inositol 1,4,5-triphosphate (IP_3). There is a transient stimulation of **diacylglycerol (DAG)**. These changes are qualitatively similar but quantitatively smaller than those normally detected in animal systems. The assays, however, involved whole pulvini, which contain vascular, collenchyma, and epidermal tissues as well as

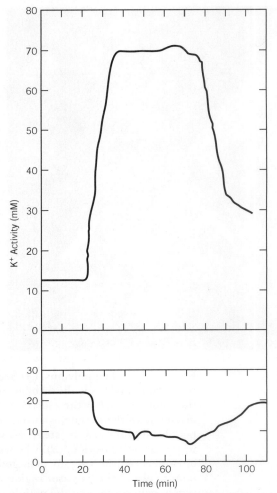

FIGURE 23.20 Changes in K^+ activity in the apoplast of extensor (upper curve) and flexor (lower curve) cells during closure of *Samanea* leaflets. Loss of K^+ from the protoplasts is followed by loss of water and turgor. Closure and opening were stimulated by dark and white light periods as indicated by the bar between the two graphs. (Adapted from Lowen, C. Z., R. L. Satter. 1989. Light promoted changes in the apoplastic K^+ activity in the *Samanea samanan* pulvinus. *Planta* 179:412–427. Figure 1. Copyright Springer-Verlag.)

the motor cells. The changes could be appropriately greater if restricted to the smaller population of motor cells. If inositol phospholipid metabolism functions in plants as it does in animals, DAG would be expected to activate a protein-kinase C (or its plant equivalent) to phosphorylate certain proteins. IP_3 would be expected to release free calcium—exogenous IP_3 does liberate calcium—although from which compartment is not known. Both the phosphorylated protein and/or transient increases in free calcium stimulate proton extrusion by activating the proton pump. The resulting electrochemical gradient energizes the uptake of K^+ and other ions, which in turn stimulates the osmotic

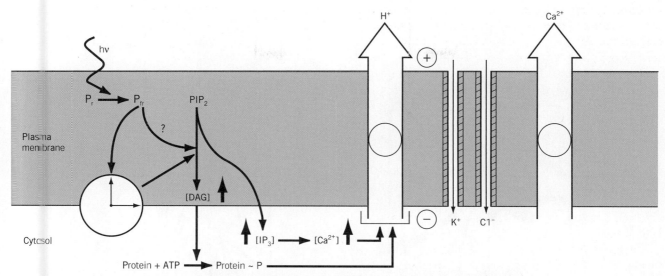

FIGURE 23.21 **A proposed model for the interaction of phytochrome, biological clocks, and the inositol triphosphate system in leaf movements of nyctinastic plants. Light, mediated by phytochrome and modulated by the endogenous clock, accelerates inositol phospholipid turnover and increases the level of the second messengers inositol-1,4,5-triphosphate (IP_3) and diacylglycerol (DAG). The second messengers stimulate a release of Ca^{2+} into the cytosol and phosphorylation of various proteins which in turn stimulate the extrusion of protons from the cell. K^+ diffuses into the cell in response to the proton motive force. An active transport pump extrudes Ca^{2+} as an aid to restoring Ca^{2+} homeostasis.**

uptake of water and motor cell swelling. The presence of a calcium pump that extrudes Ca^{2+} would help to ensure the restoration of Ca^{2+} homeostasis.

Many details of this model remain to be described, especially the function of the inositol phospholipid cycle in plants. Still, plant cells are known to contain virtually all the required components and the model is consistent with what has been observed in pulvini thus far. Significant advances are to be expected in the future, especially now that patch-clamp techniques—long a mainstay of electrophysiology research for animal cells—can be applied to plant cell protoplasts. This state-of-the-art technique has been in use for plant cells only since about 1984, but has proven invaluable for the study of ion channels.

23.3.3 SEISMONASTY IS A RESPONSE TO MECHANICAL STIMULATION

A limited number of leguminous plants that possess pulvini and exhibit nyctinastic movements also exhibit a response to mechanical stimulation. This phenomenon is known as **seismonasty**. Since seismonastic plants respond to touch, they are sometimes considered thigmonastic. However, seismonastic plants respond to

a wider variety of stimuli including shaking or wind, falling raindrops, wounding by cutting, and intense heat or burning.

The best known example of seismonastic plants is the tropical shrub *Mimosa pudica* (Figure 23.22). The survival advantage of such a response is not certain. Some have suggested that since these plants grow in arid, exposed areas where they are exposed to drying winds, folding of the leaves may be a means of reducing water loss. Others suggest that it is a means of protection from large herbivores or insects. However, one thing is clear—the response is very rapid. When the pulvinus is stimulated directly, bending begins in less than one second!

The ultimate response, leaf movement, of course involves movement of pulvini motor cells just as in nyctinastic movements. However, there are three essential characteristics of the seismonastic response that have served to focus attention on the early steps of signal transduction. The first of these is the rapidity of the response. Second, seismonasty follows the "all-or-none principle," which means that there is no obvious relationship between the intensity of the stimulus and the extent of the response. Third, excitation is propagated from the place of stimulation. The similarity of these characteristics to animal nerve transmission has given

A.   B.

FIGURE 23.22 **Seismonasty in the sensitive plant *Mimosa pudica*. The plant is shown in the open (*A*) and closed (*B*) positions. The plant on the right (*B*) was photographed about 10 seconds after closure was stimulated by a sharp tap to the stem with a pencil.**

rise to the expectation that plants may also be capable of transmitting stimuli in the form of potential changes. Indeed it has now been well established that virtually any part of the *Mimosa* plant can perceive stimuli and transmit them as electric pulses to the pulvini. Although plants do not have discrete nerve tissue, it appears that phloem sieve tubes can and do function as conduits for signal transmission. Stimulation of the petiole results in a rapid depolarization that is propagated basipetally along the sieve tube at a rate of about 2 cm s^{-1}. The unique structure of the sieve tube with its protoplasmic continuity through the sieve plates appears to be well suited for transmission of electrical signals. The appearance of the action potential is correlated with a rapid uptake of protons, suggesting that proton flux is responsible for the depolarization. When the action potential reaches the pulvinus, it appears to stimulate a rapid unloading of both K$^+$ and sugars into the apoplast. Water would follow and the resulting loss of turgor would cause collapse of the motor cells.

Other investigators have found that substances isolated from phloem sap of *Mimosa* and other species will stimulate closure of *Mimosa* pulvini when applied to the cut end of the stem. The active substance has been identified as a glycosylated derivative of gallic acid (4-0-β-d-gluco-pyranosyl-6′-sulphate)). Called "turgorin," this substance has been isolated from 14 higher plants that exhibit nyctinastic movements. It has been suggested that turgorin may give rise to action potentials in a manner similar to the animal neurotransmitter, acetylcholine.

SUMMARY

Plant movements serve to orient the plant body in space. Thus roots exhibit positive gravitropism, growing down in order to mine the soil for mineral nutrients and water. Shoots exhibit negative gravitropism and positive phototropism in order to optimize the interception of sunlight for photosynthesis.

There are several categories of plant movements. Growth movements involve cell division and elongation and are consequently irreversible. Turgor movements involve changes in turgor pressure and cell volume, and are reversible. Tropisms are directionally related to the stimulus whereas the directionality of nastic movements is inherent in the tissue and are not related to any vector in the stimulus. Nutations are rotary or helical movements that are best observed with time-lapse photography.

Under natural conditions, phototropism is a growth response to a light gradient, although in the laboratory it is usually studied by subjecting organs to unilateral light. Organs may either grow toward (positive phototropism) or away from (negative phototropism) the higher irradiance. Phototropism is a response to blue and UV-A light; mediated by a flavoprotein called phototropin, located in the plasma membrane.

The phototropic response is characterized by differential growth on the lighted and shaded sides of the responding organ. The most generally accepted theory to account for differential growth in coleoptiles and stems is the Cholodeny-Went theory. This theory proposes a lateral redistribution of auxin as it flows basipetally from the apex where it is synthesized. In the case of positive phototropism, the higher concentration of auxin flows down the shaded side of the organ, causing cells on the shaded side to elongate more rapidly than those on the lighted side.

Unlike most stimuli to which plants are exposed, gravity is omnipresent and nonvarying. There is no gravitational gradient. Gravity can be sensed only by movement of cellular structures (statoliths), which then establishes an asymmetry that is translated in terms of pressure. Although there is no direct proof, the weight of evidence indicates that statoliths are the starch-containing plastids, amyloplasts.

Sensitivity to gravity in the root is localized in the columella, a group of cells in the central core of the root cap. The primary transducer that senses the pressure of the statoliths and initiates the signal transduction chain remains unknown. Some evidence suggests it might be the endoplasmic reticulum. Another theory proposes that the sedimenting amyloplasts activate stretch-activated ion channels in the plasma membrane. As in phototropism, the gravitropic response involves differential growth that can be explained by redistribution of auxin transport. The steps between pressure sensing and auxin redistribution are unknown, but experiments indicate that pH changes, calcium ions, and inositol triphosphate may all be involved. Ultimately, the signal chain results in the relocation of auxin transport facilitators of the PIN family.

Plants exhibit a variety of nastic responses. One of the most prominent is the periodic movement of leaves known as sleep movements, or nyctinasty. Leaf movement is mediated by turgor changes in specialized motor cells located in structures called pulvini, found at the distal end of the petiole. Turgor changes are mediated by a flux of potassium ion induced by an interaction between phytochrome, biological clocks, and the inositol triphosphate system. Another nastic response is illustrated by seismonasty in the sensitive plant *Mimosa*. Seismonasty involves similar turgor changes in response to physical disturbance.

CHAPTER REVIEW

1. Define phototropism. Is phototropism restricted to unilateral lighting? In what way(s) can a light gradient be established across a plant organ such as a stem?

2. What pigment(s) function as the photoreceptor for phototropism? List the evidence that supports your conclusion.

3. Review the experimental basis for the Cholodny-Went hypothesis.

4. Shoots and roots express various levels of gravitropic response. What are the physiological advantages to be gained by such variation in response?

5. Review the statolith theory for gravitropism as it applies to roots. How is the gravitational stimulus perceived by a root and how does it respond? What is the evidence that calcium is involved in root gravitropism?

6. Describe nyctinasty. What might be the physiological significance or survival value of nyctinasty? In what ways is the seismonastic response similar to nyctinasty? In what ways is it different?

FURTHER READING

Abas, L. et al. 2006. Intracellular trafficking and proteolysis of the *Arabidopsis* auxin-efflux facilitator PIN2 are involved in root gravitropis. *Nature Cell Biology* 8:249–256.

Blancaflor, E. B., J. M. Fasano, S. Gilroy 1998. Mapping the functional roles of cap cells in the response of *Arabidopsis* primary roots to gravity. *Plant Physiology* 116:213–222.

Brown, A. H. 1993. Circumnutations: From Darwin to space flights. *Plant Physiology* 101:345–348.

Celaya, R. B., E. Liscum 2005. Phototropins and associated signaling: Providing the power of movement in higher plants. *Photochemistry and Photobiology* 81:73–80.

Christie, J. M. 2007. Phototropin blue-light receptors. *Annual Review of Plant Biology* 58:21–45.

Darwin, C. 1881. *The Power of Movement in Plants*. New York: Appleton-Century-Crofts.

Esmon, C. A. et al. 2006. A gradient of auxin and auxin-dependent transcription precedes tropic growth responses. *Proceedings of the National Academy of Sciences, USA.* 103:236–241.

Fasano, J. M., S. J. Swanson, E. B. Blancaflor, P. E. Dowd, T. Kao, S. Gilroy 2001. Changes in root cap pH are required for the gravity response of the *Arabidopsis* root. *The Plant Cell* 13:907–921.

Haga, K., M. Iino 2006. Asymmetric distribution of auxin correlates with gravitropism and phototropism but not with autostaightening (autotropism) in pea epicotyls. *Journal of Experimental Botany* 57:837–847.

Iino, M. 2006. Toward understanding the ecological functions of tropisms: Interactions among and effects of light on tropisms. *Current Opinion in Plant Biology* 9:89–93.

Jarillo, J. A. et al. 2001. Phototropin-related NPL1 controls chloroplast relocation induced by blue light. *Nature* 410:952–954.

Kimura, M., T. Kagawa 2006. Phototropin and light-signaling in phototropism. *Current Opinion in Plant Biology* 9:503–508.

Morita, M. T., M. Tasaka 2004. Gravity sensing and signaling. *Current Opinion in Plant Biology* 7:712–718.

Perbal, G., D. Driss-Ecole 2003. Mechanotransduction in gravisensing cells. *Trends in Plant Science* 8:498–504.

Satter, R. L., H. I. Gorton, T. C. Vogelmann 1990. *The Pulvinus: Motor Organ for Leaf Movement*. Rockville, MD: American Society of Plant Physiologists.

24

Measuring Time: Controlling Development by Photoperiod and Endogenous Clocks

Two hundred and fifty years ago, Carl v. Linné, better known for his development of the binomial system of nomenclature, designed a flower clock based on the opening and closing of the petals at specific but different times of the day. The plants were arranged in a circle and one could tell the time of day by simply noting which flowers were open and which were closed. It is often difficult for the layman to understand that plants can tell time without a Timex™, but many aspects of plant behavior can be interpreted in no other way. One example is the consistent flowering of various species at particular times of the year. Roses always bloom in the summer and chrysanthemums in the fall. Indeed, the flowering of many plants is so predictable from one year to the next that gardeners have for centuries incorporated them into their gardens as floral calendars, unerringly marking the progress of the seasons. In the northern latitudes, perennial plants sense the short days of autumn as a signal to induce bud dormancy, thus anticipating the unfavorable conditions of winter. The most reliable indication of the advancing season is the length of day, and an organism's capacity to measure the proportion of daylength in a 24-hour period is known as **photoperiodism**. However, photoperiodism is only one of the more outward manifestations of a far more fundamental timekeeping mechanism, known as the **biological clock**.

In this chapter we will examine

- photoperiodism; including the distinction between short-day plants, long-day plants, and other response types; the central role of the dark period; the nature of photoperiodic perception; and a discussion of current hypotheses to account for the elusive floral stimulus,

- vernalization—the low-temperature requirement for flowering in winter annuals and biennial plants,

- the biological clock, with an emphasis on endogenous rhythms with a 24-hour periodicity and their role in photoperiodic time measurement,

- the molecular genetic basis for the circadian clock and the search for the central oscillator, and

- a brief discussion of the significance of photoperiodism in nature.

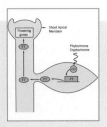

BOX 24.1
HISTORICAL PERSPECTIVES: THE DISCOVERY OF PHOTOPERIODISM

Although it had earlier been suggested that latitudinal variations in daylength contributed to plant distribution, the first efforts at controlled experimentation were conducted by a French scientist in 1912. J. Tournois found that both *Humulus* (hops) and *Cannabis* (hemp) plants flowered precociously during the winter in the greenhouse. Tournois eliminated temperature, humidity, and light intensity as environmental cues and in 1914 concluded that the changing of either daylength or nightlength was responsible for early flowering. Unfortunately, World War I intervened and Tournois did not live to continue his experiments. At the same time, H. Klebs was studying flowering of *Sempervivum funkii* (commonly known as "hens-and-chickens"). *Sempervivum* grows as a vegetative rosette in the winter greenhouse. By supplementing normal daylight with artificial light, Klebs was able to break the rosette habit, stimulate stem elongation, and induce flowering. From his experiments, Klebs concluded that length of day triggered flowering in nature. However, it remained for W. W. Garner and H. A. Allard to demonstrate the full impact of daylength on flowering and coin the term *photoperiodism*.

W. W. Garner and H. A. Allard were scientists with the U.S. Department of Agriculture near Washington, D.C. The initial focus of their work was a mutant cultivar of tobacco (*Nicotiana tabacum*), called Maryland Mammoth. In the field, Maryland Mammoth plants grew to be very tall with large leaves. Such characteristics would obviously be advantageous to the tobacco industry at the time (in the early 1920s), but breeding efforts were frustrated by the fact that the plants would not flower in the field during the normal growing season at that latitude. In the greenhouse, however, even very small plants flowered in the winter and early spring. Clearly, flowering was not simply a matter of the age of

the plants. Another problem of interest to Garner and Allard concerned flowering in soybean (*Glycine max*). When the cultivar Biloxi was sown over a 3-month period from May to August, all of the plants flowered within a 3-week period in September (Figure 24.1). The earliest seeded plants thus took 125 days to flower while those seeded last required only 58 days. Again it appeared that all plants, regardless of age, were simply awaiting some signal to initiate flowering.

Like Tournois, Garner and Allard eliminated a variety of environmental conditions (such as nutrition, temperature, and light intensity) as the "signal," coming finally (and with some reluctance) to the conclusion that flowering was controlled by the relative length of day and night. Using a crude but effective system of rolling plant benches in and out of darkened garagelike buildings at predetermined times, Garner and Allard proceeded to describe the flowering characteristics of scores of different species with respect to daylength. They went on to suggest that bird migration might also be keyed to daylength—a phenomenon that is now well documented. We now know that photoperiodic control is not limited to flowering, but is a basic regulatory component in many aspects of plant and animal behavior.

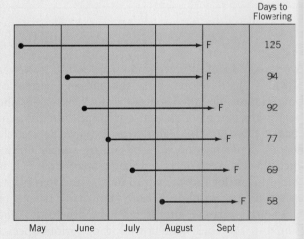

FIGURE 24.1 **September soybeans. Soybeans (*Glycine max*, cv. Biloxi) sown over a three-month period all flower within a three-week period in September.**

24.1 PHOTOPERIODISM

The switch from the vegetative state to the flowering state is arguably one of the most dramatic and mysterious events in the life of a flowering plant. Photoperiodism influences many aspects of plant development such

as tuber development, leaf fall, and dormancy, but the control of flowering by photoperiod has attracted the major share of interest. Indeed, it was the failure of a tobacco (*Nicotiana tobacum*) mutant to flower under field conditions that led to the discovery of photoperiodism (Box 24.1).

24.1.1 PHOTOPERIODIC RESPONSES MAY BE CHARACTERIZED BY A VARIETY OF RESPONSE TYPES

Photoperiodic responses fall into one of three general categories. They are: **short-day plants** (SD plants), **long-day plants** (LD plants), and daylength-indifferent or **day-neutral plants** (DNP) (Table 24.1 and Figure 24.2). Short-day plants are those that flower only, or flower earlier, in response to daylengths that are *shorter* than a certain value *within a 24-hour cycle*. Long-day plants respond to daylengths that are *longer* than a certain value, while day-neutral plants flower irrespective of daylength. Within the long- and short-day categories, we also recognize **obligate** and **facultative** requirements.[1] Plants that have an absolute requirement for a particular photoperiod before they will flower are considered obligate photoperiodic types. The common cocklebur (*Xanthium strumarium*), for example, is an *obligate short-day plant*. *Xanthium* will not flower unless it receives an appropriate short photoperiod. On the other hand, most spring cereals such as wheat (*Triticum* sp.) and rye (*Secale cereale*) are *facultative long-day plants*. Although spring cereals will eventually flower even if maintained under continuous short days, flowering is dramatically accelerated under long days. The popular research object *Arabidopsis* is also a facultative long day plant. However, the distinction between obligate and facultative response is not always hard and fast for a particular species or cultivar. Photoperiod requirement is often modified by external conditions such as temperature. A particular species may, for example, have an obligate requirement at one temperature but respond as a facultative plant at another temperature.

In addition to these three basic categories, there are a number of other response types that flower under some combination of long and short days. Various species of the genus *Bryophyllum* are, for example, **long-short-day plants** (LSD plant)—they will flower only if a certain number of short days are preceded by a certain number of long days. The reverse is true of the **short-long-day plant** (SLD plant) *Trifolium repens* (white clover). A few plants have highly specialized requirements. **Intermediate-daylength plants**, for example, flower only in response to daylengths of intermediate length but remain vegetative when the day is either too long or too short. Another type of behavior is **amphophotoperiodism**, illustrated by *Madia elegans* (tarweed). In this case, flowering is delayed under intermediate daylength (12 to 14 hours) but occurs rapidly under daylengths of 8 hours or 18 hours. There are many, often subtle, variations to the three basic response

[1] The terms obligate and facultative may be interchanged with the terms qualitative and quantitative, repectively.

TABLE 24.1 Representative plants exhibiting the principal photoperiodic response types.

Short-Day Plants	
Chenopodium rubrum	red goosefoot
Chrysanthemum sp.	chrysanthemum
Cosmos sulphureus	yellow cosmos
Euphorbia pulcherrima	poinsettia
Glycine max	soybean
Nicotiana tobacum	tobacco (Maryland Mammoth)
Perilla crispa	purple perilla
Pharbitis nil	Japanese morning glory
Xanthium strumarium	cocklebur
Long-Day Plants	
Anethum graveolens	dill
Beta vulgaris	Swiss chard
Hyoscyamus niger	black henbane
Lolium sp.	rye grass
Raphanus sativus	radish
Secale cereale	spring rye
Sinapis alba	white mustard
Spinacia oleracea	spinach
Triticum aestivum	spring wheat
Day-Neutral Plants	
Cucumis sativus	cucumber
Gomphrena globosa	globe amaranth
Helianthus annuus	sunflower
Phaseolus vulgarus	common bean
Pisum sativum	garden pea
Zea mays	corn

types, encompassing a large number of flowering plants. However, most of what is known about the physiology of photoperiodism in plants has been learned from a relatively small number of short-day and long-day plants.

24.1.2 CRITICAL DAYLENGTH DEFINES SHORT-DAY AND LONG-DAY RESPONSES

It is important to understand that the distinction between SD plant and LD plant is not based on the absolute length of day. Consider, for example, that both *Xanthium* and *Hyoscyamus niger* (black henbane) will flower with 12 to 13 hours of light per day (Figure 24.3). Yet the former is properly classified as a SD plant and the latter as a LD plant. Whether a plant is classified

SHORT DAYS

A.

B.

LONG DAYS

C.

D.

FIGURE 24.2 **Flowering response of the SD plant Japanese morning glory (*Pharbitis nil*) and the LD plant black henbane (*Hyoscyamus niger*) to short days and long days. Note the prominent flowers (arrows) in Japanese morning glory under short days and in black henbane under long days. Note also that black henbane remains as a rosette under short days. Plants of each species under both photoperiod regimes are of the same age.**

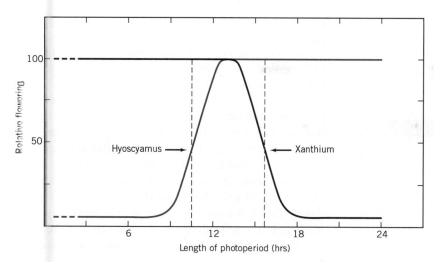

FIGURE 24.3 **A diagram to illustrate the concept of critical daylength in populations of *Xanthium strumarium* (cocklebur), a short day plant, and in *Hyoscyamus niger* (black henbane), a long day plant. Critical daylengths are indicated by the vertical dotted lines. Note that *Xanthium* flowers when the daylength is shorter than its critical daylength and *Hyoscyamus* flowers when the daylength is longer than its critical daylength.**

as a SD plant or LD plant depends instead on its behavior relative to a **critical daylength**. Plants that flower when the daylength is *shorter than some critical maximum* are classified as SD plant. Those that flower in response to daylengths *longer than a critical minimum* are classified as LD plant. Thus the critical daylength for the SD plant *Xanthium* is 15.5 hours, meaning that it will flower whenever the daylength is less than 15.5 hours out of every 24. The critical daylength for the LD plant *Hyoscyamus* is 11 hours and it will flower when the daylength exceeds that value. In the absence of further information, the actual daylength under which a plant will flower is no indication of its response type. A corollary to this observation is that, although SD plants tend to flower in the spring and fall and LD plants tend to flower in midsummer, the classification as a SD plant or LD plant is not necessarily an indication of the time of year that species will flower.

Many plants require more or less continuous exposure to the appropriate photoperiod, at least until floral primordia have been developed, in order to flower successfully. Others will proceed to flower even if, once exposed to even a single proper photoperiod, the plant is returned to unfavorable photoperiods. Such plants are said to be **induced** and the appropriate photoperiod is referred to as an **inductive treatment**. The phenomenon of induction raises intriguing, though unresolved, questions about the physiological properties of the induced state. Clearly a physiological change has taken place in induced plants and this change persists, even though no anatomical or morphological change is evident at the apex where flowers will appear.

Induction can also be experimentally useful. One of the reasons *Xanthium* has been so widely used for studies of photoperiodism is that a single short-day cycle will irreversibly lead to flowering, even in plants that are returned to long days. Such an extreme sensitivity to induction is not widespread, but it has been demonstrated in other SD plants such as Japanese

morning glory (*Pharbitis nil*), duckweed (*Lemna purpusilla*), and lambs quarters (*Chenopodium rubrum*), and in the LD plants dill (*Anethum graveolens*) and rye grass (*Lolium temulentum*).

Induction is not an all-or-none process, but can be achieved in degrees. Although *Xanthium* will respond to a single inductive cycle, the initiation of floral primordia is more rapid and more prolific if multiple cycles are given. Other plants may exhibit fractional induction—a summation of inductive photoperiods despite interruption with noninductive cycles. *Plantago lanceolata* (plantain), for example, normally requires a threshold of about 25 long days to induce flowering. Plants given only 10 long days will remain vegetative, but a schedule of 10 long days followed by 10 short days and then another 15 long days will induce flowering. The plants are apparently able to sum the long days in spite of the intervening short days. Examples of summation with up to 30 intervening noninductive periods have been reported.

24.1.3 PLANTS ACTUALLY MEASURE THE LENGTH OF THE DARK PERIOD

In their original publications, Garner and Allard suggested that plants responded to the relative lengths of day and night and coined the term *photoperiodism*, which combines the Greek roots for *light* and *duration*. Photoperiodism turns out to be a misleading term, however, because it implies that plants measure the duration of daylight. In fact, plants measure neither the relative length of day and night nor the length of the photoperiod—they measure the length of the dark period. This was elegantly demonstrated by the experiments of K. C. Hamner and J. Bonner in 1938 (Figure 24.4). Under 24-hour cycles of light and dark, *Xanthium* flowered with dark periods longer than 8.5 hours but remained vegetative on schedules of

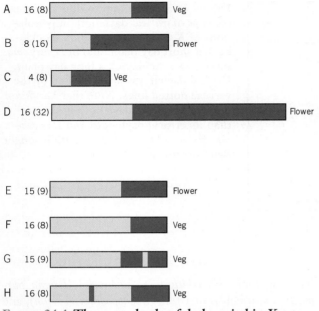

FIGURE 24.4 **The central role of dark period in** *Xanthium strumarium*, **a SD plant. The photoperiod regime is shown to the left. The number enclosed in brackets indicates the length of the dark period. Note that the plants flower whenever the dark period is uninterrupted for 9 hours or more.**

16 hours light and 8 hours darkness (Figure 24.4A,B). On schedules of 4 hours light and 8 hours darkness, plants remained vegetative even though the 4-hour photoperiod is much shorter than the 15.5-hour critical photoperiod (Figure 24.4C). On the other hand, schedules of 16 hours light and 32 hours darkness induced rapid flowering even though the photoperiod exceeded the critical daylength (Figure 24.4D).

The results of the Hamner and Bonner experiment allow two conclusions. First, the relative length of day and night is not the determining factor in photoperiodism, because the ratio of light to dark is the same in schedules B, C, and D (Figure 24.4) but with different results. Second, it is the length of the dark period that is important. The consistent feature within the experiment is that *Xanthium* will flower whenever the dark period exceeds 8.5 hours and will remain vegetative whenever the dark period is less than 8.5 hours. The critical role of the dark period was confirmed by interrupting the dark period with brief light exposures (Figure 24.4 E–H). The flowering effect of an inductive 9-hour dark period can be nullified by interrupting the dark period with a brief light-break (Figure 24.4G), but a "dark interruption" of a long light period has no effect (Figure 24.4H). Experiments with LD plants give similar results; LD plants require a dark period shorter than some critical maximum. With LD plants, a light-break in the middle of an otherwise noninductive long dark period will shorten the dark period to less than the maximum and permit flowering to occur. At this point it is clear that photoperiodism has relatively little to do with daylength per se. It is instead a response to the duration and timing of light and dark periods. Thus, the critical daylength for a SD plant actually represents the maximum length of day in a normal 24-hour regime that will allow a dark period of sufficient length. In the case of LD plants, long dark periods are inhibitory and the critical daylength is the minimum in a 24-hour regime that will keep the dark period short enough to allow flowering.

The fluence given during a light-break need not be very high to be effective. As little as one minute of incandescent light at a low fluence will prevent flowering in *Xanthium*. Even bright moonlight is sufficient to delay flowering in some SD plants. This raises an interesting possible relationship between nyctinastic leaf movements, discussed in the previous chapter, and control of flowering. It is possible, at least in some cases, that nyctinastic leaf movements could serve to reorient the leaves parallel to incident moonlight and thus reduce the impact of moonlight on the time-sensing mechanism.

24.1.4 PHYTOCHROME AND CRYPTOCHROME ARE THE PHOTORECEPTORS FOR PHOTOPERIODISM

Photoperiodism is a response to the length of a dark period, but the length of a dark period is defined by the interval between photoperiods. In other words, the length of a dark period is determined by the *timing of light-off and light-on signals*. In the case of a SD plant such as *Xanthium*, a light-break given in the middle of an otherwise inductive long dark period, as shown earlier in Figure 24.4, may be construed as a premature light-on signal that interferes with the timing process. Light-breaks are useful because they are effective with short exposures of low-fluence-rate light and can be applied to a single induced leaf. The light-break thus provides an opportunity to explore the nature of the pigment involved in photoperiodism by determining an action spectrum.

Early action spectra on several SD plants in the late 1940s indicated that red light was most effective as a light-break, with a maximum effectiveness near 650 nm. Then, at the same time that photoreversibility of seed germination was demonstrated, H. A. Borthwick and his colleagues also reported that red light inhibition of flowering in *Xanthium* was reversible with far-red. Later, similar results were demonstrated for other SD plants. Red, far-red photoreversibility of the light-break clearly implicates phytochrome in the photoperiodic timing process. The situation with LD plants, however, is not as clear. A light-break with red light in the middle of an otherwise noninductive long night should promote flowering. It does in *Hyoscyamus* and some others, but many LD plants are not sensitive to light-breaks and some are inhibited by red light. The role of phytochrome is far from clear at this point, but based on recent work with phytochrome mutants in *Arabidopsis*, it has been suggested that phyA is required to promote flowering of an LDP under certain conditions. PhyB, on the other hand, seems to inhibit flowering.

Blue light is also effective in controlling photoperiod. In *Arabidopsis*, for example, blue light is effective at promoting flowering when applied in a light-break experiment, suggesting that cryptochromes are involved as well. It is not yet clear what role cry1 may have in photoperiodic control of flowering. On the other hand, cry2 is known to promote flowering, but is believed to interact directly with the biological clock. More will be said about the role of cry2 later in this chapter, when we turn our attention to the biological clock.

24.1.5 THE PHOTOPERIODIC SIGNAL IS PERCEIVED BY THE LEAVES

The actual change from the vegetative to reproductive growth occurs, of course, in apical meristems—usually beginning at the shoot apex and appearing later in the axillary buds. Contrary to expectations, however, the photoperiodic signal is perceived not by the stem apex but by the leaves. This has been demonstrated in a variety of ways. Some of the earliest experiments were conducted by a Russian plant physiologist M. Chailakhyan. He reported flowering in *Chrysanthemum morifolium*, a SD plant, in which the apical, defoliated portion was kept on long days but the leafy portion was subjected to short days. When conditions were reversed, with the upper, defoliated portion kept on short days and the leafy portion on long days, the plants remained vegetative (Figure 24.5). Although the plants in this experiment still contained axillary buds, flowering has been successfully induced in plants from which the axillary buds have been removed. In later experiments it was shown that plants such as *Perilla* and *Xanthium* stripped of all but one leaf could be induced to flower if the remaining leaf were provided the appropriate photoperiod. Leaves may also be removed from induced plants and grafted to noninduced receptors where they will induce a flowering response. Finally, leaves need not even be attached to the plant in order to be induced. When excised leaves of *Perilla* (SD plant) were exposed to short days and grafted back to noninduced plants, the plants flowered even when maintained under long days.

The sensitivity of the leaf may vary with age. In *Chrysanthemum*, *Perilla*, and soybean (*Glycine*), the youngest fully expanded leaf was found to be most sensitive. In experiments with *Xanthium* in which the plants are stripped of all but the most sensitive leaf, it has been shown that peak sensitivity is reached during the period of most rapid expansion, when the leaf is about half its

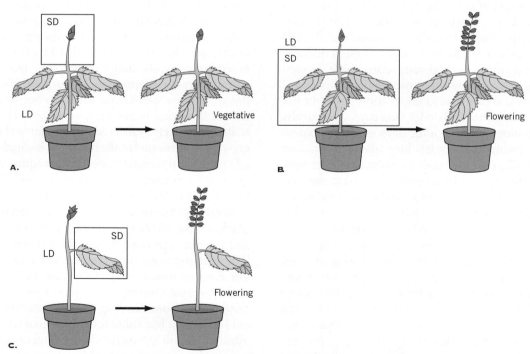

FIGURE 24.5 **The role of the leaf in perception of the photoperiodic stimulus in the short-day plant *Perilla*. (*A*) Plants remain vegetative when the shoot apex is covered to provide short days and the leaves are maintained under long days. (*B*) Plants flower when the leaves are given short days but the meristem is maintained under long days. (*C*) Flowering will occur when only a single leaf is provided short days.**

final size. These observations lead to two conclusions: first, that the leaf is independently responsible for perceiving the phototropic signal, and second, that the leaf initiates a signal chain, probably involving a diffusible floral stimulus, that communicates this information to the shoot apical meristem. Grafting experiments were also used to estimate velocity of movement of the stimulus, which for most species was found to be in the range of about 2.5 to 3.5 mm hr^{-1}.

24.1.6 CONTROL OF FLOWERING BY PHOTOPERIOD REQUIRES A TRANSMISSIBLE SIGNAL

The spatial separation between the site of perception of the photoperiodic signal (the leaves) and the site of flowering (the shoot apical meristem) logically requires a transmissible signal that carries the information from the leaf to the shoot apex. Over the past 70 years, several hypotheses have been advanced in order to explain the transmission of a floral stimulus, but the most persistent has been the hypothesis of a floral hormone.

Russian plant physiologist M. Chailakhyan was the first to suggest, in 1936, that the floral stimulus might be a hormone, which he proposed be called **florigen**. The concept of a flowering hormone was based primarily on the fact that the stimulus was transmissible across a graft union. For example, when several *Xanthium* plants are approach-grafted in sequence, all can be brought to flower if only the first is induced by short days (Figure 24.6). Members of the same family, such as the SD plant tobacco (*Nicotiana tobacum*) and the LD plant black henbane (*Hyoscyamus niger*), both in the family Solanaceae, can be grafted with relative ease. In such a partnership, *Hyoscyamus* will flower under short days if the tobacco is also maintained under short days, but not if the tobacco is maintained under long days. Conversely, tobacco will flower under long days when grafted to *Hyoscyamus* maintained under long days. A number of successful interspecific and intergeneric grafts have yielded similar results. These results have led to the conclusion that the final product of photoperiodic induction appears to be physiologically equivalent in plants of different photoperiodic classes and is probably identical to the constitutive floral stimulus in day-neutral plants.

Given the universal nature of other plant hormones, it should not be too surprising that the same floral stimulus operates in all photoperiodic classes. The major unanswered question that remains, however, is with regard to the chemical identity of the florigen. One approach toward answering this question is to prepare from flowering plants an extract that will evoke flowering in noninduced plants. Subsequent fractionation of the extract should lead to identification of the active substance. Unfortunately, although several attempts have been made, they have met with limited success. In one

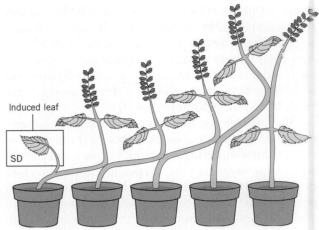

FIGURE 24.6 **Transmission of the floral stimulus in grafted plants. Several plants are "approach" grafted and the terminal plant is induced to flower. All plants will flower, indicating that the floral stimulus has been transmitted from the single induced leaf through all of the plants.**

of the most successful attempts to date, a methanol extract from freeze-dried *Xanthium* plants evoked a weak flowering response when applied to leaves of other *Xanthium* plants kept under long days. Unfortunately, few other attempts have been successful and none has been consistently repeated.

Of the several classes of plant hormones, only gibberellins have been shown consistently to evoke flowering in a wide variety of species. Repeated applications of dilute gibberellin solutions (containing principally GA$_3$) to the apex of annual *Hyoscyamus*, *Samolus pariflorus*, and *Silene armeria* (all LD plants) elicited a flowering response under short days. Does this mean that gibberellin is equivalent to florigen? The answer to this question is clearly negative. To begin with, evocation of flowering in response to gibberellin application is almost entirely restricted to LD plants that normally grow as rosettes under short days. This includes annual LD plants and biennial species that require an overwintering cold treatment before flowering as LD plants (see Chapter 25). For example, carrot (*Daucus carota*), Chinese cabbage (*Brassica pekinensis*), and biennial strains of black henbane (*Hyoscyamus niger*) all grow as rosettes and remain vegetative during the first growing season. The meristems are then subjected to an over wintering cold treatment. The following spring, the stems undergo rapid internode elongation and the plants will flower in response to long days. In the absence of any cold treatment but under long days, exogenously applied gibberellin will stimulate stem elongation and flowering. It thus appears that gibberellin will substitute for the cold requirement of biennial species or the long-day requirement of annual LD plants. But gibberellin will not evoke flowering in most SD plants (such as *Xanthium* or Biloxi soybean) kept under long days.

Molecular genetic studies with *Arabidopsis*, a facultative long-day plant, have recently cast a different light on the nature of florigen. A number of experiments conducted over the last decade have indicated a prominent role for the gene *CONSTANS* (*CO*). It was first observed that plants carrying mutant alleles of *CO* flower late under long days, but flower normally under noninductive short days. It was also found that *CO* mRNA accumulated in leaves under long days as well as in vegetative meristems and leaf initials. However, in spite of these and other observations, it was not clear whether CO stimulated flowering by acting in the leaves where the photoperiod signal is perceived or in the meristem where flowering occurs. The key proved to be experiments with transgenic *Arabidopsis* plants in which the *CO* gene was linked to a promoter from melon (*Cucumis melo*)—a promoter that is active only in the phloem companion cells in the minor veins of mature leaves. The transgenic plants expressing the *CO* gene flowered early under noninductive short-day conditions. Early flowering was also observed in transgenic *co* mutant plants that were engineered to constitutively express the *CO* gene. Finally, it was demonstrated that when scions homozygous for the *co* mutation were grafted to *CO*-expressing stock, early flowering was observed in the scions. Other experiments designed to test this idea have provided no evidence that CO itself is translocated out of the leaf cells These results strongly indicate that the *CO* gene participates in generating a phloem-mobile transmissible floral signal.

A likely candidate for the mobile floral signal is FLOWERING LOCUS T (FT), a small protein (ca. 20 kD) that is known to promote flowering in a dosage-dependent manner. The *CO* gene encodes a transcription factor that induces the expression of *FT* mRNA in leaves. The presence of RNA and peptides in phloem sap is well documented and in a recent study of *Brassica napus*, FT was one of 140 proteins identified in the phloem sap. According to one current model, CO responds to an inductive photoperiod by activating FT mRNA transcription and FT protein synthesis in leaf phloem parenchyma cells (Figure 24.7). The FT protein then moves through the phloem sieve tube elements to the shoot apical meristem where it induces a group of gene known as **floral meristem identity genes**. The role of floral identity genes will be discussed further in Chapter 25.

24.1.7 PHOTOPERIODISM NORMALLY REQUIRES A PERIOD OF HIGH FLUENCE LIGHT BEFORE OR AFTER THE DARK PERIOD

Although a SD plant will flower in response to a single long dark period, to be most effective the inductive dark period must be preceded by a period of light. The

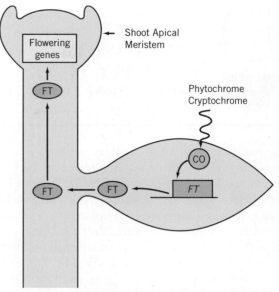

FIGURE 24.7 **A model for the long-distance transport of floral stimulus. The photoperiod is measured in the leaf phloem parenchyma cells where CO is activated by phytochrome and cryptochrome. Activated CO induces the transcription of the *FT* gene. FT protein is translocated through the phloem sieve tube cells to the shoot apical meristem. Once in the meristem, FT interacts with transcription factors to induce the transcription of flowering genes.**

function of the pre-dark light requirement is not clear since the requirements vary markedly depending on the species and conditions of the experiment. For maximum flowering with a single inductive dark period, *Xanthium* requires 8 to 12 hours of light. *Pharbitis* requires at least 6 hours. In others such as *Kalanchoe*, a few seconds of light per day were sufficient to induce flowering. Where longer light periods are necessary, it may be because photosynthetic products are required for the processes initiated in the dark period. Where very brief periods of light are effective, clearly photosynthesis cannot be involved and some other explanation for the light requirement must be sought.

In many cases it has been demonstrated that high-fluence light following the inductive dark period is also important. Again the experimental details are sketchy and the requirements seem to vary, but two explanations have been offered. One possibility is that the postinductive light period provides a stream of photosynthate that enhances translocation of the floral stimulus out of the leaf. However, some of the results are not readily explained as translocation effects. There is also evidence suggesting that the floral stimulus is subject to destruction or inactivation in the leaf if the dark period is too long. Consequently, it has been suggested that light may be required to stabilize or prevent inactivation of the floral stimulus in the leaf.

Box 24.2
HISTORICAL PERSPECTIVES: THE BIOLOGICAL CLOCK

The possibility that an internal, or endogenous, time-keeper might be involved was first raised by the French astronomer M. De Mairan in 1729. De Mairan found that leaf movements in the sensitive plant (*Mimosa*) persisted even when the plants were placed in darkness for several days. Subsequently studies by J. G. Zinn (in 1759) and A. P. De Candolle (in 1825) confirmed DeMairan's findings (see Sweeny, 1987). Curiously, DeCandolle found that under continuous light the time between maximum opening of the leaves was closer to 22 or 23 hours rather than the 24 hours under natural conditions. It would be a full century before the significance of this finding was fully appreciated!

The study of leaf movements continued to interest botanists and plant physiologists through the latter half of the eighteenth and the early nineteenth centuries. Much of the work simply confirmed the widespread occurrence of leaf movements and that they persisted under either continuous light or continuous darkness. In 1863, J. Sachs reported no correlation between leaf movements and temperature fluctuations, thus eliminating temperature as a cause. One difficulty, from an experimental perspective, was that studies of periodic phenomena required around-the-clock monitoring. In 1875, W. Pfeffer devised an apparatus for automatic and continuous recording of leaf position. Pfeffer attached the leaf, via a fine thread, to a stylus, which in turn recorded the position of the leaf on a rotating drum coated with carbon (lampblack). With some improvements, the same apparatus is occasionally used even today (Figure 24.8).

Over a period of 40 years, from 1875 to 1915, Pfeffer contributed several papers devoted to leaf movements in *Phaseolus vulgaris*, the common garden bean. At one point, he showed that plants that had lost their rhythmic leaf movements (as they will under prolonged continuous light or darkness) will regain them if exposed to a new light–dark cycle. If the new cycle is inverted with respect to natural day and night, leaf movement will also be reversed. Pfeffer concluded that persistent leaf movements under continuous light or darkness were a "learned" behavior. Others showed that regardless of any previous light–dark cycle, under continuous illumination sleep movements clearly reverted to a 24-hour oscillation. In the end, Pfeffer was forced to conclude that leaf movements were an endogenous, and probably inherited, behavior.

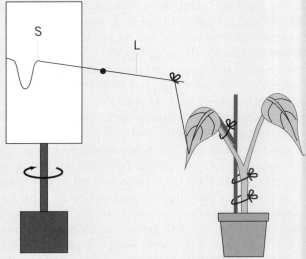

FIGURE 24.8 **A diagram illustrating the principle of the drum recording apparatus used by Bünning and others for recording leaf movements. The recording stylus (S) is attached to a finely balanced lever (L), which is in turn tied to the midvein of the leaf. As the leaf changes position, the stylus describes a tracing on the rotating drum.**

During the 1920s, improvements in the technology for maintaining constant environments, especially with respect to light and temperature, set the stage for significant advances in understanding leaf movements. One key observation was made by Rose Stoppel in Germany. Maintaining bean plants in a dark room at constant temperature, Stoppel observed that maximum night position of the leaves (i.e., near vertical orientation) occurred at the same time (between 3:00 and 4:00 A.M.) every day, exactly 24 hours apart. She reasoned that an endogenous, biological timer could be that accurate only if some environmental factor acted to, in effect, reset the clock on a daily basis. Stoppel referred to this factor as factor X.

The endogenous nature of the biological clock finally became evident through the work of two young botanists who had been given the task of determining whether subtle atmospheric factors might influence plants. E. Bünning and K. Stern became interested in Stoppel's factor X and, in order to achieve satisfactory constant temperature conditions, set up their bean plants and recording devices in Stern's potato cellar. Like Stoppel, Bünning and Stern found that the maximum night position came every 24 hours, but, surprisingly, some 7 to 8 hours later than in Stoppel's experiments. Bünning and Stern quickly recognized that the key was the very weak red light used when watering the plants and tending the recording equipment. Stoppel visited her experiments early in the morning while Bünning and Stern, because Stern's potato cellar was some distance

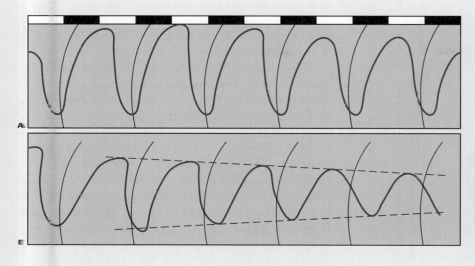

FIGURE 24.9 **Leaf movement in bean (*Phaseolus vulgaris*) is a manifestation of an endogenous rhythm generated by the biological clock. (*A*) A 6-day record of primary leaf position with alternating light and dark periods. Light and dark bars represent light and dark periods, respectively. The vertical lines indicate midnight solar time. Period length is 24 hours. (*B*) Free-running rhythmicity under continuous light. After the first period, the period length is extended to 25.7 hours. Note also that the amplitude (dashed lines) diminishes with time under continuous light. (From the data of Bünning, E. M. Tazawa 1957. *Planta* 50:107.)**

from the laboratory, made their visits in the late afternoon. Interestingly, at that time the textbook dogma was that red light had no effect on plant morphogenesis. Bünning and Stern concluded that the dogma was wrong—even weak red light apparently synchronized leaf movement so that the maximum night position

always occurred about 16 hours later. Indeed, when the red light was eliminated, the period between maximum night positions was no longer 24 hours but 25.4 hours (Figure 24.9). Bünning and Stern had identified factor X and lay the groundwork for the eventual discovery of the role of phytochrome in circadian rhythms.

Actually, most of the data fit a combination of these two ideas and the effect of the postinduction light requirement may be to quickly move the stimulus, such as FT protein, away from a site of inactivation in the leaf.

24.2 THE BIOLOGICAL CLOCK

Photoperiodism is inextricably linked to an internal time-measuring system known as the **endogenous biological clock** (Box 24.2). Another long-observed and dramatic manifestation of the biological clock is the diurnal rise and fall of leaves (known as nyctinastic, or sleep, movements) (Figure 24.10 and Chapter 23). Superficially, nyctinastic movements appear to be subject to external, or exogenous, control—namely the daily pattern of light and dark periods. However, under the right circumstances, it can be shown that the oscillations in leaf position are independent of external factors. On the other hand, photosynthetic carbon uptake describes a periodicity because it is light-driven and daylight is periodic over time. Photosynthesis is thus *diurnal*. It is active only during daylight hours and thus mirrors the daily light-dark cycle.

There are three criteria that serve to distinguish between simple diurnal phenomena and rhythms driven by an endogenous clock. (1) A clock-driven rhythm

persists under constant conditions (that is, in the absence of external cues). (2) A clock-driven rhythm can be "reset" by external signals, such as light or temperature. (3) There is no lasting effect of temperature on the timing of the clock-driven rhythm. These three criteria will be addressed in turn.

24.2.1 CLOCK-DRIVEN RHYTHMS PERSIST UNDER CONSTANT CONDITIONS

The key to an endogenous rhythm is that it persists, for at least several cycles, under constant conditions (usually constant light or constant darkness). The rhythmicity expressed under constant conditions is described as **free-running**. The time required to complete a cycle is referred to as the period (τ; tau) (Figure 24.11A). Period is conveniently described as the time from peak to peak, but it applies equally well to any two comparable points in the repeating cycle. Biological rhythms are traditionally classified according to the length of their free-running period. Thus a **circadian** rhythm has a period of approximately 24 hours (*circa*, about + *diem*, day). Bean leaf movement is a circadian rhythm because its free-running period length is about 25.4 hours. A period of about 28 days, the time between one full moon and the next, describes a **lunar** rhythm and a

FIGURE 24.10 **Circadian clock-induced movements of bean leaves. Seedlings of bean (*Phaseolus vulgaris*) are shown with the primary leaf in the horizontal day position (*A*) and vertical night position (*B*).**

period of one year is an **annual** rhythm. Also of interest are rhythms in metabolic activity with periods substantially less than 24 hours (measured in minutes or hours). These are known as **ultradian** rhythms.

The difference between the maximum and minimum, or peak and trough, of a rhythm is known as the **amplitude** (A) (Figure 24.11A). The amplitude of a free-running circadian rhythm usually diminishes with time until it eventually disappears altogether. In some cases this is probably due to declining energy reserves in prolonged darkness since the amplitude can be maintained, at least for a while, by feeding sucrose. More often, however, the clock seems to run down and an external signal is required to start up the rhythm again. The term **phase** has two slightly different but related usages. Any point of the cycle that can be identified by its relationship to the rest of the cycle can be considered a phase. The position of the peak (maximum night position of leaves, maximum flowering, etc.) is the most common reference point for phase relationships because it is usually most readily identified. In Figure 24.11A, for example, the two rhythms are displayed out of phase by approximately 6 hours. Phase may also be used to describe an arbitrary part of the cycle, such as night phase or day phase.

Discussions of endogenous circadian rhythms are sometimes complicated by the fact that two time frames are involved: **solar time** and **circadian time (CT)**. Solar time is an external time, based on a normal 24-hour day. Circadian time, on the other hand, is an internal time and is based on the free-running period. One cycle is considered to be 24 hours long, regardless of its actual length in solar time. Each hour of circadian time is therefore 1/24 of the free-running period. Thus if the free-running period is 30 hours, events that occur at 0, 15, and 30 hours of darkness will have occurred at circadian times CT:0, CT:12, and CT:24, respectively. The circadian time scale is useful in assessing phase

FIGURE 24.11 **Examples of circadian oscillations. (*A*) Two rhythms with period (τ) and amplitude (A). Although the period and amplitude for both rhythms are the same, the two are slightly out of phase. (*B*) Free-running rhythm and entrainment. A free-running rhythm with a period of 28 hours (dashed line) is entrained to 24 hours by a daily light-on signal. The open and closed bars represent light and dark conditions.**

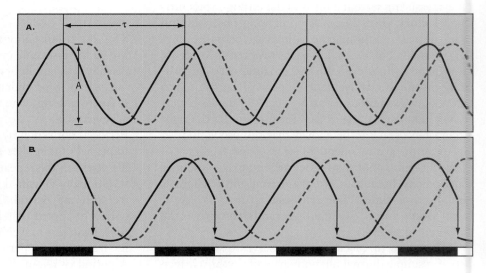

relationships within experiments or between rhythms with different periods. Finally, the phase of the free-running cycle that corresponds to day in a normal light–dark environment is known as **subjective day** and that which corresponds to normal night is the **subjective night**. In the case of sleep movements in bean, for example, the phase of the free-running cycle during which the leaves are in the horizontal position would be considered the subjective day. The phase during which the leaves are in the vertical position would be subjective night.

The rhythmic movements of bean leaves are normally coupled, or synchronized, to the solar day–night cycle (Figure 24.11B). The same coupling was evident in both Stoppel's and Bünning's experiments when the rhythm was coupled to the red light signal (Box 24.2). Such a coupling of a circadian rhythm to a regular external environmental signal is known as **entrainment**. The signal that synchronizes or entrains the rhythms is often referred to as a **zeitgeber** (Ger. *zeit*, time + *geben*, to give).

A large number of circadian rhythms have now been described in a wide range of organisms. The list includes single-celled flagellates, algae, fungi, crustacea, insects, birds, and mammals (including humans), in addition to flowering plants. A full list of known rhythms in flowering plants alone would cover several pages. An abbreviated list is provided in Table 24.2. A more extensive list, with references, is provided by Sweeny (1987).

24.2.2 LIGHT RESETS THE BIOLOGICAL CLOCK ON A DAILY BASIS

The action of the biological clock or endogenous rhythms is to ensure that certain functions occur at a particular time of day. For example, the oscillations of the clock in beans determines that the leaves rise during the day and fall at night. The period of the endogenous rhythm is fixed, but it may be "fast" or "slow" relative to the 24-hour solar period. Moreover, the daily duration of light and dark within the 24-hour solar period changes steadily throughout the season. How does the organism reconcile a nonvarying endogenous periodicity with these daily and seasonal changes in daylight? How are these rhythms kept in phase? To answer this question, we go back to the concept of entrainment. Entrainment is, in effect, a means for moving the oscillations of the clock forward or back in time on a daily basis, just as you might reset the time of an alarm clock every night before retiring. Entrainment is also useful to the experimentalist because it is the one significant way in which circadian rhythms can be manipulated in the laboratory.

Entrainment is not limited to solar periodicity. Within limits, rhythms can be entrained to light–dark cycles either shorter (18 to 20 hours) or longer (up to 30 hours or more) than 24 hours. Entrainment to extremely short or long cycles is rare. More useful information, however, can be obtained by studying whether the rhythm is moved forward or back in time and by how much. The experiment normally entails giving brief light pulses (usually 1 hour or less) at various times during an established free-running rhythm in constant darkness. The timing of the next peak is then compared with controls that have not been given a light pulse. For example, when populations of *Chenopodium* seedlings are exposed to single dark periods of various lengths, their capacity to flower fluctuates rhythmically for at least three cycles (Figure 24.12). In this case, a common light-off signal sets the rhythm in motion and the timing of the light-on signal (i.e., the length of the dark period) determines whether the plants will flower. If, however, relatively brief pulses of light are given at various times during the dark period, the pulses will reset the clock or shift the phase of the rhythm. The result is a **phase response curve** such as that shown in Figure 24.13. This curve demonstrates that a light pulse given early in the subjective night causes a delay of the first and subsequent peaks relative to the control. Somewhere near the middle of the subjective night there is a **phase jump** such that pulses given in the latter half of the subjective might cause subsequent peaks to be advanced relative to the controls. Note that light pulses given during the subjective day have very little effect on phase relationships. Similar phase response curves have been demonstrated for a variety of circadian rhythms and help to explain entrainment. It seems to be a character of the system that re-phasing is accomplished in such a way as to require the *least net displacement* of the rhythm. Thus, during the early part of the subjective night, the light pulse is apparently interpreted as a delayed light-off, or dusk, signal and the phase of the endogenous rhythm is adjusted accordingly. As the pulse arrives later, the delay is increased until at some point the pulse is now interpreted as an early light-on, or *dawn*, signal. This causes the rhythm to be advanced. Phase-shifting in this way constantly adjusts or entrains the rhythm to local

TABLE 24.2 Examples of circadian rhythmic phenomena in higher plants

Rhythm	Organism
Sleep movements	Many species
Stomatal opening	Banana, tobacco, *Vicia*
Stem growth	Tomato, *Chenopodium*
CO_2 production	Orchid flowers
Gas uptake	Dry onion seeds
Membrane potential	Spinach leaves
mRNA expression	Pea

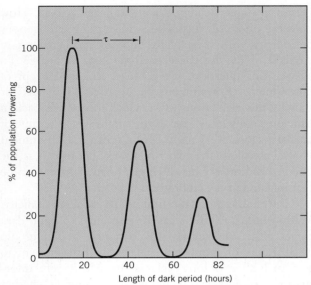

FIGURE 24.12 **Flowering in the SD plant** *Chenopodium rubrum* **responds rhythmically to the length of a single dark period. Populations of** *C. rubrum* **seedlings were exposed to a single dark period of varied length as indicated. The free-running periodicity (τ) is approximately 30 hours. The amplitude diminishes because the young seedlings are depleting their carbon supply over the extended dark period. Amplitude can be maintained by supplying the seedlings with glucose. (From King, R. W. Time measurement in photoperiodic control of flowering, Ph.D. thesis, University of Western Ontario, 1971. With permission of R. W. King.)**

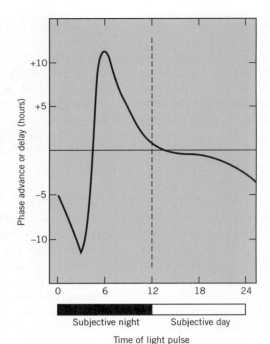

FIGURE 24.13 **Phase response curve. Light-on signals during the early part of the subjective night cause a delay in the timing of the next peak; given late in the subjective night, the result is an advance of the next peak. Light-on signals given during the subjective day have little or no effect on the phase of the rhythm. (Redrawn from King, R. W. Time measurement in photoperiodic control of flowering, Ph.D. thesis, University of Western Ontario, 1971. With permission of R. W. King.)**

solar time. The observation that similar phase response curves can be described for such different phenomena as insect pupal eclosion, bioluminescence in the dinoflagellate *Gonyaulax*, and CO_2 evolution and flowering in higher plants indicates similar properties, if not mechanisms, for the circadian clock in a variety of different organisms.

Although light plays an obvious role in resetting circadian rhythms, the photoreceptor involved is clearly not the same in all cases. Action spectra for resetting rhythms in *Gonyaulax*, the protozoan *Paramecium*, fungi, and insects all share a large peak in the blue region of the spectrum, suggesting that a flavoprotein blue-light receptor such as cry2 might be involved. In higher plants, however, both phytochromes and cryptochromes appear to be involved. For example, establishment of rhythms in dark-grown bean plants and phase-setting of leaf movement in *Samanea* both show a classic phytochrome photoreversibility with brief red and far-red light treatments. Others, however, such as the CO_2-evolution rhythm in *Bryophyllum* leaves, can be reset with red light but the effect is not reversible with far-red. The effect of far-red light in the *Bryophyllum* system is to abolish the rhythm altogether! This might also be a phytochrome effect, but these and other experiments make it clear that

photocontrol of phase-setting is not a straightforward process.

24.2.3 THE CIRCADIAN CLOCK IS TEMPERATURE-COMPENSATED

Most chemical reactions, and thus growth and other biological responses, respond to temperature with a Q_{10} near 2. This means that a 10°C increase in temperature will approximately double the rate of the process. A decrease in temperature leads to a decrease in the rate by the same amount. While such temperature sensitivity may be advantageous to the organism in some cases, the accuracy of a biological clock would be severely compromised if it were sensitive to often-random temperature fluctuations brought about by local environments. As it turns out, the period length of circadian rhythms is relatively insensitive to temperature. A classic example is again bean leaf movement. When seedlings are raised in the dark from seed, leaf movements tend to be small and unsynchronized. A single flash of light (the *zeitgeber*) initiates larger, synchronized movements.

In Bünning's experiments, the synchronized movements had a periodicity of 28.3 hours at a constant 15°C and 28 hours at 25°C. Although these data would seem to suggest that the circadian rhythm is insensitive to temperature, this is not strictly true. When seedlings were shifted from 20°C to 15°C, the initial period was 29.7 hours. Seedlings shifted from 20°C to 25°C had a period of 23.7 hours. These periods, however, lasted only for the first cycle or two—later cycles returned to periods of approximately 28 hours. Clearly the circadian rhythm is temperature-sensitive, but some mechanism quickly compensates for variations in temperature. Consequently, the Q_{10} for most circadian rhythms is near 1. Amplitude may be affected by temperature, but **temperature-compensation** is clearly a characteristic of the period for most circadian rhythms.

24.2.4 THE CIRCADIAN CLOCK IS A SIGNIFICANT COMPONENT IN PHOTOPERIODIC TIME MEASUREMENT

Endogenous rhythms of all kinds are fundamentally a question of time measurement, a concept that is not easily imagined within the framework of conventional biochemistry. Moreover, the clock is exclusively internal and, except for resetting by light and temperature, is not generally subject to manipulation from the outside. This means the clock is not amenable to traditional experimental strategies, since these normally require that the investigator be able to control or manipulate the system in some way. Another difficulty is our inability to distinguish between oscillations that are part of the timekeeping mechanism and those that are simply the "hands" responding to the output of the basic oscillator.

In 1936, E. Bünning first proposed that photoperiodism was tied to circadian rhythms. Bünning proposed that the rhythm was comprised of two phases—the **photophile**, or light-loving phase, and the **scotophile**, or dark-loving phase—which alternated about every 12 hours. According to Bünning's hypothesis, light falling on the plant during the photophile phase would promote flowering while light during the scotophile phase would inhibit flowering. In most experimental situations, when the plant is placed under continuous conditions, the photophile phase would probably be equivalent to subjective day and the scotophile phase equivalent to subjective night. In order to demonstrate rhythmicity in flowering and test Bünning's hypothesis, several novel experimental strategies have been employed. The difficulty is that, unlike leaf movement or carbon dioxide evolution, flowering is not a continuous process.

One strategy to demonstrate rhythmicity in flowering is illustrated by the SD plant *Chenopodium rubrum*, described in the previous section. Seedlings of *Chenopodium* can be induced to flower with a single dark period when the seedlings are only 4 1/2 days old. Before and after the dark period, the seedlings are maintained under continuous light. As shown in Figure 24.12, when the length of the dark period is varied, light during the first 8 to 10 hours inhibits flowering as would be expected with a SD plant. Thereafter, the capacity of the seedlings to flower as a function of the length of the dark period expresses a rhythmic pattern for at least three cycles, with a free-running period of 30 hours. With minor variations in the experimental strategy, essentially the same results have been demonstrated by other investigators using a variety of plants including cocklebur, soybean, Japanese morning glory, and duckweed.

A second example of the close dependency of photoperiodic time measurement on rhythmic phases comes from a series of elegant experiments conducted by W. S. Hillman with the aquatic plant *Lemna purpusilla* (Figure 24.14). *Lemna purpusilla* is a SD plant with a critical daylength of 12 hours. As an experimental organism, *Lemna* offers the unique advantage of growing heterotrophically in darkness when supplied with glucose. This means that long light periods can be eliminated and timing can be controlled by skeleton photoperiods—short pulses of light that serve to mark the beginning and end of dark periods. All that is required to induce flowering in *Lemna* are two 15-minute (0.25 h) light periods every 24 hours (Figure 24.15). Thus, when taken from *continuous light* to a schedule of 13 hours dark: 0.25 hours light: 10.5 hours dark: 0.25 hours light, *Lemna* will flower. On a schedule of 10.5 hours dark: 0.25 hours light: 13 hours dark:

Figure 24.14 **Duckweed (*Lemna sp.*). The genus *Lemna* is the smallest flowering plant. Each leaflike frond is about 1 to 2 mm across. Flowers arise at the point of the frond and consist of male and female parts only (no petals or sepals). There are short-day and long-day species of *Lemna*.**

FIGURE 24.15 **Skeletal photoperiods and the control of flowering in** *Lemna pur-pusilla*, **a SD plant.** *Lemna* **is arrhythmic under continuous light. The first light-off signal (arrow) starts up the rhythm. Because the first phase is the scotophile phase,** *Lemna* **interprets the first dark period, regardless of its length, as night. (*A*) If the "night" dark period exceeds the critical value, the plants will flower; if not (*B*), the plants remain vegetative. (*C*) The hypothetical scotophile and photophile phases of the rhythm.**

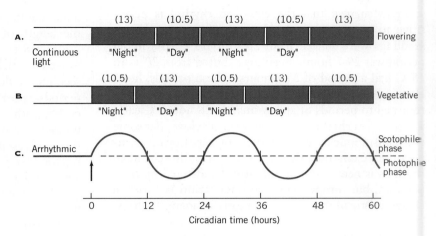

0.25 hours light, however, the plants remain vegetative. Note that the only difference between the two schedules is the length of the *first* dark period following continuous light. Even though both schedules contain a 13-hour inductive dark period, flowering is induced only when the long dark period comes first.

Clearly, *Lemna* recognizes the brief light pulses as the beginning and end of two different dark periods, but how does the plant know which is which? As with other plants under continuous light, *Lemna* becomes arrhythmic; that is, the amplitude of the rhythm damps out until the rhythm disappears. Transfer of the plants to darkness starts up the rhythm. Since flowering occurred only when the first dark period exceeded the critical night length, the first dark period (and every alternate period thereafter) must have been interpreted as *night*. This appears to be a general rule—the light-off signal following a period of continuous light starts up the rhythm and the first dark period (CT 0 to 12) is always the night phase. The second phase (CT 12 to 24) is the day phase. In subsequent experiments, Hillman confirmed that flowering occurs only when a dark period longer than the critical night length coincides with the night phase of its circadian rhythm. It is interesting to note that this interpretation is not limited to flowering plants—skeleton photoperiods elicit similar responses with respect to photoperiodic effects in insects and birds.

24.2.5 DAYLENGTH MEASUREMENT INVOLVES AN INTERACTION BETWEEN AN EXTERNAL LIGHT SIGNAL AND A CIRCADIAN RHYTHM

Several models have been proposed to explain the integration of photoperiodic time measurement with the biological clock. The model that is most consistent with both the earlier physiological studies and more recent genetic evidence is the **external coincidence model**. This model is essentially an updated version

of Bünning's scotophile and photophile hypothesis (Figure 24.16). Keep in mind that this model has been developed primarily for *Arabidopsis*, which is a facultative long-day plant. In its simplest form, the external coincidence model separates time measurement into two interacting components: the circadian clock itself and a clock-regulated daylength measuring mechanism. According to this model, the circadian clock sets a light-sensitive phase within the light–dark cycle. It does this by controlling the cyclic output of a key regulator

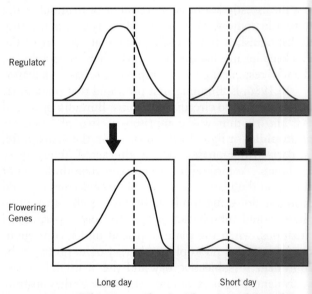

FIGURE 24.16 **The external coincidence model explains daylength measurement in long-day plants. A critical regulator gene is expressed maximally in late afternoon. Because the function and stability of the regulator protein, whose function is to turn on flowering genes, is regulated by light, the downstream expression of flowering genes will occur only under long days when the maximum expression of the regulator coincides with the presence of daylight. The vertical dashed line represents the day to night transition. The model also applies to short-day plants if it assumed that the regulator inhibits flowering.**

such that it reaches a maximum concentration in late afternoon. In a long-day plant, the regulator is ultimately responsible for turning on flowering genes in the shoot apical meristem but in order to initiate this chain of events, the regulator must first be activated by light. Consequently, flowering will be accelerated only when the late-afternoon expression of the regulator coincides with the presence of daylight, i.e., under long days. Under short days, the regulator does not reach maximum concentration until well after the beginning of the dark period. Consequently, the regulator can not be light-activated and is thus unable to initiate the cascade of events that leads to transcription of the flowering genes.

Although relatively little information is available for short-day plants, the external coincidence model would apply if it is assumed that the regulator inhibits, rather than promoting, flowering in short day plants.

24.2.5 THE CIRCADIAN CLOCK IS A NEGATIVE FEEDBACK LOOP

A basic model for the circadian clock system in plants requires three components: input pathways, a central oscillator, and output pathways (Figure 24.17). In plants, the principal input pathway is mediated by the photoreceptors phytochrome and cryptochrome, which set the phase of the oscillator in response to an external time cue (e.g., daylength). The output pathways connect the oscillator to other physiological processes whose overt rhythms reflect the timing of the central oscillator. Most output pathways are believed to be mediated by clock-controlled genes.

The strategy for the genetic dissection of the circadian clock is first to identify genes whose transcription is controlled by the clock and thus exhibits circadian oscillations. One then looks for mutants that influence the expression of the clock-controlled genes. If a mutant can be identified that influences timing at any level, the wildtype gene can be isolated and its gene product analyzed for clues to its role in the timing mechanism.

The first examples of clock-controlled genes in plants were those involved in photosynthesis, in particular those encoding the chlorophyll a/b–binding proteins found in the light-harvesting complex. Others include the small subunit of Rubisco, the enzyme Rubisco activase, and, in crassulacean acid metabolism (CAM) plants, the enzyme PEP carboxylase kinase. The chlorophyll a/b–binding proteins are encoded by a small family of genes (*CAB*). Expression of *CAB*, as measured by mRNA levels, is cyclic. *CAB* transcription begins to increase shortly before dawn and reaches a peak a few hours later. This cyclic expression of *CAB* has been exploited to develop a sensitive, rapid, and automated system for monitoring circadian rhythms at the molecular level. This has been accomplished by fusing a reporter gene,

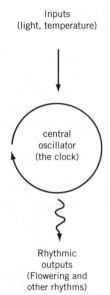

Inputs
(light, temperature)

central
oscillator
(the clock)

Rhythmic
outputs
(Flowering and
other rhythms)

FIGURE 24.17 **The three principal components of a simple circadian system include input pathways, a central oscillator, and output pathways. In plants, the input pathway(s) originate with light signals, mediated by phytochrome and cryptochrome, that entrain (or, reset) the central oscillator and may activate clock components. The oscillator is localized within individual cells and regulates the expression of clock-controlled genes, which in turn regulate overt rhythms. The output rhythms are the "hands" of the clock. Multiple-output pathways allow a single oscillator to control overt rhythms with the same periodicity but different phases.**

the bioluminescent firefly luciferase (*luc*), with the promoter for one of the chlorophyll a/b–binding protein genes, *CAB2*. Luciferase has a sufficiently short life span so that its activity will accurately reflect the activity of the promoter. Because the *CAB2* promoter is linked to the central oscillator, transgenic plants containing the resulting **CAB2::luc reporter system** will thus emit light rhythmically and these emissions can be monitored with sensitive light-monitoring equipment. Use of the *CAB2::luc* reporter system has played a large role in facilitating genetic analysis of the clock.

A number of genes closely related to the clock have now been described and the list continues to grow. Many of these were originally identified as flowering-time mutants in *Arabidopsis* and only later were found to be associated with the clock. One of the first circadian timing mutants to be identified was *toc1* (*timing of CAB*). The gene product TOC1 is localized in the nucleus where it has a role in transcription. The *toc1* mutation shortens period length for a wide range of clock-controlled processes. *TOC1* transcripts exhibit a 24 hour periodicity with a peak in the evening. Two other clock-associated genes are *CIRCADIAN CLOCK ASSOCIATED 1* (*CCA1*) and *LATE ELONGATED HYPOCOTYL* (*LHY*). CCA1 is a transcription factor that was discovered because of its capacity to bind with the *CAB* promoter in *Arabidopsis*. It appears to be a key element in the signal transduction pathway linking phytochrome with the expression of *CAB*. However, expression of *CCA1* itself cycles with a 24-hour period, peaking in the morning soon after dawn. There are several reasons for believing that *CCA1* is closely associated with the circadian oscillator. In transgenic plants that constitutively express high levels of CCA1, other circadian outputs, such as leaf movement, lose their rhythmicity and flowering is delayed. The circadian rhythms of other clock-regulated genes, including *CAB*, *CAT3* (a gene encoding the enzyme catalase), and an RNA-binding protein (*CCR2*) are also eliminated in transgenic plants that over express *CCA1*.

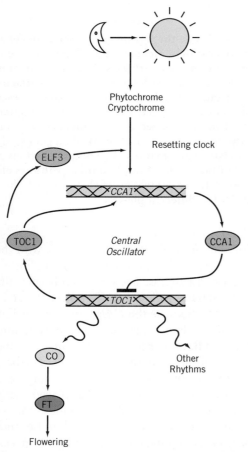

FIGURE 24.18 **A simplified molecular model of the *Arabidopsis* circadian clock. The input signal, light, is perceived by phytochrome and cryptochrome. The central oscillator is a negative feedback system. TOC1 is a positive regulator of the *CCA1* gene while CCA1 is a negative regulator of the *TOC1* gene. As TOC1 levels increase, an increase in the levels of CCA1 follows. CCA1 in turn shuts down the production of TOC1. The result is that TOC1 and CCA1 levels alternately rise and fall. ELF3, itself a product of the oscillator, blocks the input signal in the evening in order to ensure that the clock is reset by the morning light-on signal. One output of the clock is the transcription of *CONSTANS (CO)*, which initiates the flowering process.**

Another clock-associated mutant is *elf3* (*early flowering 3*). The *elf3* mutation not only advances flowering under short days, but also renders both leaf movement and *CAB* expression arrhythmic when entrained plants are shifted to continuous light. However, *CAB2::luc* activity still oscillates when entrained plants are shifted to continuous dark. ELF3 protein levels are regulated by the clock and peak in the evening. ELF3 functions to repress (or, gate) the light input pathway in the evening by binding to PHYB and inhibiting its activity. The result is that ELF3 makes the clock insensitive to light in the evening and ensures that clock is reset by the morning light-on signal.

According to the current model, CCA1 and TOC1 are part of a negative feedback loop that makes up the central oscillator (Figure 24.18). TOC1 acts as a positive regulator of the *CCA1* gene, thus stimulating its transcription and the accumulation of CCA1 protein. CCA1 protein, however, acts as a negative regulator of the *TOC1* gene, so as CCA1 accumulates, one of its actions is to repress *TOC1* transcription. In this manner, TOC1 and CCA1 levels alternately increase and decrease, or oscillate, over a 24-hour period. The oscillations are kept in phase by the dawn light-on signal received by phytochrome and cryptochrome and gated by ELF3, itself a product of the oscillator.

The circadian oscillator generates multiple output rhythms, possibly through the regulatory action of TOC1, CCA1, or related proteins on different genes. One of these certainly is the rhythmic output of CONSTANS (CO) in the leaf, which initiates the floral signaling cascade as described earlier. Other multiple outputs with different free-running periods could be explained in several ways. The clock is more complex than the simplified model shown in Figure 24.17. As the information accumulates, it is becoming increasingly clear that many additional genes and proteins are involved, forming multiple feedback loops. These additional layers of complexity could easily lead to multiple rhythms with different periodicities. Alternatively, there could be multiple oscillators within each cell, each linked to different genes, or there could be different oscillators within different cells in the same tissue.

24.3 PHOTOPERIODISM IN NATURE

Photoperiodism almost certainly reflects the need for plants to synchronize their life cycles to the time of year. Outside of the tropics, daylength is the most reliable predictor of seasonal change (Figure 24.19). Not surprisingly, photoperiodism is more important to plants in the subtropical and temperate latitudes where seasonal variations in daylength are more pronounced. But even many tropical plants respond to the small changes in daylength that occur within 5 or 10 degrees of the

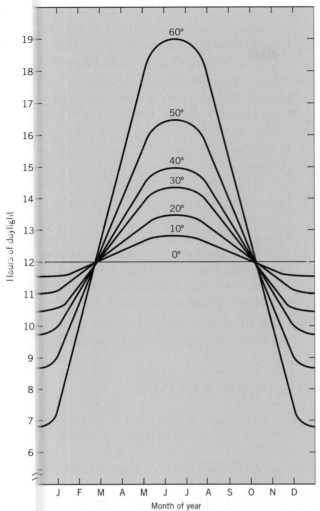

FIGURE 24.19 Daylength as function of latitude and month of year. Daylength is plotted as the time between sunrise and sunset on the 20th of each month. (Data from *The American Ephemeris and Nautical Almanac*, U.S. Naval Observatory, 1969.)

Photoperiodism also helps to ensure that plants flower in their temporal niche, reducing competition with others as well as ensuring that reproductive development is completed before the onslaught of unfavorable winter conditions. In many species, germination, for a variety of reasons, may not be uniform. If flowering relied solely on plant size, nonuniform germination would be expected to spread flowering out in time as well. To the extent that cross-pollination is required or advantageous, flowering synchronized by photoperiod would serve to ensure the maximum pollinating population or to coordinate flowering with the appearance of a particular pollinating insect.

Photoperiod and its effects on geographical distribution of plants can have a direct impact on humans, as illustrated by the case of common ragweed (*Ambrosia artemisifolia*). Ragweed is an annual SD plant, with a critical daylength of about 14.5 hours. The further north one goes, the longer the summer daylength. At the latitude of Winnepeg, Canada, (50°N), for example, the daylength exceeds 16 hours through most of June and July and doesn't drop below 14.5 hours until mid-August (Figure 24.19). Ragweed induced to flower at that time of the year would have insufficient time to flower and produce mature seed before the arrival of killing frosts in early fall. Since common ragweed can reproduce only by seed, it is abundant only in the more southerly regions of Ontario, Quebec, and the maritime provinces. It is rarely found, and then only on scattered patches of agricultural land, throughout most of western and central Canada. Hay-fever sufferers in these regions are at least spared the inconvenience of highly allergenic ragweed pollen, which plagues their neighbors to the south. This is not to suggest that photoperiod is the only factor that limits the distribution of plants such as ragweed, but it is clearly a significant part of the equation.

equator. Does this mean that the photoperiod response ties a species to one particular latitude? Probably not, since the critical photoperiod only sets the upper (for SD plant) or lower (for LD plant) limits for daylength. Beyond that, flowering and other responses to photoperiod can usually occur within fairly broad limits. Moreover, there is evidence that populations of plants are able to genetically adapt to latitude, thus giving rise to **physiological ecotypes**. In a variety of species, including *Betula* (birch), *Chenopodium* (lambs quarters), *Oxyria digyna* (mountain sorrel), and *Xanthium*, there are known ecotypes or photoperiodic races characterized by different critical daylengths. As a rule, the length of the critical day is longer as the individuals are collected at more northerly latitudes. In most cases, the critical daylength seems to key flowering to a consistent time interval before the arrival of damaging autumn frosts at that latitude.

SUMMARY

Photoperiodism is a response to the duration and timing of light and dark periods. There are three basic photoperiodic response types: short-day (SD) plants, long-day (LD) plants, and day-neutral (DN) plants. Other response types are variations on the three basic types and may be modified by environmental conditions such as temperature. A photoperiod requirement may be qualitative, in which case the requirement is absolute, or quantitative, in which case the favorable photoperiod merely hastens the response. The distinction between LD plants and SD plants is based on their response to daylengths greater than or shorter than the critical daylength. The absolute critical daylength varies from one species to another and the critical daylength for a LD plant may be shorter than the critical daylength for a SD plant.

Plants actually measure the length of a dark interval between the light-off and light-on signals. Action spectra of the light-break, which interrupts an otherwise inductive dark period, indicate that phytochrome is involved in the light signals. Photoperiodic light signals are perceived in the leaf but the response ultimately occurs elsewhere in the plant. This separation of perception and response suggests the logical necessity for a transmissible stimulus. In the case of flowering, the stimulus was proposed to be a hormone called florigen. It now appears that transmissible signal may be a small protein, FT, which is synthesized in the leaf phloem parenchyma and carried through the sieve tubes to the shoot apical meristem where it turns on the flowering genes.

Many aspects of plant development, including photoperiodism and nyctinasty, are controlled by an internal circadian clock. The circadian clock is difficult to study by traditional methods because, other than shifting the phase (equivalent to setting the clock), the clock is not readily manipulated by external influence. Three criteria that distinguish between simple periodic phenomena and clock-driven rhythms are (1) a clock-driven rhythm persists under constant conditions, (2) a clock-driven rhythm is reset or phased by environmental signals such as light and temperature, and (3) clock-driven rhythms exhibit temperature compensation.

Time measurement in photoperiodism involves an interaction between phytochrome and the endogenous biological clock. The nature of the clock and the phytochrome/clock interactions is beginning to yield to genetic studies. The use of a *CAB2::luc* reporter system has enabled identification of several clock-associated genes in plants. The central oscillator in plants is a negative feedback system similar to the oscillator previously found in insects, animals, and cyanobacteria.

Because changing daylength is the most reliable predictor of seasonable change, photoperiodism almost certainly reflects the need for plants to synchronize their life cycle to the time of year. Photoperiodism helps ensure that plants flower in their temporal niche, reducing competition with others, or that reproduction is complete before the onslaught of unfavorable winter conditions.

CHAPTER REVIEW

1. You have discovered a new plant whose photoperiod characteristics are not yet described. How would you go about determining whether this plant were a SD plant, a LD plant, or a day-neutral plant?

2. Describe the three hypotheses that have been invoked to explain the photoperiodic floral stimulus. What is the basis for each hypothesis?

3. Many metabolic processes appear to vary throughout a normal day. How would you determine whether these processes were regulated by an endogenous, circadian clock?

4. What is the physiological significance of physiological ecotypes, or photoperiodic races within a species that are characterized by different critical daylengths?

5. How does the response of *Lemna* to skeletal photoperiods lend support to Bünning's notion of a photophile and scotophile phase?

6. Distinguish between diurnal, circadian, and free-running rhythms.

7. Endogenous circadian rhythms have been described extensively, but their underlying mechanism is only now yielding to genetic information. Why is it so difficult to experimentally unlock the secrets of the circadian clock?

8. What has recent genetic evidence added to our understanding of the circadian clock?

FURTHER READING

Ayre, B. G., R. Turgeon. 2004. Graft transmission of a floral stimulant derived from *CONSTANS*. *Plant Physiology* 135:2271–2278.

Gardener, M. J. et al. 2006. How plants tell time. *Biochemical Journal* 397:15–24.

Hayama, R., G. Coupland. 2004. The molecular basis of diversity in the photoperiodic flowering responses of Arabidopsis and rice. *Plant Physiology* 135:677–684.

Hayama, R., G. Coupland. 2003. Shedding light on the circadian clock and the photoperiodic control of flowering. *Current Opinion in Plant Biology* 6:13–19.

Imaizumi, T., S. A. Kay. 2006. Photoperiod control of flowering: Not only by coincidence. *Trends in Plant Science* 11:550–558.

Hillman, W. S. 1962. *The Physiology of Flowering*. New York: Prentice-Hall.

Kondo, T., M. Ishiura. 1999. The circadian clocks of plants and cyanobacteria. *Trends in Plant Science* 4: 171–176.

Putterill, J., R. Laurie, R. Macknight. 2004. It's time to flower: Genetic control of flowering time. *Bioessays* 26:363–373.

Salisbury, F. B. 1963. *The Flowering Process*. New York: Pergamon Press.

Sweeny, B. 1987. *Rhythmic Phenomena in Plants*. 2nd ed. New York: Academic Press.

Thomas, B., D. Vince-Prue. 1997. *Photoperiodism in Plants*. 2nd ed. San Diego: Academic Press.

Wigge, P. A. et al. 2005. Integration of spatial and temporal information during floral induction in *Arabidopsis*. *Science* 309:1056–1059.

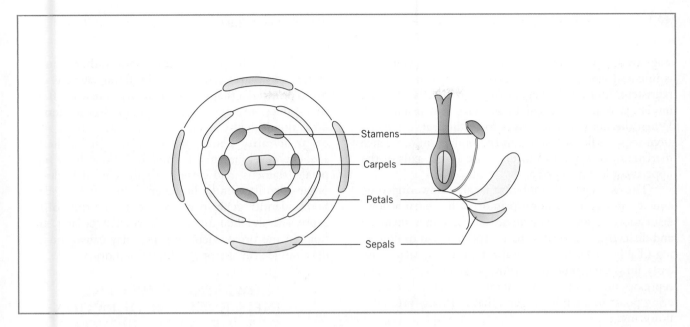

25

Flowering and Fruit Development

Flowering and fruit development have long held the interest of developmental biologists and physiologists because they represent a dramatic change in the pattern of shoot development and have significant economic implications. The switch of the shoot apical meristem from vegetative to floral organs and the subsequent development of fruit is a critical step in the developmental history of a plant and must be regulated precisely in order to ensure reproductive success. In the previous chapter it was shown that synchronization of flowering time with an environmental cue such as photoperiod is not a simple event. It is possible only through the interactions of many genetic and biochemical pathways: pathways involved in signal perception and transduction; pathways involved in the regulation of the circadian clock; and pathways involved in the development of the floral primordia.

Over the years, flowering in a number of species, including maize (*Zea mays*), petunia (*Petunia* sps.), snapdragon (*Antirrhinum* sps.), tobacco (*Nicotiana tobacum*), and annual ryegrass (*Lolium temulentum*), has been the subject of molecular and genetic studies. More recently, focus has shifted to the model plant *Arabidopsis*, where a number of genes that influence flowering have been identified and a model proposed to account for the genetic specification of floral organ initiation. In many cases, especially winter cereals and biennials, a period of low temperature can significantly alter the flowering

response. The low-temperature treatment, called vernalization, influences flowering time under long days. Finally, flowering is followed by the development of a specialized organ, the fruit, which ensure the proper environment for seed maturation and dispersal of the mature seed.

In this chapter we will examine

- the molecular genetic control of flower initiation and development in the shoot apical meristem,

- the phenomenon of vernalization and its relationship to other flowering time pathways, and

- the basic principles of fruit set and fruit development.

25.1 FLOWER INITIATION AND DEVELOPMENT INVOLVES THE SEQUENTIAL ACTION OF THREE SETS OF GENES

As noted in the previous chapter, flowering in many plants is influenced by environmental factors such as photoperiod and temperature and involves the synthesis of a mobile floral stimulus in the leaves. Other plants do not require external inputs and the signal is generated in the leaves when the plant simply reaches a minimum stage of development. In any case, flower development is initiated when that signal arrives at the shoot apical meristem. During the vegetative state, the shoot apical meristem is programmed to produce leaf primordial. When the floral signal arrives from the leaf, the meristem acquires floral identity and secondary inflorescence meristems, or floral primordia, arise in the axils of the uppermost leaf primordia.

The use of genetic mutants that alter flowering-time and the initiation of floral primordia has been a powerful tool for identifying many of the genes involved and dissecting the various pathways that lead to flowering (Table 25.1). As a general rule, mutants that cause early flowering indicate a wildtype gene that normally represses flowering while mutants that cause late flowering point to a wildtype gene that normally promotes flowering.

Extensive research with *Arabidopsis* has identified three sequential stages to the flowering process, each with its own set of genes. The first set of genes comprises the **flowering-time genes**. Flowering-time genes determine when the plant initiates flowering, either in response to the appropriate environmental signal or by monitoring the developmental state of the plant. Most mutations of the flowering-time genes cause the plants to flower later than normal, although a few will cause flowering to advance. One role of flowering-time genes is to activate the expression of **floral-identity genes**. Floral-identity genes commit undifferentiated

primordia to the production of floral rather than vegetative structures. Mutations in floral-identity genes cause primordia that would normally develop as flowers to produce structures with vegetative characteristics. The floral-identity genes in turn activate a set of **organ-identity genes** that serve to control the subsequent development of floral organs such as sepals, petals, stamens, and carpels. Expression of floral- and organ-identity genes, however, is not strictly linear. Some identity genes overlap in both the time of their expression and their function. As we shall see later, mutations in the organ-identity genes may cause abnormal development of any or all of the floral organs.

25.1.1 FLOWERING-TIME GENES INFLUENCE THE DURATION OF VEGETATIVE GROWTH

Flowering-time genes provide the connection between florigen, or the floral induction signal, and the transition to the production of floral organs. Flowering-time mutants may therefore interfere with the production of the signal in the leaf (including the timing mechanism, or circadian clock), translocation of the signal to the apex, or its activity in the apex. Most flowering-time mutants identified thus far cause plants to flower later than normal, indicating that the mutants interfere with pathways that normally promote flowering. Note that flowering time, as it is used here, refers to a developmental time

TABLE 25.1 **Some principal genes involved in flowering.**

	Gene Name	Gene product	Pathway/function
AG	*AGAMOUS*	(Unknown)	Organ identity
AP1	*APETALA 1*	(Unknown)	Floral/organ identity
AP2	*APETALA 2*	(Unknown)	Organ identity
AP3	*APETALA 3*	(Unknown)	Organ Identity
CCA1	*CIRCADIAN CLOCK ASSOCIATED 1*	Transcription factor	Circadian clock
CO	*CONSTANS*	Transcription factor	Long-day pathway
ELF3	*EARLY FLOWERING 3*	Novel protein	Circadian clock
ELF4	*EARLY FLOWERING 4*	Novel protein	Circadian clock
FCA	*FCA*	RNA-binding protein	Autonomous pathway
FLC	*FLOWERING LOCUS C*	Transcription factor	Floral repressor
FT	*FLOWERING LOCUS T*	Lipid-binding protein	Floral promoter
GA1	*GIBBERELLIC ACID 1*	(Unknown)	Gibberellic acid pathway
LD	*LUMIDEPENDENS*	Nuclear protein	Autonomous pathway
LFY	*LEAFY*	Transcription factor	Floral identity gene
TOC1	*TIMING OF CAB 1*	Transcription factor	Circadian clock
VRN1	*VERNALIZATION 1*	DNA-binding protein	Vernalization pathway
VRN2	*VERNALIZATION 2*	Repressor protein	Vernalization pathway
VRN3	*VERNALIZATION 3*	Equivalent to FT	Vernalization/LD pathway

rather than chronological time (e.g., days to flowering). For example, *Arabidopsis* is a facultative long-day plant with a critical photoperiod of 8 to 10 hours. The vegetative *Arabidopsis* plant grows as a rosette and flowering is preceded by stem elongation. Thus, under long days *Arabidopsis* flowers with 4 to 7 leaves in the rosette (about 3 weeks) but under short days flowering is delayed until 20 leaves have formed (7 to 10 weeks). Note that flowering is not related to elapsed time, but to the number of rosette leaves produced before the flowering stem appears. Late-flowering mutants simply extend vegetative growth and increase the number of leaves in the rosette before the stem elongates and the flowers develop.

It is interesting to note that, although a large number of late-flowering mutants have been described, no single *Arabidopsis* mutant that remains vegetative indefinitely has yet been identified. This fits with the general assumption that there are multiple pathways controlling flowering time with a certain amount of built-in redundancy. Redundancy provides that inactivation of genes in one pathway is at least partially compensated for by other genes or complementary pathways. In *Arabidopsis*, at least five separate, but interacting, pathways for controlling flowering time have been identified (Figure 25.1).

Several flowering-time mutants, including *fca*, *ld*, and *fae*, flower later than wildtype plants under both LD and SD conditions but remain sensitive to vernalization. Because flowering in the mutants is equally affected under both LD and SD conditions, the corresponding wildtype genes are thought to be active in an **autonomous pathway** that monitors developmental stage and initiates flowering in response to

internal developmental signals. Such a signal is commonly reflected in a minimum leaf number that must be achieved before flowering can proceed. Flowering of a second group of mutants, including *constans* (*CO*) and *gigantea* (*GI*), is delayed under LD conditions, but not under SD conditions. These mutants also show no response to low-temperature treatments, or vernalization, which is normally linked to a LD flowering response and not a SD response. As described in Chapter 24, the *CO* gene is believed to be a central component in the photoperiodic or **long-day pathway** and is responsible for promoting the mobile floral stimulus FT. Both *CO* and *GI* have been cloned and studied in some detail. It appears that *GI* operates before *CO* in the same pathway and that floral promotion under long days depends on the amount of *CO* protein and, subsequently, FT protein that is produced. The action of FT at the shoot apical meristem is at least in part mediated by a transcription factor FD. At this point it appears that the combination of FT and FD proteins is responsible for initiating flowering under long days by activating floral identity genes in the shoot apical meristem.

A single mutant identified as *ga1*, which flowers late under LD conditions and does not flower at all under SD conditions, is thought to represent a separate pathway mediated by gibberellin, the gibberellic acid or **GA pathway**. Finally, there is a small group of genes that appear to act primarily as **floral repressors**. This conclusion is based on the observation that their loss-of-function mutants [e.g., *elf3* (*early flowering 3*) and *phytochrome B* (*phyB*)] cause early flowering. Some, such as the *phyB* mutants, retain a response to photoperiod while others, such as *elf3*, do not. The loss of photoperiod response in *elf3* appears due to disruption of the circadian rhythm component of photoperiodic timing. Other early-flowering mutants, such as *emf1* and *emf2* (*embryonic flower 1-2*), flower almost immediately following germination, without forming any rosette leaves. Instead, the plants form reproductive structures on the surface of their cotyledons. Because wildtype plants eventually flower, it appears that the repressor activity of *EMF* genes must be intended to prevent precocious flowering. The activity of repressor genes must decline during development or at some point be turned off, eventually allowing one or more of the promotory pathways to take precedence. In addition to the above four genetic pathways, a possible fifth pathway that mediates low-temperature effects on flowering has been suggested by recent studies.

One of the challenges that remain is to sort out to what extent known flowering-time genes are involved in the production, transmission, and perception of the floral induction signal. In this regard, an interesting late-flowering mutation has recently been isolated from maize (*Zea mays*). Flowering in maize follows

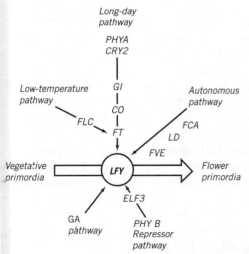

FIGURE 25.1 Five separate genetic pathways control flowering time in *Arabidopsis*. All pathways appear to converge on *LEAFY* (*LFY*), a floral-identity gene in the shoot apical meristem that mediates the switch from a vegetative meristem to a floral meristem.

an autonomous pathway and wildtype plants normally flower after producing a fixed number of leaves. Maize plants carrying the mutant *id1* (*indeterminate 1*), however, continue to produce only leaves long after the wildtype plant has produced ears and tassels. When they eventually do flower, the floral structures are aberrant. However, unlike many of the *Arabidopsis* flowering-time genes, which are expressed in the shoot apical meristem and activate floral-identity genes, *id1* is expressed only in the young leaves. This pattern suggests that the function of *ID1* may be more closely related to the synthesis of the floral stimulus (or repression of floral inhibitors) in the leaves.

25.1.2 FLORAL-IDENTITY GENES AND ORGAN-IDENTITY GENES OVERLAP IN TIME AND FUNCTION

While the principal effect of flowering-time mutants is on the duration of vegetative development, mutations in the floral-identity genes disrupt the transition of the undifferentiated primordia to floral meristems. At least four floral-identity genes have been isolated from *Arabidopsis*: *LEAFY* (*LFY*), *APETALA1* (*AP1*), *APETELA 2* (*AP2*), and *CAULIFLOWER* (*CAL*). Floral-identity genes are expressed in the apical meristem prior to the formation of floral organs but their expression is up-regulated rapidly following floral induction by long days or application of gibberellin. However, their individual roles have been difficult to study because of extensive redundancy (i.e., a loss of function due to one mutation is readily compensated by one of the other wildtype genes). The *LEAFY* gene appears to play a key role in floral meristem identity. This can be demonstrated by placing the gene under the control of a strong gene promoter (designated 35S) from the cauliflower mosaic virus. Transgenic plants that contain the 35S::*LEAFY* combination (called a construct) bypass the requirement for a floral induction signal and express the gene constitutively. In such plants, a shortened primary shoot terminates early in clusters of flowers and all secondary shoots produce flowers from the rosette. By contrast, the *lfy* mutant produces more inflorescence branches than a wildtype plant but the "flowers" consist of green, leaf-like organs. Moreover, constitutive expression of the flowering-time gene *CO* leads to a rapid activation of *LEAFY* in wildtype plants. *LEAFY* appears to have a central role in the flowering process. It is probably the principal target of the mobile flowering-time mobile stimulus FT when it arrives in the meristem. *LEAFY*, in turn, activates organ identity genes such as *APETALA1* (*AP1*).

The *Arabidopsis* flower is rather typical among advanced flowering plants, consisting of four distinct whorls of floral organs (Figure 25.2).The outermost

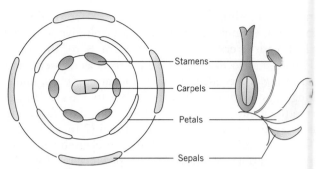

FIGURE 25.2 (*A*) The *Arabidopsis* flower consists of four distinct whorls. The outermost whorl (whorl 1) consists of four sepals, which are green and leaf-like. The next whorl (whorl 2) consists of four yellow petals. The third whorl contains six stamens, and the innermost whorl (whorl 4) contains two fused carpels at the base of the pistil.

whorl (whorl 1) consists of four sepals, which are green and leaf-like. The next whorl (whorl 2) consists of four yellow petals. The third whorl (whorl 3) contains six stamens, or male reproductive organs, and the innermost whorl (whorl 4) contains two fused carpels at the base of the female reproductive structure, the pistil. Mutations in combination with studies of temporal and spatial expression patterns have identified five genes that are involved in the determination of organ identity: *APETALA1* (*AP1*), *APETALA2* (*AP2*), *APETALA3* (*AP3*), *PISTILATA* (*PI*), and *AGAMOUS* (*AG*). Note that *AP1* and *AP2* have both been previously identified as floral identity genes as well. Mutations in the organ-identity genes generally result in the modification, displacement, or total absence of floral organs. In addition, mutations in any one of these genes generally influence the development of two adjacent floral organs.

The influence of organ-identity genes on the development of the *Arabidopsis* flower can best be understood by viewing the floral meristem as three overlapping developmental fields or fields of gene activity; designated A, B, and C. Field A includes the sepals and petals (whorls 1 and 2), field B includes the petals and stamens (whorls 2 and 3), and field C includes the stamens and central carpels (whorls 3 and 4). This view is referred to as the **ABC model** for floral organ specification, in which a particular gene or pair of genes is associated with each developmental field, but controls the identity of two adjacent whorls of organs (Figure 25.3). According to this model, expression of *AP1* alone specifies sepals; *AP1* in combination with *AP3* and *PI* specifies petals; *AP3* and *PI* in combination with *AG* specify stamens; and *AG* alone specifies the carpels. Expression of *AP2* is apparently required throughout the meristem, in part to suppress, in combination with a sixth, unknown gene (*X*), the expression of *AG* in those whorls that are destined to become sepals and petals.

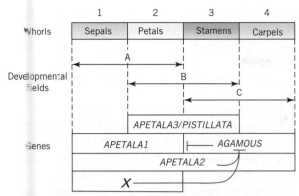

FIGURE 25.3 The ABC model for floral organ specification in *Arabidopsis*. (B) The floral meristem is visualized as being controlled by three developmental fields, identified as A, B, and C. Field A is involved in the specification of whorls 1 and 2. Field B is involved in the specification of whorls 2 and 3. Field C is involved in the specification of whorls 3 and 4. Field A alone specifies sepals, fields A and B together specify petals, fields B and C together specify stamens, and field C alone specifies carpels. Each field is associated with a specific gene or gene pair. *APETALA1* is expressed only in field A, *APETALA3* and *PISTILLATA* are expressed in field B, and *AGAMOUS* is expressed only in field C. *AGAMOUS* also represses (T-bar) the expression of *APETALA1* in field C. *APETALA2* is expressed throughout the meristem and, in conjunction with an unknown gene (*X*), represses the expression of *AGAMOUS* in field A. (After J. D. Bewley et al., 2000.)

The ABC model can be used to either predict or interpret what will happen to organ development in loss-of-function mutants for each of these genes (Figure 25.4). For example, in the *ap3* (or *pi*) mutant, developmental field B would be inoperative and the meristem would be unable to form either petals or stamens. Not only is this the case, but the whorls that would normally be filled with petals and stamens are filled instead with sepal-like and carpel-like structures. Similarly, in the *ag* mutant, stamens and carpels are not formed and all four whorls are filled with sepals or petals. No doubt this model will continue to be refined as new genes are discovered and the various genes and their products are subjected to further molecular analysis.

Although it is becoming more evident which genes are active in controlling which aspects of floral initiation and development, their biochemical and physiological function is not yet clear. However, most of the gene products that have been analyzed thus far share certain characteristics, called *motifs*, which are typical of transcription factors. Transcription factors are DNA-binding proteins that enable RNA polymerases to recognize promoters and begin transcription of the gene in eukaryotes. It thus appears that the principal function of most flowering-time, floral-identity, and organ-identity genes is to regulate other genes that may then direct synthesis of the components that actually make up the individual organs.

25.2 TEMPERATURE CAN ALTER THE FLOWERING RESPONSE TO PHOTOPERIOD

There are many examples of interactions between temperature and photoperiod, particularly with respect to flowering behavior (see Salisbury, 1963, for an extensive listing). In most cases the interaction results in relatively subtle changes in the length of the critical

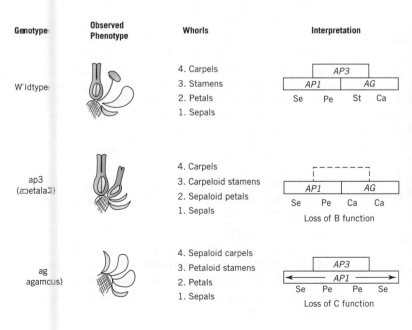

FIGURE 25.4 The ABC model helps to interpret floral-identity in loss-of-function mutants. In wildtype plants, all three developmental fields are active and produce the normal complement of sepals (Se), petals (Pe), stamens (St), and carpels (Ca). The *apetala3* mutant represents a loss of B function. This leaves *AP1* to be expressed alone in whorl 2, replacing the normal petals with sepaloid structures. Also, *AG* is expressed alone in whorl 3, producing carpeloid structures in place of normal stamens. The *agamous* mutant represents a loss of C function, leaving *AP1* to be expressed in all four whorls. The result is petaloid structures in whorl 3 and sepaloid structures in whorl 4.

photoperiod or a tendency toward daylength neutrality or an inability to flower altogether at high or low temperature extremes. There are other plants, however, for which flowering is either quantitatively or qualitatively dependent on exposure to low temperature. This phenomenon is known as **vernalization**. Vernalization is a means of preventing precocious reproductive development late in the growing season, ensuring instead that seed production does not begin until the beginning of the next growing season so that the seed will have sufficient time to reach maturity.

Vernalization refers specifically to the promotion of flowering by a period of low-temperature and should not be confused with other miscellaneous effects of low-temperature on plant development. The term itself is a translation of the Russian *yarovizatsya*; both words combining the root for spring (Russian, *yarov*; Latin, *ver*) with a suffix meaning "to make" or "become." Coined by the Russian T. D. Lysenko in the 1920s, vernalization reflects the ability of a cold treatment to make a winter cereal mimic the behavior of a spring cereal with respect to its flowering behavior. The response had actually been observed many years earlier by agriculturalists, but didn't receive critical attention of the scientific community until J. G. Gassner showed in 1918 that the cold requirement of winter cereals could be satisfied during seed germination. For his part, Lysenko received considerable notoriety for his conviction that the effect was an inheritable conversion of the winter strain to a spring strain. His position—a form of the thoroughly discredited Lamarkian doctrine of inheritance of acquired characteristics—was adopted as Soviet dogma in biology and remained so until the 1950s. The adoption of Lysenko's views as official dogma had a significant impact on Soviet biology and placed agriculture in the USSR at a severe disadvantage for decades.[1]

25.2.1 VERNALIZATION OCCURS MOST COMMONLY IN WINTER ANNUALS AND BIENNIALS

Typical winter annuals are the so-called "winter" cereals (wheat, barley, rye). "Spring" cereals are normally daylength insensitive. They are planted in the spring and come to flower and produce grain before the end of the growing season. Winter strains, however, if planted in the spring would normally fail to flower or produce mature grain within the span of a normal growing

season. Winter cereals are instead planted in the fall. They germinate and over-winter as small seedlings, resume growth in the spring, and are harvested usually about midsummer. The over-wintering cold treatment, or vernalization, renders the plants sensitive to long days.

One of the most thorough studies of vernalization and photoperiodism was carried out on the Petkus cultivar of rye (*Secale cereale*) by F. G. Gregory and O. N. Purvis, beginning in the 1930s. There are two strains of Petkus rye: a spring strain and a winter strain. The spring strain of Petkus rye is a *facultative long-day plant*. Under short days, floral initiation does not occur until after about 22 leaves have been produced, typically requiring a season of about 4.5 months. Under the appropriate long-day regime, however, flowering in the spring strain is initiated after as few as seven leaves have been produced, requiring only about two months. When sown in the spring, the winter strain is insensitive to photoperiod. The winter strain flowers equally slowly—requiring four to five months—regardless of daylength.

If seeds of the winter strain are sown in the fall, however, the germinated seedlings are subjected to an over wintering low-temperature treatment. When they resume growth in the spring, winter strain plants respond as long-day plants in the same way as the spring strain. The effect of the over wintering cold treatment can also be achieved by vernalizing the seed, that is, by holding the *germinated seed* near 1°C for several weeks. Note that the low-temperature treatment, at least in the case of winter annuals, does not alone promote early flower initiation. Rather, the effect of vernalization is to *render the seedling sensitive to photoperiod*.

Another example of vernalization is seen in biennial plants. Biennials are *monocarpic* plants that normally flower (and die) in the second season, again following an over-wintering cold treatment. Typical biennials include many varieties of sugar- and table-beet (*Beta vulgaris*), cabbages and related plants (*Brassica* sp.), carrots (*Daucus carota*) and other members of the family Umbellifereae, foxglove (*Digitalis purpurea*), and some strains of black henbane (*Hyoscyamus niger*). Biennials share with the winter annuals the property that subjecting the growing plant to a cold treatment stimulates a subsequent photoperiodic flowering response.

Biennials typically grow as a rosette, characterized by shortened internodes, in the first season (Figure 25.5). Over winter, the leaves die back but the crown, including the apical meristem, remains protected. New growth the following spring is characterized by extensive stem elongation, called **bolting**, followed by flowering. The cold requirement in biennials is qualitative (i.e., absolute). In the absence of a cold treatment many biennials can be maintained in

---

[1] The story of vernalization is a classic example of what can happen when science becomes enmeshed in political ideology. For the interested student, this unfortunate episode in the history of science has been artfully documented by D. Joravsky in his book *The Lysenko Affair* (Chicago: University of Chicago Press, 1970).

FIGURE 25.5 Vernalization and stem elongation in cabbage. Left: Cabbage plants were vernalized for six weeks at 5°C before being returned to the greenhouse. Center: Plants were sprayed weekly with a solution containing 5 × 10⁻⁷ M gibberellic acid. Right: Control plants grown at normal greenhouse temperature remain in a rosette habit. Except for the vernalization treatment, all plants were maintained in the greenhouse under a long-day (16-hour) photoperiod.

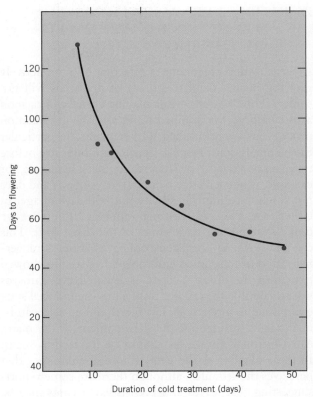

FIGURE 25.6 **Vernalization in Petkus rye (*Secale cereale*). Seeds were germinated in moist sand at 1°C for the time indicated. Cold treatments were scheduled so that all seeds were returned to the greenhouse at the same time. The number of days to flowering progressively decreased with increasing length of the cold treatment. (From Purvis, O. N., F. G. Gregory. 1937. *Annals of Botany*, N.S. 1:569–591. Copyright, The Annals of Botany Company.)**

the nonflowering rosette habit indefinitely. As a rule, vernalized plants, whether winter annuals or biennials, tend to respond as long-day flowering plants, although some biennials are daylength-indifferent following vernalization. One exception to the rule is the perennial *Chrysanthemum morifolium*, a SDP. Some varieties of *Chrysanthemum* require vernalization before responding as a quantitative SDP. As a perennial, *Chrysanthemum* normally requires vernalization on an annual basis. Many other plants such as pea (*Pisum sativum*) and spinach (*Spinacea oleracea*) can be induced to flower earlier with a cold treatment but it is not an absolute requirement.

25.2.2 THE EFFECTIVE TEMPERATURE FOR VERNALIZATION IS VARIABLE

The range of temperatures effective in vernalization varies widely depending on the species and duration of exposure. In Petkus rye, the effective range is −5°C to +15°C, with a broad optimum between +1°C and +7°C. Within these limits, vernalization is proportional to the duration of treatment. Flowering advances sharply after as little as one to two weeks' treatment at 1°C to 2°C and is maximally effective after about seven weeks at that temperature (Figure 25.6). Within the effective range, the temperature optimum is generally higher for shorter treatment periods. Presumably, a longer exposure to lower temperatures within the effective

range is required because the metabolic reactions leading to the vernalized state progress more slowly.

Like flowering, the vernalized state is more or less permanent in most species, giving rise to the concept of an **induced state**. For example, vernalized *Hyoscyamus*, a LDP, can be held under short days for up to 10 months before losing the capacity to respond to long-day treatment. On the other hand, all cold-requiring plants that have been studied are capable of being **devernalized**—vernalization can be reversed if followed immediately by a high-temperature treatment. Flowering in vernalized winter wheat, for example, can be fully nullified if the seedlings are held near 30°C for three to five days. For most plants, then, there is a "neutral" temperature where neither vernalization nor devernalization occurs. For Petkus rye the neutral temperature is about 15°C. Vernalized seeds of Petkus rye can also be devernalized by drying them for several weeks or by maintaining the seeds under anaerobic conditions for a period of time following the cold treatment.

25.2.3 THE VERNALIZATION TREATMENT IS PERCEIVED BY THE SHOOT APEX

A vernalization treatment is effective only on actively growing plants. Cold treatment of dry seeds will not suffice. Thus winter cereals may be vernalized as soon as the embryo has imbibed water and the germination process has been initiated. Other plants, in particular the biennials, must reach a certain minimum size before they can be vernalized. *Hyoscyamus* (black henbane), for example, is not sensitive before 10 days of age and does not reach maximum sensitivity until 30 days of age. In either case, the cold treatment appears to be effective only in the meristematic zones of the shoot apex. This can be shown by localized cooling treatments or vernalization of moistened embryos. Early studies showed that even the cultured apex of isolated rye embryos was susceptible to vernalization. Thus the induced state established in a relatively few meristematic cells can be maintained throughout the development of the plant. Most biennials, however, cannot be induced as seeds. In these plants it is the over-wintering stem apex that perceives the stimulus, although there are some reports suggesting that leaves and even isolated roots may be susceptible in some cases.

25.2.4 THE VERNALIZED STATE IS TRANSMISSIBLE

Experiments with isolated embryos have shown that vernalization treatments are effective only when the embryo is supplied with carbohydrate and oxygen is present, indicating that it is an energy-dependent metabolic process. Still, the nature of the induced state has eluded researchers for many years. To the extent that the meristem itself is the site of perception, any *necessity* for a transmissible hormone appears to be ruled out. A cold-induced, permanent change in the physiological or genetic state of the meristematic cells (referred to as "mitotic memory") would be self-propagating, that is, it could be passed on to daughter cells through cell division. There is some support for this argument. In plants such as Petkus rye and *Chrysanthemum*, only tissue produced in a direct cell line from the induced meristem is vernalized. If the cold treatment is localized to a single apex, it will flower, but all the buds that did not receive the cold treatment will remain vegetative.

In other experiments, especially with *Hyoscyamus*, transmission of the vernalized state across a graft union has been demonstrated. A list of successful experiments has been tabulated by A. Lang in his 1965 review (see Further Reading). These experiments are comparable to the transmission of florigen" across a graft union (Chapter 24) and result in flowering in nonvernalized receptor plants. If a vernalized *Hyoscyamus* plant is grafted to a nonvernalized plant, both will flower under long days. Transmission requires a successful (i.e., living) graft union and appears to be coordinated with the flow of photoassimilate between the donor and receptor.

Experiments such as those described above led G. Melchers to propose the existence of a transmissible vernalization stimulus called **vernalin**. Like florigen, vernalin has resisted all attempts at isolation and remains a hypothetical substance. Unfortunately, the vernalin story is to some extent clouded by interpretation. The grafting experiments all require vernalization followed by long days. They do not clearly distinguish between the transmission of "vernalin" and the possibility that the nonvernalized partner is responding instead to the floral stimulus itself (e.g., FT), which would be transmitted from the vernalized donor under long days.

Adding to the complexity of vernalization is the apparent involvement of gibberellins in the response to low temperature (see Figure 25.5). This was dramatically demonstrated by A. Lang in 1957 when he showed that repeated application of $10 \mu g$ of GA_3 to the apex would stimulate flowering in nonvernalized biennial strains of *Hyoscyamus* and several other biennials maintained under short days. No such promotion occurred in *Xanthium* and other short-day plants treated with gibberellin under noninductive long days. Subsequently it has been shown that gibberellin levels tend to increase in response to low-temperature treatments in several cold-requiring species.

The role of gibberellins is not clear, although in noninduced plants very high concentrations of the gibberellin precursor *ent*-kaurenoic acid accumulate in the shoot apex. This suggests that the cold treatment is required to complete the synthesis of gibberellins in these plants.

25.2.5 GIBBERELLIN AND VERNALIZATION OPERATE THROUGH INDEPENDENT GENETIC PATHWAYS

Results such as those described in the previous section have raised the question: Are vernalin and gibberellin equivalent? The answer, on both physiological and genetic grounds, is no. It is true that gibberellin appears to substitute for the cold requirement of some vernalizable plants and for the long-day requirement in some long-day plants, or, in the case of vernalization, both. But virtually every situation in which gibberellin has successfully substituted for low temperature or long days in promoting flowering involves bolting, or the rapid elongation of stems from the rosette vegetative state. Far less success has been achieved with gibberellins in caulescent long-day plants—those whose stems are already elongated in the vegetative state. Moreover, the developmental pattern in responsive plants differs significantly depending on whether stem elongation is stimulated by low temperature or gibberellin treatment. Following low-temperature treatment, flower buds are evident at the time stem elongation begins. Following gibberellin treatment, on the other hand, the stem first elongates to produce a vegetative shoot. Flower buds do not appear until later. These results suggest independent pathways for vernalization and gibberellins.

Recent genetic studies of flowering have confirmed that gibberellin and vernalization operate via separate genetic pathways (see Figure 25.1). When a triple mutant was constructed containing mutant alleles (*co-2*, *fca-1*, *ga1-3*) that impair each of the long-day, autonomous, and GA-dependent pathways, the mutant plants failed to flower under either long days or short days in controlled environment rooms. After 90 to 100 rosette leaves had been produced without flowering, the plants were then transferred to a long-day greenhouse. After six months, the majority of the mutant plants had died without ever flowering. However, if the triple mutant seedlings were first vernalized at 5°C for 7 weeks, all the plants flowered after approximately 50 leaves had been produced. The most straightforward interpretation of these results would be that (1) the triple mutant has an absolute requirement for vernalization and (2) vernalization promotes flowering through yet another genetic pathway that is separate from the GA-dependent, long-day, and autonomous pathways.

Both the autonomous pathway and vernalization reduce the expression of the gene *FLOWERING LOCUS C* (*FLC*). The product of this gene is a transcription factor that represses flowering. However, when the wild-type *FLC* gene is absent, control by the autonomous pathway is also eliminated but the effect of vernalization is not. Thus it appears that the autonomous pathway acts solely through controlling *FLC* expression, but vernalization is able to promote flowering through two pathways: either through suppressing *FLC* expression or through some yet-to-be-discovered *FLC*-independent mechanism.

25.2.6 THREEE GENES DETERMINE THE VERNALIZATION REQUIREMENT IN CEREALS

This brings us to the question of whether vernalin, like florigen, is a hormone. Or a better question to ask might be: what is the molecular basis for vernalization? It has actually been known for more than 30 years, primarily through plant breeding experiments, that three genes (*VERNALIZATION 1, 2*, and *3*, or *VRN1*, *VRN2*, and *VRN3*) have a major role in determining the vernalization requirements in cereal grains. These genes have now been isolated and characterized.

The *VRN1* gene encodes a transcription factor and is induced by a vernalization treatment. Furthermore,

there is a quantitative relationship between the amount of *VRN1* expressed in a vernalized plant and the amount by which flowering time is reduced under long days. *VRN2* represses flowering under long days by blocking the expression of the floral stimulus *FT*. On the other hand, *VRN2* is itself repressed by *VRN1*. Varieties of winter cereals that lack a functional copy of *VRN2* respond normally to long days without requiring a prior cold treatment. *VRN3* is the cereal equivalent (called an ortholog) of *FLOWERING LOCUS T* (*FT*) in *Arabidopsis*.

A model to illustrate how these three genes interact to control flowering in winter wheat and barley is presented in Figure 25.7. Prior to receiving a cold treatment, the winter cereals are unable to respond to long days because *FT* (or *VRN3*) expression is repressed by the presence of *VRN2*. When the seeds are sown in late summer or early autumn, the shoot apex develops vegetatively until winter, when *VRN1* expression is promoted. In the spring, growth is renewed; the low-temperature-induced expression of *VRN1* remains high; and *VRN1* suppresses any further expression of

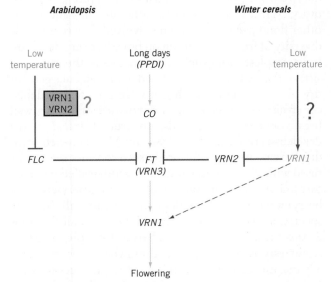

FIGURE 25.7 **A model comparing the regulation of flowering by vernalization in winter cereals and *Arabidopsis*. Long days (mediated by *PHOTOPERIOD 1* (*PPD1*) in cereals) are sensed by the *CONSTANS* gene (*CO*) which activates *FLOWERING LOCUS T* (*FT*) expression. *VRN2* prevents floral induction before winter by repressing *FT* expression. A low-temperature vernalization treatment induces expression of *VRN1*, which represses *VRN2* and allows expression of *FT* under long days. How the low temperature induces *VRN1* expression isn't known. In cereals, *VRN1* also acts as a floral meristem identity gene. In *Arabidopsis*, *FT* expression is repressed by *FLOWERING LOCUS C* (*FLC*). Flowering proceeds under long days following vernalization because the low temperature represses *FLC* expression. (Based on Trevaskis et al. 2007. *Trends in Plant Science*.)**

VRN2. FT is then allowed to be expressed and the plant now responds to long days by initiating floral primordia. *VRN2* is referred to as a floral integrator gene because, as shown in the model, its central role is to integrate the low temperature and photoperiod responses.

Flowering in the dicot *Arabidopsis* is also accelerated by vernalization and the molecular mechanism is very similar to that in the cereals. The principal difference is that in *Arabidopsis* the gene *FLOWERING LOCUS C* (*FLC*), not *VRN2*, is most directly responsible for suppressing *FT* expression and maintaining the vegetative state of the apex. The repression of *FLC*, in turn, appears to be regulated by low temperature through a complex involving *VRN1* and *VRN2* (Figure 25.7).

25.3 FRUIT SET AND DEVELOPMENT IS REGULATED BY HORMONES

The **fruit** is the final stage in the growth of the reproductive organ. Botanically a fruit is a mature or ripened ovary wall and its contents, although in some plants other floral parts may become involved. There is a wide diversity of fruits, depending on how the ovary develops. In its simplest form, such as peas or beans, the fruit consists of the seed or seeds enclosed within an enlarged but dry ovary (the pod). Such fruits are classified as dehiscent fruits—dehiscent because at maturity the ovary wall breaks open to free the seeds. The fruit of *Arabidopsis* is a dry dehiscent fruit. Maize (*Zea mays*) is a non-dehiscent dry fruit consisting of a single seed with its seed coats fused with the dry ovary wall (a structure called the pericarp). Tomato is an example of a fleshy fruit (actually a berry) with an enlarged, fleshy inner fruit wall. In some species, a structure other than the ovary wall develops as the fruit. These are called **pseudocarpic fruits**. A strawberry is one example. A strawberry "fruit" actually consists of a number of individual one-seeded fruits (called achenes) borne on the surface of an enlarged, fleshy receptacle. In many cases, it is clear that the fruit undergoes considerable cell division and cell enlargement as well as significant qualitative changes. These changes are due largely to changes in hormone content.

Fruit development, maturation, and ripening have been widely studied because of their biological significance—fruits protect the developing seed and serve as a vehicle for dispersal of the mature seed—as well as their practical importance as a significant component in human nutrition. The development, maturation, and ripening of fleshy fruits have received the bulk of the attention over the years because of problems associated with transportation, storage, and other aspects of post-harvest physiology.

25.3.1 THE DEVELOPMENT OF FLESHY FRUITS CAN BE DIVIDED INTO FIVE PHASES

Tomato (*Lycopersicon esculentum*) has become a popular model in which to study fleshy fruit development, in part because there are numerous mutants available and the plant is easily transformed. With tomato as a model, the life history of a fruit can be divided into five more-or-less distinct phases. Phase I involves the development of the ovary in preparation for fertilization and seed development and ends with the decision to either abort further development or to proceed with further cell division and cell enlargement in the ovary walls. This decision to proceed with ovary development is generally referred to as **fruit set**. In phase II, or the initial phase of fruit development, growth of the nascent fruit is due primarily to cell division in the ovary walls. The cells thus become small and dense, with very small vacuoles. During phase III, cell division effectively ceases and further growth of the fruit is mostly by cell enlargement. Once the fruit has reached its final size, it enters phase IV, or a period of ripening. In a fleshy fruit like tomato, ripening involves the development of color and flavor constituents (e.g., carotenoids, sugars, and acids) and a softening of the tissue that render the fruit attractive to animals. Tissue softening is due primarily to increased activity of enzymes such as **polygalacturonase (PG)**. PG degrades the pectic substances that are found in the middle lamella and which are responsible for cell-to-cell adhesion. Finally, in phase V, senescence sets in and the fruit begins the decay process.

As might be expected, all of the plant hormones are active at various stages during fruit development (Figure 25.8). During seed development and first and second phases of fruit development, auxins, cytokinins, and gibberellins are all present and active. Cytokinin level peaks during phase II, the period of most active cell division. Auxin level peaks in early phase III, coinciding with the initiation of cell enlargement, and then declines as the fruit reaches mature size. A second surge in auxin level occurs in the early stages of ripening, along with the appearance of significant levels of ethylene. The role of gibberellins is not well understood, but they are probably involved with cytokinins in initiating cell division and with auxin in maintaining cell enlargement. Tomato is a climacteric fruit and the burst in respiration is related to the appearance of ethylene and the qualitative changes in the fruit that represent ripening (see Chapter 21 for a discussion of ethylene).

25.3.2 FRUIT SET IS TRIGGERED BY AUXIN

Normally, successful pollination and fertilization of the ovule by sperm nuclei are required for fruit set to occur.

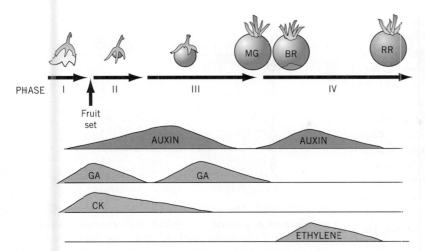

FIGURE 25.8 **Hormonal changes during the first four phases of fruit development in a tomato (*Lysopericon esculentum*). The graphs for auxin, gibberellin (GA), cytokinins (CK), and ethylene show the approximate point in development when hormone concentrations peak. Mature green (MG). Breaker (BR) refers to the stage when the first signs of color appear. Breaker marks the beginning of the ripening phase. The red ripe stage (RR) marks the end of ripening and the beginning of senescence.**

In the absence of fertilization, flowers fail to produce seeds and, in most cases, the floral parts senesce without forming fruits. The commitment to fruit set therefore appears to rely on positive growth signals generated prior to or immediately following fertilization. There are some cases, however, in which unfertilized flowers go on to produce normal, but seedless, fruits. The phenomenon of fruit development in the absence of pollination and fertilization is known as virgin fruiting or **parthenocarpy** (from *parthenos*, meaning virgin). Parthenocarpy may occur naturally due to lack of pollination or pollination that does not result in fertilization, or from pollination followed by embryo abortion. Common examples of naturally parthenocarpic fruits are bananas, navel oranges, most varieties of figs, pineapple, and some seedless grapes. Studies involving either natural or induced parthenocarpy have had a major role in helping to understand the initial phases of normal fruit development, especially fruit set.

In the mid-1930s, it was found that pollen and extracts of pollen were both a rich source of auxin and could stimulate fruit set in unpollenated plants of the family Solanaceae (e.g., tomato, peppers). Auxin induces parthenocarpy in a small number of plants in other families as well, particularly in the Cucurbitae (cucumber, pumpkin), and Citrus. Parthenocarpy is a biological curiosity—of what value is a seedless fruit—but it has significant economic implications as well. In California, for example, synthetic auxins are used to stimulate fruit set in early tomato crops, when cool night temperatures would otherwise tend to reduce fruit set. Many other fruits such as citrus and watermelons are more salable in the seedless form. Gibberellins, either alone or in combination, may also induce parthenocarpy in species such as pear and citrus.

Other than pollen, at least one source of auxin for fruit initiation is the seed. This was demonstrated in a classic study of strawberry (*Fragaria ananassa*) conducted by J. P. Nitsch in the 1950s. Since the fruit of a strawberry is the floral receptacle and the achenes are borne on the surface, it is a simple task to remove the seeds without damaging the underlying receptacle. Nitsch found that removal of the seeds prevented further development of the fruit, but supplying the fruit with auxin restored normal development. Furthermore, if some seeds were removed and others left in place, the fruit would develop only in those regions where the seeds were undisturbed (Figure 25.9). Other studies confirmed that strawberry achenes were a good source of auxin.

Molecular approaches have followed up on the classical physiological studies and confirmed that auxin from the developing ovule has a central role in fruit set and early fruit development. In general, these approaches have taken advantage of artificially induced parthenocarpic fruit development, either by (1) increasing the synthesis of auxin in the ovule or ovary, or (2) by manipulating auxin regulatory proteins such as AUX/IAA or auxin response factors (ARF). (Refer to Chapter 18 for the role of AUX/IAA and auxin response factors in auxin-regulated responses.)

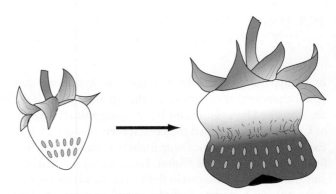

FIGURE 25.9 **Nitsch's experiment showing the effect of seed (achenes) removal on the development of strawberry fruit. All but two rows of seeds were removed from the receptacle during early development of the fruit. Subsequent fruit development was limited to the region where the seeds were left in place.**

One approach, for example, involves the *Agrobacterium tumifaciens* tumor-inducing genes *iaaM* and *iaaH*. These are the genes that encode IAA biosynthesis and cause the overproduction of IAA when *Agrobacterium* invades host cells. *IaaM* and *iaaH* can be used to construct a transgene by linking them to a promoter that restricts their expression specifically to the ovule. When this transgene construct is inserted into eggplant, tobacco, tomato, strawberry, and raspberry plants, the constitutive production of IAA induces parthenocarpic fruit development. Another gene that induces parthenocarpy is the *rolB* gene, isolated from another Agrobacterium species (*Agrobacterium rhizogenes*). The *rolB* gene, again linked to a promoter specific for the ovary and young fruit, induced parthenocarpy when it was inserted into tomato. The mechanism of *rolB* action is not known, but it appears to increase the tissue sensitivity to auxin.

In *Arabidopsis*, loss-of-function mutants for the auxin response factor gene *ARF8* induce parthenocarpic fruit development. ARF8 is a transcription factor that is expressed primarily in the ovule and surrounding tissues. Further experiments indicated that ARF8 acts as a negative regulator for fruit set, i.e., it inhibits fruit set, probably in combination with one or more AUX/IAA proteins. Expression of the *ARF8* gene is also switched off after fertilization.

Most of the data for the role of auxin in fruit set is consistent with the following model. Prior to pollination and fertilization, further development of the ovary is blocked by the presence of AUX/IAA and/or ARF repressor proteins such as ARF8. Pollination (and fertilization of the ovules) induces an increase in the level of auxin, thus leading to the auxin-mediated degradation of the AUX/IAA proteins via the 26S-proteasome pathway. This frees up the auxin-response factors to regulate auxin-responsive genes that are necessary for fruit set and subsequent fruit development.

25.3.3 RIPENING IS TRIGGERED BY ETHYLENE IN CLIMACTERIC FRUITS

In many, but not all, fleshy fruits the metabolic and visual changes that occur during the ripening process are accompanied by a significant burst in respiratory activity or CO_2 evolution, called the **climacteric** (Figure 25.10). Examples of climacteric fruits include tomato, cucurbits (cucumber and related fruits), banana, apple, peaches, and plums. Nonclimacteric fleshy fruits, which do not show the CO_2 burst, include strawberry, grape, citrus, and all nonfleshy, or dry fruits such as *Arabidopsis* or maize. The ripening process in climacteric fruits has attracted a lot of research over the years because of its economic importance and because just prior to the respiratory burst there is a significant increase in

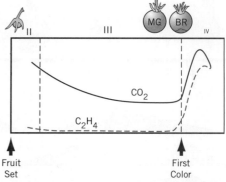

FIGURE 25.10 The pattern of CO_2 and ethylene (C_2H_4) production in tomato (*Lycopersicon esculentum*), a climacteric fruit. CO2 evolution is high during the cell division phase and steadily declines during the cell expansion and maturation phase. It then rises sharply at the breaker stage which identifies the beginning of the ripening phase. The CO2 "burst" is immediately preceded by a significant increase in the production of ethylene. (After McGlasson, W. D. 1978. In: H.O. Hultin, M. Milner (eds.), *Postharvest Biology and Biotechnology*. Westport, CT: Food and Nutrition Press.)

the production of ethylene (Figure 25.10). Moreover, ethylene synthesis is also auto-catalytic. Once ethylene production begins in one fruit, its production is stimulated in surrounding fruit—hence the old axiom that one rotten apple spoils the barrel. The role of ethylene in fruit ripening has assumed significant commercial importance. For example, tomatoes, bananas, and other climacteric fruits that have to be shipped any distance are commonly picked at the mature green stage and then ripened at their destination by gassing with ethylene (Box 25.1).

The rate limiting steps in the biosynthesis of ethylene are catalyzed by the enzymes ACC synthase (ACS) and ACC oxidase (ACO) (see Chapter 21). In tomato there are two ACS genes that are expressed in the fruit and appear to be responsible for triggering ripening. Both genes, *LeACS1A* and *LeACS4*, are under developmental control and are induced at the onset of ripening.[2] Furthermore, the induction of both genes is impaired by mutations at either the *ripening inhibitor* (*rin*) or the *nonripening* (*nor*) locus. In other words, fruits of tomato plants carrying the *rin* and *nor* mutations do not produce ethylene, do not exhibit a climacteric CO_2 burst, and do not ripen. The expression of the gene *LeACS4* is also controlled by ethylene itself and thus appears to be responsible for regulating the autocatalytic production of ethylene by a positive feedback system.

A large number of ethylene signaling components have been identified in both *Arabidopsis* and tomato,

---

[2]In gene designations such as *LeACS*, the Le identifies genes isolated specifically from tomato, *Lycopersicon esculentum*.

BOX 25.1
ETHYLENE: IT'S A GAS!

Ethylene has been unwittingly used by humans for centuries to stimulate fruit ripening. In ancient China, harvested fruit was commonly ripened in rooms where incense was being burned and in California in the early twentieth century kerosene stoves were used to ripen lemons. These practices worked because ethylene is a common combustion product. For the same reason, ethylene can still occasionally cause ethylene-mediated growth problems for modern-day greenhouse operations that heat with natural gas.

Once the role of ethylene in ripening became known, however, attention turned to methods for controlling fruit ripening by preventing or delaying the effects of ethylene production in climacteric fruits. Apples, for example, are harvested in the fall but are generally available in the markets throughout the year because of controlled atmospheric storage. Controlled atmospheric storage employs a combination of low temperature, low ambient oxygen, and high ambient CO_2. Oxygen levels are generally reduced to 1 percent to 3 percent from the normal 21 percent. CO_2 levels may be increased up to 8 percent from the normal 0.035 percent. These conditions lower both ethylene production and the respiration rate of the fruit, and thus modulate, if not prevent, the ethylene-stimulated respiratory climacteric.

including some ethylene receptors that are present only in the fruit and are strongly induced during the ripening process. The challenge now is to understand how these many components interact to form a coherent signal transduction chain that regulates a multitude of fruit-ripening genes.

SUMMARY

Flower initiation and development involves the action of at least three sets of genes. These genes regulate flowering-time (including the circadian clock), floral-meristem identity, and floral-organ identity. Flowering-time mutants may flower either later than or earlier than the wildtype. Flowering time in *Arabidopsis* is under the control of at least four different genetic pathways. The long-day pathway genes constitute the photoperiodic control system. The autonomous pathway monitors the state of vegetative development and initiates flowering in response to endogenous signals. The GA pathway involves a single gene, mediated by gibberellin. The fourth pathway involves a set of genes that normally repress flowering. Mutants in the repressor pathway flower early.

Mutations in floral-identity genes disrupt the transition to a floral meristem, while organ-identity genes control the initiation of sepals, petals, stamens, and carpels. Some floral-identity genes and organ-identity genes overlap in time of expression and function. The ABC model for floral organ specification views the meristem as three overlapping developmental fields and identifies the genes and combination of genes that specify specific floral organs. Most of the flowering-time, floral-identity, and organ-identity genes encode products that have some characteristics of transcription factors.

Vernalization is the promotion of flowering by a period of low temperature. In the case of winter annuals, such as cereals, vernalization changes the photoperiodic behavior from daylength indifference to a quantitative long-day response. Biennials typically grow as a rosette until vernalized. The flowering stem then bolts (elongates) and responds as a long-day plant. A temperature of approximately 0°C to 5°C, applied to the actively growing apex of the plant for several weeks, is required for vernalization to be effective. Temperature in vernalizable plants is perceived in the stem apex and is transmissible. The use of flowering-time mutant has shown that vernalization operates independently of the long-day, autonomous, and gibberellin-dependent genetic pathways.

Fruits are classified as dry or fleshy, depending on the extent of ovary development following fertilization. The development of fleshy fruits such as tomato and apple can be divided into five phases. Phases II through IV represent the progression from fruit set to the fully ripe fruit. Auxin is responsible for fruit set, while auxins, cytokinins, and gibberellins are involved at various times until the fruit reaches maturity. Fruit development is normally initiated by a release of auxin associated with pollination and/or fertilization of the ovule. Parthenocarpy is the development of seedless fruits in the absence of pollination or fertilization. Parthenocarpy can be induced in some fruits by the application of auxin. Phase IV, or ripening, is characterized by the development of color and flavor constituents and softening of the fruit due to a breakdown in the pectic substances that are involve in cell-to-cell adhesion. Ripening in some fleshy fruits is accompanied by a marked increase in respiratory rate and CO_2 release (the respiratory climacteric) triggered by ethylene. Ethylene release is catalyzed at least in part by ACC synthase and ACC oxidase, two rate limiting enzymes that are developmentally controlled. Ethylene initiates a signal transduction chain that turns on ripening genes.

CHAPTER REVIEW

1. How are events in the leaf connected with the conversion of a vegetative meristem to a floral meristem?

2. Distinguish between flowering-time genes, floral-identity genes, and organ-identity genes.

3. What would be the expected observed phenotype of an *Arabidopsis* flower that carries a loss-of-function mutation for the *APETALA1* gene?

4. Most plants have several pathways that control flowering-time. What might be the advantage(s) of multiple pathways?

5. How does the concept of induction apply to vernalization?

6. Gibberellin often appears to substitute for vernalization. What is the evidence that vernalization and gibberellin operate through independent genetic pathways?

7. Compare the roles of *FLOWERING LOCUS C* and *VERNALIZATION 2* in controlling flowering.

8. What is "fruit set"? What do you understand the meaning of "decision" to be as it is used with respect to fruit set?

9. What is the "climacteric"?

10. How does controlled atmospheric storage enhance the storage life of climacteric fruit?

FURTHER READING

Adams-Phillips, L., B. Cornelius, J. Giovannoni. 2004. Signal transduction systems regulating fruit ripening. *Trends in Plant Science* 9:331–338.

Bewley, J. D. et al. 2000. Reproductive development. In: Buchanan, B., W. Gruissem, R. Jones (eds.), *Biochemistry and Molecular Biology of Plants*. Rockville, MD: American Society of Plant Physiologists. (An excellent review of reproductive development.)

Coen, E. S., E. M. Meryerowitz. 1991. The war of the whorls: Genetic interactions controlling flower development. *Nature* 353:31–37.

Dennis, E. S., W. J. Peacock. 2007. Epigenetic regulation of flowering. *Current Opinion in Plant Biology* 10:520–527

Gillaspy, G., H. Ben-David, W. Gruissem. 1993. Fruits: A developmental perspective. *The Plant Cell* 5:1439–145 .

Giovannoni, J. 2004. Genetic regulation of fruit development and ripening. *Plant Cell* 16:S170–S180.

Lang, A. 1965. Physiology of flower initiation. In: W. Ruhland (ed.), *Handbuch der Pflanzenphysiolgie (Encyclopedia of Plant Physiology)*. Berlin: Springer-Verlag, XV (1) 1380–1536.

Pandolfini, T., B. Molesini, A. Spena. 2007. Molecular dissection of the role of auxin in fruit iniitation. *Trends in Plant Science* 12:327–329.

Salisbury, F. B. 1963. *The Flowering Process*. Oxford: Pergamon Press.

Trevaskis, B., M. N. Hemming, E. S. Dennis, W. J. Peacock. 2007. The molecular basis of vernalization-induced flowering in cereals. *Trends in Plant Science* 12:352–357.

Trevaskis, B., M. N. Hemming, W. J. Peacock, E. S. Dennis. 2006. HvVRN2 responds to daylength, whereas HvVRN1 is regulated by vernalization and developmental status. *Plant Physiolgy* 140:1397–1405.

Yanofsky, M. F. 1995. Floral meristems to floral organs: Genes controlling early events in Arabidopsis development. *Annual Review of Plant Physiology and Plant Molecular Biology* 46:167–188. (Currently titled *Annual Review of Plant Biology*.)

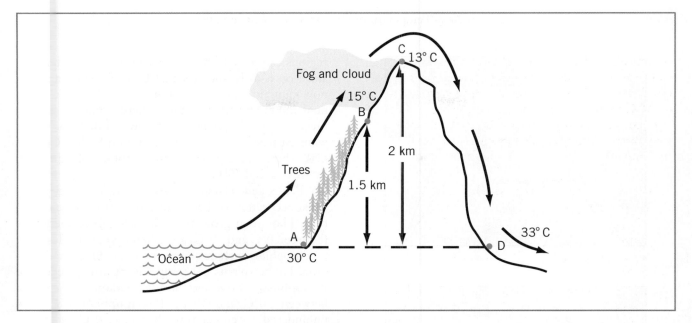

26

Temperature: Plant Development and Distribution

Plants are chemical machines and one universal characteristic of chemical machines is their sensitivity to temperature. Temperature, along with light and water, is one of the most critical factors in the physical environment of plants. This is especially so because plants, unlike homeothermic animals, are not able to maintain their tissues at a constant temperature. Environmental temperature therefore exerts a profound influence on cellular metabolism and, as a result, plant growth and their geographic distribution.

All of the chemical machinery of nature—every individual enzymatic reaction, every metabolic function, every physiological process—has temperature limits above and below which it cannot function and an optimum temperature range where it proceeds at a maximum rate. Temperature also affects the integrity of cell structure (especially the structure and properties of membranes), limits the distribution of species in space and time, and influences the direction of specific developmental events.

Temperature as an environmental stress and flowering regulator has been discussed in earlier chapters. In this chapter we will introduce some of the other ways in which temperature is known to influence plant growth, development, and distribution. Specific topics include:

- the role of temperature in perennial plants, in particular its role in bud and seed dormancy, and
- some examples of how temperature influences the geographic distribution of plants.

26.1 TEMPERATURE IN THE PLANT ENVIRONMENT

Of all the planets, the thermal environment on earth is particularly fit to give rise to and sustain life. This is because life functions in an aqueous medium and the range of temperatures encountered over most of the earth's surface generally ensures that sufficient water is maintained in the liquid state. The temperature at which biological processes can occur is generally limited by the freezing point of water on the low side and the irreversible denaturation of proteins on the high side. Between these two extremes, a plot of growth versus temperature for individual organisms assumes the shape of an asymmetric bell curve, similar to that for individual enzyme reactions or multiple enzyme-catalyzed metabolic sequences (Figure 26.1). In fact, the temperature curve for growth of an

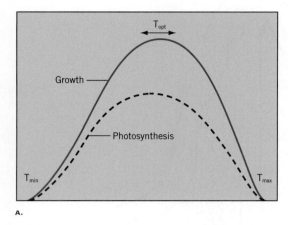

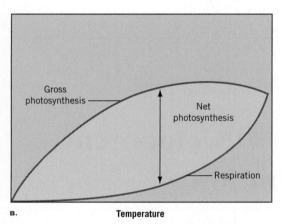

FIGURE 26.1 **Temperature and plant growth. (*A*) A schematic illustration of the three cardinal temperatures for plant growth. Typically, the pattern of the growth curve reflects the pattern of temperature effects on net photosynthesis. (*B*) Net photosynthesis is the difference between gross carbon uptake by photosynthesis and carbon evolution by respiration.**

organism effectively represents a composite of the temperature curves for photosynthesis, respiration, and other critical metabolic processes. Growth curves thus exhibit, just as do individual metabolic and enzyme reactions, the three **cardinal temperatures** (minimum, optimum, and maximum). Just as the actual values of cardinal temperatures vary between different metabolic processes, the actual values of cardinal temperatures for growth curves will vary from species to species. Assuming other factors are not limiting, these cardinal temperatures generally define the temperature range over which growth is possible. It is close to the extremes of this range that plants experience temeprature stress as described in earlier chapters.

Green plants probably first evolved in the tropical regions, not so much because of warmer temperatures (although that may have been a factor), but because the temperatures there were relatively stable. With time, plants gradually migrated into the temperate and polar regions as they adapted to wider variations in temperature on a daily and seasonal basis. Green plants are now found in regions as extreme as the Antarctic continent and the northern tundra, where temperatures over much of the year are near or below freezing, and in the warmest places on earth such as Death Valley (California), where summer temperatures commonly approach or even exceed 50°C.

Plants and related organisms may be broadly classified according to their ability to withstand temperature. Those that grow optimally at lower temperatures (between 0°C and 10°C) are called **psychrophiles**. The psychrophiles include primarily algae, fungi, and bacteria. Higher plants generally fall into the category of **mesophiles**, whose optimum temperatures lie roughly between 10°C and 30°C. **Thermophiles** will grow unhindered at temperatures between 30°C and 65°C, although there are reports of cyanobacteria growing at temperatures as high as 85°C. These temperature ranges apply to hydrated, actively growing organisms. Dehydrated organisms and organs, such as resurrection plants (*Selaginella lepidophylla*) and dry seeds with moisture contents as low as 5 percent, are able to withstand a much broader range of temperatures for extended periods of time.

Plants in nature are subjected to a complex mosaic of fluctuating air and soil temperature regimes such that it is very difficult to study the effects of temperature in a natural setting. Air temperature, for example, fluctuates widely, and often rapidly, depending on the time of day, cloud cover, season, and other factors. Soil is a major heat sink as it absorbs and stores solar energy during the day. At night, some of this heat is radiated back into the atmosphere, which both cools the soil and warms the surface. Soil temperature also varies with the soil structure, organic content, and other physical characteristics as well as slope and aspect (the direction it faces with respect to the sun).

Both air and soil temperatures have an impact on plant growth. Air temperature influences leaf temperature and therefore the rates of photosynthesis, respiration, and other metabolic reactions. On the other hand, soil temperature influences germination, root development, and nutrient uptake. For example, maize seeds will not germinate below about 10°C and the time required for germination of winter wheat increases linearly with a decrease in soil temperature below 25°C. Several investigators have shown that the uptake of nutrients such as calcium, boron, nitrogen, and phosphorous increases with increasing temperature. In cold soils, soybean roots spread out closer to the soil surface, while in warmer soils the roots grow more deeply. The stem apex (or, crown) of many grasses is situated at or near the soil surface, which means that leaf development in grasses is probably influenced more by soil temperature than by air temperature. The study of soil temperature and its influence on growth and development is of more than passing interest because soil temperature can be a significant factor in determining agricultural yield.

While the temperature in tropical climes is relatively stable, plants growing in temperate regions and closer to the poles are subject to more or less predictable variations in temperature on a daily and seasonal basis. It is perhaps not too surprising that plants have evolved ways to incorporate this information in their developmental and survival strategies. Plants use this information to ensure dormancy of buds, tubers, and seeds, and to modify their flowering behavior, all of which appear to be keys to survival over periods unfavorable to normal growth and development.

26.2 BUD DORMANCY

One aspect of development that is strongly influenced by temperature is **dormancy**. Dormancy is a term that is applied to tissues such as buds, seeds, tubers, and corms that fail to grow even though they are provided with adequate moisture and oxygen at an appropriate temperature. Dormancy is a difficult process to study because it is a progressive process, often occurring in degrees. Buds that are just entering dormancy, for example, may be stimulated to renew growth rather easily. On the other hand, buds that have developed full dormancy may require prolonged or severe treatment to break dormancy and renew growth. Dormancy in some organs may be enforced by other organs in the plant or by external factors. As a result, dormancy terminology has been quite inconsistent and can be confusing. Terms such as quiescence, rest, true dormancy, and imposed or enforced dormancy have been used by different authorities to describe various states or conditions. However, it is most convenient to group dormancy mechanisms into one of three types: **paradormancy**, in which the inhibition of growth arises from another part of the plant (e.g., apical dominance); **ecodormancy**, in which growth is imposed by limitations in the environment (e.g., lack of water); and **endodormancy**, in which the dormancy is an inherent property of the dormant structure itself (see Lang, 1987).

Bud dormancy is an example of endodormancy. A bud is a shortened, very compact terminal or axillary shoot axis in which the internodes have failed to elongate and leaves or floral parts have not enlarged. The whole is enclosed in a set of modified leaves, called **bud scales** (Figure 26.2). Bud scales serve a protective function, both insulating the bud and preventing desiccation. Many popular fall flowering bulbs are also large buds (Box 26.1).

The growth of deciduous perennial plants typical of temperate regions, where warm summers alternate with cold winters, is cyclical; such plants normally undergo

FIGURE 26.2 An axillary bud of sweet bay (*Laurus noblis*). Note the prominent bud scales that enclose and protect the bud.

a cessation of active growth on an annual basis. Both vegetative and floral buds formed in the latter part of the summer or early fall will survive over the winter in a dormant condition and do not normally renew their growth until more favorable temperature conditions return in the spring. However, it is not the low temperature itself that inhibits bud growth. Dormancy mechanisms are well under way before the cold temperatures arrive and, similar to vernalization, a period of low temperature is required before bud regrowth can occur in the spring. Just as vernalization prevents precocious flowering in the fall, bud dormancy prevents precocious bud growth, ensuring that the meristems enclosed within the bud are able to survive the adverse conditions of winter.

In order to ensure survival of the plant, dormancy mechanisms must be in place *before* the arrival of unfavorable conditions. This means that the plant must be able to *anticipate* climatic change. Mechanisms must also be in place to ensure that the buds do not break dormancy until such time that environmental conditions are appropriate to sustain normal growth and development. Premature breaking of buds during an unseasonably warm period in the winter, for example, could have serious consequences for the survival of the plant. In short, dormancy is a precisely regulated requirement for the perennial habit, cued by factors in the environment and maintained, and ultimately broken, by specific metabolic changes in the organism.

Dormancy studies have focused on three principal questions. (1) What are the environmental signals that stimulate the onset of dormancy and how are they

Box 26.1
BULBS AND CORMS

Terminal and axillary buds borne on the aerial portion of plants are not the only buds that are subject to dormancy mechanisms. A large number of plants have fleshy underground storage organs capable of carrying the plants through seasonal cold periods. Popularly referred to as bulbs, they are more accurately defined as bulbs, tubers, or corms. The true bulb (e.g., lily, hyacinth, tulip, daffodil, and onion)

is a large bud, consisting of a small, conical stem with numerous leaves surrounding one or more central meristems (Figure 26.3). The leaves are modified for food storage. In an onion, for example, these leaves are the portion of the bulb that we eat. Corms, represented by plants such as crocus and gladiolus, are solid shortened stems with numerous buds systematically arranged under a protective covering of paper-thin leaves.

The most popular spring-flowering bulbs and corms, including ornamental onion (*Allium* sps.) are generally referred to as "fall bulbs" because they are planted in the fall after entering a period of dormancy and require an extended cold period before the buds break dormancy and flower in the spring.

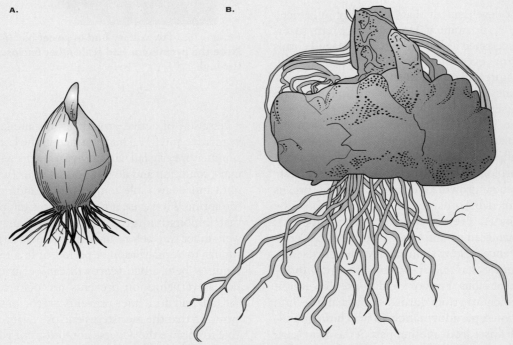

FIGURE 26.3 **Examples of fall bulbs and corms. (A) Tulip; a true bulb. The bud is enclosed by layers of clasping, fleshy storage leaves, similar to an onion. (B) Crocus; a corm. The buds are borne on a flattened stem. Both go dormant in late summer to early fall. Dormancy is broken by planting the bulbs or corms in the fall and subjecting them to cool winter ground temperatures.**

perceived? (2) What metabolic changes are responsible for the reduced activity? (3) What signals the startup of renewed growth at the appropriate time?

26.2.1 BUD DORMANCY IS INDUCED BY PHOTOPERIOD

Plants anticipate seasonal change by monitoring photoperiod. The onset of dormancy in buds is a typical short-day response, coincident with leaf fall, decreased

cambial activity, and increased capacity to withstand low temperature, or cold hardiness In temperate woody species, the short days and decreasing temperatures of late summer and autumn induce the leaf primordia to form bud scales instead of leaves. The formation of scales is followed by the induction of cold hardiness and the cessation of cell division in the meristem. Once growth has ceased and meristem has entered dormancy, the meristem becomes insensitive to any growth-promoting signals.

Like flowering, the short-day photoperiod signal that initiates the onset of dormancy is perceived in the leaves. It should not be surprising, then, that the same players that detect photoperiod for the control of flowering are also involved in controlling dormancy. For example, when the genes *CONSTANS* (*CO*) and *FLOWERING LOCUS T* (*FT*) were over-expressed in transgenic poplar trees, bud growth did not stop following exposure to short days. On the other hand, down-regulation of *FT* triggers the onset of dormancy. Thus it appears that the combination of *CO* and *FT* represents a universal photoperiodic signal module. It is not known, however, which gene(s) or gene product(s) FT interacts with to trigger bud dormancy.

Relatively little is also known about the physiological state of dormant buds except that during their formation the bud primordia accumulate storage materials such as starches, fats, and proteins. Dormant buds are further characterized by low respiratory activity, a significant loss of water, and the inability to grow even if temperature, oxygen, and water supply are adequate. There have been reports that the endogenous gibberellin levels decline at the onset of dormancy and that some buds may be released from dormancy if treated with gibberellin or cytokinins, but, otherwise, little is known about the hormonal status of dormant buds.

26.2.2 A PERIOD OF LOW TEMPERATURE IS REQUIRED TO BREAK BUD DORMANCY

Although induction of bud dormancy is coincident with decreasing temperature and short days, the principal role of temperature appears to be in breaking dormancy. Most dormant buds have a chilling requirement that must be met before the cells are capable of renewed cell division and enlargement. Studies on the chilling requirement for breaking dormancy have concentrated on commercial fruit species and deciduous ornamentals. This is because in the northern hemisphere, the chilling requirement largely determines the southerly limits of cultivation for these plants. The process is especially critical in fruit trees since the flower bud that bears fruit are initiated in the previous summer. The bud then over-winters and, having satisfied its chilling requirement, floral development continues the following spring. Apple trees, for example, will not bear fruit without a cold winter.

Temperatures near or just above freezing appear to be most effective at breaking dormancy. The amount of chilling required varies with species, cultivar, and even location of the buds on the trees. Species such as apple (*Malus pumila*), pear (*Pyrus communis*), and cherry (*Prunus* sps.) require approximately 7 to 9 weeks of exposure to temperatures below 7°C in order to overcome dormancy. Others may require up to 22 weeks

(American plum, *Prunus americana*) or as few as four to six weeks (apricot, *Prunus armeniaca*). Persimmons (*Diospyros kaki*) require only four days of low temperature and so can be grown successfully much further south than other fruit trees. The temperature in temperate regions often varies widely throughout the winter, but this generally poses no problem for dormant tissues. In most cases, buds and other dormant tissues are able to sum the periods of cold and will not renew growth until the appropriate amount of cold treatment has been accumulated.

There is wide variation in the chilling requirement of different species and ecotypes (or, genetic races) of maple. More than 12 weeks of chilling are required to break dormancy in sugar maple (*Acer saccharum*) collected in southern Canada, while those collected from the warmer regions near the southern limits of its distribution required only a few weeks of low temperature. Similar results were obtained for seedlings of red maple (*Acer rubrum*).

Very little is known about the molecular mechanisms involved in breaking bud dormancy. What is known comes largely from the study of various species of poplar (*Populus* sps.), birch (*Betula papyrifera*), and apple where recent efforts have been directed toward characterizing the expression of dormancy-responsive genes. In one species of poplar (*P. tricocarpa*), for example, a gene homologous with *FLOWRING LOCUS C* (*FLC*-like gene or *PtFLC*) has been implicated in bud dormancy. FLC is the floral suppressor that is down-regulated in *Arabidopsis* during a vernalization treatment (Chapter 25). *PtFLC* is expressed in the shoot apices of poplars grown under long days but, like FLC in *Arabidopsis*, PtFLC declines in dormant buds during the low temperature period. By analogy with vernalization, the down-regulation of *PtFLC* could be a key component in the cold-mediate release of bud dormancy in poplar. In another study, it was found that expression of the gene *KNAP2* increased during the onset of dormancy in apple buds, but was down-regulated during the breaking of dormancy. KNAP2 is a KNOTTED-like homeobox protein; a group of master control proteins that have a fundamental role in development of plants and animals alike (see Chapter 20).

26.3 SEED DORMANCY

26.3.1 NUMEROUS FACTORS INFLUENCE SEED DORMANCY

Seeds are in many respects similar to buds—they consist of a small embryonic axis (along with some storage tissues) enclosed by a series of membranes, collectively called the **seed coat**. The seed coat serves a protective function much as bud scales do. Its presence often suppresses germination by restricting the uptake of water and exchange of oxygen, it mechanically limits the expansion of the embryo and, in some cases, contains inhibitors that prevent growth of the embryo. These limitations can be removed and the germination of many seeds accelerated by mechanically disrupting or removing the seed coat, a process called **scarification**. In the laboratory, scarification may be accomplished with files or sandpaper. In nature, abrasion by sand, microbial action, or passage of the seed through animal gut will accomplish the same end. Seed coats can be very tough. Uniformity and rate of germination of morning glory (*Pharbitis nil*), cotton, and some tropical legume seeds, for example, can be improved by soaking the dry seed in concentrated sulfuric acid for up to an hour. Scarification by passage through animal gut no doubt occurs as a result of the acidic conditions in the gut.

As with buds, dormancy in seeds refers to the situation wherein the embryo fails to grow because of physiological or environmental limitations. These limitations commonly include the inability of water or oxygen to penetrate the seed coat. Seeds of some plants, particularly in the family Leguminoseae, have specialized structures that control seed moisture content. E. Hyde described a structure in seeds of lupine (*Lupinus arboreus*) that functions as a hygroscopically operated check-valve and that limits imbibition of water by the seed. Because water cannot pass through the unscarified seed coat, the only possible route of entry is through a small pore, called the **hilum** (Figure 26.4). When the water

content of the seed is higher than ambient, the hilum is open to permit the exit of water and allow the seed to dry. But when the moisture content outside the seed is higher than inside, cells surrounding the hilum swell, thus closing off the pore and preventing the uptake of water. In addition, as the seed dries out the permeability of the seed coat to water also decreases and the dormancy of the seed increases. Other seeds have pores that are blocked with a plug, called the **strophiolar plug**, which must be mechanically dislodged before water and oxygen can enter.

There is a considerable body of evidence to suggest that seed coats also interfere with gas exchange, oxygen uptake in particular. As noted above, removal of the seed coat often leads to a significant increase in respiratory consumption of oxygen. Measurements of the oxygen permeability of seed coats have been made and there is general agreement that permeability is very low in those seeds tested. However, it is not always clear that limited oxygen permeability is the primary cause of dormancy. The complexity of the situation and problems of interpretation are well illustrated by studies of the genus *Xanthium*, or cocklebur.

A cocklebur contains two seeds: an upper, dormant seed and a lower, nondormant seed. Dormancy of the upper seed can be overcome either by removing the seed coat or by subjecting the intact seed to high oxygen tension. The inference is that seed coat permeability in the dormant seed limits the supply of oxygen to the embryo and thus prevents germination. However, several other observations have cast doubt on this conclusion. There are, for example, no measurable differences between the dormant and nondormant seed with respect to the permeability of the seed coat to oxygen. Moreover, the rate of oxygen diffusion through the seed coats is more than sufficient to support measured rates of oxygen consumption by the embryos inside. Clearly, dormancy of the upper seed in *Xanthium* cannot be due to limited permeability of the seed coat to oxygen. Why then, do the upper, dormant seeds require a higher oxygen level to elicit germination? It appears that the seed coat is a barrier, not to the uptake of oxygen but to the removal of an inhibitor from the embryo. Aqueous extracts of *Xanthium* seeds have revealed the presence of two unidentified inhibitors, based on tests of the extracts in a wheat coleoptile elongation assay. The same two inhibitors are found in diffusate collected from isolated embryos placed on a moist medium, but not in diffusate from seeds surrounded by an intact seed coat. Thus germination in the dormant seed appears to be prevented by the presence of these inhibitors and the seed coat serves as a barrier that prevents those inhibitors from being leached out. The oxygen requirement can be explained by the observation that high oxygen tension reduces the quantity of an extractable inhibitor, presumably by some oxidative degradation.

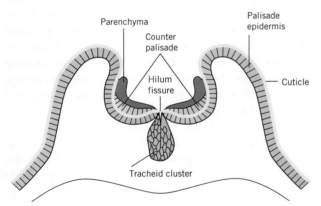

FIGURE 26.4 A cross-section through a portion of the seed of *Lupinus arboreus* (tree lupine) showing the hilum, a hygroscopic valve that regulates water loss from the seed. The counter palisade is a group of thick-walled cells lying on the outer surface of the cuticle. When the seed moisture content is higher than the moisture level in the ambient air, the hilum fissure is open and water escapes from the tracheid cluster. When the water ambient moisture level is higher than inside the seed, the counter palisade cells swell and close off the hilum fissure in order to prevent the seed from rehydrating. (From Hyde, E. O. C. 1954. *Annals of Botany* 18:241. With permission of The Annals of Botany Company.)

Even the role of inhibitors in seed dormancy is not clear. Along with hormones such as auxins and gibberellins, a large number of inhibitors have been identified in seeds, fruits, and other dispersal units. These include hormones (ABA), unsaturated lactones (coumarin), phenolic compounds (ferulic acid), various amino acids, and cyanogenic compounds (i.e., compounds that release cyanide) characteristic of apple and other seeds in the family Rosaceae (see Chapter 27). The simple presence of an inhibitor does not, however, prove its role in dormancy. The inhibitors could be localized in tissues not directly involved in growth of the embryo or otherwise sequestered so as to preclude any role in preventing germination. Evidence in support of a role for inhibitors is generally limited to leaching experiments such as that described above for *Xanthium*. In some cases, dormancy can then be restored by exposing the leached seed to the putative inhibitor. In order to clearly establish whether an inhibitor has an active role in regulating germination, it is necessary to establish whether inhibitor levels in the seed correlate with the onset and termination of dormancy. In spite of the voluminous literature relating inhibitors to dormancy, there is very little critical support for a direct role. For the present, evidence for the imposition and maintenance of dormancy by inhibitors remains largely circumstantial.

26.3.2 TEMPERATURE HAS A SIGNIFICANT IMPACT ON SEED DORMANCY

Temperature has a significant impact on termination of dormancy in many seeds. In fully imbibed seeds, both alternating and low (chilling) temperatures are known to terminate dormancy. Many seeds, even though fully hydrated, will not germinate when maintained under constant temperature. They require instead a diurnal cycle of fluctuating temperature. The required temperature differential between the high and low temperature is often not great, ranging from a few degrees to perhaps 5°C or 10°C, depending on the species. Germination of broad-leaved dock (*Rumex obtusifolia*) seeds, for example, exceeds 90 percent when the temperature differential is about 10°C and when the high temperature is given for 16 hours each day.

The reaction to alternating temperature is complex and poorly understood. In *Rumex*, alternating treatments are effective only when the high temperature is greater than 15°C. Also, when the high temperature is given for only 8 hours each day, a differential of only 5°C is required to induce 90 percent germination. Although in some cases the effect of alternating temperature appears to be localized in the embryo itself, there are many well-documented cases where the effect of alternating temperature is mechanical. It is, in effect, a form of scarification, releasing the seed from some kind of seed coat–imposed dormancy.

It has long been known that freshly shed seeds of many herbaceous and woody species have dormant embryos that can be induced to growth only by a prolonged low-temperature treatment. These include maples (*Acer* sps.), hazel (*Corylus*), and many genera in the family Rosaceae (pear, *Pyrus*; apple, *Malus*; hawthorne, *Crateagus*). Normally, following the required period of low temperature, the seeds will not germinate until temperatures are more favorable for embryo emergence and seedling development. In most cases this requirement ensures that the seed shed in late summer or fall will not germinate until spring.

The exposure to low temperature that satisfies this germination requirement is known as either pre-chilling, or **stratification**. The latter term has its origin in the horticultural practice of layering seeds in moist sand or peat moss and exposing them to low temperature for several weeks or months to induce germination. It is important that the pre-chilling requirement for release of seed dormancy not be confused with vernalization, which is a cold treatment to an already germinated seedling, as discussed earlier in the previous chapter. As with breaking of bud dormancy, temperatures near freezing but below 10°C are most effective for terminating seed dormancy. The optimum for most species is near 5°C. In a population of seeds, the effectiveness is also a function of the length of the cold treatment (Figure 26.5).

It is presumed that seeds undergo some metabolic changes during the period of low temperature, generally referred to as **after-ripening**, but the exact nature of these changes is unclear. There is some evidence for redistribution of carbon and nitrogen from the endosperm to the embryo, a decline in the inhibitor content, and a rise in gibberellin and cytokinin content.

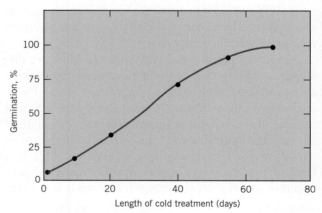

FIGURE 26.5 Breaking dormancy in apple seeds with low temperature. Moist seeds were held at 4°C for the time indicated. (Redrawn from Luckwell, L. C. 1952. *Journal of Horticultural Science* 27:53.)

Gibberellin treatments will substitute at least partially for the cold requirement in many seeds, just as they do in other cold-requiring systems.

26.4 THERMOPERIODISM IS A RESPONSE TO ALTERNATING TEMPERATURE

Growers have long recognized the beneficial effect on plant growth of lowering greenhouse temperatures during the night. This effect has been particularly well documented by the work of F. Went and his colleagues in the 1940s. Went found that tomatoes (*Lycopersicum esculentum*) grown at constant temperatures of 26°C and 18°C grew poorly and (at 26°C) failed to produce fruit. Plants maintained under alternating conditions of 26°C during the day and 18°C at night grew vigorously and produced a maximum number of fruit. In order to be effective, the day–night differential had to be synchronized with the light–dark cycle. If the temperature cycle was inverted, with the high temperature falling during the dark period, growth was even poorer than at a constant 26°C. To describe this phenomenon, F. Went coined the term **thermoperiodism**.

It is now recognized that many, but certainly not all, plants perform better under regimes with a similar temperature differential. For those that do, the effect is primarily on vegetative development, in contrast to photoperiodism where the influence is primarily on floral production. In some plants, such as potato (*Solanum*) and tobacco (*Nicotiana*), low night temperature leads to a decline in shoot–root ratio, due to preferential root growth. Stems of Begonia also respond to thermoperiodism. It has been reported that stem elongation was inhibited by low daytime temperatures alternating with high nighttime temperatures (14°C/24°C) when compared with constant daytime/nighttime temperatures (19°C/19°C). On the other hand, stem elongation was promoted by alternating high daytime temperatures with low nighttime temperatures (24°C/19°C).

Another example of the effect of temperature differentials is illustrated by floral movements in members of the Liliaceae, such as tulip (*Tulipa*) and *Crocus*. Flowers in these plants normally open during the day and close at night, but these movements are only slightly affected by light. Instead, the perianth segments or tepals[1] respond to changing temperature. This is a form of **thermonasty**, involving a differential growth response of cells on the inner (or adaxial) and outer

(or abaxial) surfaces of the tepals. The optimum temperature for growth differs by approximately 10°C between the two surfaces (Figure 26.6). The opening of the flower in response to an increase in temperature (Figure 26.6A) corresponds to a sharp but transient increase in the growth rate of cells on the inner surface (Figure 26.6B). Conversely, closure following a drop in temperature (Figure 26.6C) appears to be caused by a similar change in the growth rate of cells on the outer surface (Figure 26.6D). Other investigators have reported opening following a rise of as little as 0.2°C for *Crocus* and 1°C to 2°C for tulip. There are lower limits—*Crocus*, for example, will not open at temperatures below 8°C. Thus if the spring days are very cold, the flowers may not open at all. Thermonasty is not limited to flower parts; it has also been demonstrated for the stem angle of *Phryma leptostachya*, a perennial Asian herb.

26.5 TEMPERATURE INFLUENCES PLANT DISTRIBUTION

Temperature is thought to be one of the most important factors limiting the worldwide distribution of plants. Distribution limits often reflect temperature characteristics of major metabolic processes, especially photosynthesis. The temperature range compatible with growth of higher plants lies generally between 0°C and 45°C, although there are some plants that exceed either of these limits and within those limits temperature compatibility is very much species dependent. Various cultivars of wheat (*Triticum vulgare*), for example, will grow at temperatures from near zero to over 40°C, although the temperature optimum for growth falls in the range of 20°C to 25°C.

As a general rule, temperatures that are optimum for growth reflect the geographical region in which the species originated. Thus, plants native to warm regions either require or perform better at higher temperatures than those that originated in cooler areas of the world. The optimum for maize (*Zea mays*), a plant of tropical origin, is in the range 30°C to 35°C and it will not grow below 12°C to 15°C. Garden-cress (*Lepidium sativum*), a temperate herb, will grow at temperatures as low as 2°C but its maximum temperature for growth is 28°C.

The effects of temperature on physiology and metabolism in turn influence plant distribution, called **biogeography**. At times, temperature-related metabolic effects not only limit distribution, but have significant economic implications as well. Cotton (*Gossypium*), for example, is a southern crop in part because cool night temperatures in northern latitudes adversely affect fiber cell wall thickening, and the northern limits for maize production are very much limited by its inability to grow at lower temperatures.

---

[1]Tepal is the collective term for sepals and petals when the two share a common morphology and are indistinguishable one from the other.

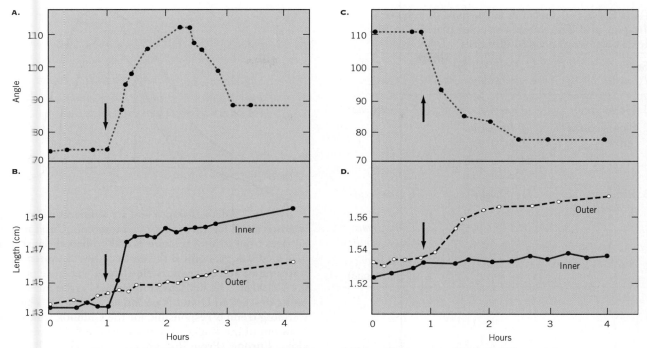

FIGURE 26.6 **The effect of 10°C temperature shifts on flower opening and differential growth in** *Tulipa flowers.* **(***A, B***) The effect of raising temperature from 7°C to 17°C. (***C, D***) The effect of lowering temperature from 20°C to 10°C. The temperature shift is indicated by the arrows. A and C show floral opening and closure, respectively. B and D show the growth of cells on the inner and outer surfaces of the tepals. (From Wood, W. M. 1953. *Journal of Experimental Botany* 4:65–77. Reprinted by permission of The Company of Biologists, Ltd.)**

However, advances in agronomy and plant breeding have moved the limits for maize steadily northward over the past several decades.

Although extensive studies of temperature effects have been conducted using controlled environment facilities, it is difficult to carry out field studies with any degree of precision. This is because the leaves and roots of plants are commonly subject to a mosaic of fluctuating temperature regimes. Leaf temperature, for example, depends not only on daily and seasonal variations in atmospheric temperature, but such factors as cloud cover, wind speed, their position in the canopy, and so forth. Root temperature depends on depth in the soil, soil moisture content, soil structure, and other physical parameters of the soil. Thus, in a natural environment, individual leaves and roots may be responding to distinctly different microclimates, each with its own unique temperature regime.

In one study, a variety of species native to either the cool coastal regions of northern California or the hot, dry desert of Death Valley were transplanted into experimental gardens in both locations. At the Death Valley site summer air temperatures commonly reach 50°C, a temperature that is lethal for many organisms. By contrast, average daily maximum temperatures at the coastal site were less than 20°C. Plants at both sites were

irrigated and fertilized so that water supply and nutrients were not limiting factors and their performance with respect to growth and survival was assessed on a regular basis. On the basis of their growth responses, the plants could be grouped into three main categories: (1) those that were unable to survive the summer months; (2) those that survived but grew slowly during the summer; and (3) those that grew most rapidly during the summer months (Table 26.1).

Virtually all of the species that are native to the cool coastal climate were unable to survive the high desert temperatures. Of the plants tested, only *Tidestromia oblongifolia*, a deciduous C4 perennial native to Death Valley, was able to thrive in the summer desert heat. Strikingly, *T. oblongifolia* was unable to survive the cool coastal temperatures. At the other extreme, *Atriplex glabriuscula*, a C3 annual native to the coastal region, thrived in the coastal garden but died in the desert in spite of ample irrigation. Two clones of the C4 species *Atriplex lentiformis* were also tested—one native to the coastal regions of southern California and one that occurs naturally in Death Valley. In terms of biomass production, the desert clone outperformed the coastal clone in the Death Valley garden; their relative performance was reversed in the coastal garden. The relative success of *A. glabriuscula* and

TABLE 26.1   **Growth responses of selected** *Atriplex* **and** *Tidestromia* **species planted in hot desert and cool coastal climates.**

Summer Growth	Death Valley Garden	Coastal Garden
1. No survival	*A glabriuscula*	*T. oblongifolia*
2. Slow summer growth	*A. lentiformis* (Coastal clone)	
	A. lentiformis (Desert clone)	
3. Rapid summer growth	*T. oblongifolia*	*A. glabriuscula*
		A. lentiformis (Coastal clone)
		A. lentiformis (Desert clone)

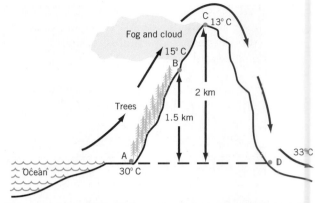

FIGURE 26.7 **Adiabatic lapse and air temperature on the windward and leeward sides of a mountain. Unsaturated air rises from point A to point B at the adiabatic lapse rate of 1°C/100 m. At point B the air becomes saturated with water vapor and cools more slowly at a wet lapse rate of 0.4°C/100 m. The dry lapse rate applies as the air warms on its descent to D and is no longer saturated. This warming trend is responsible for the Chinook winds that blow out of the Rocky Mountains or the Foehn winds of alpine Europe. (From Rosenberg, N. J. et al. 1983.** *Micro-climate: The Biological Environment.* **New York: Wiley. With permission.)**

T. oblongifolia in the two environments appeared to correlate with their capacity to assemble a competent photosynthetic apparatus. For example, the relative growth rates of the two species under laboratory controlled conditions compared favorably with the response of photosynthetic rate to temperature under the dame conditions. Both the maximum relative growth rate and maximum rate of photosynthesis occurred at approximately 25°C for *A. glabriuscula* and at approximately 40°C to 45°C for *T. oblongifolia*.

Other species are more flexible with regard to temperature. Several species, including *A. hymenelytra*, *Nerium oleander*, and the creosote bush (*Larrea divaricata*), were able to survive in both the desert and cool coastal habitats, although their growth rate was not as great as either *A. glabriuscula* or *T. oblongifolia*. Most of their growth was in fact accomplished during the spring or fall when temperatures were less extreme. In all three cases, growth at low or high temperature under controlled conditions caused an appropriate shift (by as much as 15°C) in the optimum temperature for photosynthesis. More importantly, however, there was no significant change in the maximum rate of photosynthesis, only the temperature at which the maximum rate occurred. Thus some plants exhibit a significant degree of phenotypic plasticity with respect to photosynthesis and temperature, which enables them to survive a wider range of climatic conditions. Plants restricted to one climate or another apparently do not exhibit the same degree of plasticity in their metabolic reactions.

On a worldwide basis, temperature is the most important factor affecting the relative distribution of C3 and C4 grasses. On a smaller scale, this is illustrated by the distribution of plants along an elevational gradient,

such as up a mountainside where temperature decreases with increasing altitude. This decrease in temperature with elevation is called the **adiabatic lapse**. As air rises, it expands and cools. The term *adiabatic* refers to the fact that cooling occurs without an exchange of heat. In the case of an elevational gradient, cooling as the air rises is entirely due to expansion of the air mass as the pressure decreases. The heat content and, consequently, the temperature of a unit volume of air mass are therefore lower. It is because of adiabatic lapse that the temperature gradient remains stable and the less-dense cooler air does not descend from high in the mountains to displace the warmer air in the valleys. The **adiabatic lapse rate** for dry air is constant at about 1°C/100 m elevation (Figure 26.7). The lapse rate for moist air (wet lapse rate) is more variable and lower than the dry lapse rate because condensation of water vapor releases heat.

P. W. Rundel has studied the distribution of C3 and C4 grasses along an elevational gradient in Hawaii. Rundel found a sharp transition in the distribution of the two photosynthetic types at about 1400 m. C4 grasses were predominant at warmer, drier elevations below 1400 m, while in the cooler, moist environment above 1400 m the C3 grasses were predominant. The midpoint of the transition zone is the elevation where the maximum daily temperature for the warmest month of the year is approximately 21°C. Similar distributions of C4 and C3 grasses have been reported in other elevational studies carried out in Africa and Costa Rica, and in latitudinal gradients in North America. In the latter case, the transition temperatures are slightly lower, but the principle is still valid.

SUMMARY

All living organisms can be broadly classified according to their ability to withstand temperature. Psychrophiles grow optimally at temperatures of 0°C to 10°C; mesophiles, 10°C to 30°C; and thermophiles, 30°C to 65°C. Most higher plants are mesophiles, although plants will generally survive temperatures between 0°C and 45°C. Temperature limits generally reflect the freezing point of water on the low side and denaturation of protein on the high side.

Plants also use temperature as a cue in their developmental and survival strategies. Decreasing temperature in the autumn in concert with photoperiod induces dormancy in buds, characterized by low respiratory rate and an inability to grow even if temperature, oxygen, and water supply are adequate. Most dormant buds have a chilling requirement that must be met before dormancy can be broken and growth renewed. Dormancy is also a property of many seeds, a situation in which the seed fails to germinate because of environmental and physiological limitations. Seed coats may interfere with water uptake or oxygen uptake, or may contain inhibitors that must be broken down or leached out before germination can proceed. Many seeds require alternating temperatures or a period of low temperature (pre-chilling or stratification) to break dormancy.

Some plants, such as tomato, grow poorly at constant temperature, but require alternating day–night temperatures (thermoperiodism) for optimum growth. In others, such as tulip and *Crocus*, changing temperature regulates the opening and closure of floral petals.

Temperature is a principal factor in the distribution of plants, or biogeography. Survival in extreme environments appears to be due to intrinsic differences between species in the temperature dependence of their growth responses and photosynthetic metabolism. In a study of elevational gradients up a mountainside, a sharp transition was found between C4 species (in the warmer, drier, lower elevations) and C3 species (in the cooler moister, higher elevations). It is clear that temperature stability of principal metabolic pathways is a significant determinant in plant distribution.

CHAPTER REVIEW

1. In what ways does temperature influence physiological processes? Does temperature interact with other environmental factors? If so, which ones?

2. How does temperature influence the geographical distribution of plants? What modifications might you expect to find in plants adapted to high-temperature habitats? To plants in arctic or alpine habitats?

3. Review the distinction between vernalization and stratification.

4. Dormant buds and seeds normally require an extended treatment of some sort in order to break dormancy. What is the survival value of such a requirement?

5. In what way(s) do the daily movements of floral petals, such as tulip and *Crocus*, differ from the movements of bean leaves?

FURTHER READING

Bewley, J. D., M. Black. 1994. *Seeds: Physiology of Development and Germination*. New York: Plenum.

Björkman, O., 1980. The response of photosynthesis to temperature. In: J. Grace, E. D. Ford, P. G. Jarvis (eds). *Plants and Their Atmospheric Environment*. Oxford: Blackwell Scientific Publications, pp. 273–301.

Dickinson, R. E. 1987. *The Geophysiology of Amazonia. Vegetation and Climate Interactions*. New York: Wiley.

Fitter, A. H., R. K. M. Hay. 1987. *Environmental Physiology of Plants*. 2nd ed. London: Academic Press.

Gash, J. H. C., C. A. Nobre, J. M. Roberts, R. L. Victoria. 1996. *Amazonian Deforestation and Climate*. Chichester: Wiley.

Gibson, A. C. 1996. *Structure-function Relations of Warm Desert Plants*. Berlin: Springer Verlag.

Jacobs, M. 1988. *The Tropical Rain Forest: A First Encounter*. Berlin: Springer-Verlag.

Lang. G. A. 1987. Dormancy: A new universal terminology. *HortScience* 22:817–820.

Rhode, A., R. P. Bhalerao. 2007. Plant dormancy in the perennial context. *Trends in Plant Science* 12:217–223.

Rosenberg, N. J., B. L. Blad, S. B. Verma. 1983. *Microclimate: The Biological Environment*. 2nd ed. New York: Wiley.

Rundel, P. W. 1980. The ecological distribution of C4 and C3 grasses in the Hawaiian Islands. *Oecologia* 45:354–359.

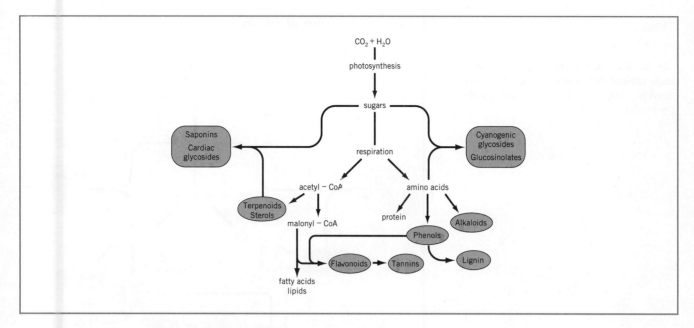

27

Secondary Metabolites

The sum of all of the chemical reactions that take place in an organism is called **metabolism**. Some aspects of metabolism, such as the metabolism of carbon and nitrogen assimilation and energy conversions, have been addressed in earlier chapters. Most of that carbon, nitrogen, and energy ends up in molecules that are common to all cells and are required for the proper functioning of cells and organisms. These molecules, e.g., lipids, proteins, nucleic acids, and carbohydrates, are called *primary metabolites* (see Appendix). Unlike animals, however, most plants divert a significant proportion of assimilated carbon and energy to the synthesis of organic molecules that may have no obvious role in normal cell function. These molecules are known as **secondary metabolites**.

In this chapter, we examine some of the broader aspects of secondary metabolites. The focus will be the biosynthesis, physiology, and ecological roles of four major classes of secondary metabolites:

* terpenes, including hormones, pigments, essential oils, steroids, and rubber,
* phenolic compounds, including coumarins, flavonoids, lignin, and tannins,
* glycosides, including saponins, cardiac glycosides, cyanogenic glycosides, and glucosinolates, and
* alkaloids.

Some secondary metabolites are also involved in defense against invading pathogens, a subject that is also addressed in this chapter.

27.1 SECONDARY METABOLITES: A.K.A NATURAL PRODUCTS

The distinction between primary and secondary metabolites is not always easily made. At the biosynthetic level, primary and secondary metabolites share many of the same intermediates and are derived from the same core metabolic pathways (Figure 27.1). In the strictest sense, however, secondary metabolites are not part of the essential molecular structure or function of the cell. Secondary metabolites generally, but not always, occur in relatively low quantities and their production may be widespread or restricted to particular families, genera, or even species. Also known as **natural products**, these novel phytochemicals were of little interest to biologists because of their apparent lack of biological significance. They were known, however, to have significant economic and medicinal value and were thus of more than a passing interest to natural products chemists. Natural products have found use in antiquity as folk remedies, soaps, and essences. They include drugs and other medicinal products,

FIGURE 27.1 **A schematic to illustrate biosynthetic relationships between principal primary and secondary metabolites (circled).**

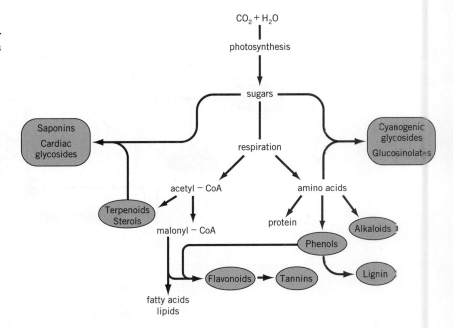

dyestuffs, feedstocks for chemical industries (gums, resins, rubber), and a variety of substances used to flavor food and drink. In recent years, however, it has become increasingly evident that many natural products do have significant ecological functions, such as protection against microbial or insect attack.

27.2 TERPENES

27.2.1 THE TERPENES ARE A CHEMICALLY AND FUNCTIONALLY DIVERSE GROUP OF MOLECULES

With nearly 15,000 structures known, terpenoids are probably the largest and most diverse class of organic compounds found in plants. As discussed earlier in Chapter 19, the unifying feature of terpenes is that they are generally lipophilic polymers based on the simple 5-carbon unit 2-methyl-1,3-butadiene, or **isoprene**, which is derived via either the mevalonic acid pathway or the MEP pathway (Chapter 19, Section 19.3).

Terpenes can be grouped into several classes, based on the number of carbon atoms (Figure 27.2). This large chemical diversity arises from the number of basic units in the chain and the various ways in which they are assembled. Formation of cyclic structures, addition of oxygen-containing functions, and conjugation with sugars or other molecules all add to the possible complexity. The name terpenoid derives from the fact that the first compounds in the group were isolated from turpentine (Ger. *terpentin*), an essential oil (chiefly pinene) distilled from the resins of several coniferous trees.

The terpene family includes hormones (gibberellins and abscisic acid): the carotenoid pigments (carotene and xanthophyll); sterols (e.g., ergosterol, sitosterol, cholesterol) and sterol derivatives (e.g., cardiac glycosides); latex (the basis for natural rubber); and many of the essential oils that give plants their distinctive odors and flavors. Cytokinin hormones and chlorophyll, although not terpenes per se, do contain terpenoid side chains. It is apparent from this list that many terpenes have significant commercial value as well as important physiological roles. Many terpenes and terpene derivatives may be considered primary metabolites. The hormones abscisic acid and gibberellin, the carotenoid and chlorophyll pigments, and sterols (steroid alcohols) all play significant roles in plant growth and development. The vast majority of terpenes, however, are secondary metabolites, many of which appear to act as toxins or feeding deterrents to herbivorous insects.

27.2.2 TERPENES ARE CONSTITUENTS OF ESSENTIAL OILS

Many plants, such as citrus, mint, *Eucalyptus*, and various herbs (sage, thyme, etc.), produce complex mixtures of alcohols, aldehydes, ketones, and terpenoids, known generally as **essential oils** (essence, as in perfume). Essential oils are responsible for the characteristic odors and flavors of these plants but they are also known to have insect-repellant properties. The terpenes and terpene derivatives found in essential oils are predominantly hemi-, mono-, and sesquiterpenes, which can be moderately to highly volatile. Several examples are shown in Figure 27.3. In most plants, the essential oils are synthesized in special glandular trichomes (hairs) on the leaf surface (Figure 27.4), although the essential oils

Number of Carbons	Class	Example
5	Hemiterpenoid	$\overset{\displaystyle COOH}{\underset{\displaystyle}{CH_3-C=CH-CH_2}}$ Tigilic acid
10	Monoterpenoid	$CH_3-\overset{CH_3}{C}=CH-(CH_2)_2-\overset{CH_3}{C}=CH-CH_2OH$ Geraniol
10	Cyclic monoterpenoid	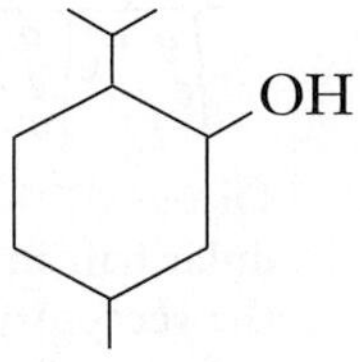 Menthol (peppermint oil)
15	Sesquiterpenoid	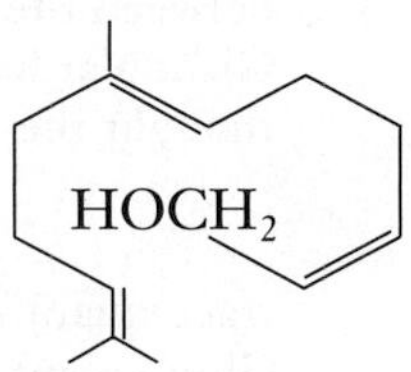 Farnesol (widespread)
20	Diterpenoid	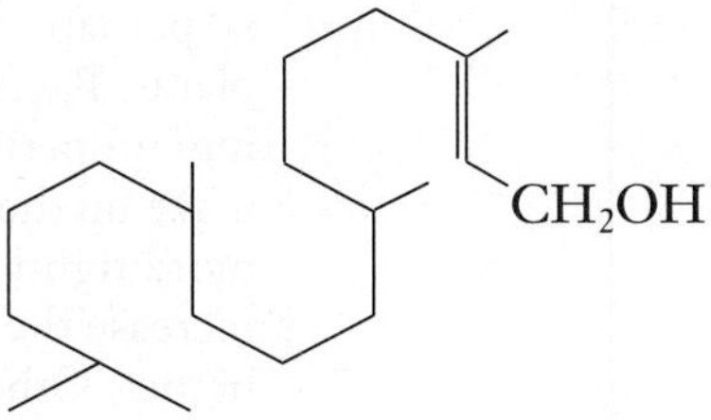 Phytol (chlorophyll)
30	Triterpenoid	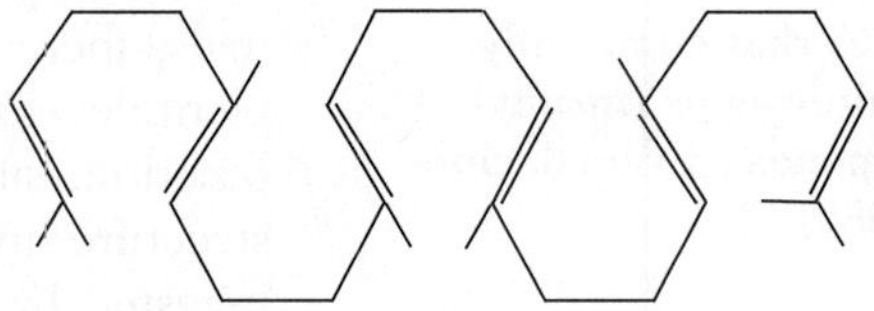 Squalene (a steroid precursor)
30	Triterpenoid	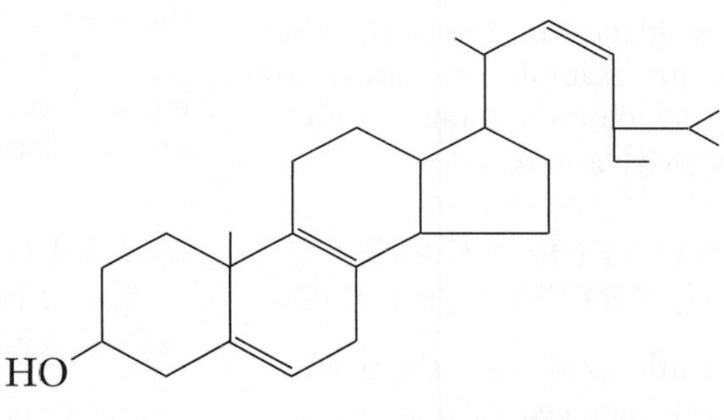 Stigmasterol (a sterol)
40	Tetraterpenoid	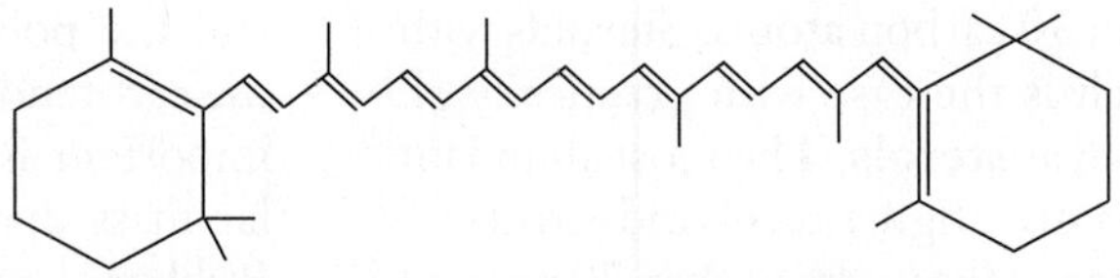 β-carotene (a carotenoid)

FIGURE 27.2 **Terpenoids are classified according to the number of carbon atoms in the basic skeleton.**

$$H_3C-CH-CH_2CH_2OH$$
$$H_3C$$

iso-Amyl Alcohol
(*Mentha, Eucalyptus*)

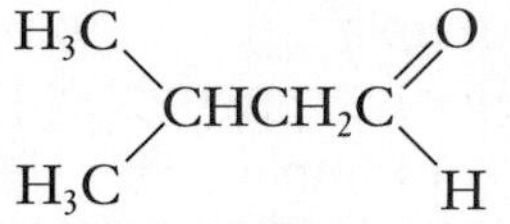

$$H_3C-CH-CH_2C-H$$
$$H_3C \qquad\qquad O$$

iso-Valeraldehyde
(*Eucalyptus*)

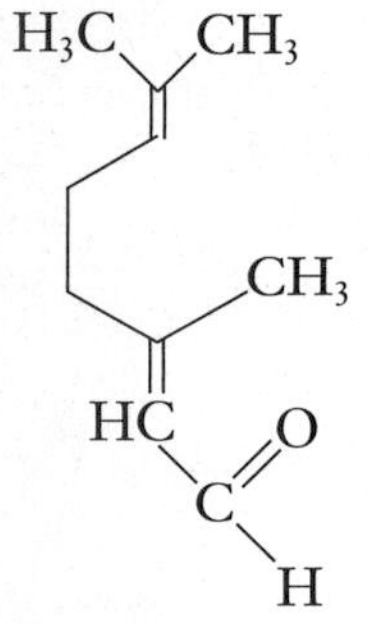

Geranial
(*Ctenium aromaticum*)
(lemon grass)

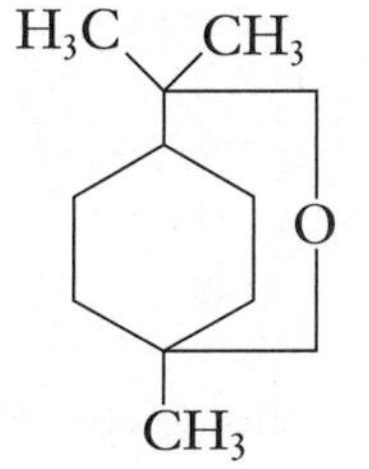

1:8 Cineole
(*Artemesia*)

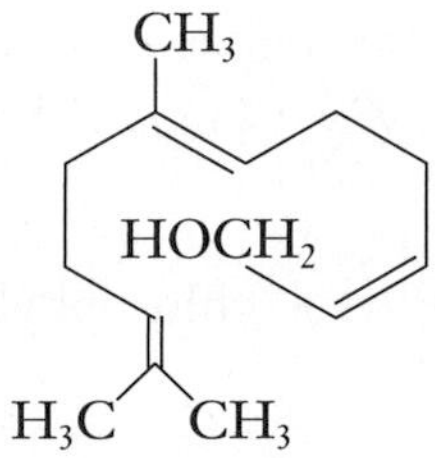

Farnesol
(widespread)

FIGURE 27.3 Representative terpenes that commonly occur in essential oils include: hemiterpenes (*iso*amyl alcohol, *iso*valeraldehyde), monoterpenes (geranial, cineole), and the sesquiterpene, farnesol.

of citrus are produced by glands in the peel. The resins of certain conifers, for example, also accumulate mixtures of terpenes, including the monoterpenes, α- and β-pinene, and myrcene (Figure 27.5).

27.2.3 STEROIDS AND STEROLS ARE TETRACYCLIC TRITERPENOIDS

Steroids and sterols are synthesized from the acyclic triterpene squalene, although they generally are modified and have fewer than 30 carbon atoms. Steroids with an alcohol group, which is the case with practically all plant steroids, are known as **sterols**. The most abundant sterols in higher plants are stigmasterol and sitosterol (Figure 27.6), which often make up more than 70 percent of the total sterols. However, plants also contain a large number of the more than 150 other sterols known to occur in nature. Plant sterols include cholesterol which, although widespread in occurrence, is present in only

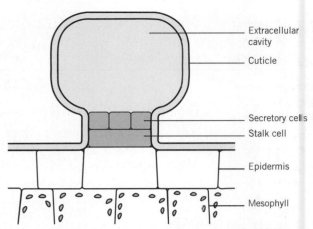

FIGURE 27.4 A schematic diagram of an epidermal glandular hair in cross-section. Essential oils are produced in the secretory cells and accumulated in a cavity that forms between the secretory cells and the overlying cuticle. Glandular hairs are found on the leaf surface, where it is thought they might serve to deter feeding by herbivores.

trace quantities. The extremely low level of cholesterol allows plant oils to be marketed as "cholesterol-free."

Sterols are constituents of plant membranes, which is perhaps their most important known function in plants. Because sterols are planar molecules, their packing properties are different from phospholipids that make up the bulk of the membrane bilayer. Sterols pack more tightly than phospholipids and therefore tend to increase the viscosity and enhance the stability of membranes. Otherwise, little is known about the function of sterols in plants. Unlike the steroid hormones in animals, there is no known hormonal role for sterols in plant development. Some sterols may have a protective function, such as the **phytoecdysones**, which have a structure similar to the insect molting hormones. When ingested by insect herbivores, phytoecdysones disrupt the insect's molting cycle. Other sterols are present as glycosides, which give rise to a number of interesting and economically significant secondary products. Steroid glycosides are discussed below.

27.2.4 POLYTERPENES INCLUDE THE CAROTENOID PIGMENTS AND NATURAL RUBBER

Larger terpenes include the tetraterpenes (40-carbon) and the polyterpenes. The principal tetraterpenes are the carotenoid family of pigments (Chapter 6). The only important isoprene derivatives with a greater molecular mass than the tetraterpenes are **rubber** and **gutta**. Rubber is a polymer consisting of up to 15,000 isopentenyl units. The polymer may be linear, as shown in Figure 27.7, or cross-linked into more complex configurations. The only difference between rubber and gutta is the configuration of the double bonds. In rubber the

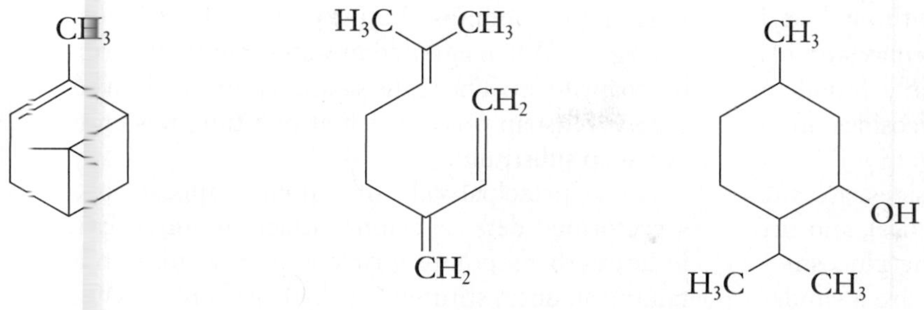

FIGURE 27.5 **Pinene, myrcene, and menthol are monoterpenes. Pinene and myrcene are found in the resins of some conifers. Menthol is the principal constituent of the essential oil of peppermint (*Mentha piperta*). Pinene also has insecticidal properties.**

α-Pinene   Myrcene   Menthol

double bonds are all *cis* configurations, while in gutta the double bonds are all *trans*.

In the plant, rubber occurs as small particles suspended in a milky-white emulsion called **latex**. Latex production is widespread in plants, with estimates ranging from a few hundred to several thousand species that produce latex in some form. The principal commercial source is *Hevea brasiliensis*, a rubber tree native to the Amazon rainforest. Others include the ornamental rubber tree (*Ficus elastica*), milkweed (*Asclepias*), and the Russian dandelion (*Taraxacum Kok-saghyz*). Latex contains about 30 to 40 percent rubber and 50 percent water. The balance is a complex mixture of resins, terpenes, proteins, and sugars. In most plants, latex is

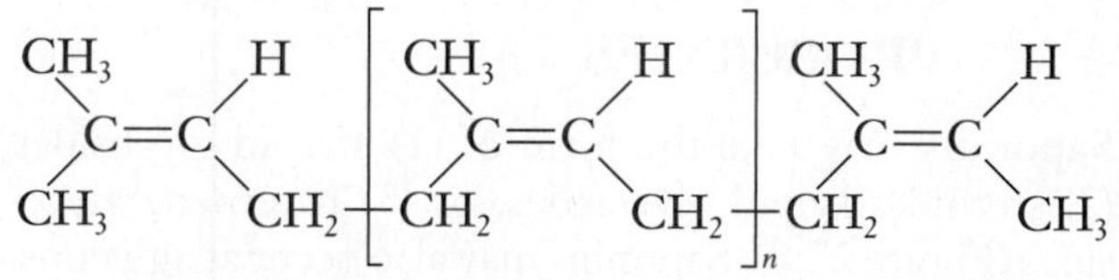

FIGURE 27.7 **Rubber is a linear polymer of isoprene units where the value of *n* may range from a few hundred to several thousand.**

produced in the phloem, accumulating in a series of long, interconnected vessels called **laticifers**.

The best-known source of gutta is a desert shrub, *Parthenium argentatum*, which grows in northern Mexico and southwestern United States. *Parthenium* (commonly known as guayule) may contain as much as 20 percent latex by weight, which is stored not in laticifers but in the vacuoles of stem and root cells. Guayule was at one time a significant commercial source of gutta for use in rubber products. However, while a single rubber tree, if properly tapped, can continue to produce for up to 30 years, guayule plants must be harvested (and, of course, replanted) annually.

Finally, there is a connection between terpenes and air pollution. Many of the essential oils, especially hemiterpenes, monoterpenes, and sesquiterpenes, are highly volatile and are given off in large quantities by plants, particularly during warm weather. Known generally as **volatile organic carbon** (**VOC**), these natural emissions from plants contribute to the formation of haze and cloud, and are involved in the formation of toxic tropospheric ozone.

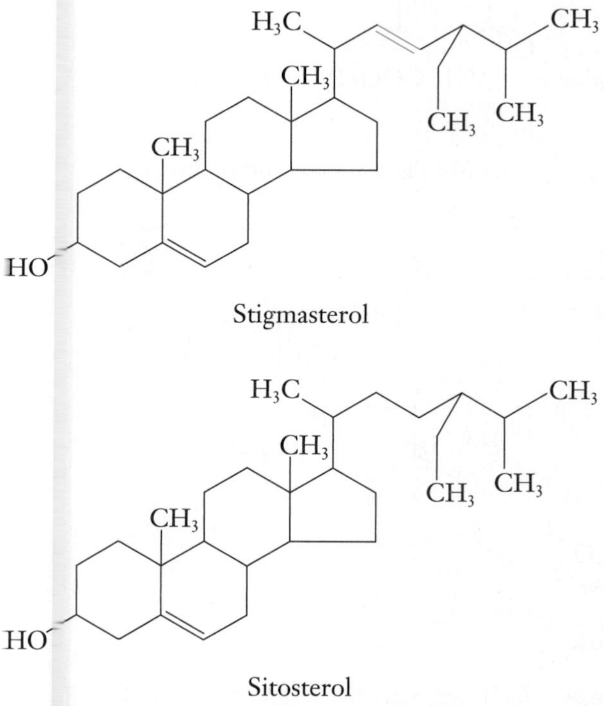

Stigmasterol

Sitosterol

FIGURE 27.6 **Stigmasterol and sitosterol differ only in the presence or absence of a double bond (highlighted). These are the two most abundant plant sterols.**

27.3 GLYCOSIDES

Some of the more interesting, if not important, derivatives synthesized by plants are glycosides. Most glycosides are thought to function as deterrents to herbivores. The term **glycoside** (Gr. *glykys*, sweet) refers to the bond formed (called a glycosidic bond) when a sugar molecule condenses with another molecule containing a hydroxyl group. Sugars may form glycosidic bonds with other sugars, such as when linked together to form polysaccharides, or with hydroxyl groups on noncarbohydrate molecules, such as steroids or amino acids. The sugar most commonly found in glycosides is glucose, although specific glycosides often contain rare sugars.

Three particularly interesting glycosides are the saponins, the cardiac glycosides (cardenolides), and the cyanogenic glycosides. A fourth family, the glucosinolates, although technically not glycosides, are a similar structure and so are included here.

27.3.1 SAPONINS ARE TERPENE GLYCOSIDES WITH DETERGENT PROPERTIES

Saponins may take the form of (1) steroid glycosides, (2) steroid-alkaloid glycosides, or (3) triterpene glycosides (Figure 27.8). Saponins may also occur as aglycones (e.g., the terpene without the sugar), which are known as **sapogenins**. In much the same way as soap, which is the sodium salt of a fatty acid, the combination of a relatively hydrophobic triterpene with a hydrophilic

sugar gives saponins the properties of a surfactant or detergent. When agitated in water, saponins form a stable soapy foam. The name saponin is in fact derived from *Saponaria* (soapwort), which at one time was employed as a soap substitute.

The principal role of saponins appears to be as a preformed defense against attack by fungi. Evidence indicates that saponins form complexes with sterols containing an unsubstituted 3-β-hydroxyl group. When the saponins react with sterols in the membranes of invading fungal hyphae, the result is a loss of membrane integrity. In a classic example of one-upmanship, however, many pathogenic fungi have developed strategies, such as the development of detoxifying enzymes, for circumventing this defense mechanism. Oat (*Avena*), for example, produces a triterpenoid saponin, **avenacin A-1**, which is localized in the root epidermal cells and effectively protects against an invasion by a fungal pathogen (*Gaeumannomyces graminis* var. *tritici*) that infects the roots of both wheat (*Triticum*) and barley (*Hordeum*). However, another strain of *G. graminis* (var. *avenae*) produces an enzyme, **avenacinase**, that detoxifies avenacin A-1 and

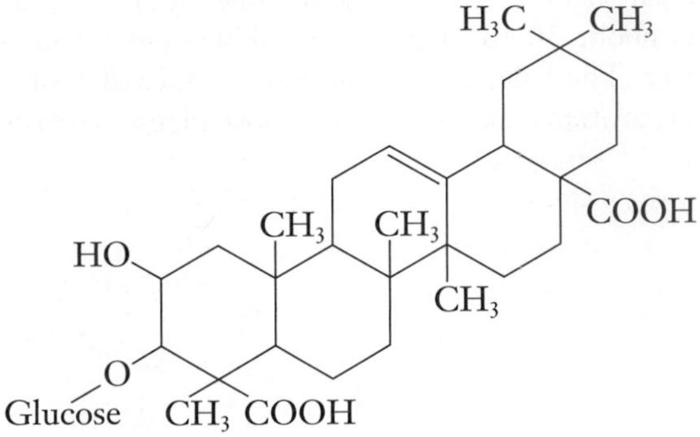

FIGURE 27.8 **Saponins are triterpenoids or steroids containing one or more sugar units. Medigenic acid glucoside is a triterpenoid saponin from alfalfa (*Medicago sativa*). Disogenin glycoside is a steroidal saponin isolated from clover (*Meliotus* spp.). The addition of a hydrophilic sugar group to a normally hydrophobic terpenoid gives saponins surfactant properties similar to soap.**

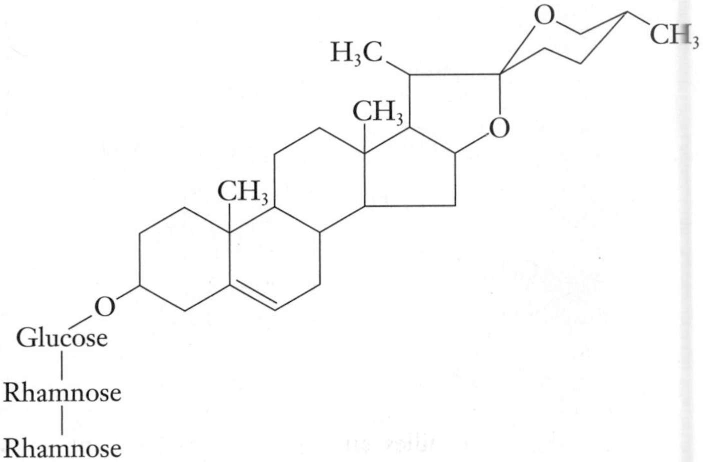

Medigenic acid glucoside

Disogenin-glycoside

a low the pathogen to invade oats as well as wheat and rye.

The effect of saponins on eukaryotic membranes is highly nonspecific and it is not clear how plants protect their own membranes against the deleterious effects of their own saponins. One possibility is that the saponins are stored in the form of a biologically inactive molecule, called a **bisdesmosidic saponin**, which has two sugar chains rather than one. When under attack the inactive form may be converted to the active **monodesmosidic** form by hydrolytic removal of the second sugar chain. Alternatively, biologically active, monodesmosidic saponins may be sequestered in vacuoles or organelles whose membranes contain a high proportion of sterols with a protected 3-β-hydroxyl position.

The effect of saponins on animals is somewhat variable. While not significantly toxic to mammals, saponins do have a bitter, acrid taste and will cause severe gastric irritation if ingested. Saponins will hemolyze red blood cells, however, if injected into the bloodstream. This action is presumably because of their detergent properties and their ability to disrupt membranes generally. On the other hand, saponins are highly toxic to fish and have been used as fish poisons. Saponins have also been implicated in reports of livestock poisoning. Alfalfa saponins, for example, can cause digestive problems and bloating in cattle. At the same time, there are reports that saponins contained in alfalfa sprouts will lower serum cholesterol levels. Commercially, saponins from the bark of *Quillaja saponaria* have been used as surfactants in photographic film, in shampoos, liquid detergents, toothpastes, and beverages (as emulsifiers). The saponin **glyscyrrhizin** from licorice (*Glyscyrrhiza glabra*) has been used in medicines and as a sweetener and flavor-enhancer in foods and cigarettes.

27.3.1 CARDIAC GLYCOSIDES ARE HIGHLY TOXIC STEROID GLYCOSIDES

The **cardiac glycosides** (or, **cardenolides**) are structurally similar to the steroid saponins and have similar detergent properties. They are distinguished from other steroid glycosides by the presence of a lactone ring (attached at C17) and the rare sugars (found almost exclusively in this group of steroids) that form the glycoside (Figure 27.9). Like the saponins, cardenolides occur naturally as either the glycoside or the aglycone (or genic).

The cardenolides have a wide distribution; they have been recorded in more than 200 species representing 55 genera and 12 families and are a principal agent in accidental poisonings of humans. Perhaps the best known is **digitalis**, a mixture of cardenolides extracted from the seeds, leaves, and roots of purple foxglove,

Digitalis purpurea or Grecian foxglove, *D. lanata*. The two principal cardenolides in digitalis are **digitoxin** and its close analog **digoxin**. *Digitalis* is also the source of a saponin, **digitonin**.

Since the late eighteenth century, digitalis has been used for its therapeutic value in treating heart conditions such as atherosclerosis. Because they disrupt the heart muscle Na^+/K^+-ATPase pumps (hence the appellation *cardiac*), cardenolides are highly toxic to vertebrates. The extreme toxicity of cardenolides has long been exploited by African hunters, who coated their arrows and spears with cardenolide-rich extract from plants. In therapeutic use, however, carefully regulated doses can both slow and strengthen the heartbeat. Unfortunately, the lethal and therapeutic doses are very close, so the therapy must be carefully monitored.

Other common sources of cardenolides are the milkweeds, *Asclepias* and *Calotropis*. These two species are known as "milkweeds" because they produce a milky-white, cardenolide-rich latex. The milkweeds are particularly interesting because they are the principal host for ovipositing monarch butterflies. The emerging larvae feed on the milkweed leaves and sequester the cardenolides without ill effect. The cardenolides are retained through metamorphosis and are present in the adult monarchs. When birds, such as blue jays,

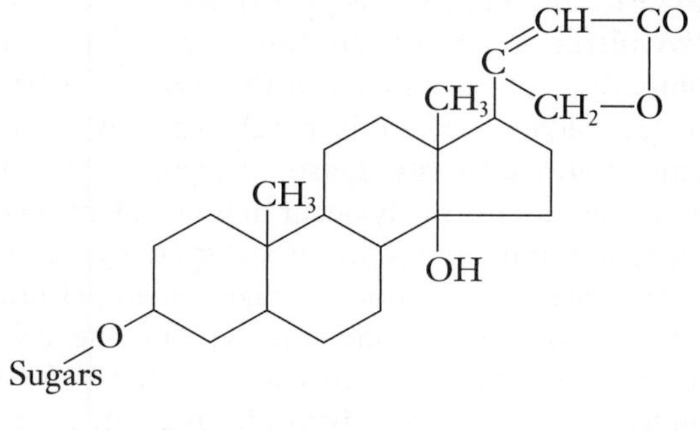

Digitoxin

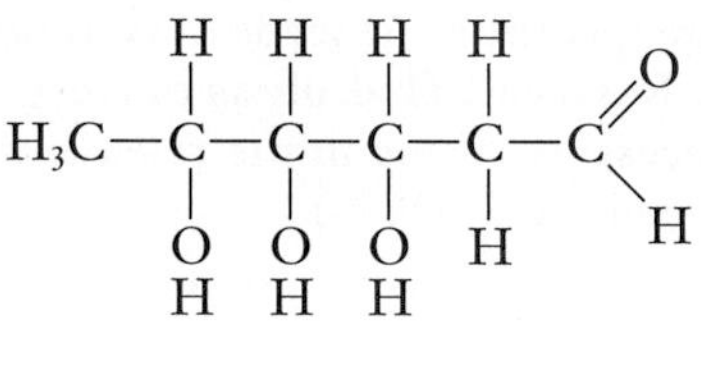

Digitoxose

FIGURE 27.9 Digitoxin, a cardiac glycoside. The sugar component of digitoxin consists of 1 molecule of glucose and 1 molecule of acetyl-digitoxose. The structure of digitoxose, one of the rare sugars found in cardiac glycosides, is shown below.

attempt to feed on monarchs, the accumulated carede-nolides induce an emetic reaction that forces the bird to vomit. The bird then wisely avoids attempting to feed on monarch larvae for some time.

27.3.3 CYANOGENIC GLYCOSIDES ARE A NATURAL SOURCE OF HYDROGEN CYANIDE

It might seem odd that plants synthesize chemicals capable of releasing deadly hydrogen cyanide or prussic acid (HCN), but more than 60 different cyanogenic compounds of plant origin have been described from more than a dozen plant families. Predominant among these are the **cyanogenic glycosides**. A common cyanogenic glycoside is **amygdalin** (Figure 27.10), which occurs in many representatives of the family Rosaceae. It is found in the seeds of apples and pears and in the bark, leaves, and seed of the stone fruits (apricot, peaches, plums, cherries). Most cyanogenic glycosides appear to be derived from one of four amino acids (phenylalanine, tyrosine, valine, and isoleucine) or from nicotinic acid. Intact cyanogenic glycosides are not themselves toxic, but when the plant is damaged by a herbivore, the glycoside undergoes an enzymatic breakdown and cyanide is released. Cyanide, a noncompetitive inhibitor of cytochrome oxidase, is acutely toxic.

The enzymatic breakdown of cyanogenic glycosides is a two-step process (Figure 27.10). First, the sugars are released by the enzyme β-**glycosidase**. The resulting hydroxynitrile is moderately unstable and will slowly decompose, liberating HCN in the process. Normally, however, decomposition is accelerated by a second enzyme, **hydroxynitrile lyase**. Enzymatic release of cyanide does not normally occur in intact plants because the enzymes and the substrate are spatially separated. In some cases, separation is maintained within the cell, but in others, the enzymes are in one cell and the cyanogenic glycosides in another. In *Sorghum*, for example, the cyanogenic glycoside **dhurrin** is synthesized and stored in epidermal cells, while the glycosidase

and lyase enzymes are found in the mesophyll cells. Only when the tissue is crushed and the contents of the two cells are mixed will cyanogenesis occur.

There is some evidence that the presence of cyanogenic glycosides deters feeding by insects and other herbivores, although most animals have the ability to detoxify small quantities of cyanide. Clearly the effectiveness of cyanogenic glycosides as a deterrent depends on many factors, such as the amount present, the rate of release of cyanide, and the ability of the animal to detoxify. The level of cyanogenic glycosides in plants is highly variable, influenced by both genetic control and environmental stress. The latter is a concern when using *Sorghum* for livestock forage. Dhurrin accumulates rapidly and can cause livestock poisoning when *Sorghum* plants are stressed by drought or frost.

Many common food plants naturally contain cyanogenic glycosides in concentration sufficiently low that they are not normally a health hazard. These include soy and other beans (Fabaceae); apples, apricots, peaches, plums, and other fruits in the family Rosaceae; and flax seed (*Linum*), which is a popular health food. One food source that contains large amounts of cyanogenic glycosides is cassava, a potato-like root of the tropical plant *Manihot esculenta*. Cassava, also known as manioc or, in North America, tapioca, is a major source of starch for millions of people in tropical countries. However, poisoning is avoided by careful preparation of the plant. This includes grinding the root and expressing the fluids, or boiling the root in several changes of water.

27.3.4 GLUCOSINOLATES ARE SULFUR-CONTAINING PRECURSORS TO MUSTARD OILS

Glucosinolates are found primarily in the mustard family (Brassicaceae) and related families in the order Capparales. They are precursors to the mustard oils, an economically important class of flavor constituents

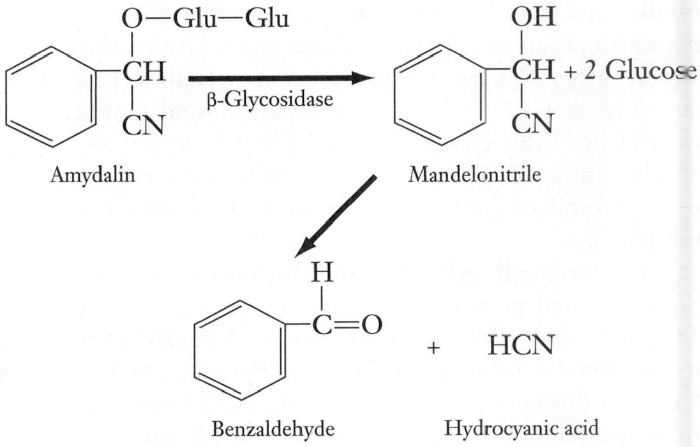

FIGURE 27.10 **Amygdalin is a cyanogenic glycoside found in large quantities in seeds of common fruits in the family Rosaceae. Hydrolysis of amygdalin is a two-step process, resulting in the release of highly toxic hydrocyanic acid (HCN).**

Amydalin   β-Glycosidase   Mandelonitrile + 2 Glucose

Benzaldehyde   Hydrocyanic acid

that gives the pungent taste to condiments such as mustards and horseradish as well as the distinctive flavor of cabbages, broccoli, and cauliflower.

All glucosinolates are **thioglucosides** (*thio*, sulfur) with the general structure shown in Figure 27.11A. The sugar is always glucose. The diversity encountered in structure and properties is due to the R group, which may range from a simple methyl group to large linear or branched chains containing aromatic or heterocyclic structures. The biological activity of glucosinolates depends primarily on their hydrolysis to **mustard oils** (Figure 27.11B). Hydrolysis of glucosinolates is catalyzed by an enzyme called myrosinase (a thioglucosidase). The hydrolysis product is unstable and immediately undergoes a rearrangement to form a thiocyanate or *iso*thiocyanate. Like the cyanogenic glycosides, glucosinolates are spatially separated from the hydrolytic enzymes so that the mustard oils are normally formed only when the cells are disrupted, allowing the enzyme and substrate to come together. As with other defense compounds, some herbivores are deterred or repelled by the presence of glucosinolates in a plant, while others have adapted to use the glucosinolates or mustard oils as attractants to stimulate feeding and ovipositing.

Glucosinolates, or rather their absence, have had a significant impact on the oilseed industry. Rape seed (principally *Brassica napus*) is a good source of vegetable oil, but its high content of glucosinolate together with high erucic acid (a 22-carbon fatty acid) gives the oil undesirable taste and poor storage properties. New strains have been bred with low glucosinolates and erucic acid. These strains, called **canola** in order to distinguish them from normal rape, are now an economically important oil source.

27.4 PHENYLPROPANOIDS

Aromatic amino acids may be directed toward either primary or secondary metabolism. Also known as **phenolics**, or **polyphenols**, **phenylpropanoids** are a large

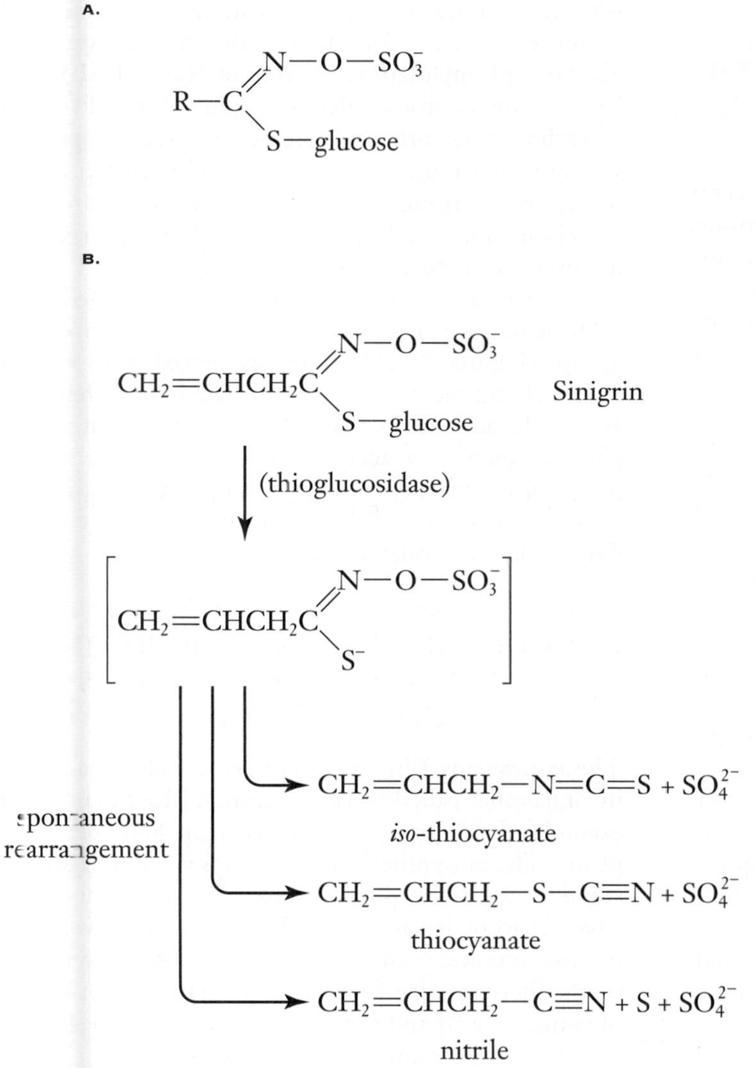

A.

B.

Sinigrin

(thioglucosidase)

iso-thiocyanate

thiocyanate

spontaneous rearrangement

nitrile

FIGURE 27.11 (*A*) All glucosinolates are thioglucosides with the same basic structure. In a thioglucoside the sugar is linked to the rest of the molecule via a sulfur atom. Variation is introduced by the composition of the R group. (*B*) Enzymatic removal of glucose from a glucosinolate creates an unstable product that spontaneously rearranges to form the pungent mustard oils in the thiocyanate, *iso*thiocyanate, or nitrile forms.

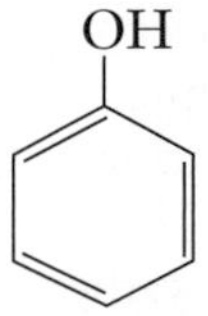

FIGURE 27.12 **Phenylpropanoids are derivatives of the simple hydroxylated aromatic ring, phenol.**

family of secondary metabolites derived from the aromatic amino acids. Phenylpropanoids are a chemically diverse family of compounds ranging from simple phenolic acids to very large and complex polymers such as tannins and lignin. Also included are the flavonoid pigments that were described earlier in Chapter 6. The basic structure is phenol, a hydroxylated aromatic ring (Figure 27.12). As with other secondary products, many phenolics appear to be involved in plant/herbivore interactions. Some (e.g., lignin) are important structural components, while others appear to be simply metabolic end-products with no obvious function.

27.4.1 SHIKIMIC ACID IS A KEY INTERMEDIATE IN THE SYNTHESIS OF BOTH AROMATIC AMINO ACIDS AND PHENYLPROPANOIDS

The biosynthesis of most phenylpropanoids begins with the aromatic amino acids phenylalanine, tyrosine, and tryptophan. These aromatic amino acids are, in turn, synthesized from phosphoenolpyruvate and erythrose-4-phosphate by a sequence of reactions known as the **shikimic acid pathway** (Figure 27.13). The shikimic acid pathway is common to bacteria, fungi, and plants, but is not found in animals. Phenylalanine and tryptophan are consequently among the 10 amino acids considered essential for animals (including humans) and represent the principal source of all aromatic molecules in animals. Tyrosine is not classified as essential because animals can synthesize it by hydroxylation of phenylalanine.

Synthesis of the aromatic amino acids begins with the condensation of one molecule of erythrose-4-P, from the pentose-phosphate respiratory pathway, with one molecule of phosphoenolpyruvate (PEP) from glycolysis. The resulting 7-carbon sugar is then cyclized and reduced to form shikimate. Shikimate is then converted to chorismate, a critical branch point that leads either to phenylalanine and tyrosine or to tryptophan. The shikimic acid pathway is an excellent example of *feedback regulation*, where the concentrations of product molecules regulate the flow of carbon through key enzymes. The first key enzyme in the pathway is the aldolase that catalyzes the initial condensation erythrose-4-P with PEP. Aldolase is inhibited by all

three end-products: tyrosine, phenylalanine, and tryptophan. In a similar manner, the conversion of chorismate to prephenate is inhibited by both phenylalanine and tyrosine, while the conversion of chorismate to anthranilate is inhibited by tryptophan.

Another step of interest in the pathway is the second step in the conversion of shikimate to 3-enolpyruvylshikimate-5-P. The enzyme **3-enolpyruvylshikimate-5-phosphate synthase (EPSPS)** is inhibited by the herbicide glyphosate (marketed commercially as RoundUp™). Plants treated with glyphosate will die because they are unable to synthesize the aromatic amino acids and their derivatives, especially protein. Thus glyphosate-treated plants effectively die of protein starvation.

27.4.2 THE SIMPLEST PHENOLIC MOLECULES ARE ESSENTIALLY DEAMINATED VERSIONS OF THE CORRESPONDING AMINO ACIDS

Synthesis of most secondary phenolic products begins with the deamination of phenylalanine to cinnamic acid (Figures 27.14, 27.15). The enzyme that catalyzes this reaction, **phenylalanine ammonia lyase (PAL)**, is a key enzyme because it effectively controls the diversion of carbon from primary metabolism, such as protein synthesis, into the production of phenylpropanoids. An alternate route, the deamination of tyrosine to p-OH-cinnamic acid, appears to be limited largely, if not exclusively, to grasses.

Cinnamic acid is readily converted to p-OH-cinnamic acid by the addition of a hydroxyl group (Figure 27.15). The sequential addition of hydroxyl and methoxy groups gives rise to caffeic acid and ferulic acid, respectively. None of these four simple phenols appear to accumulate to any extent. Their principal function appears to be as precursors to more complex derivatives such as coumarins, lignin, tannins, flavonoids, and isoflavonoids.

27.4.3 COUMARINS AND COUMARIN DERIVATIVES FUNCTION AS ANTICOAGULANTS

The **coumarins** (Figure 27.16) are a widespread family of lactones called benzopyranones. More than 1,500 examples are known, from more than 800 species of plants. The biosynthesis of coumarins is not well understood because the putative enzymes and their genes have yet to be isolated. It is most likely, however, that the two simplest forms, coumarin and 7-OH-coumarin (umbelliferone), are formed by an *ortho*-hydroxylation of cinnamic acid and p-coumaric acid, respectively, followed by ring closure (Figure 27.17).

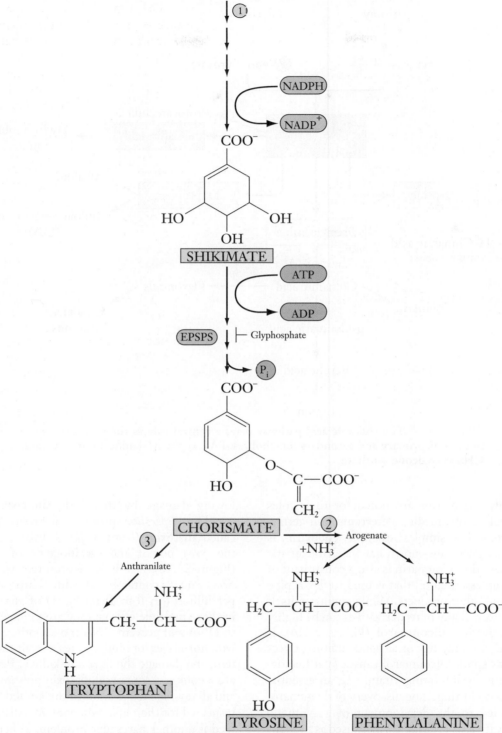

FIGURE 27.13 **The shikimic acid pathway for biosynthesis of aromatic amino acids in plants. Initial precursors are erythrose-4-P from the pentose-phosphate pathway and phosphoenolpyruvate from glycolysis. Enzymes indicated as 1, 2, and 3 are subject to feedback inhibition and thus are important regulatory enzymes in the pathway. Those enzymes are (1) 3-deoxyarabino-heptulosonate-7-phosphate synthase; (2) anthranilate synthase; (3) chorismate mutase. The enzyme EPSP synthase (EPSPS), which catalyzes the second of three reactions in the conversion of shikimate to chorismate, is inhibited by the herbicide glyphosate. Glyphosphate thus prevents the synthesis of the amino acids tyrosine and phenylalanine.**

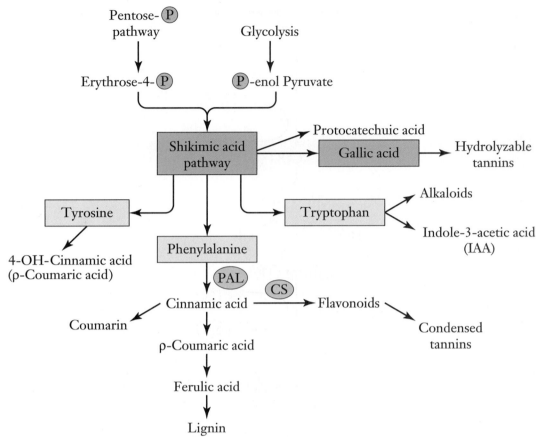

FIGURE 27.14 **The shikimic acid pathway plays a central role in the synthesis of numerous primary and secondary metabolites. PAL = phenylalanine ammonia lyase. CHS = chalcone synthase.**

As a family, coumarins are noted for their roles as antimicrobial agents, feeding deterrents, and germination inhibitors. The simplest example, coumarin, is the product that gives new-mown hay its characteristic pleasantly sweet odor. Coumarin is also a constituent of Bergamot oil, an essential oil that is used to flavor pipe tobacco, tea, and other products. While coumarin itself is only mildly toxic, many of its derivatives can be highly toxic. One derivative, **dicoumarol** (Figure 27.16), is typically found in moldy hay or silage containing sweet clover (*Meliotus* spp.). Dicoumarol causes fatal hemorrhaging in cattle by inhibiting vitamin K, an essential cofactor in blood clotting. The discovery of dicoumarol in the 1940s led to the development of a synthetic coumarin derivative, Warfarin™, widely used as a rodent poison. Scopoletin (Figure 27.16), the most prevalent coumarin in higher plants, is often present in seed coats. It is suspected of being a germination inhibitor that must be leached out of the seed coat before germination can proceed.

The most toxic coumarin derivatives are synthesized not by plants, but by fungi. The fungus *Aspergillus flavus* commonly infects foodstuffs such as livestock feed, peanuts, and maize (the latter especially fol-

lowing damage by the European corn borer). The invading *Aspergillus* produces a group of mycotoxins called aflatoxins (from *A. fla.* + toxin), believed to be the most potent and carcinogenic of natural toxins (Figure 27.16). Deaths have been recorded from ingestion of maize contaminated with as little as 6 to 15 parts per billion (or, 6 to 15 µg kg⁻¹) of aflatoxin. Aflatoxins have multiple effects: they are mutagenic; they bind to DNA and prevent RNA transcription; they compete with hormones for binding sites, impair the immune system, and damage the liver and kidney. Peanut products are a common source of aflatoxin poisoning in humans and all raw, shelled peanuts in the United States must be inspected for their aflatoxin content. Aflatoxins in cattle feed is another particular problem. When cattle ingest the toxin, they convert it to another equally toxic form and secrete it into the milk.

27.4.4 LIGNIN IS A MAJOR STRUCTURAL COMPONENT OF SECONDARY CELL WALLS

Lignin is a highly branched polymer of three simple phenolic alcohols known as **monolignols**

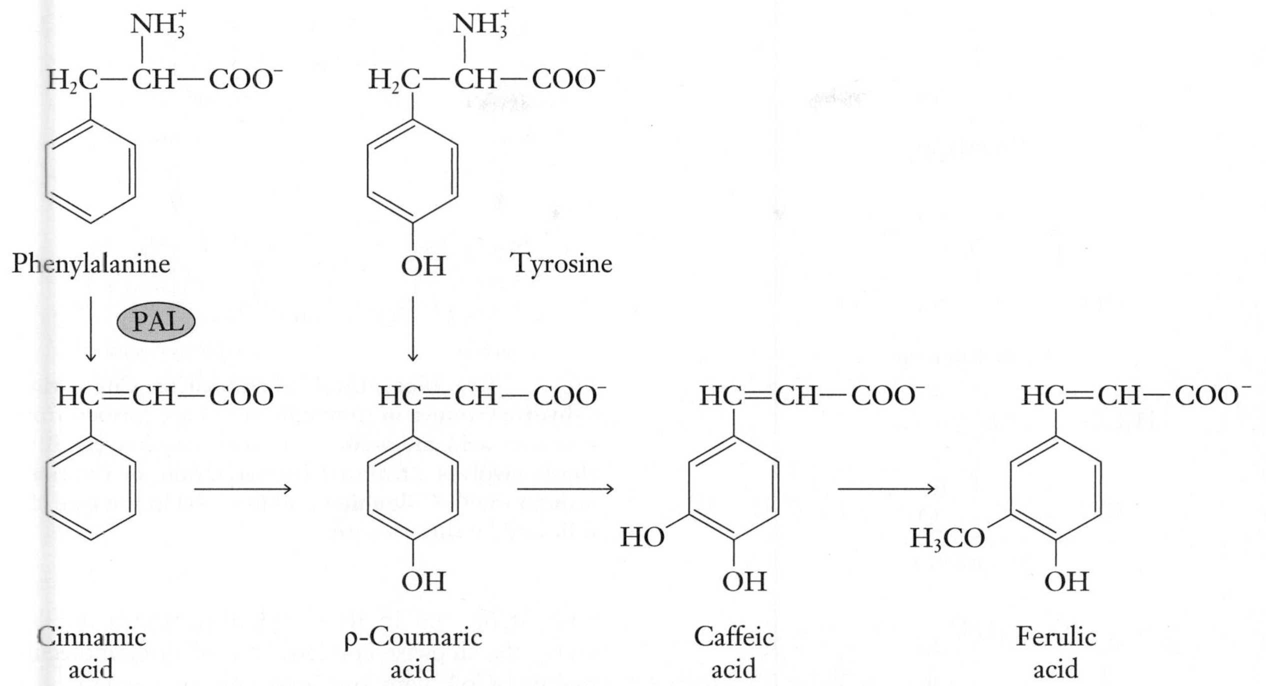

FIGURE 27.15 **Phenolic building blocks. Deamination of phenylalanine followed by hydroxylation to form *p*-coumaric acid are the first two steps in phenylpropanoid biosynthesis. In grasses, *p*-coumaric acid may be formed directly by deamination of tyrosine. Phenylalanine ammonia lyase (PAL) is a critically regulated enzyme that controls the diversion of phenylalanine from protein biosynthesis to phenylpropanoid biosynthesis.**

(Figure 27.18). Gymnosperm lignin is comprised mainly of coniferyl alcohol subunits while angiosperm lignin is a mixture of coniferyl and sinapyl alcohol subunits. The biosynthesis of lignin is not well understood, but it is believed that the alcohols are oxidized to free radicals by the ubiquitous plant enzyme, **peroxidase**. The free radicals then react spontaneously and randomly to form polymeric lignin.

Lignin is found in cell walls, especially the secondary walls of tracheary elements in the xylem. In spite of its abundance (second only to cellulose), the structure of lignin is not well understood. Lignin is a very large polymer; it is insoluble in water and most organic solvents, and impossible to extract without considerable degradation. Moreover, the three basic monomers may link together in a variety of ways to form a multibranched, three-dimensional structure. The complexity is so great that, like snowflakes, each lignin "molecule" may be unique. For the same reason that monolignols are able to form extensive linkages with each other, the lignin polymer is able to form numerous cross-links with other cell wall polymers. The result is a high degree of mechanical strength and rigidity in lignified woody stems. Lignin is in fact what makes wood, wood!

Although the principal function of lignin is structural, it has also been implicated as a defensive chemical. Lignin itself is not readily digested by herbivores and, because it is covalently linked to cellulose and cell wall

xyloglucans, its presence decreases the digestibility of these polymers as well. Also, when fungal pathogens enter host cells, they do so by enzymatically degrading the host cell wall. Several studies have shown that lignins and other phenolic derivatives accumulate at the site of fungal penetration, presumably slowing the rate of cell wall degradation.

27.4.5 FLAVONOIDS AND STILBENES HAVE PARALLEL BIOSYNTHETIC PATHWAYS

As noted earlier in Chapter 6, flavonoids include the anthocyanin pigments that serve to attract insect pollinators and the isoflavonoids that function as phytoalexins, or antibacterial and antifungal agents. Flavonoids have a role in symbiont recognition between rhizobia and host roots. It has also been demonstrated that exposure of plants to UV radiation increases the content of flavonoids, suggesting that flavonoids such as kaempferol may offer a measure of protection by screening out harmful UV-B radiation.

Flavonoids represent a very large class of phenolic derivatives (more than 4,500 representatives are known), with a variety of functions. Biochemically, however, all flavonoids share a common structure consisting of three rings, labeled A, B, and C (Figure 27.19).

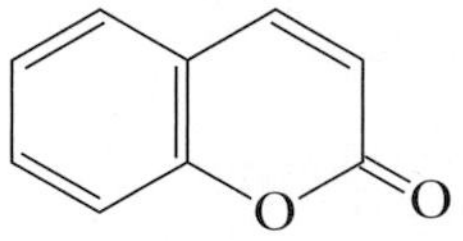

Coumarin

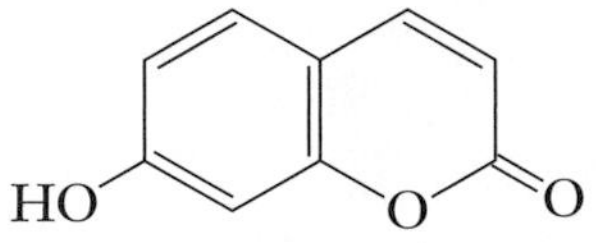

Umbelliferone

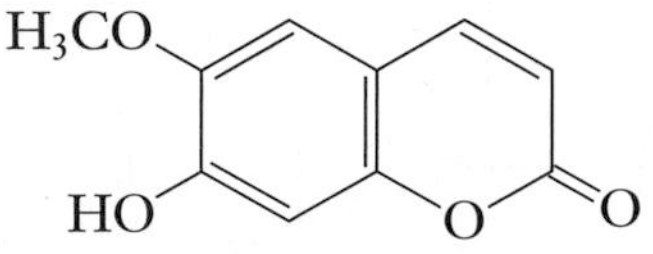

Scopoletin

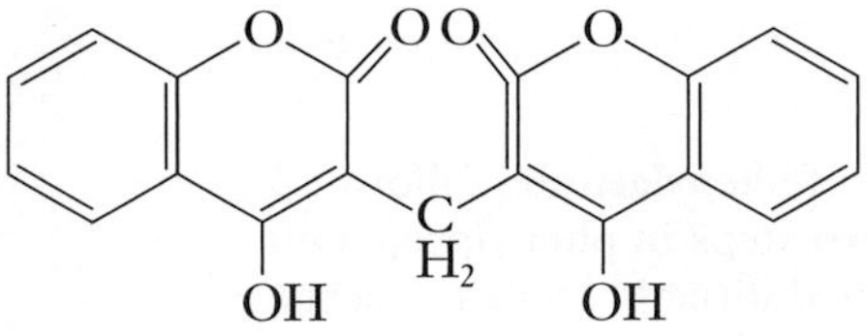

Dicoumarol

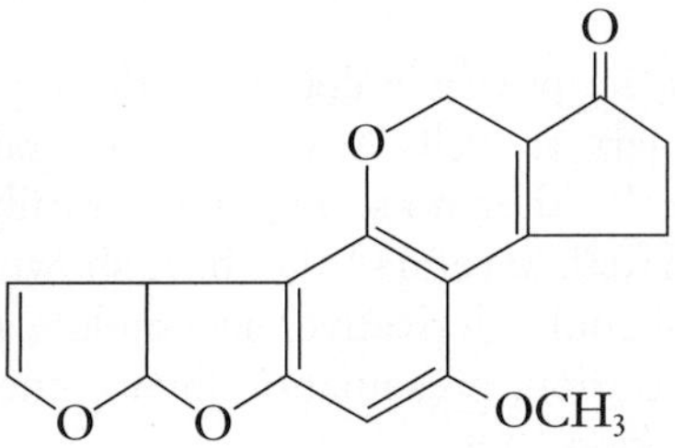

Aflatoxin B₁

Figure 27.16 **Coumarins. More than 300 coumarins have been reported from more than 70 plant families. Coumarin and 7-hydroxycoumarin (umbelliferone) are derived from cinnamic acid and *p*-coumaric acid, respectively. Scopoletin is probably the most widespread coumarin in plants. Dicoumarol is a powerful anticoagulant found in moldy hay. Dicoumarol derivatives and synthetic coumarins are commonly used to thin blood in cardiac patients. Aflatoxin B₁, a mycotoxin, is among the most carcinogenic and toxic of naturally occurring compounds.**

Cinnamic acid

p-Coumaric acid

Coumarin

7-Hydroxy coumarin

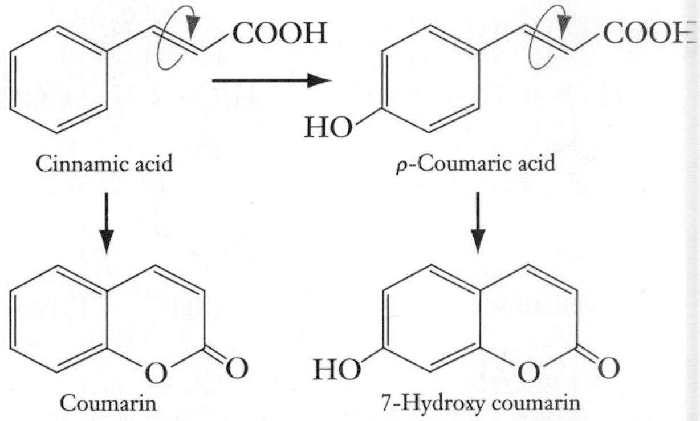

Figure 27.17 **Biosynthesis of coumarins. Coumarin and 7-hydroxycoumarin (umbelliferone) are formed from cinnamic acid and *p*-coumaric acid, respectively. Biosynthesis involves a *trans/cis*-isomerization, or rotation around the C-C double bond (arrows) in the side chain, followed by ring closure.**

committed step in flavonoid biosynthesis. CHS catalyzes the stepwise condensation of three molecules of malonyl-CoA with one molecule of *p*-coumaryl-CoA to form 4,2′,4′,6′-tetrahydroxychalcone, or, naringenin chalcone. A chalcone is a C₆—C₃—C₆ pattern wherein ring C is not yet closed. The enzyme chalcone isomerase then catalyzes the closure of ring C to form naringenin. Naringenin is the precursor to flavonols such as kaempferol and quercitin as well as the anthocyanin pigments.

Stilbenes are a group of C₆—C₂—C₆ compounds that have a chemical defense role, primarily against fungal invasions of heartwood (the distinctively colored core of the tree). Stilbenes are one of several phenylpropanoid derivatives that are continually infused into the heartwood of trees after the cells have been lignified. A defensive role is suggested by the localized deposition of stilbenes in regions of sapwood as well, when the sapwood is attacked by insects of pathogenic fungi. The role of stilbenes in this case appears to be one of preventing the spread of the invading insect or pathogen.

Stilbene biosynthesis, like the flavonoids, involves the sequential condensation of malonyl-CoA units with either cinnamoyl-CoA or *p*-coumaroyl-CoA, except that the enzyme catalyzing the reaction is **stilbene synthase**.

27.4.6 TANNINS DENATURE PROTEINS AND ADD AN ASTRINGENT TASTE TO FOODS

The name **tannin** is derived from the historic practice of using plant extracts to "tan" animal hides, that is, to convert hides to leather. Such extracts contain a mixture of chemically complex phenol derivatives that bind to, and thus denature, proteins. For years, the most common test for tannins was the capacity to precipitate

Both ring B and the 3-carbon bridge that makes up ring C are derived from the shikimic acid pathway via phenylalanine and *p*-coumaric acid. The six carbons that make up the A ring are derived from **malonic acid**, in the form of a malonyl-coenzyme A complex, **malonyl-CoA**. (Malonyl-CoA is also a principal intermediate in fatty acid synthesis.) The key enzyme is **chalcone synthase (CHS)**, which catalyzes the first

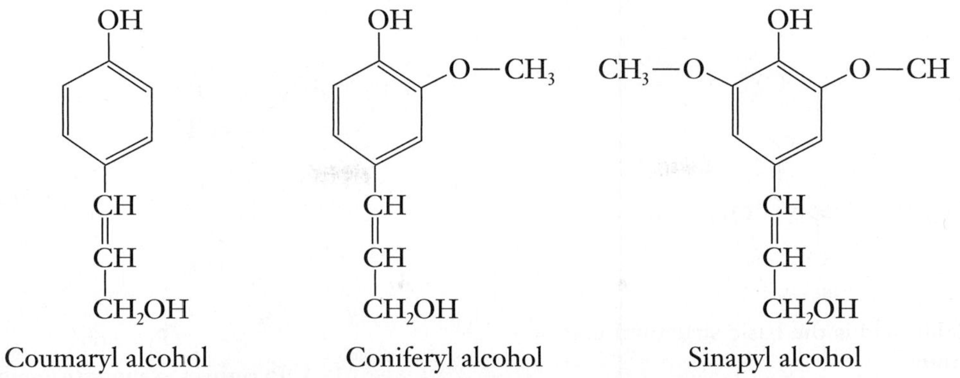

Coumaryl alcohol   Coniferyl alcohol   Sinapyl alcohol

FIGURE 27.18 **Monolignols are the principal lignin monomers. Extensive cross-linkages most commonly form between the ring alcohol group and the double-bonded carbon atoms.**

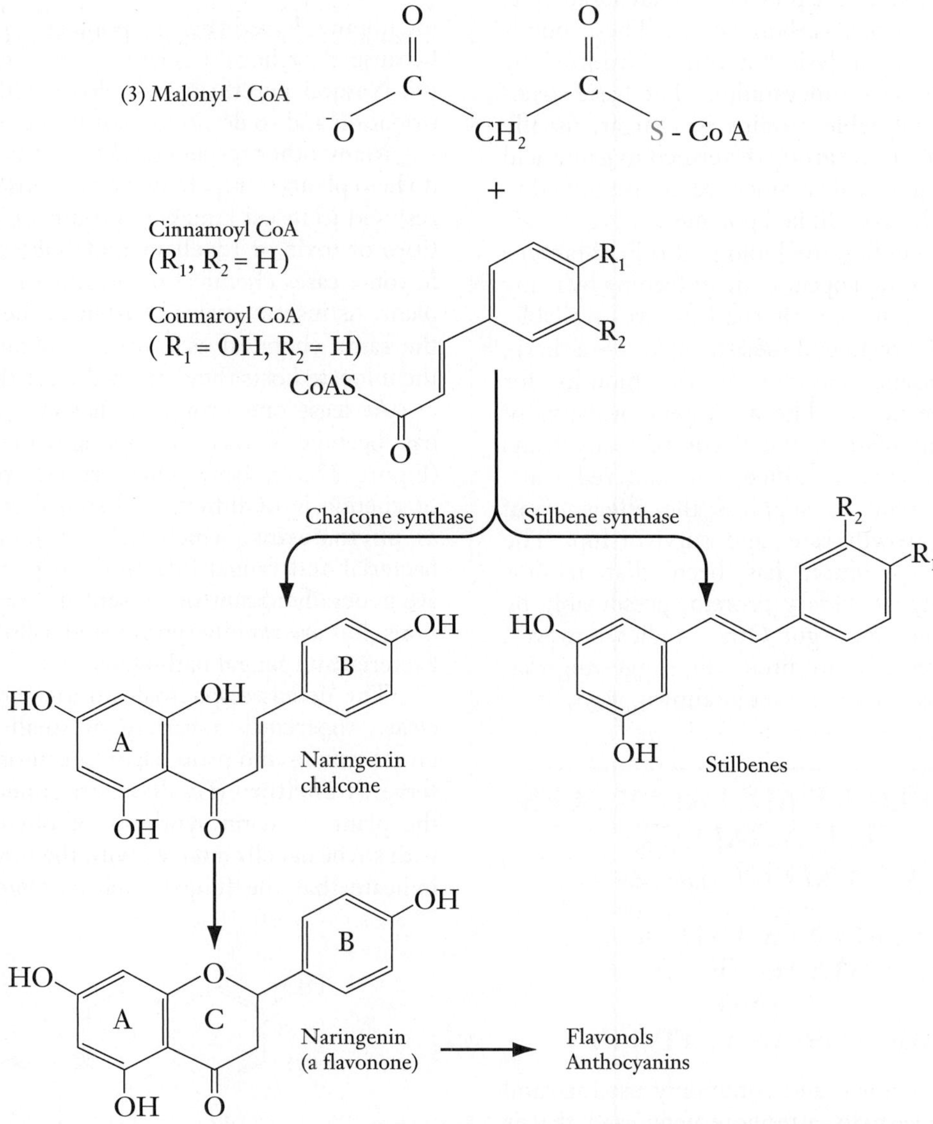

FIGURE 27.19 **The flavonoid ring structure and stilbenes are both synthesized by sequential condensations of 3 molecules of malonyl-coenzyme A (CoAS) with either cinnamoyl-coenzyme A or *p*-coumaroyl-coenzyme A. Malonyl-coenzyme A is in turn formed by the carboxylation of acetyl-CoA, the first step in the formation of long-chain fatty acids.**

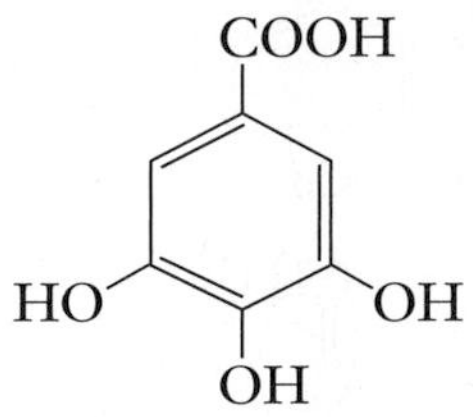

FIGURE 27.20 **Gallic acid is the basic structural unit of hydrolyzable tannins.**

FIGURE 27.21 **Pyrethrin, a monoterpenoid ester, is a commonly used organic insecticide. Pyrethrin is isolated from thedried, unexpanded flowerheads of** *Chrysanthemum cineraiifolium* **(Asteraceae).**

gelatin (a protein). Two categories of tannins are now recognized: condensed tannins and hydrolyzable tannins. Condensed tannins are polymers of flavonoid units linked by strong carbon-carbon bonds. These bonds are not subject to hydrolysis but can be oxidized by strong acid to release anthocyanidins. The basic structural unit of hydrolyzable tannins is a sugar, usually glucose, with its hydroxyl groups esterified to gallic acid (Figure 27.20). Gallic acid residues are in turn joined to form an extensively cross-linked polymer.

Like other phenolics, the biological role of tannins is not clear. Tannins do appear to deter feeding by many animals when tannin-free alternatives are available. This effect could be related to the astringency—a sharp, somewhat unpleasant sensation in the mouth—for which tannins are noted. The astringent property of tannins is a component in the flavor of many fruits as well as drinks such as coffee, tea, and red wine. As well, tannins tend to suppress the efficiency of feed utilization, growth rate, and survivorship. The conventional interpretation has been that tannins reduce digestibility of dietary protein, presumably by binding with protein in the gut. Other studies, however, have cast doubt on this interpretation, suggesting that tannins may be toxic in other, yet unknown, ways.

27.5 SECONDARY METABOLITES ARE ACTIVE AGAINST INSECTS AND DISEASE

27.5.1 SOME TERPENES AND ISOFLAVONES HAVE INSECTICIDAL AND ANTI-MICROBIAL ACTIVITY

One of the better-known and commonly used natural insect toxins is **pyrethrin**, a monoterpene ester that is produced in the ovaries of flowers of *Chrysanthemum* (or, *Pyrethrum*) *cineraiifolium* (Figure 27.21). *C. cineraiifolium* is widely cultivated in Montenegro, Japan, and eastern Africa, where the unexpanded flower heads are dried and powdered. Pyrethrin is a neurotoxin that interferes with sodium channels in the insect nerve

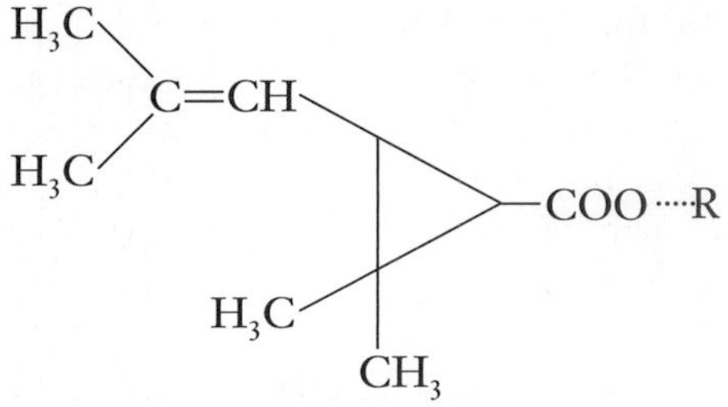

membrane. Pyrethrins are popular organic insecticides because they have a relatively low mammalian toxicity. Natural pyrethrins are also readily inactivated by oxidation and so do not persist in the environment.

Many other terpenes, while not insecticides per se, do help plants to repel invasions by insects. Many plants respond to insect attack by producing additional quantities of toxic or repellant metabolites. Unfortunately, in some cases chemical deterrents can turn against the plant. As insects evolve resistance, they are able to use the same chemical as a host-recognition cue to help them locate hosts they can feed on without ill effect.

At least one group of flavones, the **isoflavones**, has become known for its anti-microbial activities (Figure 27.22). Isoflavones are one of several classes of chemicals of differing chemical structures, known as **phytoalexins**, which help to limit the spread of bacterial and fungal infections in plants. Phytoalexins are generally absent or present in very low concentrations, but are rapidly synthesized following invasion by bacterial and fungal pathogens.

The details of phytoalexin metabolism are not yet clear. Apparently a variety of small polysaccharides, glycoproteins and proteins of fungal or bacterial origin, serve as **elicitors** (L. *elicere*, to entice) that stimulate the plant to begin synthesis of phytoalexins. Studies with soybean cells infected with the fungus *Phytophthora* indicate that the fungal elicitors trigger transcription

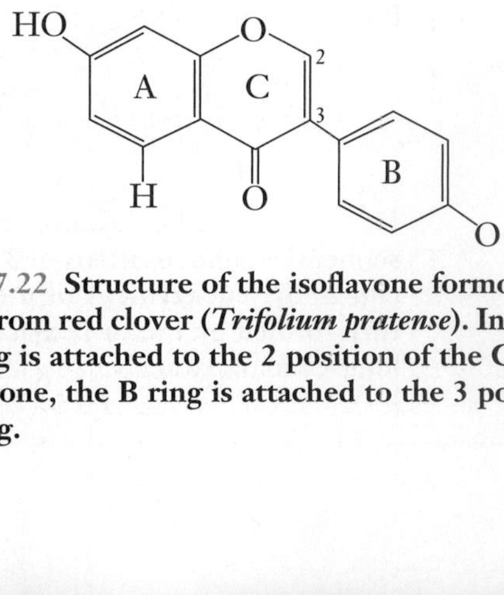

FIGURE 27.22 **Structure of the isoflavone formononetin, isolated from red clover (***Trifolium pratense***). In a flavone, the B ring is attached to the 2 position of the C ring. In an isoflavone, the B ring is attached to the 3 position of the C ring.**

of mRNA for enzymes involved in the synthesis of isoflavones. The production of phytoalexins appears to be a common defense mechanism. Isoflavones are the predominant phytoalexin in the family Leguminoseae, but other families such as Solanaceae, appear to use terpene derivatives.

27.5.2 RECOGNIZING POTENTIAL PATHOGENS

As discussed earlier in Chapter 13, plants challenged by insects or potentially pathogenic organisms also respond with a **hypersensitive reaction**, characterized by changes in the composition and properties of the cell wall and other factors, including necrotic lesions at the site of the invasion, that limit the spread of the invading pathogen. Because the hypersensitive reaction is a form of developmental response, we must assume that a signal detection and transduction chain is involved.

Attempts to explain the susceptibility of plants to infection have shown that disease has an underlying genetic basis. In fact both pathogens and potential host plants carry genes that determine whether or not disease will occur and disease will occur only when those genes are compatible. Compatibility can be explained by the **gene-for-gene model,** which predicts that resistance will occur only when the pathogenic microorganism carries a dominant allele at the **avirulence (*Avr*) locus** and host plants carry a complementary dominant allele at the **resistance (*R*) locus** (Figure 27.23). Any other combination of dominant and recessive alleles leads to a successful infection. A matching pair of dominant pathogen avirulence genes and dominant plant resistance genes initiates a hypersensitive reaction in the plant.

Although a number of avirulence genes have been isolated from both bacteria and fungi, the specific function of their products is not known. One possibility is that avirulence genes encode enzymes for the production of elicitors and resistance genes encode receptors that recognize elicitors. A variety of elicitors have been identified, most of them extracellular microbial products commonly associated with cell walls of bacteria and fungi. For example, fungal elicitors include β-glucans, chitosan (a chitin subunit),[1] and arachidonic acid (an unsaturated lipid). Other elicitors include various polysaccharides, glycoproteins, and small peptides. Even pectic fragments, resulting from initial degradation of the plant cell wall pectins, or mechanical damage are capable of eliciting a hypersensitive reaction.

Recognition of elicitors by the plant cell likely takes place at the plasma membrane. It is expected that some form of signal transduction pathway is required to relay this information to the nucleus in order to effect gene transcription. A variety of common signaling agents have been suggested, including changes in pH, and ion fluxes (especially potassium and calcium). For example, a transient uptake of Ca^{2+} (and efflux of K^+) was observed when cultured cells were challenged with a fungal elicitor. Moreover, expression of defense response genes can be regulated by regulating intracellular Ca^{2+} levels. Thus, defense responses can be activated by stimulating Ca^{2+} uptake with Ca^{2+} ionophores or inhibited by blocking Ca^{2+} channels. Other early events in elicitor-treated cells include protein phosphorylation and the production of active oxygen species (O_2^- and H_2O_2), known as the oxidative burst. The precise role of these various signals and how they interact in the signal cascade is unknown. It is a topic that is under active investigation in numerous laboratories.

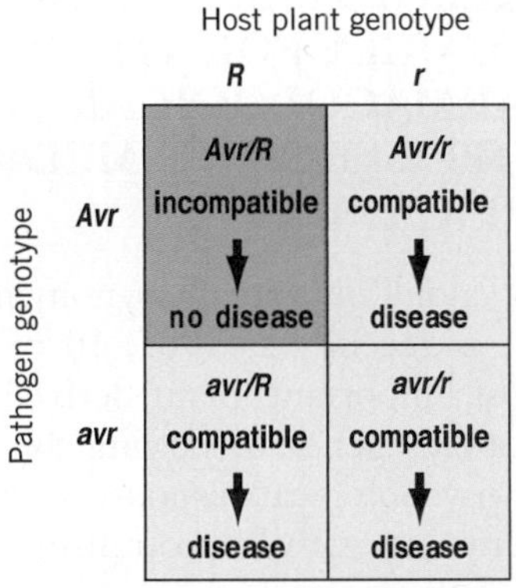

27.5.3 SALICYLIC ACID, A SHIKIMIC ACID DERIVATIVE, TRIGGERS SYSTEMIC ACQUIRED RESISTANCE

In many cases, the hypersensitive reaction leads to a general immunity against infection known as **systemic acquired resistance** or **SAR** (Chapter 13). Most of the evidence suggests that the signal for SAR is salicylic acid (SA) or its volatile methylated derivative methylsalicylate. The role of SA in SAR is indicated by several lines of evidence. For example, SAR is characterized by a rise in the levels of SA along with the activation of defense-related genes and the appearance of their products, pathogenesis-related (PR) proteins. In addition, SA

Host plant genotype

	R	*r*
Avr	**Avr/R** incompatible ↓ no disease	**Avr/r** compatible ↓ disease
avr	**avr/R** compatible ↓ disease	**avr/r** compatible ↓ disease

(Pathogen genotype)

FIGURE 27.23 **The gene-for-gene model predicts that incompatibility between a pathogen and host requires complementary dominant avirulence and resistance genes in the pathogen and host, respectively. All other genotypes are compatible and lead to invasion of the host by the pathogen or, i.e., disease.**

---

[1]The principal carbohydrate in most fungal cell walls is chitin rather than cellulose. Chitin is a polymer of N-acetylglucosamine that forms microfibrils similar to cellulose.

is synthesized from chorismate via isochorismate, a reaction catalyzed by the enzyme, but *Arabidopsis* mutants that failed to accumulate SA after SAR induction were shown to be deficient of this enzyme. Accumulation of SA in infected plants carrying mutations of the isochorismate synthase gene is no more than 5 to 10 percent of wildtype plants and systemic resistance is severely compromised.

The molecular genetics and biochemistry of SAR is slowly yielding to investigation. A central player is the regulatory protein NPR1 (NON EXPRESSOR OF PATHOGENESIS-RELATED GENES1). *NPR1* is expressed at low levels throughout the plant, but its messenger RNA levels rise some two- to threefold following pathogen infection or salicylic acid treatment. In addition, several *Arabidopsis* mutants that were not responsive to salicylic acid were found to have mutations in the *NPR1* gene. *NPR1* protein interacts with a group of transcription factors known as the TGA family of transcription factors; an interaction that requires salicylic acid and stimulates the transcription of PR genes.

Of particular interest is the recent discovery that changes in the redox status of the cell play an important role in the NPR1/TGA interaction. A change in the cellular redox potential, from an initial burst of reactive oxygen species to a more reducing environment, is induced by salicylic acid and appears to facilitate the transcription of PR genes by triggering the translocation of *NPR1* from the cytosol into the nucleus where it is free to interact with the TGA transcription factors.

27.6 JASMONATES ARE LINKED TO UBIQUITIN-RELATED PROTEIN DEGRADATION

As previously shown in Chapter 13, a plant's resistance to insects and disease is also mediated by the fatty acid derivative **jasmonic acid** and its methyl ester (**methyljasmonate**), collectively referred to as jasmonates (See Figure 13.16). Jasmonates were first recognized for their ability to promote senescence of detached barley leaf segments, but a role in disease resistance was suggested when phytoalexin biosynthesis in cell cultures was linked to jasmonic acid content. It is now known that jasmonic acid accumulates in wounded plants and in plants treated with many elicitors. Jasmonic acid has also been linked to the activation of a number of genes encoding proteins with antifungal properties.

Jasmonic acid is synthesized by the oxidation and rearrangement of linolenic acid, an unsaturated fatty acid (see Appendix). Plant membranes are a rich source of linolenic acid as a constituent fatty acid of phospholipids. It is thought that elicitors first bind with an unknown

receptor in the plasma membrane. The elicitor-receptor complex activates a membrane-bound phospholipase that releases linolenic acid. The formation of jasmonic acid then involves several steps in which the linolenic acid undergoes successive oxidizations and cyclization.

Central to jasmonate signaling is a family of transcriptional repressors (JAZ) and an F-box protein (COI1) that is structurally and functionally related to the TIR1 auxin receptor (Chapter 18). Because of this similarity, the current model for jasmonate signaling bears a close resemblance to that of auxin-mediated degradation of AUX/IAA transcriptional repressors previously described in Chapter 18. According to this model, jasmonates promote binding of JAZ repressor proteins to the E3 ubiquitin ligase (SCF^{COI1}) and their subsequent degradation via the ubiquitin-26S proteasome system, thus allowing for the activation of jasmonate-responsive genes.

27.7 ALKALOIDS

27.7.1 ALKALOIDS ARE A LARGE FAMILY OF CHEMICALLY UNRELATED MOLECULES

As a group, **alkaloids** share three principal characters: they are soluble in water, they possess at least one nitrogen atom, and they exhibit high biological activity. Often the nitrogen will accept a proton, which gives it a slightly basic, or alkaline, character in solution (hence the name *alkaloid*). Alkaloids are for the most part heterocyclic, although a few aliphatic (noncyclic) nitrogen compounds, such as **mescaline** and **colchicine**, are considered alkaloids. Altogether, some 12,000 alkaloids have been found to occur in approximately 20 percent of the species of flowering plants, mostly herbaceous dicots.

27.7.2 ALKALOIDS ARE NOTED PRIMARILY FOR THEIR PHARMACOLOGICAL PROPERTIES AND MEDICAL APPLICATIONS

The word "alkaloid" is virtually synonymous with the word "drug"; as recently as 1985, 10 of the 12 commercially most important plant-derived drugs were alkaloids. Alkaloids generate varying degrees of physiological and psychological response in humans, most often by interfering with neurotransmitters. In large doses, most alkaloids are highly toxic, but in smaller doses they may have therapeutic value.

From prehistory to the present, alkaloids or alkaloid-rich extracts have been used for a variety of pharmacological purposes, such as muscle relaxants, tranquilizers, antitussives, pain killers, poisons, and mind-altering drugs. One of the oldest known is opium, an exudate obtained from the immature seed capsule of the opium poppy (*Papaver somniferon*). The use of opium mixed with wine to induce sleep and relieve pain was noted on Sumerian tablets dating back to 2500 BC. The species name *somniferon* was chosen by Linnaeus because of its sleep-inducing properties. Opium is a latex gum containing a mixture of more than 20 different alkaloids, including **morphine, codeine**, and **papaverine** (Figure 27.24). The opium poppy has been traditionally cultivated in the Golden Crescent of the eastern Mediterranean (presently Iran, Afghanistan, and Pakistan) and in parts of southeast Asia. The genus *Papaver* contains 9 other species, many of them common garden ornamentals, but none of which produce alkaloids of medical interest. In 399 BC, the Greek philosopher Socrates was executed by consuming an extract of hemlock (*Conium* spp.) containing the alkaloid **coniine**. In the Western Hemisphere, the Aztecs and other native cultures used the peyote cactus (*Lophophora williamsii*), containing **mescaline** and a large number of other alkaloids, for its hallucinogenic properties.

Alkaloids are generally classified on the basis of the predominant ring system present in the molecule (Figure 27.25). In spite of the extensive variation in structure, however, alkaloids are generated from a limited number of simple precursors. Most alkaloids are synthesized from a few common amino acids (tyrosine, tryptophan, ornithine or argenine, and lysine). The tobacco alkaloid **nicotine** is synthesized from nicotinic acid and **caffeine** is a purine derivative.

Although a few alkaloids are found in several genera or even families, most species display their own unique, genetically determined pattern. As with other secondary metabolites, individual alkaloids may be restricted to particular organs, such as roots, leaves, or young fruit.

Berberine, however, is found in several sources, including seeds of barberry (*Berberis vulgaris*). Barberry was at one time a common horticultural hedge but is now out of fashion due to the high incidence of poisoning in young children attracted to its bright-red berries.

The quinolizidine alkaloids, such as **lupinine**, are frequently called lupine alkaloids because of their high abundance in the genus Lupinus. Although range animals are deterred from eating lupines because of their bitter taste, grazing on lupines is still a common cause of poisoning of grazing cattle. The highest concentration of alkaloids occurs in the seed, so livestock losses are generally highest in the fall. Other alkaloids that cause poisoning of livestock include **senecionine** (*Senecio*, groundsel), **lycotonine** (*Delphinium*, larkspur), **scopolamine** (*Datura stramonium*, jimson weed), and **atropine** (also known as **hyoscyamine**) from black henbane (*Hyoscyamus niger*). Agricultural crops are not without some risk to human consumption. The family Solanaceae is noted for genera such as *Datura*, *Hyoscyamus*, and *Atropa belladonna* (the deadly nightshades), all containing high amounts of toxic alkaloids. Edible members of the Solanaceae include potato (*Solanum tuberosum*) and tomato (*Lycopersicum esculentum*), which produce the steroidal alkaloids α-solanine and tomatine, respectively. The solanine alkaloids are cholinesterase inhibitors that interfere with nerve transmission. Fortunately, the alkaloid content of both potato tubers and ripe tomato fruit is well below toxic levels, although green vines have higher levels and are potentially toxic if eaten. Potato tubers that have been exposed to strong sunlight and have begun to green may also synthesize toxic levels of the α-solanine and should not be eaten.

The indole alkaloids are often referred to as terpenoid-alkaloids because, although the basic indole ring structure is derived from tryptophan, the rest of the molecule is derived from the mevalonic acid

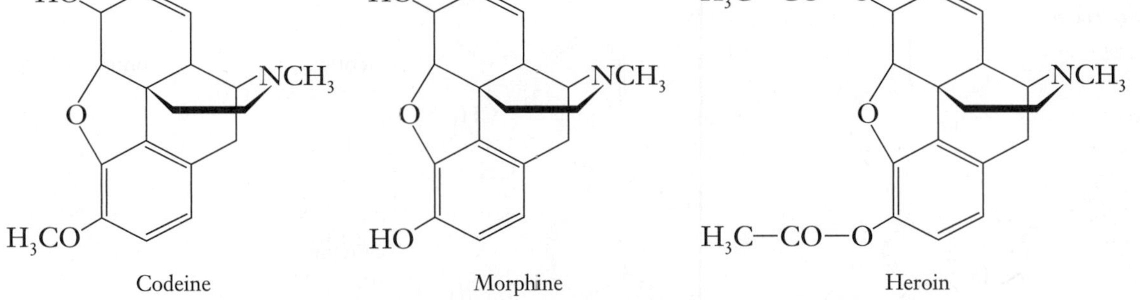

FIGURE 27.24 **Codeine, morphine, and heroin are structurally related alkaloids. Codeine and morphine are naturally occurring alkaloids isolated from the seed capsules of the opium poppy (*Papaver somniferon*). Heroin is a semisynthetic alkaloid produced by acetylation of morphine. Codeine is commonly used as a cough-suppressant and local anesthetic. Morphine is used primarily as an analgesic or pain killer. Heroin was originally synthesized by pharmaceutical chemists in the late nineteenth century as part of an effort to find a more effective alternative to morphine.**

Alkaloid Class	Example	Other Representatives
Quinoline	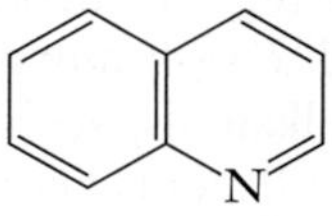 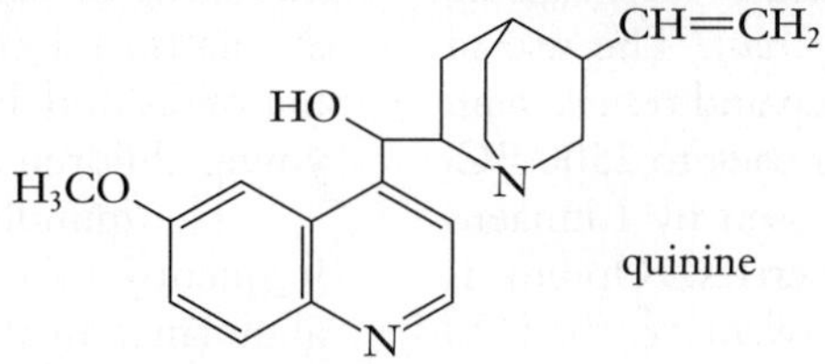 quinine	
Isoquinoline	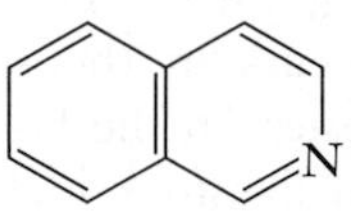 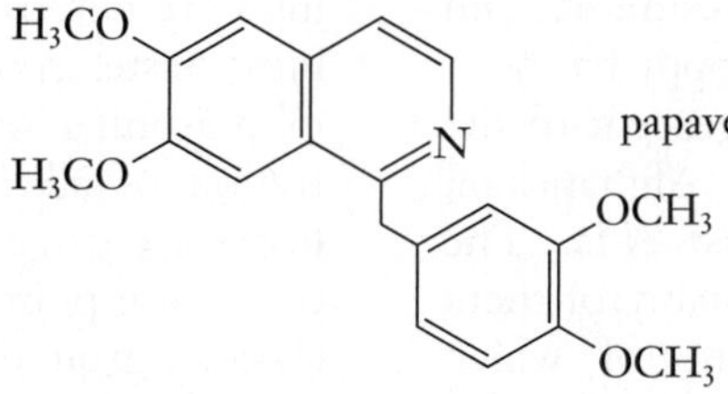 papaverine	morphine codeine berberine
Indole	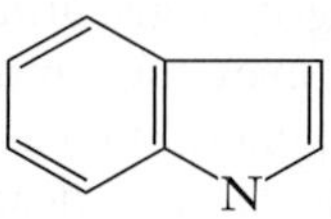 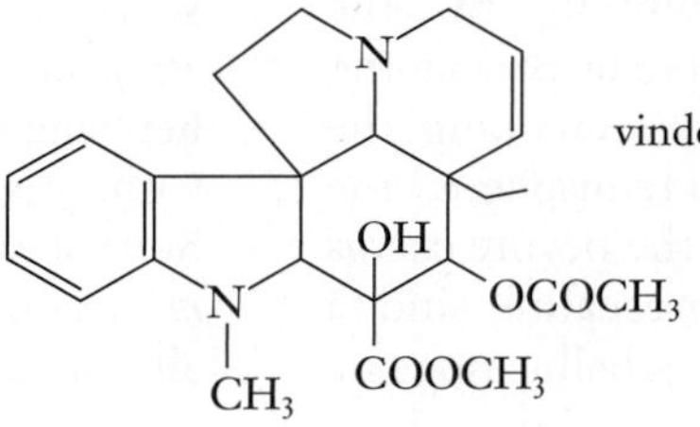 vindoline	vinblastine reserpine strychnine
Pyrrolizidene	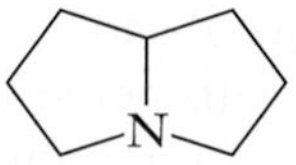 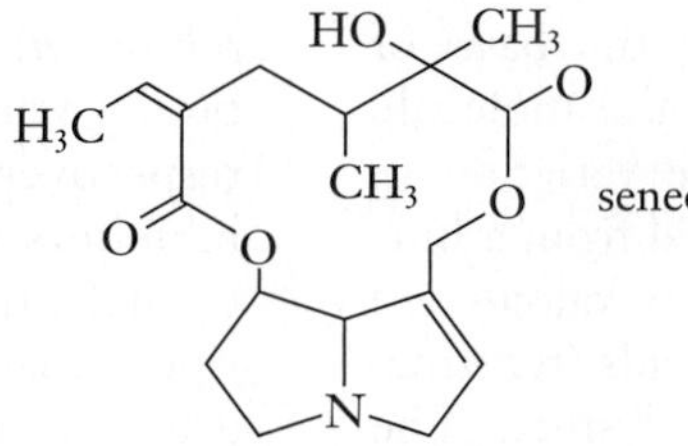 senecionine	retrorsine
Quinolizidine	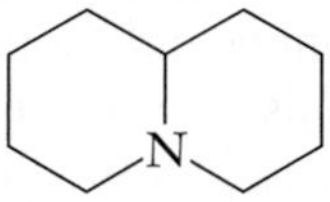 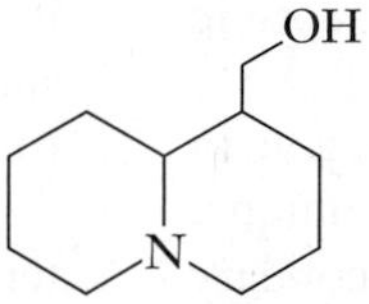 lupinine	cytisine
Tropane	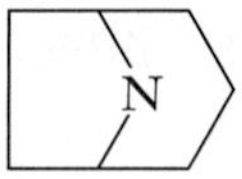 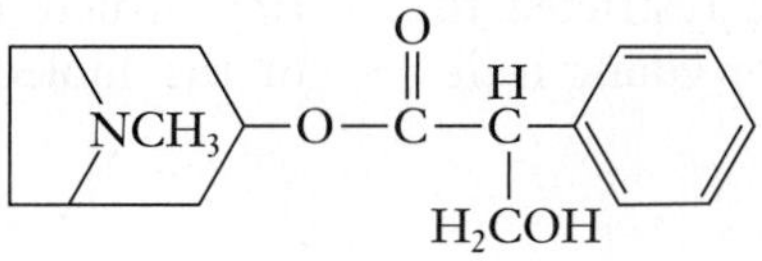 atropine	scopolamine cocaine
Piperidine	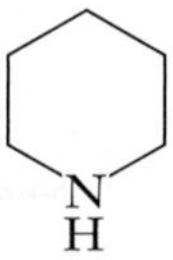 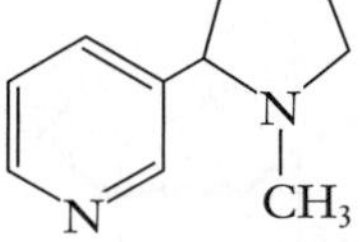 nicotine	coniine
Purine	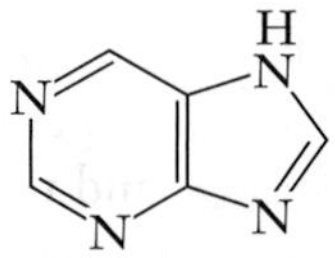 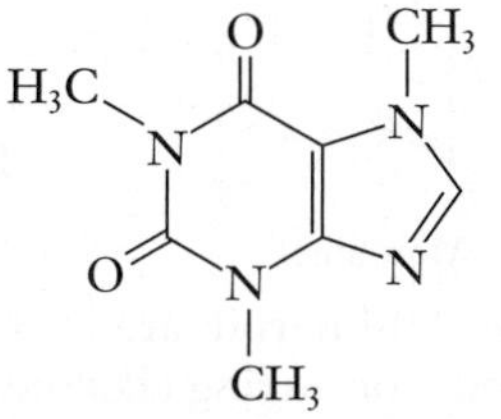 caffeine	

FIGURE 27.25 Alkaloids are a chemically diverse group of heterocyclic, nitrogen-containing compounds that are classified according to the predominant type of ring structure.

pathway. Two well-known terpenoid-indole alkaloids, **vinblastine** and **vincristine**, are produced by the Madagascar periwinkle, *Catharanthus roseus*. In the 1950s it was discovered that vinblastine and vincristine arrest cell division in metaphase by inhibiting microtubule formation. They have since been used in the treatment of Hodgkin's lymphoma, leukemia, and other forms of cancer. Unfortunately, *C. roseus* produces a complex array of indole alkaloids of which vinblastine and vincristine represent only a very small proportion (about 0.05% of leaf dry weight). This makes their extraction and purification difficult and costly. Recent efforts have led to the development of cell culture systems in order to produce higher yields at reasonable cost. Unfortunately, because alkaloid production is so often tissue-specific, production in cell culture has not been highly successful. Other well-known alkaloids such as **cocaine** (from leaves of the coca plant, *Erythroxylum coca*), **codeine** (from the opium poppy, *Papaver somniferum*), **nicotine** (tobacco), and **caffeine** (coffee beans and tea leaves) are widely used as stimulants or sedatives.

27.7.3 LIKE MANY OTHER SECONDARY METABOLITES, ALKALOIDS SERVE AS PREFORMED CHEMICAL DEFENSE MOLECULES

Whether alkaloids have any specific function in plants has been a subject of debate for many years. Some plants appear to invest a significant proportion of their resources, especially nitrogen, in a diverse array of alkaloids and yet they have no obvious function in the physiology of the plants themselves. Like other secondary products, however, there are at least two general arguments supporting a defensive role. Firstly, most alkaloids have a bitter taste, which is considered a universally repellant character for all animals, including insects. Secondly, all alkaloids are biologically active and many are significantly toxic to insects and other animals. Nicotine, for example, is a potent insect poison and one of the first insecticides used by humans.

Many alkaloids have antibiotic properties and may have a role in defense against microbial infection. Interestingly, it has been known for a long time that alkaloid concentrations are very low when measured on a whole plant basis, but, as noted earlier, may be very high in selected organs or tissues. Often the tissues that accumulate alkaloids are those most vulnerable in terms of plant fitness (young influorescences or seed pods, for example) or peripheral tissues that would be first attacked by herbivores. Thus the morphine content of whole young poppy capsules is less than 2 percent, but 25 percent or more in the latex that exudes when the capsule is wounded. Quinine accumulates in the outer bark of the tree *Cinchona officinalis* and *Rauwolfia* alkaloids are concentrated primarily in the root bark. Although alkaloids have been studied extensively for their pharmacological and medicinal value, there is much yet to be learned with respect to their physiology and chemical ecology.

SUMMARY

Although the products of plant metabolism are often designated as either primary or secondary metabolites, the distinction between the two is not easily made. Primary metabolites such as protein, lipid, carbohydrate, and nucleic acids comprise the basic metabolic machinery of all cells. Others, such as chlorophyll and lignin, are more restricted in occurrence but are equally essential to the growth and development of the organism. Secondary metabolites, on the other hand, may be found only in specific tissues or at particular stages of development and have no obvious role in the development or survival of the organism.

Three metabolic pathways that give rise to both primary and secondary metabolites are the mevalonic acid and MEP pathways and the shikimic acid pathway. The mevalonic acid and MEP pathways give rise to two 5-carbon compounds, isopentenyl pyrophosphate and dimethylallyl pyrophosphate, that form the basis for the terpenoid family. Terpenoids include primary metabolites such as phytol (a portion of the chlorophyll molecule), membrane sterols, carotenoid pigments, and the hormones gibberellin and abscisic acid. Secondary products include a range of 5-carbon hemiterpenoids through 40-carbon tetraterpenes and the polyterpenes, rubber and gutta.

The shikimic acid pathway produces the aromatic amino acids phenylalanine, tyrosine, and tryptophan—all primary metabolites required for protein synthesis. Deamination of phenylalanine to cinnamic acid, catalyzed by the enzyme phenylalanine ammonia lyase (PAL), effectively diverts carbon from primary metabolism into the synthesis of a wide range of secondary metabolites—coumarins, lignin, tannins, flavonoids, and isoflavonoids—based on simple phenolic acids.

Salicylic acid is involved in establishing the hypersensitive reaction to insect herbivory and pathogen infection. The gene-for-gene model predicts that no infection will occur when the pathogen and host plant carry dominant avirulence and response genes, respectively. An early event in this sensing/signaling pathway is the activation of defense-related genes and synthesis of their products, pathogenesis-related (PR) proteins. The plant immune response continues with the development of a general immune response known as systemic acquired resistance (SAR). SAR signaling includes the protein NPR1, which responds to a change in cellular redox potential by migrating into the nucleus where it interacts with transcription factors to stimulate the transcription of PR genes.

Other interesting and useful secondary metabolites include the saponins (terpene glycosides), cardiac glycosides, cyanogenic glycosides, and glucosinolates. The alkaloids are a heterogeneous group of nitrogenous compounds with significant pharmacological properties. Although the physiological roles of secondary metabolites are poorly understood, most are toxic to some degree and appear to serve primarily in defense against microbial infection and attack by herbivores.

it often takes a week before any signs of injury are observed. Explain this delay.

9. In what ways do plants have some control over insect herbivory?

10. How do plants respond to invasion by microbial pathogens?

CHAPTER REVIEW

1. Distinguish between primary and secondary metabolism.

2. What are terpenes? How do they originate within the plant and what functions do they serve?

3. Explain why sterols may be considered both primary and secondary metabolites.

4. What do saponins, cardenolides, and amygdalin have in common?

5. Why are amino acids such as phenylalanine and tryptophan essential in the human diet?

6. What are EPSPS and PAL? What key metabolic roles do they play?

7. Alkaloids are, for the most part, chemically unrelated. What is the basis for grouping them together?

8. How does the herbicide glyphosate kill plants? When plants are treated with glyphosate,

FURTHER READING

Cordell, G. (ed.). 1997. *The Alkaloids.* San Diego: Academic Press.

Durrant, W. E., X. Dong. 2004. Systemic acquired resistance. *Annual Review of Phytopathology* 42:185–209.

Fobert, P. R., C. Després. 2005. Redox control of systemic acquired resistance. *Current Opinion in Plant Biology* 8:373–382.

Grant, M., C. Lamb. 2006. Systemic immunity. *Current Opinion in Plant Biology* 9:414–420.

Harborne, J. B., F. A. Tomas-Barberan (eds.). 1991. *Ecological Chemistry and Biochemistry of Plant Terpenoids.* Oxford: Clarendon Press.

Klein, R. M. 1987. *The Green World: An Introduction to Plants and People.* New York: Harper & Row.

O'Kennedy, R., R. D. Thornes (eds.). 1997. *Coumarins: Biology, Application, and Mode of Action.* Chichester: Wiley.

Rosenthal, G. A., M. R. Berenbaum. 1991. *Herbivores: Their Interactions with Secondary Metabolites. Vol. 1: The Chemical Participants.* 2nd ed. San Diego: Academic Press.

Staswick, P. E. 2008. JAZing up jasmonate signaling. *Trends in Plant Sciences* 13:66–71.

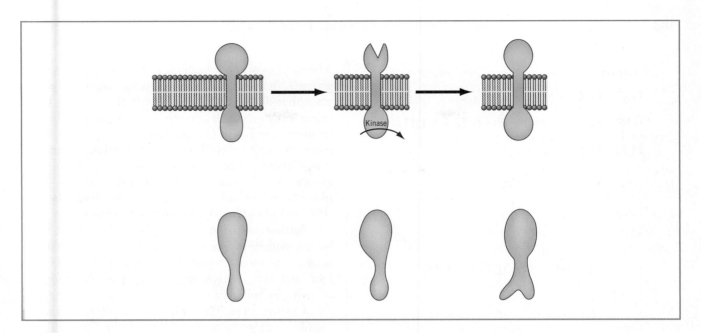

Appendix

Building Blocks: Lipids, Proteins, and Carbohydrates

I.1 LIPIDS

Lipids (Gr. *lipos*, fat) are a chemically diverse group of molecules defined primarily by their ability to dissolve readily in organic solvents, but only sparingly in water. These include fats, oils, and the phospholipids and glycolipids that make up the cellular membranes. Sterols (steroid alcohols) and molecules containing long-chain hydrocarbons such as the pigments chlorophyll and carotene, although chemically distinct from fats and oils, are also considered lipids on the basis of their solubility properties.

Fats and oils are composed of long hydrocarbon chains called **fatty acids** and the three-carbon alcohol **glycerol** (Figure I.1). Fatty acids are attached to the glycerol through an ester link between the carboxyl group of the fatty acid and a hydroxyl group on the glycerol molecule. Because one fatty acid (an **acyl** group) is esterified to each of the three glycerol carbons, a molecule of fat or oil is known as **triacylglycerol**, or a **triglyceride**. Fatty acids vary according to the length of the hydrocarbon chain (i.e., the number of carbon atoms) as well as the number of carbon–carbon double bonds and the position of those double bonds in the chain (Table I.1). Fatty acids that have no double bonds are known as saturated fatty acids, referring to the fact that all the carbon–carbon bonds are saturated with hydrogen atoms. Fatty acids with double bonds are considered unsaturated. Unsaturated fatty acids are identified not only by the number of double bonds, but also by their location in the molecule. For example, there are two forms of linolenic acid: α-linolenic acid has double bonds at the 9, 12, and 15 carbons, while in γ-linolenic the double bonds are at the 6, 9, and 12 carbons. Most fatty acids contain an even number of carbon atoms because their synthesis involves the successive addition of two-carbon acetate (CH_3—COO^-) units.

The difference between fats and oils is a matter of melting points. The melting points of fatty acids increase with chain length and the extent of saturation. Saturated fatty acids tend to be solid at room temperature and unsaturated fatty acids tend to be liquid at room temperature. Thus fats, which are normally solid at room temperature, are triglycerides composed of predominantly saturated fatty acids, while oils, normally fluid at room temperature, are composed of predominantly unsaturated fatty acids. Plant cells, especially those in seeds, contain mostly oils that are often stored in lipid bodies called **oleosomes**. Lipids may, in fact,

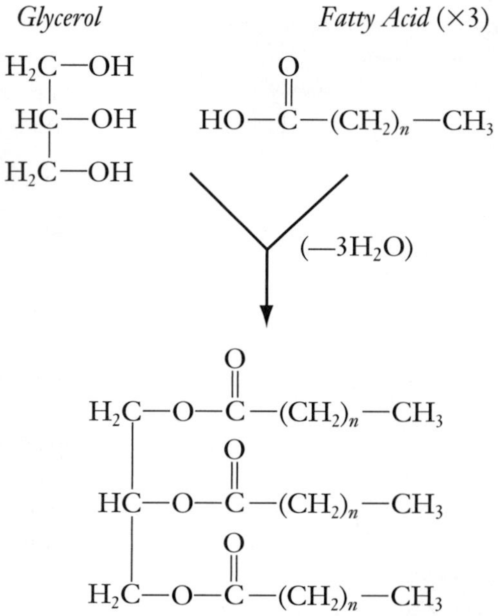

Glycerol   *Fatty Acid* ($\times 3$)

$$H_2C-OH$$
$$HC-OH \qquad HO-\overset{\displaystyle O}{\overset{\|}{C}}-(CH_2)_n-CH_3$$
$$H_2C-OH$$

$(-3H_2O)$

$$H_2C-O-\overset{\displaystyle O}{\overset{\|}{C}}-(CH_2)_n-CH_3$$
$$HC-O-\overset{\displaystyle O}{\overset{\|}{C}}-(CH_2)_n-CH_3$$
$$H_2C-O-\overset{\displaystyle O}{\overset{\|}{C}}-(CH_2)_n-CH_3$$

Triglyceride

FIGURE I.1 **A triglyceride is formed by ester linkages between three long-chain fatty acids and glycerol. The value of *n* is normally 14 to 18.**

be the dominant form of stored carbon in smaller seeds because it is an efficient form in which to store energy. Lipids are less oxidized than carbohydrate and therefore will yield more energy when oxidized during respiration. Lipid bodies are also commonly found in chloroplasts, which contain an extensive internal membrane system. These lipid bodies may represent a reservoir of lipids for membrane synthesis.

The principal class of lipids found in most cellular membranes is the **phospholipids**. Phospholipids are **diglycerides**; meaning that only *two* fatty acids (rather than three) are esterified to the glycerol molecule.

The third position is occupied instead by a phosphate group (Figure I.2). The simplest phospholipid is **phosphatidic acid (PA)**, in which R = −OH. PA is found only in small quantities in most membranes. In most membrane phospholipids, R is a small, polar molecule such as **choline, ethanolamine, serine,** or *myo*-**inositol**. Phospholipids containing these head groups are known as **phosphatidylcholine (PC), phosphatidylethanolamine (PE), phosphatidylserine (PS),** and **phosphatidylinositol (PI)**, respectively.

Further variation in phospholipids is introduced by the nature of the two fatty acids; in plant membranes, the most abundant are 16:0, 16:1, 18:0, 18:1, 18:2, and 18:3. A high proportion of unsaturated fatty acids in the membrane lipids contributes to the fluidity of membranes. This is because the carbon–carbon double bond introduces a "kink" in the fatty acid (see Figure I.2) that prevents the resulting phospholipids from packing as tightly as they do with only saturated fatty acids. Phospholipids are often described as having a charged, polar phosphate "head" and a long hydrocarbon "tail" represented by the two fatty acids. This gives the molecule a dual character in that the phosphate head is *hydrophilic* ("water-loving") and the fatty acid tail is *hydrophobic* ("water-fearing"). A molecule with both hydrophilic and hydrophobic properties is known as **amphipathic**.

Some plant membranes contain large amounts of **glycolipids**, lipids in which the head group contains one or more sugar residues. The internal membranes of the chloroplast are particularly distinctive in this regard. They contain only about 10 percent phospholipid and about 80 percent **monogalactosyl diglyceride (MGDG)** and **digalactosyl diglyceride (DGDG)**. MGDG and DGDG are similar to phospholipids except that the phosphate group is replaced with

TABLE I.1   **Some Common Biological Fatty Acids.**

Common Name	Symbol[1]	Structure	mp (°C)[2]
Saturated fatty acids			
Lauric acid	12:0	$CH_3(CH_2)_{10}COOH$	44
Myristic acid	14:0	$CH_3(CH_2)_{12}COOH$	52
Palmitic acid	16:0	$CH_3(CH_2)_{14}COOH$	63
Stearic acid	18:0	$CH_3(CH_2)_{16}COOH$	69
Unsaturated fatty acids			
Palmitoleic acid	16:1	$CH_3(CH_2)_5CH{=}CH(CH_2)_7COOH$	−0.5
Oleic acid	18:1	$CH_3(CH_2)_7CH{=}CH(CH_2)_7COOH$	13
Linoleic acid	18:2	$CH_3(CH_2)_4(CH{=}CHCH_2)_2(CH_2)_6COOH$	−9
α-Linolenic acid	18:3	$CH_3CH_2(CH{=}CHCH_2)_3(CH_2)_6COOH$	−17

[1]The number before the colon indicates the number of carbon atoms in the chain. The number following the colon indicates the number of double bonds.
[2]mp = melting point.

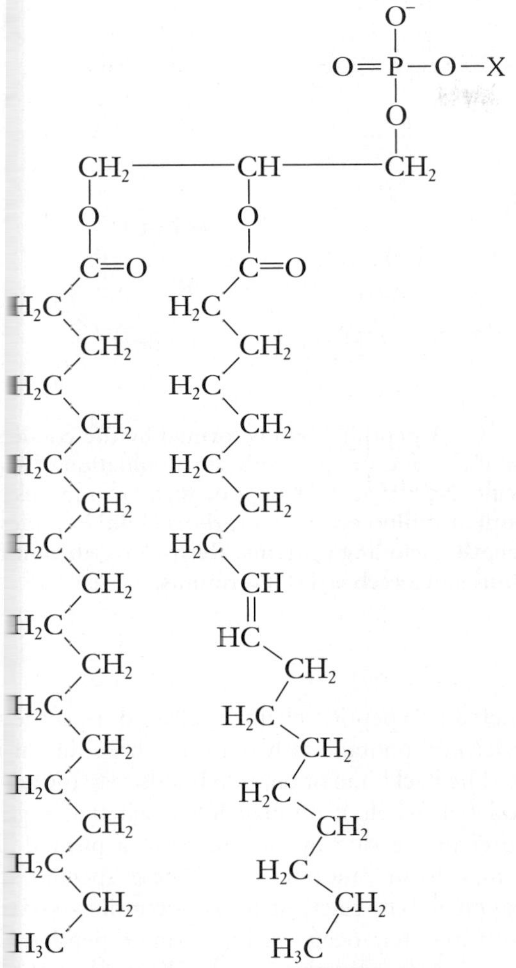

FIGURE I.2 **A phospholipid is a triglyceride with a phosphate group in place of one fatty acid. When R = OH (hydrogen), the molecule is known as phosphatidic acid. More commonly, R will represent another small polar molecule such as choline or ethanolamine.**

one or two molecules, respectively, of the sugar galactose. The remaining 10 percent of the chloroplast membrane lipid is accounted for by **sulfolipid**; in this case, a sulfur group is attached to the galactose in MGDG.

The principal sites of a fatty acid and lipid biosynthesis are the endoplasmic reticulum, the chloroplast, and the mitochondria. Chloroplasts appear to have the enzymatic machinery for synthesizing all the C_{16} and C_{18} fatty acids and the addition of galactosyl residues to the diglycerides in order to make MGDG and DGDG. Chloroplasts do not, however, appear to make phospholipids; this is accomplished in the endoplasmic reticulum, which contains all the enzymes necessary for the synthesis of PC and PE. Mitochondria are apparently unable to synthesize either PC or PE, even though these are the principal phospholipids of the mitochondrial membranes.

I.2 PROTEINS

Proteins are composed of amino acids. There are 20 "standard" amino acids, specified by the genetic code, that make up all proteins. All amino acids have the same core structure, in which a single carbon atom (the α-carbon) carries both an amino group ($-NH_2$) and a carboxylic acid group ($-COOH$) (Figure I.3). At physiological pH, both the amino group and the carboxyl group of an amino acid are ionized. Amino acids thus carry both a negative charge and a positive charge at the same time. Such molecules are known as **zwitterions**.

The structures of the 20 individual amino acids vary according to the nature of the R group, or side chain. This side chain may consist of nonpolar groups which give the amino acid hydrophobic properties, or either neutral or charged polar groups, which give the amino acid hydrophilic properties (Figure I.4).

A **peptide bond** forms when the amino group of one amino acid forms a covalent link with the carboxyl group of a second amino acid (Figure I.5). When two amino acids are linked by a single peptide bond, the resulting molecule is called a dipeptide. Three amino acids constitute a tripeptide, and so forth. A chain containing a large number of amino acids is called a **polypeptide**. Most proteins are polypeptides containing in the range of 40 to 4,000 amino acids, which allows for an extremely large number of different protein molecules. For example, a simple dipeptide may have any one of 20 choices for the first amino acid paired with any one of the same 20 choices for the second amino acid. This allows for 20^2, or 400, possible unique dipeptides. Similarly, for a protein containing 100 amino acids, there are 20^{100}, or 1.27×10^{130}, possible unique combinations.

Because the chemical properties of the side chains of individual amino acids are each different, it is clear that organisms are able to synthesize an enormous number of different proteins with an equally large range of biochemical properties. The variety of proteins is further enhanced with the addition of nonstandard amino acids, or amino acid derivatives. These amino acid derivatives are the result of specific modifications made to a standard amino acid after the peptide chain has been

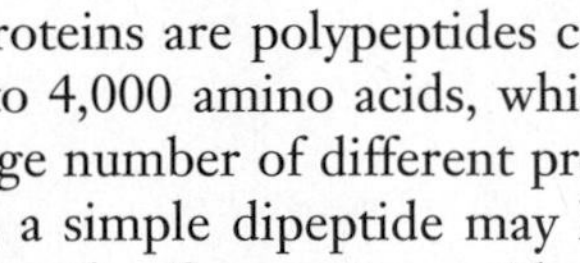

$$\overset{\displaystyle R}{\underset{\displaystyle {}^+H_3N-CH-COO^-}{|}}$$

FIGURE I.3 **All amino acids share the same basic structure. Here the amino acid is represented in the zwitterionic form, which predominates at physiological pH. There are 20 different side chains (R) that make up the 20 standard amino acids found in all proteins.**

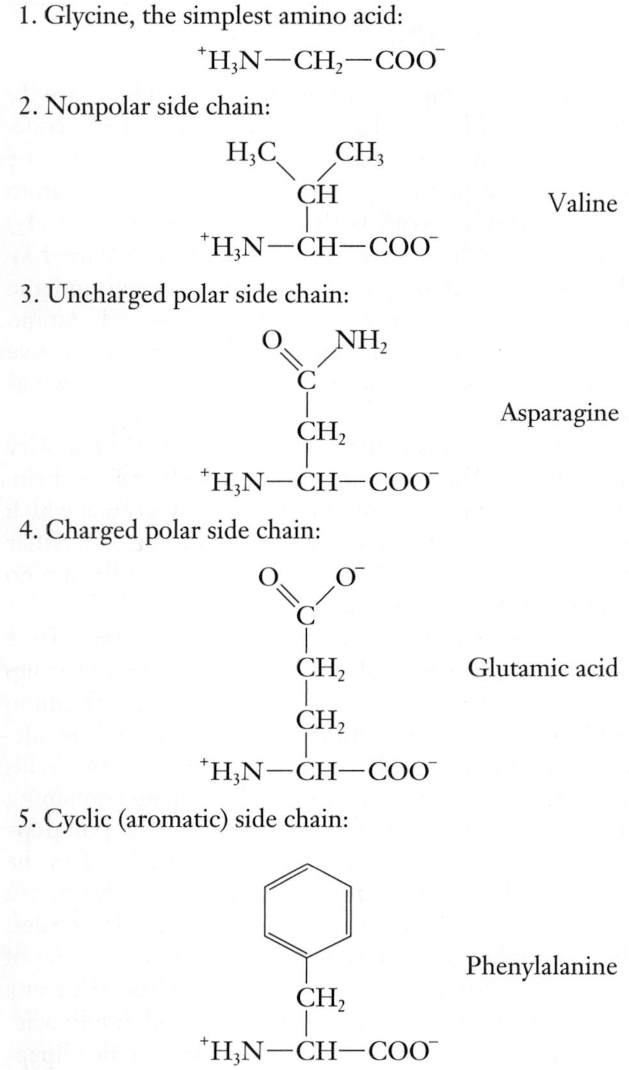

1. Glycine, the simplest amino acid:

$$^+H_3N-CH_2-COO^-$$

2. Nonpolar side chain:

Valine

3. Uncharged polar side chain:

Asparagine

4. Charged polar side chain:

Glutamic acid

5. Cyclic (aromatic) side chain:

Phenylalanine

FIGURE I.4 **Examples of amino acids representing the four principal groups of side chains. Glycine (R = hydrogen) is the simplest of the nonpolar side chains. Phenylalanine and others with cyclic side chains are known as aromatic amino acids.**

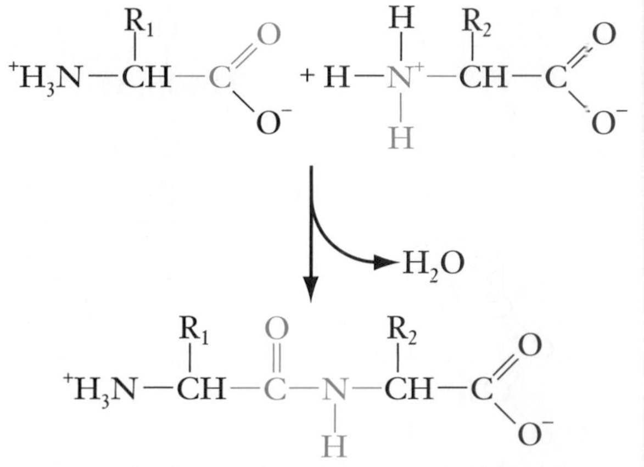

FIGURE I.5 **A peptide bond is formed by the condensation of two amino acids with the elimination of a water molecule. Note that the dipeptide formed in this example has both an amino end and a carboxyl end. Peptides of any length, including proteins, always have an amino (N) terminus and a carboxyl (C) terminus.**

synthesized. One example is **hydroxyproline**, a nonstandard amino acid that is found in certain cell wall proteins.

Protein molecules have several levels of organization, as shown in Figure I.6. The sequence of amino acids in a polypeptide is referred to as the **primary structure**. Primary structure is determined by the sequence of nucleotides in the deoxyribonucleic acid (DNA) that makes up the gene for that protein. This sequence is first transcribed into a **messenger ribonucleic acid (mRNA)**, which migrates from the nucleus into the cytoplasm where it attaches to particles called **ribosomes**. Amino acids are sequentially assembled into polypeptides on the ribosome according to the sequence of nucleotides, or message, in the mRNA. When the

completed polypeptide chain is released from the ribosome, it folds spontaneously to form a three-dimensional shape. The backbone of peptide bonds may form a helical coil (an α-helix), or may fold such that segments of the chain lie side by side to form a pleated sheet, or it may form random coils. These specific spatial arrangements are referred to as **secondary structure** and all three may occur within a single peptide chain. The number and distribution of helices, pleated sheets, and random coils in the polypeptide determine the three-dimensional configuration of the entire polypeptide, called **tertiary structure**. Tertiary structure of the protein is commonly referred to as **conformation**. Some protein molecules are composed of a single polypeptide, while others are composed of multiple polypeptide chains, called **subunits**. The arrangement of subunits in the protein molecule is referred to as **quaternary structure**. The three-dimensional structure of a protein is stabilized by a variety of noncovalent bonds including hydrogen bonds and electrostatic interactions between ionized amino and carboxyl groups on the amino acid side chains. At the tertiary level, covalent disulfide bonds (—S—S—) may form between nearby amino acids with sulfur groups in their side chains. Alternatively, neighboring amino acids with nonpolar side groups may form strong hydrophobic interactions.

Subtle differences in the composition and structure give each protein its unique characteristics and its ability to discriminate between other molecules with which it interacts. In spite of the large size and complex structure of most proteins, a change in even a single amino acid in the sequence may dramatically alter its biological properties.

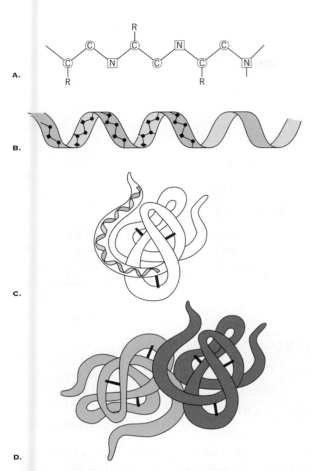

O
‖
CH   H₂COH

HCOH   C=O

H₂COH   H₂COH

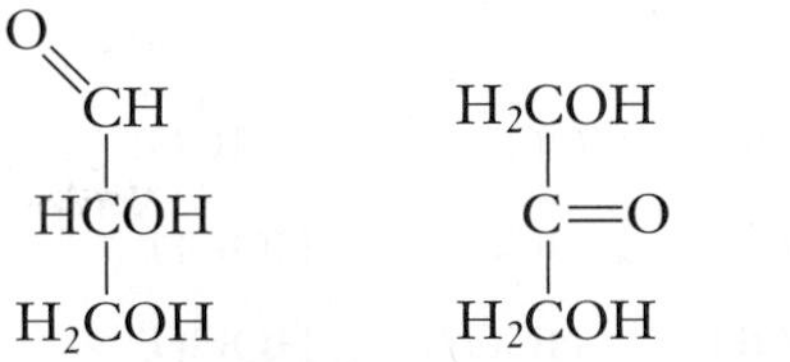

Glyceraldehyde   Dihydroxyacetone

FIGURE I.7 Glyceraldehyde and dihydroxyacetone are the two simplest carbohydrates. Glyceraldehyde is an aldo sugar, or aldose. Dihydroxyacetone is a keto sugar, or ketose.

features of monosaccharides are illustrated by the **triose** (three-carbon) sugars **glyceraldehyde** and **dihydroxyacetone** (Figure I.7). Note that both molecules contain a carbonyl oxygen (—C = O). In glyceraldehyde the carbonyl oxygen forms an aldehyde group (—CHO), while in dihydroxyacetone it forms a ketone group (—C—CO—C—). Glyceraldehyde is therefore referred to as an aldo sugar, or **aldose**, and dihydroxyacetone is referred to as a keto sugar, or **ketose**. Other common monosaccharides include the 4-carbon sugar, erythrose (a **tetrose**); the 5-carbon sugar, ribose (a **pentose**); the 6-carbon sugar, glucose (a **hexose**); and the 7-carbon sugar, sedoheptulose (a **heptose**) (Figure I.8).

The four most common hexoses found in plants are D-glucose, D-mannose, D-galactose, and D-fructose (Figure I.9). Mannose is similar to glucose except for the orientation of the hydroxyl group on the second carbon. Similarly, galactose and glucose differ only in the orientation of the hydroxyl group on carbon 4. These are examples of **stereoisomers**. Stereoisomerism arises because the four bonds of a carbon atom do not lie in a single plane as they are usually represented on paper, but form a three-dimensional tetrahedron:

When the carbon atom is attached to *four different* substituents, there are two possible configurations that are mirror images of each other. Such carbons are known as **asymmetric carbons**. Glucose, for example,

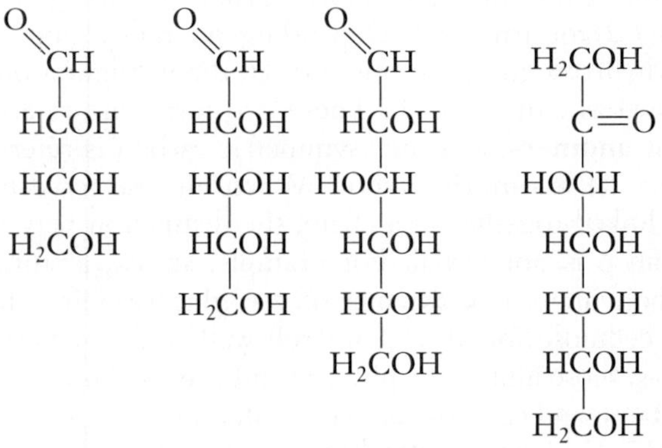

FIGURE I.6 The structural hierarchy in proteins. (A) Primary structure refers to the amino acid sequence in a peptide chain. (B) Secondary structure. The amino acid chain spontaneously forms an α-helix, a β-sheet, or a random coil. All three configurations may occur within a single peptide chain. (C) Tertiary structure is the three-dimensional shape, or conformation, of a complete folded peptide chain. (D) Quaternary structure refers to an assembly of multiple peptide chains.

I.3 CARBOHYDRATES

Carbohydrates contain carbon, hydrogen, and oxygen in the general proportions of 1:2:1. In other words, the hydrogen (H) and oxygen (O) are found in the same proportions as they are found in water, hence "carbon hydrate." The composition of carbohydrates is $(CH_2O)_n$, where $n \geq 3$. Smaller carbohydrate molecules play a central role in the energy metabolism of cells and are the principal source of carbon skeletons for almost all other organic molecules.

I.3.1 MONOSACCHARIDES

The simplest carbohydrates are known as **monosaccharides** (L. *saccharum*, sugar), because they cannot be hydrolyzed to form simpler carbohydrates. The basic

O   O   O
‖   ‖   ‖
CH   CH   CH   H₂COH

HCOH   HCOH   HCOH   C=O

HCOH   HCOH   HOCH   HOCH

H₂COH   HCOH   HCOH   HCOH

         H₂COH   HCOH   HCOH

                  H₂COH   HCOH

                          H₂COH

D-Erythrose   D-Ribose   D-Glucose   D-Sedoheptulose

FIGURE I.8 Examples of common monosaccharides with 4-, 5-, 6-, and 7-carbon atoms.

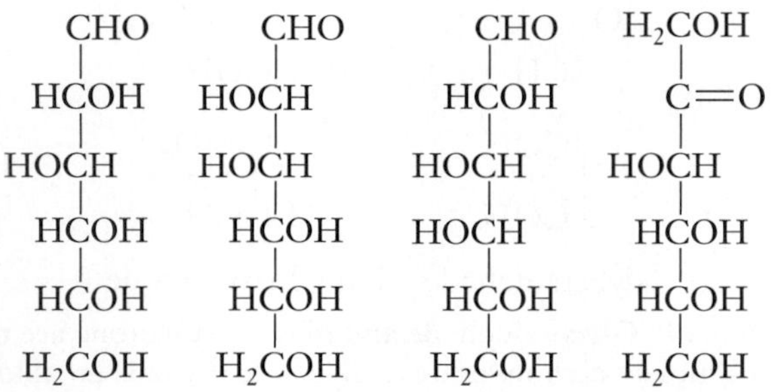

CHO HCOH HOCH HCOH HCOH H₂COH — D-Glucose

CHO HOCH HOCH HCOH HCOH H₂COH — D-Mannose

CHO HCOH HOCH HOCH HCOH H₂COH — D-Galactose

H₂COH C=O HOCH HCOH HCOH H₂COH — D-Fructose

FIGURE I.9 Example of four common hexoses found in plants. Note that the three aldo sugars, glucose, mannose, and galactose, differ only with respect to the orientation of the hydroxyl groups on the second, third, fourth, and fifth asymmetric carbon atoms.

contains four asymmetric carbon atoms (carbons 2, 3, 4, 5), which allows for up to 2^4 or16 possible sterioisomers. Mannose and galactose are just two of the 16 possible sterioisomers of glucose. Each stereoisomer has different chemical, physical, and biological properties.

Note that the names of the four sugars above contain the prefix D- (fr. L, *dextra*, right hand). This identifies the structure of the D-enantiomer, which, in solution, rotates plane polarized light to the right, or clockwise, from the point of view of the observer. By convention, the D-enantiomer is drawn with the hydroxyl group on the *highest-numbered* asymmetric carbon placed to the right of the carbon chain. If this hydroxyl is written to the left (and all other hydroxyl groups are similarly inverted), the designation is L- (fr. L, *laevus*, left). L-sugars rotate plane polarized light counterclockwise, or to the left. Most naturally occurring sugars are in the D configuration. One exception is L-galactose, a constituent of agar.

Although monosaccharides are often depicted as linear molecules, in solution the aldehyde or ketone group normally reacts intramolecularly with a hydroxyl group to form a cyclic molecule (Figure I.10). Note that ring closure generates an additional asymmetric center at the former carbonyl carbon (C1 for glucose and C2 for fructose). Depending on the position of the hydroxy group on this carbon, the configuration is known as either α or β. These two forms are referred to as **anomers**, and the asymmetric carbon is referred to as the **anomeric** carbon. When glucose molecules are linked together in a chain, the distinction between α and β is not trivial. For example, starch, a storage carbohydrate, is a polymer of α-D-glucose units while the cellulose found in plant cell walls is a polymer of β-D-glucose units. The physical and chemical properties of starch and cellulose are very different!

The aldehyde or ketone group of monosaccharides may be reduced to an alcohol group (Figure I.11). Reduction of glucose, for example, yields **sorbitol** and mannose yields **mannitol**. The sugar alcohols (sorbitol

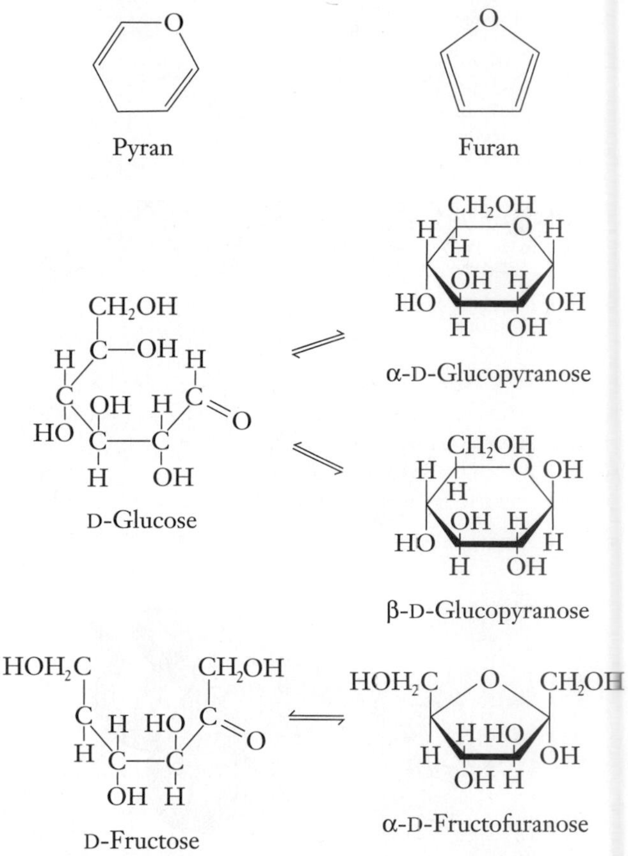

Pyran

Furan

α-D-Glucopyranose

β-D-Glucopyranose

D-Glucose

HOH₂C CH₂OH D-Fructose

α-D-Fructofuranose

FIGURE I.10 In solution, monosaccharides cyclize to form an internal lactone ring. The ring forms between the carbonyl carbon and a nonterminal hydroxyl group. A monosaccharide that forms a six-membered ring (five carbons and one oxygen) is called a *pyranose*, after pyran, the simplest molecule containing such a ring. Similarly, a monosaccharide that forms a five-membered ring (four carbons and one oxygen) is called a *furanose*, after furan. Glucose, a 6-carbon aldo sugar, forms a pyranose, while fructose, a 6-carbon keto sugar, forms a furanose.

and mannitol) frequently serve as storage carbohydrate in algae and other lower plants. Alternatively, the primary alcohol group of an aldose may be oxidized to form a carbonyl, or acid, group. The acid form of glucose is **glucuronic acid** and of galactose, **galacturonic acid**. Glucuronic and galacturonic acids are important components of noncellulosic cell wall polymers.

Monosaccharides such as glyceraldehyde and dihydroxyacetone, are known as **reducing sugars** because of the ease with which the free aldehyde group reduces mild oxidizing agents. The standard test is the reduction of silver ion (Ag+) to metallic silver using Tollens' reagent. A silver mirror lining on the surface of the reaction vessel indicates the presence of a reducing sugar.

I.3.2 POLYSACCHARIDES

Individual monosaccharides may be linked together in chains of varying length by **glycosidic bond** between the

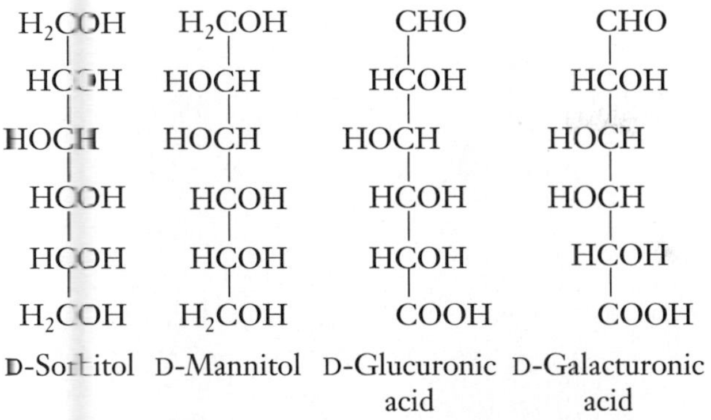

D-Sorbitol D-Mannitol D-Glucuronic D-Galacturonic
acid acid

FIGURE I.11 **Examples of alcohol and acid derivatives of some common hexose sugars. Sorbitol and mannitol are important carbon storage products, especially in the lower plants. Glucuronic acid and glalacturonic acid are important components of noncellulosic cell wall polymers.**

anomeric hydroxyl group of one sugar and a hydroxyl group of a second sugar (Figure I.12). Small carbohydrate polymers are identified by the number of sugar residues (e.g., 2 sugars = disaccharide, 3 = trisaccharide, etc.) while longer chains are referred to simply as **polysaccharides** (also known as **glycans).** The glycosidic bond is effectively analogous to the peptide bond in proteins.

The principal disaccharide found in higher plants is **sucrose**, formed by the condensation of α-D-glucose with α-D-fructose (Figure I.12). Note that the bond is formed between the number-1 carbon of glucose and the number-2 carbon of fructose. Because the bond involves the number-1 and number-2 carbons of α-sugars, it is designated as an α(1 → 2) link. Also, because both reducing groups are involved in forming the bond, sucrose is not a reducing sugar.

Polysaccharides usually consist of a single type of sugar (i.e., glucose), although polysaccharides consisting of more than one sugar are known. The two most

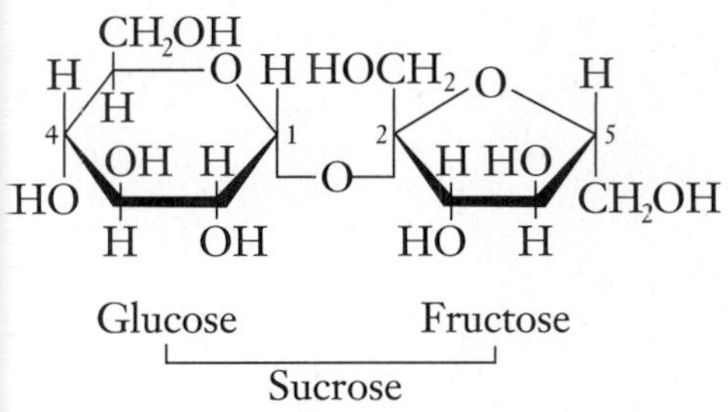

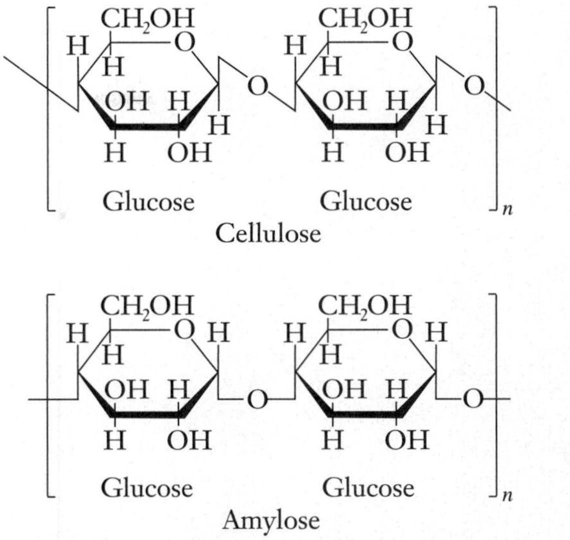

FIGURE I.13 **The structures of cellulose and amylose. Both are polymers of glucose. In cellulose, carbon-1 of the first β-D-glucose residue is linked to carbon-4 of the next β-D-glucose residue (a β(1 → 4) glycosidic bond). Amylose is a linear form of starch in which α-D-glucose residues are linked by α(1 → 4) glycosidic bonds. In either case, the value of *n* may be up to several thousand.**

common polysaccharides in higher plants are starch and cellulose. Starch is made up entirely of (1 → 4)-linked and (1 → 6)-branched α-D-glucose residues (Figure I.13). Cellulose, on the other hand, is a long, unbranched chain of (1 → 4)-linked β-D-glucose residues (Figure I.13). Because both starch and cellulose are made of glucose, they are commonly referred to as **glucans.** Starch forms a helically coiled structure used primarily for storage of carbon and energy. Cellulose is a linear molecule that may contain 3,000 or more glucose residues and is noted for its high degree of structural strength. Cellulose, which may comprise up to 80 percent of the dry weight of a plant, is the most abundant organic substance in the world and perhaps one of the most economically important. A polymer similar to cellulose is **chitin**, the principal component of the exoskeleton of some invertebrates (crustaceans, insects, spiders) and the cell walls of some fungi. Chitin is a β(1 → 4)-linked polymer of N-acetylglucosamine, which differs chemically from cellulose only in that the hydroxyl group on carbon-2 of the glucose unit is replaced by an —NHCOCH₃ group.

Another important class of polysaccharides is the **fructans.** Fructans are polymers with varying numbers of fructose molecules added to the fructose end of sucrose and are particularly common in grasses. Fructans built up of β(2 → 6) links are known as **levans**, while those with β(1 → 6) links are known as **inulins**. Inulin is an important storage carbohydrate in Jerusalem artichoke (*Helianthus tuberosum*).

Glucose Fructose

Sucrose

FIGURE I.12 **Sucrose, the principal disaccharide in plants, is composed of a molecule each of glucose and fructose. A glycosidic bond is formed between carbon-1 of the glucose molecule and carbon-2 of fructose (α 1 → 2 link.)**

Index/Glossary